REFORM WITH REASON

A report of a Joint Task Force of the Mathematical Association of America and the Association of American Colleges, "Challenges for College Mathematics: An agenda for the Next Decade," outlined some of the problems with calculus instruction. This book addresses the problems outlined in the Joint Task Force Report:

● *The **learning** problem*

If the students cannot clearly state the essential definitions and formulas *as a first step* they cannot hope to master the more subtle relationships and concepts of the course. See the CONCEPT PROBLEMS at the end of each chapter.   p. 117

Also see *Student Survival Manual*

● *The **technology** problem*

Students of the 1990s need to incorporate technology as a tool in calculus. We assume that all students have calculators, and that many may have access to computer software. There are over 130 COMPUTATIONAL WINDOWS throughout the text, but we have taken care that these windows are independent of the calculus we are trying to teach. In the text these windows are "platform-neutral," with specific keystrokes confined to the technology lab manuals.   See p. 90

Also see *Technology Manuals*

● *The **self-esteem** problem*

This book help builds students' confidence in their mathematical abilities. Hints, warnings, suggestions, and "WHAT THIS SAYS" boxes are provided to rephrase precise mathematical language into meaningful statements to help the students' understanding.   See p. 170

Also see *Student Handbook*

● *The **communication** problem*

Students must be provided intellectually stimulating introductory courses. We provide opportunities for writing essays and book reports, as well as requiring students to describe concepts in their own words in a series of "WHAT DOES THIS SAY?" questions.

See pp. 81, 228

● *The **foundation** problem*

Students must be provided intellectually stimulating introductory courses. One of the goals of this book is to encourage fluency in mathematics. The range of problems offered in a calculus book must extend from routine practice to challenging. We include past Putnam examination problems as well as problems found in current mathematical journals. We also offer a running commentary in a set of problems called SPY PROBLEMS.   See p. 67

● *The **connections** problem*

Students must connect areas of mathematics and areas of application. There are a significant number of applied problems. Even though we have provided an abundance of essential applications for the engineer and the physicist, we believe that a beginning calculus course should also include applications to a wide variety of other disciplines. In almost every problem set we have included MODELING PROBLEMS that are designed to build modeling skills and HISTORICAL QUEST problems crafted to provide connections with the development of mathematical ideas.   See p. 337

● *The **research** problem*

Students are provided opportunities for undergraduate research and independent projects. We have included group research projects, individual research questions connected with guest essays, as well as **Putnam, Journal**, and **Think Tank** problems.   See p. 206

These features are spread throughout the text; we have chosen one specific page example for each.

 **Summaries** **Where to Look for Help**

PREREQUISITES

Your algebra, trigonometry, or precalculus textbooks, if you kept them, are good places to look because you are used to the notation and presentation. If you do not have these books for reference, we have provided the necessary review in *Student Mathematics Handbook*. Also see

Properties of Absolute Value, see Table 1.1, p. 3
Exact Values of Trigonometric Functions, see Table 1.2, p. 10
Directory of Curves, see Table 1.3, p. 37
Definition of Inverse Trigonometric Functions, see Table 1.4, p. 48

TECHNOLOGY

See *Technology Manuals for Maple, Mathematica,* or *Mathlab.* We have also found the TI-92 to be of great value in working many calculus problems. Inexpensive software includes *Derive* (which is included with the TI-92), and *Converge* for graphical representation of many calculus concepts.

DIFFERENTIATION AND GRAPHING

Definition, p. 132
Derivative formulas, see page facing inside back cover
Holes, Poles, Jumps, and Continuity, see Figure 2.23, p. 107
A Summary of First Derivative and Second Derivative Tests, see Table 4.1, p. 256
Graphing strategy, see Table 4.2, p. 269
Optimization Procedure, p. 275

INTEGRATION

Definition, p. 320 (indefinite), p. 340 (definite)
Summation formulas, p. 334
Integration formulas, see inside back cover
Short integration table, see Appendix D
Integration table, see *Student Mathematics Handbook*
Volumes of Common Solids, see Table 6.1, pp. 415–416
Modeling Applications Involving Integration, see Table 6.3, pp. 461–462
Integration Strategy, see Table 7.2, pp. 499–500

INFINITE SERIES

Guidelines for Determining Convergence of Series, Table 8.1, pp. 579–580
Power Series for Elementary Functions, Table 8.2, p. 615

ANALYTIC GEOMETRY, VECTORS, AND THREE DIMENSIONS

Volume Formulas, Table 6.1, p. 415–416
Summary of Volumes of Revolution, Table 6.2, p. 429
Directory of Polar-Form Curves, Table 9.1, p. 639
Drawing lesson on sketching a prism, p. 418
Drawing lesson on plotting points in three dimensions, p. 678
Drawing lesson on drawing vertical planes, p. 679
Drawing lesson on sketching a surface, p. 681
Quadric Surfaces, Table 10.1, p. 683
Drawing lesson on lines in space, p. 706
Drawing lesson on planes and their traces, p. 711
Data on Planetary Orbits, Table 11.1, p. 754
Curvature Formulas, Table 11.2, p. 767
Summary of Velocity, Acceleration, and Curvature, Table 11.3, p. 786
Conversion formulas for coordinate systems, p. 939

FUNCTIONS OF TWO VARIABLES

Drawing lesson on drawing level curves, p. 796
Comparison of Important Integral Theorems, p. 1020

DIFFERENTIAL EQUATIONS

Introduction to differential equations, Section 5.6, p. 364
Growth and decay, see Table 7.3, p. 508
First-Order differential equations, Section 7.6, p. 503
Second-Order linear differential equations, Section 15.2, p. 1041
Summary of Strategies for Solving First-Order Differential Equations, p. 1036

Multivariable
CALCULUS
SECOND EDITION

Ceiling interior of the Saminid Mausoleum, built around 900 A.D. in Bukhara, Uzbekistan. This small cubical tomb (about 9 1/2 meters for the length of each side), built entirely of small yellow bricks, is the oldest surviving building, and probably the first, to demonstrate how to put a perfect hemisphere dome on a square base. The Iranian architects—Bukhara back then was part of a geographically larger Iranian culture—who designed this building were well versed in math, as were their later western counterparts who put up the medieval cathedrals of Europe, but these Islamic architects worked from a strong knowledge of geometry and algebra. The "squinch" above the corner of the square supports the weight of the dome—this was an innovation. Note the geometry in action: Take the midpoint of each side at the square base, and create eight line segments. Then take the midpoint of each of these eight line segments and connect adjacent midpoints at a higher level (you can see this in the photo above the squinch). So a four-sided object moves upward to one with eight sides. Repeat this process again and you have a 16-sided shape. Notice the circle of the hemisphere sits easily on the 16 sides. In summary, the square approaches the circle as a limit. In writing of another, far larger building with such a dome, built in 1088 in Isfahan, Iran, architecture critic Eric Schroeder has written: "European dome-builders never approached their skill. How ingeniously the Western builder compensated his ignorance of the mechanics of dome construction is attested by the ten chains round the base of St. Peter's, and the concealed cone which fastens the haunch of St. Paul's. But engineers could not hope to prescribe an ideally light dome of plain masonry before Newton's work on the calculus (late in the seventeenth century)." Read on, and upon completion of this text you will be qualified, at least mathematically, to do many marvelous things with the practical tool called calculus.

Multivariable
CALCULUS
SECOND EDITION

GERALD L. BRADLEY

Claremont McKenna College

KARL J. SMITH

Santa Rosa Junior College

PRENTICE HALL

Upper Saddle River, New Jersey 07458

Library of Congress has cataloged another edition as follows:

Bradley, Gerald L. (date)
 Calculus/Gerald L. Bradley, Karl J. Smith. — 2nd ed.
  p. cm.
 Includes index.
 ISBN 0-13-660135-9
 1. Calculus.   I. Smith, Karl J.   II. Title.
 QA303.B88218   1999   98-18962
 515--dc21   CIP

Acquisitions Editor: George Lobell
Editor-in-Chief: Jerome Grant
Editorial Director: Tim Bozik
Associate Editor-in-Chief, Development: Carol Trueheart
Production Editor: Barbara Mack
Senior Managing Editor: Linda Mihatov Behrens
Executive Managing Editor: Kathleen Schiaparelli
Assistant Vice President of Production and Manufacturing: David W. Riccardi
Marketing Manager: Melody Marcus
Manufacturing Buyer: Alan Fischer
Manufacturing Manager: Trudy Pisciotti
Supplements Editor/Editorial Assistant: Gale Epps
Art Director: Maureen Eide
Associate Creative Director: Amy Rosen
Director of Creative Services: Paula Maylahn
Assistant to Art Director: John Christiana
Art Manager: Gus Vibal
Art Editor: Rhoda Sidney
Interior Designer: Geri Davis, The Davis Group, Inc.
Cover Designer: Joseph Sengotta
Cover Photo: Louvre Museum and Pyramid. Tom Craig/FPG International
Photo Researcher: Beth Boyd
Photo Research Administrator: Melinda Reo
Art Studio: Monotype Composition
Historical Quest Portraits: Steven S. Nau
Copy Editor: Susan Reiland

 © 1999, 1995 by Prentice-Hall, Inc.
Simon & Schuster/A Viacom Company
Upper Saddle River, New Jersey 07458

Printed in the United States of America

10 9 8 7 6 5 4 3 2 1

ISBN 0-13-863945-0

Prentice-Hall International (UK) Limited, *London*
Prentice-Hall of Australia Pty. Limited, *Sydney*
Prentice-Hall Canada Inc., *Toronto*
Prentice-Hall Hispanoamericana, S.A., *Mexico*
Prentice-Hall of India Private Limited, *New Delhi*
Prentice-Hall of Japan, Inc., *Tokyo*
Simon & Schuster Asia Pte. Ltd., *Singapore*
Editora Prentice-Hall do Brasil, Ltda., *Rio de Janeiro*

Contents

*   **CHAPTER 1**   *Functions and Graphs*   **1**

Input value x
Square
Multiply by 5
Add 2
Output value $5x^2 + 2$

CHAPTER 2   *Limits and Continuity*   **73**

$y = |x| \sin \dfrac{1}{x}$

*The shaded Chapters 1–9 contents appear only in *Single Variable Calculus*. Chapter 10 appears in both *Single Variable Calculus* and *Multivariable Calculus*.

CHAPTER 6

Additional Applications of the Integral   405

CHAPTER 7

Methods of Integration   465

CHAPTER 8

Infinite Series   539

CHAPTER 9

Polar Coordinates and Parametric Forms   625

CHAPTER 14

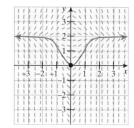

Vector Analysis   *957*

CHAPTER 15

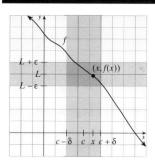

Introduction to Differential Equations   *1029*

APPENDICES

Foreword

Over the years I have seen and used a number of calculus texts. I have not always viewed "improvements" in these as real improvements, at least for my students. However, when I was asked to take a look at the manuscript for the 2nd edition of the Bradley/Smith text, I was struck by how readable it is and how basically friendly. It is by no means a "once-over-lightly" treatment of calculus, but at the same time it does not wear out the reader with excessive detail and mind-numbing rigor. This is, after all, a text intended for a course at a level that does not require all that much rigor. The explanations are careful without being pedantic and fussy, well illustrated with examples without exhausting the reader with an overabundance, and uncluttered with alternate ways of doing things.

In particular, I like the historical-biographical essays with associated projects and the calculator/computer projects, all placed where the instructor can use them or not, depending on taste or what the length of the term will allow. Some are provocative and open-ended, allowing for additional time to be spent by students who are ready for more of a challenge. In fact, this expandability would even allow the use of the text in an honors section while the text is also used for regular sections, thus allowing students to transfer in and out of honors sections during the calculus sequence.

The applications are interesting and wide-ranging. A number of them were new to me. Yet I was convinced that they would be understandable to students. In my experience with a number of recent calculus texts, some applications are so far from the experience of most of my students that they end up being mainly frustrating, rather than beneficial. Such texts also use language students don't know and ideas from other fields in which students have no background. This is not the case with Bradley/Smith.

The historical and biographical sketches, along with associated projects involving further reading and exploration, lend a humane touch to a subject this is often viewed by the unconverted as cold, mechanical, and not the work of real human beings. The authors here have done a nice job of making it clear that the calculus as we know it has been developed over many years by a variety of investigators. It is the work of some of the best scientific minds of history. And if the subject does not come easily and without help, it is not surprising. The subject is clear to us now but it is the result of centuries of human effort and ingenuity. Calculus remains one of the greatest of scientific achievements, not only a useful subject but one that forms a profound and cohesive body of knowledge. It's worth the effort it takes to master it.

Gerald L. Alexanderson
Santa Clara University

Preface

Calculus teaching is undergoing great changes, most of which will be of lasting benefit to students. We applaud teaching by using numerical, graphical, and analytical approaches to important concepts. The goal is clear student understanding of concepts, not simply the ability to "manipulate symbols." You will find this book rich with tables, graphs, and algebraic characterizations of each main concept. Our goal in writing this text was to blend the best aspects of calculus reform with the reasonable goals and methodology of traditional calculus. In incorporating so much of calculus reform, we made a deliberate effort not "to throw the baby out with the bath water." Calculus should not be a terminal course, but rather one that prepares students in engineering, science, math, and other related areas to move on to more advanced and necessary career or professional courses. Unlike some reform books, this text addresses topics such as continuity, the mean value theorem, l'Hopital's rule, parametric equations, polar coordinates, sequences, and series. In short, this text is an attempt at Reform with Reason.

The acceptance and response from our first edition has been most gratifying. In spite of the fact that many professors are reluctant to adopt a first edition textbook, we found widespread acceptance for our book and appreciate the many suggestions we received. With this second edition, we checked and rechecked the accuracy of the text material, and have taken extraordinary effort to ensure the accuracy of the answers. We incorporated many of your suggested improvements:

- This edition correctly reflects the precalculus mathematics being taught at most colleges and universities. We assume knowledge of the trigonometric functions, as well as e^x and $\ln x$. These functions are reviewed in Chapter 1. We also assume a knowledge of the conic sections and their graphs.

- It is possible to begin the course with Chapter 1 or Chapter 2 (where the calculus topics begin).

- Modeling was added as a major theme in this edition. Modeling is introduced in Section 2.1, and then appears in almost every section of the book. These applications are designed MODELING PROBLEMS.

- We have taken the introduction of differential equations seriously. Students in many allied disciplines need to use differential equations early in their studies, and consequently cannot wait for a post calculus course. In this edition, we introduce differential equations in a natural and reasonable way. Slope fields are introduced as a geometric view of antidifferentiation in Section 5.1, and then are used to introduce a graphical solution to differential equations in Section 5.6. We consider separable differential equations in Chapter 5 and first-order linear equations in Chapter 7, and demonstrate the use of both modeling a variety of applied situations. Exact and homogeneous differential equations appear in Chapter 15, along with an introduction to second-order linear equations. The "early and often" approach to differential equations is intended to illustrate their value in continuous modeling and to provide a solid foundation for further study.

- This edition offers an early presentation of transcendental functions: logarithmic, exponential, and trigonometric functions are heavily integrated into all chapters of this book (especially Chapters 1–5).

- We continue to exploit the *humanness* of mathematics, but instead of simply including *Historical Notes*, as we did in the first edition, we have transformed these notes into *problems* that lead the reader from the development of a concept to actually participating in the discovery process. The problems are designated as Historical Quest problems. The problems are not designed to be "add-on or challenge problems," but rather are intended to become an integral part of the usual assignment. The level of difficulty of Quest problems ranges from easy to difficult. An extensive selection of biographies of noted mathematicians can be found on the Internet site accompanying this text.

- We are aware that most calculus books grow "bigger and bigger" as they progress from one edition to the next. As they grow, they add "more features" and become more and more expensive for the student. We are determined that this does not happen with this book. In an effort to keep down costs, we made the decision that a full color edition, while visually appealing, does not add to *mathematical* understanding enough to justify the added cost. Consequently, this edition uses color functionally as a pointer and not as a decoration. We ask for your feedback regarding our decision.

The major issue driving calculus reform is the poor performance of students trying to master the concepts of calculus. Much of this failure can be attributed to the way most students learn mathematics in high school, which often involves stressing rote memorization over insight and understanding. On the other hand, some reform texts are perceived as spending so much time with the development of insight and understanding that students are not given enough exposure to important computational and problem-solving skills to perform well in more advanced courses. In our view, it is equally wrong to foster a situation in which the student understands too little about a lot or a lot about too little. This text aims at a middle ground by providing sound development, stimulating problems, and well-developed pedagogy within a framework of a traditional topic structure. "Think then do" is a fair summary of our approach.

Conceptual Understanding Through Verbalization

Besides developing some minimal skills in algebraic manipulation and problem solving, today's calculus text should require students to cultivate verbal skills in a mathematical setting. This is not just because real mathematics wields its words precisely and compactly, but because verbalization should help students think conceptually.

COOPERATIVE LEARNING (GROUP RESEARCH PROJECTS)   In July 1991, the National Science Foundation funded a group of instructors who met on the campus of New Mexico State University to discuss the topic, "Discovering Calculus through Student Projects." The participants concluded that encouraging students to work on significant projects in small groups acts as a counterbalance to traditional lecture methods and can serve to foster both conceptual understanding and the development of technical skills. However, many instructors still believe that mathematics can be learned only through independent work. We feel that independent work is of primary importance, but that students must also learn to work with others in group projects. After all, an individual's work in the "real world" is often done as part of a group and almost always involves solving problems for which there is no answer "in the back of the book." In response to this need, we have included challenging exercises (the Journal and Putnam problems) and a number of group research projects, each of which appears at the end of a chapter and involves intriguing questions whose mathematical content is tied loosely to the chapter just concluded. These projects have been developed and class-tested by their individual authors, to whom we are greatly indebted. Note that the complexity of these projects increases as we progress through the book, and the mathematical maturity of the student is developed.

MATHEMATICAL COMMUNICATION   We have included several opportunities for mathematical communication in terms that can be understood by nonprofessionals. The guest essays provide alternate viewpoints. The questions that follow are called MATHEMATICAL ESSAYS and are included to encourage individual writing assignments and mathematical exposition. We believe that students will benefit from individual writing and research in mathematics. Shorter problems encouraging written communication are included in the problem sets and are designed by the logo WHAT DOES THIS SAY? Another pedagogical feature is the **"What this says:"** box in which we rephrase mathematical ideas in everyday language. In the problem sets we encourage students to summarize procedures, processes, or to describe a mathematical result in everyday terms. Concept problems are found throughout the book as well as at the end of each chapter, and these problems, as well as the **mathematical essay** problems, are included to prove that mathematics is more than "working problems and getting answers." Mathematics education *must* include the communication of mathematical ideas.

Integration of Technology

COMPUTATIONAL WINDOWS   Reform is driven partly by the need to embrace the benefits technology brings to the learning of mathematics. Simply adding a lab course to the traditional calculus is possible, but this may lead to unacceptable work loads for all involved. We choose to include technology as an aid to the understanding of calculus, rather than to write a calculus course developed around the technology. While we have included over 130 windows devoted to the use of technology, we strive to keep such references "platform neutral" because specific calculators and computer programs frequently change and are better considered in separate technology manuals. The technology in the book is organized under the title "TECHNOLOGY WINDOWS" to give insight into how technological advances can be used to help understand calculus. TECHNOLOGY WINDOWS also appear in the exercisesand involve problems requiring a graphing calculator or software and computer. On the other hand, problems that are not specially designated may still use technology (for example to solve a higher-degree equation).

TECHNOLOGY LABORATORY MANUALS   For those with personal access to a computer, companion Technological Manuals (available wrapped with the text at a small charge) discuss TI graphing calculators, HP graphing calculators, *Mathematica*, *Maple*, and *MATLAB*.

SIGNIFICANT DIGITS   We have included a brief treatment in Appendix C. On occasion, we show the entire calculator or computer output of 12 digits for clarity even though such a display may exceed the requisite number of significant digits.

GREATER TEXT VISUALIZATION   Related to, but not exclusively driven by, the use of technology, is the greater use of graphs and other mathematical pictures throughout this text. Over 1,900 graphs appear—more than nearly any other calculus text. This increased visualization is intended to help develop greater student intuition. Much of this visualization appears in the wide margins to accompany the text. Its purpose is to provide explanation to supplement and/or replace that of the text prose. Also, since many tough calculus problems are often tough geometry (and algebra) problems, this increased emphasis on graphs will help students' problem-solving skills. Additional graphs are related to the student problems, including answer art.

Problem Solving

PROBLEMS   We believe that students *learn* mathematics by *doing* mathematics. Therefore, the problems and applications are perhaps the most important feature of any calculus book, and you find the problems in this book extend from routine practice to challenging. The problem sets are divided into *A* Problems (routine), *B* Problems (requiring independent thought), and *C* Problems (theory problems). In this book we also include past Putnam examination problems as well as problems found in current mathematical journals. You will find the scope and depth of the problems in this book to be extraordinary. Even though engineering and physics examples and problems play a prominent role, we include applications from a wide variety of fields such as biology, economics, ecology, psychology, and sociology. The problems have been in the development stages for over ten years and most of them have been class tested. In addition, the chapter summaries provide not only topical review, but also many miscellaneous exercises. Although the chapter reviews are typical of examinations, the miscellaneous problems are not presented as graded problems, but rather as a random list of problems loosely tied to the ideas of that chapter. In addition, there are cumulative reviews located at natural subdivision points in the text: Chapters 1-6, Chapters 7-11, and Chapters 12-15.

JOURNAL PROBLEMS   In an effort to show that "mathematicians work problems too," we have reprinted problems from leading mathematics journals. We have chosen problems which are within reason of the intended audience of this book. If students need help or hints for these problems, they can search out the original presentation and solution in the cited journal. In addition, we have included problems from the **Putnam Examinations**. These problems, which are more challenging, are offered in the miscellaneous problems at the end of various chapters and are provided to give insight into the type of problems that are asked in mathematical competitions. The annual national competition is given under the auspices of the Mathematical Association of America. The problems are designed to recognize mathematically talented college and university students.

THINK TANK PROBLEMS   It has been said that mathematical discovery is directed toward two major goals—the formulation of proofs and the construction of counterexamples. Most calculus books focus only on the first goal (the body of proofs and true statements), but we feel that some attention should be paid to the formulation of counterexamples for false statements. Throughout this book we ask the student to formulate an example satisfying certain conditions. We have designated this type of problem as a *think tank* problem.

Topics

STUDENT MATHEMATICS HANDBOOK   The content of this text adapts itself to either semester or quarter systems, and both differentiation and integration can be introduced in the first course. We begin calculus with a minimum of review. The prerequisite material which is often included in a calculus textbook has been bound separately in a companion book, *Student Mathematics Handbook*. Our handbook is offered *free of charge* with every *new* copy of the textbook. The Handbook not only includes the necessary review material and formulas, but also contains a catalog of curves and a complete integral table. We feel this is an important supplement to the textbook because we have found that the majority of errors our students make in a calculus class are not errors in calculus, but errors in basic algebra and trigonometry. A unified and complete treatment of this prerequisite material, easily referenced and keyed to the textbook, has been a valuable tool for our students taking calculus. Those portions of the text that benefit from an appropriate precalculus review are marked by the symbol (**SMH**)

SEQUENCE OF TOPICS   We resisted the temptation to label certain sections as optional, because that is a prerogative of individual instructors and schools. However, the following sections could be skipped without any difficulty: 4.7, 5.9, 6.5 (delay until Sec. 13.6), 7.8, 12.8, and 13.8. To assist instructors with the pacing of the course, we have written the material so each section reasonably can be covered in one classroom day, but to do so requires that the students read the text in order to tie together the ideas which might be discussed in a classroom setting.

PROOFS   One of the trends in the move to reform calculus is to minimize the role of a mathematical theorem. We do not agree with this aspect of reform. Precise reasoning has been, and we believe will continue to be, the backbone of good mathematics. While never sacrificing good pedagogy and student understanding, we present important results as *theorems*. We do not pretend to prove every theorem in this book; in fact we often only outline the steps one would take in proving a theorem. We refer the reader to Appendix B for certain longer proofs, or sometimes to an advanced calculus text. Why then do we include the heading "PROOF" after each theorem? It is because we want the student to know for a result to be a theorem there must be a proof. We use the heading not necessarily to give a complete proof in the text, but to give some direction to where a proof can be found, or an indication of how it can be constructed.

Supplementary Materials

Student Mathematics Handbook and Integration Table for CALCULUS by Karl J. Smith offers a review of prerequisite material, a catalog of curves, and a complete integral table. This handbook is presented free of charge along with the purchase of a new book.

A *Student Survival and Solutions Manual* by Karl J. Smith offers a running commentary of hints and suggestions to help ensure the students' success in calculus. Since this manual is written by one of the authors of this text, the solutions given in the manual complement all of the procedures and development of the textbook. The problem numbers in the book shown in color indicate the solutions included in the *Survival Manual*.

Technology Manuals by John Gresser offers computer applications keyed to the sections in the book. These manuals are identical except for the specific keystrokes on TI calculators, *Mathematica*, and *Maple*. There are similar manuals for the HP and MATLAB by Frank Hagin and Jack Cohen.

A Complete Solutions Manual by Karl J. Smith contains a brief solution for every problem in the book.

An Answer Book by Karl J. Smith contains only the answers to most problems in the book.

An Instructor's Guide offers sample tests and reviews for each chapter in the book. This guide also includes sample transparencies.

Computerized Computer Testing Program is available in both IBM and Macintosh formats.

Resources for Calculus, Volumes 1-5, A. Wayne Roberts (Project Director) available from the Mathematical Association of America, 1993:

> Dudley, Underwood (ed), *Readings for Calculus*, MAA Notes, No. 28
>
> Fraga, Robert (ed), *Calculus Problems for a New Century*, MAA Notes, No. 28
>
> Jackson, Michael B., and Ramsay, John (eds), *Problems for Student Investigation*, MAA Notes, No. 30
>
> Snow, Anita E. (ed), *Learning for Discovery: A Lab Manual for Calculus*, MAA Notes, No. 27
>
> Straffin, Philip (ed), *Applications of Calculus*, MAA Notes, No. 29

This is a valuable collection of resource materials for calculus instructors. All material may be reproduced for classroom use.

Acknowledgments

The writing and publishing of a calculus book is a tremendous undertaking. We take this responsibility very seriously because a calculus book is instrumental in transmitting knowledge from one generation to the next. We would like to thank the many people who helped us in the preparation of this book. First, we thank our editor, George Lobell, who led us masterfully through the development and publication of this book. We sincerely appreciate Donald Gecewicz, who read and critiqued each word of the manuscript. We also appreciate the work of Barbara Mack in college production, who kept us all on track, and we especially thank Susan Reiland for her meticulous attention to detail.

Of primary concern is the accuracy of the book. We had the assistance of many: Jerry Alexanderson and Mike Ecker, who read the entire manuscript and offered us many valuable suggestions; the accuracy checkers of the first edition, Ken Sydel, Diana Gerardi, Kurt Norlin, Terri Bittner, and Mary Toscano; and of the second edition, Nancy and Mary Toscano. We also would like to thank the following readers of the text for their many suggestions for improvement:

REVIEWERS OF THE SECOND EDITION:

Gerald Alexanderson, Santa Clara University; David Arterburn, New Mexico Tech; Linda A. Bolte, Eastern Washington University; Brian Borchers, New Mexico Tech; Mark Farris, Midwestern State University; Sally Fieschbeck, Rochester Institute of Technology; Mike Ecker, Pennsylvania State University, Wilkes-Barre Campus; Stuart Goldenberg, California Polytechnic Institute; Roger Jay, Tomball College; John H. Jenkins, Embry-Riddle Aeronautical University; Kathy Kepner, Paducah Community College; Daniel King, Sarah Lawrence College; Don Leftwich, Oklahoma Christian University; Ching Lu, Southern Illinois University at Edwardsville; Ann Morlet, Cleveland State University; Dena Jo Perkins, Oklahoma Christian University; Judith Reeves, California Polytechnic Institute; Jim Roznowski, Delta College; Lowell Stultz, Kalamazoo Valley Community College; Tingxiu Wang, Oakton Community College

REVIEWERS OF THE FIRST EDITION:

Neil Berger, University of Illinois at Chicago; Michael L. Berry, West Virginia Wesleyan College; Barbara H. Briggs, Tennessee Technical University; Robert Broschat, South Dakota State University; Robert D. Brown, University of Kansas; Dan Chiddix, Ricks College; Philip Crooke, Vanderbilt University; Ken Dunn, Dalhousie University; John H. Ellison, Grove City College; William P. Francis, Michigan Technological University; Harvey Greenwald, California Polytechnic San Luis Obispo; Richard Hitt, University of South Alabama; Joel W. Irish, University of Southern Maine; Clement T. Jeske, University of Wisconsin-Platteville; Lawrence Kratz, Idaho State University; Sam Lessing, Northeast Missouri University; Estela S. Llinas, University of Pittsburgh at Greensburg; Pauline Lawman, Western Kentucky University; William E. Mastrocola, Colgate University; Philip W. McCartney, Northern Kentucky University; E. D. McCune, Stephen F., Austin State University; John C. Michels, Chemeketa Community College; Pamela B. Pierce, College of Wooster; Connie Schrock, Emporia State University; Tatiana Shubin, San Jose State University; Tingxiu Wang, Oakton Community College

Gerald L. Bradley
Karl J. Smith

About the Authors

Gerald L. Bradley

Karl J. Smith

Gerald L. Bradley received his B.S. degree in mathematics from Harvey Mudd College in 1962, as a member of the second full graduating class, and his Ph.D. from the California Institute of Technology in 1966. A former NSF and Woodrow Wilson Fellow, Professor Bradley has taught at Claremont McKenna College since 1966, and twice has served as chairman of the mathematics department. His primary field of interest and research is matrix theory, and he is the author of *A Primer of Linear Algebra*, also with Prentice Hall. He has a strong interest in the development of undergraduate mathematics, and has co-authored a top-selling business calculus text as well as a basic text in finite mathematics. His personal interests include history, archaeology, and bridge, but he spends most of his spare time with his passion: writing science fiction novels.

Karl J. Smith received his B.A. and M.A. (in 1967) degrees in mathematics from UCLA. Then he moved in 1968 to northern California to teach at Santa Rosa Junior College, where he has been ever since. Along the way, he served as department chair, and he received a Ph.D. in 1979 in mathematics education at Southeastern University. A past president of the American Mathematical Association of Two-Year Colleges, Professor Smith is very active nationally in mathematics education. He was founding editor of *Western AMATYC News*, a chairperson of the committee on Mathematics Excellence, and a NSF grant reviewer. He was a recipient in 1979 of an Outstanding Young Men of America Award, in 1980 of an Outstanding Educator Award, and in 1989 of an Outstanding Teacher Award. Professor Smith is the author of several successful textbooks. In fact, over one million students have learned mathematics from his textbooks.

10 Vectors in the Plane and in Space

CONTENTS

PREVIEW

In this chapter, we focus on various algebraic and geometric aspects of vector representations. Then in Chapter 11, we see how vectors can be combined with calculus to study motion in space and other applications.

PERSPECTIVE

Suppose a child is pulling a sled by a rope across a flat field. Intuitively, we would expect quite different results if the child pulls the rope straight up than if the same effort is applied at an angle of $\pi/4$, for instance. In other words, to describe the force exerted by the child on the sled, we must specify not only a magnitude but also a direction. In general, a *scalar* quantity is one that can be described in terms of magnitude alone, whereas a *vector* quantity requires both magnitude and direction. Force, velocity, and acceleration are common vector quantities, and an important goal of this chapter is to develop methods for dealing with such quantities.

10.1 *Vectors in the Plane*

INTRODUCTION TO VECTORS

A **vector** in a plane can be thought of as a directed line segment, an "arrow" with **initial point** P and **terminal point** Q. The direction of the vector is that of the arrow, and its magnitude is represented by the arrow's length, as shown in Figure 10.1. We shall indicate such a vector by writing **PQ** in boldface type, but in your work you may write an arrow over the designated points: $\overrightarrow{PQ}$. The order of letters you write down is important: **PQ** means that the vector is from P to Q, but **QP** means that the vector is from Q to P. The first letter is the initial point and the second letter is the terminal point. We shall denote the magnitude (length) of a vector by $\|\mathbf{PQ}\|$. Two vectors are regarded as **equal** (or **equivalent**) if they have the same magnitude and the same direction, even if they do not coincide.

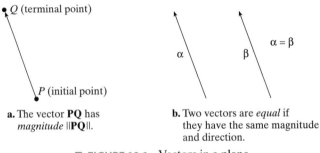

a. The vector **PQ** has *magnitude* $\|\mathbf{PQ}\|$.

b. Two vectors are *equal* if they have the same magnitude and direction.

■ **FIGURE 10.1** Vectors in a plane

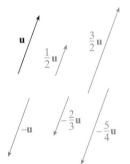

■ **FIGURE 10.2** Some multiples of the vector **u**

A vector with magnitude 0 is called a **zero** (or **null**) **vector** and is denoted by **0**. The **0** vector has no specific direction, and we shall adopt the convention of assigning it any direction that may be convenient in a particular problem.

If **v** is a vector other than **0**, then any vector **w** that is parallel to **v** is called a **scalar multiple** of **v** and satisfies $\mathbf{w} = s\mathbf{v}$ for some nonzero number s. A **scalar** quantity is one that has only magnitude and, in the context of vectors, is used to describe a real number. The scalar multiple $s\mathbf{v}$ has length $|s|$ times that of **v**; it points in the same direction as **v** if $s > 0$ and the opposite direction if $s < 0$ (see Figure 10.2). Notice that for any distinct points P and Q, $\mathbf{PQ} = -\mathbf{QP}$. For the zero vector **0**, we define $s\mathbf{0} = \mathbf{0}$ for any scalar s.

Physical experiments indicate that force and velocity vectors can be added (or **resolved**) according to a **triangular rule** displayed in Figure 10.3a, and we use this rule as our definition of vector addition. In particular, to add the vector **v** to the vector **u**, we place the end (initial point) of **v** at the tip (terminal point) of **u** and define the **sum**, also called the **resultant, u + v**, to be the vector that extends from the initial point of **u** to the terminal point of **v**.

Equivalently, **u + v** is the diagonal of the parallelogram formed with sides **u** and **v**, as shown in Figure 10.3b. The **difference u − v** is just the vector **w** that satisfies **v + w = u**, and it may be found by placing the initial points of **u** and **v** together and extending a vector from the terminal point of **v** to the terminal point of **u** (see Figure 10.3c).

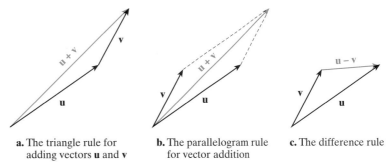

a. The triangle rule for adding vectors **u** and **v**

b. The parallelogram rule for vector addition

c. The difference rule

■ **FIGURE 10.3**   Vector addition and subtraction

A vector **OQ** with initial point at the origin O of a coordinate plane can be uniquely represented by specifying the coordinates of its terminal point Q. If Q has coordinates (a, b), we denote the vector **OQ** by $\langle a, b \rangle$, where the pointed brackets $\langle \ \rangle$ are used to distinguish the *vector* **OQ** $= \langle a, b \rangle$ from the *point* (a, b).

Technology Window

Brackets are used to represent vectors on some calculators and on most programs, such as *Mathematica*, *Maple*, *Derive*, and *Mathlab*. For example, the vector **QB** is input as $[a, b]$. This is called a *two-dimensional vector* because it has two components. In this book, we consider vectors with three or more components, and these are represented on a calculator in a similar way:

$$\mathbf{V} = [2, 3, 4] \qquad \mathbf{W} = [0, -3, \tfrac{\pi}{2}]$$

V and **W** are called *three-dimensional vectors*.

The vector **PQ** with initial point $P(c, d)$ and terminal point $Q(a, b)$ can be denoted by **PQ** $= \langle a - c, b - d \rangle$. Using analytic geometry (see Problem 55), it can be shown that **PQ** equals vector **OR** with initial point at the origin $(0, 0)$ and terminal point $R(a - c, b - d)$, as shown in Figure 10.4.

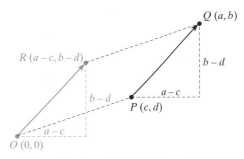

■ **FIGURE 10.4**   Given $P(c, d)$ and $Q(a, b)$, the vector **PQ** is $\langle a - c, b - d \rangle$.

Vector operations are easily represented when vectors are given in component form. In particular, we have

$$\langle a_1, b_1 \rangle = \langle a_2, b_2 \rangle \qquad \text{if and only if } a_1 = a_2 \text{ and } b_1 = b_2$$
$$k\langle a, b \rangle = \langle ka, kb \rangle \qquad \text{for constant } k$$
$$\langle a, b \rangle + \langle c, d \rangle = \langle a + c, b + d \rangle$$
$$\langle a, b \rangle - \langle c, d \rangle = \langle a - c, b - d \rangle$$

These formulas may be verified by analytic geometry. For instance, the rule for multiplication by a scalar may be obtained by using the relationships in Figure 10.5a, and Figure 10.5b can be used to obtain the rule for vector addition.

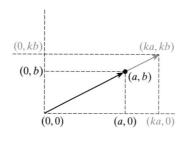
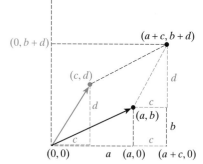

a. Multiplication by a scalar:
$k\langle a, b\rangle = \langle ka, kb\rangle$

b. Vector addition
$\langle a, b\rangle + \langle c, d\rangle = \langle a + c, b + d\rangle$

■ **FIGURE 10.5**   Vector operations

EXAMPLE 1   *Vector operations*

For the vectors $\mathbf{u} = \langle 2, -3\rangle$ and $\mathbf{v} = \langle -1, 7\rangle$ find:
a. $\mathbf{u} + \mathbf{v}$ b. $\frac{3}{4}\mathbf{u}$ c. $3\mathbf{u} - \frac{1}{2}\mathbf{v}$

Solution
a. $\mathbf{u} + \mathbf{v} = \langle 2, -3\rangle + \langle -1, 7\rangle = \langle 2 + (-1), -3 + 7\rangle = \langle 1, 4\rangle$
b. $\frac{3}{4}\mathbf{u} = \frac{3}{4}\langle 2, -3\rangle = \langle \frac{3}{4}(2), \frac{3}{4}(-3)\rangle = \langle \frac{3}{2}, \frac{-9}{4}\rangle$
c. $3\mathbf{u} - \frac{1}{2}\mathbf{v} = 3\langle 2, -3\rangle - \frac{1}{2}\langle -1, 7\rangle$

$\qquad = \langle 6, -9\rangle + \langle \frac{1}{2}, -\frac{7}{2}\rangle$ *Scalar multiplication*

$\qquad = \langle 6 + \frac{1}{2}, -9 - \frac{7}{2}\rangle$ *Add vectors.*

$\qquad = \langle \frac{13}{2}, -\frac{25}{2}\rangle$ *Simplify.*   ■

In general, an expression of the form $a\mathbf{u} + b\mathbf{v}$ is called a **linear combination** of the vectors $\mathbf{u}$ and $\mathbf{v}$. Note that if $\mathbf{u} = \langle u_1, u_2\rangle$ and $\mathbf{v} = \langle v_1, v_2\rangle$, then

$$a\mathbf{u} + b\mathbf{v} = a\langle u_1, u_2\rangle + b\langle v_1, v_2\rangle = \langle au_1 + bv_1, au_2 + bv_2\rangle$$

Vector addition and multiplication of a vector by a scalar behave much like ordinary addition and multiplication. The following theorem lists several useful properties of these operations.

═══════════════

THEOREM 10.1   *Properties of vector operations*

For any vectors $\mathbf{u}, \mathbf{v}$, and $\mathbf{w}$ in the plane and scalars s and t:

Commutativity of vector addition	$\mathbf{u} + \mathbf{v} = \mathbf{v} + \mathbf{u}$
Associativity of vector addition	$(\mathbf{u} + \mathbf{v}) + \mathbf{w} = \mathbf{u} + (\mathbf{v} + \mathbf{w})$
Associativity of scalar multiplication	$(st)\mathbf{u} = s(t\mathbf{u})$
Identity for addition	$\mathbf{u} + \mathbf{0} = \mathbf{u}$
Inverse property for addition	$\mathbf{u} + (-\mathbf{u}) = \mathbf{0}$
Distributivity laws	$(s + t)\mathbf{u} = s\mathbf{u} + t\mathbf{u}$
	$s(\mathbf{u} + \mathbf{v}) = s\mathbf{u} + s\mathbf{v}$

Proof   Each vector property can be established by using a corresponding property of real numbers. For example, to prove associativity of vector addition, let $\mathbf{u} = \langle u_1, u_2 \rangle$, $\mathbf{v} = \langle v_1, v_2 \rangle$, and $\mathbf{w} = \langle w_1, w_2 \rangle$. Then

$$
\begin{aligned}
(\mathbf{u} + \mathbf{v}) + \mathbf{w} &= (\langle u_1, u_2 \rangle + \langle v_1, v_2 \rangle) + \langle w_1, w_2 \rangle \\
&= \langle u_1 + v_1, u_2 + v_2 \rangle + \langle w_1, w_2 \rangle \\
&= \langle (u_1 + v_1) + w_1, (u_2 + v_2) + w_2 \rangle \\
&= \langle u_1 + (v_1 + w_1), u_2 + (v_2 + w_2) \rangle \\
&\qquad\qquad \textit{Associativity of addition for the real numbers} \\
&= \langle u_1, u_2 \rangle + (\langle v_1, v_2 \rangle + \langle w_1, w_2 \rangle) \\
&= \mathbf{u} + (\mathbf{v} + \mathbf{w})
\end{aligned}
$$

You are asked to prove the other six properties in the problem set.   ═

EXAMPLE 2   *Vector proof of a geometric property*

Show that the line segment joining the midpoints of two sides of a triangle is parallel to the third side and has half its length.

Solution   Consider $\triangle ABC$, and let P and Q be the midpoints of sides $\overline{AC}$ and $\overline{BC}$, respectively, as shown in Figure 10.6.

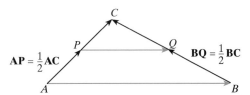

■ **FIGURE 10.6**   Vector proof of a geometric property

Given: $\mathbf{AP} = \frac{1}{2}\mathbf{AC}$ and $\mathbf{BQ} = \frac{1}{2}\mathbf{BC}$.

Proof: We want to show that $\mathbf{PQ}$ is parallel to $\mathbf{AB}$ and $\|\mathbf{PQ}\| = \frac{1}{2}\|\mathbf{AB}\|$, which means that we must establish the vector equation $\mathbf{PQ} = \frac{1}{2}\mathbf{AB}$. Toward this end, we begin by noting that $\mathbf{AB}$ can be expressed as the following vector sum:

$$
\begin{aligned}
\mathbf{AB} &= \mathbf{AP} + \mathbf{PQ} + \mathbf{QB} \\
&= \tfrac{1}{2}\mathbf{AC} + \mathbf{PQ} - \mathbf{BQ} && \mathbf{AP} = \tfrac{1}{2}\mathbf{AC} \textit{ and } \mathbf{QB} = -\mathbf{BQ} \\
&= \tfrac{1}{2}(\mathbf{AB} + \mathbf{BC}) + \mathbf{PQ} - \tfrac{1}{2}\mathbf{BC} && \mathbf{AC} = (\mathbf{AB} + \mathbf{BC}) \textit{ and } \mathbf{BQ} = \tfrac{1}{2}\mathbf{BC} \\
&= \tfrac{1}{2}\mathbf{AB} + \tfrac{1}{2}\mathbf{BC} + \mathbf{PQ} - \tfrac{1}{2}\mathbf{BC} \\
&= \tfrac{1}{2}\mathbf{AB} + \mathbf{PQ} \\
\tfrac{1}{2}\mathbf{AB} &= \mathbf{PQ} && \textit{Subtract } \tfrac{1}{2}\mathbf{AB} \textit{ from both sides.}\quad ■
\end{aligned}
$$

When a vector $\mathbf{u}$ is represented in component from $\mathbf{u} = \langle u_1, u_2 \rangle$, its length is given by the formula

$$
\|\mathbf{u}\| = \sqrt{u_1^2 + u_2^2}
$$

This is a simple application of the Pythagorean theorem, as shown in Figure 10.7. Another important relationship involving the length of vectors is the *triangle inequality*

$$
\|\mathbf{u} + \mathbf{v}\| \le \|\mathbf{u}\| + \|\mathbf{v}\|
$$

for any vectors **u** and **v**. Equality will occur precisely when **u** and **v** are multiples of one another (that is, when **u** and **v** have the same direction). To establish the inequality, we observe that **u** and **v** are two sides of a triangle in the plane, and that the third side has length $\|\mathbf{u} + \mathbf{v}\|$ and is "shorter" than the sum $\|\mathbf{u}\| + \|\mathbf{v}\|$ of the lengths of the other two sides, as shown in Figure 10.7.

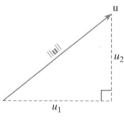
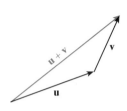

a. The vector $\mathbf{u} = (u_1, u_2)$ has length
$$\|\mathbf{u}\| = \sqrt{u_1^2 + u_2^2}$$

b. The triangle inequality
$$\|\mathbf{u} + \mathbf{v}\| \le \|\mathbf{u}\| + \|\mathbf{v}\|$$

■ **FIGURE 10.7** Geometric representation of two vector properties

EXAMPLE 3	*Speed and heading of a motorboat*

A river 4 mi wide flows south with a current of 5 mi/h. What speed and heading should a motorboat assume to travel directly across the river from east to west in 20 min?

Solution   Begin by drawing a diagram, as shown in Figure 10.8.

Let **B** be the velocity vector of the boat in the direction of the angle θ. If the river's current has velocity **C**, the given information tells us that $\|\mathbf{C}\| = 5$ mi/h and that **C** points directly south. Moreover, because the boat is to cross the river from east to west in 20 min (that is, $\frac{1}{3}$ h), its *effective velocity* after compensating for the current is a vector **V** that points west and has magnitude

$$\|\mathbf{V}\| = \frac{\text{WIDTH OF THE RIVER}}{\text{TIME OF CROSSING}} = \frac{4 \text{ mi}}{\frac{1}{3} \text{ h}} = 12 \text{ mi/h}$$

The effective velocity **V** is the resultant of **B** and **C**; that is, $\mathbf{V} = \mathbf{B} + \mathbf{C}$. Because **V** and **C** act in perpendicular directions, we can determine **B** by referring to the right triangle with sides $\|\mathbf{V}\| = 12$ and $\|\mathbf{C}\| = 5$ and hypotenuse $\|\mathbf{B}\|$. We find that

$$\|\mathbf{B}\| = \sqrt{\|\mathbf{V}\|^2 + \|\mathbf{C}\|^2} = \sqrt{12^2 + 5^2} = 13$$

The direction of the velocity vector **V** is given by the angle θ in Figure 10.8, and we find that

$$\tan \theta = \frac{5}{12} \quad \text{so that} \quad \theta = \tan^{-1}\left(\frac{5}{12}\right) \approx 0.3948$$

Thus, the boat should travel at 13 mi/h in a direction of approximately 0.3948 radian. In navigation, it is common to specify direct in degrees rather than radians; this is 22.6° north of west.   ■

A **unit vector** is simply a vector with length 1, and a **direction vector** for a given vector **v** is a unit vector **u** that points in the same direction as **v**. Such a vector can be found by dividing **v** by its length $\|\mathbf{v}\|$; that is,

$$\mathbf{u} = \frac{\mathbf{v}}{\|\mathbf{v}\|}$$

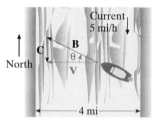

■ **FIGURE 10.8** A velocity problem

EXAMPLE 4   *Finding a direction vector*

Find a direction vector for the vector $\mathbf{v} = \langle 2, -3 \rangle$.

Solution   The vector $\mathbf{v}$ has length (magnitude) $\|\mathbf{v}\| = \sqrt{2^2 + (-3)^2} = \sqrt{13}$. Thus, the required direction vector is the unit vector

$$\mathbf{u} = \frac{\mathbf{v}}{\|\mathbf{v}\|} = \frac{\langle 2, -3 \rangle}{\sqrt{13}} = \frac{1}{\sqrt{13}} \langle 2, -3 \rangle = \left\langle \frac{2}{\sqrt{13}}, \frac{-3}{\sqrt{13}} \right\rangle$$

STANDARD REPRESENTATION OF VECTORS IN THE PLANE

The unit vectors $\mathbf{i} = \langle 1, 0 \rangle$ and $\mathbf{j} = \langle 0, 1 \rangle$ point in the directions of the positive x- and y-axes, respectively, and are called **standard basis vectors.** Any vector $\mathbf{v} = \langle v_1, v_2 \rangle$ in the plane can be expressed as a linear combination of the vectors $\mathbf{i}$ and $\mathbf{j}$, because

$$\mathbf{v} = \langle v_1, v_2 \rangle = v_1 \langle 1, 0 \rangle + v_2 \langle 0, 1 \rangle = v_1 \mathbf{i} + v_2 \mathbf{j}$$

This is called the **standard representation** of the vector $\mathbf{v}$, and it can be shown that the representation is unique in the sense that if $\mathbf{v} = a\mathbf{i} + b\mathbf{j}$, then $a = v_1$ and $b = v_2$. In this context, the scalars v_1 and v_2 are called the **horizontal and vertical components of v**, respectively. See Figure 10.9.

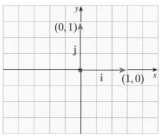
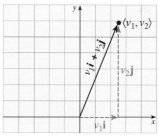

a. The standard basis vectors $\mathbf{i}$ and $\mathbf{j}$

b. Any vector $\mathbf{v} = \langle v_1, v_2 \rangle$ can be expressed uniquely as $\mathbf{v} = v_1 \mathbf{i} + v_2 \mathbf{j}$

■ **FIGURE 10.9** Standard representation of vectors in the plane

EXAMPLE 5   *Finding the standard representation of a vector*

If $\mathbf{u} = 3\mathbf{i} + 2\mathbf{j}$, $\mathbf{v} = -2\mathbf{i} + 5\mathbf{j}$, and $\mathbf{w} = \mathbf{i} - 4\mathbf{j}$, what is the standard representation of the vector $2\mathbf{u} + 5\mathbf{v} - \mathbf{w}$?

Solution   Using Theorem 10.1, we find that

$$\begin{aligned}
2\mathbf{u} + 5\mathbf{v} - \mathbf{w} &= 2(3\mathbf{i} + 2\mathbf{j}) + 5(-2\mathbf{i} + 5\mathbf{j}) - (\mathbf{i} - 4\mathbf{j}) \\
&= [2(3) + 5(-2) - 1]\mathbf{i} + [2(2) + 5(5) - (-4)]\mathbf{j} \\
&= -5\mathbf{i} + 33\mathbf{j}
\end{aligned}$$

EXAMPLE 6   *Finding standard representation of a vector connecting two points*

Find the standard representation of the vector **PQ** for the points $P(3, -4)$ and $Q(-2, 6)$.

Solution   The component form of **PQ** is

$$\mathbf{PQ} = \langle (-2) - 3, 6 - (-4) \rangle = \langle -5, 10 \rangle$$

This means that **PQ** has the standard representation $\mathbf{PQ} = -5\mathbf{i} + 10\mathbf{j}$.

> **EXAMPLE 7**   *Computing a resultant force*

Two forces $\mathbf{F}_1$ and $\mathbf{F}_2$ act on the same body. It is known that $\mathbf{F}_1$ has magnitude 3 newtons and acts in the direction of $-\mathbf{i}$, whereas $\mathbf{F}_2$ has magnitude 2 newtons and acts in the direction of the unit vector

$$\mathbf{u} = \tfrac{3}{5}\mathbf{i} - \tfrac{4}{5}\mathbf{j}$$

What additional force $\mathbf{F}_3$ must be applied to keep the body at rest?

Solution   According to the given information, we have

$$\mathbf{F}_1 = 3(-\mathbf{i}) = -3\mathbf{i} \quad \text{and} \quad \mathbf{F}_2 = 2(\tfrac{3}{5}\mathbf{i} - \tfrac{4}{5}\mathbf{j}) = \tfrac{6}{5}\mathbf{i} - \tfrac{8}{5}\mathbf{j}$$

and we want to find $\mathbf{F}_3 = a\mathbf{i} + b\mathbf{j}$ so that $\mathbf{F}_1 + \mathbf{F}_2 + \mathbf{F}_3 = \mathbf{0}$. Substituting into this vector equation, we obtain

$$(-3\mathbf{i}) + (\tfrac{6}{5}\mathbf{i} - \tfrac{8}{5}\mathbf{j}) + (a\mathbf{i} + b\mathbf{j}) = 0\mathbf{i} + 0\mathbf{j}$$

By combining terms on the left, we find that

$$(-3 + \tfrac{6}{5} + a)\mathbf{i} + (-\tfrac{8}{5} + b)\mathbf{j} = 0\mathbf{i} + 0\mathbf{j}$$

Because the standard representation is unique, we must have

$$-3 + \tfrac{6}{5} + a = 0 \quad \text{and} \quad -\tfrac{8}{5} + b = 0$$
$$a = \tfrac{9}{5} \qquad\qquad\qquad b = \tfrac{8}{5}$$

The required force is $\mathbf{F}_3 = \tfrac{9}{5}\mathbf{i} + \tfrac{8}{5}\mathbf{j}$. This is a force of magnitude

$$\|\mathbf{F}_3\| = \sqrt{\left(\frac{9}{5}\right)^2 + \left(\frac{8}{5}\right)^2} = \frac{1}{5}\sqrt{145} \text{ newtons}$$

which acts in the direction of the unit vector

$$\mathbf{v} = \frac{\mathbf{F}_3}{\|\mathbf{F}_3\|} = \frac{5}{\sqrt{145}}\left(\frac{9}{5}\mathbf{i} + \frac{8}{5}\mathbf{j}\right) = \frac{9}{\sqrt{145}}\mathbf{i} + \frac{8}{\sqrt{145}}\mathbf{j}$$ ∎

10.1 Problem Set

A *Sketch each vector given in Problems 1–4 assuming that its initial point is at the origin.*

  1. $3\mathbf{i} - 4\mathbf{j}$ **2.** $-2\mathbf{i} - 3\mathbf{j}$

  3. $-\tfrac{1}{2}\mathbf{i} + \tfrac{5}{2}\mathbf{j}$ **4.** $-2(-\mathbf{i} + 2\mathbf{j})$

The initial point P and terminal point Q of a vector are given in Problems 5–8. Sketch each vector and then write it in component form.

  5. $P(3, -1), Q(7, 2)$ **6.** $P(5, -2), Q(5, 8)$

  7. $P(3, 4), Q(-2, 4)$ **8.** $P(\tfrac{1}{2}, 6), Q(-3, -2)$

Express each vector $\mathbf{PQ}$ in Problems 9–12 in standard form and also find its length.

  9. $P(-1, -2), Q(1, -2)$ **10.** $P(5, 7), Q(6, 8)$

 11. $P(-4, -3), Q(0, -1)$ **12.** $P(3, -5), Q(2, 8)$

Find a unit vector that points in the direction of each of the vectors given in Problems 13–16.

 13. $\mathbf{i} + \mathbf{j}$ **14.** $\tfrac{1}{2}\mathbf{i} + \tfrac{1}{4}\mathbf{j}$

 15. $3\mathbf{i} - 4\mathbf{j}$ **16.** $-4\mathbf{i} + 7\mathbf{j}$

Let $\mathbf{u} = \langle -3, 4 \rangle$ and $\mathbf{v} = \langle 1, -1 \rangle$. Find scalars s and t so that the given equation in Problems 17–20 is satisfied.

 17. $s\mathbf{u} + t\mathbf{v} = \langle 6, 0 \rangle$ **18.** $s\langle 0, -3 \rangle + t\mathbf{u} = \mathbf{v}$

 19. $s\mathbf{v} + t\langle -2, 1 \rangle = \mathbf{u}$ **20.** $s\mathbf{u} + \langle 8, 11 \rangle = t\mathbf{v}$

Suppose $\mathbf{u} = 3\mathbf{i} - 4\mathbf{j}$, $\mathbf{v} = 4\mathbf{i} - 3\mathbf{j}$, and $\mathbf{w} = \mathbf{i} + \mathbf{j}$. Express each of the expressions in Problems 21–24 in standard form.

 21. $2\mathbf{u} + 3\mathbf{v} - \mathbf{w}$ **22.** $\tfrac{1}{2}(\mathbf{u} + \mathbf{v}) - \tfrac{1}{4}\mathbf{w}$

 23. $\|\mathbf{v}\|\mathbf{u} + \|\mathbf{u}\|\mathbf{v}$ **24.** $\|\mathbf{u}\|\|\mathbf{v}\|\mathbf{w}$

Find all real numbers x and y that satisfy the vector equations given in Problems 25–28.

25. $(x - y - 1)\mathbf{i} + (2x + 3y - 12)\mathbf{j} = \mathbf{0}$

26. $x\mathbf{i} - 4y^2\mathbf{j} = (5 - 3y)\mathbf{i} + (10 - 7x)\mathbf{j}$

27. $(x^2 + y^2)\mathbf{i} + y\mathbf{j} = 20\mathbf{i} + (x + 2)\mathbf{j}$

28. $(y - 1)\mathbf{i} + y\mathbf{j} = (\log x)\mathbf{i} + [\log 2 + \log(x + 4)]\mathbf{j}$

*In Problems 29–32, find a unit vector **u** with the given characteristics.*

29. **u** makes an angle of $30°$ with the positive x-axis.

30. **u** has the same direction as the vector $2\mathbf{i} - 3\mathbf{j}$.

31. **u** has the direction opposite that of $-4\mathbf{i} + \mathbf{j}$.

32. **u** has the direction of the vector from $P(-1, 5)$ to $Q(7, -3)$.

In Problems 33–36, let $\mathbf{u} = 4\mathbf{i} - \mathbf{j}$, $\mathbf{v} = \mathbf{i} + 2\mathbf{j}$, and $\mathbf{w} = -3\mathbf{i} + 4\mathbf{j}$.

33. Find a unit vector in the same direction as $\mathbf{u} + \mathbf{v}$.

34. Find a vector of length 3 with the same direction as $\mathbf{u} - 2\mathbf{v} + 2\mathbf{w}$.

35. Find the terminal point of the vector $5\mathbf{i} + 7\mathbf{j}$ if the initial point is $(-2, 3)$.

36. Find the initial point of the vector $-\mathbf{i} + 2\mathbf{j}$ if the terminal point is $(-1, -2)$.

37. a. Use vectors to find the coordinates of the midpoint of the line segment joining the points $P(-3, -8)$ and $Q(9, -2)$.

b. What point is located $\frac{5}{6}$ of the distance from P to Q?

38. If $\|\mathbf{v}\| = 3$ and $-3 \le r \le 1$, what are the possible values of $r\|\mathbf{v}\|$?

39. Show that $\mathbf{v} = (\cos \theta)\mathbf{i} + (\sin \theta)\mathbf{j}$ is a unit vector for any angle θ.

40. If **u** and **v** are nonzero vectors and

$$r = \frac{\|\mathbf{u}\|}{\|\mathbf{v}\|}$$

What is $\|r\mathbf{v}\|$?

B **41.** If **u** and **v** are nonzero vectors with $\|\mathbf{u}\| = \|\mathbf{v}\|$, does it follow that $\mathbf{u} = \mathbf{v}$? Explain.

42. If $\mathbf{u} = 2\mathbf{i} - 3\mathbf{j}$ and $\mathbf{v} = x\mathbf{i} + y\mathbf{j}$, describe the set of points in the plane whose coordinates (x, y) satisfy $\|\mathbf{v} - \mathbf{u}\| \le 2$.

43. Let $\mathbf{u}_0 = x_0\mathbf{i} + y_0\mathbf{j}$ for constants x_0 and y_0, and let $\mathbf{u} = x\mathbf{i} + y\mathbf{j}$. Describe the set of all points in the plane whose coordinates satisfy:

a. $\|\mathbf{u} - \mathbf{u}_0\| = 1$ **b.** $\|\mathbf{u} - \mathbf{u}_0\| \le 2$

44. Let $\mathbf{u} = 3\mathbf{i} - \mathbf{j}$ and $\mathbf{v} = -6\mathbf{i} + 2\mathbf{j}$. Show that there are no numbers a, b for which $a\mathbf{u} + b\mathbf{v} = 2\mathbf{i} + 5\mathbf{j}$.

45. Suppose **u** and **v** are a pair of nonzero, nonparallel vectors. Find numbers a, b, c such that $a\mathbf{u} + b(\mathbf{u} - \mathbf{v}) + c(\mathbf{u} + \mathbf{v}) = \mathbf{0}$.

46. Let $\mathbf{u} = \langle 2, 1 \rangle$ and $\mathbf{v} = \langle -3, 4 \rangle$.

a. Sketch the vector $c\mathbf{u} + (1 - c)\mathbf{v}$ for the cases where $c = 0$, $c = \frac{1}{4}$, $c = \frac{1}{2}$, $c = \frac{3}{4}$, and $c = 1$.

b. In general, if the initial point of $c\mathbf{u} + (1 - c)\mathbf{v}$ for $0 \le c \le 1$ is at the origin, where is its terminal point (x, y)?

47. Two forces $\mathbf{F}_1 = 3\mathbf{i} + 4\mathbf{j}$ and $\mathbf{F}_2 = 3\mathbf{i} - 7\mathbf{j}$ act on an object. What additional force should be applied to keep the body at rest?

48. Three forces $\mathbf{F}_1 = \mathbf{i} - 2\mathbf{j}$, $\mathbf{F}_2 = 3\mathbf{i} - 7\mathbf{j}$, and $\mathbf{F}_3 = \mathbf{i} + \mathbf{j}$ act on an object. What additional force $\mathbf{F}_4$ should be applied to keep the body at rest?

49. A river 2.1 mi wide flows south with a current of 3.1 mi/h. What speed and heading should a motorboat assume to travel across the river from east to west in 30 min?

50. Four forces act on an object: $\mathbf{F}_1$ has magnitude 10 lb and acts at an angle of $\frac{\pi}{6}$ measured counterclockwise from the positive x-axis; $\mathbf{F}_2$ has magnitude 8 lb and acts in the direction of the vector $\mathbf{j}$; $\mathbf{F}_3$ has magnitude 5 lb and acts at an angle of $4\pi/3$ measured counterclockwise from the positive x-axis. What must the fourth force $\mathbf{F}_4$ be to keep the object at rest?

C **51.** Use vector methods to show that the diagonals of a parallelogram bisect each other.

52. In a triangle, let **u**, **v**, and **w** be the vectors from each vertex to the midpoint of the opposite side. Use vector methods to show that $\mathbf{u} + \mathbf{v} + \mathbf{w} = \mathbf{0}$.

53. Two nonzero vectors **u** and **v** are said to be **linearly independent** in the plane if they are not parallel.

a. If **u** and **v** have this property and $a\mathbf{u} = b\mathbf{v}$ for constants a, b, show that $a = b = 0$.

b. Show that the standard representation of a vector is unique. That is, if the vector **u** has the representation $\mathbf{u} = a_1\mathbf{i} + b_1\mathbf{j}$ and $\mathbf{u} = a_2\mathbf{i} + b_2\mathbf{j}$ is another such representation, then $a_1 = a_2$ and $b_1 = b_2$.

54. Prove that the medians of a triangle intersect at a single point by completing the following argument.

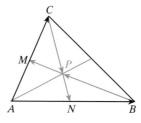

a. Let M and N be the midpoints of sides $\overline{AC}$ and $\overline{AB}$, respectively. Show that

$$\mathbf{CN} = \tfrac{1}{2}\mathbf{AB} - \mathbf{AC} \quad \text{and} \quad \mathbf{BM} = \tfrac{1}{2}\mathbf{AC} - \mathbf{AB}$$

b. Let P be the point where medians $\overline{BM}$ and $\overline{CN}$ intersect, and let r, s be constants such that

$$\mathbf{CP} = r\,\mathbf{CN} \quad \text{and}$$

$$\mathbf{BP} = s\,\mathbf{BM}$$

Note that **CP** + **PB** = **CB**. Use this relationship to prove that $r = s = \dfrac{2}{3}$. Explain why this shows that any pair of medians meet at a point located $\dfrac{2}{3}$ the distance from each vertex to the midpoint of the opposite side. Why does this show that *all three* medians meet at a single point?

c. The *centroid* of a triangle is the point where the medians meet. Show that a triangle with vertices $A(x_1, y_1)$, $B(x_2, y_2)$, $C(x_3, y_3)$, has centroid with coordinates

$$P\left(\frac{x_1 + x_2 + x_3}{3}, \frac{y_1 + y_2 + y_3}{3}\right)$$

55. Show that the vector with initial point $P(c, d)$ and terminal point $Q(a, b)$ has the component form **PQ** = $\langle a - c, b - d\rangle$. See Figure 10.4.

56. Prove the following parts of Theorem 10.1.
   a. Commutativity
   b. Identity
   c. Inverse property of addition
   d. The distributive law
      $s(\mathbf{u} + \mathbf{v}) = s\mathbf{u} + s\mathbf{v}$

57. Let $A, B, C,$ and D be any four points in the plane. If M and N are midpoints of $\overline{AC}$ and $\overline{BD}$, show that

$$\overrightarrow{MN} = \tfrac{1}{4}(\overrightarrow{AB} + \overrightarrow{AD} + \overrightarrow{CB} + \overrightarrow{CD})$$

58. Let $A, B, C,$ and D be vertices of a quadrilateral, and let $M, N, P,$ and Q be the midpoints of the sides $\overline{AB}, \overline{BC}, \overline{CD},$ and $\overline{AD}$, respectively. Use vector methods to show that $MNPQ$ is a parallelogram.

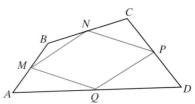

59. **SPY PROBLEM**   The Spy finally locates Blohardt's submarine (Problem 60, Section 9.4) and forces it to surface, but inside he finds only a note attached to the periscope. There are map coordinates, a crude drawing, and these words:

"From the snowman, pace off the distance to the woodpile, turn left, pace off an equal distance, then drive a stake. From the snowman again, pace off the distance to the flagpole, turn right, pace off an equal distance and drive a second stake. Dig halfway between the stakes."

A bloody letter "P" marks the place on the map between the stakes. "Purity!" gasps the Spy. Gripped by trepidation, he arrives at the location indicated by the map coordinates. There is a woodpile and a flagpole, but the snowman has melted! How can the Spy find where to dig?

10.2   Quadric Surfaces and Graphing in Three Dimensions

IN THIS SECTION   **three-dimensional coordinate system, graphs in** $\mathbb{R}^3$ **(planes, spheres, cylinders, quadric surfaces)**

THREE-DIMENSIONAL COORDINATE SYSTEM

Our next goal is to see how analytic geometry and vector methods can be applied in space. We have already considered ordered pairs and a two-dimensional coordinate system. We shall denote this two-dimensional system by $\mathbb{R}^2$. Because we exist in a three-dimensional world, it is also important to consider a three-dimensional system. We call this *three-space* and denote it by $\mathbb{R}^3$. We introduce a coordinate system to three-space by choosing three mutually perpendicular axes to serve as a frame of reference. The orientation of our reference system will be *right-handed* in the sense that if you stand at the origin with your right arm along the positive x-axis and your left arm along the positive y-axis, as shown in Figure 10.10, your head will then point in the direction of the positive z-axis.

To orient yourself to a three-dimensional coordinate system, think of the x-axis and y-axis as lying in the plane of the floor and the z-axis as a line perpendicular to the floor. All the graphs that we have drawn in the first nine chapters of this book would now be drawn on the floor. If you orient yourself in a room (your classroom, for

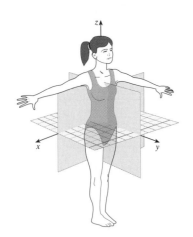

■ **FIGURE 10.10**   A "right-handed" rectangular coordinate system for $\mathbb{R}^3$

example) as shown in Figure 10.11, you may notice some important planes. Assume that the room is 25 ft × 30 ft × 8 ft and fix the origin at a front corner (where the board hangs).

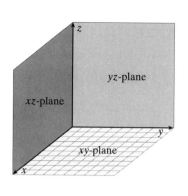

■ **FIGURE 10.11**   A typical classroom; assume the dimensions are 25 ft by 30 ft with an 8-foot ceiling.

Floor: ***xy*-plane;** equation is $z = 0$.
Ceiling: plane parallel to the *xy*-plane; equation is $z = 8$.
Front wall: ***yz*-plane;** equation is $x = 0$.
Back wall: plane parallel to the *yz*-plane; equation is $x = 30$.
Left wall: ***xz*-plane;** equation is $y = 0$.
Right wall: plane parallel to the *xz*-plane; equation is $y = 25$.

The *xy*-, *xz*-, and *yz*-planes are called the **coordinate planes.** Points in $\mathbb{R}^3$ are located by their position in relation to the three coordinate planes and are given appropriate coordinates. Specifically, the point P is assigned coordinates (a, b, c) to indicate that it is a, b, and c units, respectively, from the *yz*-, *xz*-, and *xy*-planes. Name the coordinates of several objects in your classroom (or in Figure 10.11).

| **EXAMPLE 1** | *Points in three dimensions* |

Graph the following ordered triplets:
a. $(10, 20, 10)$ b. $(-12, 6, 12)$ c. $(-12, -18, 6)$ d. $(20, -10, 18)$

Solution

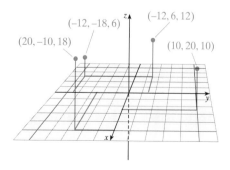

In Example 1, we measured distances in the *x*-, *y*-, and *z*-directions. We will, however, also need to measure distances between points in $\mathbb{R}^3$. The formula for distance in $\mathbb{R}^2$ easily extends to $\mathbb{R}^3$ (see Problem 52).

Distance Formula

The distance $|P_1P_2|$ between $P_1(x_1, y_1, z_1)$ and $P_2(x_2, y_2, z_2)$ is

$$|P_1P_2| = \sqrt{(x_2 - x_1)^2 + (y_2 - y_1)^2 + (z_2 - z_1)^2}$$

For instance, the distance between $(10, 20, 10)$ and $(-12, 6, 12)$ is

$$d = \sqrt{(-12 - 10)^2 + (6 - 20)^2 + (12 - 10)^2} = \sqrt{684} = 6\sqrt{19}$$

GRAPHS IN $\mathbb{R}^3$

The **graph of an equation** in $\mathbb{R}^3$ is the collection of all points (x, y, z) whose coordinates satisfy a given equation. This graph is called a **surface.** You are not expected to spend a great deal of time graphing three-dimensional surfaces, but the drawing lessons in this section should help. You may also have access to a computer program to help you look at graphs in three dimensions. We will discuss lines and planes more thoroughly in Section 10.5, but it is worthwhile to begin with a brief introduction to certain surfaces in $\mathbb{R}^3$.

Assignment:
Plot: $P(3, 4, -5)$

DRAWING LESSON 2: PLOTTING POINTS	
a. Sketch x-axis and y-axis, adding tickmarks. Outline the xy-plane.	**b.** Sketch z-axis, adding tickmarks. Use dashed segments for hidden parts.
c. Plot x-distance and y-distance; darken segments from each along gridlines. Colored pencil or highlighter may help you visualize the figure.	**d.** Plot z-distance, using the unit size from the z-axis. Lightly sketch a grid on the xy-plane, using tickmarks as guides.

Planes   We shall obtain equations for planes in space after we discuss vectors in space. However, in beginning to visualize objects in $\mathbb{R}^3$, we do not want to ignore planes, because they are so common (for example, the walls, ceiling, and floor in Figure 10.11). In Section 10.5, we will show that the graph of $ax + by + cz = d$ is a **plane** if $a, b, c,$ and d are real numbers (not all zero).

DRAWING LESSON 3: DRAWING VERTICAL PLANES

Assignment:
To drawing lesson
2, add planes
$x = 2$ and $y = 0$.

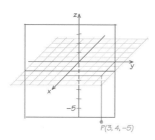

a. Draw a segment of the line $x = 2$ on the *xy*-plane. Through each endpoint, draw a segment parallel to the *z*-axis. Then connect the endpoints.

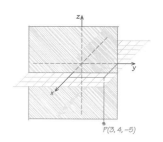

b. Shade the plane $x = 2$ where it is not hidden by the *xy*-plane. Erase hidden parts of both planes, and use your eraser to dash hidden parts of the axes.

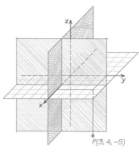

c. Follow the same procedure to draw and shade the plane $y = 0$. Draw the intersection of the two planes.

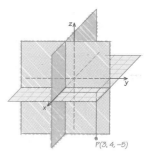

d. Use colored pencils or highlighters to distinguish individual planes.

EXAMPLE 2   *Graphing planes*

Graph the planes defined by the given equations.
a. $x + 3y + 2z = 6$
b. $y + z = 5$
c. $x = 4$

Solution   To graph a plane, find some ordered triplets satisfying the equation. The best ones to use are often those that fall on a coordinate axis (the intercepts).
a. Let $x = 0$ and $y = 0$; then $z = 3$; plot the point $(0, 0, 3)$.
   Let $x = 0$ and $z = 0$; then $y = 2$; plot the point $(0, 2, 0)$.
   Let $y = 0$ and $z = 0$; then $x = 6$; plot the point $(6, 0, 0)$.
   Use these points to draw the intersection lines (called **trace lines**) of the plane you are graphing with each of the coordinate planes. The result is shown in Figure 10.12a.
b. When one of the variables is missing from an equation of a plane, then that plane is parallel to the axis corresponding to the missing variable; thus, $y + z = 5$ is parallel to the *x*-axis. Draw the line $y + z = 5$ on the *yz*-plane, and then complete the plane, as shown in Figure 10.12b.
c. When two variables are missing, then the plane is parallel to the coordinate plane determined by the missing numbers. In this case, $x = 4$ is parallel to the *yz*-plane, as shown in Figure 10.12c.

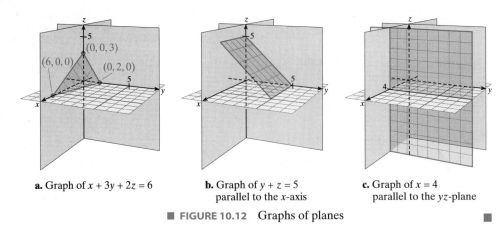

a. Graph of $x + 3y + 2z = 6$

b. Graph of $y + z = 5$ parallel to the x-axis

c. Graph of $x = 4$ parallel to the yz-plane

■ **FIGURE 10.12**   Graphs of planes   ■

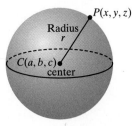

■ **FIGURE 10.13**   Graph of a sphere with center (a, b, c), radius r

Spheres   A **sphere** is defined as the collection of all points located a fixed distance (the **radius**) from a fixed point (the **center**). In particular, if $P(x, y, z)$ is a point on the sphere with radius r and center $C(a, b, c)$, then the distance from C to P is r. Thus,

$$r = \sqrt{(x - a)^2 + (y - b)^2 + (z - c)^2}$$

If you square both sides of this equation, you can see that it is equivalent to the equation of a sphere displayed in the following box. Conversely, if the point (x, y, z) satisfies an equation of this form, it must lie on a sphere with center (a, b, c) and radius r.

Equation of a Sphere

The graph of the equation

$$(x - a)^2 + (y - b)^2 + (z - c)^2 = r^2$$

is a sphere with center (a, b, c) and radius r, and any sphere has an equation of this form. This is called the **standard form of the equation of a sphere** (or simply *standard-form sphere*).

EXAMPLE 3   *Center and radius of a sphere from a given equation*

Show that the graph of the equation $x^2 + y^2 + z^2 + 4x - 6y - 3 = 0$ is a sphere, and find its center and radius.

Solution   By completing the square in both variables x and y, we have

$$(x^2 + 4x) + (y^2 - 6y) + z^2 = 3$$
$$(x^2 + 4x + 2^2) + [y^2 - 6y + (-3)^2] + z^2 = 3 + 4 + 9$$
$$(x + 2)^2 + (y - 3)^2 + z^2 = 16$$

Comparing this equation with the standard form, we see that it is the equation of a sphere with center $(-2, 3, 0)$ and radius 4.   ■

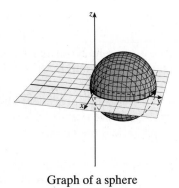

Graph of a sphere

DRAWING LESSON 4: SKETCHING A SURFACE

ASSIGNMENT:

Graph $z = \dfrac{1}{1 + x^2 + y^2}$

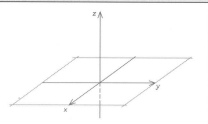

a. Draw the xy-plane in three dimensions, adding the z-axis.

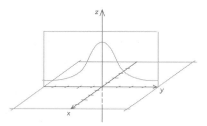

b. Draw a trace in one of the coordinate planes (in this case, the plane $x = 0$). If necessary, adjust the z-scale to show the trace more clearly.

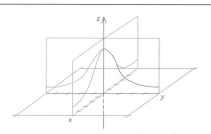

c. Draw a trace in another coordinate plane (in this case, the plane $y = 0$).

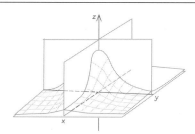

d. Erase all hidden lines. Draw several additonal trace curves to reveal the contours of the surface.

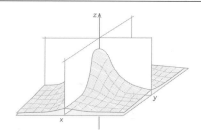

e. Erase all hidden lines. Use highlighters or pencils to color the surface and the xy-plane.

Cylinders   A **cross section** of a surface in space is a curve obtained by intersecting the surface with a plane. If parallel planes intersect a given surface in congruent cross-sectional curves, the surface is called a **cylinder.** We define a cylinder with *principal cross sections C and generating line L* to be the surface obtained by moving lines parallel to L along the boundary of the curve C, as shown in Figure 10.14. In this context, the curve C is called a **directrix** of the cylinder, and L is the **generatrix.**

We shall deal primarily with cylinders in which the directrix is a conic section and the generatrix L is one of the coordinate axes. Such a cylinder is often named for the type of conic section in its prinicipal cross sections, and is described by an equation involving only two of the variables x, y, z. In this case, the generating line L is parallel to

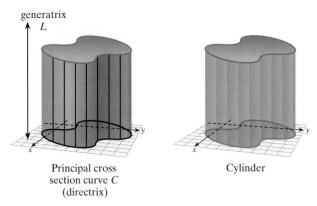

■ FIGURE 10.14   A cylinder with directrix C and generatrix L

the coordinate axis of the missing variable. Thus,

$$x^2 + y^2 = 5 \qquad \text{is a \textbf{circular cylinder} with } L \text{ parallel to the } z\text{-axis}$$
$$y^2 - z^2 = 9 \qquad \text{is a \textbf{hyperbolic cylinder} with } L \text{ parallel to the } x\text{-axis}$$
$$x^2 + 2z^2 = 25 \qquad \text{is an \textbf{elliptic cylinder} with } L \text{ parallel to the } y\text{-axis.}$$

The graphs of these cylinders are shown in Figure 10.15.

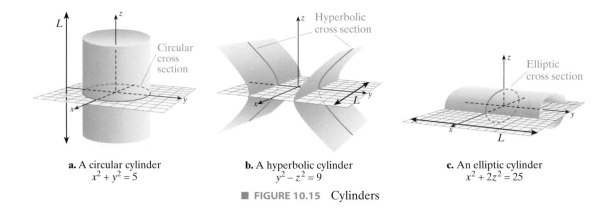

a. A circular cylinder
$x^2 + y^2 = 5$

b. A hyperbolic cylinder
$y^2 - z^2 = 9$

c. An elliptic cylinder
$x^2 + 2z^2 = 25$

■ FIGURE 10.15   Cylinders

Quadric Surfaces   Spheres and elliptic, parabolic, and hyperbolic cylinders are examples of **quadric surfaces.** In general, such a surface is the graph of an equation of the form

$$Ax^2 + By^2 + Cz^2 + Dxy + Exz + Fyz + Gx + Hy + Iz + J = 0$$

Quadric surfaces may be thought of as the generalizations of the conic sections in $\mathbb{R}^3$. The **trace** of a curve is found by setting one of the variables equal to a constant and then graphing the resulting curve. If $x = k$ (k is a constant), the resulting curve is drawn in the plane $x = k$, which is parallel to the yz-plane; and if $z = k$, the curve is drawn in the plane $z = k$, parallel to the xy-plane. Table 10.1 shows the quadric surfaces.

■ **TABLE 10.1   Quadric Surfaces**

Surface	Description	Surface	Description
Elliptic cone 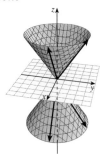	The trace in the xy-plane is a point; in planes parallel to the xy-plane, it is an ellipse. Traces in the xz- and yz-planes are intersecting lines; in planes parallel to these, they are hyperbolas: $$z^2 = \frac{x^2}{a^2} + \frac{y^2}{b^2}$$	*Elliptic paraboloid* 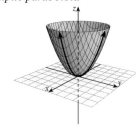	The trace in the xy-plane is a point; in planes parallel to the xy-plane, it is an ellipse. Traces in the xz- and yz-planes are parabolas. $$z = \frac{x^2}{a^2} + \frac{y^2}{b^2}$$
Hyperboloid of one sheet 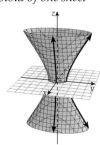	The trace in the xy-plane is an ellipse; in the xz- and yz-planes, the traces are hyperbolas. $$\frac{x^2}{a^2} + \frac{y^2}{b^2} - \frac{z^2}{c^2} = 1$$	*Hyperboloid of two sheets* 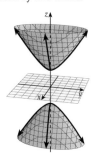	There is no trace in the xy-plane. In planes parallel to the xy-plane that intersect the surface, the traces are ellipses. Traces in the xz- and yz-planes are hyperbolas. $$\frac{x^2}{a^2} + \frac{y^2}{b^2} - \frac{z^2}{c^2} = -1$$
Ellipsoid 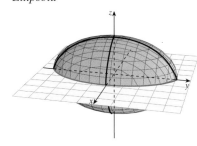	The traces in the coordinate planes are ellipses. $$\frac{x^2}{a^2} + \frac{y^2}{b^2} + \frac{z^2}{c^2} = 1$$	*Hyperbolic paraboloid,* also called a *saddle* 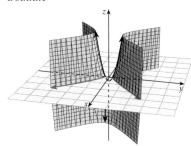	The trace in the xy-plane is two intersecting lines; in planes parallel to the xy-plane, the traces are hyperbolas. Traces in the xz- and yz-planes are parabolas. $$z = \frac{y^2}{b^2} - \frac{x^2}{a^2}$$
Sphere 	The sphere is a special kind of ellipsoid for which $a = b = c = r$. $$x^2 + y^2 + z^2 = r^2$$		

| | EXAMPLE 4 | *Identifying and sketching a quadric surface* |

Identify and sketch the surface with equation $9x^2 - 16y^2 + 144z = 0$.

Solution   Look at Table 10.1 on page 683 and note that the equation is second degree in x and y but first degree in z. This means it is an elliptic paraboloid or a hyperbolic paraboloid. Solve the equation for z:

$$9x^2 - 16y^2 + 144z = 0$$
$$144z = 16y^2 - 9x^2$$
$$z = \frac{y^2}{9} - \frac{x^2}{16}$$

We recognize this as a hyperbolic paraboloid.

Next, we take cross sections of $9x^2 - 16y^2 + 144z = 0$:

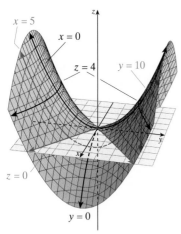

Cross Section	Chosen Value	Equation of Trace	Description of Trace
xy-plane	$z = 0$	$\frac{y^2}{9} - \frac{x^2}{16} = 0$	Two intersecting lines
parallel to xy-plane	$z = 4$	$\frac{y^2}{36} - \frac{x^2}{64} = 1$	Hyperbola
xz-plane	$y = 0$	$z = -\frac{x^2}{16}$	Parabola opens down
parallel to xz-plane	$y = 10$	$z - \frac{100}{9} = -\frac{x^2}{16}$	Parabola opens down
yz-plane	$x = 0$	$z = \frac{y^2}{9}$	Parabola opens up
parallel to yz-plane	$x = 5$	$z + \frac{25}{16} = \frac{y^2}{9}$	Parabola opens up

■ **FIGURE 10.16** Graph of $9x^2 - 16y^2 + 144z = 0$

These traces are also shown in Figure 10.16.   ■

10.2 Problem Set

Ⓐ *In Problems 1–4, plot the points P and Q in $\mathbb{R}^3$, and find the distance $|\overrightarrow{PQ}|$.*

1. $P(3, -4, 5), Q(1, 5, -3)$ **2.** $P(0, 3, 0), Q(-2, 5, -7)$
3. $P(-3, -5, 8), Q(3, 6, -7)$ **4.** $P(0, 5, -3), Q(2, -1, 0)$

In Problems 5–8, find the standard-form equation of the sphere with the given center C and radius r.

5. $C(0, 0, 0), r = 1$ **6.** $C(-3, 5, 7), r = 2$
7. $C(0, 4, -5), r = 3$ **8.** $C(-2, 3, -1), r = \sqrt{5}$

Find the center and radius of each sphere whose equations are given in Problems 9–12.

9. $x^2 + y^2 + z^2 - 2y + 2z - 2 = 0$
10. $x^2 + y^2 + z^2 + 4x - 2z - 8 = 0$

11. $x^2 + y^2 + z^2 - 6x + 2y - 2z + 10 = 0$
12. $x^2 + y^2 + z^2 - 2x - 4y + 8z + 17 = 0$

In Problems 13–22, match the equation with its graph (A–L).

13. $x^2 = z^2 + y^2$ **14.** $z^2 = \frac{x^2}{4} + \frac{y^2}{9}$

15. $\frac{x^2}{2} - \frac{y^2}{4} + \frac{z^2}{9} = 1$ **16.** $\frac{x^2}{9} + \frac{y^2}{16} - \frac{z^2}{4} = 1$

17. $x^2 + y^2 + z^2 = 9$ **18.** $y = x^2 + z^2$

19. $x = \frac{y^2}{25} + \frac{z^2}{16}$ **20.** $y = \frac{z^2}{4} - \frac{x^2}{9}$

21. $y^2 + z^2 - x^2 = -1$ **22.** $\frac{x^2}{4} - \frac{y^2}{9} + \frac{z^2}{9} = -1$

A.

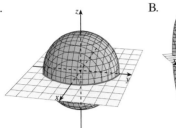

B.

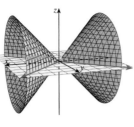

J.

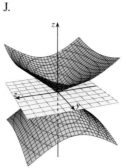

K.

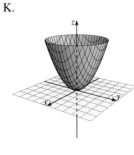

C.

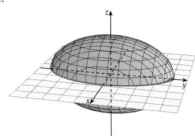

L.

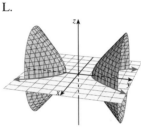

D.

E.

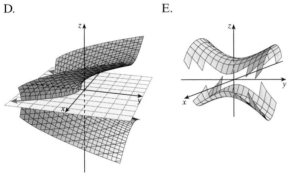

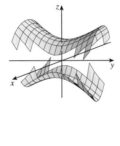

F.

G.

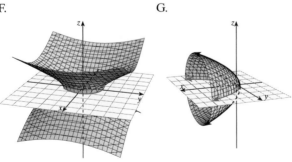

H.

I.

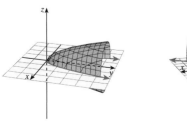

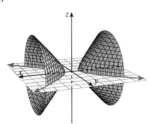

B *The vertices A, B, and C of a triangle in space are given in Problems 23–26. Find the lengths of the sides of the triangle and determine whether it is a right triangle, an isosceles triangle, both, or neither.*

23. $A(3, -1, 0)$, $B(7, 1, 4)$, $C(1, 3, 4)$

24. $A(1, 1, 1)$, $B(3, 3, 2)$, $C(3, -3, 5)$

25. $A(1, 2, 3)$, $B(-3, 2, 4)$, $C(1, -4, 3)$

26. $A(2, 4, 3)$, $B(-3, 2, -4)$, $C(-6, 8, -10)$

27. ■ **What Does This Say?** Describe a procedure for sketching a quadric surface.

28. ■ **What Does This Say?** Describe a procedure for identifying a quadric surface by looking at its equation.

In Problems 29–40, sketch the graph of each equation in $\mathbb{R}^3$.

29. $2x + y + 3z = 6$ 30. $x = 4$

31. $x + 2y + 5z = 10$ **32.** $x + y + z = 1$

33. $3x - 2y - z = 12$ 34. $y = z^2$

35. $x = -1$ **36.** $y^2 + z^2 = 1$

37. $z = e^y$ 38. $z = \ln x$

39. $x + z = 1$ 40. $z = y^{-1}$

In Problems 41–48, identify the quadric surface and describe the traces. Sketch the graph.

41. $9x^2 + 4y^2 + z^2 = 1$ **42.** $\dfrac{x^2}{4} + y^2 + \dfrac{z^2}{9} = 1$

43. $\dfrac{x^2}{4} + \dfrac{y^2}{9} - z^2 = 1$ 44. $\dfrac{x^2}{9} - y^2 - z^2 = 1$

45. $z = x^2 + \dfrac{y^2}{4}$ 46. $z = \dfrac{x^2}{9} - \dfrac{y^2}{16}$

47. $x^2 + 2y^2 = 9z^2$

48. $z^2 = 1 + \dfrac{x^2}{9} + \dfrac{y^2}{4}$

49. Find an equation for a sphere, given that the endpoints of a diameter of the sphere are $(1, 2, -3)$ and $(-2, 3, 3)$.

Ⓒ 50. Find the point P that lies $\frac{2}{3}$ of the distance from the point $A(-1, 3, 9)$ to the midpoint of the line segment joining points $B(-2, 3, 7)$ and $C(4, 1, -3)$.

51. Let $P(3, 2, -1)$, $Q(-2, 1, c)$, and $R(c, 1, 0)$ be points in $\mathbb{R}^3$. For what values of c (if any) is PQR a right triangle?

52. Derive the formula for the distance between $P_1(x_1, y_1, z_1)$ and $P_2(x_2, y_2, z_2)$. (*Hint:* Project the segment $\overline{P_1P_2}$ onto the xy-plane, and then use the Pythagorean theorem.)

10.3   The Dot Product

IN THIS SECTION   vectors in $\mathbb{R}^3$, definition of dot product, angle between vectors, projections, work as a dot product

VECTORS IN $\mathbb{R}^3$

A vector in $\mathbb{R}^3$ may be thought of as a directed line segment (an "arrow") in space. The vector $\mathbf{P_1P_2}$ with initial point $P_1(x_1, y_1, z_1)$ and terminal point $P_2(x_2, y_2, z_2)$ has the component form

$$\mathbf{P_1P_2} = \langle x_2 - x_1, y_2 - y_1, z_2 - z_1 \rangle$$

Vector addition and multiplication of a vector by a scalar are defined for vectors in $\mathbb{R}^3$ in essentially the same way as these operations were defined for vectors in $\mathbb{R}^2$. In addition, the properties of vector algebra listed in Theorem 10.1 of Section 10.1 apply to vectors in $\mathbb{R}^3$ as well as to those in $\mathbb{R}^2$.

For example, we observed that each vector in $\mathbb{R}^2$ can be expressed as a unique linear combination of the standard basis vectors $\mathbf{i}$ and $\mathbf{j}$. This representation can be extended to vectors in $\mathbb{R}^3$ by adding a vector $\mathbf{k}$ defined to be the unit vector in the direction of the positive z-axis. In component form, we have in $\mathbb{R}^3$,

$$\mathbf{i} = \langle 1, 0, 0 \rangle \qquad \mathbf{j} = \langle 0, 1, 0 \rangle \qquad \mathbf{k} = \langle 0, 0, 1 \rangle$$

We call these the **standard basis vectors** in $\mathbb{R}^3$. The **standard representation** of the vector with initial point at the origin O and terminal point $Q(a_1, a_2, a_3)$ is $\mathbf{OQ} = a_1\mathbf{i} + a_2\mathbf{j} + a_3\mathbf{k}$, as shown in Figure 10.17b.

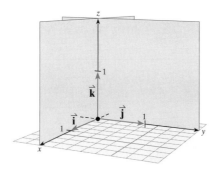

a. The standard basis vectors in $\mathbb{R}^3$

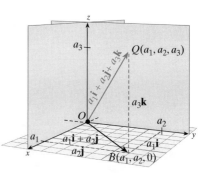

b. The vector from O to $Q(a_1, a_2, a_3)$ is $\mathbf{OQ} = a_1\mathbf{i} + a_2\mathbf{j} + a_3\mathbf{k}$.

■ **FIGURE 10.17**   Standard representation of vectors in $\mathbb{R}^3$

The vector **PQ** with initial point $P(x_0, y_0, z_0)$ and terminal point $Q(x_1, y_1, z_1)$ has the standard representation

$$\mathbf{PQ} = (x_1 - x_0)\mathbf{i} + (y_1 - y_0)\mathbf{j} + (z_1 - z_0)\mathbf{k}$$

EXAMPLE 1   *Standard representation of a vector in* $\mathbb{R}^3$

Find the standard representation of the vector **PQ** with initial point $P(-1, 2, 2)$ and terminal point $Q(3, -2, 4)$.

Solution   We have

$$\mathbf{PQ} = [3 - (-1)]\mathbf{i} + [-2 - 2]\mathbf{j} + [4 - 2]\mathbf{k} = 4\mathbf{i} - 4\mathbf{j} + 2\mathbf{k}$$ ∎

By referring to Figure 10.17b, we can also derive a formula for the length of a vector, which in turn can be used to establish the distance formula (stated in the previous section) between any two points in $\mathbb{R}^3$. Specifically, note that $\triangle OBQ$ in Figure 10.17b is a right triangle with hypotenuse $\|\mathbf{OQ}\|$ and legs $\|\mathbf{BQ}\| = |a_3|$ and $\|\mathbf{OB}\| = \sqrt{a_1^2 + a_2^2}$; by applying the Pythagorean theorem, we conclude that the vector $\mathbf{OQ} = a_1\mathbf{i} + a_2\mathbf{j} + a_3\mathbf{k}$ has length

$$\|\mathbf{OQ}\| = \sqrt{\|\mathbf{OB}\|^2 + \|\mathbf{BQ}\|^2} = \sqrt{(a_1^2 + a_2^2) + a_3^2} = \sqrt{a_1^2 + a_2^2 + a_3^2}$$

Moreover, if $A(a_1, a_2, a_3)$ and $B(b_1, b_2, b_3)$ are any two points in $\mathbb{R}^3$, the distance between them is the length of the vector **AB**. We find

$$\mathbf{AB} = \mathbf{OB} - \mathbf{OA} = (b_1 - a_1)\mathbf{i} + (b_2 - a_2)\mathbf{j} + (b_3 - a_3)\mathbf{k}$$

so that the distance between A and B is given by

$$\|\mathbf{AB}\| = \sqrt{(b_1 - a_1)^2 + (b_2 - a_2)^2 + (b_3 - a_3)^2}$$

Magnitude of a Vector

The **magnitude,** or length, of the vector $\mathbf{v} = a_1\mathbf{i} + a_2\mathbf{j} + a_3\mathbf{k}$ is

$$\|\mathbf{v}\| = \sqrt{a_1^2 + a_2^2 + a_3^2}$$

EXAMPLE 2   *Magnitude of a vector*

Find the magnitude of the vector $\mathbf{v} = 2\mathbf{i} - 3\mathbf{j} + 5\mathbf{k}$ and the distance between the points $A(1, -1, -4)$ and $B(-2, 3, 8)$.

Solution $\|\mathbf{v}\| = \sqrt{2^2 + (-3)^2 + 5^2} = \sqrt{38}$ and

$$\|\overline{AB}\| = \sqrt{(-2 - 1)^2 + [3 - (-1)]^2 + [8 - (-4)]^2} = 13$$ ∎

As in $\mathbb{R}^2$, if **v** is a given nonzero vector in $\mathbb{R}^3$, then a unit vector **u** that points in the same direction as **v** is

$$\mathbf{u} = \frac{\mathbf{v}}{\|\mathbf{v}\|}$$

EXAMPLE 3   *Finding a direction vector*

Find a unit vector that points in the direction of the vector **PQ** from $P(-1, 2, 5)$ to $Q(0, -3, 7)$.

Solution $\quad$ $\mathbf{PQ} = [0 - (-1)]\mathbf{i} + [-3 - 2]\mathbf{j} + [7 - 5]\mathbf{k} = \mathbf{i} - 5\mathbf{j} + 2\mathbf{k}$

$$\|\mathbf{PQ}\| = \sqrt{1^2 + (-5)^2 + 2^2} = \sqrt{30}$$

Thus,

$$\mathbf{u} = \frac{\mathbf{PQ}}{\|\mathbf{PQ}\|} = \frac{\mathbf{i} - 5\mathbf{j} + 2\mathbf{k}}{\sqrt{30}} = \frac{1}{\sqrt{30}}\mathbf{i} - \frac{5}{\sqrt{30}}\mathbf{j} + \frac{2}{\sqrt{30}}\mathbf{k}$$

■

As in $\mathbb{R}^2$, two vectors in $\mathbb{R}^3$ are **parallel** if they are multiples of one another. Nonzero vectors $\mathbf{u}$ and $\mathbf{v}$ are parallel if and only if $\mathbf{u} = s\mathbf{v}$ for some nonzero scalar s.

> **EXAMPLE 4** $\qquad$ *Parallel vectors*

A vector $\mathbf{PQ}$ has initial point $P(1, 0, -3)$ and length 3. Find Q so that $\mathbf{PQ}$ is parallel to $\mathbf{v} = 2\mathbf{i} - 3\mathbf{j} + 6\mathbf{k}$.

Solution $\quad$ Let Q have coordinates (a_1, a_2, a_3). Then

$$\mathbf{PQ} = [a_1 - 1]\mathbf{i} + [a_2 - 0]\mathbf{j} + [a_3 - (-3)]\mathbf{k} = (a_1 - 1)\mathbf{i} + a_2\mathbf{j} + (a_3 + 3)\mathbf{k}$$

Because $\mathbf{PQ}$ is parallel to $\mathbf{v}$, we have $\mathbf{PQ} = s\mathbf{v}$ for some scalar s; that is,

$$(a_1 - 1)\mathbf{i} + a_2\mathbf{j} + (a_3 + 3)\mathbf{k} = s(2\mathbf{i} - 3\mathbf{j} + 6\mathbf{k})$$

Thanks to the uniqueness of the standard representation, this implies that

$$a_1 - 1 = 2s \qquad a_2 = -3s \qquad a_3 + 3 = 6s$$
$$a_1 = 2s + 1 \qquad\qquad\qquad\qquad a_3 = 6s - 3$$

Because $\mathbf{PQ}$ has length 3, we have

$$3 = \sqrt{(a_1 - 1)^2 + a_2^2 + (a_3 + 3)^2}$$
$$= \sqrt{[(2s + 1) - 1]^2 + (-3s)^2 + [(6s - 3) + 3]^2}$$
$$= \sqrt{4s^2 + 9s^2 + 36s^2} = \sqrt{49s^2} = 7|s|$$

Thus, $s = \pm\frac{3}{7}$ and so

$$a_1 = 2(\pm\tfrac{3}{7}) + 1 \qquad a_2 = -3(\pm\tfrac{3}{7}) \qquad a_3 = 6(\pm\tfrac{3}{7}) - 3$$
$$= \tfrac{13}{7}, \tfrac{1}{7} \qquad\qquad = -\tfrac{9}{7}, \tfrac{9}{7} \qquad\qquad = -\tfrac{3}{7}, -\tfrac{39}{7}$$

There are two points that satisfy the conditions for the required terminal point Q: $(\tfrac{13}{7}, -\tfrac{9}{7}, -\tfrac{3}{7})$ and $(\tfrac{1}{7}, \tfrac{9}{7}, -\tfrac{39}{7})$. ■

DEFINITION OF DOT PRODUCT

The **dot (scalar) product** and the **cross (vector) product** are two important vector operations. We shall examine the cross product in the next section. The dot product is also known as a scalar product because it is a product of vectors that gives a scalar (that is, real number) as a result. Sometimes the dot product is called the **inner product.**

> **Dot Product**
>
> The **dot product** of vectors $\mathbf{v} = a_1\mathbf{i} + a_2\mathbf{j} + a_3\mathbf{k}$ and $\mathbf{w} = b_1\mathbf{i} + b_2\mathbf{j} + b_3\mathbf{k}$ is the scalar denoted by $\mathbf{v} \cdot \mathbf{w}$ and given by
>
> $$\mathbf{v} \cdot \mathbf{w} = a_1b_1 + a_2b_2 + a_3b_3$$

The dot product of two vectors $\mathbf{v} = a_1\mathbf{i} + a_2\mathbf{j}$ and $\mathbf{w} = b_1\mathbf{i} + b_2\mathbf{j}$ in a plane is given by a similar formula with $a_3 = b_3 = 0$, namely,

$$\mathbf{v} \cdot \mathbf{w} = a_1 b_1 + a_2 b_2$$

EXAMPLE 5   *Dot product*

Find the dot product $\mathbf{v} = -3\mathbf{i} + 2\mathbf{j} + \mathbf{k}$ and $\mathbf{w} = 4\mathbf{i} - \mathbf{j} + 2\mathbf{k}$.

Solution $\mathbf{v} \cdot \mathbf{w} = -3(4) + 2(-1) + 1(2) = -12$ ∎

EXAMPLE 6   *Dot product in component form*

If $\mathbf{v} = \langle 4, -1, 3 \rangle$ and $\mathbf{w} = \langle -1, -2, 5 \rangle$, find the dot product, $\mathbf{v} \cdot \mathbf{w}$.

Solution $\mathbf{v} \cdot \mathbf{w} = 4(-1) + (-1)(-2) + 3(5) = 13$ ∎

Before we can apply the dot product to geometric and physical problems, we need to know how it behaves algebraically. A number of important general properties of the dot product are listed in the following theorem.

THEOREM 10.2   *Properties of the dot product*

If $\mathbf{u}$, $\mathbf{v}$, and $\mathbf{w}$ are vectors in $\mathbb{R}^2$ or $\mathbb{R}^3$ and c is a scalar, then:

Magnitude of a vector	$\mathbf{v} \cdot \mathbf{v} = \|\mathbf{v}\|^2$
Zero product	$\mathbf{0} \cdot \mathbf{v} = 0$
Commutativity	$\mathbf{v} \cdot \mathbf{w} = \mathbf{w} \cdot \mathbf{v}$
Product of a multiple	$c(\mathbf{v} \cdot \mathbf{w}) = (c\mathbf{v}) \cdot \mathbf{w} = \mathbf{v} \cdot (c\mathbf{w})$
Distributivity	$\mathbf{u} \cdot (\mathbf{v} + \mathbf{w}) = \mathbf{u} \cdot \mathbf{v} + \mathbf{u} \cdot \mathbf{w}$

Proof   Let $\mathbf{u} = a_1\mathbf{i} + a_2\mathbf{j} + a_3\mathbf{k}$, $\mathbf{v} = b_1\mathbf{i} + b_2\mathbf{j} + b_3\mathbf{k}$, and $\mathbf{w} = c_1\mathbf{i} + c_2\mathbf{j} + c_3\mathbf{k}$.

Magnitude of a vector

$$\|\mathbf{v}\|^2 = (\sqrt{a_1^2 + a_2^2 + a_3^2})^2 = a_1^2 + a_2^2 + a_3^2 = \mathbf{v} \cdot \mathbf{v}$$

Zero product, commutativity, and **scalar multiple** can be established in a similar fashion.

Distributivity

$$\mathbf{u} \cdot (\mathbf{v} + \mathbf{w})$$
$$= (a_1\mathbf{i} + a_2\mathbf{j} + a_3\mathbf{k}) \cdot [(b_1 + c_1)\mathbf{i} + (b_2 + c_2)\mathbf{j} + (b_3 + c_3)\mathbf{k}]$$
$$= a_1(b_1 + c_1) + a_2(b_2 + c_2) + a_3(b_3 + c_3)$$
$$= a_1 b_1 + a_1 c_1 + a_2 b_2 + a_2 c_2 + a_3 b_3 + a_3 c_3$$

and,

$$\mathbf{u} \cdot \mathbf{v} + \mathbf{u} \cdot \mathbf{w} = (a_1 b_1 + a_2 b_2 + a_3 b_3) + (a_1 c_1 + a_2 c_2 + a_3 c_3)$$
$$= a_1 b_1 + a_1 c_1 + a_2 b_2 + a_2 c_2 + a_3 b_3 + a_3 c_3$$

Thus, $\mathbf{u} \cdot (\mathbf{v} + \mathbf{w}) = \mathbf{u} \cdot \mathbf{v} + \mathbf{u} \cdot \mathbf{w}$.   ⹀

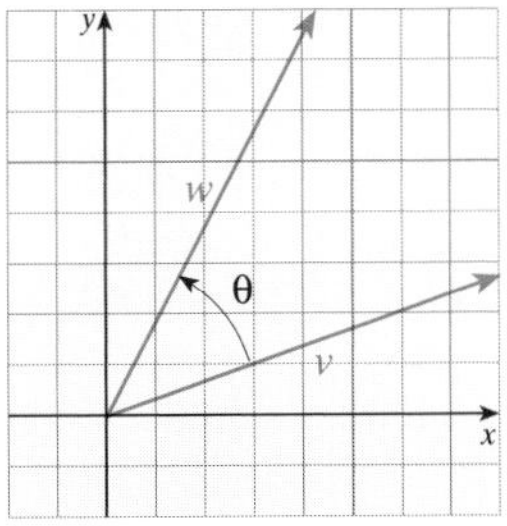

■ **FIGURE 10.18**   The angle between two vectors

ANGLE BETWEEN VECTORS

The angle between two nonzero vectors $\mathbf{v}$ and $\mathbf{w}$ is defined to be the angle θ with $0 \leq \theta \leq \pi$ that is formed when the vectors are in standard position (initial points at the origin), as shown in Figure 10.18.

The angle between vectors plays an important role in certain applications and may be computed by using the following formula involving the dot product.

==========

THEOREM 10.3   *Angle between two vectors*

If θ is the angle between the nonzero vectors $\mathbf{v}$ and $\mathbf{w}$, then

$$\cos \theta = \frac{\mathbf{v} \cdot \mathbf{w}}{\|\mathbf{v}\| \|\mathbf{w}\|}$$

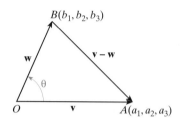

$B(b_1, b_2, b_3)$

$\mathbf{w}$ $\mathbf{v} - \mathbf{w}$

θ

O $\mathbf{v}$ $A(a_1, a_2, a_3)$

■ **FIGURE 10.19** Finding an angle between two vectors

Proof   Suppose $\mathbf{v} = a_1\mathbf{i} + a_2\mathbf{j} + a_3\mathbf{k}$ and $\mathbf{w} = b_1\mathbf{i} + b_2\mathbf{j} + b_3\mathbf{k}$, and consider $\triangle AOB$ with vertices at the origin O and the points $A(a_1, a_2, a_3)$ and $B(b_1, b_2, b_3)$, as shown in Figure 10.19.

Note that sides $\overline{OA}$ and $\overline{OB}$ have lengths

$$\|\mathbf{v}\| = \sqrt{a_1^2 + a_2^2 + a_3^2} \quad \text{and} \quad \|\mathbf{w}\| = \sqrt{b_1^2 + b_2^2 + b_3^2}$$

respectively, and that side $\overline{AB}$ has length

$$\|\mathbf{v} - \mathbf{w}\| = \sqrt{(a_1 - b_1)^2 + (a_2 - b_2)^2 + (a_3 - b_3)^2}$$

Next, we use the law of cosines:

$$a^2 = b^2 + c^2 - 2bc \cos \theta \qquad \text{Law of cosines}$$
$$\|\mathbf{v} - \mathbf{w}\|^2 = \|\mathbf{v}\|^2 + \|\mathbf{w}\|^2 - 2\|\mathbf{v}\| \|\mathbf{w}\| \cos \theta \qquad \text{See Figure 10.19.}$$
$$\cos \theta = \frac{\|\mathbf{v}\|^2 + \|\mathbf{w}\|^2 - \|\mathbf{v} - \mathbf{w}\|^2}{2\|\mathbf{v}\| \|\mathbf{w}\|} \qquad \text{Solve for } \cos \theta.$$
$$= \frac{a_1^2 + a_2^2 + a_3^2 + b_1^2 + b_2^2 + b_3^2 - [(a_1 - b_1)^2 + (a_2 - b_2)^2 + (a_3 - b_3)^2]}{2\|\mathbf{v}\| \|\mathbf{w}\|}$$
$$= \frac{2a_1b_1 + 2a_2b_2 + 2a_3b_3}{2\|\mathbf{v}\| \|\mathbf{w}\|} = \frac{\mathbf{v} \cdot \mathbf{w}}{\|\mathbf{v}\| \|\mathbf{w}\|}$$

==========

EXAMPLE 7   *Angle between two given vectors*

Let $\triangle ABC$ be the triangle with vertices $A(1, 1, 8)$, $B(4, -3, -4)$, and $C(-3, 1, 5)$. Find the angle formed at A.

Solution   Draw $\triangle ABC$ and label the angle formed at A as α, as shown in Figure 10.20. The angle α is the angle between vectors $\mathbf{AB}$ and $\mathbf{AC}$, where

$$\mathbf{AB} = (4 - 1)\mathbf{i} + (-3 - 1)\mathbf{j} + (-4 - 8)\mathbf{k} = 3\mathbf{i} - 4\mathbf{j} - 12\mathbf{k}$$
$$\mathbf{AC} = (-3 - 1)\mathbf{i} + (1 - 1)\mathbf{j} + (5 - 8)\mathbf{k} = -4\mathbf{i} - 3\mathbf{k}$$

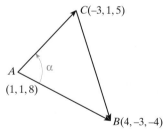

$C(-3, 1, 5)$

A α
$(1, 1, 8)$

$B(4, -3, -4)$

■ **FIGURE 10.20** Find an angle of a triangle

Thus,

$$\cos \alpha = \frac{\mathbf{AB} \cdot \mathbf{AC}}{\|\mathbf{AB}\| \|\mathbf{AC}\|} = \frac{3(-4) + (-4)(0) + (-12)(-3)}{\sqrt{3^2 + (-4)^2 + (-12)^2} \sqrt{(-4)^2 + (-3)^2}}$$
$$= \frac{24}{\sqrt{169} \sqrt{25}} = \frac{24}{65}$$

and the required angle is $\alpha = \cos^{-1}\left(\frac{24}{65}\right) \approx 1.19$.   ■

The formula for the angle between vectors is often used in conjunction with the dot product. If we multiply both sides of the formula by $\|\mathbf{v}\| \|\mathbf{w}\|$, we obtain the following alternate form for the dot product formula.

Geometric Formula for the Dot Product

$$\mathbf{v} \cdot \mathbf{w} = \|\mathbf{v}\| \|\mathbf{w}\| \cos \theta$$

where θ is the angle ($0 \leq \theta \leq \pi$) between the vectors $\mathbf{v}$ and $\mathbf{w}$.

Two vectors are said to be **perpendicular,** or **orthogonal,** if the angle between them is $\theta = \pi/2$. The following theorem provides a useful criterion for orthogonality.

THEOREM 10.4   *The orthogonal vector theorem*

Nonzero vectors $\mathbf{v}$ and $\mathbf{w}$ are **orthogonal** if and only if

$$\mathbf{v} \cdot \mathbf{w} = 0$$

Proof   If the vectors are orthogonal, then the angle between them is $\frac{\pi}{2}$; therefore,

$$\mathbf{v} \cdot \mathbf{w} = \|\mathbf{v}\| \|\mathbf{w}\| \cos \frac{\pi}{2} = 0$$

Conversely, if $\mathbf{v} \cdot \mathbf{w} = 0$, and $\mathbf{v}$ and $\mathbf{w}$ are nonzero vectors, then $\cos \theta = 0$, so that $\theta = \frac{\pi}{2}$ (since $0 \leq \theta \leq \pi$) and the vectors are orthogonal.

■   *What This Says:*   The orthogonal vector theorem allows you to show that two vectors are orthogonal by finding the dot product of the vectors and showing this dot product is 0. On the other hand, if you know the vectors are orthogonal, then it follows that the dot product of the vectors is 0.

EXAMPLE 8   *Orthogonal vectors*

Determine which (if any) pairs of the following vectors are orthogonal:
$$\mathbf{u} = 3\mathbf{i} + 7\mathbf{j} - 2\mathbf{k} \qquad \mathbf{v} = 5\mathbf{i} - 3\mathbf{j} - 3\mathbf{k} \qquad \mathbf{w} = \mathbf{j} - \mathbf{k}$$

Solution

$\mathbf{u} \cdot \mathbf{v} = 3(5) + 7(-3) + (-2)(-3) = 0$; $\mathbf{u}$ and $\mathbf{v}$ are orthogonal.

$\mathbf{u} \cdot \mathbf{w} = 3(0) + 7(1) + (-2)(-1) = 9$; $\mathbf{u}$ and $\mathbf{w}$ are not orthogonal.

$\mathbf{v} \cdot \mathbf{w} = 5(0) + (-3)(1) + (-3)(-1) = 0$; $\mathbf{v}$ and $\mathbf{w}$ are orthogonal.   ■

PROJECTIONS

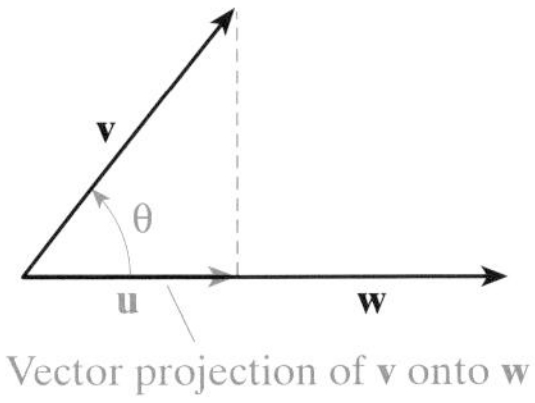

Let $\mathbf{v}$ and $\mathbf{w}$ be two vectors in $\mathbb{R}^2$ drawn so that they have a common initial point, as shown in Figure 10.21.* If we drop a perpendicular from the tip of $\mathbf{v}$ to the line determined by $\mathbf{w}$, we determine a vector called the **vector projection of $\mathbf{v}$ onto $\mathbf{w}$**, which we have labeled $\mathbf{u}$ in Figure 10.21.

Vector projection of **v** onto **w**

■ **FIGURE 10.21**   Projection of **v** onto **w**

---

*Even though Figure 10.21 is drawn in $\mathbb{R}^2$, the projection formula applies to $\mathbb{R}^3$ as well.

To find a formula for the vector projection, note that $\mathbf{u} = t\mathbf{w}$ for some scalar t and that $\mathbf{v} - t\mathbf{w}$ is orthogonal to $\mathbf{w}$. Thus,

$$(v - t\mathbf{w}) \cdot \mathbf{w} = 0$$
$$\mathbf{v} \cdot \mathbf{w} = t(\mathbf{w} \cdot \mathbf{w})$$
$$t = \frac{\mathbf{v} \cdot \mathbf{w}}{\mathbf{w} \cdot \mathbf{w}}$$

and the vector projection is

$$\mathbf{u} = \left(\frac{\mathbf{v} \cdot \mathbf{w}}{\mathbf{w} \cdot \mathbf{w}} \right)\mathbf{w}$$

WARNING Note that $\left(\dfrac{\mathbf{v} \cdot \mathbf{w}}{\mathbf{w} \cdot \mathbf{w}} \right)\mathbf{w}$ is not the same as $\left(\dfrac{\mathbf{v}}{\mathbf{w}} \right)\mathbf{w}$; you cannot "cancel" the vector $\mathbf{w}$. Remember, $\mathbf{v} \cdot \mathbf{w}$ and $\mathbf{w} \cdot \mathbf{w}$ are numbers, whereas $\mathbf{v}/\mathbf{w}$ is not defined. ⬅

The length of the vector projection is called the **scalar projection of v onto w** (also called the **component of v along w**) and may be computed by the formula

$$\|\mathbf{u}\| = \left| \frac{\mathbf{v} \cdot \mathbf{w}}{\mathbf{w} \cdot \mathbf{w}} \right| \|\mathbf{w}\| = \frac{|\mathbf{v} \cdot \mathbf{w}|}{\|\mathbf{w}\|}$$

To summarize:

> **Scalar and Vector Projections of v onto w**
>
> **Scalar projection** of $\mathbf{v}$ onto $\mathbf{w}$: $\left| \dfrac{\mathbf{v} \cdot \mathbf{w}}{\|\mathbf{w}\|} \right|$ a number
>
> **Vector projection** of $\mathbf{v}$ in the direction of $\mathbf{w}$: $\left(\dfrac{\mathbf{v} \cdot \mathbf{w}}{\mathbf{w} \cdot \mathbf{w}} \right)\mathbf{w}$ a vector

In $\mathbb{R}^3$ it is important to note that the $\mathbf{i}$, $\mathbf{j}$, and $\mathbf{k}$ components of any vector $\mathbf{v}$ are the scalar projections of $\mathbf{v}$ onto the appropriate basis vectors, as shown in Figure 10.22.

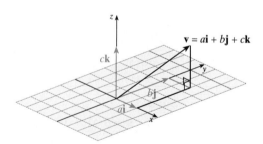

■ **FIGURE 10.22** The $\mathbf{i}$, $\mathbf{j}$, and $\mathbf{k}$ components of $\mathbf{v} = a\mathbf{i} + b\mathbf{j} + c\mathbf{k}$ are projections of $\mathbf{v}$ onto $\mathbf{i}$, $\mathbf{j}$, and $\mathbf{k}$.

EXAMPLE 9   *Scalar and vector projections*

Find the scalar and vector projections of $\mathbf{v} = 2\mathbf{i} - 3\mathbf{j} + 5\mathbf{k}$ onto $\mathbf{w} = 2\mathbf{i} - 2\mathbf{j} + \mathbf{k}$.

Solution   The vector projection of $\mathbf{v}$ onto $\mathbf{w}$ is

$$\left(\frac{\mathbf{v} \cdot \mathbf{w}}{\mathbf{w} \cdot \mathbf{w}} \right)\mathbf{w} = \left(\frac{2(2) + (-3)(-2) + 5(1)}{2^2 + (-2)^2 + 1^2} \right)(2\mathbf{i} - 2\mathbf{j} + \mathbf{k})$$

$$= \frac{15}{9}(2\mathbf{i} - 2\mathbf{j} + \mathbf{k}) = \frac{10}{3}\mathbf{i} - \frac{10}{3}\mathbf{j} + \frac{5}{3}\mathbf{k}$$

To find the scalar projection, we can find the length of the vector projection or we can use the scalar projection formula (which is usually easier than finding the length directly):

$$\text{Scalar projection of } \mathbf{v} \text{ onto } \mathbf{w} = \left| \frac{\mathbf{v} \cdot \mathbf{w}}{\|\mathbf{w}\|} \right| = \left| \frac{15}{3} \right| = 5 \qquad \blacksquare$$

WORK AS A DOT PRODUCT

In Section 6.5, we noted that when a constant force $\mathbf{F}$ acts in the direction of motion of an object moving from P to Q along a straight line, it produces work W where

$$W = (\text{MAGNITUDE OF } \mathbf{F}) (\text{DISTANCE MOVED}) = \|\mathbf{F}\| \|\mathbf{PQ}\|$$

The force remains constant but acts in a direction that makes an angle θ with the direction of motion (see Figure 10.23a). Experiments indicate that the work produced is

$$W = (\text{SCALAR COMPONENT OF } \mathbf{F} \text{ ALONG } \mathbf{PQ}) (\text{DISTANCE MOVED})$$
$$= (\left|\mathbf{F}\right| \cos \theta) \|\mathbf{PQ}\| = \mathbf{F} \cdot \mathbf{PQ}$$

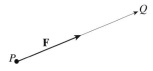

a. If the force $\mathbf{F}$ acts along the line of motion then $W = \|\mathbf{F}\| \|\mathbf{PQ}\|$.

b. If $\mathbf{F}$ acts at a nonzero angle with the line of motion, then $W = \|\text{proj. of } \mathbf{F} \text{ onto } \mathbf{PQ}\| \|\mathbf{PQ}\|$.

■ **FIGURE 10.23** Work W as a dot product

Work as a Dot Product

An object that moves along a line with displacement $\mathbf{PQ}$ against a constant force $\mathbf{F}$ performs

$$W = \mathbf{F} \cdot \mathbf{PQ}$$

units of work.

EXAMPLE 10   *Work performed by a constant force*

A boat sails north aided by a wind blowing in a direction of N30°E (30° east of north) with magnitude 500 lb. How much work is performed by the wind as the boat moves 100 ft?

Solution   The wind force is $\|\mathbf{F}\| = 500$ lb, acting in a direction $\theta = 30°$, as shown in Figure 10.24. The displacement direction is $\mathbf{PQ} = 100\mathbf{j}$, so $\|\mathbf{PQ}\| = 100$ ft. Thus, $\mathbf{F} = 500 \cos 60°\mathbf{i} + 500 \sin 60°\mathbf{j} = 250\mathbf{i} + 250\sqrt{3}\mathbf{j}$. Thus, the work performed is

$$W = \mathbf{F} \cdot \mathbf{PQ} = 100(250\sqrt{3}) = 25{,}000\sqrt{3}$$

Thus, the work is approximately 43,300 ft-lb.   ■

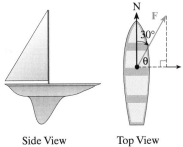

Side View   Top View
■ **FIGURE 10.24** Work performed on a sail

10.3 Problem Set

A **1.** ■ **What Does This Say?** Discuss how to find a dot product, and describe an application of dot product.

  2. ■ **What Does This Say?** What does it mean for vectors to be orthogonal? What is meant by the orthogonal projection of a vector **v** on another vector **w**?

Find the standard representation of the vector **PQ**, *and then find* $\|\mathbf{PQ}\|$ *in Problems 3–6.*

  3. $P(1, -1, 3), Q(-1, 1, 4)$

  4. $P(0, 2, 3), Q(2, 3, 0)$

  5. $P(1, 1, 1), Q(-3, -3, -3)$

  6. $P(3, 0, -4), Q(0, -4, 3)$

Find the dot product **v** · **w** *in Problems 7–10.*

  7. $\mathbf{v} = \langle 3, -2, 4 \rangle; \mathbf{w} = \langle 2, -1, -6 \rangle$

  8. $\mathbf{v} = \langle 2, -6, 0 \rangle; \mathbf{w} = \langle 0, -3, 7 \rangle$

  9. $\mathbf{v} = 2\mathbf{i} + 3\mathbf{j} - \mathbf{k}; \mathbf{w} = -3\mathbf{i} + 5\mathbf{j} + 4\mathbf{k}$

  10. $\mathbf{v} = 3\mathbf{i} - \mathbf{j}; \mathbf{w} = 2\mathbf{i} + 5\mathbf{j}$

State whether the given pairs of vectors in Problems 11–14 are orthogonal.

  11. $\mathbf{v} = \mathbf{i}; \mathbf{w} = \mathbf{k}$

  12. $\mathbf{v} = \mathbf{j}; \mathbf{w} = -\mathbf{k}$

  13. $\mathbf{v} = 3\mathbf{i} - 2\mathbf{j}; \mathbf{w} = 6\mathbf{i} + 9\mathbf{j}$

  14. $\mathbf{v} = 4\mathbf{i} - 5\mathbf{j} + \mathbf{k}; \mathbf{w} = 8\mathbf{i} + 10\mathbf{j} - 2\mathbf{k}$

Evaluate the expressions given in Problems 15–18.

  15. $\|\mathbf{i} + \mathbf{j} + \mathbf{k}\|$

  16. $\|\mathbf{i} - \mathbf{j} + \mathbf{k}\|$

  17. $\|2\mathbf{i} + \mathbf{j} - 3\mathbf{k}\|^2$

  18. $\|2(\mathbf{i} - \mathbf{j} + \mathbf{k}) - 3(2\mathbf{i} + \mathbf{j} - \mathbf{k})\|^2$

Let $\mathbf{v} = \mathbf{i} - 2\mathbf{j} + 2\mathbf{k}$ *and* $\mathbf{w} = 2\mathbf{i} + 4\mathbf{j} - \mathbf{k}$; *and find the vector or scalar requested in Problems 19–22.*

  19. $2\|\mathbf{v}\| - 3\|\mathbf{w}\|$

  20. $\|\mathbf{v}\|\mathbf{w}$

  21. $\|2\mathbf{v} - 3\mathbf{w}\|$

  22. $\|\mathbf{v} - \mathbf{w}\|(\mathbf{v} + \mathbf{w})$

Determine whether each vector in Problems 23–26 is parallel to $\mathbf{u} = 2\mathbf{i} - 3\mathbf{j} + 5\mathbf{k}$.

  23. $\mathbf{v} = \langle 4, -6, 10 \rangle$

  24. $\mathbf{v} = \langle -2, 6, -10 \rangle$

  25. $\mathbf{v} = \langle 1, -\frac{3}{2}, 2 \rangle$

  26. $\mathbf{v} = \langle -1, \frac{3}{2}, -\frac{5}{2} \rangle$

Let $\mathbf{v} = 3\mathbf{i} - 2\mathbf{j} + \mathbf{k}$ *and* $\mathbf{w} = \mathbf{i} + \mathbf{j} - \mathbf{k}$. *Evaluate the expressions in Problems 27–30.*

  27. $(\mathbf{v} + \mathbf{w}) \cdot (\mathbf{v} - \mathbf{w})$

  28. $(\mathbf{v} \cdot \mathbf{w})\mathbf{w}$

  29. $(\|\mathbf{v}\|\mathbf{w}) \cdot (\|\mathbf{w}\|\mathbf{v})$

  30. $\dfrac{2\mathbf{v} + 3\mathbf{w}}{\|3\mathbf{v} + 2\mathbf{w}\|}$

Find the angle between the vectors given in Problems 31–34. Round to the nearest degree.

  31. $\mathbf{v} = \mathbf{i} + \mathbf{j} + \mathbf{k}; \mathbf{w} = \mathbf{i} - \mathbf{j} + \mathbf{k}$

  32. $\mathbf{v} = 2\mathbf{i} + \mathbf{k}; \mathbf{w} = \mathbf{j} - 3\mathbf{k}$

  33. $\mathbf{v} = 2\mathbf{j} + \mathbf{k}; \mathbf{w} = \mathbf{i} - 2\mathbf{k}$

  34. $\mathbf{v} = 4\mathbf{i} - \mathbf{j} + \mathbf{k}; \mathbf{w} = 2\mathbf{i} + 3\mathbf{j} + 5\mathbf{k}$

Find the vector and scalar projections of **v** *onto* **w** *in Problems 35–38.*

  35. $\mathbf{v} = \mathbf{i} + \mathbf{j} + \mathbf{k}; \mathbf{w} = 2\mathbf{k}$

  36. $\mathbf{v} = \mathbf{i} + 2\mathbf{k}; \mathbf{w} = -3\mathbf{j}$

  37. $\mathbf{v} = 2\mathbf{i} - 3\mathbf{j}; \mathbf{w} = 2\mathbf{j} - 3\mathbf{k}$

  38. $\mathbf{v} = \mathbf{i} + \mathbf{j} - 2\mathbf{k}; \mathbf{w} = \mathbf{i} + \mathbf{j} + \mathbf{k}$

  39. Find two distinct unit vectors orthogonal to $\mathbf{v} = \mathbf{i} + \mathbf{j} - \mathbf{k}; \mathbf{w} = -\mathbf{i} + \mathbf{j} + \mathbf{k}$

  40. Find two distinct unit vectors orthogonal to $\mathbf{v} = 2\mathbf{i} + \mathbf{j} + 2\mathbf{k}; \mathbf{w} = -\mathbf{i} + 2\mathbf{j} - \mathbf{k}$.

  41. Find a unit vector that points in the direction opposite to $\mathbf{v} = 2\mathbf{i} + 3\mathbf{j} - 2\mathbf{k}$.

  42. Find a vector that points in the same direction as $\mathbf{v} = \mathbf{i} + 2\mathbf{j} - \mathbf{k}$ and has one-third its length.

  43. Find x, y, and z that solve
  $$x(\mathbf{i} + \mathbf{j} + \mathbf{k}) + y(\mathbf{i} - \mathbf{j} + 2\mathbf{k}) + z(\mathbf{i} + \mathbf{k}) = 2\mathbf{i} + \mathbf{k}$$

  44. Find x, y, and z that solve
  $$x(\mathbf{i} - \mathbf{k}) + y(\mathbf{j} + \mathbf{k}) + z(\mathbf{i} - \mathbf{j}) = 5\mathbf{i} - \mathbf{k}$$

  45. Find a number a that guarantees that the vectors $3\mathbf{i} - 2\mathbf{j} + \mathbf{k}$ and $2\mathbf{i} + a\mathbf{j} - 2a\mathbf{k}$ will be orthogonal.

  46. Find x if the vectors $\mathbf{v} = 3\mathbf{i} - x\mathbf{j} + 2\mathbf{k}$ and $\mathbf{w} = x\mathbf{i} + \mathbf{j} - 2\mathbf{k}$ are to be orthogonal.

B **47.** Find the angles between the vector $2\mathbf{i} + \mathbf{j} - \mathbf{k}$ and each of the coordinate axes. The cosines of these angles (to the nearest degree) are known as the **direction cosines.**

  48. Find the cosine of the angle between the vectors $\mathbf{v} = \mathbf{i} - \mathbf{j} + 2\mathbf{k}$ and $\mathbf{w} = 2\mathbf{i} + \mathbf{j} - \mathbf{k}$. Then find the vector projection of **v** onto **w**.

  49. Let $\mathbf{v} = 4\mathbf{i} - \mathbf{j} + \mathbf{k}$ and $\mathbf{w} = 2\mathbf{i} + 3\mathbf{j} - \mathbf{k}$. Find:
  a. $\mathbf{v} \cdot \mathbf{w}$
  b. $\cos\theta$, where θ is the angle between **v** and **w**
  c. a scalar s such that **v** is orthogonal to $\mathbf{v} - s\mathbf{w}$
  d. a scalar s such that $s\mathbf{v} + \mathbf{w}$ is orthogonal to **w**

  50. Let $\mathbf{v} = 2\mathbf{i} - 3\mathbf{j} + 6\mathbf{k}$ and $\mathbf{w} = 4\mathbf{i} + 3\mathbf{k}$. Find:
  a. $\mathbf{v} \cdot \mathbf{w}$
  b. $\cos\theta$, where θ is the angle between **v** and **w**
  c. a scalar s such that **v** is the orthogonal to $\mathbf{v} - s\mathbf{w}$
  d. a scalar t such that $\mathbf{v} + t\mathbf{w}$ is orthogonal to **w**

  51. Find the scalar component of the force $\mathbf{F} = 4\mathbf{i} - 2\mathbf{j} + 3\mathbf{k}$ in the direction of the vector $\mathbf{v} = \mathbf{i} - \mathbf{j} + 2\mathbf{k}$.

52. Find the work done by the constant force $\mathbf{F} = 2\mathbf{i} + 3\mathbf{j} + \mathbf{k}$ when it moves a particle along the line from $P(1, 0, -1)$ to $Q(3, 1, 2)$.

53. Find the work performed when a force $\mathbf{F} = \frac{6}{7}\mathbf{i} - \frac{2}{7}\mathbf{j} + \frac{6}{7}\mathbf{k}$ is applied to an object moving along the line from $P(-3, -5, 4)$ to $Q(4, 9, 11)$.

54. Fred and his son Sam are pulling a heavy log along flat horizontal ground by ropes attached to the front of the log. The ropes are 8 ft long. Fred holds his rope 2 ft above the log and 1 ft to the side, and Sam holds his end 1 ft above the log and 1 ft to the opposite side, as shown in Figure 10.25.

■ **FIGURE 10.25**   Problem 54

If Fred exerts a force of 30 lb and Sam exerts a force of 20 lb, what is the resultant force on the log?

55. Find the force required to keep a 5,000-lb van from rolling downhill if it is parked on a 10° slope.

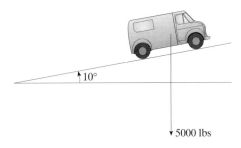

56. A block of ice is dragged 20 ft across a floor, using a force of 50 lb. Find the work done if the direction of the force is inclined θ to the horizontal, where
  a. $\theta = \frac{\pi}{3}$ **b.** $\theta = \frac{\pi}{4}$

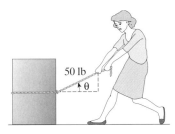

57. Suppose that the wind is blowing with a 1,000-lb magnitude force $\mathbf{F}$ in the direction of N60°W behind a boat's sail. How much work does the wind perform in moving the boat in a northerly direction a distance of 50 ft? Give your answer in foot-pounds.

Technology Window

Orthogonality of functions   *One of the big discoveries in modern mathematics was that it is very fruitful to extend some of the geometric notions of this chapter to functions—for example, taking the dot product of two functions or asking whether two functions are orthogonal, or attempting to project a function onto a second function (or onto a "plane" of functions). You are introduced to these important concepts in Problems 58 and 59.*

58. We say that two functions f and g are **orthogonal** on $[a, b]$ if

$$\int_a^b f(x)g(x)\, dx = 0$$

  a. Show that the two functions x^2 and $x^3 - 5x$ are orthogonal on $[-b, b]$ for any positive b.
  b. For positive integers k and $n \neq k$, show that $\sin kx$ and $\sin nx$ are orthogonal on the interval $[-\pi, \pi]$. That is, the *family* of functions $\sin x, \sin 2x, \sin 3x, \ldots$ are *mutually orthogonal* on $[-\pi, \pi]$. You may need the product-to-sum identity (identity 38 from the *Student Mathematics Handbook*):

$$2\sin \alpha \sin \beta = \cos(\alpha - \beta) - \cos(\alpha + \beta)$$

  c. Make a careful sketch of $\sin x$ and $\sin 2x$ on $[-\pi, \pi]$ and explain why the integral of their product is 0.

59. Historical Quest

Jean Baptiste Joseph Fourier began his studies by training for the priesthood. In spite of this first calling, his interest in mathematics remained intense, and in 1789 he wrote, "Yesterday was my 21st birthday, and at that age Newton and Pascal had already acquired many claims to immortality." He did not take his religious vows but continued his mathematical research while teaching. Fourier was renowned as an outstanding teacher, but it was not until around 1804–1807 that he did his important mathematical work on the theory of heat. During that time, he made the remarkable observation that most important odd functions [that is, $f(-x) = -f(x)$] can be approximated by

JEAN FOURIER
1768–1830

$$f(x) \approx b_1 \sin x + b_2 \sin 2x + \cdots + b_n \sin nx = S_n(x)$$

Technology Window

This formula reminds us of Taylor approximations; in fact, some call $S_n(x)$ a *trigonometric polynomial*. However, this expansion is physically more meaningful because it breaks $f(x)$ down into frequency components, where these components can represent things like pitch in sound.

a. Assume for now that $f(x)$ is actually *equal* to $S_n(x)$. To find the coefficients b_k, for $k = 1, 2, \ldots, n$, multiply the above trigonometric polynomial by $\sin kx$ and then integrate over $[-\pi, \pi]$. Show that

$$\int_{-\pi}^{\pi} \sin^2 kx \, dx = \pi$$

$$b_k = \frac{1}{\pi} \int_{-\pi}^{\pi} f(x) \sin kx \, dx$$

$$= \frac{2}{\pi} \int_{0}^{\pi} f(x) \sin kx \, dx$$

b. Why, in the evaluation of b_k, is it necessary to integrate over $[0, \pi]$ only?

c. Now let us do a sine (Fourier) expansion of the simple function $f(x) = x$. Using the work from part **a**, show that

$$x \approx 2 \sum_{k=1}^{\infty} \frac{(-1)^{k+1}}{k} \sin(kx)$$

d. Use the Fourier series in part **c** to evaluate the alternating harmonic series

$$\sum_{k=1}^{\infty} \frac{(-1)^{k+1}}{2k - 1} = 1 - \frac{1}{3} + \frac{1}{5} - \cdots$$

 60. Let $u_0 = a\mathbf{i} + b\mathbf{j} + c\mathbf{k}$ and let $\mathbf{u} = x\mathbf{i} + y\mathbf{j} + z\mathbf{k}$. Describe the set of points in $\mathbb{R}^3$ defined by

$$\|\mathbf{u}_0 - \mathbf{u}\| < r$$

where $r > 0$ and a, b, c are constants.

61. Suppose the vectors $\mathbf{v}$ and $\mathbf{w}$ are sides of a triangle with area $25\sqrt{3}$. Find $\mathbf{v} \cdot \mathbf{w}$.

62. Find the angle (to the nearest degree) between the diagonal of a cube and a diagonal of one of its faces.

63. If $\mathbf{v} \cdot \mathbf{v} = 0$, what can you conclude about $\mathbf{v}$?

64. **THINK TANK PROBLEM**   Let $\mathbf{u}$ and $\mathbf{v}$ be nonzero vectors, and let θ be the acute angle between $\mathbf{u}$ and $\mathbf{v}$. What can be said about the vector

$$\mathbf{B} = \|\mathbf{v}\|\mathbf{u} + \|\mathbf{u}\|\mathbf{v}$$

Prove your conjecture.

65. Use vector methods to prove that an angle inscribed in a semicircle must be a right angle.

66. a. Show that $(\mathbf{v} + \mathbf{w}) \cdot (\mathbf{v} + \mathbf{w})$
   $= \|\mathbf{v}\|^2 + \|\mathbf{w}\|^2 + 2(\mathbf{v} \cdot \mathbf{w})$.

b. Use part **a** to prove the **triangle inequality:**

$$\|\mathbf{v} + \mathbf{w}\| \leq \|\mathbf{v}\| + \|\mathbf{w}\|$$

Hint: Note that

$$\|\mathbf{v} + \mathbf{w}\|^2 = (\mathbf{v} + \mathbf{w}) \cdot (\mathbf{v} + \mathbf{w})$$

67. The **Cauchy–Schwarz** inequality in $\mathbb{R}^3$ states that for any vectors $\mathbf{v}$ and $\mathbf{w}$,

$$|\mathbf{v} \cdot \mathbf{w}| \leq \|\mathbf{v}\| \|\mathbf{w}\|$$

a. Prove the Cauchy–Schwarz inequality. (*Hint:* Use the formula for the angle between vectors.)

b. Show that equality in the Cauchy–Schwarz inequality occurs if and only if $\mathbf{v} = t\mathbf{w}$ for some scalar t.

c. Use the Cauchy–Schwarz inequality to prove the **triangle inequality:**

$$\|\mathbf{v} + \mathbf{w}\| \leq \|\mathbf{v}\| + \|\mathbf{w}\|$$

10.4   The Cross Product

IN THIS SECTION   definition and basic properties of the cross product, geometric interpretation of cross product, area and volume, torque

DEFINITION AND BASIC PROPERTIES OF THE CROSS PRODUCT

 We begin by defining the *cross product* in terms of a determinant. If you need to review the basic properties of determinants, consult the *Student Mathematics Handbook*.

Cross Product

If $\mathbf{v} = a_1\mathbf{i} + a_2\mathbf{j} + a_3\mathbf{k}$ and $\mathbf{w} = b_1\mathbf{i} + b_2\mathbf{j} + b_3\mathbf{k}$, the **cross product,** written $\mathbf{v} \times \mathbf{w}$, is the vector

$$\mathbf{v} \times \mathbf{w} = (a_2b_3 - a_3b_2)\mathbf{i} + (a_3b_1 - a_1b_3)\mathbf{j} + (a_1b_2 - a_2b_1)\mathbf{k}$$

These terms can be obtained by using a determinant

$$\mathbf{v} \times \mathbf{w} = \begin{vmatrix} \mathbf{i} & \mathbf{j} & \mathbf{k} \\ a_1 & a_2 & a_3 \\ b_1 & b_2 & b_3 \end{vmatrix}$$

To verify the determinant formula for the cross product, we expand about the first row:

$$\mathbf{v} \times \mathbf{w} = \begin{vmatrix} \mathbf{i} & \mathbf{j} & \mathbf{k} \\ a_1 & a_2 & a_3 \\ b_1 & b_2 & b_3 \end{vmatrix} = \begin{vmatrix} a_2 & a_3 \\ b_2 & b_3 \end{vmatrix}\mathbf{i} - \begin{vmatrix} a_1 & a_3 \\ b_1 & b_3 \end{vmatrix}\mathbf{j} + \begin{vmatrix} a_1 & a_2 \\ b_1 & b_2 \end{vmatrix}\mathbf{k}$$

j is in row 1, column 2, so do not forget negative sign here.

$$= (a_2 b_3 - a_3 b_2)\mathbf{i} - (a_1 b_3 - a_3 b_1)\mathbf{j} + (a_1 b_2 - a_2 b_1)\mathbf{k}$$
$$= (a_2 b_3 - a_3 b_2)\mathbf{i} + (a_3 b_1 - a_1 b_3)\mathbf{j} + (a_1 b_2 - a_2 b_1)\mathbf{k}$$

EXAMPLE 1   *Cross product*

Find $\mathbf{v} \times \mathbf{w}$, where $\mathbf{v} = 2\mathbf{i} - \mathbf{j} + 3\mathbf{k}$ and $\mathbf{w} = 7\mathbf{j} - 4\mathbf{k}$.

Solution

$$\mathbf{v} \times \mathbf{w} = \begin{vmatrix} \mathbf{i} & \mathbf{j} & \mathbf{k} \\ 2 & -1 & 3 \\ 0 & 7 & -4 \end{vmatrix}$$

Do not forget minus here (j is negative position).

$$= [(-1)(-4) - 3(7)]\mathbf{i} - [2(-4) - 0(3)]\mathbf{j} + [2(7) - 0(-1)]\mathbf{k}$$
$$= -17\mathbf{i} + 8\mathbf{j} + 14\mathbf{k}$$

 See Problem 66, p. 36 of handbook.

Technology Window

Mathematical programs such as *Mathematica*, *Maple*, *Derive*, and *Mathlab* all do vector operations, including dot and cross product. Recall from a previous Technology Window that brackets are sometimes used to input vectors, with the components separated by commas, as in $[a, b, c]$.

The graphic at the right shows the dot product and the cross product of the vectors in Example 1. Notice that the output shows a number, -19, for the dot product, and a vector, $[-17, 8, 14]$, for the cross product. In vector notation,

$$[-17, 8, 14] = -17\mathbf{i} + 8\mathbf{j} + 14\mathbf{k}$$

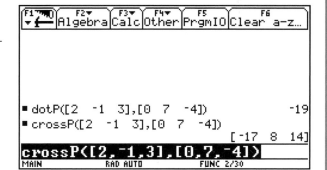

Properties of determinants can also be used to establish properties of the cross product. For instance, the following computation shows that the cross product is *not* commutative. This property is sometimes called **anticommutativity.**

$$\mathbf{v} \times \mathbf{w} = \begin{vmatrix} \mathbf{i} & \mathbf{j} & \mathbf{k} \\ a_1 & a_2 & a_3 \\ b_1 & b_2 & b_3 \end{vmatrix} = - \begin{vmatrix} \mathbf{i} & \mathbf{j} & \mathbf{k} \\ b_1 & b_2 & b_3 \\ a_1 & a_2 & a_3 \end{vmatrix} = -(\mathbf{w} \times \mathbf{v})$$

This and other properties are listed in the following theorem.

THEOREM 10.5   *Properties of the cross product*

If **u**, **v**, and **w** are vectors in $\mathbb{R}^3$ and s and t are scalars, then:

Scalar distributivity $(s\mathbf{v}) \times (t\mathbf{w}) = st(\mathbf{v} \times \mathbf{w})$

Distributivity for cross product over addition

$$\mathbf{u} \times (\mathbf{v} + \mathbf{w}) = (\mathbf{u} \times \mathbf{v}) + (\mathbf{u} \times \mathbf{w})$$
$$(\mathbf{u} + \mathbf{v}) \times \mathbf{w} = (\mathbf{u} \times \mathbf{w}) + (\mathbf{v} \times \mathbf{w})$$

Anticommutativity $\mathbf{v} \times \mathbf{w} = -(\mathbf{w} \times \mathbf{v})$

Product of a multiple $\mathbf{v} \times s\mathbf{v} = 0$; in particular, $\mathbf{v} \times \mathbf{v} = \mathbf{0}$

Zero product $\mathbf{v} \times \mathbf{0} = \mathbf{0} \times \mathbf{v} = \mathbf{0}$

Lagrange's identity $\|\mathbf{v} \times \mathbf{w}\|^2 = \|\mathbf{v}\|^2\|\mathbf{w}\|^2 - (\mathbf{v} \cdot \mathbf{w})^2$

cab-bac formula $\mathbf{a} \times (\mathbf{b} \times \mathbf{c}) = (\mathbf{c} \cdot \mathbf{a})\mathbf{b} - (\mathbf{b} \cdot \mathbf{a})\mathbf{c}$

Proof   Let $\mathbf{u} = a_1\mathbf{i} + a_2\mathbf{j} + a_3\mathbf{k}$, $\mathbf{v} = b_1\mathbf{i} + b_2\mathbf{j} + b_3\mathbf{k}$, and $\mathbf{w} = c_1\mathbf{i} + c_2\mathbf{j} + c_3\mathbf{k}$.

Scalar distributivity and **vector distributivity** are proved by using the definition of cross product and the corresponding properties of real numbers.

Anticommutativity was proved in the paragraph preceding this theorem.

Product of a multiple

$$\mathbf{v} \times s\mathbf{v} = \begin{vmatrix} \mathbf{i} & \mathbf{j} & \mathbf{k} \\ b_1 & b_2 & b_3 \\ sb_1 & sb_2 & sb_3 \end{vmatrix} = s\begin{vmatrix} \mathbf{i} & \mathbf{j} & \mathbf{k} \\ b_1 & b_2 & b_3 \\ b_1 & b_2 & b_3 \end{vmatrix} = 0$$

A determinant with two identical rows is 0.

Zero product is obvious.

Lagrange's identity

$$\begin{aligned}\|\mathbf{v} \times \mathbf{w}\|^2 &= (b_2c_3 - b_3c_2)^2 + (b_3c_1 - b_1c_3)^2 + (b_1c_2 - b_2c_1)^2 \\ &= (b_1^2 + b_2^2 + b_3^2)(c_1^2 + c_2^2 + c_3^2) - (b_1c_1 + b_2c_2 + b_3c_3)^2 \\ &= \|\mathbf{v}\|^2\|\mathbf{w}\|^2 - (\mathbf{v} \cdot \mathbf{w})^2\end{aligned}$$

====

GEOMETRIC INTERPRETATION OF THE CROSS PRODUCT

The following theorem shows that the vector $(\mathbf{v} \times \mathbf{w})$ is orthogonal to both the vectors **v** and **w**. See Figure 10.26.

THEOREM 10.6   *Orthogonality property of the cross product*

If **v** and **w** are nonzero vectors in $\mathbb{R}^3$ that are not multiples of one another, then

$$\mathbf{v} \times \mathbf{w} \text{ is orthogonal to both } \mathbf{v} \text{ and } \mathbf{w}.$$

Proof   We focus on the case where $\mathbf{v} \times \mathbf{w}$ is a nonzero vector. We will show that $(\mathbf{v} \times \mathbf{w})$ is orthogonal to **v** and leave the proof that $\mathbf{v} \times \mathbf{w}$ is orthogonal to **w** as an exercise. Let $\mathbf{v} = a_1\mathbf{i} + a_2\mathbf{j} + a_3\mathbf{k}$ and $\mathbf{w} = b_1\mathbf{i} + b_2\mathbf{j} + b_3\mathbf{k}$. Then

$$\mathbf{v} \times \mathbf{w} = \begin{vmatrix} \mathbf{i} & \mathbf{j} & \mathbf{k} \\ a_1 & a_2 & a_3 \\ b_1 & b_2 & b_3 \end{vmatrix} = (a_2b_3 - a_3b_2)\mathbf{i} - (a_1b_3 - a_3b_1)\mathbf{j} + (a_1b_2 - a_2b_1)\mathbf{k}$$

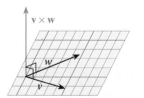

■ **FIGURE 10.26** Vector product of two vectors

To show that this vector is orthogonal to **v**, we find $\mathbf{v} \cdot (\mathbf{v} \times \mathbf{w})$:

$$\mathbf{v} \cdot (\mathbf{v} \times \mathbf{w}) = a_1(a_2b_3 - a_3b_2) - a_2(a_1b_3 - a_3b_1) + a_3(a_1b_2 - a_2b_1)$$
$$= a_1a_2b_3 - a_1a_3b_2 - a_1a_2b_3 + a_2a_3b_1 + a_1a_3b_2 - a_2a_3b_1 = 0 \quad \blacksquare$$

Because both $\mathbf{v} \times \mathbf{w}$ and $\mathbf{w} \times \mathbf{v}$ are orthogonal to the plane determined by **v** and **w**, and because $(\mathbf{v} \times \mathbf{w}) = -(\mathbf{w} \times \mathbf{v})$, we see that one points up from the given plane and the other points down. To see which is which, we use the **right-hand rule** described in Figure 10.27.

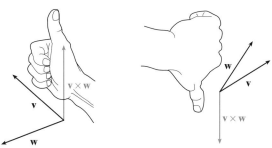

■ **FIGURE 10.27**   The right-hand rule: If you place the palm of your right hand along **v** and curl your fingers towards **w** to cover the smaller angle between **v** and **w**, then your thumb points in the direction of $\mathbf{v} \times \mathbf{w}$.

EXAMPLE 2   *Right-hand rule*

Use the right-hand rule to verify each of the following cross products.

$$\mathbf{i} \times \mathbf{j} = \mathbf{k} \qquad \mathbf{j} \times \mathbf{i} = -\mathbf{k} \qquad \mathbf{i} \times \mathbf{i} = \mathbf{0}$$
$$\mathbf{i} \times \mathbf{k} = -\mathbf{j} \qquad \mathbf{k} \times \mathbf{i} = \mathbf{j} \qquad \mathbf{j} \times \mathbf{j} = \mathbf{0}$$
$$\mathbf{j} \times \mathbf{k} = \mathbf{i} \qquad \mathbf{k} \times \mathbf{j} = -\mathbf{i} \qquad \mathbf{k} \times \mathbf{k} = \mathbf{0}$$

Solution $\mathbf{i} \times \mathbf{j} = \begin{vmatrix} \mathbf{i} & \mathbf{j} & \mathbf{k} \\ 1 & 0 & 0 \\ 0 & 1 & 0 \end{vmatrix} = \mathbf{k}$

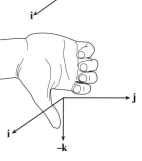

Place the palm of your right hand along **i** and curl your fingers toward **j**. The answer is **k**.

$$\mathbf{j} \times \mathbf{i} = \begin{vmatrix} \mathbf{i} & \mathbf{j} & \mathbf{k} \\ 0 & 1 & 0 \\ 1 & 0 & 1 \end{vmatrix} = -\mathbf{k}$$

Place the palm of your right hand along **j** and curl your fingers toward **i**. The answer is $-\mathbf{k}$.

$$\mathbf{i} \times \mathbf{i} = \begin{vmatrix} \mathbf{i} & \mathbf{j} & \mathbf{k} \\ 1 & 0 & 0 \\ 1 & 0 & 0 \end{vmatrix} = \mathbf{0}$$

The other parts are left for you to verify.   ■

EXAMPLE 3   *A vector orthogonal to two given vectors*

Find a nonzero vector that is orthogonal to both $\mathbf{v} = -2\mathbf{i} + 3\mathbf{j} - 7\mathbf{k}$ and $\mathbf{w} = 5\mathbf{i} + 9\mathbf{k}$.

Solution The cross product $\mathbf{v} \times \mathbf{w}$ is orthogonal to both $\mathbf{v}$ and $\mathbf{w}$.

$$\mathbf{v} \times \mathbf{w} = \begin{vmatrix} \mathbf{i} & \mathbf{j} & \mathbf{k} \\ -2 & 3 & -7 \\ 5 & 0 & 9 \end{vmatrix} = (27 + 0)\mathbf{i} - (-18 + 35)\mathbf{j} + (0 - 15)\mathbf{k}$$

$$= 27\mathbf{i} - 17\mathbf{j} - 15\mathbf{k} \qquad \blacksquare$$

In Section 10.3, we showed that the dot product satisfies $(\mathbf{v} \cdot \mathbf{w}) = \|\mathbf{v}\|\|\mathbf{w}\|\cos \theta$, where θ is the angle between $\mathbf{v}$ and $\mathbf{w}$. The following theorem establishes a similar result for the cross product.

THEOREM 10.7 *Geometric interpretation of cross product*

If $\mathbf{v}$ and $\mathbf{w}$ are nonzero vectors in $\mathbb{R}^3$ with θ the angle between $\mathbf{v}$ and $\mathbf{w}$ ($0 \le \theta \le \pi$), then

$$\mathbf{v} \times \mathbf{w} = (\|\mathbf{v}\|\|\mathbf{w}\|\sin \theta)\mathbf{n}$$

where θ is the angle between $\mathbf{v}$ and $\mathbf{w}$ ($0 \le \theta \le \pi$) and $\mathbf{n}$ is a unit normal vector determined by the right-hand rule (see Figure 10.27).

Proof We first show that $\|\mathbf{v} \times \mathbf{w}\| = \|\mathbf{v}\|\|\mathbf{w}\|\,|\sin \theta|$ and then focus on the direction.

$$\|\mathbf{v} \times \mathbf{w}\|^2 = \|\mathbf{v}\|^2\|\mathbf{w}\|^2 - [\|\mathbf{v}\|\|\mathbf{w}\|\cos \theta]^2 \qquad \textit{Lagrange's identity}$$

$$= \|\mathbf{v}\|^2\|\mathbf{w}\|^2 - \|\mathbf{v}\|^2\|\mathbf{w}\|^2 \cos^2\theta$$

$$= \|\mathbf{v}\|^2\|\mathbf{w}\|^2 (1 - \cos^2\theta)$$

$$= \|\mathbf{v}\|^2\|\mathbf{w}\|^2\sin^2\theta$$

Thus, $\|\mathbf{v} \times \mathbf{w}\| = \|\mathbf{v}\|\|\mathbf{w}\|\,|\sin \theta| = \|\mathbf{v}\|\|\mathbf{w}\|\sin \theta$ *Because $\sin \theta \ge 0$*

We already know that $\mathbf{v} \times \mathbf{w}$ is orthogonal to both $\mathbf{v}$ and $\mathbf{w}$ (Theorem 10.6), so it either points up from the plane, or down from the plane. The fact that $\mathbf{n}$ is determined by the right-hand rule follows from the results in Example 2. $\quad =\!=$

> ■ **What This Says:** Theorem 10.7 provides a *coordinate-free* way of computing the cross product. This is important to physicists and engineers because it enables them to choose whichever coordinate system is most convenient to compute cross products that arise in applications.

AREA AND VOLUME

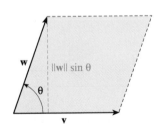

■ **FIGURE 10.28** A geometric interpretation of $\|\mathbf{v} \times \mathbf{w}\|$

There is another interpretation for Theorem 10.7—namely, $\|\mathbf{v} \times \mathbf{w}\|$ is equal to the area of a parallelogram having $\mathbf{v}$ and $\mathbf{w}$ as adjacent sides, as shown in Figure 10.28. To see this, note that because $h = \|\mathbf{w}\|\sin \theta$, we have

$$\text{AREA} = (\text{BASE})(\text{HEIGHT}) = \|\mathbf{v}\|(\|\mathbf{w}\|\sin \theta) = \|\mathbf{v} \times \mathbf{w}\|$$

Here is an example that uses this formula. You might also note that even though $\mathbf{v} \times \mathbf{w} \ne \mathbf{w} \times \mathbf{v}$, it is true that $\|\mathbf{v} \times \mathbf{w}\| = \|\mathbf{w} \times \mathbf{v}\|$.

EXAMPLE 4 *Area of a triangle*

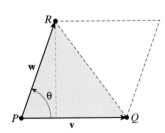

■ **FIGURE 10.29** Area of a triangle by cross product

Find the area of the triangle with vertices $P(-2, 4, 5)$, $Q(0, 7, -4)$, and $R(-1, 5, 0)$.

Solution Draw this triangle as shown in Figure 10.29. Then $\triangle PQR$ has half the area of the parallelogram determined by the vectors $\mathbf{PQ}$ and $\mathbf{PR}$; that is, the triangle has area

$$A = \tfrac{1}{2}\|\mathbf{PQ} \times \mathbf{PR}\|$$

First find

$$\mathbf{PQ} = (0 + 2)\mathbf{i} + (7 - 4)\mathbf{j} + (-4 - 5)\mathbf{k} = 2\mathbf{i} + 3\mathbf{j} - 9\mathbf{k}$$

$$\mathbf{PR} = (-1 + 2)\mathbf{i} + (5 - 4)\mathbf{j} + (0 - 5)\mathbf{k} = \mathbf{i} + \mathbf{j} - 5\mathbf{k}$$

and compute the cross product:

$$\begin{aligned}
\mathbf{PQ} \times \mathbf{PR} &= \begin{vmatrix} \mathbf{i} & \mathbf{j} & \mathbf{k} \\ 2 & 3 & -9 \\ 1 & 1 & -5 \end{vmatrix} \\
&= (-15 + 9)\mathbf{i} - (-10 + 9)\mathbf{j} + (2 - 3)\mathbf{k} \\
&= -6\mathbf{i} + \mathbf{j} - \mathbf{k}
\end{aligned}$$

Thus, the triangle has area

$$\begin{aligned}
A &= \tfrac{1}{2}\|\mathbf{PQ} \times \mathbf{PR}\| \\
&= \tfrac{1}{2}\sqrt{(-6)^2 + 1^2 + (-1)^2} \\
&= \tfrac{1}{2}\sqrt{38}
\end{aligned}$$

The cross product can also be used to compute the volume of a parallelepiped in $\mathbb{R}^3$. Consider the parallelepiped determined by three nonzero vectors $\mathbf{u}$, $\mathbf{v}$, and $\mathbf{w}$ that do not all lie in the same plane, as shown in Figure 10.30.

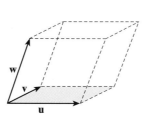

a. The parallelepiped determined by $\mathbf{u}$, $\mathbf{v}$, and $\mathbf{w}$

b. The parallelepiped has volume $V = Ah = |(\mathbf{u} \times \mathbf{v}) \cdot \mathbf{w}|.$

■ **FIGURE 10.30**   Computing volume with the triple scalar product

It is known from solid geometry that this parallelogram has volume $V = Ah$, where A is the area of the face determined by $\mathbf{u}$ and $\mathbf{v}$, and h is the altitude from the tip of $\mathbf{w}$ to this face (see Figure 10.30b).

The face determined by $\mathbf{u}$ and $\mathbf{v}$ is a parallelogram with area $A = \|\mathbf{u} \times \mathbf{v}\|$, and we know that the cross-product vector $\mathbf{u} \times \mathbf{v}$ is perpendicular to both $\mathbf{u}$ and $\mathbf{v}$ and hence to the face determined by $\mathbf{u}$ and $\mathbf{v}$. From the geometric formula for the dot product, we have

$$(\mathbf{u} \times \mathbf{v}) \cdot \mathbf{w} = \|\mathbf{u} \times \mathbf{v}\|\|\mathbf{w}\|\cos\theta$$

where θ is the angle between $\mathbf{u} \times \mathbf{v}$ and $\mathbf{w}$. Thus, the parallelepiped has altitude

$$h = \left|\|\mathbf{w}\|\cos\theta\right| = \left|\frac{(\mathbf{u} \times \mathbf{v}) \cdot \mathbf{w}}{\|\mathbf{u} \times \mathbf{v}\|}\right|$$

and the volume is given as

$$V = Ah = \|\mathbf{u} \times \mathbf{v}\|\left|\frac{(\mathbf{u} \times \mathbf{v}) \cdot \mathbf{w}}{\|\mathbf{u} \times \mathbf{v}\|}\right| = \left|(\mathbf{u} \times \mathbf{v}) \cdot \mathbf{w}\right|$$

The combined operation $(\mathbf{u} \times \mathbf{v}) \cdot \mathbf{w}$ is called the **triple scalar product** of $\mathbf{u}$, $\mathbf{v}$, and $\mathbf{w}$. These observations are summarized in the following box.

> **Volume Interpretation of the Triple Scalar Product**
>
> Let $\mathbf{u}, \mathbf{v}$, and $\mathbf{w}$ be nonzero vectors that do not all lie in the same plane. Then the parallelepiped determined by these vectors has volume
> $$V = \left|(\mathbf{u} \times \mathbf{v}) \cdot \mathbf{w}\right|$$

EXAMPLE 5   *Volume of a parallelepiped*

Find the volume of the parallelepiped determined by the vectors

$$\mathbf{u} = \mathbf{i} - 2\mathbf{j} + 3\mathbf{k} \qquad \mathbf{v} = -4\mathbf{i} + 7\mathbf{j} - 11\mathbf{k} \qquad \mathbf{w} = 5\mathbf{i} + 9\mathbf{j} - \mathbf{k}$$

Solution   We first find the cross product

$$\mathbf{u} \times \mathbf{v} = \begin{vmatrix} \mathbf{i} & \mathbf{j} & \mathbf{k} \\ 1 & -2 & 3 \\ -4 & 7 & -11 \end{vmatrix} = (22 - 21)\mathbf{i} - (-11 + 12)\mathbf{j} + (7 - 8)\mathbf{k}$$
$$= \mathbf{i} - \mathbf{j} - \mathbf{k}$$

Thus,

$$V = \left|(\mathbf{u} \times \mathbf{v}) \cdot \mathbf{w}\right| = \left|(\mathbf{i} - \mathbf{j} - \mathbf{k}) \cdot (5\mathbf{i} + 9\mathbf{j} - \mathbf{k})\right| = \left|5 - 9 + 1\right| = 3$$

The volume of the parallelepiped is 3 cubic units.   ■

THEOREM 10.8   *Determinant form for a triple scalar product*

If $\mathbf{u} = a_1\mathbf{i} + a_2\mathbf{j} + a_3\mathbf{k}$, $\mathbf{v} = b_1\mathbf{i} + b_2\mathbf{j} + b_3\mathbf{k}$, and $\mathbf{w} = c_1\mathbf{i} + c_2\mathbf{j} + c_3\mathbf{k}$, then the triple scalar product can be found by evaluating the determinant

$$(\mathbf{u} \times \mathbf{v}) \cdot \mathbf{w} = \begin{vmatrix} a_1 & a_2 & a_3 \\ b_1 & b_2 & b_3 \\ c_1 & c_2 & c_3 \end{vmatrix}$$

Proof   The proof follows by expanding the determinant (see Problem 52).   ═

We can use Theorem 10.8 to rework Example 5:

$$(\mathbf{u} \times \mathbf{v}) \cdot \mathbf{w} = \begin{vmatrix} 1 & -2 & 3 \\ -4 & 7 & -11 \\ 5 & 9 & -1 \end{vmatrix} = -3$$

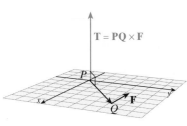

SMH   *See Problem 65, p. 36 of the handbook.*

Thus, the volume is $\left|-3\right| = 3$, as computed directly in Example 5.

WARNING   In the problem set, we have included several exercises involving the triple scalar product and the triple vector product $\mathbf{u} \times \mathbf{v} \times \mathbf{w}$. In case you wonder why we neglect the product $(\mathbf{u} \cdot \mathbf{v}) \times \mathbf{w}$, notice that such a product makes no sense, because $\mathbf{u} \cdot \mathbf{v}$ is a *scalar* and the cross product is an operation involving only vectors. Thus, the product $\mathbf{u} \cdot \mathbf{v} \times \mathbf{w}$ must mean $\mathbf{u} \cdot (\mathbf{v} \times \mathbf{w})$.   ←

TORQUE

A useful physical application of the cross product involves **torque.** Suppose the force $\mathbf{F}$ is applied to the point Q. Then the torque of $\mathbf{F}$ around P is defined as the cross product of the "arm" vector $\mathbf{PQ}$ with the force $\mathbf{F}$, as shown in Figure 10.31.

■ **FIGURE 10.31** Torque given by a cross product

Thus, the torque, **T**, of **F** at Q about P is

$$\mathbf{T} = \mathbf{PQ} \times \mathbf{F}$$

The magnitude of the torque, $\|\mathbf{T}\|$, provides a measure of the tendency of the vector arm **PQ** to rotate counterclockwise about an axis perpendicular to the plane determined by **PQ** and **F** (see Figure 10.31).

EXAMPLE 6   *Torque on the hinge of a door*

Figure 10.32 shows a half-open door that is 3 ft wide. A horizontal force of 30 lb is applied at the edge of the door. Find the torque of the force about the hinge on the door.

Solution   We represent the force by $\mathbf{F} = -30\mathbf{i}$ (see Figure 10.32). Because the door is half open, it makes an angle of $\frac{\pi}{4}$ with the horizontal, and we can represent the "arm" **PQ** by the vector

$$\mathbf{PQ} = 3\left(\cos\frac{\pi}{4}\mathbf{i} + \sin\frac{\pi}{4}\mathbf{j}\right) = 3\left(\frac{\sqrt{2}}{2}\mathbf{i} + \frac{\sqrt{2}}{2}\mathbf{j}\right) = \frac{3\sqrt{2}}{2}\mathbf{i} + \frac{3\sqrt{2}}{2}\mathbf{j}$$

The torque can now be found:

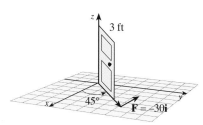

$$\mathbf{T} = \mathbf{PQ} \times \mathbf{F} = \begin{vmatrix} \mathbf{i} & \mathbf{j} & \mathbf{k} \\ 3\sqrt{2}/2 & 3\sqrt{2}/2 & 0 \\ -30 & 0 & 0 \end{vmatrix} = 45\sqrt{2}\,\mathbf{k}$$

■ **FIGURE 10.32**   The torque of a force applied at the edge of a door about the hinges

The magnitude of the torque ($45\sqrt{2}$ lb-ft) is a measure of the tendency of the door to rotate about its hinges.   ■

10.4 Problem Set

Ⓐ *Find* **v** × **w** *for the vectors given in Problems* 1–10.

1. $\mathbf{v} = \mathbf{i}; \mathbf{w} = \mathbf{j}$ **2.** $\mathbf{v} = \mathbf{k}; \mathbf{w} = \mathbf{k}$
3. $\mathbf{v} = 3\mathbf{i} + 2\mathbf{k}; \mathbf{w} = 2\mathbf{i} + \mathbf{j}$
4. $\mathbf{v} = \mathbf{i} - 3\mathbf{j}; \mathbf{w} = \mathbf{i} + 5\mathbf{k}$
5. $\mathbf{v} = 3\mathbf{i} - 2\mathbf{j} + 4\mathbf{k}; \mathbf{w} = \mathbf{i} + 4\mathbf{j} - 7\mathbf{k}$
6. $\mathbf{v} = 5\mathbf{i} - \mathbf{j} + 2\mathbf{k}; \mathbf{w} = 2\mathbf{i} + \mathbf{j} - 3\mathbf{k}$
7. $\mathbf{v} = 3\mathbf{i} - \mathbf{j} + 2\mathbf{k}; \mathbf{w} = 2\mathbf{i} + 3\mathbf{j} - 4\mathbf{k}$
8. $\mathbf{v} = -\mathbf{j} + 4\mathbf{k}; \mathbf{w} = 5\mathbf{i} + 6\mathbf{k}$
9. $\mathbf{v} = \mathbf{i} - 6\mathbf{j} + 10\mathbf{k}; \mathbf{w} = -\mathbf{i} + 5\mathbf{j} - 6\mathbf{k}$
10. $\mathbf{v} = \cos\theta\,\mathbf{i} + \sin\theta\mathbf{j}; \mathbf{w} = -\sin\theta\,\mathbf{i} + \cos\theta\,\mathbf{j}$

Find $\sin\theta$ *where* θ *is the angle between* **v** *and* **w** *in Problems* 11–16.

11. $\mathbf{v} = \mathbf{i} + \mathbf{k}; \mathbf{w} = \mathbf{i} + \mathbf{j}$
12. $\mathbf{v} = \mathbf{i} + \mathbf{j}; \mathbf{w} = \mathbf{i} + \mathbf{j} + \mathbf{k}$
13. $\mathbf{v} = \mathbf{j} + \mathbf{k}; \mathbf{w} = \mathbf{i} + \mathbf{k}$
14. $\mathbf{v} = \mathbf{i} + \mathbf{j}; \mathbf{w} = \mathbf{j} + \mathbf{k}$
15. $\mathbf{v} = \mathbf{i} + 2\mathbf{j} + 3\mathbf{k}; \mathbf{w} = 4\mathbf{i} + 5\mathbf{j} + 6\mathbf{k}$
16. $\mathbf{v} = \cos\theta\,\mathbf{i} - \sin\theta\mathbf{j}; \mathbf{w} = \sin\theta\,\mathbf{i} - \cos\theta\,\mathbf{j}$

Find a unit vector that is orthogonal to both **v** *and* **w** *in Problems* 17–20.

17. $\mathbf{v} = 2\mathbf{i} + \mathbf{k}; \mathbf{w} = \mathbf{i} - \mathbf{j} - \mathbf{k}$
18. $\mathbf{v} = \mathbf{j} - 3\mathbf{k}; \mathbf{w} = -\mathbf{i} + \mathbf{j} + \mathbf{k}$
19. $\mathbf{v} = \mathbf{i} + \mathbf{j} + \mathbf{k}; \mathbf{w} = 3\mathbf{i} + 12\mathbf{j} - 4\mathbf{k}$
20. $\mathbf{v} = 2\mathbf{i} - 2\mathbf{j} + \mathbf{k}; \mathbf{w} = 4\mathbf{i} + 2\mathbf{j} - 3\mathbf{k}$

Find the area of the parallelogram determined by the vectors in Problems 21–24.

21. $3\mathbf{i} + 4\mathbf{j}$ and $\mathbf{i} + \mathbf{j} - \mathbf{k}$
22. $2\mathbf{i} - \mathbf{j} + 2\mathbf{k}$ and $4\mathbf{i} - 3\mathbf{j}$
23. $4\mathbf{i} - \mathbf{j} + \mathbf{k}$ and $2\mathbf{i} + 3\mathbf{j} - \mathbf{k}$
24. $2\mathbf{i} + 3\mathbf{k}$ and $2\mathbf{j} - 3\mathbf{k}$

Find the area of △*PQR in Problems* 25–28.

25. $P(0, 1, 1), Q(1, 1, 0), R(1, 0, 1)$
26. $P(1, 0, 0), Q(2, 1, -1), R(0, 1, -2)$
27. $P(1, 2, 3), Q(2, 3, 1), R(3, 1, 2)$
28. $P(-1, -1, -1), Q(1, -1, -1), R(-1, 1, -1)$

Determine whether each product in Problems 29–31 is a scalar or a vector or does not exist. Explain your reasoning.

29. a. $\mathbf{u} \times (\mathbf{v} \cdot \mathbf{w})$

  b. $\mathbf{u} \cdot (\mathbf{v} \times \mathbf{w})$

30. a. $\mathbf{u} \times (\mathbf{v} \times \mathbf{w})$

  b. $\mathbf{u} \cdot (\mathbf{v} \cdot \mathbf{w})$

31. a. $(\mathbf{u} \times \mathbf{v}) \cdot (\mathbf{u} \times \mathbf{w})$

  b. $(\mathbf{u} \times \mathbf{v}) \times (\mathbf{u} \times \mathbf{w})$

In Problems 32–35, find the volume of the parallelepiped determined by vectors $\mathbf{u}$, $\mathbf{v}$, and $\mathbf{w}$.

32. $\mathbf{u} = \mathbf{i} + \mathbf{j}$; $\mathbf{v} = \mathbf{j} + 2\mathbf{k}$; $\mathbf{w} = 3\mathbf{k}$

33. $\mathbf{u} = \mathbf{j} + \mathbf{k}$; $\mathbf{v} = 2\mathbf{i} + \mathbf{j} + 2\mathbf{k}$; $\mathbf{w} = 5\mathbf{i}$

34. $\mathbf{u} = \mathbf{i} + \mathbf{j} + \mathbf{k}$; $\mathbf{v} = \mathbf{i} - \mathbf{j} - \mathbf{k}$; $\mathbf{w} = 2\mathbf{i} + 3\mathbf{k}$

B **35.** $\mathbf{u} = 2\mathbf{i} + \mathbf{j} - \mathbf{k}$; $\mathbf{v} = 3\mathbf{i} + \mathbf{k}$; $\mathbf{w} = \mathbf{j} + \mathbf{k}$

36. ■ **What Does This Say?** Contrast dot and cross products of vectors, including a discussion of some of their properties.

37. ■ **What Does This Say?** What is the right-hand rule?

38. ■ **What Does This Say?** Give a geometric interpretation of $|(\mathbf{u} \times \mathbf{v}) \cdot \mathbf{w}|$.

39. **THINK TANK PROBLEM**

  a. If $\mathbf{u} \times \mathbf{w} = \mathbf{v} \times \mathbf{w}$, does it follow that $\mathbf{u} = \mathbf{v}$?

  b. If $\mathbf{u} \cdot \mathbf{w} = \mathbf{v} \cdot \mathbf{w}$, does it follow that $\mathbf{u} = \mathbf{v}$?

  c. If both $\mathbf{u} \times \mathbf{w} = \mathbf{v} \times \mathbf{w}$, and $\mathbf{u} \cdot \mathbf{w} = \mathbf{v} \cdot \mathbf{w}$, does it follow that $\mathbf{u} = \mathbf{v}$?

40. Find a number s that guarantees that the vectors $\mathbf{i}, \mathbf{i} + \mathbf{j} + \mathbf{k}$, and $\mathbf{i} + 2\mathbf{j} + s\mathbf{k}$ will all be parallel to the same plane.

41. Find a number t that guarantees that the vectors $\mathbf{i} + \mathbf{j}, 2\mathbf{i} - \mathbf{j} + \mathbf{k}$, and $\mathbf{i} + \mathbf{j} + t\mathbf{k}$ will all be parallel to the same plane.

42. Find the angle between the vector $2\mathbf{i} - \mathbf{j} + \mathbf{k}$ and the plane determined by the points $P(1, -2, 3), Q(-1, 2, 3)$, and $R(1, 2, -3)$.

43. Let $\mathbf{u} = \mathbf{i} + \mathbf{j}, \mathbf{v} = 2\mathbf{i} - \mathbf{j} + \mathbf{k}$, and $\mathbf{w} = 3\mathbf{i}$. Compute $(\mathbf{u} \times \mathbf{v}) \times \mathbf{w}$ and $\mathbf{u} \times (\mathbf{v} \times \mathbf{w})$. What does this say about the associativity of cross product?

44. Show that $(a\mathbf{u}) \times (b\mathbf{v}) = ab(\mathbf{u} \times \mathbf{v})$ for scalars a and b.

45. For a given vector $\mathbf{v}$ in $\mathbb{R}^3$, find all vectors $\mathbf{w}$ such that $\mathbf{v} \times \mathbf{w} = \mathbf{w}$.

46. What can be said about nonzero vectors $\mathbf{v}$ and $\mathbf{w}$ if $\mathbf{v} \cdot \mathbf{w} = \mathbf{0}$? What can be said if $\mathbf{w} \times \mathbf{v} = \mathbf{0}$?

47. One end of a 2-ft lever pivots about the origin in the yz-plane, as shown in Figure 10.33.

  If a vertical force of 40 lb is applied at the end of the lever, what is the torque of the lever about the pivot point (the origin) when the lever makes an angle of $30°$ with the xy-plane?

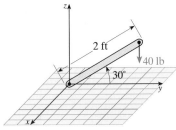

■ **FIGURE 10.33**   Finding the torque

48. A 40-lb child sits on a seesaw, 3 ft from the fulcrum. What torque is exerted when the child is 2 ft above the horizontal? What is the maximum torque exerted by the child?

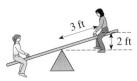

■ **FIGURE 10.34**   Seesaw torque

49. a. Show that the vectors $\mathbf{u}, \mathbf{v}$, and $\mathbf{w}$ are coplanar (all in the same plane) if
$$\mathbf{u} \cdot (\mathbf{v} \times \mathbf{w}) = 0 \quad \text{or} \quad (\mathbf{u} \times \mathbf{v}) \cdot \mathbf{w} = 0$$

  b. Are the vectors $\mathbf{u} = \mathbf{i} + 3\mathbf{j} + \mathbf{k}, \mathbf{v} = 2\mathbf{i} - \mathbf{j} - \mathbf{k}$, and $\mathbf{w} = 7\mathbf{j} + 3\mathbf{k}$ coplanar?

50. Show that the triangle with vertices $(x_1, y_1), (x_2, y_2)$, (x_3, y_3) has area $A = \frac{1}{2}D$, where
$$D = \begin{vmatrix} x_1 & y_1 & 1 \\ x_2 & y_2 & 1 \\ x_3 & y_3 & 1 \end{vmatrix}$$

51. Using the properties of determinants, show that
$$\mathbf{u} \cdot (\mathbf{v} \times \mathbf{w}) = (\mathbf{u} \times \mathbf{v}) \cdot \mathbf{w}$$
for any vectors $\mathbf{u}, \mathbf{v}$, and $\mathbf{w}$.

C **52.** Prove the determinant formula for evaluating a triple scalar product.

53. Let A, B, C, and D be four points that do not lie in the same plane. It can be shown that the volume of the tetrahedron with vertices A, B, C, and D satisfies
$$\begin{bmatrix} \text{VOLUME OF} \\ \text{TETRAHEDRON } ABCD \end{bmatrix} = \frac{1}{3} \begin{pmatrix} \text{AREA OF} \\ \triangle ABC \end{pmatrix} \begin{pmatrix} \text{ALTITUDE FROM} \\ D \text{ TO } \triangle ABC \end{pmatrix}$$

Show that the volume is given by
$$V = \frac{1}{6} |(\mathbf{AB} \times \mathbf{AC}) \cdot \mathbf{AD}|$$

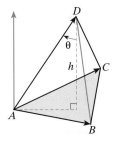

54. Show that if **u**, **v**, and **w** are vectors in $\mathbb{R}^3$ with **u** + **v** + **w** = **0**, then

$$\mathbf{u} \times \mathbf{v} = \mathbf{v} \times \mathbf{w} = \mathbf{w} \times \mathbf{u}$$

55. Suppose **u**, **v**, and **w** are nonzero vectors in $\mathbb{R}^3$ with **u** × **v** = **w** and **u** · **v** = 0. Show that

$$\mathbf{v} = s(\mathbf{w} \times \mathbf{u}) \quad \text{and} \quad \mathbf{u} = t(\mathbf{v} \times \mathbf{w})$$

for scalars s and t.

56. Show that

$$(c\mathbf{u}) \times (d\mathbf{v}) = cd(\mathbf{u} \times \mathbf{v})$$

57. Show that

$$\tan \theta = \frac{\|\mathbf{v} \times \mathbf{w}\|}{\mathbf{v} \cdot \mathbf{w}}$$

where θ $(0 \leq \theta < \frac{\pi}{2})$ is the angle between **v** and **w**.

58. Let **u**, **v**, and **w** be nonzero vectors in $\mathbb{R}^3$ that do not all lie in the same plane. Show that

$$\left| \mathbf{u} \cdot (\mathbf{v} \times \mathbf{w}) \right| = \left| \mathbf{v} \cdot (\mathbf{u} \times \mathbf{w}) \right|$$

What other triple scalar products involving **u**, **v**, and **w** have the same absolute values?

59. Let **a**, **b**, and **c** be vectors in space. Prove the "cab – bac" formula.

$$\mathbf{a} \times (\mathbf{b} \times \mathbf{c}) = (\mathbf{c} \cdot \mathbf{a})\mathbf{b} - (\mathbf{b} \cdot \mathbf{a})\mathbf{c}$$

Establish the validity of the equations in Problems 60–63 for arbitrary vectors **u**, **v**, **w**, *and* **z** *in* $\mathbb{R}^3$. *You may use the result of Problem 59.*

60. $(\mathbf{u} \times \mathbf{v}) \times (\mathbf{w} \times \mathbf{z}) = (\mathbf{u} \cdot \mathbf{w} \times \mathbf{z})\mathbf{v} - (\mathbf{v} \cdot \mathbf{w} \times \mathbf{z})\mathbf{u}$

61. $\mathbf{u} \times (\mathbf{v} \times \mathbf{w}) + \mathbf{v} \times (\mathbf{w} \times \mathbf{u}) + \mathbf{w} \times (\mathbf{u} \times \mathbf{v}) = \mathbf{0}$

62. $\mathbf{u} \times \mathbf{v} = (\mathbf{u} \cdot \mathbf{v} \cdot \mathbf{i})\mathbf{i} + (\mathbf{u} \cdot \mathbf{v} \cdot \mathbf{j})\mathbf{j} + (\mathbf{u} \cdot \mathbf{v} \times \mathbf{k})\mathbf{k}$

63. $\mathbf{u} \times [\mathbf{u} \times (\mathbf{u} \times \mathbf{v})] \cdot \mathbf{w} = -\|\mathbf{u}\|^2 \mathbf{u} \cdot \mathbf{v} \times \mathbf{w}$

64. Show that if vectors **OP**, **OQ**, **OR**, and **OS** lie in the same plane, then

$$(\mathbf{OP} \times \mathbf{OQ}) \times (\mathbf{OR} \times \mathbf{OS}) = \mathbf{0}$$

10.5 Lines and Planes in Space

IN THIS SECTION   lines in $\mathbb{R}^3$, direction cosines, planes in $\mathbb{R}^3$

LINES IN $\mathbb{R}^3$

As in the plane, a line in space is completely determined once we know one of its points and its direction. We used the concept of slope to measure the direction of a line in the plane, but in space, it is more convenient to specify direction with vectors.

Suppose L is a line in space whose location is determined by the vector **v** = $A\mathbf{i} + B\mathbf{j} + C\mathbf{k}$ and also suppose L contains $Q(x_0, y_0, z_0)$. We say that L is **aligned with v**, as shown in Figure 10.35.

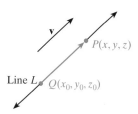

■ **FIGURE 10.35** If L is aligned with **v** and contains Q, then P is on L whenever **QP** = t**v**.

We also say that the line has **direction numbers** A, B, and C and denote these direction numbers by $[A, B, C]$. The vector **v** is called the direction vector of the line L. If $P(x, y, z)$ is any point on L, then the vector **QP** is parallel to **v** and must satisfy the vector equation **QP** = t**v** for some number t. If we introduce coordinates and use the standard representation, we can rewrite this vector equation as

$$(x - x_0)\mathbf{i} + (y - y_0)\mathbf{j} + (z - z_0)\mathbf{k} = t[A\mathbf{i} + B\mathbf{j} + C\mathbf{k}]$$

By equating components on both sides of this equation, we find that the coordinates of P must satisfy the linear system

$$x - x_0 = tA \qquad y - y_0 = tB \qquad z - z_0 = tC$$

where t is a real number.

Parametric Form of a Line in $\mathbb{R}^3$

If L is a line that contains the point (x_0, y_0, z_0) and is aligned with the vector $\mathbf{v} = A\mathbf{i} + B\mathbf{j} + C\mathbf{k}$, then the point (x, y, z) is on L if and only if its coordinates satisfy

$$x = x_0 + tA \qquad y = y_0 + tB \qquad z = z_0 + tC$$

for some number t.

WARNING t will be a different value for each point (x, y, z) on L.

Turning things around, if we are given the equation of a line with direction numbers $[A, B, C]$, then $\mathbf{v} = A\mathbf{i} + B\mathbf{j} + C\mathbf{k}$ is the **vector aligned with L.**

DRAWING LESSON 5: DRAWING A LINE IN SPACE

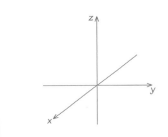

a. Draw the three coordinate axes.

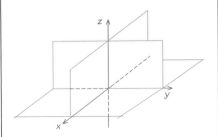

b. Draw the three coordinate planes.

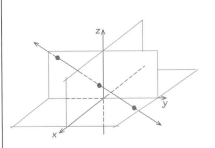

c. Plot points where the line intersects each coordinate plane. Used dashed lines for hidden parts.

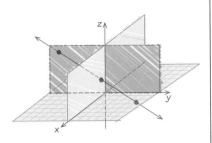

d. Use highlighters or pencils to color the planes to add depth to the figure.

EXAMPLE 1   *Parametric equations of a line in space*

Find the parametric equations for the line that contains the point $(3, 1, 4)$ and is aligned with the vector $\mathbf{v} = -\mathbf{i} + \mathbf{j} - 2\mathbf{k}$. Find where this line passes through the coordinate planes and sketch the line.

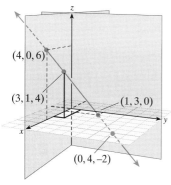

FIGURE 10.36 Graph of the line $x = 3 - t, y = 1 + t, z = 4 - 2t$

Solution   The direction numbers are $[-1, 1, -2]$ and $x_0 = 3$, $y_0 = 1$, $z_0 = 4$, so the line has the parametric form

$$x = 3 - t \qquad y = 1 + t \qquad z = 4 - 2t$$

This line will intersect the xy-plane when $z = 0$;

$$0 = 4 - 2t \quad \text{implies} \quad t = 2$$

If $t = 2$, then $x = 3 - 2 = 1$ and $y = 1 + 2 = 3$. This is the point $(1, 3, 0)$. Similarly, the line intersects the xz-plane at $(4, 0, 6)$ and the yz-plane at $(0, 4, -2)$. Plot these points and draw the line, as shown in Figure 10.36.   ■

In the special case where none of the direction numbers A, B, or C is 0, we can solve each of the parametric-form equations for t to obtain the following **symmetric equations** for a line.

Symmetric Form of a Line in $\mathbb{R}^3$

If L is a line that contains the point (x_0, y_0, z_0) and is aligned with the vector $\mathbf{v} = A\mathbf{i} + B\mathbf{j} + C\mathbf{k}$ (A, B, and C nonzero numbers), then the point (x, y, z) is on L if and only if its coordinates satisfy

$$\frac{x - x_0}{A} = \frac{y - y_0}{B} = \frac{z - z_0}{C}$$

| EXAMPLE 2 | *Symmetric form of the equation of a line in space* |

Find symmetric equations for the line L through the points $P(-1, 3, 7)$ and $Q(4, 2, -1)$. Find the points of intersection with the coordinate planes and sketch the line.

Solution   The required line passes through P and is aligned with the vector

$$\mathbf{PQ} = [4 - (-1)]\mathbf{i} + [2 - 3]\mathbf{j} + [-1 - 7]\mathbf{k} = 5\mathbf{i} - \mathbf{j} - 8\mathbf{k}$$

Thus, the direction numbers of the line are $[5, -1, -8]$, and we can choose either P or Q as (x_0, y_0, z_0). Choosing P, we obtain

$$\frac{x + 1}{5} = \frac{y - 3}{-1} = \frac{z - 7}{-8}$$

Next, we find points of intersection with the coordinate planes. For the xy-plane $(z = 0)$, we have

$$\frac{x + 1}{5} = \frac{0 - 7}{-8} \quad \text{and} \quad \frac{y - 3}{-1} = \frac{0 - 7}{-8}$$

$$x = \frac{27}{8} \qquad\qquad y = \frac{17}{8}$$

The point of intersection of the line with the xy-plane is $\left(\frac{27}{8}, \frac{17}{8}, 0\right)$. Similarly, the other intersections are

$$xz\text{-plane:} \quad (14, 0, -17) \qquad yz\text{-plane:} \quad \left(0, \frac{14}{5}, \frac{27}{5}\right)$$

The graph is shown in Figure 10.37.   ■

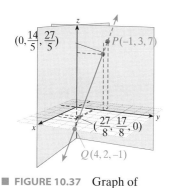

FIGURE 10.37 Graph of $\dfrac{x + 1}{5} = \dfrac{y - 3}{-1} = \dfrac{z - 7}{-8}$

Recall that two lines in $\mathbb{R}^2$ must intersect if their slopes are different (because they cannot be parallel). However, two lines in $\mathbb{R}^3$ may have different direction numbers and still not intersect because there is enough "room" in space for the lines to lie in parallel planes but be aligned with vectors that are not parallel. In this case, the lines are said to be **skew.** The three different situations that can occur (ignoring the trivial case where the lines coincide) are shown in Figure 10.38.

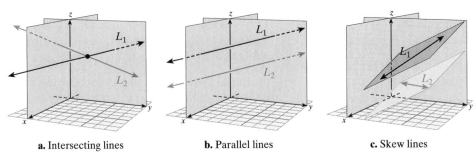

a. Intersecting lines   **b.** Parallel lines   **c.** Skew lines

■ **FIGURE 10.38**   Lines in space may intersect, be parallel, or skew.

EXAMPLE 3   *Skew lines in space*

Determine whether the following lines intersect, are parallel, or are skew.

$$L_1: \quad \frac{x-1}{2} = \frac{y+1}{1} = \frac{z-2}{4} \qquad \text{and} \qquad L_2: \quad \frac{x+2}{4} = \frac{y}{-3} = \frac{z+1}{1}$$

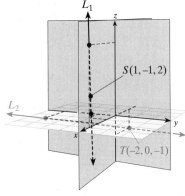

L_1

$S(1,-1,2)$

L_2

$T(-2,0,-1)$

Graph of L_1 and L_2. Note that $S(1, -1, 2)$ lies on L_1 and $T(-2, 0, -1)$ lies on L_2.

Solution   Note that L_1 has direction numbers [2, 1, 4] (that is, L_1 is aligned with $2\mathbf{i} + \mathbf{j} + 4\mathbf{k}$) and L_2 has direction numbers [4, −3, 1]. If we solve

$$\langle 2, 1, 4 \rangle = t\langle 4, -3, 1 \rangle$$

for t, we find no possible solution for any value of t. This implies that the lines are not parallel.

Next we determine whether the lines intersect or are skew. Note that $S(1, -1, 2)$ lies on L_1 and $T(-2, 0, -1)$ lies on L_2. The lines intersect if and only if there is a point P that lies on both lines. To determine this, we write the equations of the lines in parametric form. We use a different parameter for each line because it is possible for two lines to arrive at the same point "at different times" (values of the parameter):

$$L_1: x = 1 + 2s \qquad y = -1 + s \qquad z = 2 + 4s$$
$$L_2: x = -2 + 4t \qquad y = -3t \qquad z = -1 + t$$

The lines intersect if there are numbers s and t for which

$$x = 1 + 2s = -2 + 4t$$
$$y = -1 + s = -3t$$
$$z = 2 + 4s = -1 + t$$

This is equivalent to the system of linear equations

$$\begin{cases} 2s - 4t = -3 \\ s + 3t = 1 \\ 4s - t = -3 \end{cases}$$

Any solution of this system must correspond to a point of intersection of L_1 and L_2, and if no solution exists, then L_1 and L_2 are skew. Because this is a system of three equations with two unknowns, we first solve the first two equations simultaneously to find $s = -\frac{1}{2}, t = \frac{1}{2}$. Because $(s, t) = (-\frac{1}{2}, \frac{1}{2})$ does not satisfy the third equation, it follows that L_1 and L_2 do not intersect, so they must be skew. ∎

EXAMPLE 4   *Intersecting lines*

Show that the lines

$$L_1: \quad \frac{x-1}{2} = \frac{y+1}{1} = \frac{z-2}{4} \quad \text{and} \quad L_2: \quad \frac{x+2}{4} = \frac{y}{-3} = \frac{z-\frac{1}{2}}{-1}$$

intersect and find the point of intersection.

Solution L_1 has direction numbers $[2, 1, 4]$ and L_2 has direction numbers $[4, -3, -1]$. Because there is no t for which $[2, 1, 4] = t[4, -3, -1]$, the lines are not parallel. Express the lines in parametric form:

$$L_1: \quad x = 1 + 2s \qquad y = -1 + s \qquad z = 2 + 4s$$
$$L_2: \quad x = -2 + 4t \qquad y = -3t \qquad z = \tfrac{1}{2} - t$$

At an intersection point we must have

$$1 + 2s = -2 + 4t \quad \text{or} \quad 2s - 4t = -3$$
$$-1 + s = -3t \quad \text{or} \quad s + 3t = 1$$
$$2 + 4s = \tfrac{1}{2} - t \quad \text{or} \quad 4s + t = -\tfrac{3}{2}$$

Solving the first two equations simultaneously, we find $s = -\frac{1}{2}, t = \frac{1}{2}$. This solution satisfies the third equation, namely,

$$4(-\tfrac{1}{2}) + \tfrac{1}{2} = -\tfrac{3}{2}$$

To find the coordinates of the point of intersection, substitute $s = -\frac{1}{2}$ into the parametric-form equations for L_1 (or substitute $t = \frac{1}{2}$ into L_2) to obtain

$$x_0 = 1 + 2(-\tfrac{1}{2}) = 0$$
$$y_0 = -1 + (-\tfrac{1}{2}) = -\tfrac{3}{2}$$
$$z_0 = 2 + 4(-\tfrac{1}{2}) = 0$$

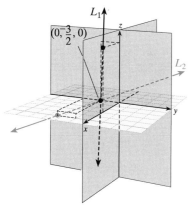

Graph of L_1 and L_2

Thus, the lines intersect at $P(0, -\frac{3}{2}, 0)$. ∎

DIRECTION COSINES

Besides using direction numbers, the direction of a nonzero vector $\mathbf{v} = a_1\mathbf{i} + a_2\mathbf{j} + a_3\mathbf{k}$ can be measured in terms of angles $\alpha, \beta,$ and γ between $\mathbf{v}$ and the coordinate axes, as shown in Figure 10.39. These angles are called the **direction angles** of $\mathbf{v}$ and their cosines are known as the **direction cosines** of $\mathbf{v}$.

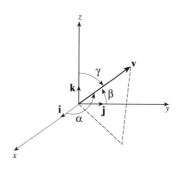

■ **FIGURE 10.39**   Direction cosines of the vector $\mathbf{v}$

Note that because $\mathbf{i}$ is a unit vector represented by $\langle 1, 0, 0 \rangle$, we have

$$\mathbf{v} \cdot \mathbf{i} = (a_1 \cdot 1) + (a_2 \cdot 0) + (a_3 \cdot 0) = a_1$$

Therefore,

$$a_1 = \mathbf{v} \cdot \mathbf{i} = \|\mathbf{v}\| \|\mathbf{i}\| \cos \alpha \quad \text{so that} \quad \cos \alpha = \frac{a_1}{\|\mathbf{v}\|}$$

Similar formulas hold for the other direction cosines. If $\mathbf{u}$ is a unit vector in the same direction as $\mathbf{v}$, we have

$$\mathbf{u} = \frac{\mathbf{v}}{\|\mathbf{v}\|} = \frac{a_1}{\|\mathbf{v}\|}\mathbf{i} + \frac{a_2}{\|\mathbf{v}\|}\mathbf{j} + \frac{a_3}{\|\mathbf{v}\|}\mathbf{k} = \cos \alpha \, \mathbf{i} + \cos \beta \, \mathbf{j} + \cos \gamma \, \mathbf{k}$$

$$\|\mathbf{u}\|^2 = \left(\frac{a_1}{\|\mathbf{v}\|}\right)^2 + \left(\frac{a_2}{\|\mathbf{v}\|}\right)^2 + \left(\frac{a_3}{\|\mathbf{v}\|}\right)^2 = \cos^2\alpha + \cos^2\beta + \cos^2\gamma = 1$$

EXAMPLE 5   *Direction angles and direction cosines*

Find the direction cosines and direction angles (to the nearest degree) of the vector $\mathbf{v} = -2\mathbf{i} + 3\mathbf{j} + 5\mathbf{k}$, and verify the formula $\cos^2\alpha + \cos^2\beta + \cos^2\gamma = 1$.

Solution   We find $\|\mathbf{v}\| = \sqrt{(-2)^2 + 3^2 + 5^2} = \sqrt{38}$. Therefore,

$$\cos \alpha = \frac{a_1}{\|\mathbf{v}\|} = \frac{-2}{\sqrt{38}} \approx -0.3244428 \qquad \alpha \approx \cos^{-1}(-0.324428) \approx 109°$$

$$\cos \beta = \frac{a_2}{\|\mathbf{v}\|} = \frac{3}{\sqrt{38}} \approx 0.4866642 \qquad \beta \approx \cos^{-1}(0.4866642) \approx 61°$$

$$\cos \gamma = \frac{a_3}{\|\mathbf{v}\|} = \frac{5}{\sqrt{38}} \approx 0.8111071 \qquad \gamma \approx \cos^{-1}(0.8111071) \approx 36°$$

$$\text{and} \quad \cos^2\alpha + \cos^2\beta + \cos^2\gamma = \left(\frac{-2}{\sqrt{38}}\right)^2 + \left(\frac{3}{\sqrt{38}}\right)^2 + \left(\frac{5}{\sqrt{38}}\right)^2 = 1 \qquad ■$$

$\gamma = 36°$ ◇ $\beta = 61°$

$\alpha = 109°$

Graph of $\mathbf{v} = -2\mathbf{i} + 3\mathbf{j} + 5\mathbf{k}$

P_0

■ **FIGURE 10.40**   A plane may be described by specifying one of its points and a normal vector $\mathbf{N}$.

PLANES IN $\mathbb{R}^3$

Planes in space can also be characterized by vector methods. In particular, any plane is completely determined once we know one of its points and its orientation—that is, the "direction" it faces. A common way to specify the direction of a plane is by means of a vector $\mathbf{N}$ that is orthogonal to every vector in the plane, as shown in Figure 10.40. Such a vector in called a **normal** to the plane.

> ■ *What This Says:*   To "get a handle" on a plane $Ax + By + Cz + D = 0$, we use the plane's normal vector. We will see in this section that the direction numbers of the normal vector are proportional to A, B, C.

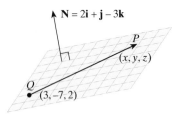

$\mathbf{N} = 2\mathbf{i} + \mathbf{j} - 3\mathbf{k}$

P
(x, y, z)

Q
$(3, -7, 2)$

Graph of plane in Example 6

EXAMPLE 6   *Obtain the equation for a plane*

Find an equation for the plane that contains the point $Q(3, -7, 2)$ and is normal to the vector $\mathbf{N} = 2\mathbf{i} + \mathbf{j} - 3\mathbf{k}$.

Solution   The normal vector $\mathbf{N}$ is orthogonal to every vector in the plane. In particular, if $P(x, y, z)$ is any point in the plane, then $\mathbf{N}$ must be orthogonal to the vector

$$\mathbf{QP} = (x - 3)\mathbf{i} + (y + 7)\mathbf{j} + (z - 2)\mathbf{k}$$

Because the dot (or scalar) product of two orthogonal vectors is 0, we have

$$\mathbf{N} \cdot \mathbf{QP} = 2(x - 3) + (1)(y + 7) + (-3)(z - 2) = 0$$
$$2x - 6 + y + 7 - 3z + 6 = 0$$
$$2x + y - 3z + 7 = 0$$

Therefore, $2x + y - 3z + 7 = 0$ is the equation of the plane.   ■

By generalizing the approach illustrated in Example 6, we can show that the plane that contains the point (x_0, y_0, z_0) and has normal vector $\mathbf{N} = A\mathbf{i} + B\mathbf{j} + C\mathbf{k}$ must have the Cartesian equation

$$A(x - x_0) + B(y - y_0) + C(z - z_0) = 0$$

This is called the **point-normal form** of the equation of a plane. By rearranging terms, we can rewrite this equation in the form $Ax + By + Cz + D = 0$. This is called the **standard form** of the equation of a plane. The numbers $[A, B, C]$ are called **attitude numbers** of the plane (see Figure 10.41).

Notice from Figure 10.41 that *attitude numbers of a plane are the same as direction numbers of a normal line*. This means that you can find normal vectors to a plane by *inspecting the equation of the plane*.

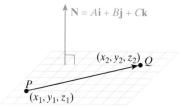

■ **FIGURE 10.41** The graph of a plane with attitude numbers $[A, B, C]$

Assignment:
Sketch
$3x + 2y + 6z = 18$

DRAWING LESSON 6: DRAWING A PLANE IN SPACE

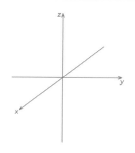

a. Draw the three coordinate axes.

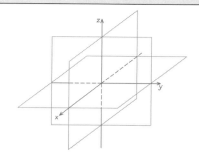

b. Draw the three coordinate planes.

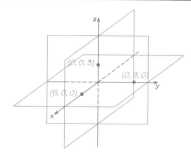

Let $x = 0$, $z = 0$, and find y; for this example, $y = 9$
Let $y = 0$, $x = 0$, and find z; for this example, $z = 3$
Let $z = 0$, $y = 0$, and find x; for this example, $x = 6$

c. Plot the points where the plane intersects the coordinate axes.

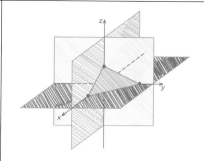

d. Connect the points plotted in part c and shade the part of the plane in the first octant.

e. If desired, extend the plane into the other octants. Use highlighters or pencils to color the planes to add depth to the figure.

| EXAMPLE 7 | *Relationship between normal vectors and planes* |

Find normal vectors to the planes

a. $5x + 7y - 3z = 0$ b. $x - 5y + \sqrt{2}z = 6$ c. $3x - 7z = 10$

Solution
a. A normal to the plane $5x + 7y - 3z = 0$ is $\mathbf{N} = 5\mathbf{i} + 7\mathbf{j} - 3\mathbf{k}$.
b. For $x - 5y + \sqrt{2}z = 6$, the normal is $\mathbf{N} = \mathbf{i} - 5\mathbf{j} + \sqrt{2}\mathbf{k}$.
c. For $3x - 7z = 10$, it is $\mathbf{N} = 3\mathbf{i} - 7\mathbf{k}$.   ∎

Forms for the Equation of a Plane

A plane with normal $\mathbf{N} = A\mathbf{i} + B\mathbf{j} + C\mathbf{k}$ that contains the point (x_0, y_0, z_0) has the following equations:

Point-normal form: $A(x - x_0) + B(y - y_0) + C(z - z_0) = 0$
Standard form: $Ax + By + Cz + D = 0$

for some constants $A, B, C,$ and D.

THEOREM 10.9   *The normal of a given plane*

Let $A, B, C,$ and D be constants with $A, B,$ and C not all zero. Then the graph of the equation

$$Ax + By + Cz + D = 0$$

is the equation of a plane with normal vector $\mathbf{N} = A\mathbf{i} + B\mathbf{j} + C\mathbf{k}$.

Proof   Suppose $A \neq 0$. Then the equation $Ax + By + Cz + D = 0$ can be written as

$$A\left[x + \frac{D}{A}\right] + By + Cz = 0$$

which is the point-normal form of the plane that passes through the point $(-D/A, 0, 0)$ with a normal $\mathbf{N} = A\mathbf{i} + B\mathbf{j} + C\mathbf{k}$. A similar argument applies if $A = 0$ and either B or C is not zero.   ═

| EXAMPLE 8 | *Equation of a line orthogonal to a given plane* |

Find an equation of the line that passes through the point $Q(2, -1, 3)$ and is orthogonal to the plane $3x - 7y + 5z + 55 = 0$. Where does the line intersect the plane?

Solution   By *inspection* of the equation of the plane, we see that $\mathbf{N} = 3\mathbf{i} - 7\mathbf{j} + 5\mathbf{k}$ is a normal vector. Because the required line is also orthogonal to the plane, it must be parallel to $\mathbf{N}$. Thus, the line contains the point $Q(2, -1, 3)$ and has direction numbers $[3, -7, 5]$, so its equation is

$$\frac{x - 2}{3} = \frac{y + 1}{-7} = \frac{z - 3}{5}$$

To find the point where this line intersects the plane, we first set the three fractions equal to t and rewrite the line in parametric form:

$$x = 2 + 3t, \quad y = -1 - 7t, \quad \text{and} \quad z = 3 + 5t$$

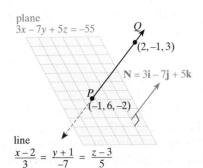

plane
$3x - 7y + 5z = -55$
Q
$(2, -1, 3)$
$\mathbf{N} = 3\mathbf{i} - 7\mathbf{j} + 5\mathbf{k}$
P
$(-1, 6, -2)$

line
$\dfrac{x-2}{3} = \dfrac{y+1}{-7} = \dfrac{z-3}{5}$

Now, substitute into the equation of the plane:

$$3(2 + 3t) - 7(-1 - 7t) + 5(3 + 5t) = -55$$
$$6 + 9t + 7 + 49t + 15 + 25t = -55$$
$$83t = -83$$
$$t = -1$$

Then, the point of intersection is found by substituting $t = -1$ for $x, y,$ and z:

$$x = 2 + 3(-1) = -1$$
$$y = -1 - 7(-1) = 6$$
$$z = 3 + 5(-1) = -2$$

The point of intersection is $(-1, 6, -2)$. ∎

> **EXAMPLE 9**   *Equation of a plane containing three given points*

Find the standard-form equation of a plane containing $P(-1, 2, 1)$, $Q(0, -3, 2)$, and $R(1, 1, -4)$.

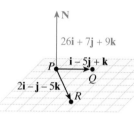

Solution   Because a normal $\mathbf{N}$ to the required plane is orthogonal to the vectors $\mathbf{PQ}$ and $\mathbf{PR}$, we find $\mathbf{N}$ by computing the cross product $\mathbf{N} = \mathbf{PQ} \times \mathbf{PR}$.

$$\mathbf{PQ} = (0 + 1)\mathbf{i} + (-3 - 2)\mathbf{j} + (2 - 1)\mathbf{k} = \mathbf{i} - 5\mathbf{j} + \mathbf{k}$$
$$\mathbf{PR} = (1 + 1)\mathbf{i} + (1 - 2)\mathbf{j} + (-4 - 1)\mathbf{k} = 2\mathbf{i} - \mathbf{j} - 5\mathbf{k}$$

$$\mathbf{N} = \mathbf{PQ} \times \mathbf{PR} = \begin{vmatrix} \mathbf{i} & \mathbf{j} & \mathbf{k} \\ 1 & -5 & 1 \\ 2 & -1 & -5 \end{vmatrix}$$
$$= (25 + 1)\mathbf{i} - (-5 - 2)\mathbf{j} + (-1 + 10)\mathbf{k}$$
$$= 26\mathbf{i} + 7\mathbf{j} + 9\mathbf{k}$$

We can now find the equation of the plane using this normal vector and any point in the plane. We will use the point P:

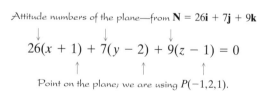

Attitude numbers of the plane—from $\mathbf{N} = 26\mathbf{i} + 7\mathbf{j} + 9\mathbf{k}$

$$26(x + 1) + 7(y - 2) + 9(z - 1) = 0$$

Point on the plane; we are using $P(-1, 2, 1)$.

Thus, the equation of the plane is

$$26x + 26 + 7y - 14 + 9z - 9 = 0$$
$$26x + 7y + 9z + 3 = 0$$ ∎

> **EXAMPLE 10**   *Equation of a line parallel to the intersection of two given planes*

Find the equation of a line passing through $(-1, 2, 3)$ that is parallel to the line of intersection of the planes $3x - 2y + z = 4$ and $x + 2y + 3z = 5$.

Solution   By inspection, we see that the normals to the given planes are $\mathbf{N}_1 = 3\mathbf{i} - 2\mathbf{j} + \mathbf{k}$ and $\mathbf{N}_2 = \mathbf{i} + 2\mathbf{j} + 3\mathbf{k}$. The desired line is perpendicular to both of these

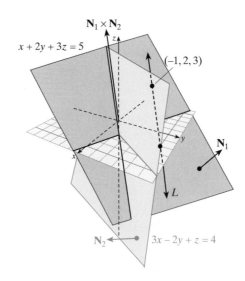

normals, so the aligned vector is found by computing the cross product:

$$\mathbf{N}_1 \times \mathbf{N}_2 = \begin{vmatrix} \mathbf{i} & \mathbf{j} & \mathbf{k} \\ 3 & -2 & 1 \\ 1 & 2 & 3 \end{vmatrix} = (-6 - 2)\mathbf{i} - (9 - 1)\mathbf{j} + (6 + 2)\mathbf{k}$$
$$= -8\mathbf{i} - 8\mathbf{j} + 8\mathbf{k}$$

The direction of this vector is $\langle -8, -8, 8 \rangle = -8\langle 1, 1, -1 \rangle$, so the equation of the desired line is

$$\frac{x + 1}{1} = \frac{y - 2}{1} = \frac{z - 3}{-1} \qquad \blacksquare$$

Example 10 can also be used to find the equation of the line of intersection of the two planes. Instead of using the given point $(-1, 2, 3)$, you will first need to find a point in the intersection and then proceed, using the steps of Example 10. We conclude this section by finding the equation of a plane containing two given (nonparallel) lines.

| EXAMPLE 11 | *Equation of a plane containing two intersecting lines* |

Find the standard-form equation of the plane determined by the intersecting lines

$$\frac{x - 2}{3} = \frac{y + 5}{-2} = \frac{z + 1}{4} \quad \text{and} \quad \frac{x + 1}{2} = \frac{y}{-1} = \frac{z - 16}{5}$$

Solution   Proceeding as in Example 4, we find that the lines intersect at $(-19, 9, -29)$. The aligned vectors for these two lines are $\mathbf{v}_1 = 3\mathbf{i} - 2\mathbf{j} + 4\mathbf{k}$ and $\mathbf{v}_2 = 2\mathbf{i} - \mathbf{j} + 5\mathbf{k}$. The normal to the desired plane is orthogonal to both $\mathbf{v}_1$ and $\mathbf{v}_2$, so we take the normal to be the cross product:

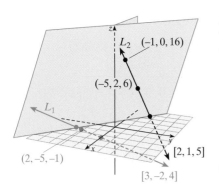

$$\mathbf{N} = \mathbf{v}_1 \times \mathbf{v}_2 = \begin{vmatrix} \mathbf{i} & \mathbf{j} & \mathbf{k} \\ 3 & -2 & 4 \\ 2 & -1 & 5 \end{vmatrix} = (-10 + 4)\mathbf{i} - (15 - 8)\mathbf{j} + (-3 + 4)\mathbf{k}$$
$$= -6\mathbf{i} - 7\mathbf{j} + \mathbf{k}$$

The point of intersection $P(-19, 9, -29)$ is certainly in the plane, as are $(2, -5, -1)$ and $(-1, 0, 16)$. We use $(2, -5, -1)$ to obtain

$$-6(x - 2) - 7(y + 5) + 1(z + 1) = 0$$
$$6x + 7y - z + 22 = 0 \qquad \blacksquare$$

10.5 Problem Set

(A) 1. ■ **What Does This Say?** Contrast the parametric and symmetric forms of the equation of a line.

2. ■ **What Does This Say?** Describe the relationship between normal vectors and planes.

Write each equation for a plane given in Problems 3–6 in standard form.

3. $4(x + 1) - 2(y + 1) + 6(z - 2) = 0$
4. $5(x - 2) - 3(y + 2) + 4(z + 3) = 0$
5. $-3(x - 4) + 2(y + 1) - 2(z + 1) = 0$
6. $-2(x + 1) + 4(y - 3) - 8z = 0$

Find the parametric and symmetric equations for the line(s) passing through the given points with the properties described in Problems 7–16.

7. $(1, -1, -2)$; parallel to $3\mathbf{i} - 2\mathbf{j} + 5\mathbf{k}$
8. $(1, 0, -1)$; parallel to $3\mathbf{i} + 4\mathbf{j}$
9. $(1, -1, 2)$; through $(2, 1, 3)$
10. $(2, 2, 3)$; through $(1, 3, -1)$
11. $(1, -3, 6)$; parallel to $\dfrac{x - 5}{1} = \dfrac{y + 2}{-3} = \dfrac{z}{-5}$
12. $(1, -1, 2)$; parallel to $\dfrac{x + 3}{4} = \dfrac{y - 2}{5} = \dfrac{z + 5}{1}$
13. $(0, 4, -3)$; parallel to $\dfrac{2x - 1}{22} = \dfrac{y + 2}{-6} = \dfrac{z - 1}{10}$
14. $(1, 0, -4)$; parallel to $x = -2 + 3t, y = 4 + t, z = 2 + 2t$
15. $(3, -1, 0)$; parallel to the xy-plane and the yz-plane
16. $(-1, 1, 6)$; perpendicular to $3x + y - 2z = 5$

Find the points of intersection of each line in Problems 17–20 with each of the coordinate planes.

17. $\dfrac{x - 4}{4} = \dfrac{y + 3}{3} = \dfrac{z + 2}{1}$
18. $\dfrac{x + 1}{1} = \dfrac{y + 2}{2} = \dfrac{z - 6}{3}$
19. $x = 6 - 2t, y = 1 + t, z = 3t$
20. $x = 6 + 3t, y = 2 - t, z = 2t$

In Problems 21–26, tell whether the two lines intersect, are parallel, are skew, or coincide. If they intersect, give the point of intersection.

21. $\dfrac{x - 4}{2} = \dfrac{y - 6}{-3} = \dfrac{z + 2}{5}$; $\dfrac{x}{4} = \dfrac{y + 2}{-6} = \dfrac{z - 3}{10}$
22. $x = 4 - 2t, y = 6t, z = 7 - 4t$;
    $x = 5 + t, y = 1 - 3t, z = -3 + 2t$
23. $x = 3 + 3t, y = 1 - 4t, z = -4 - 7t$;
    $x = -3t, y = 5 + 4t, z = 3 + 7t$
24. $x = 2 - 4t, y = 1 + t, z = \frac{1}{2} + 5t$;
    $x = 3t, y = -2 - t, z = 4 - 2t$
25. $\dfrac{x - 3}{2} = \dfrac{y - 1}{-1} = \dfrac{z - 4}{1}$;
    $\dfrac{x + 2}{3} = \dfrac{y - 3}{-1} = \dfrac{z - 2}{1}$
26. $\dfrac{x + 1}{2} = \dfrac{y - 3}{-1} = \dfrac{z - 2}{1}$;
    $\dfrac{x + 1}{2} = \dfrac{y + 1}{3} = \dfrac{z - 3}{-4}$

Find the direction cosines and the direction angles for the vectors given in Problems 27–32.

27. $\mathbf{v} = 2\mathbf{i} - 3\mathbf{j} - 5\mathbf{k}$
28. $\mathbf{v} = 3\mathbf{i} - 2\mathbf{k}$
29. $\mathbf{v} = 5\mathbf{i} - 4\mathbf{j} + 3\mathbf{k}$
30. $\mathbf{v} = \mathbf{j} - 5\mathbf{k}$
31. $\mathbf{v} = \mathbf{i} - 3\mathbf{j} + 9\mathbf{k}$
32. $\mathbf{v} = \mathbf{i} - \mathbf{j} + 3\mathbf{k}$

Find an equation for the plane that contains the point P and has the normal vector N given in Problems 33–38.

33. $P(-1, 3, 5)$; $\mathbf{N} = 2\mathbf{i} + 4\mathbf{j} - 3\mathbf{k}$
34. $P(0, -7, 1)$; $\mathbf{N} = -\mathbf{i} + \mathbf{k}$
35. $P(0, -3, 0)$; $\mathbf{N} = -2\mathbf{j} + 3\mathbf{k}$
36. $P(1, 1, -1)$; $\mathbf{N} = -\mathbf{i} - 2\mathbf{j} + 3\mathbf{k}$
37. $P(0, 0, 0)$; $\mathbf{N} = \mathbf{k}$
38. $P(0, 0, 0)$; $\mathbf{N} = \mathbf{i}$

39. Find two unit vectors parallel to the line
$$\dfrac{x - 3}{4} = \dfrac{y - 1}{2} = \dfrac{z + 1}{1}$$

40. Find two unit vectors parallel to the line
$$\dfrac{x - 1}{2} = \dfrac{y + 2}{4} = \dfrac{z + 5}{1}$$

41. Find two unit vectors perpendicular to the plane $2x + 4y - 3z = 4$.

42. Find two unit vectors perpendicular to the plane $5x - 3y + 2z = 15$.

(B) 43. Show that the vector $3\mathbf{i} - 4\mathbf{j} + \mathbf{k}$ is orthogonal to the line that passes through the points $P(0, 0, 1)$ and $Q(2, 1, -1)$.

44. Show that the vector $7\mathbf{i} + 4\mathbf{j} + 3\mathbf{k}$ is orthogonal to the line passing through the points $P(-2, 2, 7)$ and $Q(3, -3, 2)$.

45. Find two unit vectors that are parallel to the line of intersection of the planes $x + y = 1$ and $x - 2z = 3$.

46. Find two unit vectors that are parallel to the line of intersection of the planes $x + y + z = 3$ and $x - y + z = 1$.

47. Find an equation for the plane that passes through $P(1, -1, 2)$ and is normal to $\mathbf{PQ}$ where Q is $Q(2, 1, 3)$.

48. Find an equation for the line that passes through the point $(1, -5, 3)$ and is orthogonal to the plane $2x - 3y + z = 1$.

49. Find an equation for the plane that contains the point $(2, 1, -1)$ and is orthogonal to the line
$$\frac{x - 3}{3} = \frac{y + 1}{5} = \frac{z}{2}$$

50. Find a plane that passes through the point $(1, 2, -1)$ and is parallel to the plane $2x - y + 3z = 1$.

51. Show that the line
$$\frac{x - 1}{2} = \frac{y + 1}{3} = \frac{z - 2}{4}$$
is parallel to the plane $x - 2y + z = 6$.

52. Find the point where the line
$$\frac{x - 1}{2} = \frac{y + 1}{-1} = \frac{z}{3}$$
intersects the plane $3x + 2y - z = 5$.

53. The *angle* between two planes is defined to be the acute angle between their normal vectors. Find the angle between the planes $2x + y - 4z = 3$ and $x - y + z = 2$, rounded to the nearest degree.

54. Find the equation of the line that passes through the point $P(2, 3, 1)$ and is parallel to the line of intersection of the planes $x + 2y - 3z = 4$ and $x - 2y + z = 0$.

55. Find the equation of the line that passes through the point $P(0, 1, -1)$ and is parallel to the line of intersection of the planes $2x + y - 2z = 5$ and $3x - 6y - 2z = 7$.

56. Find a vector that is parallel to the line of intersection of the planes $2x + 3y = 0$ and $3x - y + z = 1$.

57. Find the equation of the line of intersection of the planes $3x + y - z = 5$ and $x - 6y - 2z = 10$.

58. Find the equation of the line of intersection of the planes $2x - y + z = 8$ and $x + y - z = 5$.

59. Let $\mathbf{v} = 2\mathbf{i} + \mathbf{j}$ and $\mathbf{w} = 2\mathbf{i} - \mathbf{j} = 3\mathbf{k}$. Find the direction cosines and the direction angles of $\mathbf{v} \times \mathbf{w}$.

60. Find the direction cosines of a vector determined by the line of intersection of the planes $x + y + z = 3$ and $2x + 3y - z = 4$.

Ⓒ 61. What can be said about the lines
$$\frac{x - x_0}{a_1} = \frac{y - y_0}{b_1} = \frac{z - z_0}{c_1}$$
and
$$\frac{x - x_0}{a_2} = \frac{y - y_0}{b_2} = \frac{z - z_0}{c_2}$$
in the case where $a_1 a_2 + b_1 b_2 + c_1 c_2 = 0$?

62. In Figure 10.42, $\mathbf{N}$ is normal to the plane P and L is a line that intersects P. Assume $\mathbf{N} = a\mathbf{i} + b\mathbf{j} + c\mathbf{k}$ and that L is given by
$$x = x_0 + At, \quad y = y_0 + Bt, \quad z = z_0 + Ct$$
  a. Find $\cos\theta$ for the angle θ between L and the plane P.
  b. Find the angle between the plane $x + y + z = 10$ and the line
$$\frac{x - 1}{2} = \frac{y + 3}{3} = \frac{z - 2}{-1}$$

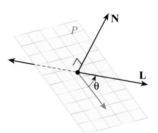

■ FIGURE 10.42   Problem 62

63. Show that a plane with x-intercept a, y-intercept b, and z-intercept c has the equation
$$\frac{x}{a} + \frac{y}{b} + \frac{z}{c} = 1$$
assuming $a, b,$ and c are all nonzero.

64. Suppose planes p_1 and p_2 intersect. If $\mathbf{v}_1$ and $\mathbf{w}_1$ are vectors on p_1, and $\mathbf{v}_2$ and $\mathbf{w}_2$ are on plane p_2, then show that
$$(\mathbf{v}_1 \times \mathbf{w}_1) \times (\mathbf{v}_2 \times \mathbf{w}_2)$$
is aligned with the line of intersection of the planes.

10.6   *Vector Methods for Measuring Distance in* $\mathbb{R}^3$

IN THIS SECTION   **distance from a point to a plane, distance from a point to a line**   ■

DISTANCE FROM A POINT TO A PLANE

To prepare for deriving a formula for the distance from a point to a plane, we will consider a simpler case, namely, the distance from a point to a line in $\mathbb{R}^2$. Let L be any given line and P any given point not on L. We wish to find the distance from P to L—that is, the perpendicular distance d, as shown in Figure 10.43.

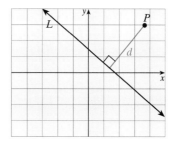

■ **FIGURE 10.43**   Distance from P to L

If L is a vertical line, then the distance from P to L is easy to find (why?). If L is not vertical, then we let Q be any point on the line and $\mathbf{N}$ be normal to L. Since Q can be any point on the line, we choose a convenient point, say the y-intercept (see Figure 10.44).

The distance we seek is seen to be the scalar projection of the vector $\mathbf{QP}$ onto $\mathbf{N}$. Thus,

$$d = \left| \frac{\mathbf{QP} \cdot \mathbf{N}}{\|\mathbf{N}\|} \right| = \frac{\left|\mathbf{QP} \cdot \mathbf{N}\right|}{\|\mathbf{N}\|}$$

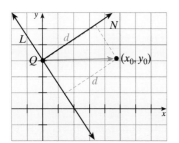

■ **FIGURE 10.44**   Procedure for finding the distance from a point to a line

In particular, we will now apply this formula to find the distance from the point $P(x_0, y_0)$ to the line

$$Ax + By + C = 0$$

Because we have chosen Q to be the y-intercept, $y = -C/B$ (because L is not vertical, $B \neq 0$). Then $\mathbf{QP} = (x_0 - 0)\mathbf{i} + (y_0 + C/B)\mathbf{j}$. It can be shown that the normal to the line $Ax + By + C = 0$ is $\mathbf{N} = A\mathbf{i} + B\mathbf{j}$. Then

$$d = \left| \frac{\mathbf{QP} \cdot \mathbf{N}}{\|\mathbf{N}\|} \right| = \frac{\left|Ax_0 + By_0 + C\right|}{\sqrt{A^2 + B^2}}$$

EXAMPLE 1   *Distance from a point to a line in $\mathbb{R}^2$*

Find the distance from the point $(5, -3)$ to the line $4x + 3y - 15 = 0$:
a. as a scalar projection   b. by using the formula

Solution

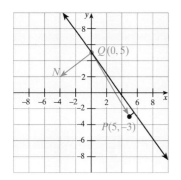

a. Let P be the point $(5, -3)$ and Q be the y-intercept $(0, 5)$ of the line. Then
   $\mathbf{QP} = 5\mathbf{i} - 8\mathbf{j}$ and $\mathbf{N} = 4\mathbf{i} + 3\mathbf{j}$

$$d = \left| \frac{\mathbf{QP} \cdot \mathbf{N}}{\|\mathbf{N}\|} \right| = \left| \frac{20 - 24}{\sqrt{16 + 9}} \right| = \left| \frac{-4}{5} \right| = \frac{4}{5}$$

b. Note $A = 4, B = 3, C = -15, x_0 = 5$, and $y_0 = -3$ so that

$$d = \frac{Ax_0 + By_0 + C}{\sqrt{A^2 + B^2}} = \left| \frac{4(5) + 3(-3) - 15}{\sqrt{4^2 + 3^2}} \right| = \left| \frac{-4}{5} \right| = \frac{4}{5}$$ ■

A projection is also used to obtain the following formula for the distance from a point to a plane.

THEOREM 10.10   *Distance from a point to a plane in $\mathbb{R}^3$*

The distance from the point (x_0, y_0, z_0) to the plane $Ax + By + Cz + D = 0$ is given by

$$d = \frac{\left|Ax_0 + By_0 + Cz_0 + D\right|}{\sqrt{A^2 + B^2 + C^2}}$$

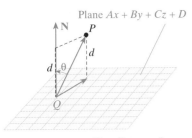

Plane $Ax + By + Cz + D$

■ **FIGURE 10.45**   The distance from a point to a plane in $\mathbb{R}^3$

If Q is any point in the given plane, the required distance is found by projecting the vector $\mathbf{QP}$ onto a normal $\mathbf{N}$ for the plane. Thus, the distance from the point to the plane is given by

$$d = \|\mathbf{QP}\|\,|\cos\theta| = \frac{\|\mathbf{QP}\|\,\|\mathbf{N}\|\,|\cos\theta|}{\|\mathbf{N}\|} = \frac{|\mathbf{QP}\cdot\mathbf{N}|}{\|\mathbf{N}\|}$$

where θ is the (acute) angle between $\mathbf{QP}$ and $\mathbf{N}$ (see Figure 10.45).

Suppose P has coordinates (x_0, y_0, z_0) and the given plane has the standard form $Ax + By + Cz + D = 0$. Then $\mathbf{N} = A\mathbf{i} + B\mathbf{j} + C\mathbf{k}$ is a normal to this plane, and if $Q(x_1, y_1, z_1)$ is any particular point in the plane, we have

$$\mathbf{QP} = (x_0 - x_1)\mathbf{i} + (y_0 - y_1)\mathbf{j} + (z_0 - z)\mathbf{k}$$

The dot product of $\mathbf{QP}$ with $\mathbf{N}$ is given by

$$\begin{aligned}
\mathbf{QP}\cdot\mathbf{N} &= (x_0 - x_1)A + (y_0 - y_1)B + (z_0 - z_1)C\\
&= (Ax_0 + By_0 + Cz_0) - (Ax_1 + By_1 + Cz_1)\\
&= Ax_0 + By_0 + Cz_0 - (-D) \quad \text{Because } Ax_1 + By_1 + Cz_1 + D = 0
\end{aligned}$$

Because the normal vector has length $\|\mathbf{N}\| = \sqrt{A^2 + B^2 + C^2}$, we can substitute into the formula to obtain

$$d = \left|\frac{\mathbf{QP}\cdot\mathbf{N}}{\|\mathbf{N}\|}\right| = \frac{|Ax_0 + By_0 + Cz_0 + D|}{\sqrt{A^2 + B^2 + C^2}}$$

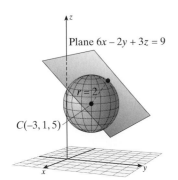

Plane $6x - 2y + 3z = 9$

$r = 2$

$C(-3, 1, 5)$

■ **FIGURE 10.46**   The sphere with center $C(-3, 1, 5)$ and radius 2 is tangent to the plane $6x - 2y + 3z = 9$.

EXAMPLE 2   *Equation of a sphere given a tangent plane*

Find an equation for the sphere with center $C(-3, 1, 5)$ that is tangent to the plane $6x - 2y + 3x = 9$.

Solution   The radius r of the sphere is the distance from the center C to the given tangent plane, as shown in Figure 10.46.

$$r = \left|\frac{6(-3) + (-2)(1) + 3(5) - 9}{\sqrt{6^2 + (-2)^2 + 3^2}}\right| = \left|\frac{-14}{7}\right| = 2$$

Therefore, an equation of the sphere is

$$(x + 3)^2 + (y - 1)^2 + (z - 5)^2 = 2^2 \qquad ■$$

Vector methods can also be used to derive a formula for the distance between two skew lines L_1 and L_2 in $\mathbb{R}^3$.

THEOREM 10.11   *Distance between skew lines in $\mathbb{R}^3$*

Assume L_1 and L_2 are skew lines containing the points P_1 and P_2 and are aligned with the vectors $\mathbf{v}_1$ and $\mathbf{v}_2$, respectively. Then the distance between the lines is

$$d = \left|\frac{(\mathbf{v}_1 \times \mathbf{v}_2)\cdot\mathbf{P_1P_2}}{\|\mathbf{v}_1 \times \mathbf{v}_2\|}\right| = \left|\frac{\mathbf{N}\cdot\mathbf{P_1P_2}}{\|\mathbf{N}\|}\right|$$

Proof   You are asked to prove this theorem in Problem 32. Notice that the distance d between L_1 and L_2 (see Figure 10.47) is the same as the distance between two parallel planes containing the lines. Because $\mathbf{N} = \mathbf{v}_1 \times \mathbf{v}_2$ is normal to both planes, it follows that the required distance d is a scalar multiple of $\|\mathbf{v}_1 \times \mathbf{v}_2\|$.   =

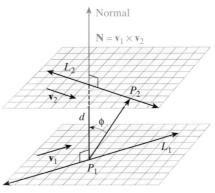

■ **FIGURE 10.47** Distance between lines

DISTANCE FROM A POINT TO A LINE

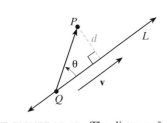

■ **FIGURE 10.48** The distance from a point to a line in $\mathbb{R}^3$

Next, we shall derive a formula for the distance from a point P to a line L in $\mathbb{R}^3$. Let Q be a point on L and let $\mathbf{v}$ be a vector aligned with L. Then, as shown in Figure 10.48, the distance from P to L is given by

$$d = \|\mathbf{QP}\| \left|\sin \theta\right|$$

where θ is the acute angle between $\mathbf{v}$ and the vector $\mathbf{QP}$.

This reminds us of the cross product $\mathbf{v} \times \mathbf{QP}$, and because

$$\|\mathbf{v} \times \mathbf{QP}\| = \|\mathbf{v}\| \|\mathbf{QP}\| \left|\sin \theta\right|$$

we have

$$d = \|\mathbf{QP}\| \left|\sin \theta\right| = \frac{\|\mathbf{v} \times \mathbf{QP}\|}{\|\mathbf{v}\|}$$

THEOREM 10.12   *Distance from a point to a line*

The distance from the point P to the line L is given by the formula

$$d = \frac{\|\mathbf{v} \times \mathbf{QP}\|}{\|\mathbf{v}\|}$$

where $\mathbf{v}$ is a vector aligned with L and Q is any point on L.

Proof   A sketch of the proof precedes the statement of the theorem.   =

EXAMPLE 3   *Distance from a point to a line*

Find the distance from the point $P(3, -8, 1)$ to the line

$$\frac{x - 3}{3} = \frac{y + 7}{-1} = \frac{z + 2}{5}$$

Solution   We need to find a point Q on the line. We see that $Q(3, -7, -2)$ is on the line and that

$$\mathbf{QP} = -\mathbf{j} + 3\mathbf{k}$$

The vector $\mathbf{v}$ aligned with L is $\mathbf{v} = 3\mathbf{i} - \mathbf{j} + 5\mathbf{k}$. We now find

$$\mathbf{v} \times \mathbf{QP} = \begin{vmatrix} \mathbf{i} & \mathbf{j} & \mathbf{k} \\ 3 & -1 & 5 \\ 0 & -1 & 3 \end{vmatrix} = (-3 + 5)\mathbf{i} - (9 - 0)\mathbf{j} + (-3 + 0)\mathbf{k}$$

$$= 2\mathbf{i} - 9\mathbf{j} - 3\mathbf{k}$$

Finally,

$$d = \frac{\|\mathbf{v} \times \mathbf{QP}\|}{\|\mathbf{v}\|} = \frac{\sqrt{(2)^2 + (-9)^2 + (-3)^2}}{\sqrt{3^2 + (-1)^2 + 5^2}} = \frac{\sqrt{94}}{\sqrt{35}} \approx 1.64$$ ∎

10.6 Problem Set

A *Find the distance between the point and the line in Problems 1–6.*

1. $(4, 5); 3x - 4y + 8 = 0$

2. $(9, -3); 3x - 4y + 8 = 0$

3. $(4, -3); 12x + 5y - 2 = 0$

4. $(1, -6); x - 3y + 15 = 0$

5. $(8, 14); x - 3y + 15 = 0$

6. $(4, 5); 2x - 5y = 0$

Find the distance between the point and the plane given in Problems 7–12.

7. $P(1, 0, -1); x + y - z = 1$

8. $P(0, 0, 0); 2x - 3y + 5z = 10$

9. $P(1, 1, -1); x - y + 2z = 4$

10. $P(2, 1, -2); 3x - 4y + z = -1$

11. $P(a, -a, 2a); 2ax - y + az = 4a, a \neq 0$

12. $P(a, 2a, 3a); 3x - 2y + z = -1/a, a \neq 0$

Find the distance from the point $(-1, 2, 1)$ to each plane given in Problems 13–16.

13. the plane through the points $A(0, 0, 0)$, $B(1, 2, 4)$, and $C(-2, -1, 1)$

14. the plane through the point $(1, 0, 1)$ with normal vector $2\mathbf{i} - \mathbf{j} + 2\mathbf{k}$

15. the plane through the point $(-3, 5, 1)$ with normal vector $3\mathbf{i} + \mathbf{j} + 5\mathbf{k}$

16. the plane through the points $A(-1, 1, 1)$, $B(4, 3, 7)$, and $C(3, -1, 0)$

Find the distance from the point P to the line L in Problems 17–22.

17. $P(1, 0, -1); \dfrac{x - 2}{3} = \dfrac{y + 1}{1} = \dfrac{z - 1}{2}$

18. $P(1, 0, 1); \dfrac{x}{3} = \dfrac{y - 1}{2} = \dfrac{z}{1}$

19. $P(1, -2, 2); \dfrac{x}{1} = \dfrac{2y}{1} = \dfrac{z}{-1}$

20. $P(0, 1, -1); \dfrac{2x - 1}{2} = \dfrac{y}{2} = \dfrac{z}{-1}$

21. $P(a, 0, -a); \dfrac{x + a}{2} = \dfrac{y - a}{1} = \dfrac{z - a}{2}; a \neq 0$

22. $P\left(0, a, \dfrac{a}{2}\right); \dfrac{x - a}{1} = \dfrac{y}{1} = \dfrac{z + 4a}{1}$

B **23.** Find the equation of the sphere with center $C(-2, 3, 7)$ that is tangent to the plane $2x + 3y - 6z = 5$.

24. a. Show that the line

$$\frac{x - 1}{3} = \frac{y}{-2} = \frac{z + 1}{1}$$

is parallel to the plane $x + 2y + z = 1$.

b. Find the distance from the line to the plane in part **a**.

25. Three of the four vertices of a parallelogram in $\mathbb{R}^3$ are $Q(-1, 3, 5)$, $R(6, -3, 2)$, and $S(2, 4, -3)$. What is the area of the parallelogram?

26. Find an equation for the set of all points $P(x, y, z)$ such that the distance from P to the point $P_0(-1, 2, 4)$ is the same as the distance from P to the plane $2x - 5y + 3z = 7$. (Do not expand binomials.)

27. Find an equation for the set of all points $P(x, y, z)$ such that the distance from P to the line

$$\frac{x - 1}{4} = \frac{y + 1}{-1} = \frac{z}{3}$$

is 5. (Do not expand trinomials.)

Find the (perpendicular) distance between the lines given in Problems 28–31.

28. $\dfrac{x + 1}{3} = \dfrac{y - 2}{-2} = \dfrac{z - 1}{1}$ and $\dfrac{x - 2}{5} = \dfrac{y + 1}{1} = \dfrac{z}{3}$

29. $x = 2 - t, y = 5 + 2t, z = 3t$ and $x = 2t, y = -1 - t, z = 1 + 2t$

30. $\dfrac{x+1}{1} = \dfrac{y-3}{2} = \dfrac{z+2}{3}$ and the line

passing through $(1, 3, -2)$ and $(0, 1, -1)$

31. $x = -1 + t$, $y = -2t$, $z = 3$ and the line passing through $(0, -1, 2)$ and $(1, -2, 3)$

C 32. Prove Theorem 10.11 (distance between skew lines in $\mathbb{R}^3$).

33. The planes $Ax + By + Cz = D_1$ and $Ax + By + Cz = D_2$ are parallel.

    a. Is the distance between the planes $\left|D_1 - D_2\right|$? If not, what is the correct formula?

b. Find the distance between the parallel planes

$$x + y + 2z = 2 \quad \text{and} \quad x + y + 2z = 4$$

34. Show that the planes

$$A_1x + B_1y + C_1z + D_1 = 0$$

and

$$A_2x + B_2y + C_2z + D_2 = 0$$

are mutually orthogonal if and only if

$$A_1A_2 + B_1B_2 + C_1C_2 = 0$$

Chapter 10 Review

Proficiency Examination

Concept Problems

1. What is a vector and what is a scalar?
2. Give both an algebraic and a geometric interpretation of multiplication of a vector by a scalar.
3. What is the parallelogram rule for vector sums?
4. State each of the following properties of vector operations:

    a. commutativity of vector addition
    b. associativity of vector addition
    c. identity for vector addition
    d. inverse property for vector addition
    e. magnitude of a vector for dot product
    f. commutativity for dot product
    g. dot product of a scalar multiple
    h. distributivity for dot product over addition
    i. cross product of the zero vector
    j. anticommutativity for cross product
    k. distributivity for cross product over addition

5. If $\mathbf{u}$ and $\mathbf{v}$ are parallel vectors, what can be said about $\mathbf{u} \times \mathbf{v}$?
6. How do you find the length of a vector? What is a unit vector?
7. State the triangle inequality.
8. What are the standard basis vectors in $\mathbb{R}^3$?
9. What is the standard-form equation of a sphere?
10. What is a cylinder?
11. What is the distance between two points in $\mathbb{R}^3$?
12. How do you find a unit vector $\mathbf{u}$ in the direction of a given vector $\mathbf{v}$?
13. What is Lagrange's identity?
14. Define dot product.

15. What is the formula for the angle between two vectors?
16. What is meant by *orthogonal* vectors? What is the algebraic condition for orthogonality?
17. What is a vector projection? Give a formula for finding the vector projection of $\mathbf{v}$ onto $\mathbf{w}$.
18. What is a scalar projection? Give a formula for finding the scalar projection of $\mathbf{v}$ onto $\mathbf{w}$.
19. What is the vector formula for work?
20. Define cross product.
21. What is the right-hand rule for a coordinate system?
22. a. Give a geometric interpretation of cross product.
    b. What is the formula for magnitude of a cross product?
23. What is the determinant form for the triple scalar product?
24. How do you find the volume of a parallelepiped?
25. a. What is the parametric form of a line in $\mathbb{R}^3$?
    b. What is the symmetric form of a line in $\mathbb{R}^3$?
26. What are the direction cosines of a vector in $\mathbb{R}^3$?
27. What is the normal vector of a plane in $\mathbb{R}^3$?
28. What is the point-normal form of a plane in $\mathbb{R}^3$?
29. What is the standard form of a plane in $\mathbb{R}^3$?
30. What is the formula for the distance from a point to a plane?
31. What is the formula for the distance from a point to a line?

Practice Problems

32. Given $\mathbf{v} = 2\mathbf{i} - 3\mathbf{j} + \mathbf{k}$, $\mathbf{w} = 3\mathbf{i} - 2\mathbf{j}$. Find each of the following vectors.
    a. $2\mathbf{v} + 3\mathbf{w}$
    b. $\|\mathbf{v}\|^2 - \|\mathbf{w}\|^2$
    c. vector projection of $\mathbf{v}$ onto $\mathbf{w}$

d. scalar projection of **w** onto **v**

e. **v · w**

f. **v × w**

33. Given $\mathbf{u} = 2\mathbf{i} - 3\mathbf{j} + \mathbf{k}$, $\mathbf{v} = \mathbf{i} + \mathbf{j} - 2\mathbf{k}$, and $\mathbf{w} = 3\mathbf{i} + 5\mathbf{k}$. In each of the following cases, either perform the indicated computation or explain why it is not defined.

   a. $(\mathbf{u} \times \mathbf{v}) \cdot \mathbf{w}$ b. $(\mathbf{u} \cdot \mathbf{v}) \times \mathbf{w}$

   c. $(\mathbf{u} \times \mathbf{v}) \times \mathbf{w}$ d. $(\mathbf{u} \cdot \mathbf{v}) \cdot \mathbf{w}$

Find the equations for the lines and planes in Problems 34–37.

34. the line through the points $P(-1, 4, -3)$ and $Q(0, -2, 1)$

35. the plane that contains the point $P(1, 1, 3)$ and is normal to the vector $\mathbf{v} = 2\mathbf{i} + 3\mathbf{k}$

36. the line of intersection of the planes $2x + 3y + z = 2$ and $y - 3z = 5$

37. the plane that contains the points $P(0, 2, -1)$, $Q(1, -3, 5)$, and $R(3, 0, -2)$

38. Find the direction cosines and the direction angles of the vector $\mathbf{u} = -2\mathbf{i} + 3\mathbf{j} + \mathbf{k}$. Round to the nearest degree.

39. In each case, determine whether the lines intersect, are parallel, or are skew. If they intersect, find the point of intersection.

   a. $x = 2t - 3$, $y = 4 - t$, $z = 2t$; and
$$\frac{x + 2}{3} = \frac{y - 3}{5}; z = 3$$

b. $\dfrac{x - 7}{5} = \dfrac{y - 6}{4} = \dfrac{z - 8}{5}$; and
$$\frac{x - 8}{6} = \frac{y - 6}{4} = \frac{z - 9}{6}$$

40. Let $\mathbf{u} = 2\mathbf{i} + \mathbf{j}$, $\mathbf{v} = \mathbf{i} - \mathbf{j} - \mathbf{k}$, and $\mathbf{w} = 3\mathbf{i} + 5\mathbf{k}$.

   a. Find the volume of the parallelepiped determined by these vectors.

   b. Find a positive number A that guarantees that the tetrahedron determined by $A\mathbf{u}$, $A\mathbf{v}$, and $\mathbf{w}$ has volume that is twice the volume of the original tetrahedron.

41. Find the distance from $P(-1, 1, 4)$ to $2x + 5y - z = 3$.

42. Find the distance between the skew lines $x = t$, $y = 2t$, $z = 3t - 1$ and $x = 1 - t$, $y = t + 2$, $z = t$.

43. Find the distance from the point $P(4, 5, 0)$ to the line
$$\frac{x - 2}{3} = \frac{y}{5} = \frac{z + 1}{-1}$$

44. An airplane flies at 200 mi/h parallel to the ground at an altitude of 10,000 ft. If the plane flies due south and the wind is blowing toward the northeast at 50 mi/h, what is the ground speed of the plane (that is, effective speed)?

45. A girl pulls a sled 50 ft on level ground with a rope inclined at an angle of 30° with the horizontal (the ground). If she applies 3 lb of tension to the rope, how much work is performed on the sled?

Supplementary Problems

1. A triangle in $\mathbb{R}^3$ has vertices $A(0, 2, -1)$, $B(1, 1, 3)$, and $C(1, 0, -4)$.

   a. Find the perimeter of the triangle.

   b. Find the area of the triangle.

   c. Find the three vertex angles of the triangle. (Round to the nearest degree.)

   d. Find a number p such that the points A, B, C, and $D(p, p, 0)$ form a tetrahedron of volume $V = 100$ cubic units.

Find equations, in both parametric and symmetric forms, of the lines described in Problems 2–3. Find two additional points on each line.

2. passing through $A(1, -2, 3)$, $B(4, -1, 2)$

3. passing through $P(1, 4, 0)$, with direction numbers $[2, 0, 1]$

4. Find the equation of the line passing through $P(3, 4, -1)$ and parallel to the line of intersection of the planes
$$x + 2y + 2z + 5 = 0 \text{ and } 2x + y - 3z - 6 = 0.$$

Find the equation of the plane satisfying the conditions given in Problems 5–12.

5. the xy-plane

6. the plane parallel to the xz-plane passing through $(4, 3, 7)$

7. the plane through $(1, -3, 4)$ with attitude numbers $[3, 4, -1]$

8. the plane through $(-1, 4, 5)$ and orthogonal to a line with direction numbers $[4, 4, -3]$

9. the plane through $(4, -3, 2)$ and parallel to the plane $5x - 2y + 3z - 10 = 0$

10. the plane containing
$$\frac{x - 3}{4} = \frac{z - 1}{2}; y = -2; \text{ and } \frac{x - 3}{3} = \frac{y + 2}{1} = \frac{z - 1}{-2}$$

11. the plane passing through $P(4, 1, 3)$, $Q(-4, 2, 1)$, and $R(1, 0, 2)$

12. the plane passing through $(4, -1, 2)$ and parallel to the line
$$\frac{x + 2}{3} = \frac{y - 2}{-1} = \frac{z + 1}{2} \text{ and } \frac{x - 2}{1} = \frac{y - 3}{2} = \frac{z - 4}{3}$$

13. If **u** and **v** are orthogonal unit vectors, show that $(\mathbf{u} \times \mathbf{v}) \times \mathbf{u} = \mathbf{v}$. What is $(\mathbf{u} \times \mathbf{v}) \times \mathbf{v}$?

14. Show that three planes where normals **u**, **v**, **w** satisfies $\mathbf{u} \times \mathbf{v} \cdot \mathbf{w} \neq 0$ intersect in exactly one point.

15. Show that $\mathbf{u} \times (\mathbf{v} \times \mathbf{w}) = (\mathbf{u} \times \mathbf{v}) \times \mathbf{w}$ if and only if $\mathbf{v} \times (\mathbf{w} \times \mathbf{u}) = \mathbf{0}$.

16. Given the vectors $\mathbf{v} = 3\mathbf{i} - 2\mathbf{j} + \mathbf{k}$ and $\mathbf{w} = 4\mathbf{i} + \mathbf{j} - 3\mathbf{k}$, find $\|\mathbf{v}\|$, $\mathbf{v} - \mathbf{w}$, and $2\mathbf{v} + 3\mathbf{w}$.

17. Given $\mathbf{u} = \mathbf{i} - \mathbf{j} + \mathbf{k}$, $\mathbf{v} = 3\mathbf{i} - 2\mathbf{j} + 5\mathbf{k}$, and $\mathbf{w} = \mathbf{i} + \mathbf{j} - \mathbf{k}$, find $(\mathbf{u} - \mathbf{v}) \cdot \mathbf{w}$ and $(2\mathbf{u} + \mathbf{v}) \times (\mathbf{u} - \mathbf{w})$.

18. Given the vectors $\mathbf{v} = 4\mathbf{i} + 2\mathbf{j} + \mathbf{k}$, $\mathbf{w} = 2\mathbf{i} + \mathbf{j} - 5\mathbf{k}$. Find
    a. $5\mathbf{v} - 3\mathbf{w}$
    b. $\|2\mathbf{v} - \mathbf{w}\|$
    c. vector projection of $\mathbf{v}$ onto $\mathbf{w}$
    d. scalar projection of $\mathbf{w}$ onto $\mathbf{v}$

19. Find the direction cosines for $\mathbf{v} = (2\mathbf{i} + \mathbf{j}) \times (\mathbf{i} + \mathbf{j} - 3\mathbf{k})$.

20. Find two unit vectors that are parallel to the line
$$\frac{x}{6} = \frac{y}{2} = \frac{z - 1}{6}$$

21. Find the (acute) angle, rounded to the nearest degree, between the intersecting lines
$$\frac{x - 1}{3} = \frac{y - 3}{-1} = \frac{z + 5}{2} \text{ and } \frac{x - 1}{2} = \frac{y - 3}{-1} = \frac{z + 5}{-2}$$

22. Find the area of the parallelogram determined by $3\mathbf{i} - 4\mathbf{j}$ and $-\mathbf{i} - \mathbf{j} + \mathbf{k}$.

23. Find the center and the radius of the sphere
$$4x^2 + 4y^2 + 4z^2 + 12y - 4z + 1 = 0$$

24. Find the points of intersection of the line $x = 6 + 3t$, $y = 10 - 2t$, $z = 5t$ with each of the coordinate planes.

25. Find the point of intersection of the planes $3x - y + 4z = 15$, $2x + y - 3z - 1 = 0$, and $x + 3y + 5z - 2 = 0$

26. Find the equation of the plane determined by the intersecting lines
$$\frac{x + 3}{3} = \frac{y}{-2} = \frac{z - 7}{6} \text{ and } \frac{x + 6}{1} = \frac{y + 5}{-3} = \frac{z - 1}{2}$$

27. Find an equation for the set of all points that are equidistant from the planes $3x - 4y + 12z = 6$ and $4x + 3z = 7$.

28. Vertices B and C of $\triangle ABC$ lie along the line
$$\frac{x + 2}{2} = \frac{y - 1}{1} = \frac{z}{4}$$
Find the area of the triangle given that A has coordinates $(1, -1, 2)$ and disk $\overline{BC}$ has length 5.

29. Find the work done by the constant force $\mathbf{F} = 5\mathbf{i} + 4\mathbf{j} + \mathbf{k}$ in moving a particle along the line from $P(2, 1, -1)$ to $Q(4, 1, 2)$.

30. How much work does it take to move a container 25 m along a horizontal loading platform onto a truck using a constant force of 100 newtons at an angle of $\frac{\pi}{6}$ from the horizontal?

31. Find a formula for the surface area of the tetrahedron determined by vectors $\mathbf{u}$, $\mathbf{v}$, and $\mathbf{w}$. Assume the vectors do not all lie in the same plane.

32. Suppose $\mathbf{v}$ and $\mathbf{w}$ are nonzero vectors. Show that $\|\mathbf{v}\|\mathbf{w} + \|\mathbf{w}\|\mathbf{v}$ and $\|\mathbf{v}\|\mathbf{w} - \|\mathbf{w}\|\mathbf{v}$ are orthogonal vectors.

33. Let $\mathbf{v} = \cos\theta \mathbf{i} + \sin\theta \mathbf{j}$ and $\mathbf{w} = \cos\phi \mathbf{i} + \sin\phi \mathbf{j}$. Find $\mathbf{v} \times \mathbf{w}$. Interpret this cross product geometrically, and use it to derive a well-known trigonometric identity.

34. Find the three vertex angles (rounded to the nearest degree) of the triangle whose vertices are $(1, -2, 3)$, $(-1, 2, -3)$, and $(2, 1, -3)$.

35. Find a relationship between the numbers a_1, b_1, and c_1 so that the angle between the vectors $\mathbf{v} = a_1\mathbf{i} + b_1\mathbf{j} + c_1\mathbf{k}$ and $\mathbf{i} - 2\mathbf{j}$ is the same as the angle between $\mathbf{v}$ and $2\mathbf{i} + \mathbf{k}$.

36. Find an equation of the plane that passes through $(a, 0, 0)$, $(0, a, 0)$, and $(0, 0, a)$.

37. Find the area of the triangle with vertices $A(0, -1, 2)$, $B(1, 2, -1)$, and $C(3, -1, 2)$.

38. Find an equation for the set of all points equidistant from $D(0, 0, 6)$ and the xy-plane.

39. Find the area of the triangle determined by the vectors $\mathbf{v} = \mathbf{i} - \mathbf{j} + \mathbf{k}$ and $\mathbf{w} = 2\mathbf{i} + \mathbf{j} - 2\mathbf{k}$.

40. Find an equation for the plane that passes through the origin and is parallel to the vectors $\mathbf{v} = \mathbf{i} - 2\mathbf{j} + 3\mathbf{k}$ and $\mathbf{w} = -\mathbf{i} + \mathbf{j} + 2\mathbf{k}$.

41. Find an equation for the plane that passes through the origin and whose normal vector is parallel to the line of intersection of the planes $2x - y + z = 4$ and $x + 3y - z = 2$.

42. Find a number A such that the planes $2Ax + 3y + z = 1$ and $x - Ay + 3z = 5$ are orthogonal.

43. The lines L_1 and L_2 are aligned with the vectors $\mathbf{v}_1 = \mathbf{i} - \mathbf{j}$ and $\mathbf{v}_2 = \mathbf{i} - \mathbf{j} + 2\mathbf{k}$, respectively. Find an equation for the line L that passes through the point $(-1, 2, 0)$ and is orthogonal to both L_1 and L_2.

44. A parallelepiped is determined by the vectors $\mathbf{u} = \mathbf{i} - \mathbf{j} + \mathbf{k}$, $\mathbf{v} = \mathbf{i} + 2\mathbf{j} - \mathbf{k}$, and $\mathbf{w} = 2\mathbf{i} + \mathbf{j} + \mathbf{k}$. Find the altitude from the tip of $\mathbf{w}$ to the side determined by $\mathbf{u}$ and $\mathbf{v}$.

45. In Chapter 11, we show that $\mathbf{T} = \mathbf{i} + 2x\mathbf{j}$ is a vector in the direction of the tangent line at each point $P(x, x^2)$ on the parabola $y = x^2$. Find a unit vector normal to the parabola at the point $(3, 9)$.

46. For any nonzero vectors $\mathbf{u}$, $\mathbf{v}$, and $\mathbf{w}$, show that the vector $(\mathbf{u} \times \mathbf{v}) \times (\mathbf{u} \times \mathbf{w})$ is parallel to $\mathbf{u}$.

47. Let $\mathbf{v} = a\mathbf{i} + b\mathbf{j} + c\mathbf{k}$ and $\mathbf{w} = A\mathbf{i} + B\mathbf{j} + C\mathbf{k}$, where a, b, c, A, B, and C are constants. Describe the set of vectors $\mathbf{v} + t\mathbf{w}$, where t is any scalar.

48. Use vectors to show that the sum of the squares of the lengths of the sides of a parallelogram equals the sum of the squares of the lengths of the diagonals.

49. The vectors **u**, **v**, and **w** are said to be *linearly independent* in $\mathbb{R}^3$ if the only solution to the equation $a\mathbf{u} + b\mathbf{v} + c\mathbf{w} = \mathbf{0}$ is $a = b = c = 0$. Otherwise, the vectors are *linearly dependent*. Determine whether the vectors $\mathbf{u} = -\mathbf{i} + 2\mathbf{k}, \mathbf{v} = 2\mathbf{i} - \mathbf{j} + 3\mathbf{k}, \mathbf{w} = \mathbf{i} + 3\mathbf{j} - 2\mathbf{k}$ are linearly independent or dependent.

50. Figure 10.49 shows a parallelogram $ABCD$. If M is the midpoint of side $\overline{AB}$, show that the line $\overline{CM}$ intersects diagonal $\overline{BD}$ at a point P located one-third of the distance from B to D by completing the following steps.

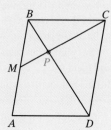

■ **FIGURE 10.49** Parallelogram $ABCD$

a. Let a and b be scalars such that $\mathbf{MP} = a\mathbf{MC}$ and $\mathbf{BP} = b\mathbf{BD}$. Show that
$$\tfrac{1}{2}\mathbf{AB} + b[\mathbf{AD} - \mathbf{AB}] = a[\tfrac{1}{2}\mathbf{AB} + \mathbf{AD}]$$

b. Use the fact that $\mathbf{AB}$ and $\mathbf{AD}$ are linearly independent (see Problem 49) to show that
$$\tfrac{1}{2} - b - \tfrac{1}{2}a = 0 \text{ and } a - b = 0$$

Solve this system of equations to show that P has the required location.

51. Show that $\mathbf{u} = a_1\mathbf{i} + a_2\mathbf{j} + a_3\mathbf{k}, \mathbf{v} = b_1\mathbf{i} + b_2\mathbf{j} + b_3\mathbf{k}$ and $\mathbf{w} = c_1\mathbf{i} + c_2\mathbf{j} + c_3\mathbf{k}$ are linearly dependent (see Problem 49), if and only if
$$\begin{vmatrix} a_1 & a_2 & a_3 \\ b_1 & b_2 & b_3 \\ c_1 & c_2 & c_3 \end{vmatrix} = 0$$

52. Show that
$$\begin{vmatrix} \mathbf{u}_1 \cdot \mathbf{v}_1 & \mathbf{u}_1 \cdot \mathbf{v}_2 \\ \mathbf{u}_2 \cdot \mathbf{v}_1 & \mathbf{u}_2 \cdot \mathbf{v}_2 \end{vmatrix} = (\mathbf{u}_1 \times \mathbf{u}_2) \cdot (\mathbf{v}_1 \times \mathbf{v}_2)$$

Hint: See Problem 59, Section 10.4.

53. Let $A(-2, 3, 7), B(1, 5, -3), C(2, 8, -1)$ be the vertices of a triangle in $\mathbb{R}^3$. What are the coordinates of the point M where the medians of the triangle meet (the centroid)?

54. The medians of a triangle meet at a point (the centroid) located two-thirds of the distance from each vertex to the midpoint of the opposite side. Generalize this result by showing that the four lines that join each vertex of a tetrahedron to the centroid of the opposite face meet at a point located three-fourths of the distance from the vertex to the centroid.

55. A triangle in $\mathbb{R}^3$ is determined by the vectors **v** and **w** as shown in Figure 10.50. Show that the traingle has area
$$A = \tfrac{1}{2}\sqrt{\|\mathbf{v}\|^2\|\mathbf{w}\|^2 - (\mathbf{v} \cdot \mathbf{w})^2}$$

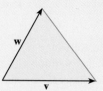

■ **FIGURE 10.50** Area of a triangle

56. **THINK TANK PROBLEM** In Figure 10.51, $\triangle ABC$ is equilateral and the points M, N, and O are located so that
$$\mathbf{AM} = \tfrac{1}{3}\mathbf{AB} \qquad \mathbf{BN} = \tfrac{1}{3}\mathbf{BC} \qquad \mathbf{CO} = \tfrac{1}{3}\mathbf{CA}$$

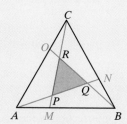

■ **FIGURE 10.51** Problem 56

It can be shown that $\triangle PQR$ is also equilateral, and $\|\mathbf{PM}\| = \|\mathbf{QN}\| = \|\mathbf{RO}\|$ and $\|\mathbf{AP}\| = \|\mathbf{BQ}\| = \|\mathbf{CR}\|$.

a. Show that $\mathbf{PQ} = \tfrac{3}{7}\mathbf{AN}$, then show that $\|\mathbf{AN}\|^2 = \tfrac{7}{9}\|\mathbf{AB}\|^2$. (*Hint:* Use the law of cosines.)

b. Show that $\triangle PQR$ has area $\tfrac{1}{7}$ that of $\triangle ABC$.

c. Do you think the same result would hold if $\triangle ABC$ were not equilateral? Investigate your conjecture.

57. **Gram-Schmidt orthogonalization process** Let **u**, **v**, and **w** be nonzero vectors in $\mathbb{R}^3$ that do not lie on the same plane. Define vectors $\boldsymbol{\alpha}$ and $\boldsymbol{\beta}$ as follows:
$$\boldsymbol{\alpha} = \mathbf{v} - \left[\frac{\mathbf{v} \cdot \mathbf{u}}{\|\mathbf{u}\|^2}\right]\mathbf{u} \quad \text{and} \quad \boldsymbol{\beta} = \mathbf{w} - \left[\frac{\mathbf{w} \cdot \mathbf{u}}{\|\mathbf{u}\|^2}\right]\mathbf{u} - \left[\frac{\mathbf{w} \cdot \boldsymbol{\alpha}}{\|\boldsymbol{\alpha}\|^2}\right]\boldsymbol{\alpha}$$

a. Show that $\mathbf{u}, \boldsymbol{\alpha}, \boldsymbol{\beta}$ are mutually orthogonal (any pair is orthogonal).

b. If $\boldsymbol{\gamma}$ is any vector in $\mathbb{R}^3$, show that
$$\boldsymbol{\gamma} = \left[\frac{\boldsymbol{\gamma} \cdot \mathbf{u}}{\|\mathbf{u}\|^2}\right]\mathbf{u} + \left[\frac{\boldsymbol{\gamma} \cdot \boldsymbol{\alpha}}{\|\boldsymbol{\alpha}\|^2}\right]\boldsymbol{\alpha} + \left[\frac{\boldsymbol{\gamma} \cdot \boldsymbol{\beta}}{\|\boldsymbol{\beta}\|^2}\right]\boldsymbol{\beta}$$

58. a. In Figure 10.52, $ABCD$ is a rectangle, with M the midpoint of side $\overline{CD}$ and $\mathbf{AR}_k = \dfrac{1}{2k + 1}\mathbf{AB}$. If P is the intersection of $\overline{AM}$ and $\overline{CR}_k$, and R_{k+1} is the foot of

the perpendicular drawn from P to $\overline{AB}$, show that

$$\mathbf{AR}_{k+1} = \frac{1}{2k+3}\mathbf{AB}$$

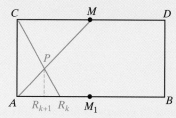

■ **FIGURE 10.52**   Problem 58

b. It is easy to subdivide a line segment line $\overline{AB}$ in half, or fourths, or eighths, etc. However, dividing it into thirds or fifths, … is more difficult. Use the result obtained in part **a** to describe a procedure for subdividing $\overline{AB}$ into an odd number of equal parts.

59. **PUTNAM EXAMINATION PROBLEM**   Find the equations of two straight lines, each of which cuts all four of the following lines:

$$L_1: x = 1, y = 0 \qquad L_2: y = 1, z = 0$$
$$L_3: z = 1, x = 0 \qquad L_4: x = y = -6z$$

60. **PUTNAM EXAMINATION PROBLEM**   Find the equation of the smallest sphere that is tangent to both the lines

$$L_1: x = t + 1, y = 2t + 4, z = -3t + 5 \text{ and}$$
$$L_2: x = 4t - 12, y = t + 8, z = t + 17$$

61. **PUTNAM EXAMINATION PROBLEM**   The hands of an accurate clock have lengths 3 cm and 4 cm. Find the distance between the tips of the hands when the distance is increasing most rapidly.

Star Trek

This project is to be done in groups of three or four students. Each group will submit a single written report.

The starship *Enterprise* has been captured by the evil Romulans and is being held in orbit by a Romulan tractor beam. The orbit is elliptical with the planet Romula at one focus of the ellipse. Repeated efforts to escape have been futile and have almost exhausted the fuel supplies. Morale is low and food reserves are dwindling.

In searching the ship's log, Lieutenant Commander Data discovers that the *Enterprise* had been captured long ago by a Romulan tractor beam and had escaped. The key to that escape was to fire the ship's thrusters at exactly the right position in the orbit. Captain Picard gives the command to feed the required information into the computer to find that position. But, alas, a Romulan virus has rendered the computer useless for this task. Everyone turns to you and asks for your help in solving the problem.

Here is what Data discovered. If F represents the focus of the ellipse and P is the position of the ship on the ellipse, then the vector $\overrightarrow{FP}$ can be written as a sum $\mathbf{T} + \mathbf{N}$, where $\mathbf{T}$ is tangent to the ellipse and $\mathbf{N}$ is normal to the ellipse (not necessarily unit vectors). The thrusters must be fired when the ratio $\|\mathbf{T}\|/\|\mathbf{N}\|$ is equal to the eccentricity of the ellipse.

Your mission is to save the starship from the evil Romulans.

*Time is said to have only **one dimension**, and space to have **three dimensions**... the mathematical **quaternion** partakes of **both** of these elements; in technical language it may be said to be "time plus space," or "space plus time"; and in the sense it has, or at least involves a reference to, **four dimensions**...*

W. R. HAMILTON
GRAVES' LIFE OF
HAMILTON (NEW YORK,
1882–1889), VOL. 3, P. 635.

MAA Notes 17 (1991): "Priming the Calculus Pump: Innovations and Resources," by Marcus S. Cohen, Edward D. Gaughan, R. Arthur Knoebel, Douglas S. Kurtz, and David J. Pengelley.

11 Vector-Valued Functions

CONTENTS

PREVIEW

The marriage of calculus and vector methods forms what is called *vector calculus*. The key to using vector calculus is the concept of a *vector-valued function*. In this chapter, we introduce such functions and examine some of their properties. We shall see that vector-valued functions behave much like the *scalar-valued functions* studied earlier in this text.

PERSPECTIVE

A car travels down a curved road at a constant speed of 55 mi/h. What additional information do we need about the car to determine whether it will stay on the road or skid off as it rounds a particular curve? Can we modify the road (say, by banking) so an average-sized car can travel at moderate speeds without skidding? How does a highway department decide what warning sign to install on a particular curve? A soldier fires a howitzer whose muzzle speed and angle of elevation are known. If the shell overshoots its target by 40 yd, how should the angle of elevation be changed to ensure a hit on the next shot? If a satellite is 20,000 mi above the earth, how fast must it travel to remain stationary above a particular point on the equator? These and other similar questions can be answered using vector calculus.

11.1 *Introduction to Vector Functions*

vector-valued functions, operations with vector functions, limits and continuity

VECTOR-VALUED FUNCTIONS

In Section 9.4, we described a *plane curve* using parametric equations

$$x = f(t) \quad \text{and} \quad y = g(t)$$

where f and g are continuous functions of t on some interval. We extend this definition to three dimensions. A **space curve** is the set of all ordered triples $[f_1(t), f_2(t), f_3(t)]$ satisfying the parametric equations $x = f_1(t)$, $y = f_2(t)$, $z = f_3(t)$, where f_1, f_2, and f_3 are continuous functions of t on some domain D.

The concept of a vector-valued function is fundamental to the ideas we plan to explore. Here is a definition of this concept.

> **Vector-Valued Function**
>
> A **vector-valued function** (or, simply, a **vector function**) **F** with *domain D* assigns to each scalar t in the set D a unique vector $\mathbf{F}(t)$. The set of all vectors **v** of the form $\mathbf{v} = \mathbf{F}(t)$ for t in D is the *range* of **F**. In this text, we shall be concerned with vector functions whose range is in $\mathbb{R}^2$ or $\mathbb{R}^3$. That is,
>
> $$\mathbf{F}(t) = f_1(t)\mathbf{i} + f_2(t)\mathbf{j} \qquad \text{in } \mathbb{R}^2 \text{ (plane)}$$
> $$\mathbf{F}(t) = f_1(t)\mathbf{i} + f_2(t)\mathbf{j} + f_3(t)\mathbf{k} \quad \text{in } \mathbb{R}^3 \text{ (space)}$$
>
> where f_1, f_2, and f_3 are real-valued (**scalar-valued**) functions of the real number t defined on the domain set D. In this context, f_1, f_2, and f_3 are called the **components** of **F**.

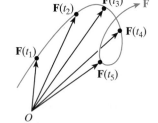

■ **FIGURE 11.1** The graph of the vector function $\mathbf{F}(t)$ is traced out by the terminal point of $\mathbf{F}(t)$ as t varies over D.

Let **F** be a vector function, and suppose the initial point of the vector $\mathbf{F}(t)$ is at the origin. The graph of **F** is the curve traced out by the terminal point of the vector $\mathbf{F}(t)$ as t varies over the domain set D, as shown in Figure 11.1.

EXAMPLE 1   *Graph of a vector function*

Sketch the graph of the vector function

$$\mathbf{F}(t) = (3 - t)\mathbf{i} + (2t)\mathbf{j} + (-4 + 3t)\mathbf{k}$$

for all t.

Solution   The graph is the collection of all points (x, y, z) with

$$x = 3 - t \qquad y = 2t \qquad z = -4 + 3t$$

for all t. We recognize these as the parametric equations for the line in $\mathbb{R}^3$ that contains the point $P_0(3, 0, -4)$ and is aligned with the vector

$$\mathbf{v} = -\mathbf{i} + 2\mathbf{j} + 3\mathbf{k}$$

as shown in Figure 11.2a.

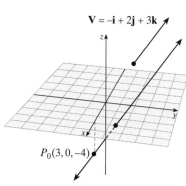

■ **FIGURE 11.2a** The graph of $\mathbf{F}(t) = (3 - t)\mathbf{i} + (2t)\mathbf{j} + (3t - 4)\mathbf{k}$

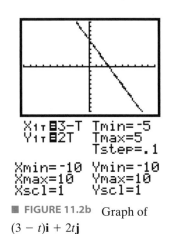

■ **FIGURE 11.2b**   Graph of $(3 - t)\mathbf{i} + 2t\mathbf{j}$

If you have access to technology, you can draw vector functions in three dimensions quite easily. However, if you have only a graphing calculator, you may be limited to graphics in two dimensions. In order to visualize this vector function, draw a graph (in parametric form) as shown in Figure 11.2b. Note that the line passes through $(3, 0)$. The direction of the line (in $\mathbb{R}^2$) is $\langle -1, 2 \rangle$ which means a run of -1 with a rise of 2. From *this* graph imagine the same line in the plane $z = -4$ (because the line in $\mathbb{R}^3$ passes through $(3, 0, -4)$) and *then* imagine a change in the direction of z to be three units (because the direction is $\langle -1, 2, 3 \rangle$). ■

EXAMPLE 2   *Graph of a circular helix*

Sketch the graph of the vector function

$$\mathbf{F}(t) = (2 \sin t)\mathbf{i} - (2 \cos t)\mathbf{j} + (3t)\mathbf{k}$$

Solution   The graph of $\mathbf{F}$ is the collection of all points (x, y, z) in $\mathbb{R}^3$ whose coordinates satisfy

$$x = 2 \sin t \qquad y = -2 \cos t \qquad z = 3t \qquad \text{for all } t$$

The first two components satisfy

$$x^2 + y^2 = (2 \sin t)^2 + (-2 \cos t)^2 = 4(\sin^2 t + \cos^2 t) = 4$$

which means that the graph lies on the surface of the right circular cylinder with radius 2, whose axis of symmetry is the z-axis, as shown in Figure 11.3. We also know that as t increases, the z-coordinate of the point $P(x, y, z)$ on the graph of $\mathbf{F}$ increases according to the formula $z = 3t$, which means that the point (x, y, z) on the graph rises in a spiral on the surface of the cylinder $x^2 + y^2 = 4$. The point on the graph of $\mathbf{F}$ that corresponds to $t = 0$ is $(0, -2, 0)$, and the points that correspond to $t = \frac{\pi}{2}$ and $t = \pi$ are $(2, 0, \frac{3\pi}{2})$ and $(0, 2, 3\pi)$, respectively. Thus, the graph spirals upward counterclockwise (as viewed from above). The graph, which is known as a **right circular helix,** is shown in Figure 11.3a. In order to visualize this helix using a graphing calculator, draw the circle as shown in Figure 11.3b, and *imagine* the rise in the z-direction in the manner of a helix. ■

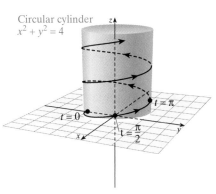

■ **FIGURE 11.3a**   The graph of $\mathbf{F}(t) = (2 \sin t)\mathbf{i} - (2 \cos t)\mathbf{j} + (3t)\mathbf{k}$

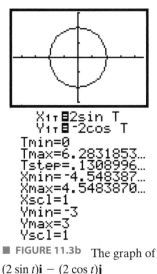

■ **FIGURE 11.3b**   The graph of $(2 \sin t)\mathbf{i} - (2 \cos t)\mathbf{j}$

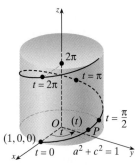

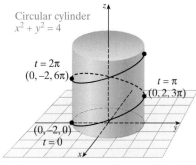

a. A computer-generated image of the helix in Example 2: $\mathbf{F}(t) = (2 \sin t)\mathbf{i} - (2 \cos t)\mathbf{j} + 3t\mathbf{k}$

b. A computer-generated image of the helix: $\mathbf{F}(t) = (\cos 2t)\mathbf{i} + (\sin 2t)\mathbf{j} + 0.2t\mathbf{k}$

■ **FIGURE 11.4**   Examples of helixes

A well-known example of a helix is the DNA (deoxyribonucleic acid) molecule, which has a structure consisting of two intertwined helixes, as shown in Figure 11.4. Some other computer-generated helixes are also shown.

Examples 1 and 2 illustrate how the graph of a vector function

$$\mathbf{F}(t) = f_1(t)\mathbf{i} + f_2(t)\mathbf{j} + f_3(t)\mathbf{k}$$

can be obtained by examining the parametric equations

$$x = f_1(t) \qquad y = f_2(t) \qquad z = f_3(t)$$

In Example 3, we turn things around and find a vector function whose graph is a given curve.

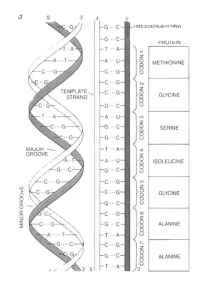

The double helix of the DNA was discovered in 1953 by James Watson and Francis Crick.

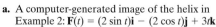

EXAMPLE 3   *Find a vector function*

Find a vector function $\mathbf{F}$ whose graph is the curve of intersection of the hemisphere $z = \sqrt{4 - x^2 - y^2}$ and the parabolic cylinder $y = x^2$, as shown in Figure 11.5.

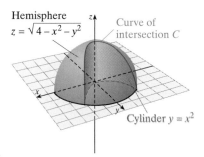

■ **FIGURE 11.5**   The curve of intersection of the hemisphere and the cylinder

Solution   Finding the parametric representation is sometimes called **parametrizing the curve.** There are several ways this can be done, but the natural choice is to let $x = t$. Then $y = t^2$ (from the equation of the parabola), and by substituting into the equation for the hemisphere, we find

$$z = \sqrt{4 - x^2 - y^2} = \sqrt{4 - (t)^2 - (t^2)^2} = \sqrt{4 - t^2 - t^4}$$

We can now state a formula for a vector function of the given graph:

$$\mathbf{F}(t) = t\mathbf{i} + t^2\mathbf{j} + \sqrt{4 - t^2 - t^4}\,\mathbf{k}$$ ∎

OPERATIONS WITH VECTOR FUNCTIONS

It follows from the definition of vector operations that vector functions can be added, subtracted, multiplied by a scalar function, and multiplied together. We summarize these operations in the following box.

Vector Function Operations

Let $\mathbf{F}$ and $\mathbf{G}$ be vector functions of the real variable t, and let $f(t)$ be a scalar function. Then, $\mathbf{F} + \mathbf{G}, \mathbf{F} - \mathbf{G}, f\mathbf{F}$, and $\mathbf{F} \times \mathbf{G}$ are vector functions, and $\mathbf{F} \cdot \mathbf{G}$ is a scalar function. These operations are summarized as follows:

Vector functions:

$$(\mathbf{F} + \mathbf{G})(t) = \mathbf{F}(t) + \mathbf{G}(t) \qquad (\mathbf{F} - \mathbf{G})(t) = \mathbf{F}(t) - \mathbf{G}(t)$$
$$(f\mathbf{F})(t) = f(t)\mathbf{F}(t) \qquad (\mathbf{F} \times \mathbf{G})(t) = \mathbf{F}(t) \times \mathbf{G}(t)$$

Scalar function: $\quad (\mathbf{F} \cdot \mathbf{G})(t) = \mathbf{F}(t) \cdot \mathbf{G}(t)$

EXAMPLE 4   *Vector function operations*

Let $\mathbf{F}(t) = t^2\mathbf{i} + t\mathbf{j} - (\sin t)\mathbf{k}$ and $\mathbf{G}(t) = t\mathbf{i} + \dfrac{1}{t}\mathbf{j} + 5\mathbf{k}$. Find

a. $(\mathbf{F} + \mathbf{G})(t)$ b. $(e^t\mathbf{F})(t)$ c. $(\mathbf{F} \times \mathbf{G})(t)$ d. $(\mathbf{F} \cdot \mathbf{G})(t)$

Solution

a. $(\mathbf{F} + \mathbf{G})(t) = \mathbf{F}(t) + \mathbf{G}(t)$

$$= [t^2\mathbf{i} + t\mathbf{j} - (\sin t)\mathbf{k}] + \left[t\mathbf{i} + \frac{1}{t}\mathbf{j} + 5\mathbf{k}\right]$$

$$= (t^2 + t)\mathbf{i} + (t + t^{-1})\mathbf{j} + (5 - \sin t)\mathbf{k}$$

b. $(e^t\mathbf{F})(t) = e^t\mathbf{F}(t) = e^t t^2\mathbf{i} + e^t t\mathbf{j} - (e^t \sin t)\mathbf{k}$

c. $(\mathbf{F} \times \mathbf{G})(t) = \mathbf{F}(t) \times \mathbf{G}(t)$

$$= [t^2\mathbf{i} + t\mathbf{j} - (\sin t)\mathbf{k}] \times \left[t\mathbf{i} + \frac{1}{t}\mathbf{j} + 5\mathbf{k}\right]$$

$$= \begin{vmatrix} \mathbf{i} & \mathbf{j} & \mathbf{k} \\ t^2 & t & -\sin t \\ t & \dfrac{1}{t} & 5 \end{vmatrix}$$

$$= \left[5t + \frac{\sin t}{t}\right]\mathbf{i} - [5t^2 + t\sin t]\mathbf{j} + [t - t^2]\mathbf{k}$$

d. $(\mathbf{F} \cdot \mathbf{G})(t) = \mathbf{F}(t) \cdot \mathbf{G}(t)$

$$= [t^2\mathbf{i} + t\mathbf{j} - (\sin t)\mathbf{k}] \cdot \left[t\mathbf{i} + \frac{1}{t}\mathbf{j} + 5\mathbf{k}\right]$$

$$= t^3 + 1 - 5\sin t$$ ∎

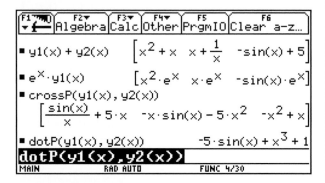

Technology Window

Representing a vector function using technology is generally the same whether you are using a calculator or a software package. Square brackets are used to denote vector functions. For Example 4, we denote **F** as the function $y1(x)$, and **G** as $y2(x)$:

$$y1(x) = [x^2, x, -\sin x]$$

$$y2(x) = [x, 1/x, 5]$$

A sample of these input values is shown. Next, we work through each of the parts of Example 4.

Notice that if the answer is a vector, it is shown in brackets (as in parts **a, b,** and **c**). If the answer is a scalar (as in part **d**), the answer is not shown in brackets.

LIMITS AND CONTINUITY

Limit of a Vector Function

Suppose the components f_1, f_2, f_3 of the vector function

$$\mathbf{F}(t) = f_1(t)\mathbf{i} + f_2(t)\mathbf{j} + f_3(t)\mathbf{k}$$

all have finite limits as $t \to t_0$, where t_0 is any number or $\pm\infty$. Then the **limit** of $\mathbf{F}(t)$ as $t \to t_0$ is the vector

$$\lim_{t \to t_0} \mathbf{F}(t) = \left[\lim_{t \to t_0} f_1(t)\right]\mathbf{i} + \left[\lim_{t \to t_0} f_2(t)\right]\mathbf{j} + \left[\lim_{t \to t_0} f_3(t)\right]\mathbf{k}$$

EXAMPLE 5   *Limit of a vector function*

Find $\lim_{t \to 2} \mathbf{F}(t)$, where $\mathbf{F}(t) = (t^2 - 3)\mathbf{i} + e^t\mathbf{i} + (\sin \pi t)\mathbf{k}$.

Solution $\lim_{t \to 2} \mathbf{F}(t) = \left[\lim_{t \to 2}(t^2 - 3)\right]\mathbf{i} + \left[\lim_{t \to 2}(e^t)\right]\mathbf{j} + \left[\lim_{t \to 2}(\sin \pi t)\right]\mathbf{k}$

$$= 1\mathbf{i} + e^2\mathbf{j} + (\sin 2\pi)\mathbf{k}$$

$$= \mathbf{i} + e^2\mathbf{j}$$

For the most part, vector limits behave like scalar limits. The following theorem contains some useful general properties of such limits.

THEOREM 11.1   *Rules for vector limits*

If the vector functions **F** and **G** are functions of a real variable t and $h(t)$ is a scalar function such that all three functions have finite limits as $t \to t_0$, then

Limit of a sum $\lim\limits_{t \to t_0} [\mathbf{F}(t) + \mathbf{G}(t)] = \lim\limits_{t \to t_0} \mathbf{F}(t) + \lim\limits_{t \to t_0} \mathbf{G}(t)$

Limit of a difference $\lim\limits_{t \to t_0} [\mathbf{F}(t) - \mathbf{G}(t)] = \lim\limits_{t \to t_0} \mathbf{F}(t) - \lim\limits_{t \to t_0} \mathbf{G}(t)$

Limit of a scalar multiple $\lim\limits_{t \to t_0} [h(t)\mathbf{F}(t)] = \left[\lim\limits_{t \to t_0} h(t)\right]\left[\lim\limits_{t \to t_0} \mathbf{F}(t)\right]$

Limit of a dot product $\lim\limits_{t \to t_0} [\mathbf{F}(t) \cdot \mathbf{G}(t)] = \left[\lim\limits_{t \to t_0} \mathbf{F}(t)\right] \cdot \left[\lim\limits_{t \to t_0} \mathbf{G}(t)\right]$

Limit of a cross product $\lim\limits_{t \to t_0} [\mathbf{F}(t) \times \mathbf{G}(t)] = \left[\lim\limits_{t \to t_0} \mathbf{F}(t)\right] \times \left[\lim\limits_{t \to t_0} \mathbf{G}(t)\right]$

These limit formulas are also valid as $t \to +\infty$ or as $t \to -\infty$, assuming all have finite limits.

Proof   We shall establish the formula for the limit of a dot product and leave the rest of the proof as an exercise. Let

$$\mathbf{F}(t) = f_1(t)\mathbf{i} + f_2(t)\mathbf{j} + f_3(t)\mathbf{k} \quad \text{and} \quad \mathbf{G}(t) = g_1(t)\mathbf{i} + g_2(t)\mathbf{j} + g_3(t)\mathbf{k}$$

Apply the limit of a vector function along with the sum rule and product rule for scalar limits to write:

$$\lim_{t \to t_0} [\mathbf{F}(t) \cdot \mathbf{G}(t)] = \lim_{t \to t_0} [f_1(t)g_1(t) + f_2(t)g_2(t) + f_3(t)g_3(t)]$$

$$= \left[\lim_{t \to t_0} f_1(t)\right]\left[\lim_{t \to t_0} g_1(t)\right] + \left[\lim_{t \to t_0} f_2(t)\right]\left[\lim_{t \to t_0} g_2(t)\right] + \left[\lim_{t \to t_0} f_3(t)\right]\left[\lim_{t \to t_0} g_3(t)\right]$$

$$= \left(\left[\lim_{t \to t_0} f_1(t)\right]\mathbf{i} + \left[\lim_{t \to t_0} f_2(t)\right]\mathbf{j} + \left[\lim_{t \to t_0} f_3(t)\right]\mathbf{k}\right) \cdot \left(\left[\lim_{t \to t_0} g_1(t)\right]\mathbf{i} + \left[\lim_{t \to t_0} g_2(t)\right]\mathbf{j} + \left[\lim_{t \to t_0} g_3(t)\right]\mathbf{k}\right)$$

$$= \left[\lim_{t \to t_0} \mathbf{F}(t)\right] \cdot \left[\lim_{t \to t_0} \mathbf{G}(t)\right]$$

EXAMPLE 6   *Limit of a cross product of vector functions*

Show that $\lim\limits_{t \to 1} [\mathbf{F}(t) \times \mathbf{G}(t)] = \left[\lim\limits_{t \to 1} \mathbf{F}(t)\right] \times \left[\lim\limits_{t \to 1} \mathbf{G}(t)\right]$ for the vector functions

$$\mathbf{F}(t) = t\mathbf{i} + (1 - t)\mathbf{j} + t^2\mathbf{k} \quad \text{and} \quad \mathbf{G}(t) = e^t\mathbf{i} - (3 + e^t)\mathbf{k}$$

Solution

$$\mathbf{F}(t) \times \mathbf{G}(t) = \begin{vmatrix} \mathbf{i} & \mathbf{j} & \mathbf{k} \\ t & 1 - t & t^2 \\ e^t & 0 & -(3 + e^t) \end{vmatrix}$$

$$= [(1 - t)(-3 - e^t) - 0]\mathbf{i} - [-t(3 + e^t) - t^2 e^t]\mathbf{j} + [0 - e^t(1 - t)]\mathbf{k}$$

$$= (te^t + 3t - e^t - 3)\mathbf{i} + (t^2 e^t + te^t + 3t)\mathbf{j} + (te^t - e^t)\mathbf{k}$$

Thus, the limit of the cross product is

$$\lim_{t \to 1} [\mathbf{F}(t) \times \mathbf{G}(t)] = \lim_{t \to 1} [te^t + 3t - e^t - 3]\mathbf{i} + \lim_{t \to 1} [t^2 e^t + te^t + 3t]\mathbf{j} + \lim_{t \to 1} [te^t - e^t]\mathbf{k}$$

$$= (e + 3 - e - 3)\mathbf{i} + (e + e + 3)\mathbf{j} + (e - e)\mathbf{k} = (2e + 3)\mathbf{j}$$

Now we find the cross product of the limits:

$$\lim_{t \to 1} \mathbf{F}(t) = \left[\lim_{t \to 1} t\right]\mathbf{i} + \left[\lim_{t \to 1} (1 - t)\right]\mathbf{j} + \left[\lim_{t \to 1} t^2\right]\mathbf{k} = \mathbf{i} + \mathbf{k}$$

$$\lim_{t \to 1} \mathbf{G}(t) = \left[\lim_{t \to 1} e^t\right]\mathbf{i} + \left[\lim_{t \to 1} (-3 - e^t)\right]\mathbf{k} = e\mathbf{i} + (-3 - e)\mathbf{k}$$

Thus,

$$\left[\lim_{t\to 1} \mathbf{F}(t)\right] \times \left[\lim_{t\to 1} \mathbf{G}(t)\right] = \begin{vmatrix} \mathbf{i} & \mathbf{j} & \mathbf{k} \\ 1 & 0 & 1 \\ e & 0 & -3-e \end{vmatrix}$$

$$= (0-0)\mathbf{i} - (-3-e-e)\mathbf{j} + (0-0)\mathbf{k} = (3+2e)\mathbf{j}$$

We see $\lim_{t\to 1}[\mathbf{F}(t) \times \mathbf{G}(t)] = \left[\lim_{t\to 1}\mathbf{F}(t)\right] \times \left[\lim_{t\to 1}\mathbf{G}(t)\right]$. ∎

Continuity of a Vector Function

A vector function $\mathbf{F}(t)$ is said to be **continuous** at t_0 if t_0 is in the domain of $\mathbf{F}$ and $\lim_{t\to t_0}\mathbf{F}(t) = \mathbf{F}(t_0)$.

> ■ *What This Says:*   This is the same as requiring each component of $\mathbf{F}(t)$ to be continuous at t_0. That is,
>
> $$\mathbf{F}(t) = f_1(t)\mathbf{i} + f_2(t)\mathbf{j} + f_3(t)\mathbf{k}$$
>
> is continuous at t_0 when t_0 is in the domain of the component functions $f_1(t), f_2(t),$ and $f_3(t)$ and
>
> $$\lim_{t\to t_0} f_1(t) = f_1(t_0) \qquad \lim_{t\to t_0} f_2(t) = f_2(t_0) \qquad \lim_{t\to t_0} f_3(t) = f_3(t_0)$$

The rules for vector limits listed in Theorem 11.1 can be used to derive general properties of vector continuity.

EXAMPLE 7   *Continuity of a vector function*

For what values of t is $\mathbf{F}(t) = (\sin t)\mathbf{i} + (1-t)^{-1}\mathbf{j} + (\ln t)\mathbf{k}$ continuous?

Solution   The vector function $\mathbf{F}$ is continuous where its component functions

$$f_1(t) = \sin t \qquad f_2(t) = (1-t)^{-1} \qquad f_3(t) = \ln t$$

are continuous. The function f_1 is continuous for all t; f_2 is continuous where $1-t \neq 0$ (that is, where $t \neq 1$); f_3 is continuous for $t > 0$. Thus, $\mathbf{F}$ is continuous when t is a positive number other than 1—that is, $t > 0, t \neq 1$. ∎

11.1 Problem Set

Ⓐ *Find the domain for the vector functions given in Problems 1–8.*

1. $\mathbf{F}(t) = 2t\mathbf{i} - 3t\mathbf{j} + \dfrac{1}{t}\mathbf{k}$

2. $\mathbf{F}(t) = (1-t)\mathbf{i} + \sqrt{t}\mathbf{j} - \dfrac{1}{t-2}\mathbf{k}$

3. $\mathbf{F}(t) = (\sin t)\mathbf{i} + (\cos t)\mathbf{j} + (\tan t)\mathbf{k}$

4. $\mathbf{F}(t) = (\cos t)\mathbf{i} - (\cot t)\mathbf{j} + (\csc t)\mathbf{k}$

5. $h(t)\mathbf{F}(t)$ where $h(t) = \sin t$ and

$$\mathbf{F}(t) = \frac{1}{\cos t}\mathbf{i} + \frac{1}{\sin t}\mathbf{j} + \frac{1}{\tan t}\mathbf{k}$$

6. $\mathbf{F}(t) + \mathbf{G}(t)$, where

$$\mathbf{F}(t) = 3t\mathbf{j} + t^{-1}\mathbf{k} \quad \text{and}$$
$$\mathbf{G}(t) = 5t\mathbf{i} + \sqrt{10-t}\mathbf{j}$$

7. $\mathbf{F}(t) - \mathbf{G}(t)$, where

$$\mathbf{F}(t) = (\ln t)\mathbf{i} + 3t\mathbf{j} - t^2\mathbf{k} \quad \text{and}$$
$$\mathbf{G}(t) = \mathbf{i} + 5t\mathbf{j} - t^2\mathbf{k}$$

8. $\mathbf{F}(t) \times \mathbf{G}(t)$ where

$$\mathbf{F}(t) = t^2\mathbf{i} - t\mathbf{j} + 2t\,\mathbf{k} \quad \text{and}$$
$$\mathbf{G}(t) = \frac{1}{t+2}\mathbf{i} + (t+4)\mathbf{j} - \sqrt{-t}\,\mathbf{k}$$

Describe the graph of the vector functions given in Problems 9–20 or sketch a graph in $\mathbb{R}^3$. A graph in $\mathbb{R}^2$ may help with your description.

9. $\mathbf{F}(t) = 2t\mathbf{i} + t^2\mathbf{j}$

10. $\mathbf{G}(t) = (1 - t)\mathbf{i} + \dfrac{1}{t}\mathbf{j}$

11. $\mathbf{G}(t) = (\sin t)\mathbf{i} - (\cos t)\mathbf{j}$

12. $\mathbf{F}(t) = (2 \cos t)\mathbf{i} + (\sin t)\mathbf{j}$

13. $\mathbf{F}(t) = t\mathbf{i} - 4\mathbf{k}$

14. $\mathbf{G}(t) = e^t\mathbf{j} + t\mathbf{k}$

15. $\mathbf{F}(t) = (\cos t)\mathbf{i} + (\sin t)\mathbf{j} + t\mathbf{k}$

16. $\mathbf{F}(t) = e^t\mathbf{i} + e^t\mathbf{j} + e^{-t}\mathbf{k}$

17. $\mathbf{G}(t) = (1 - t)\mathbf{i} + t^2\mathbf{j} + t\mathbf{k}$

18. $\mathbf{F}(t) = \left(\dfrac{\sqrt{2}}{2}\sin t\right)\mathbf{i} + \left(\dfrac{\sqrt{2}}{2}\sin t\right)\mathbf{j} + (\cos^2 t)\mathbf{k}$

19. $\mathbf{F}(t) = t\mathbf{i} + (t^2 + 1)\mathbf{j} + t^2\mathbf{k}$

20. $\mathbf{G}(t) = (2 \sin t)\mathbf{i} + (2 \cos t)\mathbf{j} + 3\mathbf{k}$

B 21. ■ **What Does This Say?** Discuss the concept of a vector-valued function.

22. ■ **What Does This Say?** Discuss the concept of limit of a vector-valued function.

23. Show that the curve given by
$$\mathbf{R}(t) = (2 \sin t)\mathbf{i} + (2 \sin t)\mathbf{j} + (\sqrt{8} \cos t)\mathbf{k}$$
lies on a sphere centered at the origin.

24. Sketch the graph of the curve given by
$$\mathbf{R}(t) = (2 \cos t)\mathbf{i} + (\sin t)\mathbf{j} - 2t\,\mathbf{k}$$

Find a vector function $\mathbf{F}$ whose graph is the curve given in Problems 25–30.

25. $y = x^2; z = 2$

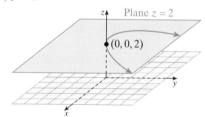

26. $x^2 + y^2 = 4; z = -1$

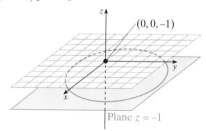

27. $x = 2t, y = 1 - t, z = \sin t$

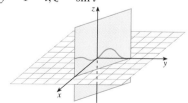

28. $\dfrac{x - 2}{3} = \dfrac{y - 1}{2} = \dfrac{z}{4}$

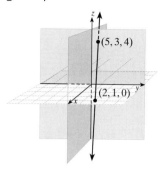

29. The curve of intersection of the hemisphere $z = \sqrt{9 - x^2 - y^2}$ and the parabolic cylinder $x = y^2$

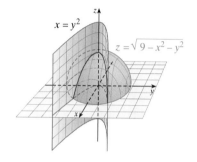

30. The line of intersection of the planes $2x + y + 3z = 6$ and $x - y - z = 1$

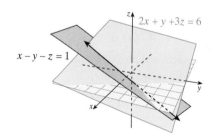

Perform the operations indicated in Problems 31–42 with
$$\mathbf{F}(t) = 2t\mathbf{i} - 5\mathbf{j} + t^2\mathbf{k}, \quad \mathbf{G}(t) = (1 - t)\mathbf{i} + \dfrac{1}{t}\mathbf{k},$$
$$\mathbf{H}(t) = (\sin t)\mathbf{i} + e^t\mathbf{j}$$

31. $2\mathbf{F}(t) - 3\mathbf{G}(t)$

32. $t^2\mathbf{F}(t) - 3\mathbf{H}(t)$

33. $\mathbf{F}(t) \cdot \mathbf{G}(t)$

34. $\mathbf{F}(t) \cdot \mathbf{H}(t)$

35. $\mathbf{G}(t) \cdot \mathbf{H}(t)$

36. $\mathbf{F}(t) \times \mathbf{G}(t)$

37. $\mathbf{F}(t) \times \mathbf{H}(t)$

38. $\mathbf{G}(t) \times \mathbf{H}(t)$

39. $2e^t\mathbf{F}(t) + t\mathbf{G}(t) + 10\mathbf{H}(t)$

40. $\mathbf{F}(t) \cdot [\mathbf{H}(t) \times \mathbf{G}(t)]$

41. $\mathbf{G}(t) \cdot [\mathbf{H}(t) \times \mathbf{F}(t)]$

42. $\mathbf{H}(t) \cdot [\mathbf{G}(t) \times \mathbf{F}(t)]$

Find each limit indicated in Problems 43–50.

43. $\lim\limits_{t \to 1} [2t\mathbf{i} - 3\mathbf{j} + e^t\mathbf{k}]$

44. $\lim\limits_{t \to 1} [3t\mathbf{i} + e^{2t}\mathbf{j} + (\sin \pi t)\mathbf{k}]$

45. $\lim\limits_{t \to 0} \left[\dfrac{(\sin t)\mathbf{i} - t\mathbf{k}}{t^2 + t - 1} \right]$

46. $\lim\limits_{t \to 1} \left[\dfrac{t^3 - 1}{t - 1}\mathbf{i} + \dfrac{t^2 - 3t + 2}{t^2 + t - 2}\mathbf{j} + (t^2 + 1)e^{t-1}\mathbf{k} \right]$

47. $\lim\limits_{t \to 0} \left[\dfrac{te^t}{1 - e^t}\mathbf{i} + \dfrac{e^{t-1}}{\cos t}\mathbf{j} \right]$

48. $\lim\limits_{t \to 0} \left[\dfrac{\sin t}{t}\mathbf{i} + \dfrac{1 - \cos t}{t}\mathbf{j} + e^{1-t}\mathbf{k} \right]$

49. $\lim\limits_{t \to 0^+} \left[\dfrac{\sin 3t}{\sin 2t}\mathbf{i} + \dfrac{\ln(\sin t)}{\ln(\tan t)}\mathbf{j} + (t \ln t)\mathbf{k} \right]$

50. $\lim\limits_{t \to 2} \left[(2\mathbf{i} - t\mathbf{j} + e^t\mathbf{k}) \times (t^2\mathbf{i} + 4\sin t\,\mathbf{j}) \right]$

Determine all values of t for which the vector function given in Problems 51–56 is continuous.

51. $\mathbf{F}(t) = t\mathbf{i} + 3\mathbf{j} - (1 - t)\mathbf{k}$

52. $\mathbf{G}(t) = t\mathbf{i} - \dfrac{1}{t}\mathbf{k}$

53. $\mathbf{G}(t) = \dfrac{\mathbf{i} + 2\mathbf{j}}{t^2 + t}$

54. $\mathbf{F}(t) = (e^t \sin t)\mathbf{i} + (e^t \cos t)\mathbf{k}$

55. $\mathbf{F}(t) = e^t \left[t\mathbf{i} + \dfrac{1}{t}\mathbf{j} + 3\mathbf{k} \right]$

56. $\mathbf{G}(t) = \dfrac{\mathbf{u}}{\|\mathbf{u}\|}$ where $\mathbf{u} = t\mathbf{i} + \sqrt{t}\mathbf{j}$

57. The graph of
$$\mathbf{R} = t\mathbf{i} - \left(\dfrac{1 - t}{t} \right)\mathbf{j} + \left(\dfrac{1 - t^2}{t} \right)\mathbf{k}$$
lies in a certain plane. What is the equation of that plane?

58. How many revolutions are made by the circular helix
$$\mathbf{R} = (2 \sin t)\mathbf{i} + (2 \cos t)\mathbf{j} + \tfrac{5}{8}t\mathbf{k}$$
in a vertical distance of 8 units?

59. Given the vector functions
$$\mathbf{F}(t) = t\mathbf{i} + t^2\mathbf{j} + t^3\mathbf{k} \quad \text{and} \quad \mathbf{G}(t) = \dfrac{1}{t}\mathbf{i} - e^t\mathbf{j}$$

directly verify each of the following limit formulas (that is, without using Theorem 11.1).

a. $\lim\limits_{t \to 0} e^t\mathbf{F}(t) = \left[\lim\limits_{t \to 0} e^t \right]\left[\lim\limits_{t \to 0} \mathbf{F}(t) \right]$

b. $\lim\limits_{t \to 1} \mathbf{F}(t) \cdot \mathbf{G}(t) = \left[\lim\limits_{t \to 1} \mathbf{F}(t) \right] \cdot \left[\lim\limits_{t \to 1} \mathbf{G}(t) \right]$

c. $\lim\limits_{t \to 1} [\mathbf{F}(t) \times \mathbf{G}(t)] = \left[\lim\limits_{t \to 1} \mathbf{F}(t) \right] \times \left[\lim\limits_{t \to 1} \mathbf{G}(t) \right]$

60. If $\mathbf{H}(t)$ is a vector function, we define the **difference operator $\Delta\mathbf{H}$** by the formula
$$\Delta\mathbf{H} = \mathbf{H}(t + \Delta t) - \mathbf{H}(t)$$
where Δt is a change in the parameter t. (*Note:* Usually, $|\Delta t|$ is a small number.) If $\mathbf{F}(t)$ and $\mathbf{G}(t)$ are vector functions, show that
$$\Delta(\mathbf{F} \times \mathbf{G})(t) = \mathbf{F}(t + \Delta t) \times \Delta\mathbf{G}(t) + \Delta\mathbf{F}(t) \times \mathbf{G}(t)$$
Hint: Note that
$$\Delta(\mathbf{F} \times \mathbf{G})(t) = \mathbf{F}(t + \Delta t) \times \mathbf{G}(t + \Delta t)$$
$$- \mathbf{F}(t + \Delta t) \times \mathbf{G}(t) + \mathbf{F}(t + \Delta t) \times \mathbf{G}(t)$$
$$- \mathbf{F}(t) \times \mathbf{G}(t)$$

61. Prove the limit of a sum rule of Theorem 11.1. (The difference rule is proved similarly.)

62. a. Prove the limit of a scalar multiple rule of Theorem 11.1.
    b. Prove the limit of a cross product rule of Theorem 11.1.

63. **THINK TANK PROBLEM**   Let $\mathbf{F}(t)$ and $\mathbf{G}(t)$ be vector functions that are continuous at t_0, and let $h(t)$ be a scalar function continuous at t_0. In each of the following cases, either show that the given function is continuous at t_0 or provide a counterexample to show that it is not continuous.
    a. $3\mathbf{F}(t) + 5\mathbf{G}(t)$ b. $\mathbf{F}(t) \cdot \mathbf{G}(t)$
    c. $h(t)\mathbf{F}(t)$ d. $\mathbf{F}(t) \times \mathbf{G}(t)$

11.2   Differentiation and Integration of Vector Functions

IN THIS SECTION   vector derivatives, tangent vectors, properties of vector derivatives, the modeling motion of an object in space, vector integrals

VECTOR DERIVATIVES

In Chapter 3, we defined the derivative of the (scalar) function f to be the limit as $\Delta x \to 0$ of the difference quotient $\Delta f / \Delta x$. We call this the **scalar derivative** to distinguish it from derivatives involving vector functions. The **difference quotient of a vector function F** is the vector expression
$$\dfrac{\Delta\mathbf{F}}{\Delta t} = \dfrac{\mathbf{F}(t + \Delta t) - \mathbf{F}(t)}{\Delta t}$$
and we define the derivative of $\mathbf{F}$ as follows.

> ### Derivative of a Vector Function
>
> The **derivative** of the vector function $\mathbf{F}$ is the vector function $\mathbf{F}'$ determined by the limit
>
> $$\mathbf{F}'(t) = \lim_{\Delta t \to 0} \frac{\Delta \mathbf{F}}{\Delta t} = \lim_{\Delta t \to 0} \frac{\mathbf{F}(t + \Delta t) - \mathbf{F}(t)}{\Delta t}$$
>
> wherever this limit exists. In the Leibniz notation, the derivative of $\mathbf{F}(t)$ is denoted by $\dfrac{d\mathbf{F}}{dt}$.

We say that the vector function $\mathbf{F}$ is **differentiable** at $t = t_0$ if $\mathbf{F}'(t)$ is defined at t_0.
The following theorem establishes a convenient method for computing the derivative of a vector function.

THEOREM 11.2   *Derivative of a vector function*

The vector function $\mathbf{F}(t) = f_1(t)\mathbf{i} + f_2(t)\mathbf{j} + f_3(t)\mathbf{k}$ is differentiable whenever the component functions $f_1, f_2,$ and f_3 are all differentiable, and in this case

$$\mathbf{F}'(t) = f_1'(t)\mathbf{i} + f_2'(t)\mathbf{j} + f_3'(t)\mathbf{k}$$

Proof   We use the definition of the derivative, along with rules of vector limits (Theorem 11.1), and the fact that the scalar derivatives $f_1'(t), f_2'(t),$ and $f_3'(t)$ and exist.

$$\mathbf{F}'(t) = \lim_{\Delta t \to 0} \frac{\mathbf{F}(t + \Delta t) - \mathbf{F}(t)}{\Delta t}$$

$$= \lim_{\Delta t \to 0} \frac{[f_1(t + \Delta t)\mathbf{i} + f_2(t + \Delta t)\mathbf{j} + f_3(t + \Delta t)]\mathbf{k} - [f_1(t)\mathbf{i} + f_2(t)\mathbf{j} + f_3(t)\mathbf{k}]}{\Delta t}$$

$$= \left[\lim_{\Delta t \to 0} \frac{f_1(t + \Delta t) - f_1(t)}{\Delta t}\right]\mathbf{i} + \left[\lim_{\Delta t \to 0} \frac{f_2(t + \Delta t) - f_2(t)}{\Delta t}\right]\mathbf{j} + \left[\lim_{\Delta t \to 0} \frac{f_3(t + \Delta t) - f_3(t)}{\Delta t}\right]\mathbf{k}$$

$$= f_1'(t)\mathbf{i} + f_2'(t)\mathbf{j} + f_3'(t)\mathbf{k}$$

EXAMPLE 1   *Differentiability of a vector function*

For what values of t is $\mathbf{G}(t) = |t|\mathbf{i} + (\cos t)\mathbf{j} + (t - 5)\mathbf{k}$ differentiable?

Solution   The component functions $\cos t$ and $t - 5$ are differentiable for all t, but $|t|$ is not differentiable at $t = 0$. Thus, the vector function $\mathbf{G}$ is differentiable for all $t \neq 0$.   ∎

EXAMPLE 2   *Derivative of a vector function*

Find the derivative of the vector function

$$\mathbf{F}(t) = e^t\mathbf{i} + (\sin t)\mathbf{j} + (t^3 + 5t)\mathbf{k}$$

Solution

$$\mathbf{F}'(t) = (e^t)'\mathbf{i} + (\sin t)'\mathbf{j} + (t^3 + 5t)'\mathbf{k} = e^t\mathbf{i} + (\cos t)\mathbf{j} + (3t^2 + 5)\mathbf{k}$$ ∎

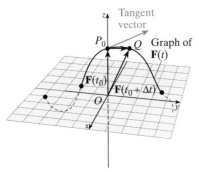

■ **FIGURE 11.6** The difference quotient is a multiple of the secant line vector.

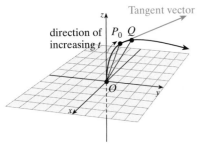

■ **FIGURE 11.7** As $\Delta t \to 0$, the vector **PQ** and hence the difference quotient $\Delta \mathbf{F}/\Delta t$ approach the tangent vector at P_0.

TANGENT VECTORS

Recall that the scalar derivative $f'(x_0)$ gives the slope of the tangent line to the graph of f at the point where $x = x_0$ and thus provides a measure of the graph's direction at that point. Our first goal is to extend this interpretation by showing how the vector derivative can be used to find tangent vectors to curves in space.

Let t be a number in the domain of the vector function $\mathbf{F}(t)$, and let P_0 be the point on the graph of $\mathbf{F}$ that corresponds to t_0, as shown in Figure 11.6. Then for any positive number Δt, the difference quotient

$$\frac{\Delta \mathbf{F}}{\Delta t} = \frac{\mathbf{F}(t_0 + \Delta t) - \mathbf{F}(t_0)}{\Delta t}$$

is a vector that points in the same direction as the secant vector

$$\mathbf{P_0 Q} = \mathbf{F}(t_0 + \Delta t) - \mathbf{F}(t_0),$$

where Q is the point on the graph of $\mathbf{F}$ that corresponds to $t = t_0 + \Delta t$ (see Figure 11.6). If we choose $\Delta t < 0$, the difference quotient $\Delta \mathbf{F}/\Delta t$ points in the *opposite* direction to that of the secant vector $\mathbf{P_0 Q}$.

Suppose the difference quotient $\Delta \mathbf{F}/\Delta t$ has a limit as $\Delta t \to 0$ and that

$$\lim_{\Delta t \to 0} \frac{\Delta \mathbf{F}}{\Delta t} \neq \mathbf{0}$$

Then, as $\Delta t \to 0$, the direction of the secant $\mathbf{P_0 Q}$—and hence that of the difference quotient $\Delta \mathbf{F}/\Delta t$—will approach the direction of the tangent vector at P_0, as shown in Figure 11.7. Thus, we expect the tangent vector at P_0 to be the limit vector

$$\lim_{\Delta t \to 0} \frac{\Delta \mathbf{F}}{\Delta t}$$

which we recognize as the vector derivative $\mathbf{F}'(t_0)$. These observations lead to the following interpretation.

> ### Tangent Vector
>
> Suppose $\mathbf{F}(t)$ is differentiable at t_0 and that $\mathbf{F}'(t_0) \neq \mathbf{0}$. Then $\mathbf{F}'(t_0)$ is a **tangent vector** to the graph of $\mathbf{F}(t)$ at the point where $t = t_0$ and points in the direction of increasing t.

EXAMPLE 3 *Finding a tangent vector*

Find a tangent vector at the point P_0 where $t = 0.2$ on the graph of the vector function

$$\mathbf{F}(t) = e^{2t}\mathbf{i} + (t^2 - t)\mathbf{j} + (\ln t)\mathbf{k}$$

What is the equation of the tangent line at P_0?

Solution Because $\mathbf{F}(0.2) = e^{0.4}\mathbf{i} - 0.16\mathbf{j} + (\ln 0.2)\mathbf{k}$, the point of tangency P_0 has coordinates $(e^{0.4}, -0.16, \ln 0.2) \approx (1.5, -0.16, -1.6)$. The derivative of $\mathbf{F}$ is

$$\mathbf{F}'(t) = 2e^{2t}\mathbf{i} + (2t - 1)\mathbf{j} + t^{-1}\mathbf{k}$$

The required tangent vector is found by evaluating $\mathbf{F}'(t)$ at $t = 0.2$:

$$\mathbf{F}'(0.2) = 2e^{0.4}\mathbf{i} + (-0.6)\mathbf{j} + 5\mathbf{k}$$

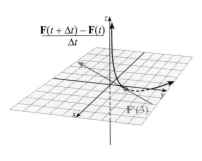

In two dimensions, we have:

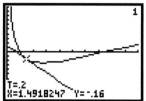

```
X₁ᴛ▪e^(2T)
Y₁ᴛ▪T²-T
X₂ᴛ▪e^.4+(2e^.4)
T
Y₂ᴛ▪-.16+(-.6)T
Tmin=-5
Tmax=5
Tstep=.1
Xmin=0   Ymin=-1
Xmax=10  Ymax=1
Xscl=1   Yscl=.1
```

Once again, we can turn to a graph in $\mathbb{R}^2$ to help us visualize this tangent vector. If you have a graphing calculator, you might wish to pay special attention to the parametric-form input for both the vector function and its tangent vector.

Once we have the point of tangency and a tangent vector, we can find an equation for the tangent line to the graph by using the parametric form for a line. In this case, the tangent line contains $P_0(e^6, 6, \ln 3)$ and is aligned with the tangent vector $\mathbf{v} = 2e^{0.4}\mathbf{i} - 0.6\mathbf{j} + 5\mathbf{k}$, so that the tangent line at P_0 has the parametric form

$$x = e^6 + 2e^6 t \qquad y = 6 + 5t \qquad z = \ln 3 + \tfrac{1}{3}t$$

where t is a real number (scalar).   ∎

We have seen that the graph of $\mathbf{F}$ will have a tangent vector $\mathbf{F}'(t_0)$ at the point P_0, where $t = t_0$ if $\mathbf{F}'(t_0)$ exists and $\mathbf{F}'(t_0) \neq \mathbf{0}$. If we also require the derivative $\mathbf{F}'$ to be continuous at t_0, then the tangent vector at each point of the graph of $\mathbf{F}$ near P_0 will be close to the tangent vector at P_0. In this case, the graph has a *continuously turning tangent* at P_0, and we say that it is *smooth* there.

> ### Smooth Curve
>
> The graph of the vector function defined by $\mathbf{F}(t)$ is said to be **smooth** on any interval of t where $\mathbf{F}'$ is continuous and $\mathbf{F}'(t) \neq \mathbf{0}$. The graph is **piecewise smooth** on an interval that can be subdivided into a finite number of subintervals on which $\mathbf{F}$ is smooth.

As indicated in Figure 11.8, a smooth graph will have no sharply angled points (cusps). For this reason, vector functions with smooth graphs are "nicely behaved" and play an especially important role in vector calculus.

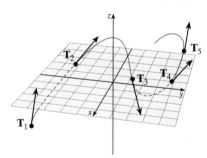
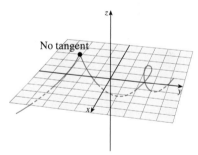

a. A curve that is smooth has a continuously turning tangent.

b. A curve that is not smooth can have "sharp" points. Note that this graph is piecewise smooth.

■ **FIGURE 11.8**   Smooth and piecewise smooth curves

PROPERTIES OF VECTOR DERIVATIVES

Higher derivatives of a vector function $\mathbf{F}$ are obtained by successively differentiating the components of $\mathbf{F}(t) = f_1(t)\mathbf{i} + f_2(t)\mathbf{j} + f_3(t)\mathbf{k}$. For instance, the **second derivative** of $\mathbf{F}$ is the vector function

$$\mathbf{F}''(t) = [\mathbf{F}'(t)]' = f_1''(t)\mathbf{i} + f_2''(t)\mathbf{j} + f_3''(t)\mathbf{k}$$

whereas the **third derivative** $\mathbf{F}'''(t)$ is the derivative of $\mathbf{F}''(t)$, and so on. In the Leibniz notation, the second vector derivative of $\mathbf{F}$ with respect to t is denoted by $\dfrac{d^2\mathbf{F}}{dt^2}$, and the third derivative, by $\dfrac{d^3\mathbf{F}}{dt^3}$.

EXAMPLE 4   *Higher derivatives of a vector function*

Find the second and third derivatives of the vector function

$$\mathbf{F}(t) = e^{2t}\mathbf{i} + (1 - t^2)\mathbf{j} + (\cos 2t)\mathbf{k}$$

Solution

$$\mathbf{F}'(t) = 2e^{2t}\mathbf{i} + (-2t)\mathbf{j} + (-2 \sin 2t)\mathbf{k}$$
$$\mathbf{F}''(t) = 4e^{2t}\mathbf{i} - 2\mathbf{j} - (4 \cos 2t)\mathbf{k}$$
$$\mathbf{F}'''(t) = 8e^{2t}\mathbf{i} + (8 \sin 2t)\mathbf{k}$$ ∎

Several rules for computing derivatives of vector functions are listed in the following theorem.

THEOREM 11.3   *Rules for differentiating vector functions*

If the vector functions $\mathbf{F}$ and $\mathbf{G}$ and the scalar function h are differentiable at t, then so are $a\mathbf{F} + b\mathbf{G}, h\mathbf{F}, \mathbf{F} \cdot \mathbf{G}$, and $\mathbf{F} \times \mathbf{G}$, and

Linearity rule	$(a\mathbf{F} + b\mathbf{G})'(t) = a\mathbf{F}'(t) + b\mathbf{G}'(t)$ for constants a, b
Scalar multiple	$(h\mathbf{F})'(t) = h'(t)\mathbf{F}(t) + h(t)\mathbf{F}'(t)$
Dot product rule	$(\mathbf{F} \cdot \mathbf{G})'(t) = (\mathbf{F}' \cdot \mathbf{G})(t) + (\mathbf{F} \cdot \mathbf{G}')(t)$
Cross product rule*	$(\mathbf{F} \times \mathbf{G})'(t) = (\mathbf{F}' \times \mathbf{G})(t) + (\mathbf{F} \times \mathbf{G}')(t)$
Chain rule	$[\mathbf{F}(h(t))]' = h'(t)\mathbf{F}'(h(t))$

Proof   We shall prove the linearity rule and leave the rest of the proof as exercises. Note that if $\mathbf{F}$ and $\mathbf{G}$ are vector functions differentiable at t, and a, b are constants, then the linear combination $a\mathbf{F} + b\mathbf{G}$ has the difference quotient

$$\frac{\Delta(a\mathbf{F} + b\mathbf{G})}{\Delta t} = \frac{a\Delta\mathbf{F}}{\Delta t} + \frac{b\Delta\mathbf{G}}{\Delta t}$$

We can now find the derivative:

$$\begin{aligned}
(a\mathbf{F} + b\mathbf{G})'(t) &= \lim_{\Delta t \to 0}\left[\frac{\Delta(a\mathbf{F} + b\mathbf{G})}{\Delta t}\right]\\
&= \lim_{\Delta t \to 0}\left[\frac{a\Delta\mathbf{F}}{\Delta t} + \frac{b\Delta\mathbf{G}}{\Delta t}\right]\\
&= a\lim_{\Delta t \to 0}\frac{\Delta\mathbf{F}}{\Delta t} + b\lim_{\Delta t \to 0}\frac{\Delta\mathbf{G}}{\Delta t}\\
&= a\mathbf{F}'(t) + b\mathbf{G}'(t)
\end{aligned}$$ ⹂

EXAMPLE 5   *Derivative of a cross product*

Let $\mathbf{F}(t) = \mathbf{i} + t\mathbf{j} + t^2\mathbf{k}$ and $\mathbf{G}(t) = t\mathbf{i} + e^t\mathbf{j} + 3\mathbf{k}$. Verify that

$$(\mathbf{F} \times \mathbf{G})'(t) = (\mathbf{F}' \times \mathbf{G})(t) + (\mathbf{F} \times \mathbf{G}')(t)$$

Solution   First find the derivative of the cross product:

$$(\mathbf{F} \times \mathbf{G})(t) = \begin{vmatrix} \mathbf{i} & \mathbf{j} & \mathbf{k} \\ 1 & t & t^2 \\ t & e^t & 3 \end{vmatrix} = (3t - t^2e^t)\mathbf{i} - (3 - t^3)\mathbf{j} + (e^t - t^2)\mathbf{k}$$

so that $(\mathbf{F} \times \mathbf{G})'(t) = (3 - 2te^t - t^2e^t)\mathbf{i} + (3t^2)\mathbf{j} + (e^t - 2t)\mathbf{k}$.

---

*The order of the factors is important in the cross product rule, because the cross product of vectors is not commutative.

Next, find $(\mathbf{F}' \times \mathbf{G})(t) + (\mathbf{F} \times \mathbf{G}')(t)$ by first finding the derivatives of $\mathbf{F}$ and $\mathbf{G}$ and then the appropriate cross products:

$$\mathbf{F}'(t) = \mathbf{j} + 2t\mathbf{k} \quad \text{and} \quad \mathbf{G}'(t) = \mathbf{i} + e^t\mathbf{j}$$

$$(\mathbf{F}' \times \mathbf{G})(t) = \begin{vmatrix} \mathbf{i} & \mathbf{j} & \mathbf{k} \\ 0 & 1 & 2t \\ t & e^t & 3 \end{vmatrix} = (3 - 2te^t)\mathbf{i} - (-2t^2)\mathbf{j} + (-t)\mathbf{k}$$

$$(\mathbf{F} \times \mathbf{G}')(t) = \begin{vmatrix} \mathbf{i} & \mathbf{j} & \mathbf{k} \\ 1 & t & t^2 \\ 1 & e^t & 0 \end{vmatrix} = (-t^2e^t)\mathbf{i} - (-t^2)\mathbf{j} + (e^t - t)\mathbf{k}$$

Finally, add these vector functions:

$$(\mathbf{F}' \times \mathbf{G})(t) + (\mathbf{F} \times \mathbf{G}')(t) = [(3 - 2te^t) - t^2e^t]\mathbf{i} + [2t^2 + t^2]\mathbf{j} + [-t + e^t - t]\mathbf{k}$$
$$= (3 - 2te^t - t^2e^t)\mathbf{i} + (3t^2)\mathbf{j} + (e^t - 2t)\mathbf{k}$$

Thus, $(\mathbf{F} \times \mathbf{G})' = (\mathbf{F}' \times \mathbf{G}) + (\mathbf{F} \times \mathbf{G}')$. ∎

EXAMPLE 6   *Derivatives of vector function expressions*

Let $\mathbf{F}(t) = \mathbf{i} + e^t\mathbf{j} + t^2\mathbf{k}$ and $\mathbf{G}(t) = 3t^2\mathbf{i} + e^{-t}\mathbf{j} - 2t\mathbf{k}$. Find

a. $\dfrac{d}{dt}[2\mathbf{F}(t) + t^3\mathbf{G}(t)]$ b. $\dfrac{d}{dt}[\mathbf{F}(t) \cdot \mathbf{G}(t)]$

Solution   Begin by finding $\dfrac{d\mathbf{F}}{dt}$ and $\dfrac{d\mathbf{G}}{dt}$:

$$\frac{d\mathbf{F}}{dt} = e^t\mathbf{j} + 2t\mathbf{k} \quad \text{and} \quad \frac{d\mathbf{G}}{dt} = 6t\mathbf{i} - e^{-t}\mathbf{j} - 2\mathbf{k}$$

a. $\dfrac{d}{dt}[2\mathbf{F}(t) + t^3\mathbf{G}(t)] = 2\dfrac{d\mathbf{F}}{dt} + \left[t^3\dfrac{d\mathbf{G}}{dt} + \dfrac{d}{dt}(t^3)\,\mathbf{G}(t)\right]$

$= 2(e^t\mathbf{j} + 2t\mathbf{k}) + t^3(6t\mathbf{i} - e^{-t}\mathbf{j} - 2\mathbf{k}) + 3t^2(3t^2\mathbf{i} + e^{-t}\mathbf{j} - 2t\mathbf{k})$

$= (6t^4 + 9t^4)\mathbf{i} + (2e^t - t^3e^{-t} + 3t^2e^{-t})\mathbf{j} + (4t - 2t^3 - 6t^3)\mathbf{k}$

$= 15t^4\mathbf{i} + [2e^t + e^{-t}t^2(3 - t)]\mathbf{j} - 4t(2t^2 - 1)\mathbf{k}$

b. $\dfrac{d}{dt}[\mathbf{F}(t) \cdot \mathbf{G}(t)] = \dfrac{d\mathbf{F}}{dt} \cdot \mathbf{G}(t) + \mathbf{F}(t) \cdot \dfrac{d\mathbf{G}}{dt}$

$= (e^t\mathbf{j} + 2t\mathbf{k}) \cdot (3t^2\mathbf{i} + e^{-t}\mathbf{j} - 2t\mathbf{k}) + (\mathbf{i} + e^t\mathbf{j} + t^2\mathbf{k}) \cdot (6t\mathbf{i} - e^{-t}\mathbf{j} - 2\mathbf{k})$

$= [0(3t^2) + e^t(e^{-t}) + 2t(-2t)] + [1(6t) + e^t(-e^{-t}) + t^2(-2)]$

$= (1 - 4t^2) + (6t - 1 - 2t^2)$

$= -6t^2 + 6t$ ∎

We shall use Theorem 11.3 in both applied and theoretical problems. In the following theorem, we establish an important geometric property of vector functions.

THEOREM 11.4   *Orthogonality of a function of constant length and its derivative*

If the nonzero vector function $\mathbf{F}(t)$ is differentiable and has constant length, then $\mathbf{F}(t)$ is orthogonal to the derivative vector $\mathbf{F}'(t)$.

Proof   We are given that $\|\mathbf{F}(t)\| = r$ for some constant r and all t. To prove that $\mathbf{F}$ and $\mathbf{F}'$ are orthogonal, we shall show that $\mathbf{F}(t) \cdot \mathbf{F}'(t) = 0$ for all t. First, note that

$$r^2 = \|\mathbf{F}(t)\|^2 = \mathbf{F}(t) \cdot \mathbf{F}(t)$$

for all t. We take the derivative of both sides (remember that the derivative of r^2 is zero because it is a constant).

$$[r^2]' = [\mathbf{F}(t) \cdot \mathbf{F}(t)]'$$
$$0 = \mathbf{F}'(t) \cdot \mathbf{F}(t) + \mathbf{F}(t) \cdot \mathbf{F}'(t)$$
$$0 = 2\mathbf{F}(t) \cdot \mathbf{F}'(t)$$
$$0 = \mathbf{F}(t) \cdot \mathbf{F}'(t)$$

Thus, $\mathbf{F}$ and $\mathbf{F}'$ are orthogonal.   ▄▄

You may recall from plane geometry that the tangent line to a circle is always perpendicular to the radial line from the center of a circle to the point of tangency (Figure 11.9). Theorem 11.4 can be used to extend this property to spheres in $\mathbb{R}^3$. Suppose P is a point on the sphere

$$x^2 + y^2 + z^2 = r^2$$

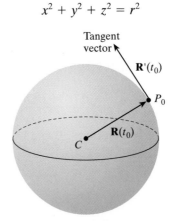

■ **FIGURE 11.9**   The tangent $\mathbf{R}'(t_0)$ at P_0 is orthogonal to the radius vector $\mathbf{R}(t_0)$ from the center of the sphere to P_0.

Let $\mathbf{R}(t) = x(t)\mathbf{i} + y(t)\mathbf{j} + z(t)\mathbf{k}$ be a vector function whose graph lies entirely on the surface of the sphere and contains the point P_0. Suppose P_0 corresponds to $t = t_0$. Then $\mathbf{R}(t_0)$ is the vector from the center of the sphere to P_0, and $\mathbf{R}'(t_0)$ is a tangent vector to the graph of $\mathbf{R}(t)$ at P_0, which means that $\mathbf{R}'(t_0)$ is also tangent to the sphere at P_0. Because $\mathbf{R}(t)$ has constant length,

$$\|\mathbf{R}(t)\| = \sqrt{x^2(t) + y^2(t) + z^2(t)} = r \quad \text{for all } t$$

it follows from Theorem 11.4 that $\mathbf{R}$ is orthogonal to $\mathbf{R}'$. Thus, the tangent vector $\mathbf{R}'(t_0)$ at P_0 is orthogonal to the vector $\mathbf{R}(t_0)$ from the center of the sphere to the point of tangency.

MODELING THE MOTION OF AN OBJECT IN SPACE

Recall from Chapter 3 that the derivative of an object's position with respect to time is the velocity, and the derivative of the velocity is the acceleration. We frequently know the acceleration and can use integration to find the velocity and the position. We will now express these concepts in terms of vector functions.

Position Vector, Velocity, and Acceleration

An object that moves in such a way that its position at time t is given by the vector function $\mathbf{R}(t)$ is said to have

Position vector $\mathbf{R}(t)$ and

Velocity $\mathbf{V} = \dfrac{d\mathbf{R}}{dt} = \mathbf{R}'(t)$

At any time t,

the **speed** is $\|\mathbf{V}\|$, the magnitude of velocity,

the **direction of motion** is $\dfrac{\mathbf{V}}{\|\mathbf{V}\|}$,   and

the **acceleration vector** is the derivative of the velocity:

$$\mathbf{A} = \frac{d\mathbf{V}}{dt} = \frac{d^2\mathbf{R}}{dt^2}$$

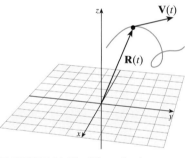

■ **FIGURE 11.10** The velocity vector is tangent to the trajectory of the object's motion.

The graph of the position vector $\mathbf{R}(t)$ is called the **trajectory** of the object's motion. According to the results obtained earlier in this section, the velocity $\mathbf{V}(t) = \mathbf{R}'(t)$ is a tangent vector to the trajectory at any point where $\mathbf{V}(t)$ exists and $\mathbf{V}(t) \neq 0$ (see Figure 11.10). The *direction of motion* is given by the unit vector $\mathbf{V}/\|\mathbf{V}\|$. Appropriately, whenever $\mathbf{V}(t) = \mathbf{0}$, the object is said to be **stationary.**

In practice, the position vector is often represented in the form

$$\mathbf{R}(t) = f_1(t)\mathbf{i} + f_2(t)\mathbf{j} + f_3(t)\mathbf{k}$$

and, in this case, the velocity and the acceleration vectors are given by

$$\mathbf{V}(t) = \mathbf{R}'(t) = f_1'(t)\mathbf{i} + f_2'(t)\mathbf{j} + f_3'(t)\mathbf{k}$$

and

$$\mathbf{A}(t) = \mathbf{V}'(t) = \mathbf{R}''(t) = f_1''(t)\mathbf{i} + f_2''(t)\mathbf{j} + f_3''(t)\mathbf{k}$$

EXAMPLE 7   *Speed and direction of a particle*

A particle's position at time t is determined by the vector

$$\mathbf{R}(t) = (\cos t)\mathbf{i} + (\sin t)\mathbf{j} + t^3\mathbf{k}$$

Analyze the particle's motion. In particular, find the particle's velocity, speed, acceleration, and direction of motion at time $t = 2$.

Solution $\mathbf{V} = \dfrac{d\mathbf{R}}{dt} = (-\sin t)\mathbf{i} + (\cos t)\mathbf{j} + 3t^2\mathbf{k}$

$$\mathbf{A} = \frac{d\mathbf{V}}{dt} = (-\cos t)\mathbf{i} - (\sin t)\mathbf{j} + 6t\mathbf{k}$$

The velocity at time $t = 2$ is $\mathbf{V}(2)$:

$$(-\sin 2)\mathbf{i} + (\cos 2)\mathbf{j} + 3(2)^2\mathbf{k} \approx -0.91\mathbf{i} - 0.42\mathbf{j} + 12\mathbf{k}$$

The acceleration at $t = 2$ is

$$(-\cos 2)\mathbf{i} - (\sin 2)\mathbf{j} + 6(2)\mathbf{k} \approx 0.42\mathbf{i} - 0.91\mathbf{j} + 12\mathbf{k}$$

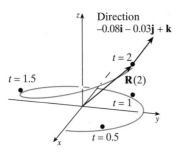

The speed is

$$\|\mathbf{V}\| = \sqrt{(-\sin t)^2 + (\cos t)^2 + (3t^2)^2}$$
$$= \sqrt{1 + 9t^4}$$

At time $t = 2$, the speed is $\|\mathbf{V}(2)\| = \sqrt{1 + 144} \approx 12.04$.

The direction of motion is

$$\frac{\mathbf{V}}{\|\mathbf{V}\|} = \frac{1}{\sqrt{145}}[(-\sin t)\mathbf{i} + (\cos t)\mathbf{j} + 3t^2\mathbf{k}]$$

At time $t = 2$, the direction is $\mathbf{V}(2)/\|\mathbf{V}(2)\|$:

$$\frac{1}{\sqrt{145}}[(-\sin 2)\mathbf{i} + (\cos 2)\mathbf{j} + 12\mathbf{k}] \approx -0.08\mathbf{i} - 0.03\mathbf{j} + \mathbf{k}$$ ∎

WARNING   Notice $\mathbf{V} = \|\mathbf{V}\| \cdot \dfrac{\mathbf{V}}{\|\mathbf{V}\|}$

$= (\text{SPEED}) \cdot (\text{DIRECTION})$.

Technology Window

To represent a velocity vector, it may seem that we could use the *nDer* feature available on many calculators. If we attempt to take the derivative of a position vector with respect to a parameter t, we find that calculator feature does not seem to accept a vector-valued function. However, we can use the *evalF* feature to define our own numerical derivative using the following formula:

$$\mathbf{V} = \frac{\mathbf{P}(t + \delta) - \mathbf{P}(t - \delta)}{2\delta} \text{ after defining a position vector } \mathbf{P} = [x, y, z]$$

That is, input: $\mathbf{V} = (\text{evalF}(P, t, t + \delta) - \text{evalF}(P, t, t - \delta))/(2\delta)$ where the Greek letter δ is a tolerance variable available on many models of calculator. A tolerance of $\delta = 0.001$ seems to work well in most applications.

Once we have defined a velocity vector, we can easily define other vectors:

Speed is $\|\mathbf{V}\|$, so we use a calculator by: $\mathbf{S} = \text{norm}(\mathbf{V})$.

Direction of motion is $\dfrac{\mathbf{V}}{\|\mathbf{V}\|}$, so we use a calculator by: $\mathbf{T} = \text{unitV}(\mathbf{V})$

VECTOR INTEGRALS

Since vector limits and derivatives are performed in a componentwise fashion, it should come as no surprise that vector integration is performed in the same way.

Indefinite Integral and Definite Integral

Let $\mathbf{F}(t) = f_1(t)\mathbf{i} + f_2(t)\mathbf{j} + f_3(t)\mathbf{k}$, where $f_1, f_2,$ and f_3 are continuous on the closed interval $a \le t \le b$. Then the **indefinite integral** of $\mathbf{F}(t)$ is the vector function

$$\int \mathbf{F}(t)\, dt = \left[\int f_1(t)\, dt\right]\mathbf{i} + \left[\int f_2(t)\, dt\right]\mathbf{j} + \left[\int f_3(t)\, dt\right]\mathbf{k} + \mathbf{C}$$

where $\mathbf{C}$ is an arbitrary constant vector. The **definite integral** of $\mathbf{F}(t)$ on $a \le t \le b$ is the vector

$$\int_a^b \mathbf{F}(t)\, dt = \left[\int_a^b f_1(t)\, dt\right]\mathbf{i} + \left[\int_a^b f_2(t)\, dt\right]\mathbf{j} + \left[\int_a^b f_3(t)\, dt\right]\mathbf{k}$$

EXAMPLE 8   *Integral of a vector function*

Find $\int_0^\pi [t\mathbf{i} + 3\mathbf{j} - (\sin t)\mathbf{k}]\,dt$.

Solution $\int_0^\pi [t\mathbf{i} + 3\mathbf{j} - (\sin t)\mathbf{k}]\,dt = \left[\int_0^\pi t\,dt\right]\mathbf{i} + \left[\int_0^\pi 3\,dt\right]\mathbf{j} - \left[\int_0^\pi \sin t\,dt\right]\mathbf{k}$

$$= \left(\frac{t^2}{2}\right)\Big|_0^\pi \mathbf{i} + (3t)\Big|_0^\pi \mathbf{j} - (-\cos t)\Big|_0^\pi \mathbf{k}$$

$$= \tfrac{1}{2}\pi^2\mathbf{i} + 3\pi\mathbf{j} + (\cos\pi - \cos 0)]\mathbf{k}$$

$$= \tfrac{1}{2}\pi^2\mathbf{i} + 3\pi\mathbf{j} - 2\mathbf{k}$$

EXAMPLE 9   *Position of a particle given its velocity*

The velocity of a particle moving in space is

$$\mathbf{V}(t) = e^t\mathbf{i} + t^2\mathbf{j} + (\cos 2t)\mathbf{k}$$

Find the particle's position as a function of t if $\mathbf{R}(0) = 2\mathbf{i} + \mathbf{j} - \mathbf{k}$.

Solution   We need to solve the initial value problem that consists of:

The differential equation $\mathbf{V}(t) = \dfrac{d\mathbf{R}}{dt} = e^t\mathbf{i} + t^2\mathbf{j} + (\cos 2t)\mathbf{k}$

The initial condition $\mathbf{R}(0) = 2\mathbf{i} + \mathbf{j} - \mathbf{k}$

We integrate both sides of the differential equation with respect to t:

$$\int d\mathbf{R} = \left[\int e^t\,dt\right]\mathbf{i} + \left[\int t^2\,dt\right]\mathbf{j} + \left[\int \cos 2t\,dt\right]\mathbf{k}$$

$$\mathbf{R}(t) = (e^t + C_1)\mathbf{i} + \left(\frac{1}{3}t^3 + C_2\right)\mathbf{j} + \left(\frac{1}{2}\sin 2t + C_3\right)\mathbf{k}$$

$$= e^t\mathbf{i} + \frac{1}{3}t^3\mathbf{j} + \left(\frac{1}{2}\sin 2t\right)\mathbf{k} + \underbrace{C_1\mathbf{i} + C_2\mathbf{j} + C_3\mathbf{k}}_{\mathbf{C}}$$

Now use the initial condition to find $\mathbf{C}$:

$$\underbrace{2\mathbf{i} + \mathbf{j} - \mathbf{k}}_{\mathbf{R}(0)} = e^0\mathbf{i} + \frac{1}{3}(0)^3\mathbf{j} + \left(\frac{1}{2}\sin 0\right)\mathbf{k} + \mathbf{C} = \mathbf{i} + \mathbf{C}$$

so

$$\mathbf{i} + \mathbf{j} - \mathbf{k} = \mathbf{C}$$

Thus, the particle's position at any time t is

$$\mathbf{R}(t) = e^t\mathbf{i} + \frac{1}{3}t^3\mathbf{j} + \left(\frac{1}{2}\sin 2t\right)\mathbf{k} + \mathbf{i} + \mathbf{j} - \mathbf{k}$$

$$= (e^t + 1)\mathbf{i} + \left(\frac{1}{3}t^3 + 1\right)\mathbf{j} + \left(\frac{1}{2}\sin 2t - 1\right)\mathbf{k}$$

11.2 Problem Set

Ⓐ *Find the vector derivative* **F′** *in Problems 1–4.*

1. $\mathbf{F}(t) = t\mathbf{i} + t^2\mathbf{j} + (t + t^3)\mathbf{k}$
2. $\mathbf{F}(s) = (s\mathbf{i} + s^2\mathbf{j} + s^2\mathbf{k}) + (2s^2\mathbf{i} - s\mathbf{j} + 3\mathbf{k})$
3. $\mathbf{F}(s) = (\ln s)[s\mathbf{i} + 5\mathbf{j} - e^s\mathbf{k}]$
4. $\mathbf{F}(\theta) = (\cos \theta)[\mathbf{i} + (\tan \theta)\mathbf{j} + 3\mathbf{k}]$

Find **F′** *and* **F″** *for the vector functions given in Problems 5–8.*

5. $\mathbf{F}(t) = t^2\mathbf{i} + t^{-1}\mathbf{j} + e^{2t}\mathbf{k}$
6. $\mathbf{F}(s) = (1 - 2s^2)\mathbf{i} + (s \cos s)\mathbf{j} - s\mathbf{k}$
7. $\mathbf{F}(s) = (\sin s)\mathbf{i} + (\cos s)\mathbf{j} + s^2\mathbf{k}$
8. $\mathbf{F}(\theta) = (\sin^2 \theta)\mathbf{i} + (\cos 2\theta)\mathbf{j} + \theta^2\mathbf{k}$

Differentiate the scalar functions in Problems 9–12.

9. $f(x) = [x\mathbf{i} + (x + 1)\mathbf{j}] \cdot [(2x)\mathbf{i} - (3x^2)\mathbf{j}]$
10. $f(x) = \langle \cos x, x, -x \rangle \cdot \langle \sec x, -x^2, 2x \rangle$
11. $g(x) = \| \langle \sin x, -2x, \cos x \rangle \|$
12. $f(x) = \| [x\mathbf{i} + x^2\mathbf{j} - 2\mathbf{k}] + [(1 - x)\mathbf{i} - e^x\mathbf{j}] \|$

In Problems 13–18, **R** *is the position vector for a particle in space at time t. Find the particle's velocity and acceleration vector and then find the speed and direction of motion for the given value of t.*

13. $\mathbf{R}(t) = t\mathbf{i} + t^2\mathbf{j} + 2t\mathbf{k}$ at $t = 1$
14. $\mathbf{R}(t) = (1 - 2t)\mathbf{i} - t^2\mathbf{j} + e^t\mathbf{k}$ at $t = 0$
15. $\mathbf{R}(t) = (\cos t)\mathbf{i} + (\sin t)\mathbf{j} + 3t\mathbf{k}$ at $t = \frac{\pi}{4}$
16. $\mathbf{R}(t) = (2 \cos t)\mathbf{i} + t^2\mathbf{j} + (2 \sin t)\mathbf{k}$ at $t = \frac{\pi}{2}$
17. $\mathbf{R}(t) = e^t\mathbf{i} + e^{-t}\mathbf{j} + e^{2t}\mathbf{k}$ at $t = \ln 2$
18. $\mathbf{R}(t) = (\ln t)\mathbf{i} + \frac{1}{2}t^3\mathbf{j} - t\mathbf{k}$ at $t = 1$

Find the tangent vector to the graph of the given vector function **F** *at the points indicated in Problems 19–24.*

19. $\mathbf{F}(t) = t^2\mathbf{i} + 2t\mathbf{j} + (t^3 + t^2)\mathbf{k}$;
    $t = 0, t = 1, t = -1$
20. $\mathbf{F}(t) = t^{-3}\mathbf{i} + t^{-2}\mathbf{j} + t^{-1}\mathbf{k}$;
    $t = 1, t = -1$
21. $\mathbf{F}(t) = \dfrac{t\mathbf{i} + t^2\mathbf{j} + t^3\mathbf{k}}{1 + 2t}$; $t = 0, t = 2$
22. $\mathbf{F}(t) = t^2\mathbf{i} + (\cos t)\mathbf{j} + (t^2 \cos t)\mathbf{k}$;
    $t = 0, t = \frac{\pi}{2}$
23. $\mathbf{F}(t) = (\sin t)\mathbf{i} + (\cos t)\mathbf{j} + at\mathbf{k}$;
    $t = \frac{\pi}{2}, t = \pi$
24. $\mathbf{F}(t) = (e^t \sin \pi t)\mathbf{i} + (e^t \cos \pi t)\mathbf{j} + (\sin \pi t + \cos \pi t)\mathbf{k}$;
    $t = 0, t = 1, t = 2$

Find the indefinite vector integrals in Problems 25–30.

25. $\int [t\mathbf{i} - e^{3t}\mathbf{j} + 3\mathbf{k}]\, dt$
26. $\int [(\cos t)\mathbf{i} + (\sin t)\mathbf{j} - (2t)\mathbf{k}]\, dt$
27. $\int [(\ln t)\mathbf{i} - t\mathbf{j} + 3\mathbf{k})]\, dt$
28. $\int e^{-t}[3\mathbf{i} + t\mathbf{j} + (\sin t)\mathbf{k}]\, dt$
29. $\int [(t \ln t)\mathbf{i} - \sin(1 - t)\mathbf{j} + t\mathbf{k}]\, dt$
30. $\int [(\sinh t)\mathbf{i} - 3\mathbf{j} + (\cosh t)\mathbf{k}]\, dt$

Find the position vector **R**(t) *given the velocity* **V**(t) *and the initial position* **R**(0) *in Problems 31–34.*

31. $\mathbf{V}(t) = t^2\mathbf{i} - e^{2t}\mathbf{j} + \sqrt{t}\,\mathbf{k}; \mathbf{R}(0) = \mathbf{i} + 4\mathbf{j} - \mathbf{k}$
32. $\mathbf{V}(t) = t\mathbf{i} - \sqrt[3]{t}\mathbf{j} + e^t\mathbf{k}; \mathbf{R}(0) = \mathbf{i} - 2\mathbf{j} + \mathbf{k}$
33. $\mathbf{V}(t) = 2\sqrt{t}\mathbf{i} + (\cos t)\mathbf{j}; \mathbf{R}(0) = \mathbf{i} + \mathbf{j}$
34. $\mathbf{V}(t) = -3t\mathbf{i} + (\sin^2 t)\mathbf{j} + (\cos^2 t)\mathbf{k}; R(0) = \mathbf{j}$

Find the position vector **R**(t) *and* **V**(t) *given the acceleration* **A**(t) *and initial position and velocity* **R**(0) *and* **V**(t) *in Problems 35–36.*

35. $\mathbf{A}(t) = (\cos t)\mathbf{i} - (t \sin t)\mathbf{k}$;
    $\mathbf{R}(0) = \mathbf{i} - 2\mathbf{j} + \mathbf{k}; \mathbf{V}(0) = 2\mathbf{i} + 3\mathbf{k}$
36. $\mathbf{A}(t) = t^2\mathbf{i} - 2\sqrt{t}\mathbf{j} + e^{3t}\mathbf{k}$;
    $\mathbf{R}(0) = 2\mathbf{i} + \mathbf{j} - \mathbf{k}; \mathbf{V}(0) = \mathbf{i} - \mathbf{j} - 2\mathbf{k}$
37. The velocity of a particle moving in space is

$$\mathbf{V}(t) = e^t\mathbf{i} + t^2\mathbf{j}$$

    Find the particle's position as a function of t if $\mathbf{R}(0) = \mathbf{i} - \mathbf{j}$.
38. The velocity of a particle moving in space is

$$\mathbf{V}(t) = (t \cos t)\mathbf{i} + e^{2t}\mathbf{k}$$

    Find the particle's position as a function of t if $\mathbf{R}(0) = \mathbf{i} + 2\mathbf{k}$.
39. The acceleration of a particle moving in space is

$$\mathbf{A}(t) = 24t^2\mathbf{i} + 4\mathbf{j}$$

    Find the particle's position as a function of t if $\mathbf{R}(0) = \mathbf{i} + 2\mathbf{j}$ and $\mathbf{V}(0) = \mathbf{0}$.

Ⓑ 40. ■ **What Does This Say?** Describe the concept of a smooth curve.
41. ■ **What Does This Say?** Compare or contrast the notions of speed and velocity of a vector function.

Let $\mathbf{v} = 2\mathbf{i} - \mathbf{j} + 5\mathbf{k}$ *and* $\mathbf{w} = \mathbf{i} + 2\mathbf{j} - 3\mathbf{k}$. *Find the derivatives in Problems 42–45.*

42. $\dfrac{d}{dt}(\mathbf{v} + t\mathbf{w})$
43. $\dfrac{d^2}{dt^2}(\mathbf{v} \cdot t^4\mathbf{w})$
44. $\dfrac{d^2}{dt^2}(t\|\mathbf{v}\| + t^2\|\mathbf{w}\|)$
45. $\dfrac{d}{dt}(t\mathbf{v} \times t^2\mathbf{w})$

In Problems 46–47, verify the indicated equation for the vector functions

$$\mathbf{F}(t) = (3 + t^2)\mathbf{i} - (\cos 3t)\mathbf{j} + t^{-1}\mathbf{k} \quad and$$
$$\mathbf{G}(t) = \sin(2 - t)\mathbf{i} - e^{2t}\mathbf{k}$$

46. $(3\mathbf{F} - 2\mathbf{G})'(t) = 3\mathbf{F}'(t) - 2\mathbf{G}'(t)$

47. $(\mathbf{F} \cdot \mathbf{G})'(t) = (\mathbf{F}' \cdot \mathbf{G})(t) + (\mathbf{F} \cdot \mathbf{G}')(t)$

In Problems 48–49, find a value of a that satisfies the given equation.

48. $\displaystyle\int_0^a \left[(t\sqrt{1 + t^2})\mathbf{i} + \left(\frac{1}{1 + t^2}\right)\mathbf{j} \right] dt$

$$= \tfrac{1}{3}(2\sqrt{2} - 1)\mathbf{i} + \tfrac{\pi}{4}\mathbf{j}$$

49. $\displaystyle\int_0^{2a} [(\cos t)\mathbf{i} + (\sin t)\mathbf{j} + (\sin t \cos t)\mathbf{k}] \, dt$

$$= \mathbf{i} + \mathbf{j} + \tfrac{1}{2}\mathbf{k}$$

In Problems 50–51, show that $\mathbf{F}(t)$ and $\mathbf{F}''(t)$ are parallel for all t with constant k.

50. $\mathbf{F}(t) = e^{kt}\mathbf{i} + e^{-kt}\mathbf{j}$

C 51. $\mathbf{F}(t) = (\cos kt)\mathbf{i} + (\sin kt)\mathbf{j}$

52. Let $\mathbf{F}(t) = u(t)\mathbf{i} + v(t)\mathbf{j} + w(t)\mathbf{k}$, where u, v, and w are differentiable scalar functions of t. Show that

$$\mathbf{F}(t) \cdot \mathbf{F}'(t) = \|\mathbf{F}(t)\|(\|\mathbf{F}(t)\|)'$$

53. Let $\mathbf{F}(t)$ be a differentiable vector function and let $h(t)$ be a differentiable scalar function of t. Show that

$$[h(t)\mathbf{F}(t)]' = h(t)\mathbf{F}'(t) + h'(t)\mathbf{F}(t)$$

54. If $\mathbf{F}$ and $\mathbf{G}$ are differentiable vector functions of t, prove

$$(\mathbf{F} \cdot \mathbf{G})'(t) = (\mathbf{F}' \cdot \mathbf{G})(t) + (\mathbf{F} \cdot \mathbf{G}')(t)$$

55. If $\mathbf{F}$ and $\mathbf{G}$ are differentiable vector functions of t, prove

$$(\mathbf{F} \times \mathbf{G})'(t) = (\mathbf{F}' \times \mathbf{G})(t) + (\mathbf{F} \times \mathbf{G}')(t)$$

56. If $\mathbf{F}$ is a differentiable vector function such that $\mathbf{F}(t) \neq \mathbf{0}$, show that

$$\frac{d}{dt}\left(\frac{\mathbf{F}(t)}{\|\mathbf{F}(t)\|}\right) = \frac{\mathbf{F}'(t)}{\|\mathbf{F}(t)\|} - \frac{[\mathbf{F}(t) \cdot \mathbf{F}'(t)]\mathbf{F}(t)}{\|\mathbf{F}(t)\|^3}$$

57. Show that

$$\frac{d}{dt}\|\mathbf{R}(t)\| = \frac{1}{\|\mathbf{R}\|}\mathbf{R} \cdot \frac{d\mathbf{R}}{dt}$$

58. Find a formula for

$$\frac{d}{dt}[\mathbf{F} \cdot (\mathbf{G} \times \mathbf{H})]$$

59. Find a formula for

$$\frac{d}{dt}[\mathbf{F} \times (\mathbf{G} \times \mathbf{H})]$$

60. If $\mathbf{G} = \mathbf{F} \cdot (\mathbf{F}' \times \mathbf{F}'')$, what is $\mathbf{G}'$?

11.3 *Modeling Ballistics and Planetary Motion*

IN THIS SECTION modeling the motion of a projectile in a vacuum, Kepler's second law

MODELING THE MOTION OF A PROJECTILE IN A VACUUM

In general, it can be quite difficult to analyze the motion of a projectile, but the problem becomes manageable if we assume that the acceleration due to gravity is constant and that the projectile travels in a vacuum. A model of the motion in the real world (with air resistance, for example) based on these assumptions is fairly realistic as long as the projectile is reasonably heavy, travels at relatively low speed, and stays close to the surface of the earth.

Figure 11.11 shows a projectile that is fired from a point P and travels in a vertical plane coordinatized so that P is directly above the origin O and the impact point I is on the x-axis at ground level. We let s_0 be the height of P above O, v_0 be the initial speed of the projectile (the **muzzle speed**), and α be the angle of elevation of the "gun" that fires the projectile.

Because our projectile is to travel in a vacuum, we shall assume that the only force acting on the projectile at any given time is the force due to gravity. This force $\mathbf{F}$ is directed downward. The magnitude of this force is the projectile's weight. If m is the mass of the projectile and g is the "free-fall" acceleration due to gravity (g is approximately 32 ft/s^2 or 9.8 m/s^2), then the magnitude of $\mathbf{F}$ is mg. Also, according to **Newton's second law of motion,** the sum of all the forces acting on the projectile must

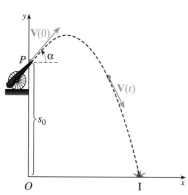

FIGURE 11.11 The motion of a projectile

equal $m\mathbf{A}(t)$, where $\mathbf{A}(t)$ is the acceleration of the projectile at time t. Because $\mathbf{F} = -m g\mathbf{j}$ is the only force acting on the projectile, it follows that

$$-m g\mathbf{j} = \mathbf{F} = m\mathbf{A}(t)$$
$$\mathbf{A}(t) = -g\mathbf{j} \quad \text{for all } t \geq 0$$

The velocity $\mathbf{V}(t)$ of the projectile can now be obtained by integrating $\mathbf{A}(t)$. Specifically, we have

$$\mathbf{V}(t) = \int \mathbf{A}(t)\, dt = \int (-g\mathbf{j})\, dt = -gt\mathbf{j} + \mathbf{C}_1$$

where $\mathbf{C}_1 = \mathbf{V}(0)$ is the velocity when $t = 0$ (that is, the *initial velocity*). Because the projectile is fired with initial speed v_0 at an angle of elevation α, the initial velocity must be

$$\mathbf{C}_1 = \mathbf{V}(0) = (v_0 \cos \alpha)\mathbf{i} + (v_0 \sin \alpha)\mathbf{j}$$

(as shown in the detail of Figure 11.11), and by substitution we find

$$\mathbf{V}(t) = -gt\mathbf{j} + (v_0 \cos \alpha)\mathbf{i} + (v_0 \sin \alpha)\mathbf{j}$$
$$= (v_0 \cos \alpha)\mathbf{i} + (v_0 \sin \alpha - gt)\mathbf{j}$$

Next, by integrating $\mathbf{V}$ with respect to t, we find that the projectile has displacement

$$\mathbf{R}(t) = \int \mathbf{V}(t)\, dt = \int [(v_0 \cos \alpha)\mathbf{i} + (v_0 \sin \alpha - gt)\mathbf{j}]\, dt$$
$$= (v_0 \cos \alpha)t\mathbf{i} + [(v_0 \sin \alpha)t - \tfrac{1}{2}gt^2]\mathbf{j} + \mathbf{C}_2$$

where $\mathbf{C}_2 = \mathbf{R}(0)$ is the position at time $t = 0$. Because the projectile begins its flight at P, the initial position is $\mathbf{R}(0) = s_0\mathbf{j}$, and by substituting this expression for $\mathbf{C}_2$, we find that the position at time t is

$$\mathbf{R}(t) = [(v_0 \cos \alpha)t]\mathbf{i} + [(v_0 \sin \alpha)t - \tfrac{1}{2}gt^2 + s_0]\mathbf{j}$$

Specifically, $\mathbf{R}(t)$ gives the position of the projectile at any time t after it is fired and before it hits the ground. If we let $x(t)$ and $y(t)$ denote the horizontal and vertical components of $\mathbf{R}(t)$, respectively, we can make the following statement.

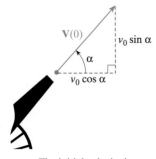

The initial velocity is
$\mathbf{v}(0) = (v_0 \cos \alpha)\mathbf{i} + (v_0 \sin \alpha)\mathbf{j}$.

Detail of Figure 11.11

Motion of a Projectile in a Vacuum

Consider a projectile that travels in a vacuum in a coordinate plane, with the x-axis along level ground. If the projectile is fired from a height of s_0 with initial speed v_0 and angle of elevation α, then at time $t\,(t \geq 0)$ it will be at the point $(x(t), y(t))$, where

$$x(t) = (v_0 \cos \alpha)t \quad \text{and} \quad y(t) = -\tfrac{1}{2}gt^2 + (v_0 \sin \alpha)t + s_0$$

The parametric equations for the motion of a projectile in a vacuum provide useful general information about a projectile's motion. For instance, note that if $\alpha \neq 90°$, we can eliminate the parameter t by solving the first equation, obtaining

$$t = \frac{x}{v_0 \cos \alpha}$$

By substituting into the second equation, we find

$$y = -\frac{1}{2}g\left[\frac{x}{v_0\cos\alpha}\right]^2 + (v_0\sin\alpha)\left[\frac{x}{v_0\cos\alpha}\right] + s_0$$

$$= \left[\frac{-g}{2(v_0\cos\alpha)^2}\right]x^2 + (\tan\alpha)x + s_0$$

This is the Cartesian equation for the trajectory of the projectile, and because the equation has the general form $y = ax^2 + bx + c$, with $a < 0$, the trajectory must be part of a downward-opening parabola. That is, *if a projectile is not fired vertically (so that* $\tan\alpha$ *is defined), its trajectory will be part of a downward-opening parabolic arc,* as shown in Figure 11.12.

■ **FIGURE 11.12** The motion of a projectile. The trajectory is part of a parabolic arc

The *time of flight* T_f of a projectile is the elapsed time between launch and impact, and the *range* is the total horizontal distance R traveled by the projectile during its flight. Because $y = 0$ at impact, it follows that T_f must satisfy the quadratic equation

$$-\tfrac{1}{2}gT_f^2 + (v_0\sin\alpha)T_f + s_0 = 0$$

Then, because the flight begins at $x = 0$ when $t = 0$ and ends at $x = R_f$ when $t = T_f$, the range R_f can be computed by evaluating $x(t)$ at $t = T_f$. That is,

$$R_f = (v_0\cos\alpha)T_f$$

EXAMPLE 1   *Modeling the path of a projectile*

A boy standing at the edge of a cliff throws a ball upward at a 30° angle with an initial speed of 64 ft/s. Suppose that when the ball leaves the boy's hand, it is 48 ft above the ground at the base of the cliff.

a. What are the time of flight of the ball and its range?
b. What are the velocity of the ball and its speed at impact?
c. What is the highest point reached by the ball during its flight?

Solution   We know that $g = 32$ ft/s², and we are given $s_0 = 48$ ft, $v_0 = 64$ ft/s, and $\alpha = 30°$. We can now write the parametric equations for the parabola:

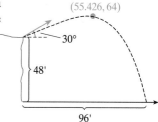

$$x(t) = (v_0\cos\alpha)t = (64\cos 30°)t = 32\sqrt{3}\,t$$
$$y(t) = -\tfrac{1}{2}gt^2 + (v_0\sin\alpha)t + s_0$$
$$= -\tfrac{1}{2}(32)t^2 + (64\sin 30°)t + 48$$
$$= -16t^2 + 32t + 48$$

a. The ball hits the ground when $y = 0$, and by solving the equation $-16t^2 + 32t + 48 = 0$ for $t \geq 0$, we find that $t = 3$, so the time of flight is $T_f = 3$ seconds. The range is

$$x(3) = 32\sqrt{3}(3) \approx 166.27688$$

That is, the ball hits the ground about 166 ft from the base of the cliff.

b. We find that $x'(t) = 32\sqrt{3}$ and $y'(t) = -32t + 32$, so the velocity at time t is

$$\mathbf{V}(t) = 32\sqrt{3}\,\mathbf{i} + (-32t + 32)\mathbf{j}$$

Thus, at impact (when $t = 3$), the velocity is $\mathbf{V}(3) = 32\sqrt{3}\,\mathbf{i} - 64\,\mathbf{j}$, and its speed is

$$\|\mathbf{V}(3)\| = \sqrt{(32\sqrt{3})^2 + (-64)^2} \approx 84.664042$$

That is, the speed at impact is about 85 ft/s.

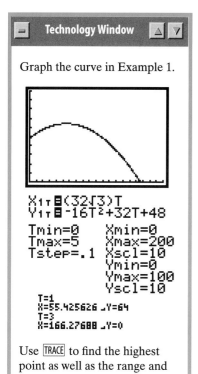

Technology Window

Graph the curve in Example 1.

```
X₁ᵀ ▤ (32√3)T
Y₁ᵀ ▤ -16T²+32T+48

Tmin=0    Xmin=0
Tmax=5    Xmax=200
Tstep=.1  Xscl=10
          Ymin=0
          Ymax=100
          Yscl=10

T=1
X=55.425626  ⌐Y=64
T=3
X=166.27688  ⌐Y=0
```

Use TRACE to find the highest point as well as the range and time of flight.

c. The ball attains its maximum height when the upward (vertical) component of its velocity $\mathbf{V}(t)$ is 0—that is, when $y'(t) = 0$. Solving the equation $-32t + 32 = 0$, we find that this occurs when $t = 1$. Therefore, the maximum height attained by the ball is

$$y_m = y(1) = -16(1)^2 + 32(1) + 48 = 64$$
$$x_m = x(1) = 32\sqrt{3}(1) \approx 55.425626$$

and the highest point reached by the ball has coordinates (rounded to the nearest unit) of $(55, 64)$. ∎

In general, if the projectile is fired from ground level (so $s_0 = 0$), the time of flight T_f satisfies

$$-\tfrac{1}{2}gT_f^2 + (v_0 \sin \alpha)T_f = 0$$
$$T_f = \frac{2}{g}v_0 \sin \alpha$$

Thus, the range R_f is

$$R_f = x(T_f) = (v_0 \cos \alpha)\left(\frac{2v_0 \sin \alpha}{g}\right) = \frac{v_0^2}{g}(2 \sin \alpha \cos \alpha) = \frac{v_0^2}{g} \sin 2\alpha$$

Notice that for a given initial speed v_0, the range assumes its largest value when $\sin 2\alpha = 1$, that is, when $\alpha = 45° = \frac{\pi}{4}$. To summarize:

Time of Flight and Range When Fired from Ground Level

A projectile fired from *ground level* has **time of flight** T_f and **range** R_f given by the equations

$$T_f = \frac{2}{g}v_0 \sin \alpha \quad \text{and} \quad R_f = \frac{v_0^2}{g} \sin 2\alpha$$

The maximal range is $R_m = \frac{v_0^2}{g}$, and it occurs when $\alpha = \frac{\pi}{4}$.

EXAMPLE 2   *Flight time and range for the motion of a projectile*

A projectile is fired from ground level at an angle of $40°$ with muzzle speed 110 ft/s. Find the time of flight and the range.

Solution   With $g = 32$ ft/s^2, we see that the flight time T_f is

$$T_f = \frac{2}{g}v_0 \sin \alpha = \frac{2}{32}(110)(\sin 40°) \approx 4.4191648$$

The range R_m is

$$R_m = \frac{v_0^2}{g} \sin 2\alpha = \frac{(110)^2}{32}(\sin 80°) \approx 372.38043$$

That is, the projectile travels about 372 ft horizontally before it hits the ground, and the flight takes a little more than 4 sec. ∎

KEPLER'S SECOND LAW

In the 17th century, the German astronomer Johannes Kepler (1571–1630) formulated three useful laws for describing planetary motion. The guest essay at the end of this chapter discusses the discovery of these laws.

> **Kepler's Laws**
>
> 1. The planets move about the sun in elliptical orbits, with the sun at one focus.
> 2. The radius vector joining a planet to the sun sweeps over equal areas in equal intervals of time.
> 3. The square of the time of one complete revolution of a planet about its orbit is proportional to the cube of the orbit's semimajor axis.

We will use vector methods to prove Kepler's second law. The other two laws can be established similarly.

We begin by introducing some useful notation involving polar coordinates. Let $\mathbf{u}_r$ and $\mathbf{u}_\theta$ denote unit vectors along the radial axis and orthogonal to that axis, respectively, as shown in Figure 11.13. Then, in terms of the unit Cartesian vectors $\mathbf{i}$ and $\mathbf{j}$, we have

$$\mathbf{u}_r = (\cos\theta)\mathbf{i} + (\sin\theta)\mathbf{j} \quad \text{and} \quad \mathbf{u}_\theta = (-\sin\theta)\mathbf{i} + (\cos\theta)\mathbf{j}$$

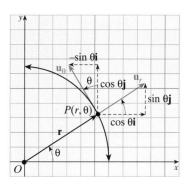

■ **FIGURE 11.13**   Describing the motion of a particle along a curve

The derivatives $\dfrac{d\mathbf{u}_r}{d\theta}$ and $\dfrac{d\mathbf{u}_\theta}{d\theta}$ satisfy

$$\frac{d\mathbf{u}_r}{d\theta} = (-\sin\theta)\mathbf{i} + (\cos\theta)\mathbf{j} = \mathbf{u}_\theta$$

$$\frac{d\mathbf{u}_\theta}{d\theta} = (-\cos\theta)\mathbf{i} + (-\sin\theta)\mathbf{j} = -\mathbf{u}_r$$

Now, suppose the sun S is at the origin (pole) of a polar coordinate system, and consider the motion of a body B (planet, comet, artificial satellite) about S. The radial vector $\mathbf{R} = \mathbf{SB}$ can be expressed as

$$\mathbf{R} = r\mathbf{u}_r = (r\cos\theta)\mathbf{i} + (r\sin\theta)\mathbf{j}$$

To date, there are more than 4,000 earth satellites.

where $r = \|\mathbf{R}\|$, and the velocity $\mathbf{V}$ satisfies

$$
\begin{aligned}
\mathbf{V} = \frac{d\mathbf{R}}{dt} &= \frac{dr}{dt}\mathbf{u}_r + r\frac{d\mathbf{u}_r}{dt} && \textit{Derivative of a scalar multiple} \\
&= \frac{dr}{dt}\mathbf{u}_r + r\frac{d\mathbf{u}_r}{d\theta}\frac{d\theta}{dt} && \textit{Chain rule} \\
&= \frac{dr}{dt}\mathbf{u}_r + r\frac{d\theta}{dt}\mathbf{u}_\theta
\end{aligned}
$$

You can find a similar formula for acceleration (see Problem 40). We summarize these formulas in the following box.

Polar Formulas for Velocity and Acceleration

$$
\mathbf{V}(t) = \frac{dr}{dt}\mathbf{u}_r + r\frac{d\theta}{dt}\mathbf{u}_\theta
$$

$$
\mathbf{A}(t) = \frac{d\mathbf{V}(t)}{dt^2} = \frac{d^2\mathbf{R}}{dt} = \left[\frac{d^2r}{dt^2} - r\left(\frac{d\theta}{dt}\right)^2\right]\mathbf{u}_r + \left[r\frac{d^2\theta}{dt^2} + 2\frac{dr}{dt}\frac{d\theta}{dt}\right]\mathbf{u}_\theta
$$

EXAMPLE 3   *Find the velocity of a moving body*

The position vector of a moving body is $\mathbf{R}(t) = 2t\mathbf{i} - t^2\mathbf{j}$ for $t \geq 0$. Express $\mathbf{R}$ and the velocity vector $\mathbf{V}(t)$ in terms of $\mathbf{u}_r$ and $\mathbf{u}_\theta$.

Solution   We note that on the trajectory, $x = 2t$, $y = -t^2$, so $\mathbf{R}$ has length

$$
r = \|\mathbf{R}(t)\| = \sqrt{(2t)^2 + (-t^2)^2} = \sqrt{4t^2 + t^4} = t\sqrt{t^2 + 4}
$$

and $\mathbf{R} = r\mathbf{u}_r = t\sqrt{t^2 + 4}\,\mathbf{u}_r$. Because $\mathbf{V}(t) = \frac{dr}{dt}\mathbf{u}_r + r\frac{d\theta}{dt}\mathbf{u}_\theta$, we need $\frac{dr}{dt}$ and $\frac{d\theta}{dt}$. We find that

$$
\frac{dr}{dt} = \sqrt{t^2 + 4} + t\left(\frac{1}{2}\right)(t^2 + 4)^{-1/2}(2t) = \frac{2t^2 + 4}{\sqrt{t^2 + 4}}
$$

and because the polar angle satisfies $\theta = \tan^{-1}\left(\frac{y}{x}\right)$ for all points (x, y) on the curve (except where $x = 0$), we have

$$
\theta = \tan^{-1}\left(\frac{-t^2}{2t}\right) = \tan^{-1}\left(-\frac{t}{2}\right)
$$

$$
\frac{d\theta}{dt} = \frac{1}{1 + \left(-\frac{t}{2}\right)^2}\left(-\frac{1}{2}\right) = \frac{-2}{t^2 + 4}
$$

Thus,

$$
\mathbf{V}(t) = \frac{2t^2 + 4}{\sqrt{t^2 + 4}}\mathbf{u}_r + t\sqrt{t^2 + 4}\left(\frac{-2}{t^2 + 4}\right)\mathbf{u}_\theta = \frac{(2t^2 + 4)\mathbf{u}_r - 2t\,\mathbf{u}_\theta}{\sqrt{t^2 + 4}}
$$

We now have the tools we need to establish Kepler's second law.

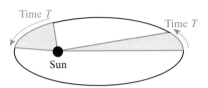

Time T Time T

Sun

■ **FIGURE 11.14** The radial line sweeps out equal area in equal time.

THEOREM 11.5   *Kepler's second law*

The radius vector from the sun to a planet in its orbit sweeps over equal areas in equal intervals of time.

Proof   The situation described by Kepler's second law is illustrated in Figure 11.14. We shall assume that the only force acting on a planet is the gravitational attraction of the sun. According to the universal law of gravitation, the force of attraction is given by

$$\mathbf{F} = -G\frac{mM}{r^2}\mathbf{u}_r$$

where G is a physical constant, and m and M are the masses of the planet and the sun, respectively. Because this is the only force acting on the planet, Newton's second law of motion tells us that $\mathbf{F} = m\mathbf{A}$, where $\mathbf{A}$ is the acceleration of the planet in its orbit. By equating these two expressions for the force $\mathbf{F}$, we find

$$m\mathbf{A} = -G\frac{mM}{r^2}\mathbf{u}_r$$

$$\mathbf{A} = \frac{-GM}{r^2}\mathbf{u}_r$$

This says that *the acceleration of a planet in its orbit has only a radial component.* Note how this conclusion depends on our assumption that the only force on the planet is the sun's gravitational attraction, which acts along radial lines from the sun to the planet.

This means that the $\mathbf{u}_\theta$ component of the planet's acceleration is 0, and by examining the polar formula for acceleration, we see that this condition is equivalent to the differential equation

$$\left[r\frac{d^2\theta}{dt^2} + 2\frac{dr}{dt}\frac{d\theta}{dt} \right] = 0$$

If we set $z = \dfrac{d\theta}{dt}$, we obtain $\dfrac{d^2\theta}{dt^2} = \dfrac{dz}{dt}$, so that

$$r\frac{dz}{dt} + 2z\frac{dr}{dt} = 0$$

$$z^{-1}\frac{dz}{dt} = -2r^{-1}\frac{dr}{dt}$$

$$\int z^{-1}\,dz = \int (-2r^{-1})dr$$

$$\ln|z| + C_1 = -2\ln|r| + C_2$$

$$\ln z = \ln(Cr^{-2}) \qquad \text{since } r > 0 \text{ and } z > 0$$

$$z = Cr^{-2}$$

$$\frac{d\theta}{dt} = Cr^{-2}$$

$$r^2\frac{d\theta}{dt} = C$$

Now, let $[t_1, t_2]$ and $[t_3, t_4]$ be two time intervals of equal length, so $t_2 - t_1 = t_4 - t_3$. According to Theorem 9.5, the area swept out in the time period $[t_1, t_2]$ (see Figure 11.15) is

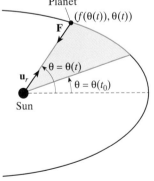

Planet
$(f(\theta(t)), \theta(t))$
F

$\mathbf{u}_r$ $\theta = \theta(t)$

$\theta = \theta(t_0)$

Sun

■ **FIGURE 11.15** The radial line sweeps out area at the rate

$$\frac{dA}{dt} = \frac{1}{2}r^2\frac{d\theta}{dt}$$

$$S_1 = \int_{t_1}^{t_2} \frac{1}{2}r^2\,d\theta = \int_{t_1}^{t_2} \frac{1}{2}\left[r^2\frac{d\theta}{dt} \right]dt = \int_{t_1}^{t_2} \frac{1}{2}C\,dt \qquad \text{since } r^2\frac{d\theta}{dt} = C$$

Thus, $S_1 = \frac{1}{2} C(t_2 - t_1)$. Similarly, the area swept out in time period $[t_3, t_4]$ is $S_2 = \frac{1}{2} C(t_4 - t_3)$, and we have

$$S_1 = \tfrac{1}{2} C(t_2 - t_1) = \tfrac{1}{2} C(t_4 - t_3) = S_2$$

so equal area is swept out in equal time, as claimed by Kepler.

An object is said to move in a **central force field** if it is subject to a single force that is always directed toward a particular point. Our derivation of Kepler's second law is based on the assumption that planetary motion occurs in a central force field. In particular, the formula

$$\mathbf{A} = \frac{-GM}{r^2} \mathbf{u}_r$$

shows that a planet's acceleration has only a radial component, and this turns out to be true for an object moving in a plane in any central force field.

Much of what we know about the bodies in our solar system is a result of the mathematics associated with Kepler's laws. Table 11.1 provides some of these data.

TABLE 11.1

Data on Planetary Orbits

Planet	Mean Orbit (in millions of miles)	Eccentricity	Period (in days)	Comparative Mass (Earth = 1.00)
Mercury	36.0	0.205635	88.0	0.06
Venus	67.3	0.006761	224.7	0.81
Earth	**93.0**	**0.016678**	**365.265**	**1.00**
				Actual mass: 5.975 × 10²⁴ kg
Mars	141.7	0.093455	687.0	0.12
Jupiter	484.0	0.048207	4,332.1	317.83
Saturn	887.0	0.055328	10,825.9	95.16
Uranus	1,787.0	0.047694	30,676.1	14.5
Neptune	2,797.0	0.010034	59,911.1	17.2
Pluto	3,675.0	0.248646	90,824.2	0.0025

11.3   Problem Set

Ⓐ In Problems 1–8, find the time of flight T_f (to the nearest tenth second) and the range R_f (to the nearest unit) of a projectile fired from ground level at the given angle α with the indicated initial speed v_0. Assume that $g = 32$ ft/s² or $g = 9.8$ m/s².

1. $\alpha = 35°, v_0 = 128$ ft/s   2. $\alpha = 45°, v_0 = 80$ ft/s
3. $\alpha = 48.5°, v_0 = 850$ m/s   4. $\alpha = 43.5°, v_0 = 185$ m/s
5. $\alpha = 23.74°, v_0 = 23.3$ m/s   6. $\alpha = 31.04°, v_0 = 38.14$ m/s
7. $\alpha = 14.11°, v_0 = 100$ ft/s   8. $\alpha = 78.09°, v_0 = 88$ ft/s

In Problems 9–14, an object moves along the given curve in the plane (described in either polar or parametric form). Find its velocity and acceleration in terms of the unit polar vectors $\mathbf{u}_r$ and $\mathbf{u}_\theta$.

9. $x = 2t, y = t$ 10. $x = \sin t, y = \cos t$

11. $r = \sin \theta, \theta = 2t$
12. $r = e^{-\theta}, \theta = 1 - t$
13. $r = 5(1 + \cos \theta), \theta = 2t + 1$
14. $r = \dfrac{1}{1 - \cos \theta}, \theta = t$

Ⓑ 15. A shell fired from ground level at an angle of 45° hits the ground 2,000 m away. What is the muzzle speed of the shell?

16. A gun at ground level has muzzle speed of 300 ft/s. What angle of elevation (to the nearest degree) should be used to hit an object 1,500 ft away?

17. At what angle (to the nearest tenth of a degree) should a projectile be fired from ground level if its muzzle speed is 167.1 ft/s and the desired range is 600 ft?

18. A shell is fired at ground level with a muzzle speed of 280 ft/s and at an elevation of 45° from ground level.
  a. Find the maximum height attained by the shell.
  b. Find the time of flight and the range of the shell.
  c. Find the velocity and speed of the shell at impact.

19. Jeff Bagwell hits a baseball at a 30° angle with a speed of 90 ft/s. If the ball is 4 ft above ground level when it is hit, what is the maximum height reached by the ball? How far will it travel from home plate before it lands? If it just barely clears a 5-foot wall in the outfield, how far (to the nearest foot) is the wall from home plate?

20. A baseball hit in the Astrodome at a 24° angle from 3 ft above the ground just goes over the 9-ft fence 400 ft from home plate. About how fast was the ball traveling, and how long did it take the ball to reach the wall?

21. One of the shortest possible home runs in major league baseball is over the right field fence at the Kingdome in Seattle, where the distance is 312 ft and the fence is 8 ft high. How long will it take for a ball to leave the playing field at that point if it is struck $3\frac{1}{2}$ ft above the ground at a 32° angle, and we assume that the ball just barely clears the fence?

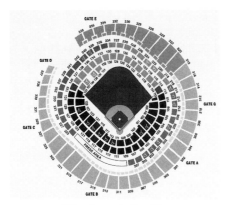

Kingdome in Seattle

22. Steve Young, a quarterback for the San Francisco '49ers, throws a pass at a 45° angle from a height of 6.5 ft with a speed of 50 ft/s. Receiver Jerry Rice races straight downfield on a fly pattern at a constant speed of 32 ft/s and catches the ball at a height of 6 ft. If Steve and Jerry both line up on the 50-yard line at the start of the play, how far does Steve fade back from the line of scrimmage before he releases the ball?

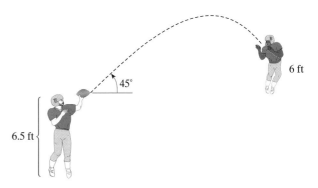

23. A golf ball is hit from the tee to a green with an initial speed of 125 ft/s at an angle of elevation of 45°. How long will it take for the ball to hit the green?

24. Basketball player Shaquille O'Neal attempts a shot while standing 20 ft from the basket. If "The Shaq" shoots from a height of 7 ft and the ball reaches a maximum height of 12 ft before passing through the 10-ft-high basket, what is the initial speed of the ball?

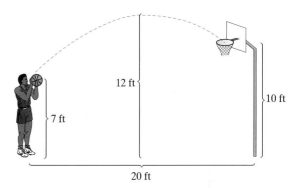

25. If a shotputter throws a shot from a height of 5 ft with an angle of 46° and initial speed of 25 ft/s, what is the horizontal distance of the throw?

26. **MODELING PROBLEM**   In 1974, Evel Knievel attempted a skycycle ride across the Snake River, and at the time there was a great deal of hype about "will he make it?" If the angle of the launching ramp was 45° and if the horizontal distance the skycycle needed to travel was 4,700 ft, at what speed did Evel have to leave the ramp to make it across the Snake River Canyon?

Assume that the opposite edges of the canyon are at the same height.

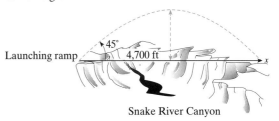

Snake River Canyon

27. A particle moves along the polar path (r, θ), where $r(t) = 3 + 2 \sin t$, $\theta(t) = t^3$. Find $\mathbf{V}(t)$ and $\mathbf{A}(t)$ in terms of $\mathbf{u}_r$ and $\mathbf{u}_\theta$.

In Problems 28–31, consider a particle moving on a circular path of radius a described by the equation $\mathbf{R}(t) = (a \cos \omega t)\mathbf{i} + (a \sin \omega t)\mathbf{j}$, *where* $\omega = \dfrac{d\theta}{dt}$ *is the constant angular velocity.*

28. Find the velocity vector and show that it is orthogonal to $\mathbf{R}(t)$.

29. Find the speed of the particle.

30. Find the acceleration vector and show that its direction is always toward the center of the circle.

31. Find the magnitude of the acceleration vector.

32. MODELING PROBLEM A 3-oz paddleball attached to a string is swung in a circular path with a 1-ft radius. If the string will break under a force of 2 lb, find the maximum speed the ball can attain without breaking the string. (*Hint:* Note that 3 oz is 3/16 lb.)

33. SPY PROBLEM The Spy figures out where to dig (Problem 59 of Section 10.1) and unearths a human-size box. With shaking hands he opens the lid, and to his relief, finds not Purity, but her favorite scarf and a cell phone. He picks it up and hears Purity's voice screaming for help. Her pleas are quickly muffled and Blohardt takes over. "Do what you're told or the next box will be heavier!" he threatens. "You're going to help me eliminate my competition, starting with Scélérat." The Spy is still peeved with Scélérat for icing his friend, Siggy Leiter, so he agrees. He's told that the Frenchman and his gang are holed up in a bunker on the side of a hill inclined at an angle of 15° to the horizontal, as shown in Figure 11.16. The Spy returns to his helicopter and flies toward the bunker traveling at 200 ft/s at an altitude of 10,000 ft. Just as the helicopter flies over the base of the hill, the Spy sights a ventilation hole in the top of the bunker at an angle of 20°. He decides to drop a canister of knockout gas into the bunker through the hole. How long should he wait before releasing the canister?

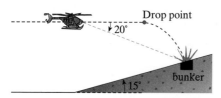

■ **FIGURE 11.16** Spy and the bunker target

34. A child, running along level ground at the top of a 30-ft-high vertical cliff at a speed of 15 ft/s, throws a rock over the cliff into the sea below. Suppose the child's arm is 3 ft above the ground and her arm speed is 25 ft/s. If the rock is released 10 ft from the edge of the cliff at an angle of 30°, how long does it take for the rock to hit the water? How far from the base of the cliff does it hit?

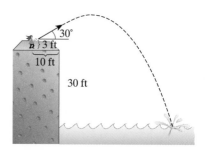

35. Find two angles of elevation α_1 and α_2 (to the nearest degree) so that a shell fired at ground level with muzzle speed of 650 ft/s will hit a target 6,000 ft away (also at ground level).

36. A gun is fired with muzzle speed $v_0 = 550$ ft/s at an angle of $\alpha = 22.0°$. The shell overshoots the target by 50 ft. At what angle should a second shot be fired with the same muzzle speed to hit the target?

37. A gun is fired with muzzle speed $v_0 = 700$ ft/s at an angle of $\alpha = 25°$. It overshoots the target by 60 ft. The target moves away from the gun at a constant speed of 10 ft/s. If the gunner takes 30 sec to reload, at what angle should a second shot be fired with the same muzzle speed to hit the target?

38. A shell is fired from ground level with muzzle speed of 750 ft/s at an angle of 25°. An enemy gun 20,000 ft away fires a shot 2 seconds later and the shells collide 50 ft above ground. What are the muzzle speed v_0 and angle of elevation α of the second gun?

C 39. Suppose a shell is fired with muzzle speed v_0 at an angle α from a height s_0 above level ground.
   a. Show that the range of R_f must satisfy

$$g(\sec^2 \alpha)\, R_f^2 - 2v_0^2 (\tan \alpha) R_f - 2v_0^2 s_0 = 0$$

   b. Show that the maximum range R_m occurs at angle α_m where

$$R_m \tan \alpha_m = \frac{v_0^2}{g}$$

   c. Show that

$$R_m = \frac{v_0}{g} \sqrt{v_0^2 + 2gs_0} \quad \text{and} \quad \alpha_m = \tan^{-1}\left(\frac{v_0}{\sqrt{v_0^2 + 2gs_0}}\right)$$

40. Use the formula

$$\mathbf{V}(t) = \frac{dr}{dt}\mathbf{u}_r + r\frac{d\theta}{dt}\mathbf{u}_\theta$$

to derive the formula for polar acceleration:

$$\mathbf{A}(t) = \left[\frac{d^2r}{dt^2} - r\left(\frac{d\theta}{dt}\right)^2\right]\mathbf{u}_r + \left[r\frac{d^2\theta}{dt^2} + 2\frac{dr}{dt}\frac{d\theta}{dt}\right]\mathbf{u}_\theta$$

Hint: Begin by using the chain rule to show

$$\frac{d\mathbf{u}_r}{dt} = \mathbf{u}_\theta\frac{d\theta}{dt} \quad \text{and} \quad \frac{d\mathbf{u}_\theta}{dt} = -\mathbf{u}_r\frac{d\theta}{dt}$$

41. Using the acceleration formula given in Problem 40, show that the force $\mathbf{F}$ on an object is given by

$$\mathbf{F}(t) = m\mathbf{A}(t) = F_r\mathbf{u}_r + F_\theta\mathbf{u}_\theta$$

where

$$\mathbf{F}_r(t) = m\frac{d^2r}{dt^2} - mr\left(\frac{d\theta}{dt}\right)^2$$

and

$$F_\theta(t) = mr\frac{d^2\theta}{dt^2} + 2m\frac{dr}{dt}\frac{d\theta}{dt}$$

42. Using Problem 41, show that

$$rF_\theta(t) = \frac{d}{dt}\left(mr^2\frac{d\theta}{dt}\right)$$

43. Use Problem 42 to show that if $F_\theta(t) = 0$, then

$$mr^2\frac{d\theta}{dt}$$

is constant.

11.4 Unit Tangent and Normal Vectors; Curvature

IN THIS SECTION unit tangent and principal unit normal vectors, arc length as a parameter, curvature

UNIT TANGENT AND PRINCIPAL UNIT NORMAL VECTORS

From Section 11.2, we know that if $\mathbf{R}(t)$ is a vector function with a smooth graph, then the derivative $\mathbf{R}'(t)$ is a tangent vector to the graph at the point corresponding to $\mathbf{R}(t)$. Because the graph is smooth, we have $\mathbf{R}'(t) \neq \mathbf{0}$, and a unit tangent vector $\mathbf{T}(t)$ can be obtained by dividing $\mathbf{R}'(t)$ by its length, that is,

$$\mathbf{T}(t) = \frac{\mathbf{R}'(t)}{\|\mathbf{R}'(t)\|}$$

A unit tangent vector at a point P of a curve is shown in Figure 11.17.

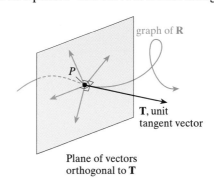

graph of $\mathbf{R}$

P

$\mathbf{T}$, unit tangent vector

Plane of vectors orthogonal to $\mathbf{T}$

■ **FIGURE 11.17** Unit tangent vector $\mathbf{T}$ and vectors orthogonal to $\mathbf{T}$ on a given curve

Based on our experience with normal lines to curves in the plane, we would expect a normal vector to be orthogonal to the tangent vector $\mathbf{T}(t)$ at each point on the graph of $\mathbf{R}$. To obtain such a vector, we recall (Theorem 11.4) that any vector function $\mathbf{R}(t)$ with constant length is always orthogonal to its derivative $\mathbf{R}'(t)$. Therefore, because $\mathbf{T}(t)$ has length 1 for all t, it follows that $\mathbf{T}(t)$ is orthogonal to $\mathbf{T}'(t)$ for all t, so that

$$\mathbf{N}(t) = \frac{\mathbf{T}'(t)}{\|\mathbf{T}'(t)\|}$$

is a unit normal vector. Actually, there are infinitely many vectors orthogonal to **T**(t)—see Figure 11.17—and we shall refer to **N**(t) as the **principal unit normal** whenever it is necessary to distinguish it from other normal vectors.

> ### Unit Tangent Vector and Principal Unit Normal Vector
>
> If **R**(t) is a vector function that defines a smooth graph, then at each point a unit tangent is
>
> $$\mathbf{T}(t) = \frac{\mathbf{R}'(t)}{\|\mathbf{R}'(t)\|}$$
>
> and the principal unit normal vector is
>
> $$\mathbf{N}(t) = \frac{\mathbf{T}'(t)}{\|\mathbf{T}'(t)\|}$$

> ■ *What This Says:* For each number t_0 in the domain of **R**(t), we have constructed a pair of unit vectors, **T**(t_0) and **N**(t_0), with **T** tangent to the graph of **R** at $t = t_0$ and with **N** orthogonal to **T** at $t = t_0$.

EXAMPLE 1   *Unit tangent and principal unit normal*

Find the unit tangent vector **T**(t) and the principal unit normal vector **N**(t) at each point on the graph of the vector function

$$\mathbf{R}(t) = e^t\mathbf{i} + e^{-t}\mathbf{j} + \sqrt{2}\,t\,\mathbf{k}$$

In particular, find **T**(1) and **N**(1).

Solution   The derivative of **R**(t) with respect to t is the vector function $\mathbf{R}'(t) = e^t\mathbf{i} - e^{-t}\mathbf{j} + \sqrt{2}\,\mathbf{k}$, which has length

$$\|\mathbf{R}'(t)\| = \sqrt{(e^t)^2 + (-e^{-t})^2 + (\sqrt{2})^2}$$

$$= \sqrt{e^{2t} + e^{-2t} + 2} = \sqrt{(e^t + e^{-t})^2} = e^t + e^{-t}$$

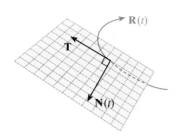

Thus,

$$\mathbf{T}(t) = \frac{\mathbf{R}'(t)}{\|\mathbf{R}'(t)\|} = \frac{e^t\mathbf{i} - e^{-t}\mathbf{j} + \sqrt{2}\,\mathbf{k}}{e^t + e^{-t}}$$

$$= \left(\frac{e^t}{e^t + e^{-t}}\right)\mathbf{i} + \left(\frac{-e^{-t}}{e^t + e^{-t}}\right)\mathbf{j} + \left(\frac{\sqrt{2}}{e^t + e^{-t}}\right)\mathbf{k}$$

So

$$\mathbf{T}(1) = \frac{e\mathbf{i} - e^{-1}\mathbf{j} + \sqrt{2}\,\mathbf{k}}{e + e^{-1}} \approx 0.88\mathbf{i} - 0.12\mathbf{j} + 0.46\mathbf{k}$$

To compute the principal unit normal **N**, we first find **T**′(t):

$$\mathbf{T}'(t) = \frac{(e^t + e^{-t})\,e^t - e^t(e^t - e^{-t})}{(e^t + e^{-t})^2}\mathbf{i} + \frac{(e^t + e^{-t})(e^{-t}) + e^{-t}(e^t - e^{-t})}{(e^t + e^{-t})^2}\mathbf{j}$$

$$- \sqrt{2}\,(e^t + e^{-t})^{-2}\,(e^t - e^{-t})\mathbf{k}$$

$$= \frac{1}{(e^t + e^{-t})^2}\,[(e^{2t} + 1 - e^{2t} + 1)\mathbf{i} + 2\mathbf{j} - \sqrt{2}(e^t - e^{-t})\mathbf{k}]$$

$$= \frac{2\mathbf{i} + 2\mathbf{j} - \sqrt{2}(e^t - e^{-t})\mathbf{k}}{(e^t + e^{-t})^2}$$

and

$$\|\mathbf{T}'(t)\| = \frac{\sqrt{2^2 + 2^2 + 2(e^t - e^{-t})^2}}{(e^t + e^{-t})^2} = \frac{\sqrt{2e^{2t} + 4 + 2e^{-2t}}}{(e^t + e^{-t})^2}$$

$$= \frac{\sqrt{2(e^t + e^{-t})^2}}{(e^t + e^{-t})^2} = \frac{\sqrt{2}}{e^t + e^{-t}}$$

So, the unit normal is

$$\mathbf{N}(t) = \frac{\mathbf{T}'(t)}{\|\mathbf{T}'(t)\|} = \frac{\dfrac{2\mathbf{i} + 2\mathbf{j} - \sqrt{2}(e^t - e^{-t})\mathbf{k}}{(e^t + e^{-t})^2}}{\dfrac{\sqrt{2}}{e^t + e^{-t}}}$$

$$= \frac{2\mathbf{i} + 2\mathbf{j} - \sqrt{2}(e^t - e^{-t})\mathbf{k}}{\sqrt{2}(e^t + e^{-t})}$$

$$= \frac{\sqrt{2}\mathbf{i} + \sqrt{2}\mathbf{j} - (e^t - e^{-t})\mathbf{k}}{(e^t + e^{-t})}$$

In particular,

$$\mathbf{N}(1) = \frac{2\mathbf{i} + 2\mathbf{j} - \sqrt{2}(e - e^{-1})\mathbf{k}}{\sqrt{2}(e + e^{-1})} \approx 0.46\mathbf{i} + 0.46\mathbf{j} - 0.76\mathbf{k}$$

Remember, because $\mathbf{T}$ and $\mathbf{N}$ are to be orthogonal, we can check our work by noting that

$$\mathbf{T} \cdot \mathbf{N} \approx (0.88\mathbf{i} - 0.12\mathbf{j} + 0.46\mathbf{k}) \cdot (0.46\mathbf{i} + 0.46\mathbf{j} - 0.76\mathbf{k}) = 0 \qquad \blacksquare$$

Technology Window

We continue to remind you that if you use some software packages, your answers may take on slightly different, yet equivalent, forms. For Example 1 we obtain

$$\mathbf{T}'(t) = \frac{2e^{2t}}{(e^{2t} + 1)^2}\mathbf{i} + \frac{2e^{2t}}{(e^{2t} + 1)^2}\mathbf{j} + \frac{\sqrt{2}e^t(1 - e^t)(e^t + 1)}{(e^{2t} + 1)^2}\mathbf{k}$$

Continuing with the software program, we find

$$\|\mathbf{T}'(t)\| = \frac{\sqrt{2}e^t}{e^{2t} + 1}$$

Finally,

$$\mathbf{N}(t) = \frac{\mathbf{T}'(t)}{\|\mathbf{T}'(t)\|} = \frac{\sqrt{2}e^t}{e^{2t} + 1}\mathbf{i} + \frac{\sqrt{2}e^t}{e^{2t} + 1}\mathbf{j} + \left(\frac{1 - e^{2t}}{e^{2t} + 1}\right)\mathbf{k}$$

$$\mathbf{N}(1) = 0.458243571484\mathbf{i} + 0.458243571484\mathbf{j} - 0.761594155955\mathbf{k}$$

Checking $\mathbf{T} \cdot \mathbf{N}$ for $t = 1$, we obtain $-1.49398426240 \; 10^{-6}$, which is scientific notation for a number very close (but not equal to) zero. Note that the rounded check in Example 1 showed that this was exactly 0, but that result was, by coincidence, due to rounding. The best you can do with most approximate results is to show that they are approximately equal.

ARC LENGTH AS A PARAMETER

In Section 11.2, we found that time t is the most natural parameter for studying motion along a curve. However, if we are primarily interested in the geometric features of a curve, it may be more convenient to use arc length s as a parameter.

> ### Arc Length Function
>
> Let $\mathbf{R}(t)$ be a vector function whose graph is a smooth curve C on the closed interval $[a, b]$. Then the **arc length function** on $[a, b]$ is defined by
>
> $$s(t) = \int_a^t \|\mathbf{R}'(u)\| \, du \qquad a \le t \le b$$

Notice that $s(t)$ increases in the direction of increasing t. That is, arc length s increases in the same direction as the parameter t along the curve C (see Figure 11.18).

In the case where $\mathbf{R}(t) = x(t)\mathbf{i} + y(t)\mathbf{j} + z(t)\mathbf{k}$, we have

$$\mathbf{R}'(u) = x'(u)\mathbf{i} + y'(u)\mathbf{j} + z'(u)\mathbf{k}$$

and

$$\|\mathbf{R}'(u)\| = \sqrt{[x'(u)]^2 + [y'(u)]^2 + [z'(u)]^2}$$

Thus, on the interval $[a, b]$, the trajectory of $\mathbf{R}(t)$ has arc length

$$L = s(b) = \int_a^b \|\mathbf{R}'(u)\| \, du = \int_a^b \sqrt{[x'(u)]^2 + [y'(u)]^2 + [z'(u)]^2} \, du$$

This gives a formula for the arc length of a curve in space that is a direct generalization of the formula obtained for the arc length of a planar curve in Chapter 9 (Theorem 9.4).

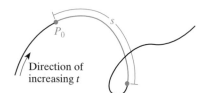

■ **FIGURE 11.18** Arc length s increases in the same direction as t along the graph of $\mathbf{R}$.

THEOREM 11.6   *Arc length of a space curve*

If C is a smooth curve defined by $\mathbf{R}(t) = x(t)\mathbf{i} + y(t)\mathbf{j} + z(t)\mathbf{k}$ on an interval $[a, b]$, then the arc length of C is given by

$$s = \int_a^b \|\mathbf{R}'(t)\| \, dt = \int_a^b \sqrt{[x'(t)]^2 + [y'(t)]^2 + [z'(t)]^2} \, dt$$

Proof   An outline of the proof precedes the statement of this theorem.   ⹀

EXAMPLE 2   *Finding the arc length of a space curve*

Find the arc length of the curve defined by

$$\mathbf{R}(t) = 12t\mathbf{i} + (5 \cos t)\mathbf{j} + (3 - 5 \sin t)\mathbf{k}$$

from $t = 0$ to $t = 2$.

Solution   We have $x(t) = 12t$, $y(t) = 5 \cos t$, and $z(t) = 3 - 5 \sin t$, so that $x'(t) = 12$, $y'(t) = -5 \sin t$, and $z'(t) = -5 \cos t$. We now use Theorem 11.6:

$$s = \int_a^b \sqrt{[x'(t)]^2 + [y'(t)]^2 + [z'(t)]^2} \, dt$$

$$= \int_0^2 \sqrt{12^2 + (-5 \sin t)^2 + (-5 \cos t)^2} \, dt = \int_0^2 13 \, dt = 26 \qquad ■$$

By applying the second fundamental theorem of calculus, we can differentiate the arc length function on $[a, b]$ to obtain

$$\frac{ds}{dt} = \|\mathbf{R}'(t)\|$$

Recall from Section 11.2 that the *speed* of an object moving on its trajectory is given by $\|\mathbf{V}(t)\| = \|\mathbf{R}'(t)\|$, where $\mathbf{V}(t)$ is the velocity vector of the object. Thus, the speed, in terms of arc length, is

$$\|\mathbf{V}(t)\| = \|\mathbf{R}'(t)\| = \frac{ds}{dt}$$

For future reference, we summarize this result in the following theorem.

THEOREM 11.7   *Speed in terms of arc length*

Suppose an object moves with displacement $\mathbf{R}(t)$, where $\mathbf{R}'(t)$ is continuous on the interval $[a, b]$. Then the object has speed

$$\|\mathbf{V}(t)\| = \|\mathbf{R}'(t)\| = \frac{ds}{dt} \qquad \text{for } a \leq t \leq b$$

Proof   An outline of the proof precedes the statement of the theorem.   ═══

EXAMPLE 3   *Speed and distance traveled by an object moving in space*

An object moves with displacement

$$\mathbf{R}(t) = e^t\mathbf{i} + (\sqrt{2}\, t + 3)\mathbf{j} + e^{-t}\mathbf{k}$$

Find the speed of the object at time t and compute the distance the object travels between times $t = 0$ and $t = 1$.

Solution   By differentiating $\mathbf{R}(t)$ with respect to t, we find the velocity:

$$\mathbf{V}(t) = \mathbf{R}'(t) = e^t\mathbf{i} + \sqrt{2}\,\mathbf{j} - e^{-t}\mathbf{k}$$

Therefore, the speed at time t is

$$\begin{aligned}
\frac{ds}{dt} = \|\mathbf{R}'(t)\| &= \sqrt{(e^t)^2 + (\sqrt{2})^2 + (-e^{-t})^2} \\
&= \sqrt{e^{2t} + 2 + e^{-2t}} = \sqrt{(e^t + e^{-t})^2} = e^t + e^{-t}
\end{aligned}$$

The distance traveled by the object between times $t = 0$ and $t = 1$ is the arc length and is given by

$$s = \int_0^1 (e^t + e^{-t})\, dt = [e^t - e^{-t}]\big|_0^1 = e - e^{-1} - 1 + 1 \approx 2.3504024$$

COMMENT   Incidentally, notice that the object is at the point $P(1, 3, 1)$ when $t = 0$ and is at $Q(e, \sqrt{2} + 3, e^{-1})$ when $t = 1$. The straight-line distance between P and Q is

$$d = \sqrt{(e - 1)^2 + (\sqrt{2} + 3 - 3)^2 + (e^{-1} - 1)^2} \approx 2.3134539$$

which, as might be expected, is slightly less than the distance measured from P to Q along the trajectory.   ■

When $\mathbf{R}$ is represented as $\mathbf{R}(s)$ in terms of the arc length parameter s, the unit tangent vector $\mathbf{T}$ can be represented as $\mathbf{T} = d\mathbf{R}/ds$. To see why this is true, recall from

Theorem 11.7 that $\|\mathbf{R}'(t)\| = ds/dt$, and because

$$\mathbf{R}'(t) = \frac{d\mathbf{R}}{dt}$$

it follows that

$$\mathbf{T} = \frac{\mathbf{R}'(t)}{\|\mathbf{R}'(t)\|} = \frac{\dfrac{d\mathbf{R}}{dt}}{\dfrac{ds}{dt}} = \frac{d\mathbf{R}}{ds}$$

Note that because s increases with t, we have $\dfrac{ds}{dt} > 0$, so the unit tangent vector $\mathbf{T} = \dfrac{d\mathbf{R}}{ds}$ must point in the direction of increasing s.

Next, to obtain a formula for the principal unit normal vector

$$\mathbf{N} = \frac{\mathbf{T}'(t)}{\|\mathbf{T}'(t)\|}$$

in terms of the arc length parameter s, we first note that

$$\mathbf{T}'(t) = \frac{d\mathbf{T}}{dt} = \frac{d\mathbf{T}}{ds}\frac{ds}{dt}$$

Because $\dfrac{ds}{dt} > 0$, it follows that the vector derivatives $\dfrac{d\mathbf{T}}{dt}$ and $\dfrac{d\mathbf{T}}{ds}$ point in the same direction, and $\mathbf{N}$ can be computed by finding $\dfrac{d\mathbf{T}}{ds}$ and dividing by its length. The scalar function $\kappa = \left\|\dfrac{d\mathbf{T}}{ds}\right\|$ is called the **curvature** of the graph (defined below).

Thus, we have

$$\mathbf{N} = \frac{\dfrac{d\mathbf{T}}{ds}}{\left\|\dfrac{d\mathbf{T}}{ds}\right\|} = \frac{1}{\kappa}\frac{d\mathbf{T}}{ds}$$

We summarize these observations in the following box.

Formulas for Unit Tangent and Principal Unit Normal in Terms of the Arc Length Parameter

If $\mathbf{R}(t)$ has a smooth graph and is represented as $\mathbf{R}(s)$ in terms of the arc length parameter s, then the unit tangent $\mathbf{T}$ and the principal normal $\mathbf{N}$ at each point satisfy

$$\mathbf{T} = \frac{d\mathbf{R}}{ds} \quad \text{and} \quad \mathbf{N} = \frac{1}{\kappa}\frac{d\mathbf{T}}{ds}$$

where $\kappa = \left\|\dfrac{d\mathbf{T}}{ds}\right\|$ is a scalar function called the **curvature** of the graph.

CURVATURE

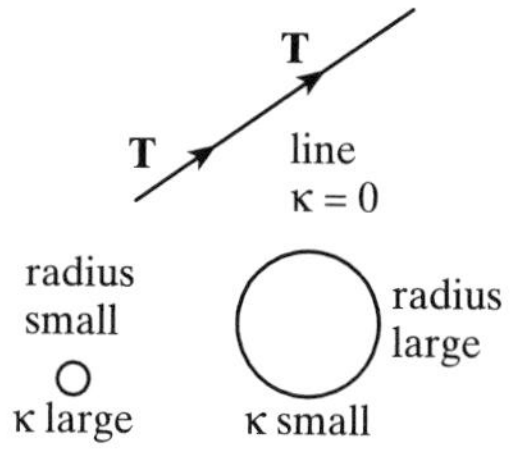

The concept of curvature defined in the above box provides a measure of the "bend" in a trajectory. In $\mathbb{R}^2$, a straight line has no bend, so it is not surprising to find it has curvature 0. The "bend" in a circle of radius r is the same at each point and is measured by the reciprocal of the radius; that is, $\kappa = 1/r$. Thus, a circle with small radius bends sharply and its curvature κ is, indeed, large, while just the opposite is true for a circle with large radius.

More generally, the curvature at a point P on a curve C may be thought of as the curvature of the circle that "best approximates" the shape of C "near" P. This best approximating circle, called the **osculating** (*kissing*) **circle,** is tangent to C at P and has radius $\rho = 1/\kappa$. For example, in the problem set (Problem 54), you are asked to show that the curvature of a function $y = f(x)$ is given by

$$\kappa = \frac{|y''|}{(1 + y'^2)^{3/2}}$$

at each point $P(x, y)$. Applying this formula to the parabola $y = x^2$, we note $y' = 2x$, $y'' = 2$, so

$$\kappa = \frac{2}{(1 + 4x^2)^{3/2}}$$

At the origin $(0, 0)$, $\kappa = 2$, while at $(1, 1)$ it is $\kappa(1) = 2/5^{3/2}$. The corresponding osculating circles are shown in Figure 11.19.

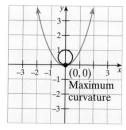

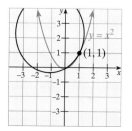

a. Maximum curvature at $(0, 0)$; smallest osculating circle

b. Curvature at $(1, 1)$ is $2/5^{3/2}$; larger osculating circle shown

■ **FIGURE 11.19**   Curvature on a parabola

For curves in $\mathbb{R}^3$, the situation is not quite as easy to visualize because the "bend" of a curve is not confined to a single plane, but can occur in all directions.

To compute the curvature κ, we can first find the unit tangent vector **T** and then substitute into the formula

$$\kappa = \left\| \frac{d\mathbf{T}}{ds} \right\| = \frac{\left\| \dfrac{d\mathbf{T}}{dt} \right\|}{\dfrac{ds}{dt}}$$

For instance, in Example 1 we found $\mathbf{R}(t) = e^t\mathbf{i} + e^{-t}\mathbf{j} + \sqrt{2}t\mathbf{k}$, and $\mathbf{T}(t) = \dfrac{e^t\mathbf{i} - e^{-t}\mathbf{j} + \sqrt{2}\mathbf{k}}{e^t + e^{-t}}$, so that

$$\frac{ds}{dt} = \left\| \mathbf{R}'(t) \right\| = e^t + e^{-t} \quad \text{and} \quad \left\| \frac{d\mathbf{T}}{dt} \right\| = \frac{\sqrt{2}}{e^t + e^{-t}}$$

Margin figure:

T **T**   line $\kappa = 0$

radius small   radius large

κ large κ small

Curvature for
$\mathbf{R}(t) = e^t\mathbf{i} + e^{-t}\mathbf{j} + \sqrt{2}t\,\mathbf{k}$

t	κ
0.0	0.35355339059
0.5	0.27805126251
1.0	0.14848335244
1.5	0.06388944489
2.0	0.02497883868
2.5	0.00940177191
3.0	0.00348817089
3.5	0.00128724714
4.0	0.00047409766
4.5	0.00017448475
5.0	0.00006419937

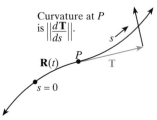

Thus, the curvature in this case is the scalar function

$$\kappa = \frac{\left\|\dfrac{d\mathbf{T}}{dt}\right\|}{\dfrac{ds}{dt}} = \frac{\dfrac{\sqrt{2}}{e^t + e^{-t}}}{e^t + e^{-t}} = \frac{\sqrt{2}}{(e^t + e^{-t})^2}$$

It can be shown (see Problem 27) that a line has curvature 0. Because the curvature is nonnegative, smaller curvature indicates a gentler curve. For Example 1, we see that if $t = 0$, the curvature is about 0.35, whereas if $t = 5$ the curvature is 0.0000642 (almost flat). Some curvature calculations for this curve are shown in the table in the margin (which was generated using a spreadsheet program).

In the case where the graph of $\mathbf{R}(t)$ lies entirely in a plane, the curvature $\kappa(t)$ has a nice geometric interpretation. Note that in this case, the unit tangent vector $\mathbf{T}$ at a point P can be expressed as

$$\mathbf{T} = (\cos \phi)\mathbf{i} + (\sin \phi)\mathbf{j}$$

where ϕ is the angle of inclination of the tangent line at P, as shown in Figure 11.20a. Then, differentiating both sides of the equation for $\mathbf{T}$, we have

$$\frac{d\mathbf{T}}{ds} = \left(-\sin \phi \frac{d\phi}{ds}\right)\mathbf{i} + \left(\cos \phi \frac{d\phi}{ds}\right)\mathbf{j}$$

and because the curvature is the length of this vector, we find

$$\kappa = \left\|\frac{d\mathbf{T}}{ds}\right\| = \left|\frac{d\phi}{ds}\right|\sqrt{(-\sin \phi)^2 + (\cos \phi)^2} = \left|\frac{d\phi}{ds}\right|$$

This characterization of curvature is demonstrated in Figure 11.20b.

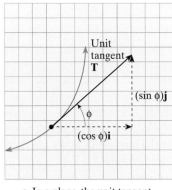

a. In a plane, the unit tangent vector $\mathbf{T} = (\cos \phi)\mathbf{i} + (\sin \phi)\mathbf{j}$.

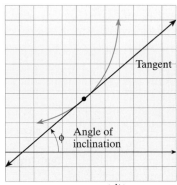

b. The curvature $\kappa = \left|\dfrac{d\phi}{ds}\right|$ measures the rate at which the curve bends away from the tangent.

■ **FIGURE 11.20**   The curvature of a graph

By looking at the formula $\kappa = \left|\dfrac{d\phi}{ds}\right|$, we see that at each point P on a planar curve, the curvature κ measures the rate at which the curve bends away from the tangent line.

The preceding curvature formula is not particularly easy to use, and because the graph C of the vector function is frequently defined by $\mathbf{R}(t)$, it is often useful to have the following formula for κ in terms of $\mathbf{R}$:

Derivative Formula for Curvature

The curvature of a graph C defined by a vector function $\mathbf{R}(t)$ can be found by

$$\kappa = \frac{\|\mathbf{R}' \times \mathbf{R}''\|}{\|\mathbf{R}'\|^3}$$

The derivation of this formula is outlined in Problem 52 and is illustrated in Example 4.

EXAMPLE 4   *Curvature of a helix*

Given $\mathbf{R}(t) = (a \cos t)\mathbf{i} + (a \sin t)\mathbf{j} + bt\mathbf{k}$ with a and b both nonnegative, express the curvature of $\mathbf{R}(t)$ in terms of a and b ($a > b$).

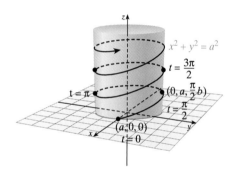

Solution

$$\mathbf{R}'(t) = (-a \sin t)\mathbf{i} + (a \cos t)\mathbf{j} + b\mathbf{k}$$
$$\mathbf{R}''(t) = (-a \cos t)\mathbf{i} + (-a \sin t)\mathbf{j}$$
$$\mathbf{R}' \times \mathbf{R}'' = \begin{vmatrix} \mathbf{i} & \mathbf{j} & \mathbf{k} \\ -a \sin t & a \cos t & b \\ -a \cos t & -a \sin t & 0 \end{vmatrix} = (ab \sin t)\mathbf{i} - (ab \cos t)\mathbf{j} + a^2\mathbf{k}$$

We now need to find the magnitude of this vector, as well as that of $\mathbf{R}'$:

$$\begin{aligned} \|\mathbf{R}' \times \mathbf{R}''\| &= \sqrt{(ab \sin t)^2 + (ab \cos t)^2 + a^4} \\ &= \sqrt{a^2b^2(\sin^2 t + \cos^2 t) + a^4} \\ &= \sqrt{a^2b^2 + a^4} \end{aligned}$$

and

$$\|\mathbf{R}'\| = \sqrt{a^2\cos^2 t + a^2\sin^2 t + b^2} = \sqrt{a^2 + b^2}$$

Then we have

$$\kappa = \frac{\|\mathbf{R}' \times \mathbf{R}''\|}{\|\mathbf{R}'\|^3} = \frac{\sqrt{a^2b^2 + a^4}}{(\sqrt{a^2 + b^2})^3} = \frac{a\sqrt{a^2 + b^2}}{(a^2 + b^2)\sqrt{a^2 + b^2}} = \frac{a}{a^2 + b^2}$$

If we increase b, then for a fixed a the curvature decreases. If $b = 0$, the helix reduces to a circle of radius a. If $a = 0$, the helix flattens out along the z-axis. ∎

In the plane, where a curve is given by $\mathbf{R}(t) = x(t)\mathbf{i} + y(t)\mathbf{j}$, the curvature may be computed by the formula

$$\kappa = \frac{|x'y'' - y'x''|}{[(x')^2 + (y')^2]^{3/2}}$$

A derivation of this formula is outlined in Problem 53. We will use this formula to show that a circle has constant curvature.

EXAMPLE 5   *Curvature of a circle*

Show that a circle of radius r has curvature $\kappa = \dfrac{1}{r}$ at each point.

Solution   Suppose a circle has radius r and center (h, k). Then its Cartesian equation is $(x - h)^2 + (y - k)^2 = r^2$, and we can obtain a parametrization by setting

$$x - h = r\cos t \quad \text{and} \quad y - k = r\sin t$$

In other words, the circle is the graph of the vector function $\mathbf{R}(t) = x(t)\mathbf{i} + y(t)\mathbf{j}$, where

$$x(t) = h + r\cos t \quad \text{and} \quad y(t) = k + r\sin t$$

Differentiating with respect to t, we obtain

$$x'(t) = -r\sin t \qquad y'(t) = r\cos t$$
$$x''(t) = -r\cos t \qquad y''(t) = -r\sin t$$

and by substituting these into the curvature formula, we find

$$\kappa = \frac{|x'y'' - y'x''|}{[(x')^2 + (y')^2]^{3/2}} = \frac{(-r\sin t)(-r\sin t) - (r\cos t)(-r\cos t)}{[(-r\sin t)^2 + (r\cos t)^2]^{3/2}}$$
$$= \frac{r^2\sin^2 t + r^2\cos^2 t}{[r^2(\sin^2 t + \cos^2 t)]^{3/2}} = \frac{r^2}{r^3} = \frac{1}{r} \qquad \blacksquare$$

The osculating circle may be interpreted as the circle whose shape best fits the shape of the graph of $\mathbf{R}(t)$ in the "vicinity" of P (see Figure 11.21). In the special case where the graph of $\mathbf{R}(t)$ is a circle C, the radius of curvature is the radius of C, and the osculating circle coincides with C.

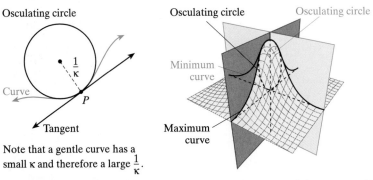

Osculating circle

Curve

$\frac{1}{\kappa}$

P

Tangent

Note that a gentle curve has a small κ and therefore a large $\frac{1}{\kappa}$.

Osculating circle

Osculating circle

Minimum curve

Maximum curve

■ **FIGURE 11.21**   The osculating circle best fits the shape of the curve at P

We conclude this section by providing a summary of various formulas for computing curvature (Table 11.2).

■ TABLE 11.2
Curvature Formulas

Type	Given Information	Formula to Use	Where Derived
Arc length parameter	$\mathbf{R}(s)$	$\left\| \dfrac{d\mathbf{T}}{ds} \right\|$	Page 763
Vector derivative form	$\mathbf{R}(t)$	$\dfrac{\|\mathbf{R}' \times \mathbf{R}''\|}{\|\mathbf{R}'\|^3}$	Page 765, Problem 52
Parametric equation form	$x = x(t), y = y(t)$	$\dfrac{\|x'y'' - y'x''\|}{[(x')^2 + (y')^2]^{3/2}}$	Problem 53
Functional form	$y = f(x)$	$\dfrac{\|y''\|}{[1 + (y')^2]^{3/2}}$	Problem 54
Polar form	$r = f(\theta)$	$\dfrac{\|r^2 + 2r'^2 - rr''\|}{(r^2 + r'^2)^{3/2}}$	Problem 55

11.4   Problem Set

Ⓐ *In Problems 1–8, find the unit tangent vector* $\mathbf{T}(t)$ *and the principal unit normal vector* $\mathbf{N}(t)$ *for the curve given by* $\mathbf{R}(t)$.

1. $\mathbf{R}(t) = t^2\mathbf{i} + t^3\mathbf{j}, t \neq 0$
2. $\mathbf{R}(t) = t^2\mathbf{i} + \sqrt{t}\,\mathbf{j}, t > 0$
3. $\mathbf{R}(t) = (e^t \cos t)\mathbf{i} + (e^t \sin t)\mathbf{j}$
4. $\mathbf{R}(t) = (t \cos t)\mathbf{i} + (t \sin t)\mathbf{j}$
5. $\mathbf{R}(t) = (\cos t)\mathbf{i} + (\sin t)\mathbf{j} + t\mathbf{k}$
6. $\mathbf{R}(t) = (\sin t)\mathbf{i} - (\cos t)\mathbf{j} + t\mathbf{k}$
7. $\mathbf{R}(t) = (\ln t)\mathbf{i} + t^2\mathbf{k}$
8. $\mathbf{R}(t) = (e^{-t}\sin t)\mathbf{i} + e^{-t}\mathbf{j} + (e^{-t}\cos t)\mathbf{k}$

In Problems 9–14, find the length of the given curve over the given interval.

9. $\mathbf{R}(t) = 2t\mathbf{i} + t\mathbf{j}$, over $[0, 4]$
10. $\mathbf{R}(t) = t\mathbf{i} + 3t\mathbf{j}$, over $[0, 4]$
11. $\mathbf{R}(t) = 3t\mathbf{i} + (3 \cos t)\mathbf{j} + (3 \sin t)\mathbf{k}$, over $[0, \frac{\pi}{2}]$
12. $\mathbf{R}(t) = t\mathbf{i} + 2t\mathbf{j} + 3t\mathbf{k}$, over $[0, 2]$
13. $\mathbf{R}(t) = (4 \cos t)\mathbf{i} + (4 \sin t)\mathbf{j} + 5t\mathbf{k}$, over $[0, \pi]$
14. $\mathbf{R}(t) = (\cos^3 t)\mathbf{i} + (\cos^2 t)\mathbf{k}$, over $[0, \frac{\pi}{2}]$

Find the curvature of the plane curves at the points indicated in Problems 15–26.

15. $y = 4x - 2$ at $x = 2$
16. $y = mx + b$ at $x = a$
17. $y = x - \frac{1}{9}x^2$, at $x = 3$
18. $y = 2x^2 + 1$, at $x = 1$
19. $y = ax^2 + bx$, at $x = c$
20. $y = x + x^{-1}$, at $x = 1$
21. $y = \sqrt{4 - x^2}$, at $x = 1$
22. $y = \sqrt{r^2 - x^2}$, at $x = 0$
23. $y = \sin x$, at $x = \frac{\pi}{2}$
24. $y = \cos x$, at $x = \frac{\pi}{4}$
25. $y = \ln x$, at $x = 1$
26. $y = e^x$, at $x = 0$

Ⓑ 27. Let $\mathbf{u}$ and $\mathbf{v}$ be constant, nonzero vectors. Show that the line given by $\mathbf{R}(t) = \mathbf{u} + \mathbf{v}t$ has curvature 0 at each point.

28. Find the unit tangent $\mathbf{T}$ and principal normal $\mathbf{N}$ for $\mathbf{R}(t) = (\cosh t)\mathbf{i} + (\sinh t)\mathbf{j}$ at the point where $t = 0$.

29. Find the unit tangent $\mathbf{T}$ and principal normal $\mathbf{N}$ for $\mathbf{R}(t) = [\ln(\sin t)]\mathbf{i} + [\ln(\cos t)]\mathbf{j}$ at the point where $t = \frac{\pi}{3}$.

30. A curve C in the plane is given parametrically by $x = 32t$, $y = 16t^2 - 4$.
    a. Sketch the graph of the curve.
    b. Find the unit tangent vector when $t = 3$.
    c. Find the radius of curvature of the point P on C where $t = 3$.

31. For the curve given by

$$\mathbf{R}(t) = (\sin t)\mathbf{i} + (\cos t)\mathbf{j} + t\mathbf{k}$$

    a. Find a unit tangent vector $\mathbf{T}$ at the point on the curve where $t = \pi$.
    b. Find the curvature when $t = \pi$.
    c. Find the length of the curve from $t = 0$ to $t = \pi$.

32. Let C be the curve given by

$$\mathbf{R}(t) = (t - \sin t)\mathbf{i} + (1 - \cos t)\mathbf{j} + (4 \sin \tfrac{1}{2})\mathbf{k}$$

    a. Find the unit tangent vector $\mathbf{T}(t)$ to C.
    b. Find $\dfrac{d\mathbf{T}}{ds}$ and the curvature $\kappa(t)$.

33. Find the point (or points) where the ellipse $9x^2 + 4y^2 = 36$ has maximum curvature.

34. Find the maximum curvature on the curve $y = e^{2x}$.

35. Find the radius of curvature at each relative extremum of the graph of $y = x^6 - 3x^2$.

36. Find the curvature of the curve given by $x = t - \sin t$, $y = 1 - \cos t$. Sketch the curve on $0 \le t \le 2\pi$ and sketch the osculating circle at the point where $t = \frac{\pi}{2}$.

37. The tangent line at a point P on a curve C in space is the line that passes through P and is aligned with the tangent vector **T** to C at P. Find parametric equations for the tangent line to $\mathbf{R}(t) = 2t\mathbf{i} - t\mathbf{j} + t^2\mathbf{k}$ at the point where $t = 1$.

38. The tangent line at a point P on a curve C in space is the line that passes through P and is aligned with the tangent vector **T** to C at P. Find parametric equations for the tangent line to $\mathbf{R}(t) = e^t\mathbf{i} - 3\mathbf{j} + (1 - t)\mathbf{k}$ at the point where $t = 0$.

There are many different formulas for finding curvature, as summarized in Table 11.2. In Problems 39–46, find the curvature of each of the given curves using the indicated formula.

39. $\mathbf{R}(t) = t\mathbf{i} + t^2\mathbf{j} + t^3\mathbf{k}$; vector derivative form

40. $\mathbf{R}(t) = (t - \cos t)\mathbf{i} + (\sin t)\mathbf{j} + 3\mathbf{k}$; vector derivative form

41. $y = x^2$; functional form

42. $y = x^3$; functional form

43. $y = x^{-1}, x > 0$; functional form

44. $y = \sin x$; functional form

45. the spiral $r = e^\theta$; polar form

46. the cardioid $r = 1 + \cos \theta$, for $0 \le \theta \le 2\pi$; polar form

47. A *pestus houseflyus* is observed to zip around a room in such a way that at time t its position with respect to the nose of an observer is given by the vector function

$$\mathbf{R}(t) = t\mathbf{u} + t^2\mathbf{v} + 2\left(\frac{2}{3}t\right)^{3/2}(\mathbf{u} \times \mathbf{v})$$

where **u** and **v** are unit vectors separated by an angle of $60°$. Compute the fly's speed and find how long it takes to move a distance of 20 units along its path (starting from the nose).

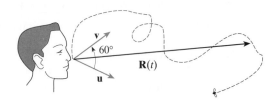

48. ■ **What Does This Say?** Describe what is meant by using arc length as a parameter.

49. ■ **What Does This Say?** Discuss the various curvature formulas given in Table 11.2.

50. **Projections onto a plane in $\mathbb{R}^3$**   A very important notion in applied mathematics is that of projecting a vector onto a subspace of some kind, for example, a plane. Later in your career, the "vectors" can also be functions in which a function is approximated by "nice" functions. In this problem, our work is limited to $\mathbb{R}^3$.

a. Suppose that **u** and **v** are unit vectors and that they are orthogonal. Show that the "linear combinations" $a\mathbf{u} + b\mathbf{v}$ generate a plane, through $(0, 0, 0)$, as a and b take on all real values.

b. Let **w** be a third vector, not on the plane described in part **a.** We want to project **w** onto the plane; we write $\mathbf{w} = \mathbf{p} + \mathbf{n} = (a\mathbf{u} + b\mathbf{v}) + \mathbf{n}$, where **p** is the desired projection. To find a, take the dot product of **w** and **u** and show that **p** is orthogonal to **n** (hence **p** is the orthogonal projection).

c. For the three following vectors, find the orthogonal projection of **w** onto the plane generated by **u** and **v**:

$$\mathbf{u} = \frac{1}{\sqrt{3}}(\mathbf{i} + \mathbf{j} + \mathbf{k})$$

$$\mathbf{v} = \frac{1}{\sqrt{6}}(2\mathbf{i} - \mathbf{j} - \mathbf{k})$$

$$\mathbf{w} = \mathbf{j} + 2\mathbf{k}$$

d. A projection is especially simple if we have $\mathbf{u} = \mathbf{i}$ and $\mathbf{v} = \mathbf{j}$. Explain the situation here for an arbitrary vector $\mathbf{w} = a\mathbf{i} + b\mathbf{j} + c\mathbf{k}$.

e. Show that for a given vector $\mathbf{w} = a\mathbf{i} + b\mathbf{j} + c\mathbf{k}$, its slope relative to the xy-plane is $m = \dfrac{c}{\sqrt{a^2 + b^2}}$.

51. This problem will make some use of Problem 50, and addresses arc length of a curve in $\mathbb{R}^3$ and the work involved in moving along its path. Suppose a mining company has built a road up a hill modeled by the following parametric equations $(0 \le t \le 2\pi)$:

$$x(t) = e^{-t/3} \cos 3t$$

$$y(t) = e^{-t/3} \sin 3t$$

$$z(t) = \frac{13t}{t^2 + 40}$$

a. Generate a rough graph of this curve. Then somehow (use your imagination!) get a decent estimate of its length, say, within 10%.

b. Use your computer to compute its length. Then compare with your estimate from part **a.**

Technology Window

c. Let $\mathbf{T}(t)$ denote the tangent to the curve at the point and assume the force

$$F(t) = \frac{8z'(t)}{\sqrt{[x'(t)]^2 + [y'(t)]^2}}$$

Compute the work required to move a cart from the bottom to the top of the hill. (Recall that work $= \int F \, ds$.)

52. Let $\mathbf{R}(t)$ be a smooth vector function.
   a. Differentiate $\mathbf{R}' = \|\mathbf{R}'\|\mathbf{T}$ to show that

$$\mathbf{R}'' = \|\mathbf{R}'\|'\mathbf{T} + \kappa\|\mathbf{R}'\|^2\mathbf{N}$$

   b. Show that

$$\mathbf{R}' \times \mathbf{R}'' = \kappa\|\mathbf{R}'\|^3(\mathbf{T} \times \mathbf{N})$$

   c. Conclude that

$$\kappa = \frac{\|\mathbf{R}' \times \mathbf{R}''\|}{\|\mathbf{R}'\|^3}$$

53. Use the formula in Problem 52c to verify the formula

$$\kappa = \frac{|x'y'' - y'x''|}{[(x')^2 + (y')^2]^{3/2}}$$

   Hint: Let $\mathbf{R}(t) = x(t)\mathbf{i} + y(t)\mathbf{j}$.

54. A curve in the plane is given by $y = f(x)$. Show that the functional form for curvature is given by

$$\frac{|f''(x)|}{\{1 + [f'(x)]^2\}^{3/2}}$$

   (Hint: Use the formula in Problem 53 with $\mathbf{R}(x) = x\mathbf{i} + f(x)\mathbf{j}$.)

55. Let f be a twice differentiable function. Show that, in polar coordinates, the curvature of the curve given by $r = f(\theta)$ satisfies

$$\kappa(\theta) = \frac{|r^2 + 2r'^2 - rr''|}{(r^2 + r'^2)^{3/2}}$$

   (Hint: Use the formula in Problem 52 with $x = f(\theta)\cos\theta, y = f(\theta)\sin\theta$.)

56. Use the formula in Problem 52 to show that the curve C given by $\mathbf{R}(t)$ has curvature

$$\kappa = \frac{[\|\mathbf{V}\|^2\|\mathbf{A}\|^2 - \mathbf{V} \cdot \mathbf{A}]^{1/2}}{\|\mathbf{V}\|^3}$$

   Hint: Use the identity

$$\|\mathbf{u} \times \mathbf{v}\|^2 = (\|\mathbf{u}\| \, \|\mathbf{v}\|)^2 - (\mathbf{u} \cdot \mathbf{v})^2$$

57. Let $P(a, b)$ be a point on the graph C of the vector function $\mathbf{R}(t)$.

a. Describe a general procedure for finding an equation for the osculating circle to C at P.
b. Find an equation for the osculating circle at the point $P(32, 12)$ on the curve C defined by

$$x = 32t, \quad y = 16t^2 - 4$$

58. If $\mathbf{T}$ and $\mathbf{N}$ are the unit tangent and normal vectors, respectively, on the trajectory of a moving body, then the cross product vector $\mathbf{B} = \mathbf{T} \times \mathbf{N}$ is called the unit **binormal** of the trajectory. Three planes determined by $\mathbf{T}, \mathbf{N},$ and $\mathbf{B}$ are shown in Figure 11.22.

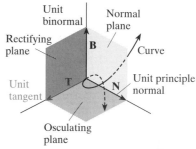

■ **FIGURE 11.22**   Three planes determined by $\mathbf{T}, \mathbf{N},$ and $\mathbf{B}$

a. Show that $\mathbf{T}$ is orthogonal to $\frac{d\mathbf{B}}{ds}$.
   (Hint: Differentiate $\mathbf{B} \cdot \mathbf{T}$.)
b. Show that $\mathbf{B}$ is orthogonal to $\frac{d\mathbf{B}}{ds}$.
   (Hint: Differentiate $\mathbf{B} \cdot \mathbf{B}$.)
c. Show that $\frac{d\mathbf{B}}{ds} = -\tau\mathbf{N}$ for some constant τ. *(Note:* τ is called the **torsion** of the trajectory.)

59. Prove the Frenet-Serret formulas:

$$\frac{d\mathbf{T}}{ds} = \kappa\mathbf{N}$$

$$\frac{d\mathbf{N}}{ds} = -\kappa\mathbf{T} + \tau\mathbf{B}$$

$$\frac{d\mathbf{B}}{ds} = -\tau\mathbf{N}$$

where κ is the curvature and $\tau = \tau(s)$ is a scalar function called the torsion, which provides a measure of the amount of twisting at each point on the trajectory.

60. Show that the torsion may be computed by the formula

$$\tau = \frac{[\mathbf{R}'(t) \times \mathbf{R}''(t)] \cdot \mathbf{R}'''(t)}{\|\mathbf{R}'(t) \times \mathbf{R}''(t)\|^2}$$

Use this formula to find the torsion for the helix

$$\mathbf{R}(t) = (a\cos t)\mathbf{i} + (a\sin t)\mathbf{j} + (bt)\mathbf{k}$$

61. A highway has an exit ramp that begins at the origin and follows the curve $y = \frac{1}{32}x^{5/2}$ to the point $(4, 1)$. Then it follows the shape of the osculating circle at $(4, 1)$ until the point where $y = 3$. What is the total length of the exit ramp?

11.5 *Tangential and Normal Components of Acceleration*

COMPONENTS OF ACCELERATION

When a body is caused to accelerate or brakes are applied, it is of interest to know how much of the acceleration acts in the direction of the body's motion, as indicated by the unit tangent vector **T**. This question is answered by the following theorem.

THEOREM 11.8   *Tangential and normal components of acceleration*

An object moving along a smooth curve (with $\mathbf{T}' \neq \mathbf{0}$) has velocity **V** and acceleration **A**, where

$$\mathbf{V} = \left(\frac{ds}{dt}\right)\mathbf{T} \quad \text{and} \quad \mathbf{A} = \left(\frac{d^2s}{dt^2}\right)\mathbf{T} + \kappa\left(\frac{ds}{dt}\right)^2\mathbf{N}$$

and s is the arc length along the trajectory.

Proof   An object moving with displacement **R** has unit tangent $\mathbf{T} = \dfrac{d\mathbf{R}}{ds}$ and unit normal $\mathbf{N} = \dfrac{1}{\kappa}\dfrac{d\mathbf{T}}{ds}$. We use the chain rule to write the velocity vector **V** as follows:

$$\mathbf{V} = \frac{d\mathbf{R}}{dt} = \frac{d\mathbf{R}}{ds}\frac{ds}{dt} = \mathbf{T}\frac{ds}{dt}$$

Differentiate both sides of this equation with respect to t and substitute $\dfrac{d\mathbf{T}}{ds} = \kappa\mathbf{N}$:

$$\mathbf{A} = \frac{d\mathbf{V}}{dt} \qquad\qquad \text{Definition of } \mathbf{A}$$

$$= \frac{d}{dt}\left[\mathbf{T}\frac{ds}{dt}\right] \qquad\qquad \mathbf{V} = \left(\frac{ds}{dt}\right)\mathbf{T}$$

$$= \frac{d^2s}{dt^2}\mathbf{T} + \frac{ds}{dt}\frac{d\mathbf{T}}{dt} \qquad\qquad \text{Product rule}$$

$$= \frac{d^2s}{dt^2}\mathbf{T} + \frac{ds}{dt}\left[\frac{d\mathbf{T}}{ds}\frac{ds}{dt}\right] \qquad\qquad \text{Chain rule}$$

$$= \frac{d^2s}{dt^2}\mathbf{T} + \left(\frac{ds}{dt}\right)^2\frac{d\mathbf{T}}{ds}$$

$$= \frac{d^2s}{dt^2}\mathbf{T} + \left(\frac{ds}{dt}\right)^2(\kappa\mathbf{N}) \qquad \frac{d\mathbf{T}}{ds} = \kappa\mathbf{N} \qquad =$$

■ *What This Says:*   At each point on the trajectory of a moving object, the velocity **V** points in the direction of the unit tangent **T**, but the acceleration **A** has both a tangential and a normal component. The trajectory may twist and turn, but the acceleration is always in the plane determined by **T** and the unit normal **N**.

The two components of acceleration have special names.

Tangential and Normal Components of Acceleration

The acceleration **A** of a moving object can be written as

$$\mathbf{A} = A_T\mathbf{T} + A_N\mathbf{N}$$

where

$$A_T = \frac{d^2s}{dt^2} \qquad \text{is the \textbf{tangential component}.}$$

$$A_N = \kappa\left(\frac{ds}{dt}\right)^2 \qquad \text{is the \textbf{normal component}.}$$

The tangential and normal components are shown in Figure 11.23.

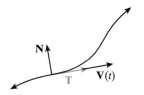

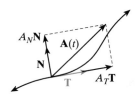

a. Velocity has only a nonzero tangential component.

b. Acceleration may have both tangential and normal component.

■ **FIGURE 11.23**   Components of velocity and acceleration

It is usually fairly easy to find A_T, but it may be more difficult to find A_N, because computing the curvature κ is often a messy process. Fortunately, there is a way to compute A_N without first finding κ. We expand the dot product of $\mathbf{A} = A_T\mathbf{T} + A_N\mathbf{N}$ with itself and use the fact that **T** and **N** are orthogonal unit vectors:

$$
\begin{aligned}
\|\mathbf{A}\|^2 &= \mathbf{A} \cdot \mathbf{A} \\
&= (A_T\mathbf{T} + A_N\mathbf{N}) \cdot (A_T\mathbf{T} + A_N\mathbf{N}) \qquad \mathbf{A} = A_T\mathbf{T} + A_N\mathbf{N} \\
&= A_T^2(\mathbf{T} \cdot \mathbf{T}) + 2A_TA_N(\mathbf{N} \cdot \mathbf{T}) + A_N^2(\mathbf{N} \cdot \mathbf{N}) \\
&= A_T^2(1) + 2A_TA_N(0) + A_N^2(1) \qquad \mathbf{T} \cdot \mathbf{T} = 1, \mathbf{N} \cdot \mathbf{T} = 0, \text{ and } \mathbf{N} \cdot \mathbf{N} = 1 \\
&= A_T^2 + A_N^2
\end{aligned}
$$

Thus, once we know **A** and A_T, we can compute A_N by applying the following formula.

Computation of Normal Component

The normal component A_N can be found using the formula

$$A_N = \sqrt{\|\mathbf{A}\|^2 - A_T^2}$$

EXAMPLE 1   *Finding tangential and normal components of acceleration*

Find the tangential and normal components of an object that moves with displacement

$$\mathbf{R}(t) = \langle t^3, t^2, t\rangle$$

Solution

$$\mathbf{V} = \frac{d\mathbf{R}}{dt} = \langle 3t^2, 2t, 1 \rangle \quad \text{and} \quad \mathbf{A} = \frac{d\mathbf{V}}{dt} = \langle 6t, 2, 0 \rangle$$

$$\frac{ds}{dt} = \|\mathbf{V}\| = \sqrt{(3t^2)^2 + (2t)^2 + (1)^2} = \sqrt{9t^4 + 4t^2 + 1}$$

$$A_{\mathrm{T}} = \frac{d^2s}{dt^2} = \frac{1}{2}(9t^4 + 4t^2 + 1)^{-1/2}(36t^3 + 8t) = \frac{18t^3 + 4t}{\sqrt{9t^4 + 4t^2 + 1}}$$

This is the tangential component of acceleration.

$$A_{\mathrm{N}} = \sqrt{\|\mathbf{A}\|^2 - A_{\mathrm{T}}^2} = \sqrt{\left[\sqrt{36t^2 + 4}\right]^2 - \left[\frac{18t^3 + 4t}{\sqrt{9t^4 + 4t^2 + 1}}\right]^2}$$

$$= \sqrt{4(9t^2 + 1) - \frac{4t^2(9t^2 + 2)^2}{9t^4 + 4t^2 + 1}}$$

$$= \sqrt{\frac{36t^4 + 36t^2 + 4}{9t^4 + 4t^2 + 1}} \qquad \text{There are several simplification steps that are not shown.}$$

$$= 2\sqrt{\frac{9t^4 + 9t^2 + 1}{9t^4 + 4t^2 + 1}}$$

This is the normal component of acceleration.   ■

EXAMPLE 2   *Finding tangential and normal components on a helix*

An object moves along the helix with position vector

$$\mathbf{R}(t) = (\cos t)\mathbf{i} + (\sin t)\mathbf{j} + t\mathbf{k}$$

Find the tangential and normal components of acceleration.

Solution

$$\mathbf{V} = \frac{d\mathbf{R}}{dt} = (-\sin t)\mathbf{i} + (\cos t)\mathbf{j} + \mathbf{k} \quad \text{and} \quad \mathbf{A} = \frac{d\mathbf{V}}{dt} = (-\cos t)\mathbf{i} + (-\sin t)\mathbf{j}$$

$$\frac{ds}{dt} = \|\mathbf{V}\| = \sqrt{\sin^2 t + \cos^2 t + 1} = \sqrt{2} \quad \text{and} \quad A_{\mathrm{T}} = \frac{d^2s}{dt^2} = 0$$

$$A_{\mathrm{N}} = \sqrt{\|\mathbf{A}\|^2 - A_{\mathrm{T}}^2} = \sqrt{\left(\sqrt{\cos^2 t + \sin^2 t}\right)^2 - 0^2} = 1$$

The tangential and normal components of acceleration are 0 and 1, respectively. This means that the acceleration satisfies

$$\mathbf{A} = A_{\mathrm{T}}\mathbf{T} + A_{\mathrm{N}}\mathbf{N} = (0)\mathbf{T} + (1)\mathbf{N} = \mathbf{N}$$

That is, the acceleration vector is the principal unit normal $\mathbf{N}$, and the acceleration is always normal to the trajectory of the uniform helix.   ■

APPLICATIONS

Now that we know how to compute the tangential and normal components of acceleration, A_{T} and A_{N}, we shall examine some applications. First, according to Newton's second law of motion, the total force acting on a moving object of mass m satisfies $\mathbf{F} = m\mathbf{A}$, where $\mathbf{A}$ is the acceleration of the object. Because $\mathbf{A} = A_{\mathrm{T}}\mathbf{T} + A_{\mathrm{N}}\mathbf{N}$, we have

$$\mathbf{F} = m\mathbf{A} = (mA_{\mathrm{T}})\mathbf{T} + (mA_{\mathrm{N}})\mathbf{N} = F_{\mathrm{T}}\mathbf{T} + F_{\mathrm{N}}\mathbf{N}$$

where

$$F_{\mathrm{T}} = m\frac{d^2s}{dt^2} \quad \text{and} \quad F_{\mathrm{N}} = m\kappa\left(\frac{ds}{dt}\right)^2$$

For instance, experience leads us to expect a car to skid if it makes a sharp turn at moderate speed or even a gradual turn at high speed. Mathematically, a "sharp turn" occurs when the radius of curvature $\rho = 1/\kappa$ is small (that is, when κ is large), and "high speed" means that ds/dt is large. In either case,

$$\mathbf{F}_{\mathrm{N}} = m\kappa\left(\frac{ds}{dt}\right)^2$$

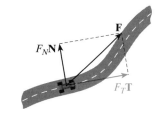

■ **FIGURE 11.24**   Tendency to skid

will be relatively large, and the car will stay on the road (its trajectory) only if there is a correspondingly large frictional force between the tires of the car and the surface of the road (see Figure 11.24).

When forces, masses, and accelerations are involved, it is customary to express the mass m in slugs of an object whose weight W has been given in pounds.* From the definition of weight,

$$m = \frac{W}{g}$$

where g is the acceleration of gravity ($g \approx 32$ ft/s²).

EXAMPLE 3   *Modeling Application: Tendency of a vehicle to skid*

A car weighing 2,700 lb makes a turn on a flat road while traveling at 56 ft/s (about 38 mi/h). If the radius of the turn is 21 ft, how much frictional force is required to keep the car from skidding?

Solution   The required frictional force is $F_{\mathrm{N}} = m\kappa\left(\dfrac{ds}{dt}\right)^2$, where m is the mass of the car and κ is the curvature of the road. We know that

$$\frac{ds}{dt} = 56 \text{ ft/s}$$

and because the car weighs $W = 2{,}700$ lb, its mass is $m = \dfrac{W}{g} = \dfrac{2{,}700}{32} \approx 84.38$ slugs. Because the turn radius is 21 ft, we have $\kappa = \dfrac{1}{21}$, so that

$$\mathbf{F}_{\mathrm{N}} = \left(\frac{2{,}700}{32}\text{ slugs}\right)\left(\frac{1}{21\text{ ft}}\right)\left(56\,\frac{\text{ft}}{\text{s}}\right)^2 = 12{,}600\,\frac{\text{lb}\cdot\text{s}^2}{\text{ft}}\frac{1}{\text{ft}}\frac{\text{ft}^2}{\text{s}^2} = 12{,}600 \text{ lb} \quad ■$$

There are certain important applications in which an object moves along its trajectory with constant speed ds/dt, and when this occurs, the acceleration $\mathbf{A}$ can have only a normal component, because $d^2s/dt^2 = 0$.

═══════════

THEOREM 11.9   *Acceleration of an object with constant speed*

The acceleration of an object moving with constant speed is always orthogonal to the direction of motion.

───────

*A slug is a unit of measurement defined as the unit of mass that receives an acceleration of 1 ft/s² when a force of 1 lb is applied to it. That is, 1 slug $= \dfrac{1 \text{ lb}}{1 \text{ ft/s}^2} = \dfrac{\text{lb s}^2}{\text{ft}}$.

Proof   Notice that we really do not need to know anything about components of acceleration to prove this result. Saying that the object has constant speed means that $\|\mathbf{R}'(t)\|$ is constant, and by Theorem 11.4, we conclude that $\mathbf{R}'(t)$ is orthogonal to its derivative $\mathbf{R}''(t) = \mathbf{A}(t)$. But $\mathbf{R}'(t)$ points in the direction of the object's motion along its trajectory, which means that the acceleration $\mathbf{A}$ is orthogonal to the direction of motion.   ▬

As an application of this result, note that when an object moves with constant speed v_0 along a circular path of radius R (so that $\kappa = 1/R$), its acceleration is directed toward the center of the path and has magnitude

$$A_N = \kappa \left(\frac{ds}{dt}\right)^2 = \frac{1}{R} v_0^2$$

(See Figure 11.25.)

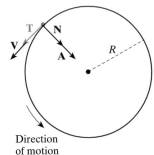

If the speed is v_0 and the path has radius R, the acceleration is

$$\mathbf{A} = \frac{v_0^2}{R} \mathbf{N}$$

Direction of motion

■ **FIGURE 11.25**   An object moving with constant speed on a circular path

EXAMPLE 4   *Modeling Application: Period of a satellite*

An artificial satellite travels at constant speed in a stable circular orbit 20,000 km above the earth's surface. How long does it take for the satellite to make one complete circuit of the earth?

Solution   Let m denote the satellite's mass and v denote its speed. We shall assume that the earth is a sphere of radius 6,440 km, so the curvature of the path is $\kappa = 1/R$, where

$$R = \underbrace{6,440}_{\text{Radius of earth}} + \overbrace{20,000}^{\text{Height}} = 26,440 \text{ km}$$

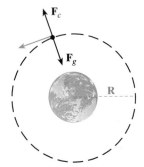

■ **FIGURE 11.26**   A satellite in a stable orbit; centripetal force $\mathbf{F}_g$ equals the force $\mathbf{F}_c$ due to gravity.

is the distance of the satellite from the center of the earth. The satellite maintains a stable orbit when the force mv^2/R produced by its centripetal acceleration equals the force F_g due to gravity. (See Figure 11.26.) According to Newton's law of universal gravitation, $\|\mathbf{F}_c\| = GmM/R^2$, where M is the mass of the earth and G is the *gravitational constant*. Thus, for stability we must have

$$\frac{mv^2}{R} = \frac{GmM}{R^2}$$

so $v = \sqrt{GM/R}$. Experiments indicate that $GM = 398,600 \text{ km}^3/\text{s}^2$, and by substituting $R = 26,440$, we obtain

$$v = \sqrt{\frac{GM}{R}} = \sqrt{\frac{398,600}{26,440}} \approx 3.88273653$$

For this example, we see the speed of the satellite is approximately 3.883 km/s.

Finally, suppose T is the time required for the satellite to make one complete circuit of the earth (called the **period** of the satellite). In each period, the satellite travels a distance equal to the circumference of a circle of radius $R = 26{,}440$ km, and because it travels at v km/s, we must have $vT = 2\pi R$, so that

$$T = \frac{2\pi R}{v} \approx \frac{2\pi(26{,}440 \text{ km})}{3.88273653} \approx 42{,}786.16852 \text{ seconds}$$

or approximately 713 minutes (11 h 53 min). ∎

11.5 Problem Set

Ⓐ *In Problems 1–8, $\mathbf{R}(t)$ is the position vector of a moving object. Find the tangential and normal components of the object's acceleration.*

1. $\mathbf{R}(t) = t\mathbf{i} + t^2\mathbf{j}$ **2.** $\mathbf{R}(t) = t\mathbf{i} + e^t\mathbf{j}$

3. $\mathbf{R}(t) = \langle t \sin t, t \cos t \rangle$ **4.** $\mathbf{R}(t) = \langle 3 \cos t, 2 \sin t \rangle$

5. $\mathbf{R}(t) = \langle t, t^2, t \rangle$ **6.** $\mathbf{R}(t) = \langle 4 \cos t, 0, \sin t \rangle$

7. $\mathbf{R}(t) = (\sin t)\mathbf{i} + (\cos t)\mathbf{j} + (\sin t)\mathbf{k}$

8. $\mathbf{R}(t) = \left(\frac{5}{13} \cos t\right)\mathbf{i} + \frac{12}{13}(1 - \cos t)\mathbf{j} + \sin t\mathbf{k}$

In Problems 9–12, the velocity $\mathbf{V}$ and acceleration $\mathbf{A}$ of a moving object are given at a certain instant. Find $\mathbf{T}$, $\mathbf{N}$, and A_T, A_N at this instant.

9. $\mathbf{V} = \langle 1, -3 \rangle$; $\mathbf{A} = \langle 2, 5 \rangle$

10. $\mathbf{V} = -\mathbf{i} + 7\mathbf{j}$; $\mathbf{A} = 4\mathbf{i} + 5\mathbf{j}$

11. $\mathbf{V} = 2\mathbf{i} + 3\mathbf{j} - \mathbf{k}$; $\mathbf{A} = -\mathbf{i} - 5\mathbf{j} + 2\mathbf{k}$

12. $\mathbf{V} = \langle 5, -1, 2 \rangle$; $\mathbf{A} = \langle 1, 0, -7 \rangle$

In Problems 13–16, the speed $\|\mathbf{V}\|$ of a moving object is given. Find A_T, the tangential component of acceleration, at the indicated time.

13. $\|\mathbf{V}\| = \sqrt{5t^2 + 3}$; $t = 1$

14. $\|\mathbf{V}\| = \sqrt{t^2 + t + 1}$; $t = 3$

15. $\|\mathbf{V}\| = \sqrt{\sin^2 t + \cos(2t)}$; $t = 0$

16. $\|\mathbf{V}\| = \sqrt{e^{-t} + t^4}$; $t = 0$

17. Find the maximum and minimum speeds of a particle whose position vector is

$$\mathbf{R}(t) = (4 \sin 2t)\mathbf{i} - (3 \cos 2t)\mathbf{j}$$

18. Where on the trajectory of

$$\mathbf{R}(t) = (2t^2 - 5t)\mathbf{i} + (5t + 2)\mathbf{j} + 4t^2\mathbf{k}$$

is the speed minimized? What is the minimum speed?

Ⓑ **19.** The position of an object at time t is (x, y), where $x = 1 + \cos 2t$, $y = \sin 2t$. Find the velocity, the acceleration, and the normal and tangential components of acceleration of the object at time t.

20. An object moves with constant angular velocity ω around the circle $x^2 + y^2 = r^2$ in the xy-plane. The position vector is

$$\mathbf{R}(t) = (r \cos \omega t)\mathbf{i} + (r \sin \omega t)\mathbf{j}$$

a. Find the tangential and normal components of acceleration.

b. Show that the curvature is $\kappa = 1/r$ at each point on the circle.

21. Find the tangential and normal components of the acceleration of an object that moves along the parabolic path $y^2 = 4x^2$ at the instant the speed is $ds/dt = 20$.

MODELING PROBLEMS   *In Problems 22–25, set up an appropriate model to answer the given questions. Be sure to state your assumptions.*

22. A pail attached to a rope 1 yd long is swung at the rate of 1 rev/s. Find the tangential and normal components of the pail's acceleration. Assume the rope is swung in a level plane.

23. A boy holds onto a pail of water weighing 2 lb and swings it in a vertical circle with a radius of 3 ft. If the pail travels at ω rpm, what is the pressure of the water on the bottom of the pail at the highest and lowest points of the swing? What is the *smallest* value of ω required to keep the water from spilling from the pail? Assume the pail is held by a handle so that its bottom is straight up when it is at its highest point.

24. A car weighing 2,700 lb (about 1.35 tons) moves along the elliptic path $900x^2 + 400y^2 = 1$. If the car travels at the constant speed 45 mi/h, how much frictional force is required to keep it from skidding as it turns the "corner" at $\left(\frac{1}{30}, 0\right)$? What about the corner $\left(0, \frac{1}{20}\right)$?

25. Curved sections of road such as expressway offramps are often banked to protect against skidding. Consider a road with a circular curve of radius 150 feet, and assume the magnitude of the force of static friction $\mathbf{F}_s$ (the force resisting the tendency to skid) is proportional to the car's weight; that is $\|\mathbf{F}_s\| = \mu W$, where μ is a constant called the *coefficient of static friction*. For this problem we assume $\mu = 0.47$.

a. First, suppose the road is not banked. What is the largest speed $\mathbf{v}_\mathrm{m}$ that a 3,500 lb car can travel around the curve without skidding? Would the answer be different for a smaller car, say one weighing 2,000 lbs?

b. Now suppose the curve is banked at 17° and answer the questions in part **a.**

c. Suppose the highway engineers want the maximum safe speed at 50 mph. At what angle should the road be banked?

26. What is the smallest radius that should be used for a circular highway curve if the normal component of the acceleration of a car traveling at 45 mi/h is not to exceed 2.4 ft/s²?

27. A Ferris wheel with radius 15 ft rotates in a vertical plane at ω rpm. What is the maximum value of ω for a Ferris wheel carrying a person of weight W? (That is, the largest ω so the passenger is not "thrown off.")

Use the formulas in Problem 35 to find A_T and A_N for the given position vector $\mathbf{R}(t)$ in Problems 28–31.

28. $\mathbf{R}(t) = t^3\mathbf{i} + t^2\mathbf{j} + t\mathbf{k}$ 29. $\mathbf{R}(t) = t\mathbf{i} + 2t\mathbf{j} + t^2\mathbf{k}$

30. $\mathbf{R}(t) = (\cos t)\mathbf{i} + (\sin t)\mathbf{j} + \mathbf{k}$

31. $\mathbf{R}(t) = (e^t \cos t)\mathbf{i} + (e^t \sin t)\mathbf{j} + e^t\mathbf{k}$

C 32. Let **T** be the unit tangent vector, **N** the principal unit normal, and **B** the unit binormal vector (see Problem 58, Section 11.4) to a given curve C. Show that

$$\frac{d\mathbf{B}}{ds} = \mathbf{T} \times \frac{d\mathbf{N}}{ds}$$

33. **MODELING PROBLEM**   An amusement park ride consists of a large (25-ft radius), flat, horizontal wheel. Customers board the wheel while it is stationary and try to stay on as long as possible as it begins to rotate. The purpose of this problem is to discover how fast the wheel can rotate without losing its passengers.

a. Suppose the wheel rotates at ω revolutions per minute and that a volunteer weighing W lb sits 15 feet from the center of the wheel. Find $\mathbf{F}_N = m\mathbf{A}_N$, where $m = W/g$ is the volunteer's mass. This is the force tending to push the volunteer off the wheel.

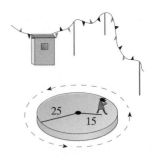

b. If the frictional force of the volunteer in part **a** to stay on the wheel is $0.12W$, find the largest value of ω that will allow the volunteer to stay in place. Wait, we did not tell you the volunteer's weight. Does it matter?

34. **MODELING PROBLEM**   A curve in a railroad track has the shape of the parabola $x = y^2/120$. If a train is loaded so that its scalar normal component of acceleration cannot exceed 30 units/s², what is its maximum possible speed as it rounds the curve at $(0, 0)$?

35. Suppose the position vector of a moving object is $\mathbf{R}(t)$. Show that the tangential and normal components of the object's acceleration may be computed by the formulas

$$A_T = \frac{\mathbf{R}' \cdot \mathbf{R}''}{\|\mathbf{R}'\|} \quad \text{and} \quad A_N = \frac{\|\mathbf{R}' \times \mathbf{R}''\|}{\|\mathbf{R}'\|}$$

36. Suppose an object moves in the plane along the curve $y = f(x)$. Use the formulas obtained in Theorem 11.8 to show that

$$A_T = \frac{f'(x)f''(x)}{\sqrt{1 + [f'(x)]^2}}; \quad \text{and} \quad A_N = \frac{|f''(x)|}{\sqrt{1 + [f'(x)]^2}}$$

37. If the graph of the function of f has an inflection point at $x = a$ and $f''(a)$ exists, show that the graph of f has curvature 0 at $(a, f(a))$.

38. A projectile is fired from ground level with an angle of elevation α and muzzle speed v_0. Find formulas for the tangential and normal components of the projectile's acceleration at time t. What are A_T and A_N at the time the projectile is at its maximum height?

39. An object connected to a string of length r is spun counterclockwise in a circular path in a horizontal plane. Let ω be the constant angular velocity of the object.

a. Show that the displacement vector of the object is

$$\mathbf{R}(t) = (r \cos \omega t)\mathbf{i} + (r \sin \omega t)\mathbf{j}$$

b. Find the normal component of the object's acceleration.

c. If the angular velocity ω is doubled, what is the effect on the normal component of acceleration? What is the effect on **A** if ω is unchanged but the length of the string is doubled? What happens to A_N if ω is doubled and r is halved?

40. An artificial satellite travels at constant speed in a stable circular orbit R km above the earth's surface (as in Example 4).
   a. Show that the satellite's period is given by the formula

   $$T = 2\pi \frac{(R + R_e)^{3/2}}{\sqrt{GM}}$$

   where R_e is the radius of the earth.
   b. A satellite is said to be in **geosynchronous orbit** if it completes one orbit every sidereal day (23 hours,

56 minutes). How high above the earth should the satellite be to achieve such an orbit? Assume the radius of the earth is approximately 6,440 km.
   c. What is the speed of a satellite in geosynchronous orbit?

41. MODELING PROBLEM   For Mars the following facts are known (all in relation to Earth):

diameter	0.533
length of day	1.029
mass	0.1074
gravity(g)	0.3776

Use these facts to determine how high above the surface of Mars a satellite must be to achieve synchronous orbit. How fast would such a satellite be traveling?

Chapter 11   Review

Proficiency Examination

Concept Problems

1. What is a vector-valued function?

2. What are the components of a vector function?

3. Describe what is meant by the graph of a vector function.

4. Define the limit of a vector function.

5. Define the derivative of a vector function.

6. Define the integral of a vector function.

7. What is a smooth curve?

8. Complete the following rules for differentiating vector functions:
   a. Linearity rule
   b. Scalar multiple rule
   c. Dot product rule
   d. Cross product rule
   e. Chain rule

9. State the theorem about the orthogonality of a derivative of constant length.

10. What are the position, velocity, and acceleration vectors?

11. What is the speed of a particle moving on a curve C?

12. What are the formulas for the motion of a projectile in a vacuum?

13. What are the formulas for time of flight and range of a projectile?

14. State Kepler's laws.

15. What are the unit polar vectors, $\mathbf{u}_r$ and $\mathbf{u}_\theta$?

16. What are the unit tangent and normal vectors?

17. What is speed in terms of arc length?

18. Give a formula for the arc length of a space curve.

19. What is the curvature of a graph?

20. What is the formula for curvature in terms of the velocity and acceleration vectors?

21. What are the formulas for unit tangent and principal unit normal in terms of the arc length parameter?

22. What is the radius of curvature?

23. What are the tangential and normal components of acceleration?

Practice Problems

24. Sketch the graph of $\mathbf{R}(t) = (3 \cos t)\mathbf{i} + (3 \sin t)\mathbf{j} + t\mathbf{k}$, and find the length of this curve from $t = 0$ to $t = 2\pi$.

25. If $\mathbf{F}(t) = \left[\dfrac{t}{1 + t}, \dfrac{\sin t}{t}, \cos t\right]$, find $\mathbf{F}'(t)$ and $\mathbf{F}''(t)$.

26. Evaluate $\displaystyle\int_1^2 \langle 3t, 0, 3\rangle \times \langle 0, \ln t, -t^2\rangle \, dt$.

27. Find a vector function $\mathbf{F}$ such that $\mathbf{F}''(t) = \langle e^t, -t^2, 3\rangle$ and $\mathbf{F}(0) = \langle 1, -2, 0\rangle, \mathbf{F}'(0) = \langle 0, 0, 3\rangle$.

28. Find the velocity $\mathbf{V}$, the speed $\dfrac{ds}{dt}$, and the acceleration $\mathbf{A}$ for the body with position vector $\mathbf{R}(t) = t\mathbf{i} + 2t\mathbf{j} + te^t\mathbf{k}$.

29. Find $\mathbf{T}, \mathbf{N}, A_T$ and A_N (the tangential and normal components of acceleration), and κ (curvature) for an object with position vector $\mathbf{R}(t) = t^2\mathbf{i} + 3t\mathbf{j} - 3t\mathbf{k}$.

30. A projectile is fired from ground level with initial velocity 50 ft/s at an angle of elevation of $\alpha = 30°$.
   a. What is the maximum height reached by the projectile?
   b. What are the time of flight and the range?

Supplementary Problems

Find the vector limits in Problems 1–6.

1. $\lim\limits_{t\to 0}\left[t^2\mathbf{i} - 3te^t\mathbf{j} + \dfrac{\sin 2t}{t}\mathbf{k}\right]$ **2.** $\lim\limits_{t\to 0}\left[\dfrac{\mathbf{i} + t\mathbf{j} - e^{-t}\mathbf{k}}{1 - t}\right]$

3. $\lim\limits_{t\to 0}\langle t, 0, 5\rangle \cdot \langle \sin t, 3t, -(1 - t)\rangle$

4. $\lim\limits_{t\to \pi}\langle 1 + t, -3, 0\rangle \times \langle 0, t^2, \cos t\rangle$

5. $\lim\limits_{t\to 0}\left[\left(1 + \dfrac{1}{t}\right)^t\mathbf{i} - \left(\dfrac{\sin t}{t}\right)\mathbf{j} - t\mathbf{k}\right]$

6. $\lim\limits_{t\to \infty}\left[\left(\dfrac{1 - \cos t}{t}\right)\mathbf{i} + 4\mathbf{j} + \left(1 + \dfrac{3}{t}\right)^t\mathbf{k}\right]$

Find $\mathbf{F}'(t)$ and $\mathbf{F}''(t)$ for the vector functions in Problems 7–12.

7. $\mathbf{F}(t) = te^t\mathbf{i} + t^2\mathbf{j}$ **8.** $\mathbf{F}(t) = (t\ln 2t)\mathbf{i} + t^{3/2}\mathbf{k}$

9. $\mathbf{F}(t) = \langle 2t^{-1}, -2t, te^{-t}\rangle$

10. $\mathbf{F}(t) = \langle t, -(1 - t), 0\rangle \times \langle 0, t^2, e^{-t}\rangle$

11. $\mathbf{F}(t) = (t^2 + e^{at})\mathbf{i} + (te^{-at})\mathbf{j} + (e^{at+1})\mathbf{k}$, *a* a constant

12. $\mathbf{F}(t) = (1 - t)^{-1}\mathbf{i} + (\sin 2t)\mathbf{j} + (\cos^2 t)\mathbf{k}$

Sketch the graph of the vector functions in Problems 13–15.

13. $\mathbf{F}(t) = te^t\mathbf{i} + t^2\mathbf{j}$ **14.** $\mathbf{F}(t) = t^2\mathbf{i} - 3t\mathbf{j}$

15. $\mathbf{F}(t) = (1 - \cos t)\mathbf{i} + (\sin t)\mathbf{k}$

Describe the graph of the vector functions in Problems 16–18.

16. $\mathbf{F}(t) = \langle 3\cos t, 3\sin t, t\rangle$ **17.** $\mathbf{F}(t) = \langle 2\sin t, 2\cos t, 5t\rangle$

18. $\mathbf{F}(t) = 2t^2\mathbf{i} + (1 - t)\mathbf{j} + 3\mathbf{k}$

Let $\mathbf{F}(t) = f_1(t)\mathbf{i} + f_2(t)\mathbf{j} + f_3(t)\mathbf{k}$ and find the indicated derivative in terms of $\mathbf{F}'$ in Problems 19–24.

19. $\dfrac{d}{dt}[\mathbf{F}(t) \cdot \mathbf{F}(t)]$ **20.** $\dfrac{d}{dt}[\|\mathbf{F}(t)\|\mathbf{F}(t)]$ **21.** $\dfrac{d}{dt}\left[\dfrac{\mathbf{F}(t)}{\|\mathbf{F}(t)\|}\right]$

22. $\dfrac{d}{dt}\|\mathbf{F}(t)\|$ **23.** $\dfrac{d}{dt}[\mathbf{F}(t) \times \mathbf{F}(t)]$ **24.** $\dfrac{d}{dt}\mathbf{F}(e^t)$

Evaluate the definite and indefinite vector integrals in Problems 25–30.

25. $\displaystyle\int_{-1}^{1} (e^{-t}\mathbf{i} + t^3\mathbf{j} + 3\mathbf{k})\, dt$

26. $\displaystyle\int_{1}^{2} [(1 - t)\mathbf{i} - t^{-1}\mathbf{j} + e^t\mathbf{k}]\, dt$

27. $\displaystyle\int [te^t\mathbf{i} - (\sin 2t)\mathbf{j} + t^2\mathbf{k}]\, dt$

28. $\displaystyle\int e^{2t}[2\mathbf{i} - t\mathbf{j} + (\sin t)\mathbf{k}]\, dt$ **29.** $\displaystyle\int t[e^t\mathbf{i} + (\ln t)\mathbf{j} + 3\mathbf{k}]\, dt$

30. $\displaystyle\int [e^t\mathbf{i} + 2\mathbf{j} - t\mathbf{k}] \cdot [e^{-t}\mathbf{i} - t\mathbf{j}]\, dt$

In Problems 31–34, $\mathbf{R}$ is the position vector of a moving body. Find the velocity $\mathbf{V}$, the speed ds/dt, and the acceleration $\mathbf{A}$.

31. $\mathbf{R}(t) = t\mathbf{i} + (3 - t)\mathbf{j} + 2\mathbf{k}$

32. $\mathbf{R}(t) = (\sin 2t)\mathbf{i} + 2\mathbf{j} - (\cos 2t)\mathbf{k}$

33. $\mathbf{R}(t) = \langle t\sin t, te^{-t}, -(1 - t)\rangle$

34. $\mathbf{R}(t) = \langle \ln t, e^t, -\tan t\rangle$

Find $\mathbf{T}$ and $\mathbf{N}$ for the vector functions given in Problems 35–38.

35. $\mathbf{R}(t) = t\mathbf{i} - t^2\mathbf{j}$ **36.** $\mathbf{R}(t) = (3\cos t)\mathbf{i} - (3\sin t)\mathbf{j}$

37. $\mathbf{R}(t) = \langle 4\cos t, -3t, 4\sin t\rangle$

38. $\mathbf{R}(t) = \langle e^t\sin t, e^t, e^t\cos t\rangle$

In Problems 39–42, find the tangential and normal components of acceleration, and the curvature of a moving object with position $\mathbf{R}(t)$.

39. $\mathbf{R}(t) = t^2\mathbf{i} + 2t\mathbf{j} + e^t\mathbf{k}$

40. $\mathbf{R}(t) = t^2\mathbf{i} - 2t\mathbf{j} + (t^2 - t)\mathbf{k}$

41. $\mathbf{R}(t) = (4\sin t)\mathbf{i} + (4\cos t)\mathbf{j} + 4t\mathbf{k}$

42. $\mathbf{R}(t) = (a\sin 3t)\mathbf{i} + (a + a\cos 3t)\mathbf{j} + (3a\sin t)\mathbf{k}$, for constant $a \neq 0$

A polar curve C is given by $r = f(\theta)$. If $f''(\theta)$ exists (see Problem 54, Section 11.4), then the curvature can be found by the formula

$$\kappa = \dfrac{|r^2 + 2(r')^2 - rr''|}{[r^2 + (r')^2]^{3/2}}$$

Find the curvature at the given point on each of the polar curves given in Problems 43–48.

43. $r = 4\cos\theta$, where $\theta = \dfrac{\pi}{3}$ **44.** $r = \theta^2$, where $\theta = 2$

45. $r = e^{-\theta}$, where $\theta = 1$

46. $r = 1 + \cos\theta$, where $\theta = \dfrac{\pi}{2}$

47. $r = 4\cos 3\theta$, where $\theta = \dfrac{\pi}{6}$

48. $r = 1 - 2\sin\theta$, where $\theta = \dfrac{\pi}{4}$

49. For what values of t is the following vector function continuous?

$$\mathbf{F}(t) = (2t - 1)\mathbf{i} + \left(\dfrac{t^2 - 1}{t - 1}\right)\mathbf{j} + 4\mathbf{k}$$

50. Find the length of the graph of the vector function

$$\mathbf{R}(t) = \left(\dfrac{t^2 - 2}{2}\right)\mathbf{i} + \dfrac{(2t + 1)^{3/2}}{3}\mathbf{j}$$

from $t = 0$ to $t = 6$.

51. Find a vector function $\mathbf{F}$ such that $\mathbf{F}(0) = \mathbf{F}'(0) = \mathbf{i}$, $\mathbf{F}''(0) = 2\mathbf{j} + \mathbf{k}$, and

$$\mathbf{F}'''(t) = (\cos t)\mathbf{i} + (\sin t)\mathbf{j} + \dfrac{t}{\pi}\mathbf{k}$$

52. An object moves in space with acceleration $\mathbf{A}(t) = \langle -t, 2, 2 - t\rangle$. When $t = 0$, it is known that the object is at the point $(1, 0, 0)$ and that it has velocity $\mathbf{V}(0) = \langle 2, -4, 0\rangle$.
 a. Find the velocity $\mathbf{V}(t)$ and the position $\mathbf{R}(t)$.
 b. What are the speed and location of the object when $t = 1$?

c. When is the object stationary and what is its position at that time?

53. The position vector for a curve is given in terms of arc length s by $\mathbf{R}(s) = \left\langle a \cos \dfrac{s}{a}, a \sin \dfrac{s}{a}, 2s \right\rangle$ for $0 \leq s \leq 2\pi a$, $a \neq 0$. Find the unit tangent vector $\mathbf{T}(s)$ and the principal unit normal $\mathbf{N}(s)$.

54. Find the radius of curvature of the curve given by $y = 1 + \sin x$ at the points where x is
    a. $\frac{\pi}{6}$ b. $\frac{\pi}{4}$ c. $\frac{3\pi}{2}$

55. Find the radius of curvature of the ellipse given by $\mathbf{R}(t) = \langle a \cos t, b \sin t \rangle$ where $a > 0, b > 0, a \neq b$, and $0 \leq t \leq 2\pi$ at the points where $t = 0$ and $\pi/2$.

56. **MODELING PROBLEM** A stunt pilot flying at an altitude of 4,000 ft with a speed of 180 mi/hr drops a weighted marker, attempting to hit a target on the ground below, as shown in Figure 11.27. How far away from the target (measured horizontally) should the pilot be when she releases the marker? You may neglect air resistance.

4000 ft

■ **FIGURE 11.27** Problem 56

57. The position of an object moving in space is given by
    $$\mathbf{R}(t) = (e^{-t}\cos t)\mathbf{i} + (e^{-t}\sin t)\mathbf{j} + e^{-t}\mathbf{k}$$
    a. Find the velocity, speed, and acceleration of the object at arbitrary time t.
    b. Determine the curvature of the trajectory at time t.

58. Find the point or points on the curve $y = e^{ax} (a > 0)$ where the radius of curvature is maximized.

59. **MODELING PROBLEM** A car weighing 3,000 lb travels at a constant speed of 60 mi/h on a flat road and then makes a circular turn on an interchange. If the radius of

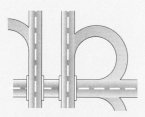

the turn is 40 ft, what frictional force is needed to keep the car from skidding?

60. The position vector of an object in space is
    $$\mathbf{R}(t) = (a \cos \omega t)\mathbf{i} + (a \sin \omega t)\mathbf{j} + \omega^2 t \mathbf{k}$$
    Find ω so that the sum of the object's tangential and normal components of acceleration equal half its speed.

61. A particle moves along a path given in parametric form where $r(t) = 1 + \cos at$ and $\theta(t) = e^{-at}$ (for positive constant a). Find the velocity and acceleration of the particle in terms of the unit polar vectors $\mathbf{u}_r$ and $\mathbf{u}_\theta$.

62. A nozzle discharges a stream of water with an initial velocity $v_0 = 50$ ft/s into the end of a horizontal pipe of inside diameter $d = 5$ ft. What is the maximum horizontal distance that the stream can reach?

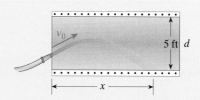

63. Sketch the graph of the vector function
    $$\mathbf{R}(t) = \left(\frac{3t}{1 + t^3}\right)\mathbf{i} + \left(\frac{3t^2}{1 + t^3}\right)\mathbf{j}$$
    then find parametric equations for the tangent line at the point where $t = 2$. This curve is called the *folium of Descartes.*

Technology Window

64. **MODELING PROBLEM** A fireman stands 5.5 m from the front of a burning building 15 m high. His fire hose discharges water from a height of 1.2 m at an angle of 62°, as shown in the figure. Use a spreadsheet or a computer program to determine the height h where the stream of water strikes the building for values of v_0 varying from 6 m/s to 26 m/s at intervals of 1 m/s.

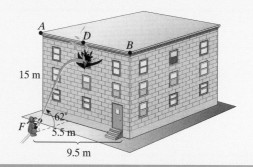

65. A DNA molecule has the shape of a double helix (see Figure 11.4). The radius of each helix is about 10^{-8} μm. Each helix rises about 3×10^{-8} μm during each complete turn and there are about 3×10^8 complete turns. Estimate the length of each helix.

66. Show that the tangential component of acceleration of a moving object is 0 if the object has constant speed. Is the converse statement also true? That is, if $A_T = 0$, can we conclude that the speed is constant?

67. The path of a particle P is an Archimedean spiral. The motion of the particle is described by the polar coordinates $r = 10t$ and $\theta = 2\pi t$, where r is expressed in inches and t is in seconds. Determine the velocity of the particle (in terms of $\mathbf{u}_r$ and $\mathbf{u}_\theta$) when
   a. $t = 0$ b. $t = 0.25$ s

68. **SPY PROBLEM** After dropping the knockout gas into Scélérat's bunker (Problem 33, Section 11.3), the Spy lands and, as expected, finds a bunker full of sleeping thugs—but no Scélérat! He races through the back door of the bunker and finds himself in a ski resort. By the time he locates Scélérat, the French fiend is skiing down the mountain. The Spy grabs a pair of skis and runs over to a conveniently located ski jump. His keen mind quickly determines that the slope of the mountain is 17° and the angle at the lip of the jump is 10°, as shown in Figure 11.28. Suppose Scélérat is 150 ft down the mountainside skiing at 75 ft/s when the Spy launches himself from the end of the ski jump with a speed of 85 ft/s. How close will he be to Scélérat when he lands? If the Spy becomes airborne at noon and he maintains his landing speed, when (if ever) does he catch Scélérat?

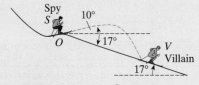

■ **FIGURE 11.28** Spy scene

In Problems 69–70, find $\mathbf{F}'(x)$.

69. $\mathbf{F}(x) = \displaystyle\int_1^x [(\sin t)\mathbf{i} - (\cos 2t)\mathbf{j} + e^{-t}\mathbf{k}]\, dt$

70. $\mathbf{F}(x) = \displaystyle\int_1^{2x} [t^2\mathbf{i} + (\sec e^{-t})\mathbf{j} - (\tan e^{2t})\mathbf{k}]\, dt$

In Problems 71–74, solve the initial value problems for $\mathbf{F}$ *as a vector function of* t.

71. differential equation: $\mathbf{F}'(t) = t\mathbf{i} + t\mathbf{j} - t\mathbf{k}$
    initial condition: $\mathbf{F}(0) = \mathbf{i} + 2\mathbf{j} - 3\mathbf{k}$

72. differential equation:
    $$\mathbf{F}'(t) = \frac{3}{2}\sqrt{t+1}\,\mathbf{i} + (t+1)^{-1}\mathbf{j} + e^t\mathbf{k}$$
    initial condition: $\mathbf{F}(0) = \mathbf{j} - 3\mathbf{k}$

73. differential equation:
    $$\mathbf{F}'(t) = (\sin 2t)\mathbf{i} + (e^t\cos t)\mathbf{j} - \left(\frac{3}{t+1}\right)\mathbf{k}$$
    initial condition: $\mathbf{F}(0) = \mathbf{i} - 3\mathbf{k}$

74. differential equation: $\dfrac{d^2\mathbf{F}}{dt^2} = -32\mathbf{j}$
    initial conditions: $\mathbf{F} = 50\mathbf{j}$ and $\dfrac{d\mathbf{F}}{dt} = 5\mathbf{i} + 5\mathbf{k}$ at $t = 0$

75. **PUTNAM EXAMINATION PROBLEM** A shell strikes an airplane flying at a height h above the ground. It is known that the shell was fired from a gun on the ground with muzzle speed v_0, but the position of the gun and its angle of elevation are both unknown. Deduce that the gun is situated within a circle whose center lies directly below the airplane and whose radius is
    $$\frac{v_0}{g}\sqrt{v_0^2 - 2gh}$$
    Neglect resistance of the atmosphere.

76. **PUTNAM EXAMINATION PROBLEM** A coast artillery gun can fire at any angle of elevation between 0° and 90° in a fixed vertical plane. If air resistance is neglected and the muzzle speed is constant ($v = v_0$), determine the set H of points in the plane that can be hit. Consider only those points above the horizontal.

77. **PUTNAM EXAMINATION PROBLEM** A particle moves on a circle with center O, starting from rest at a point P and coming to rest again at a point Q, without coming to rest at any intermediate point. Prove that the acceleration vector of the particle does not vanish at any point between P and Q, and that, at some point R between P and Q, the acceleration vector points in along the radius $\overline{RO}$.

The Stimulation of Science

Howard Eves was born in Paterson, New Jersey, in 1911 and is professor emeritus of mathematics at the University of Maine. This guest essay first appeared in *Great Moments in Mathematics Before 1650*. Professor Eves reminds us, "It must be remembered that a *moment* in history is sometimes an inspired flash and sometimes an evolution extending over a long period of time." Howard Eves's lectures (from which this guest essay was taken) are so renowned that each college and university at which he taught awarded him, at one time or another, every available honor for distinguished teaching. Howard Eves is a prolific and successful textbook author with a real love for the history of mathematics. He now lives with his wife at the retirement retreat in Lubec, Maine.

The die is cast; I have written my book; it will be read either in the present age or by posterity, it matters not which; it may well await a reader, since God has waited six thousand years for an interpreter of his words.

JOHANNES KEPLER
FROM JAMES R. NEWMAN, *THE WORLD OF MATHEMATICS*, *VOLUME I* (NEW YORK: SIMON AND SCHUSTER, 1956), P. 220.

The mighty Antaeus was the giant son of Neptune (god of the sea) and Ge (goddess of the earth), and his strength was invincible so long as he remained in contact with his Mother Earth. Strangers who came to his country were forced to wrestle to the death with him, and so it chanced one day that Hercules and Antaeus came to grips with one another. But Hercules, aware of the source of Antaeus' great strength, lifted and held the giant from the earth and crushed him in the air.

There is a parable here for mathematicians. For just as Antaeus was born of and nurtured by his Mother Earth, history has shown us that all significant and lasting mathematics is born of and nurtured by the real world. As in the case of Antaeus, so long as mathematics maintains its contact with the real world, it will remain powerful. But should it be lifted too long from the solid ground of its birth into the filmy air of pure abstraction, it runs the risk of weakening. It must of necessity return, at least occasionally, to the real world for renewed strength.

Such a rejuvenation of mathematics occurred in the seventeenth century, following discoveries made by two eminent mathematician–scientists—Galileo Galilei (1564–1642) and Johannes Kepler (1571–1630). Galileo, through a sequence of experiments started before his 25th birthday, discovered a number of basic facts concerning the motion of bodies in the earth's gravitational field, and Kepler, by 1619, had deduced all three of his famous laws of planetary motion. These achievements proved to be so influential on the development of so much of subsequent mathematics that they must be ranked as two of the GREAT MOMENTS IN MATHEMATICS. Galileo's discoveries led to the creation of the modern science of dynamics and Kepler's to the creation of modern celestial mechanics; and each of these studies, in turn, required, for their development, the creation of a new mathematical tool—the calculus—capable of dealing with change, flux, and motion.

Galileo was born in Pisa in 1564 as the son of an impoverished Florentine nobleman. After a disinterested start as a medical student, Galileo obtained parental permission to change his studies to science and mathematics, fields in which he possessed a strong natural talent. While still a medical student at the University of Pisa, he made his historically famous observation that the great pendulous lamp in the cathedral there oscillated to and fro with a period independent of the size of the arc of oscillation.*

Later, he showed that the period of a pendulum is also independent of the weight of the pendulum's bob. When he was 25, he accepted an appointment as professor of mathematics at the University of Pisa. It was during this appointment that he is alleged to have performed experiments from the leaning tower of Pisa, showing that, contrary to the teaching of Aristotle, heavy bodies do not fall faster than light ones. By rolling balls down inclined planes, he arrived at the law that the distance a body falls is proportional to the square of the time of falling, in accordance with the now-familiar formula $s = \frac{1}{2}gt^2$.

Unpleasant local controversies caused Galileo to resign his chair at Pisa in 1591, and the following year he accepted a professorship in mathematics at the University of Padua, where there reigned an atmosphere more friendly to scientific pursuits. Here at Padua, for nearly 18 years, Galileo continued his experiments and his teaching, achieving a widespread fame. While at Padua, he heard of the discovery, in about 1607, of the telescope by the Dutch lens-grinder Johann Lipersheim, and he set about making instruments of his own, producing a telescope with a magnifying power of more than 30 diameters. With this telescope he observed sunspots (contradicting Aristotle's assertion that the sun is without blemish), saw mountains on the moon, and noticed the phases of Venus, Saturn's rings, and the four bright satellites of Jupiter (all three of these lending credence to the Copernican theory of the solar system). Galileo discoveries roused the opposition of the Church, and finally, in the year 1633, he was summoned to appear before the Inquisition, and there forced to recant his scientific findings. Not many years later the great scientist became blind. He died, a prisoner in his own home, in 1642, the year Isaac Newton was born.

Johannes Kepler was born near Stuttgart, Germany, in 1571 and commenced his studies at the University of Tübingen with the intention of becoming a Lutheran minister. Like Galileo, he found his first choice of an occupation far less congenial than his deep interest in science, particularly astronomy, and he accordingly changed his plans. In 1594, when in his early twenties, he accepted a lectureship at the University of Gräz in Austria. Five years later, he became assistant to the famous Danish-Swedish astronomer Tycho Brahe, who had moved to Prague to serve as the court astronomer to Kaiser Rudolph II. Shortly after, in 1601, Brahe suddenly died, and Kepler inherited both his master's position and his vast collection of very accurate data on the positions of the planets as they moved about the sky. With amazing perseverance, Kepler set out to find, from Brahe's enormous mass of observational data, just how the planets move in space.

It has often been remarked that almost any problem can be solved if one but continuously worries over it and works at it a sufficiently long time. As Thomas Edison said, genius is 1% inspiration and 99% perspiration. Perhaps nowhere in the history of science is this more clearly demonstrated than in Kepler's incredible pertinacity in solving the problem of the motion of the planets about the sun. Thoroughly convinced of the Copernican theory that the planets revolve in orbits about the central sun, Kepler strenuously sought to determine the nature and position of those orbits and the man-

---

*This is only approximately true, the approximation being very close in the case of small amplitudes of oscillation.

ner in which the planets travel in their orbits. With Brahe's great set of observational recordings at hand, the problem became this: to obtain a pattern of motion of the planets that would exactly agree with Brahe's observations. So dependable were Brahe's recordings that any solution that should differ from Brahe's observed positions by even as little as a quarter of the moon's apparent diameter must be discarded as incorrect. Kepler had, then, first to guess with his *imagination* some plausible solution and then, with painful *perseverance,* to endure the mountain of tedious calculations needed to confirm or reject his guess. He made hundreds of fruitless attempts and performed reams and reams of calculations, laboring with undiminished zeal and patience for many years. Finally he solved his problem, in the form of his three famous laws of planetary motion, the first two around 1609 and the third one 10 years later in 1619:

I. The planets move about the sun in elliptical orbits with the sun at one focus.

II. The radius vector joining a planet to the sun sweeps over equal areas in equal intervals of time.

III. The square of the time of one complete revolution of a planet about its orbit is proportional to the cube of the orbit's semimajor axis.

The empirical discovery of these laws from Brahe's mass of data constitutes one of the most remarkable inductions ever made in science. With justifiable pride, Kepler prefaced his *Harmony of the Worlds* of 1619 with the following outburst:

> *I am writing a book for my contemporaries or—it does not matter—for posterity. It may be that my book will wait for a hundred years for a reader. Has not God waited 6000 years for an observer?*

Kepler's laws of planetary motion are landmarks in the history of astronomy and mathematics, for in the effort to justify them, Isaac Newton was led to create modern celestial mechanics. It is very interesting that 1800 years after the Greeks had developed the properties of the conic sections there should occur such an illuminating practical application of them. *One never knows when a piece of pure mathematics may receive an unexpected application.*

In order to compute the areas involved in his second law, Kepler had to resort to a crude form of the integral calculus, making him one of the precursors of that calculus. Also, in his *Stereometria doliorum vinorum* (*Solid Geometry of Wine Barrels,* 1615), he applied crude integration procedures to the finding of the volumes of 93 different solids obtained by revolving arcs of conic sections about axes in their planes. Among these solids were the torus and two that he called *the apple* and *the lemon,* these latter being obtained by revolving a major and a minor arc, respectively, of a circle about the arc's chord as an axis. Kepler's interest in these matters arose when he observed some of the poor methods in use by the wine gaugers of the time. (See Group Research Project, page 315.)

It was Kepler who introduced the word *focus* (Latin for "hearth") into the geometry of conic sections. He approximated the perimeter of an ellipse of semiaxes a and b by the formula $\pi(a + b)$. He also laid down a so-called *principle of continuity,* which postulates the existence in a plane of certain ideal points and an ideal line having many of the properties of ordinary points and lines, lying at infinity. Thus he explained that a straight line can be considered as closed at infinity and that a parabola may be regarded as the limiting case of either an ellipse or a hyperbola in which one of the foci has retreated to infinity. The ideas were extended by later geometers.

Kepler was a confirmed Pythagorean, with the result that his work is often a blend of the fancifully mystical and the carefully scientific. It is sad that his personal life was made almost unendurable by a multiplicity of worldly misfortunes. An infection from smallpox when he was four years old left his eyesight much impaired. In addition to his general lifelong weakness, he spent a joyless youth; his marriage was a constant source of unhappiness; his favorite child died of smallpox; his wife went mad and died; he was expelled from his lectureship at the University of Gräz when the city fell to the Catholics; his mother was charged and imprisoned for witchcraft, and for almost a year he desperately tried to save her from the torture chamber; he himself very narrowly escaped condemnation for heterodoxy; and his stipend was always in arrears. One report says that his second marriage was even less fortunate than his first, although he took the precaution to analyze carefully the merits and demerits of 11 women before choosing the wrong one. He was forced to augment his income by casting horoscopes, and he died of a fever in 1630 at the age of 59 while on a journey to try to collect some of his long overdue salary.

Mathematical Essays

1. Write a 500-word essay on the life and mathematics of Galileo Galilei.

2. **HISTORICAL QUEST** For this Quest, explain the remark in Galileo's *Discorsi e dimostrazioni matematiche intorno a due nuove scienze* of 1638 that "neither is the number of squares less than the totality of all numbers, nor the latter greater than the former."

GALILEO GALILEI
1564–1642

3. **HISTORICAL QUEST** In 1638, Galileo published his ideas about dynamics in his book *Discorsi e dimostrazioni matematiche intorno à due nuove scienze.* In this book, he considers the following problem:

   Suppose the larger circle of Figure 11.29 has made one revolution in rolling along the straight line from A to B, so that $|AB|$ is equal to the circumference of the large circle. Then the small circle, fixed to the large one, has also made one revolution, so that $|CD|$ is equal to the circumference of the small circle. It follows that *the two circles have equal circumferences.*

■ **FIGURE 11.29** Aristotle's wheel

   This paradox had been earlier described by Aristotle and is therefore sometimes referred to as *Aristotle's wheel.* Can you explain what is going on?

4. **HISTORICAL QUEST** In Historical Quest 39 Section 6.3, we outlined a procedure by Johannes Kepler for finding the volume of a torus.

   This Quest asks you to use Kepler's reasoning to find the area of an ellipse.

   a. Kepler divided a circle of radius r and circumference C into a large number of very thin sectors. By regarding each sector as a thin isosceles triangle of altitude r and

JOHANNES KEPLER
1571–1630

base equal to the arc of the sector, he heuristically arrived at the formula $A = \frac{1}{2} r C$ for the area of a circle. Show how this was done.

   b. Use Kepler's reasoning to obtain a volume V of a sphere of radius r and surface area S as $V = \frac{1}{3} r S$.

   c. Use Kepler's reasoning to show that an ellipse of semimajor axis a and semiminor axis b has area $A = \pi ab$.

5. Prove Kepler's first law. You may search the literature for help with this proof.

6. Prove Kepler's third law. You may search the literature for help with this proof.

7. Where is a planet in its orbit when its speed is the greatest? Support your response.

8. Two hypothetical planets are moving about the sun in elliptical orbits having equal major axes. The minor axis of one, however, is half that of the other. How do the periods of the two planets compare? Support your response.

9. **HISTORICAL QUEST** The three laws of Kepler described in this section forever changed the way we view the universe, but it was not Kepler who correctly proved these laws. Isaac Newton (Historical Quest #2, Section 2.5) proved these laws from the inverse-square law of gravitation.

   In Section 11.3, we proved Kepler's second law. Write a 500-word essay on the life and mathematics of Johannes Kepler, and include, as part of your essay, where Kepler made his mistake in his proof of his second law.

10. **Book Report** "Science is that body of knowledge that describes, defines, and where possible, explains the universe . . . we think of the history of science as a history of men. . . . [History] is the story of thousands of people who contributed to the knowledge and theories that constituted the science of their eras and made the 'great leaps' possible. Many of these people were women." So begins a history of women in science entitled *Hypatia's Heritage* by Margaret Alic (Boston: Beacon Press, 1986). Read this book and write a book report.

11. Make up a word problem involving the calculus of vector-valued functions. Send your problem to:
    Bradley and Smith
    Prentice Hall Publishing Company
    1 Lake Street
    Upper Saddle River, NJ 07458
    The best ones submitted will appear in the next edition (along with credit to the problem poser).

Chapters 7–11 Cumulative Review

■ TABLE 11.3 **Summary of Velocity, Acceleration, and Curvature**

CURVES	$\mathbb{R}^2$ (plane)	$\mathbf{R}(t) = x(t)\mathbf{i} + y(t)\mathbf{j}$
Position vector	$\mathbb{R}^3$ (space)	$\mathbf{R}(t) = x(t)\mathbf{i} + y(t)\mathbf{j} + z(t)\mathbf{k}$

The graph is called the **trajectory** of the object's motion.

Velocity vector

$$\mathbf{V}(t) = \frac{d\mathbf{R}}{dt} = \mathbf{R}'(t)$$

$$\|\mathbf{V}\| = \frac{ds}{dt} \qquad \text{This is called the \textbf{speed}.}$$

$$\frac{\mathbf{V}}{\|\mathbf{V}\|} \qquad \text{This is the direction of } \mathbf{V} \text{ or } \textit{direction of motion.}$$

Acceleration vector

$$\mathbf{A}(t) = \frac{d^2\mathbf{R}}{dt^2} = \frac{d\mathbf{V}}{dt} = A_T\mathbf{T} + A_N\mathbf{N}$$

where

$$A_T = \frac{d^2s}{dt^2} = \mathbf{A} \cdot \mathbf{T} = \frac{\mathbf{V} \cdot \mathbf{A}}{\|\mathbf{V}\|}$$

This is the **tangential component.**

$$A_N = \kappa\left(\frac{ds}{dt}\right)^2 = \frac{\|\mathbf{V} \times \mathbf{A}\|}{\|\mathbf{V}\|} = \sqrt{\|\mathbf{A}\|^2 - A_T^2}$$

This is the **normal component.**

Unit tangent and normal vectors

$$\mathbf{T} = \frac{\mathbf{V}}{\|\mathbf{V}\|} = \frac{\mathbf{R}'(t)}{\|\mathbf{R}'(t)\|} = \frac{d\mathbf{R}}{ds}$$

This is the **unit tangent vector** in the *direction of motion.*

$$\mathbf{N} = \frac{\mathbf{T}'(t)}{\|\mathbf{T}'(t)\|} = \frac{1}{\kappa}\frac{d\mathbf{T}}{ds}$$

This is the **principal unit normal vector.**

Curvature

$$\kappa = \left\|\frac{d\mathbf{T}}{ds}\right\| = \frac{\left\|\dfrac{d\mathbf{T}}{dt}\right\|}{\dfrac{ds}{dt}} = \frac{\|\mathbf{R}' \times \mathbf{R}''\|}{\|\mathbf{R}'\|^3}; \text{if } y = f(x) \text{ then}$$

$$\kappa = \frac{|x'y'' - y'x''|}{[(x')^2 + (y')^2]^{3/2}}$$

Torsion

$$\tau = \left\|\frac{d\mathbf{B}}{ds}\right\|$$

Cumulative Review Problems for Chapters 7–11

1. ■ **What Does This Say?** In your own words, outline a procedure for integrating a given function.

2. ■ **What Does This Say?** In your own words, outline a procedure for deciding whether a given series converges or diverges.

3. ■ **What Does This Say?** What is a quadric surface? In your own words, discuss classifying and sketching quadric surfaces.

4. ■ **What Does This Say?** What is a vector? What is a vector function? In your own words, discuss what is meant by vector calculus.

5. a. If $y = x\cosh^{-1} x$, find y'.

   b. If $x\sinh^{-1} y + y\tanh^{-1} x = 0$, find $\dfrac{dy}{dx}$.

Find the integrals in Problems 6–14.

6. $\displaystyle\int x \ln \sqrt[3]{x}\, dx$

7. $\displaystyle\int \sin^2 x \cos^3 x\, dx$

8. $\displaystyle\int \tan^2 x \sec x\, dx$

9. $\displaystyle\int x\sqrt{16 - x}\, dx$

10. $\displaystyle\int \dfrac{\cosh x\, dx}{1 + \sinh^2 x}$

11. $\displaystyle\int \dfrac{dx}{1 + \cos x}$

12. $\displaystyle\int \dfrac{dx}{x^2 (x^2 + 5)}$

13. $\displaystyle\int \dfrac{dx}{\sqrt{x} - \sqrt[3]{x}}$

14. $\displaystyle\int \dfrac{dx}{\sqrt{2x - x^2}}$

15. Find the equation of the line through $(-1, 2, 5)$ and perpendicular to the plane $2x - 3y + z = 11$.

16. Find the equation of the plane satisfying the given conditions.

   a. passing through $(5, 1, 2), (3, 1, -2)$, and $(3, 2, 5)$

   b. passing through $(-2, -1, 4)$ and perpendicular to the line $\dfrac{x}{2} = \dfrac{y + 1}{5} = \dfrac{3 - z}{2}$

Test the series in Problems 17–23 for convergence.

17. $\displaystyle\sum_{k=0}^{\infty} \dfrac{k^3}{k^4 + 2}$

18. $\displaystyle\sum_{k=1}^{\infty} \dfrac{1}{k \cdot 4^k}$

19. $\displaystyle\sum_{k=2}^{\infty} \dfrac{1}{k \ln k}$

20. $\displaystyle\sum_{k=0}^{\infty} \dfrac{3k^2 - 7k + 2}{(2k - 1)(k + 3)}$

21. $\displaystyle\sum_{k=1}^{\infty} \dfrac{k!}{2^k \cdot k}$

22. $\displaystyle\sum_{k=0}^{\infty} \dfrac{(-1)^{k+1} k}{k^2 + k - 1}$

23. $1 + \frac{1}{8} - \frac{1}{27} - \frac{1}{64} + \frac{1}{125} + \frac{1}{216} - - + + \cdots$

24. In each case, find the sum of the convergent series.

   a. $\displaystyle\sum_{k=1}^{\infty} \dfrac{2^{k-1}}{5^{k+3}}$

   b. $\displaystyle\sum_{k=1}^{\infty} \dfrac{1}{(3k - 1)(3k + 2)}$

25. In each case, determine whether the given improper integral converges, and if it does, find its value.

   a. $\displaystyle\int_{1}^{\infty} x^2 e^{-x}\, dx$

   b. $\displaystyle\int_{0}^{2} \dfrac{dx}{\sqrt{4 - x^2}}$

26. Find $\mathbf{v} \cdot \mathbf{w}$ and $\mathbf{v} \times \mathbf{w}$ for the vectors $\mathbf{v} = 3\mathbf{i} - 2\mathbf{j} + 5\mathbf{k}$ and $\mathbf{w} = \mathbf{i} - 3\mathbf{j} - \mathbf{k}$.

27. Find $\mathbf{F}'$ and $\mathbf{F}''$ for $\mathbf{F}(t) = 2t\mathbf{i} + e^{-3t}\mathbf{j} + t^4\mathbf{k}$.

28. Find $\displaystyle\int [e^t\mathbf{i} - \mathbf{j} - t\mathbf{k}] \cdot [e^{-t}\mathbf{i} + t\mathbf{j} - \mathbf{k}]\, dt$.

29. Find $\mathbf{T}$ and $\mathbf{N}$ for $\mathbf{R}(t) = 2(\sin 2t)\mathbf{i} + (2 + 2\cos 2t)\mathbf{j} + 6t\mathbf{k}$.

30. Show that the alternating series
$$S = \sum_{k=1}^{\infty} \dfrac{(-1)^{k+1}}{\sqrt{k}}$$
converges and determine what N must be to guarantee that the partial sum
$$S_N = \sum_{k=1}^{N} \dfrac{(-1)^{k+1}}{\sqrt{k}}$$
approximates the entire sum S with four-decimal-place accuracy.

31. a. Find the Maclaurin series representation for $f(x) = x^2 e^{-x^2}$.

   b. Use the representation in Part **a** to approximate
$$\int_{0}^{1} x^2 e^{-x^2}\, dx$$
with three-decimal-place accuracy.

32. Let $P(x) = 7 - 3(x - 4) + 5(x - 4)^2 - 2(x - 4)^3 + 6(x - 4)^4$ be the fourth-degree Taylor polynomial for the function f about 4.

   a. Find $f(4)$ and $f'''(4)$.

   b. Use the third-degree Taylor polynomial for f' about $x = 4$ to approximate $f'(4.2)$.

   c. Write a fifth-degree Taylor polynomial for
$$F(x) = \int_{4}^{x} f(t)\, dt \text{ about } 4.$$

   d. Can $f(5)$ be determined from the given information? Why or why not?

33. Solve the first-order linear differential equation
$$\dfrac{dy}{dx} + 2y = x^2$$
subject to the condition $y = 2$ when $x = 0$.

34. During the time period from $t = 0$ to $t = 6$ seconds, a particle moves along the path given by $x(t) = 3 \cos (\pi t)$, $y(t) = 5 \sin (\pi t)$.

   a. What is the position of the particle when $t = 1.5$?

   b. Graph the path of the particle from $t = 0$ to $t = 6$. Show direction.

   c. How many times does the particle pass through the point found in part **a**?

d. Find the velocity vector for the particle at any time t.

e. What is the distance traveled by the particle from $t = 1.25$ to $t = 1.75$?

35. Find the area of the region inside the circle $r = 4$ and to the right of the line $r = 2 \sec \theta$.

36. Find an equation in terms of x and y for the line tangent to the curve given by

$$x = t^2 - 2t - 1, \quad y = t^4 - 4t^2 + 2$$

at the point where $t = 1$.

37. A particle moves in space with position vector

$$\mathbf{R}(t) = (\sin t)\mathbf{i} - (\cos t)\mathbf{j} + \mathbf{k}$$

a. Find the velocity and acceleration vectors for the particle and find its speed.

b. Find the Cartesian equation for the particle's trajectory.

c. Find the curvature κ and the tangential and normal components of acceleration for the particle's motion.

38. MODELING PROBLEM   At what speed must a satellite travel to maintain a circular orbit 1,000 mi above the surface of the earth? Assume the earth is a sphere of radius 4,000 mi and that GM in Newton's law of universal gravitation is approximately $9.56 \times 10^4 \, \text{mi}^3/\text{s}^2$.

39. MODELING PROBLEM   A calculus text (other than the one you are now reading) is hurled downward from the top of a 120-ft-high building at an angle of 30° from the horizontal. Assume that the initial speed of the book is 8 ft/s. How far from the base of the building will the text land?

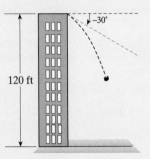

40. To close a sliding door, a person pulls on a rope with a constant force of 50 lb at a constant angle of 60°. Find the work done in moving the door 12 ft to its closed position.

CONTENTS

12

Partial Differentiation

PREVIEW

The goal of this chapter is to extend the methods of single-variable differential calculus to functions of several variables. The vector methods developed in Chapters 10 and 11 play an important role in our work, and indeed, we shall find that the closest analogue of the single-variable derivative in higher dimensions is the vector function called the *gradient*. In physics, the gradient is the rate at which a variable quantity, such as temperature or pressure, changes in value. In this chapter, we will define the *gradient of a function,* which is a vector whose components along the axes are related to the rate at which the function changes in the direction of the given component. We conclude this chapter by solving rate and optimization problems involving functions of several variables.

PERSPECTIVE

In many practical situations, the value of one quantity may depend on the values of two or more others. For example, the amount of water in a reservoir may depend on the amount of rainfall and on the amount of water consumed by local residents. The current in an electrical circuit varies with the electromotive force, the capacitance, the resistance, and the impedance in the circuit. The flow of blood from an artery into a small capillary depends on the diameter of the capillary and the pressure in both the artery and the capillary. The output of a factory may depend on the amount of capital invested in the plant and on the size of the labor force. We will analyze such situations using functions of several variables.

12.1 Functions of Several Variables

IN THIS SECTION   basic concepts; level curves and surfaces; open, closed, and bounded sets; graphs of functions of two variables

In the real world, physical quantities often depend on two or more variables (see the Perspective box at the beginning of this chapter). For example, we might be concerned with the temperature on a metal plate at various points at time t. Locations on the plate are designated as ordered pairs (x, y), so that the temperature T could be considered as a function of two location variables, x and y, as well as a time variable, t. Extending the notation for a function of a single variable, we might write this as $T(x, y, t)$.

BASIC CONCEPTS

We begin our investigation of functions of several variables by introducing notation and terminology and examining a few basic concepts.

> **Function of Two Variables**
>
> A **function of two variables** is a rule f that assigns to each ordered pair (x, y) in a set D a unique number $f(x, y)$. The set D is called the **domain** of the function, and the corresponding values of $f(x, y)$ constitute the **range** of f.

Functions of three or more variables can be defined in a similar fashion. For simplicity, we shall focus most of our attention on functions of two or three variables.

When dealing with a function of two variables f, we may write $z = f(x, y)$ and refer to x and y as the **independent variables** and to z as the **dependent variable.** Often, the functional "rule" will be given as a formula, and unless otherwise stated, *we shall assume that the domain is the largest set of points in the plane (or in space) for which the functional formula is defined and real valued.* These definitions and conventions are illustrated in Example 1 for a function of two variables.

EXAMPLE 1   *Evaluating a function of two variables and finding the domain and range*

Let $f(x, y) = \sqrt{1 - x + y}$.
a. Evaluate $f(2, 1), f(-4, 3)$, and $f(2t, t^2)$.
b. Describe the domain and range of f.

Solution
a. $f(2, 1) = \sqrt{1 - 2 + 1} = 0$

$f(-4, 3) = \sqrt{1 - (-4) + 3} = \sqrt{8} = 2\sqrt{2}$

$f(2t, t^2) = \sqrt{1 - (2t) + (t^2)} = \sqrt{(t - 1)^2} = |t - 1|$

b. The domain of f is the set of all ordered pairs (x, y) for which $\sqrt{1 - x + y}$ is defined. We must have $1 - x + y \geq 0$ or, equivalently, $y \geq x - 1$, in order for the square root to be defined. Thus, the domain of f is the shaded set shown in Figure 12.1.

Because $z = f(x, y) = \sqrt{1 - x + y}$, we see that z must be nonnegative, and the range of f is all $z \geq 0$.

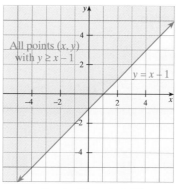

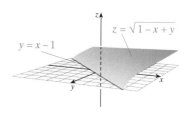

a. Graph of domain in $\mathbb{R}^2$.

b. Graph of $z = \sqrt{1 - x + y}$ in $\mathbb{R}^3$ over it's domain. Notice that the xy-plane is now a horizontal plane.

■ **FIGURE 12.1**   The domain of $f(x, y) = \sqrt{1 - x + y}$ ■

Functions of several variables can be combined just like functions of a single variable.

Operations with Functions of Two Variables

If $f(x, y)$ and $g(x, y)$ are functions of two variables with domain D, then

Sum $(f + g)(x, y) = f(x, y) + g(x, y)$

Difference $(f - g)(x, y) = f(x, y) - g(x, y)$

Product $(fg)(x, y) = f(x, y) \, g(x, y)$

Quotient $\left(\dfrac{f}{g}\right)(x, y) = \dfrac{f(x, y)}{g(x, y)}$ $g(x, y) \neq 0$

A **polynomial function in x and y** is a sum of functions of the form

$$Cx^m y^n$$

with nonnegative integers m and n, and C a constant; for instance,

$$3x^5 y^3 - 7x^2 y + 2x - 3y + 11$$

is a polynomial in x and y. A **rational function** is a quotient of two polynomial functions. Similar notation and terminology apply to functions of three or more variables.

LEVEL CURVES AND SURFACES

By analogy with the single-variable case, we define the **graph of the function $f(x, y)$** to be the collection of all 3-tuples (ordered triplets) (x, y, z) such that (x, y) is in the domain of f and $z = f(x, y)$. The graph of $f(x, y)$ is a surface in $\mathbb{R}^3$ whose projection onto the xy-plane is the domain D.

It is usually not easy to sketch the graph of a function of two variables. One way to proceed is illustrated in Figure 12.2.

Notice that when the plane $z = C$ intersects the surface $z = f(x, y)$, the result is the space curve with the equation $f(x, y) = C$. Such an intersection is called the **trace**

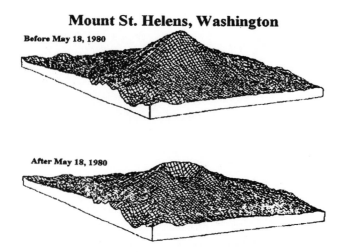

Mount St. Helens, Washington

Before May 18, 1980

After May 18, 1980

Mount St. Helens map courtesy of Bill Lennox, Humboldt State University

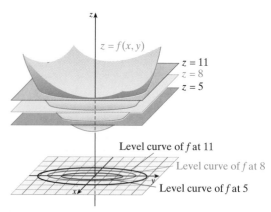

$z = f(x, y)$

$z = 11$
$z = 8$
$z = 5$

Level curve of f at 11
Level curve of f at 8
Level curve of f at 5

■ **FIGURE 12.2**   Graph of a function of two variables

Computer programs for three-dimensional sketching are common; see the *Technology Manual*. The Mt. St. Helens simulation (above) was done using a function of two variables.

of the graph of f in the plane $z = C$. The set of points (x, y) in the xy-plane that satisfy $f(x, y) = C$ is called the **level curve** of f at C, and an entire family of level curves is generated as C varies over the range of f. We can think of a level curve as a "slice" of the surface at a particular location. By sketching members of this family on the xy-plane, we obtain a useful topographical map of the surface $z = f(x, y)$. Because these level curves are used to show the shape of a surface (a mountain, for example), they are sometimes called **contour curves**.

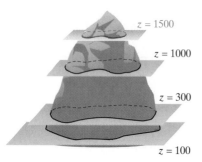

$z = 1500$
$z = 1000$
$z = 300$
$z = 100$

a. The surface $z = f(x, y)$ as a mountain

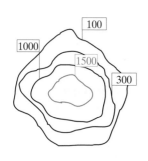

100
1000
1500
300

b. Level curves yield a topographic map of $z = f(x, y)$.

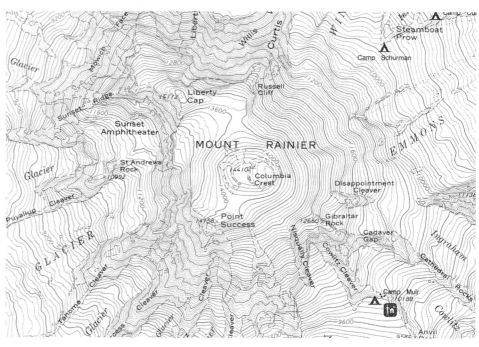

c. Topographic map of Mount Rainier

■ **FIGURE 12.3**   Level curves of a surface

For instance, imagine that the surface $z = f(x, y)$ is a "mountain," and that we wish to draw a two-dimensional "profile" of its shape. To draw such a profile, we indicate the paths of constant elevation by sketching the family of level curves in the plane and pinning a "flag" to each curve to show the elevation to which it corresponds, as shown in Figure 12.3b. Notice that regions in the map where paths are crowded together correspond to the steeper portions of the mountain. An actual topographical map of Mount Rainier is shown in Figure 12.3c.

You probably have seen level curves on the weather report in the newspaper or on the evening news, where level curves of equal temperature are called **isotherms** (see Figure 12.4). Other common uses of level curves show lines of equal pressure (called **isobars**) or those representing equal electric potential (called **equipotential lines**).

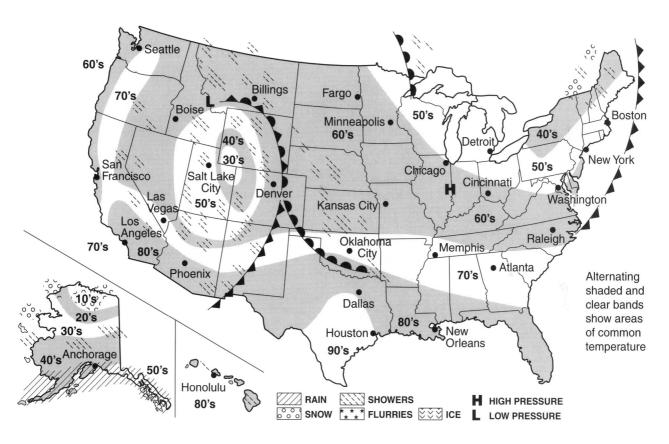

■ **FIGURE 12.4** Isotherms

| EXAMPLE 2 | *Level curves* |

Sketch some level curves of the function $f(x, y) = 10 - x^2 - y^2$.

Solution A computer graph of this curve is the surface shown in Figure 12.5a. To see the relationship between the surface and the level curves, we graph a simplified graph showing contour curves for $z = 1, 6,$ and 9, as shown in Figure 12.5b. Finally, the corresponding level curves are shown in Figure 12.5c.

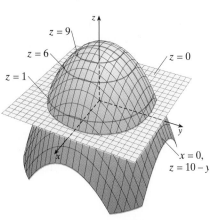

a. Computer generated graph of surface

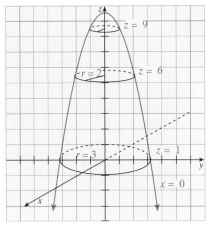

b. Simplified graph showing contour curves

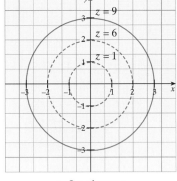

c. Level curves

■ **FIGURE 12.5**   Graph of $z = 10 - x^2 - y^2$ ■

OPEN, CLOSED, AND BOUNDED SETS

Most functions of a single variable have domains that can be described in terms of intervals. However, the domains of functions of several variables are often more complicated regions that require special terminology.

First, a point $P_0(x_0, y_0)$ is said to be an **interior point** of a set S in the plane if some open disk centered at P_0 is contained entirely within S. If S is the empty set, or if every point of S is an interior point, then S is called an **open set**. The point P_0 is called a **boundary point** of S (see Figure 12.6a) if every open disk centered at P_0 contains both points that belong to S and points that do not. The collection of all boundary points of S is called the **boundary** of S, and S is said to be **closed** if it contains its boundary; that is, if every boundary point of S is also a point of S (Figure 12.6b). If S has no boundary points, then it certainly contains its boundary, and is therefore closed. Finally, a set S in the plane is said to be **bounded** (Figure 12.6c) if it can be contained in a circle (or a rectangle).

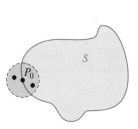

a. P_0 is a boundary point.

b. A closed set contains its boundary points.

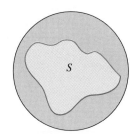

c. S is bounded.

■ **FIGURE 12.6**   Closed and bounded sets

In $\mathbb{R}^3$, the same definitions apply with the term "sphere" replacing "disk." Thus, an interior point P_0 of a set S in $\mathbb{R}^3$ has the property that some open sphere centered at P_0 is contained entirely within S, and S is open if all its points are interior point. Similar definitions hold for boundary points and closed sets. A bounded set in $\mathbb{R}^3$ is one that can be contained in a sphere (or a box).

GRAPHS OF FUNCTIONS OF TWO VARIABLES

The level curves of a function $f(x, y)$ provide information about the cross sections of the surface $z = f(x, y)$ perpendicular to the z-axis. However, a more complete picture of the surface can often be obtained by examining cross-sections in other directions as well. This procedure is used to graph a function in Example 3.

EXAMPLE 3   *Level curves*

Use the level curves of the function $f(x, y) = x^2 + y^2$ to sketch the graph of f.

Solution   The level curve $x^2 + y^2 = 0$ (that is, $C = 0$) is the point $(0, 0)$, and for $C > 0$, the level curve $x^2 + y^2 = C$ is the circle with center $(0, 0)$ and radius $\sqrt{C}$ (Figure 12.7a). There are no points (x, y) that satisfy $x^2 + y^2 = C$ for $C < 0$.

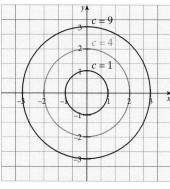

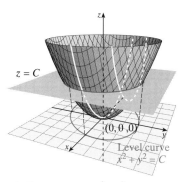

a. Level curves of $f(x, y) = x^2 + y^2$ are circles in the form $x^2 + y^2 = C$.

b. The surface $z = x^2 + y^2$. Cross sections perpendicular to the x- and y-axes shown in white are parabolas.

■ **FIGURE 12.7**   The graph of the function $f(x, y) = x^2 + y^2$

We can gain additional information about the appearance of the surface by examining cross sections perpendicular to the other two principal directions. Cross-sectional planes perpendicular to the x-axis have the form $x = A$ and intersect the surface $z = x^2 + y^2$ in parabolas of the form $z = A^2 + y^2$. Similarly, cross-sectional planes perpendicular to the y-axis have the form $y = B$ and intersect the surface in parabolas of the form $z = x^2 + B^2$.

To summarize, the surface $z = x^2 + y^2$ has cross sections that are circular in planes perpendicular to the z-axis and parabolic in the other two principal directions. For this reason, the surface is called a **circular paraboloid** or a **paraboloid of revolution**. The graph of the surface is shown in Figure 12.7b.   ■

The concept of level curve can be generalized to apply to functions of more than two variables. In particular, if f is a function of the n variables $x_1, x_2, \ldots, x_n$ and C is a number in the range of f, then the solution set of the equation $f(x_1, x_2, \ldots, x_n) = C$ is a region of n-space called the **level surface** of f at C. The level surfaces of a function of three variables are surfaces in $\mathbb{R}^3$ and can provide some insight into the nature of the function, but level surfaces of functions of four or more variables are very difficult to visualize.

DRAWING LESSON 7: SKETCHING SURFACES WITH LEVEL CURVES

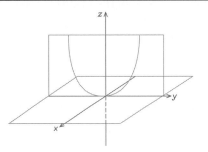

a. Draw the three coordinate axes and the xy-plane. Draw a trace of the surface in the yz-plane.

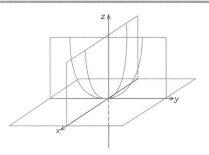

b. Draw a trace of the surface in the xz-plane.

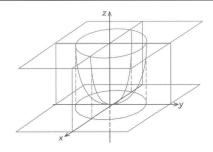

c. Draw a trace in the plane z = constant. Drop dashed segment from the intercepts of this trace down to the xy-plane. Then draw the level curve in the xy-plane. Finish outlining the surface.

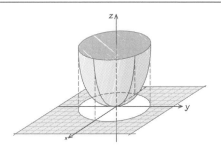

d. Lightly erase all planes except the xy-plne. Use highlighters or colored pencils to shade the surface.

EXAMPLE 4   *Isothermal surface*

Suppose a region of R is heated so that its temperature T at each point (x, y, z) is given by $T(x, y, z) = 100 - x^2 - y^2 - z^2$ degrees Celsius. Describe the isothermal surfaces for $T > 0$.

Solution   The isothermal surfaces are given by $T(x, y, z) = k$ for constant k; that is, $x^2 + y^2 + z^2 = 100 - k$. If $100 - k > 0$, the graph of $x^2 + y^2 + z^2 = 100 - k$ is a sphere of radius $\sqrt{100 - k}$ and center $(0, 0, 0)$. When $k = 100$, the graph is a single

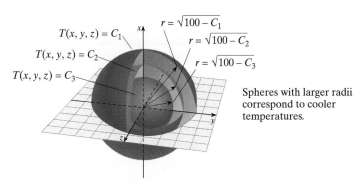

Spheres with larger radii correspond to cooler temperatures.

■ **FIGURE 12.8**   Isothermal surfaces for $T(x, y, z) = 100 - x^2 - y^2 - z^2$

point (the origin), and $T(0, 0, 0) = 100\ ^\circ\text{C}$. As the temperature drops, the constant k gets smaller, and the radius $\sqrt{100 - k}$ of the sphere gets larger. Hence, the isothermal surfaces are spheres, and the larger the radius, the cooler the surface. This situation is illustrated in Figure 12.8. ∎

As you can see by the examples in this section, you will need to recall the graphs of quadric surfaces. It will also help if you can recognize the surface by looking at its equation. What you will need to remember is summarized in Table 10.1, page 683.

Technology Window

Computer software is now available for graphing functions of two variables. Software packages such as *Mathematica, Maple,* and *Derive* will do very sophisticated graphs in three dimensions. Software usually allows a given graph to be visualized from different viewpoints by rotating a surface until a desired viewpoint is found.

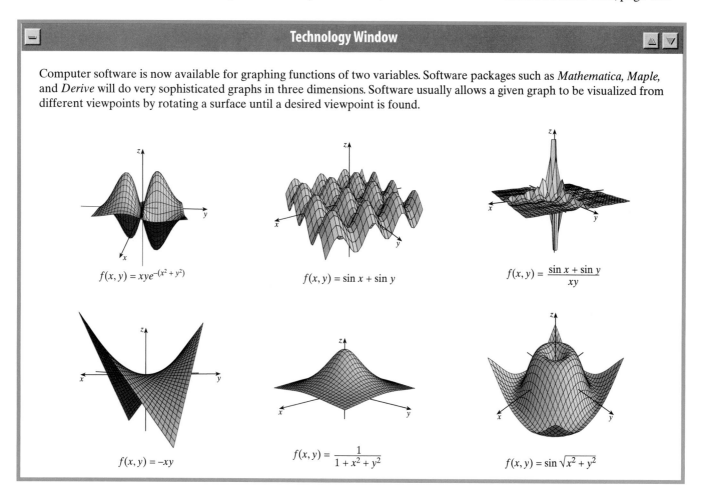

$f(x, y) = xye^{-(x^2 + y^2)}$

$f(x, y) = \sin x + \sin y$

$f(x, y) = \dfrac{\sin x + \sin y}{xy}$

$f(x, y) = -xy$

$f(x, y) = \dfrac{1}{1 + x^2 + y^2}$

$f(x, y) = \sin \sqrt{x^2 + y^2}$

12.1 Problem Set

A **1.** Let $f(x, y) = x^2y + xy^2$. If t is a real number, find:
  a. $f(0, 0)$
  b. $f(-1, 0)$
  c. $f(0, -1)$
  d. $f(1, 1)$
  e. $f(2, 4)$
  f. $f(t, t)$
  g. $f(t, t^2)$
  h. $f(1 - t, t)$

  a. $f(0, 1)$
  b. $f(5, 5)$
  c. $f(6, 1)$
  d. $f(1, 2)$
  e. $f(t, t)$
  f. $f(5t, t)$
  g. $f(t, 2t)$
  h. $f(1 + t, t)$

2. Let $f(x, y) = \left(1 - \dfrac{x}{y}\right)^2$. If t is a nonzero real number, find:

3. Let $f(x, y, z) = x^2ye^{2x} + (x + y - z)^2$. Find
  a. $f(0, 0, 0)$
  b. $f(1, -1, 1)$
  c. $f(-1, 1, -1)$
  d. $\dfrac{d}{dx} f(x, x, x)$
  e. $\dfrac{d}{dy} f(1, y, 1)$
  f. $\dfrac{d}{dz} f(1, 1, z^2)$

4. Let $f(x, y, z) = x \sin y + y \cos z$. Find
a. $f(0, 0, 0)$ b. $f(1, \frac{\pi}{2}, \pi)$ c. $f(1, \pi, \frac{\pi}{2})$
d. $\dfrac{d}{dx} f(x, x, x)$ e. $\dfrac{d}{dx} f(x, 2x, 3x)$ f. $\dfrac{d}{dy} f(y, y, 0)$

5. ■ **What Does This Say?** Discuss the notion of a function of two variables. Your discussion should include additional examples of a function of two variables.

6. ■ **What Does This Say?** How do you think that a function of three variables differs from a function of two variables?

7. ■ **What Does This Say?** Discuss level curves and amplify with some examples (different from those in this text).

Find the domain and range for each function given in Problems 8–17.

8. $f(x, y) = \sqrt{x - y}$ **9.** $f(x, y) = \dfrac{1}{\sqrt{x - y}}$

10. $f(u, v) = \sqrt{uv}$ **11.** $f(x, y) = \sqrt{\dfrac{y}{x}}$

12. $f(x, y) = \ln(y - x)$ **13.** $f(u, v) = \sqrt{u} \sin v$

14. $f(x, y) = \sqrt{(x + 3)^2 + (y - 1)^2}$

15. $f(x, y) = e^{(x+1)/(y-2)}$ **16.** $f(x, y) = \dfrac{1}{\sqrt{x^2 - y^2}}$

17. $f(x, y) = \dfrac{1}{\sqrt{9 - x^2 - y^2}}$

Sketch some level curves $f(x, y) = C$ for $C > 0$ of the functions given in Problems 18–23.

18. $f(x, y) = 2x - 3y$ **19.** $f(x, y) = x^2 - y^2$
20. $f(x, y) = x^3 - y$ **21.** $g(x, y) = x^2 - y$
22. $h(u, v) = u^2 + \dfrac{v^2}{4}$ **23.** $f(x, t) = \dfrac{x}{t}$

In Problems 24–29, sketch the level surface $f(x, y, z) = C$ for the given value of C.

24. $f(x, y, z) = y^2 + z^2$ for $C = 1$
25. $f(x, y, z) = x^2 + z^2$ for $C = 1$
26. $f(x, y, z) = x + y - z$ for $C = 1$
27. $f(x, y, z) = x + y - z$ for $C = 0$
28. $f(x, y, z) = (x + 1)^2 + (y - 2)^2 + (z - 3)^2$ for $C = 4$
29. $f(x, y, z) = 2x^2 + 2y^2 - z$ for $C = 1$

In Problems 30–37, describe the traces of the given quadric surface with each coordinate plane. You might wish to review Section 10.2 (page 676). Sketch the graph of the quadric and identify it.

30. $9x^2 + 4y^2 + z^2 = 1$ **31.** $\dfrac{x^2}{4} + y^2 + \dfrac{z^2}{9} = 1$

32. $\dfrac{x^2}{4} + \dfrac{y^2}{9} - z^2 = 1$ **33.** $\dfrac{x^2}{9} - y^2 - z^2 = 1$

34. $z = x^2 + \dfrac{y^2}{4}$ **35.** $z = \dfrac{x^2}{9} - \dfrac{y^2}{16}$

36. $x^2 + 2y^2 = 9z^2$ **37.** $z^2 = 1 + \dfrac{x^2}{9} + \dfrac{y^2}{4}$

Match each family of level curves given in Problems 38–43 with one of the surfaces labeled A–F.

38. $f(x, y) = x^2 - y^2$ **39.** $f(x, y) = e^{1-x^2+y^2}$

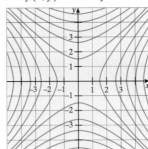
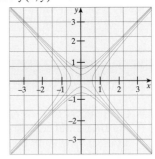

40. $f(x, y) = \dfrac{1}{x^2 + y^2}$ **41.** $f(x, y) = \sin\left(\dfrac{x^2 + y^2}{2}\right)$

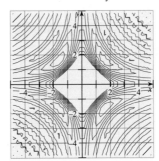
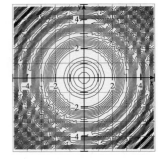

42. $f(x, y) = \dfrac{\cos xy}{x^2 + y^2}$ **43.** $f(x, y) = \sin\sqrt{x^2 + y^2}$

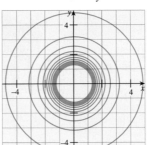
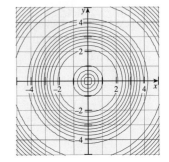

A.
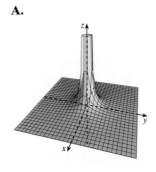

B.
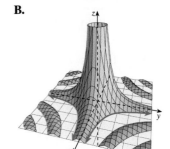

C.

D.

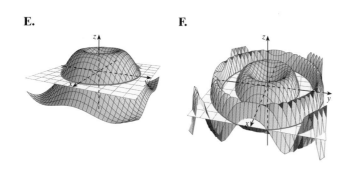

E.

F.

B *Sketch the graph of each function given in Problems* 44–55.

44. $f(x, y) = -4$

45. $f(x, y) = x$

46. $f(x, y) = y^2 + 1$

47. $f(x, y) = x^3 - 1$

48. $f(x, y) = 2x - 3y$

49. $f(x, y) = x^2 - y$

50. $f(x, y) = 2x^2 + y^2$

51. $f(x, y) = x^2 - y^2$

52. $f(x, y) = \dfrac{x}{y}$

53. $f(x, y) = \sqrt{x + y}$

54. $f(x, y) = x^2 + y^2 + 2$

55. $f(x, y) = \sqrt{1 - x^2 - y^2}$

56. The *lens equation* in optics states that

$$\frac{1}{d_0} + \frac{1}{d_i} = \frac{1}{L}$$

where d_0 is the distance of an object from a thin, spherical lens, d_i is the distance of its image on the other side of the lens, and L is the *focal length* of the lens. Describe the level curves for the constant focal length.

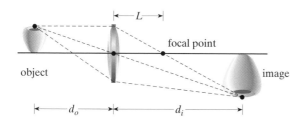

57. The EZGRO agricultural company estimates that when $100x$ worker-hours of labor are employed on y acres of land, the number of bushels of wheat produced is

$f(x, y) = Ax^a y^b$, where A, a, and b are nonnegative constants. Suppose the company decides to double the production factors x and y. Determine how this decision affects the production of wheat in each of these cases:

a. $a + b > 1$ b. $a + b < 1$ c. $a + b = 1$

58. Suppose that when x machines and y worker-hours are used each day, a certain factory will produce $Q(x, y) = 10xy$ mobile phones. Describe the relationship between the "inputs" x and y that result in an "output" of 1,000 phones each day. *Note:* You are finding a level curve of the production function Q.

C **59. MODELING PROBLEM**   At a certain factory, the daily output is modeled by $Q = CK^r L^{1-r}$ units, where K denotes capital investment, L is the size of the labor force, and C and r are constants, with $0 < r < 1$ (this is called a *Cobb–Douglas production function*). What happens to Q if K and L are both doubled? What if both are tripled?

60. Sketch the graph of $f(x, y) = xy$ in the first octant (x, y, and z are all positive).

61. The ideal gas law says that $PV = kT$, where P is the pressure of a confined gas, V is its volume, T is its temperature (Kelvin), and k is a constant. Express P as a function of V and T, and describe the level curves of this function.

62. A publishing house has found that in a certain city each of its salespeople will sell approximately

$$\frac{r^2}{2{,}000p} + \frac{s^2}{100} - s \text{ units}$$

per month, at a price of p dollars/unit, where s denotes the total number of salespeople employed and r is the amount of money spent each month on local advertising. Express the total revenue R as a function of p, r, and s.

63. MODELING PROBLEM   A manufacturer with exclusive rights to a sophisticated new industrial machine is planning to sell a limited number of the machines to both foreign and domestic firms. The price that the manufacturer can expect to receive for the machines will depend on the number of machines made available. It is estimated that if the manufacturer supplies x machines to the domestic market and y machines to the foreign market, the machines will sell for

$$60 - \frac{x}{5} + \frac{y}{20}$$

thousand dollars apiece at home and

$$50 - \frac{x}{10} + \frac{y}{20}$$

thousand dollars apiece abroad. Express the revenue R as a function of x and y. Describe the curves of constant revenue.

12.2 Limits and Continuity

IN THIS SECTION limit of a function of two variables, properties of limits, continuity

LIMIT OF A FUNCTION OF TWO VARIABLES

x	y	$f(x) = \dfrac{x}{x-y}$
0.99	2.01	−0.97059
0.999	2.001	−0.99701
1.001	2.0001	−1.00190
1.01	1.9999	−1.02031
0.99	1.99	−0.99000

As with single-variable functions, we need limits to discuss continuity, derivatives, slopes, and rates of change of functions of several variables. The single-variable limit can be extended naturally to functions of several variables. However, when we say that $f(x, y)$ approaches the number L as (x, y) approaches the point (a, b), written $(x, y) \to (a, b)$, we must remember that the approach to (a, b) can be from *any* direction, not just the left or right. The table in the margin suggests that no matter how $(x, y) \to (1, 2)$, the expression $f(x, y) = x/(x - y)$ approaches $L = -1$.

In Chapter 2, we informally defined the limit statement $\lim\limits_{x \to c} f(x) = L$ to mean that $f(x)$ can be made arbitrarily close to L by choosing x sufficiently close (but not equal) to c. The analogous informal definition for a function of two variables is now given.

Limit of a Function of Two Variables (Informal Definition)

The notation

$$\lim_{(x, y) \to (x_0, y_0)} f(x, y) = L$$

means that the functional values $f(x, y)$ can be made arbitrarily close to L by choosing a point (x, y) sufficiently close (but not equal) to the point (x_0, y_0).

What This Says: Formally, $\lim\limits_{(x, y) \to (x_0, y_0)} f(x, y) = L$ means that for each given number $\epsilon > 0$, there exists a number $\delta > 0$ so that $|f(x) - L| < \epsilon$ whenever (x, y) lies in a punctured disk $0 \leq \sqrt{(x - x_0)^2 + (y - y_0)^2} \leq \delta$ centered at $P_0(x_0, y_0)$. This definition is illustrated in Figure 12.9a.

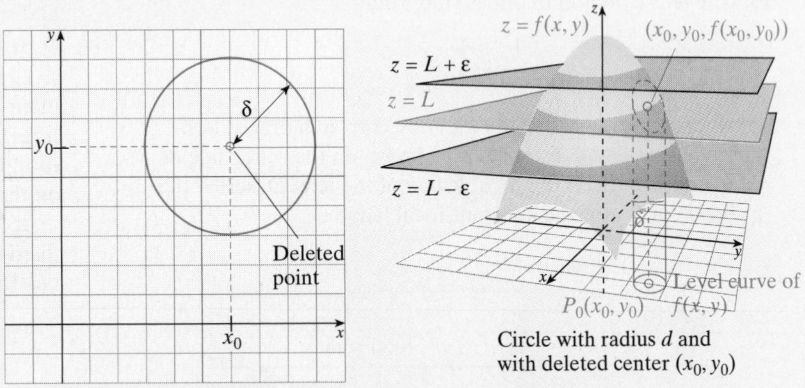

a. A punctured disk **b.** $\lim\limits_{(x, y) \to (x_0, y_0)} f(x, y) = L$

■ **FIGURE 12.9** Limit of a function of two variables

When considering the limit of a function of a single variable, we need to examine the approach of x to c from two directions (the left- and right-hand limits). However, for a function of two variables, we write $(x, y) \to (x_0, y_0)$ to mean that the point (x, y) is allowed to approach (x_0, y_0) along *any* curve that passes through (x_0, y_0). If

$$\lim_{(x, y) \to (x_0, y_0)} f(x, y)$$

is not the same for *all* approaches, or **paths,** then *the limit does not exist.*

> **EXAMPLE 1**   *Limit of a function of two variables*

a. Show

$$\lim_{(x, y) \to (0, 0)} \frac{2xy}{x^2 + y^2}$$

does not exist by evaluating this limit along the x-axis ($y = 0$), the y-axis ($x = 0$), and along the line $y = x$.

b. Evaluate

$$\lim_{(x, y) \to (1, 2)} 2xy$$

along the paths $y = 2$, $x = 1$, and $y = 2x$, and show that these values are all the same.

Solution

a. First note that the denominator is zero at $(0, 0)$ so $f(0, 0)$ is not defined. If we approach the origin along the x-axis (where $y = 0$) we find that

$$\frac{2xy}{x^2 + y^2} = \frac{2x(0)}{x^2 + 0^2} = 0 \text{ as } (x, y) \to (0, 0) \text{ along } y = 0 \text{ (and } x \neq 0)$$

We find a similar result if we approach the origin along the y-axis (where $x = 0$); see Figure 12.10.

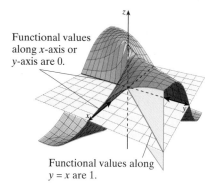

Functional values along x-axis or y-axis are 0.

Functional values along $y = x$ are 1.

■ **FIGURE 12.10**   Graph of $y = \dfrac{2xy}{x^2 + y^2}$ and limits as $(x, y) \to (0, 0)$

However, along the line $y = x$, the functional values are

$$f(x, y) = f(x, x) = \frac{2x^2}{x^2 + x^2} = 1 \quad \text{for } x \neq 0$$

so that $\dfrac{2xy}{x^2 + y^2} \to 1$ as $(x, y) \to (0, 0)$ along $y = x$

Because $f(x, y)$ tends toward different numbers as $(x, y) \to (0, 0)$ along different curves, it follows that f has no limit at the origin.

b. If we approach the point $(1, 2)$ along the line $y = 2$ (see Figure 12.11a) we have

$$2xy = 4x \rightarrow 4 \text{ as } (x, y) \rightarrow (1, 2) \text{ along } y = 2$$

Similarly, along the line $x = 1$ (Figure 12.11b)

$$2xy = 2y \rightarrow 4 \text{ as } (x, y) \rightarrow (1, 2) \text{ along } x = 1$$

Finally, along the line $y = 2x$ (Figure 12.11c)

$$2xy = 2x(2x) = 4x^2 \rightarrow 4 \text{ as } (x, y) \rightarrow (1, 2) \text{ along } y = 2x$$

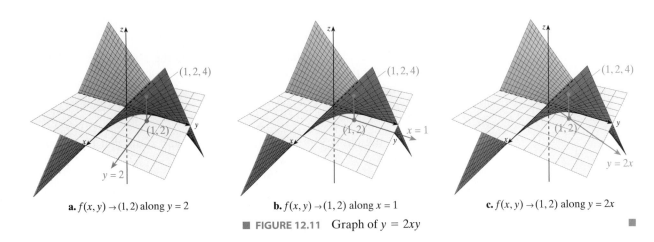

a. $f(x, y) \rightarrow (1, 2)$ along $y = 2$ **b.** $f(x, y) \rightarrow (1, 2)$ along $x = 1$ **c.** $f(x, y) \rightarrow (1, 2)$ along $y = 2x$

■ **FIGURE 12.11**   Graph of $y = 2xy$ ■

WARNING ▶ Note from Example 1a that the two-path procedure is enough to show that a *limit does NOT exist*. However, just showing that the limits along different paths in Example 1b are the same is not enough to conclude that a limit *DOES exist*, because we have not evaluated the limit along *every* path. ⚊

We now shall develop some rules that *will* allow us to conclude that a limit exists as well as help us to find the limit.

PROPERTIES OF LIMITS

The process of finding the limits shown in Example 1 is not very satisfactory. Fortunately, the various rules for manipulating limits of a function of one variable all have counterparts for limits of two variables.

Basic Formulas and Rules for Limits of a Function of Two Variables

Suppose $\lim\limits_{(x, y) \to (x_0, y_0)} f(x, y) = L$ and $\lim\limits_{(x, y) \to (x_0, y_0)} g(x, y) = M$. Then:

Scalar multiple rule $\lim\limits_{(x, y) \to (x_0, y_0)} [af](x, y) = aL$ for constant a

Sum rule $\lim\limits_{(x, y) \to (x_0, y_0)} [f + g](x, y) = L + M$

Product rule $\lim\limits_{(x, y) \to (x_0, y_0)} [fg](x, y) = LM$

Quotient rule $\lim\limits_{(x, y) \to (x_0, y_0)} \left[\dfrac{f}{g}\right](x, y) = \dfrac{L}{M}$ if $M \neq 0$

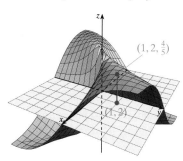

a. Graph of $z = x^2 + xy + y^2$

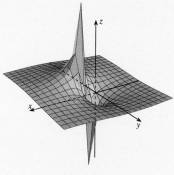

b. Graph of $f(x, y) = \dfrac{2xy}{x^2 + y^2}$

■ **FIGURE 12.12**   Graphs showing limits

| EXAMPLE 2 | *Finding limits using properties of limits* |

Assuming each limit exists, evaluate:

a. $\displaystyle\lim_{(x,\,y)\to(3,\,-4)} (x^2 + xy + y^2)$ b. $\displaystyle\lim_{(x,\,y)\to(1,\,2)} \frac{2xy}{x^2 + y^2}$

Solution

a.
$$\lim_{(x,\,y)\to(3,\,-4)} (x^2 + xy + y^2) = (3)^2 + (3)(-4) + (-4)^2 = 13$$

A graph is shown in Figure 12.12a.

b.
$$\lim_{(x,\,y)\to(1,\,2)} \frac{2xy}{x^2 + y^2} = \frac{\displaystyle\lim_{(x,\,y)\to(1,\,2)} 2xy}{\displaystyle\lim_{(x,\,y)\to(1,\,2)} (x^2 + y^2)}$$

$$= \frac{2(1)(2)}{1^2 + 2^2} = \frac{4}{5}$$

The graph is shown in Figure 12.12b.   ■

CONTINUITY

Recall that a function of a single variable x is continuous at $x = c$ if

1. $f(c)$ is defined;   **2.** $\displaystyle\lim_{x\to c} f(x)$ exists;   **3.** $\displaystyle\lim_{x\to c} f(x) = f(c)$.

Using the definition of the limit of a function of two variables, we can now define the continuity of a function of two variables analogously.

> ### Continuity of a Function of Two Variables
>
> The function $f(x, y)$ is **continuous** at the point (x_0, y_0) if and only if
>
> **1.** $f(x_0, y_0)$ is defined;
>
> **2.** $\displaystyle\lim_{(x,\,y)\to(x_0,\,y_0)} f(x, y)$ exists;
>
> **3.** $\displaystyle\lim_{(x,\,y)\to(x_0,\,y_0)} f(x, y) = f(x_0, y_0)$.
>
> Also, f is **continuous on a set S** in its domain if it is continuous at each point in S.

> ■ *What This Says:*   The function f is continuous at (x_0, y_0) if the functional value of $f(x, y)$ is close to $f(x_0, y_0)$ whenever (x, y) is sufficiently close to (x_0, y_0). Geometrically, this means that f is continuous if the surface $z = f(x, y)$ has no "gaps" or "holes."

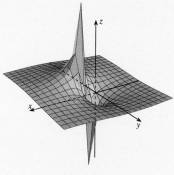

a. $f(x, y) = \dfrac{x - y}{x^2 + y^2}$
is discontinuous at $(0, 0)$; a hole

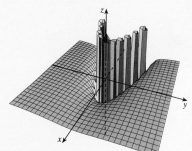

b. $f(x, y) = \dfrac{1}{y - x^2}$
is discontinuous on the parabola $y - x^2$; a gap

The basic properties of limits can be used to show that if f and g are both continuous on the set S, then so are the sum $f + g$, the multiple af, the product fg, the quotient f/g (whenever $g \neq 0$), and the root $\sqrt[n]{f}$ wherever it is defined. Also, if F is a function of two variables continuous at (x_0, y_0) and G is a function of one variable that is continuous at $F(x_0, y_0)$, it can be shown that the composite function $G \circ F$ is continuous at (x_0, y_0).

Many common functions of two variables are continuous wherever they are defined. For instance, a polynomial in two variables, such as $x^3y^2 + 3xy^3 - 7x + 2$, is continuous throughout the plane, and a rational function in two variables is continuous wherever the denominator polynomial is not zero. In this course, all of the "standard" functions of two or more variables are continuous over their domains.

> **EXAMPLE 3**   *Testing for continuity*

Test the continuity of the functions whose graphs are shown here, namely,

a. $f(x, y) = \dfrac{x - y}{x^2 + y^2}$ b. $f(x, y) = \dfrac{1}{y - x^2}$

Solution
a. Because $x - y$ and $x^2 + y^2$ are both polynomial functions in x and y, f is a rational function and the only place where it might not be continuous is where the denominator is zero. Because $x^2 + y^2 = 0$ only where both x and y are 0, the only possible point of discontinuity is at $(0, 0)$. However $f(0, 0)$ is not defined, so f is discontinuous at $(0, 0)$.
b. Once again we need to check when the denominator is 0:

$$y - x^2 = 0 \quad \text{when} \quad y = x^2$$

We can, therefore, conclude that the function is continuous at all points except those lying on the parabola $y = x^2$.   ∎

We conclude with an example in which the definition of limit is used to determine continuity.

> **EXAMPLE 4**   *Continuity using the limit definition*

Show that f is continuous at $(0, 0)$, where

$$f(x, y) = \begin{cases} y \sin \dfrac{1}{x}, & x \neq 0 \\ 0 & x = 0 \end{cases}$$

Solution   The graph is shown in Figure 12.13. It appears to be continuous. To prove continuity at $(0, 0)$, we must show that for any $\epsilon > 0$, there exists a $\delta > 0$ such that

$$\left| f(x, y) - f(0, 0) \right| = \left| y \sin \frac{1}{x} \right| < \epsilon \quad \text{whenever} \quad 0 < x^2 + y^2 < \delta^2$$

(We use $x^2 + y^2$ and δ^2 here instead of $\sqrt{x^2 + y^2}$ and δ for convenience.) Note that $\left| y \sin \dfrac{1}{x} \right| \leq |y|$ for all $x \neq 0$, because $\left| \sin \dfrac{1}{x} \right| \leq 1$ for $x \neq 0$. If (x, y) lies in the disk $x^2 + y^2 < \delta^2$, then the points $(0, y)$ that satisfy $y^2 < \delta^2$ lie in the same disk (let $x = 0$ in $x^2 + y^2 < \delta^2$). In other words, points satisfying $|y| < \delta$ lie in the disk, and if we let $\delta = \epsilon$ it follows that

$$\left| f(x, y) - f(0, 0) \right| \leq |y| < \delta = \epsilon \quad \text{whenever} \quad |y| < \delta$$ ∎

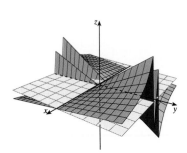

Front view

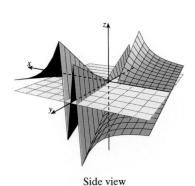

Side view

■ **FIGURE 12.13** Graph of f in Example 4

12.2 Problem Set

A **1.** ■ **What Does This Say?** Describe the notion of a limit of function of two variables.

2. ■ **What Does This Say?** Describe the basic formulas and rules for limits of a function of two variables.

In Problems 3–20, find the given limit, assuming it exists.

3. $\lim\limits_{(x, y)\to(-1, 0)} (xy^2 + x^3y + 5)$

4. $\lim\limits_{(x, y)\to(0, 0)} (5x^2 - 2xy + y^2 + 3)$

5. $\lim\limits_{(x, y)\to(1, 3)} \dfrac{x + y}{x - y}$

6. $\lim\limits_{(x, y)\to(3, 4)} \dfrac{x - y}{\sqrt{x^2 + y^2}}$

7. $\lim\limits_{(x, y)\to(1, 0)} e^{xy}$

8. $\lim\limits_{(x, y)\to(1, 0)} (x + y)e^{xy}$

9. $\lim\limits_{(x, y)\to(0, 1)} [e^{x^2+x} \ln(ey^2)]$

10. $\lim\limits_{(x, y)\to(e, 0)} \ln(x^2 + y^2)$

11. $\lim\limits_{(x, y)\to(0, 0)} \dfrac{x^2 - 2xy + y^2}{x - y}$

12. $\lim\limits_{(x, y)\to(1, 2)} \dfrac{(x^2 - 1)(y^2 - 4)}{(x - 1)(y - 2)}$

13. $\lim\limits_{(x, y)\to(0, 0)} \dfrac{e^x\tan^{-1}y}{y}$

14. $\lim\limits_{(x, y)\to(0, 0)} \dfrac{x^2y^2}{x^2 + y^2}$

15. $\lim\limits_{(x, y)\to(0, 0)} \dfrac{\sin(x + y)}{x + y}$

16. $\lim\limits_{(x, y)\to(0, 0)} (\sin x + \cos y)$

17. $\lim\limits_{(x, y)\to(5, 5)} \dfrac{x^4 - y^4}{x^2 - y^2}$

18. $\lim\limits_{(x, y)\to(a, a)} \dfrac{x^4 - y^4}{x^2 - y^2};$ *a* is a constant

19. $\lim\limits_{(x, y)\to(2, 1)} \dfrac{x^2 - 4y^2}{x - 2y}$

20. $\lim\limits_{(x, y)\to(0, 0)} \dfrac{x^2 + y^2}{\sqrt{x^2 + y^2 + 4} - 2}$

In Problems 21–29, evaluate the indicated limit, if it exists.

21. $\lim\limits_{(x, y)\to(2, 1)} (xy^2 + x^3y)$

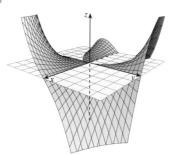

22. $\lim\limits_{(x, y)\to(1, 2)} (5x^2 - 2xy + y^2)$

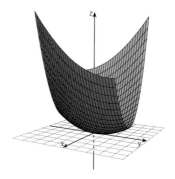

23. $\lim\limits_{(x, y)\to(0, 0)} \dfrac{x + y}{x - y}$

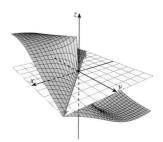

24. $\lim\limits_{(x, y)\to(0, 0)} \dfrac{x - y}{\sqrt{x^2 + y^2}}$

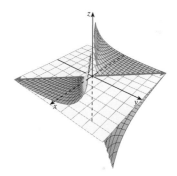

25. $\lim\limits_{(x, y)\to(0, 0)} e^{xy}$

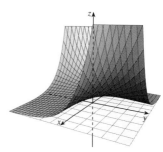

26. $\displaystyle\lim_{(x, y)\to(1, 1)} (x + y)e^{xy}$

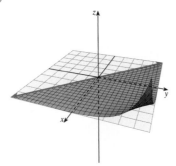

27. $\displaystyle\lim_{(x, y)\to(0, 0)} (\sin x - \cos y)$

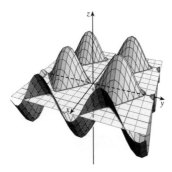

28. $\displaystyle\lim_{(x, y)\to(0, 0)} \left[1 - \frac{\sin(x^2 + y^2)}{x^2 + y^2}\right]$

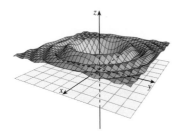

29. $\displaystyle\lim_{(x, y)\to(0, 0)} \frac{1 - \cos(x^2 + y^2)}{x^2 + y^2}$

Ⓑ *In Problems* 30–33, *show that* $\displaystyle\lim_{(x, y)\to(0, 0)} f(x, y)$ *does not exist.*

30. $f(x, y) = \dfrac{x^4y^4}{(x^2 + y^4)^3}$

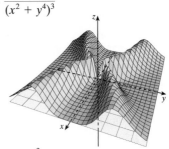

31. $f(x, y) = \dfrac{x - y^2}{x^2 + y^2}$

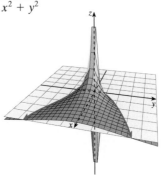

32. $f(x, y) = \dfrac{x^2 + y}{x^2 + y^2}$

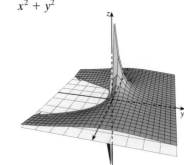

33. $f(x, y) = \dfrac{x^2y^2}{x^4 + y^4}$

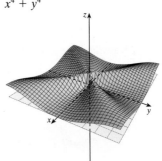

34. Let f be the functions defined by

$$f(x, y) = \begin{cases} \dfrac{xy^2}{x^2 + y^4} & \text{for } (x, y) \neq (0, 0) \\ 0 & \text{for } (x, y) = (0, 0) \end{cases}$$

Is f continuous at $(0, 0)$? Explain.

35. Let f be the functions defined by

$$f(x, y) = \begin{cases} \dfrac{xy^3}{x^2 + y^6} & \text{for } (x, y) \neq (0, 0) \\ 0 & \text{for } (x, y) = (0, 0) \end{cases}$$

Is f continuous at $(0, 0)$? Explain.

36. Let f be the function defined by

$$f(x, y) = \frac{x^2 - y^2}{x^2 + y^2} \quad \text{for } (x, y) \neq (0, 0)$$

a. Find $\displaystyle\lim_{(x, y)\to(2, 1)} f(x, y)$.

b. Prove that f has no limit at $(0, 0)$ by showing that $f(x, y)$ tends toward different numbers as $(x, y) \to (0, 0)$ along each coordinate axis.

37. Let f be the function defined by

$$f(x, y) = \frac{x^2 + 2y^2}{x^2 + y^2} \quad \text{for } (x, y) \neq (0, 0)$$

a. Find $\displaystyle\lim_{(x, y)\to(3, 1)} f(x, y)$.

b. Prove that f has no limit at $(0, 0)$ by showing that $f(x, y)$ tends toward different numbers as $(x, y) \to (0, 0)$ along each coordinate axis.

38. Given that the function

$$f(x, y) = \begin{cases} \dfrac{x^3 + y^3}{x^2 + y^2} & \text{for } (x, y) \neq (0, 0) \\ A & \text{for } (x, y) = (0, 0) \end{cases}$$

is continuous at the origin, what is A?

39. Given that the function

$$f(x, y) = \begin{cases} \dfrac{3x^3 - 3y^3}{x^2 - y^2} & \text{for } x^2 \neq y^2 \\ B & \text{otherwise} \end{cases}$$

is continuous at the origin, what is B?

40. Let

$$f(x, y) = \begin{cases} \dfrac{x^2 y^2}{x^2 + y^2} & \text{for } (x, y) \neq (0, 0) \\ 0 & \text{for } (x, y) = (0, 0) \end{cases}$$

Given that $f(x, y)$ has a limiting value at $(0, 0)$, is it continuous there?

41. Assuming that the limit exists, show that

$$\lim_{(x, y, z)\to(0, 0, 0)} \frac{xyz}{x^2 + y^2 + z^2} = 0$$

C *Use the $\delta-\epsilon$ definition of limit to verify the limit statements given in Problems 42–45.*

42. $\displaystyle\lim_{(x, y)\to(0, 0)} (2x^2 + 3y^2) = 0$ **43.** $\displaystyle\lim_{(x, y)\to(0, 0)} (x + y^2) = 0$

44. $\displaystyle\lim_{(x, y)\to(0, 0)} \frac{x^2 - y^2}{x + y} = 0$ **45.** $\displaystyle\lim_{(x, y)\to(1, -1)} \frac{x^2 - y^2}{x + y} = 2$

46. Prove that if f is continuous and $f(a, b) > 0$, then there exists a δ-neighborhood about (a, b) such that $f(x, y) > 0$ for every point (x, y) in the neighborhood.

47. Prove the scalar multiple rule:

$$\lim_{(x, y)\to(x_0, y_0)} [af](x, y) = a \lim_{(x, y)\to(x_0, y_0)} f(x, y)$$

48. Prove the sum rule:

$$\lim_{(x, y)\to(x_0, y_0)} [f + g](x, y) = L + M$$

where $L = \left[\displaystyle\lim_{(x, y)\to(x_0, y_0)} f(x, y) \right]$

and $M = \left[\displaystyle\lim_{(x, y)\to(x_0, y_0)} g(x, y) \right]$.

49. A function of two variables $f(x, y)$ may be continuous in each separate variable at $x = x_0$ and $y = y_0$ without being itself continuous at (x_0, y_0). Let $f(x, y)$ be defined by

$$f(x, y) = \begin{cases} \dfrac{xy}{x^2 + y^2} & \text{for } (x, y) \neq (0, 0) \\ 0 & \text{at } (0, 0) \end{cases}$$

Let $g(x) = f(x, 0)$ and $h(y) = f(0, y)$. Show that both $g(x)$ and $h(y)$ are continuous at 0, but that $f(x, y)$ is not continuous at $(0, 0)$.

12.3 Partial Derivatives

IN THIS SECTION   **partial differentiation, partial derivative as a slope, partial derivative as a rate, higher partial derivatives**

PARTIAL DIFFERENTIATION

The process of differentiating a function of several variables with respect to one of its variables while keeping the other variable(s) fixed is called **partial differentiation,** and the resulting derivative is a **partial derivative** of the function.

Recall that the derivative of a function of a single variable f is defined to be the limit of a difference quotient, namely,

$$f'(x) = \lim_{\Delta x \to 0} \frac{f(x + \Delta x) - f(x)}{\Delta x}$$

Partial derivatives with respect to x or y are defined similarly.

Partial Derivatives of a Function of Two Variables

If $z = f(x, y)$, then the **(first) partial derivatives** of f with respect to x and y are the functions f_x and f_y, respectively, defined by

$$f_x(x, y) = \lim_{\Delta x \to 0} \frac{f(x + \Delta x, y) - f(x, y)}{\Delta x}$$

$$f_y(x, y) = \lim_{\Delta y \to 0} \frac{f(x, y + \Delta y) - f(x, y)}{\Delta y}$$

provided the limits exist.

■ **What This Says:** For the partial differentiation of a function of two variables, $z = f(x, y)$, we find the partial derivative with respect to x by regarding y as constant while differentiating the function with respect to x. Similarly, the partial derivative with respect to y is found by regarding x as constant while differentiating with respect to y.

EXAMPLE 1   *Partial derivatives*

If $f(x, y) = x^3y + x^2y^2$, find:
a. f_x b. f_y

Solution

a. For f_x, hold y constant and find the derivative with respect to x:

$$f_x(x, y) = 3x^2y + 2xy^2$$

b. For f_y, hold x constant and find the derivative with respect to y:

$$f_y(x, y) = x^3 + 2x^2y$$ ■

Technology Window

Finding partial derivatives using technology is a natural extension of the way you have been finding other derivatives. The general format for most calculators and computer programs is the same: *derivative operator, function, variable of differentiation.* Look at one example that compares the partial derivatives with respect to x and y from Example 1.

a.

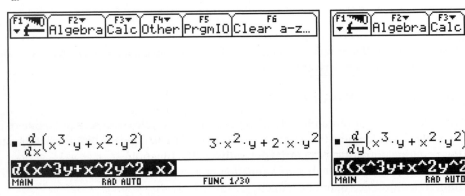

b.

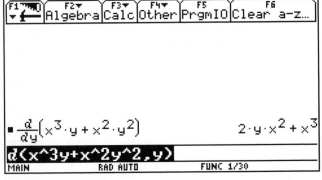

Several different symbols are used to denote partial derivatives, as indicated in the following box.

Alternative Notation for First Partial Derivatives

For $z = f(x, y)$, the partial derivatives f_x and f_y are denoted by

$$f_x(x, y) = \frac{\partial f}{\partial x} = \frac{\partial z}{\partial x} = \frac{\partial}{\partial x} f(x, y) = z_x = D_x(f)$$

and

$$f_y(x, y) = \frac{\partial f}{\partial y} = \frac{\partial z}{\partial y} = \frac{\partial}{\partial y} f(x, y) = z_y = D_y(f)$$

The values of the partial derivatives of $f(x, y)$ at the point (a, b) are denoted by

$$\left.\frac{\partial f}{\partial x}\right|_{(a, b)} = f_x(a, b) \quad \text{and} \quad \left.\frac{\partial f}{\partial y}\right|_{(a, b)} = f_y(a, b)$$

EXAMPLE 2   *Finding and evaluating a partial derivative*

Let $z = x^2 \sin(3x + y^3)$.

a. Evaluate $\dfrac{\partial z}{\partial x}$ at $\left(\frac{\pi}{3}, 0\right)$.   b. Evaluate z_y at $(1, 1)$.

Solution

a. $\dfrac{\partial z}{\partial x} = 2x \sin(3x + y^3) + x^2 \cos(3x + y^3) \cdot (3) = 2x \sin(3x + y^3) + 3x^2\cos(3x + y^3)$

Thus,

$$\left.\frac{\partial z}{\partial x}\right|_{(\pi/3, 0)} = 2\left(\frac{\pi}{3}\right) \sin \pi + 3\left(\frac{\pi}{3}\right)^2 \cos \pi = \frac{2\pi}{3}(0) + \frac{\pi^2}{3}(-1) = -\frac{\pi^2}{3}$$

b. $z_y = x^2\cos(3x + y^3) \cdot (3y^2) = 3x^2y^2\cos(3x + y^3)$, so that

$$z_y(1, 1) = 3(1)^2(1)^2\cos(3 + 1) = 3 \cos 4 \qquad \blacksquare$$

Technology Window

To evaluate a partial derivative at a point using technology, you must evaluate the partial derivative first for the active variable (with respect to the differentiation) using the *evaluate* feature of most derivative routines.

EXAMPLE 3 *Partial derivative of a function of three variables*

Let $f(x, y, z) = x^2 + 2xy^2 + yz^3$; find:

a. f_x b. f_y c. f_z

Solution

a. For f_x, think of f as a function of x alone with y and z treated as constants:

$$f_x(x, y, z) = 2x + 2y^2$$

b. $f_y(x, y, z) = 4xy + z^3$ c. $f_z(x, y, z) = 3yz^2$ ∎

EXAMPLE 4 *Partial derivative of an implicitly defined function*

Let z be defined implicitly as a function of x and y by the equation

$$x^2z + yz^3 = x$$

Find $\partial z/\partial x$ and $\partial z/\partial y$.

Solution Differentiate implicitly with respect to x, treating y as a constant:

$$2xz + x^2\frac{\partial z}{\partial x} + 3z^2y\frac{\partial z}{\partial x} = 1$$

so that

$$\frac{\partial z}{\partial x} = \frac{1 - 2xz}{x^2 + 3z^2y}$$

Similarly, holding x constant and differentiating implicitly with respect to y, we find

$$x^2\frac{\partial z}{\partial y} + z^3 + 3z^2y\frac{\partial z}{\partial y} = 0$$

so that

$$\frac{\partial z}{\partial y} = \frac{-z^3}{x^2 + 3z^2y}$$ ∎

The following example uses implicit differentiation to find the slope of the tangent line to a level curve at a particular point.

EXAMPLE 5 *Slope of a level curve*

A certain level curve of the surface $z = x^2 + 3xy + y^2$ contains the point $P(1, 1)$. Find the slope of the tangent line to this curve at P.

Solution The required level curve has the form $F(x, y) = C$, where

$$F(x, y) = x^2 + 3xy + y^2$$

We find the slope, $\dfrac{dy}{dx}$, implicitly:

$$x^2 + 3xy + y^2 = C$$

$$\frac{d}{dx}(x^2 + 3xy + y^2) = \frac{d}{dx}(C)$$

$$2x + 3y + 3x\frac{dy}{dx} + 2y\frac{dy}{dx} = 0$$

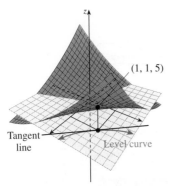

The slope of the line tangent to a level curve of a surface

$$(3x + 2y)\frac{dy}{dx} = -2x - 3y$$

$$\frac{dy}{dx} = \frac{-(2x + 3y)}{3x + 2y}$$

This provides the slope at each point on the level curve where $3x + 2y \neq 0$. In particular, at the point where $x = 1$ and $y = 1$, we have

$$\frac{dy}{dx}\bigg|_{(1,1)} = \frac{-[2(1) + 3(1)]}{3(1) + 2(1)} = \frac{-5}{5} = -1$$

so that the required line has slope -1.   ∎

PARTIAL DERIVATIVE AS A SLOPE

A useful geometric interpretation of partial derivatives is indicated in Figure 12.14. In Figure 12.14a, the plane $y = y_0$ intersects the surface $z = f(x, y)$ in a curve C parallel to the xz-plane. That is, C is the trace of the surface in the plane $y = y_0$. An equation for this curve is $z = f(x, y_0)$, and because y_0 is fixed, the function depends only on x. Thus, we can compute the slope of the tangent line to C at the point $P(x_0, y_0, z_0)$ in the plane $y = y_0$ by differentiating $f(x, y_0)$ with respect to x and evaluating the derivative at $x = x_0$. That is, the slope is $f_x(x_0, y_0)$, the value of the partial derivative f_x at (x_0, y_0). There is a similar interpretation for $f_y(x_0, y_0)$, as shown in Figure 12.14b.

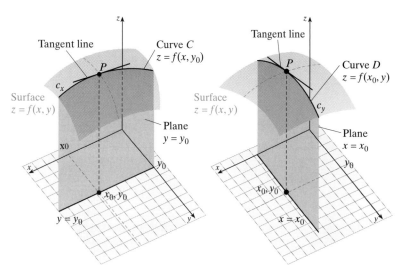

a. The tangent line in the plane $y = y_0$ to the curve C at the point P has slope $f_x(x_0, y_0)$.

b. The tangent line in the plane $x = x_0$ to the curve D at the point P has slope $f_y(x_0, y_0)$.

■ **FIGURE 12.14**   Slope interpretation of the partial derivative

---

Partial Derivative as the Slope of a Tangent Line

The line parallel to the xz-plane and tangent to the surface $z = f(x, y)$ at the point $P_0(x_0, y_0, z_0)$ has slope $f_x(x_0, y_0)$. Likewise, the tangent line to the surface at P_0 that is parallel to the yz-plane has slope $f_y(x_0, y_0)$.

| EXAMPLE 6 | *Slope of a line parallel to the xz-plane* |

Find the slope of the line that is parallel to the xz-plane and tangent to the surface $z = x\sqrt{x + y}$ at the point $P(1, 3, 2)$.

Solution   If $f(x, y) = x\sqrt{x + y} = x(x + y)^{1/2}$, then the required slope is $f_x(1, 3)$.

$$f_x(x, y) = x\left(\frac{1}{2}\right)(x + y)^{-1/2}(1 + 0) + (1)(x + y)^{1/2} = \frac{x}{2\sqrt{x + y}} + \sqrt{x + y}$$

Thus, $f_x(1, 3) = \dfrac{1}{2\sqrt{1 + 3}} + \sqrt{1 + 3} = \dfrac{9}{4}$.   ■

PARTIAL DERIVATIVE AS A RATE

The derivative of a function of one variable can be interpreted as a rate of change, and the analogous interpretation of partial derivative may be described as follows.

> **Partial Derivatives as Rates of Change**
>
> As the point (x, y) moves from the fixed point $P_0(x_0, y_0)$, the function $f(x, y)$ changes at a rate given by $f_x(x_0, y_0)$ in the direction of the positive x-axis and by $f_y(x_0, y_0)$ in the direction of the positive y-axis.

| EXAMPLE 7 | *Partial derivatives as a rate of change* |

In an electrical circuit with electromotive force (EMF) of E volts and resistance R ohms, the current is $I = E/R$ amperes. Find the partial derivatives $\partial I/\partial E$ and $\partial I/\partial R$ at the instant when $E = 120$ and $R = 15$, and interpret these derivatives as rates.

Solution   Since $I = ER^{-1}$, we have

$$\frac{\partial I}{\partial E} = R^{-1} \quad \text{and} \quad \frac{\partial I}{\partial R} = -ER^{-2}$$

and thus, when $E = 120$ and $R = 15$, we find that

$$\frac{\partial I}{\partial E} = 15^{-1} \approx 0.0667 \quad \text{and} \quad \frac{\partial I}{\partial R} = -(120)(15)^{-2} \approx -0.5333$$

This means that if the resistance is fixed at 15 ohms, the current is increasing (because the slope is positive) with respect to voltage at the rate of 0.0667 ampere per volt when the EMF is 120 volts. Likewise, with the same fixed EMF, the current is decreasing (because the slope is negative) with respect to resistance at the rate of 0.5333 ampere per ohm when the resistance is 15 ohms.   ■

HIGHER PARTIAL DERIVATIVES

The partial derivative of a function is a function, so it is possible to take the partial derivative of a partial derivative. This is very much like taking the second derivative of a function of one variable if we take two consecutive partial derivatives with respect to the same variable, and the resulting derivative is called the **second partial derivative** with respect to that variable. However, we can also take the partial derivative with respect to one variable and then take another partial derivative with respect to a different variable, producing what is called a **mixed partial derivative.** The higher partial derivatives for a function of two variables $f(x, y)$ are denoted as indictaed in the following box.

Higher Partial Derivatives

Given $z = f(x, y)$
Second partial derivatives

$$\frac{\partial^2 f}{\partial x^2} = \frac{\partial}{\partial x}\left(\frac{\partial f}{\partial x}\right) = (f_x)_x = f_{xx}$$

$$\frac{\partial^2 f}{\partial y^2} = \frac{\partial}{\partial y}\left(\frac{\partial f}{\partial y}\right) = (f_y)_y = f_{yy}$$

Mixed partial derivatives

$$\frac{\partial^2 f}{\partial x \partial y} = \frac{\partial}{\partial x}\left(\frac{\partial f}{\partial y}\right) = (f_y)_x = f_{yx}$$

$$\frac{\partial^2 f}{\partial y \partial x} = \frac{\partial}{\partial y}\left(\frac{\partial f}{\partial x}\right) = (f_x)_y = f_{xy}$$

EXAMPLE 8   *Higher partial derivatives of a function of two variables*

For $z = f(x, y) = 5x^2 - 2xy + 3y^3$, find the requested higher partial derivatives.

a. $\dfrac{\partial^2 z}{\partial x \partial y}$ b. $\dfrac{\partial^2 f}{\partial y \partial x}$ c. $\dfrac{\partial^2 z}{\partial x^2}$ d. $f_{xy}(3, 2)$

Solution   Part of what this example is illustrating is the variety of notation that can be used for higher partial derivatives.

a. First differentiate with respect to y; then differentiate with respect to x:

$$\frac{\partial z}{\partial y} = -2x + 9y^2$$

$$\frac{\partial^2 z}{\partial x \partial y} = \frac{\partial}{\partial x}\left(\frac{\partial z}{\partial y}\right) = \frac{\partial}{\partial x}(-2x + 9y^2) = -2$$

b. Differentiate first with respect to x and then with respect to y:

$$\frac{\partial f}{\partial x} = 10x - 2y$$

$$\frac{\partial^2 f}{\partial y \partial x} = \frac{\partial}{\partial y}\left(\frac{\partial f}{\partial x}\right) = \frac{\partial}{\partial y}(10x - 2y) = -2$$

c. Differentiate with respect to x twice:

$$\frac{\partial^2 z}{\partial x^2} = \frac{\partial}{\partial x}\left(\frac{\partial z}{\partial x}\right) = \frac{\partial}{\partial x}(10x - 2y) = 10$$

d. Evaluate the mixed partial found in part **b** at the point $(3, 2)$:

$$f_{xy}(3, 2) = -2$$ ■

Notice from parts **a** and **b** of Example 8 that $\dfrac{\partial^2 z}{\partial x \partial y} = \dfrac{\partial^2 z}{\partial y \partial x}$. This equality of mixed partials does not hold for all functions, but for most functions we shall encounter, it will be true. The following theorem provides sufficient conditions for this equality to occur.

THEOREM 12.1   *Equality of mixed partials*

If the function $f(x, y)$ has mixed partial derivatives f_{xy} and f_{yx} that are continuous in an open disk containing (x_0, y_0), then

$$f_{yx}(x_0, y_0) = f_{xy}(x_0, y_0)$$

Proof   This proof requires methods of advanced calculus and is omitted in this text.

EXAMPLE 9   *Partial derivatives of functions of two variables*

Find $f_{xy}, f_{yx}, f_{xx},$ and f_{xxy}, where $f(x, y) = x^2 y e^y$.

Solution   We have the first partial derivatives

$$f_x = 2xye^y \qquad f_y = x^2 e^y + x^2 y e^y$$

The mixed partial derivatives (which must be the same by the previous theorem) are

$$f_{xy} = (f_x)_y = 2xe^y + 2xye^y \qquad f_{yx} = (f_y)_x = 2xe^y + 2xye^y$$

Finally, we compute the second and higher partial derivatives:

$$f_{xx} = (f_x)_x = 2ye^y \quad \text{and} \quad f_{xxy} = (f_{xx})_y = 2e^y + 2ye^y \qquad \blacksquare$$

EXAMPLE 10   *Verifying that a function satisfies the heat equation*

Verify that $T(x, t) = e^{-t} \cos \dfrac{x}{c}$ satisfies the heat equation, $\dfrac{\partial T}{\partial t} = c^2 \dfrac{\partial^2 T}{\partial x^2}$.

Solution $\dfrac{\partial T}{\partial t} = -e^{-t} \cos \dfrac{x}{c}$ and $\dfrac{\partial^2 T}{\partial x^2} = \dfrac{\partial}{\partial x}\left(-\dfrac{1}{c} e^{-t} \sin \dfrac{x}{c}\right) = -\dfrac{1}{c^2} e^{-t} \cos \dfrac{x}{c}$

WARNING → The heat equation describes the distribution of temperature in an insulated rod. The constant c is called the *diffusivity* of the material in the rod. ←

Thus, T satisfies the heat equation $\dfrac{\partial T}{\partial t} = c^2 \dfrac{\partial^2 T}{\partial x^2}$. $\qquad \blacksquare$

Analogous definitions can be made for functions of more than two variables. For example,

$$f_{zzz} = \frac{\partial^3 f}{\partial z^3} = \frac{\partial}{\partial z}\left[\frac{\partial}{\partial z}\left(\frac{\partial f}{\partial z}\right)\right] \quad \text{or} \quad f_{xyz} = \frac{\partial^3 f}{\partial z \partial y \partial x} = \frac{\partial}{\partial z}\left[\frac{\partial}{\partial y}\left(\frac{\partial f}{\partial x}\right)\right]$$

EXAMPLE 11   *Higher partial derivatives of a function of several variables*

By direct calculation, show that $f_{xyz} = f_{yzx} = f_{zyx}$ for the function $f(x, y, z) = xyz + x^2 y^3 z^4$.

Solution   Find the first partials:

$$f_x(x, y, z) = yz + 2xy^3 z^4; \quad f_y(x, y, z) = xz + 3x^2 y^2 z^4; \quad f_z(x, y, z) = xy + 4x^2 y^3 z^3$$

Next, the required mixed partials:

$$f_{xy}(x, y, z) = (yz + 2xy^3 z^4)_y = z + 6xy^2 z^4$$
$$f_{yz}(x, y, z) = (xz + 3x^2 y^2 z^4)_z = x + 12x^2 y^2 z^3$$
$$f_{zy}(x, y, z) = (xy + 4x^2 y^3 z^3)_y = x + 12x^2 y^2 z^3$$

Finally, the higher mixed partials:

$$f_{xyz}(x, y, z) = (z + 6xy^2 z^4)_z = 1 + 24xy^2 z^3$$
$$f_{yzx}(x, y, z) = (x + 12x^2 y^2 z^3)_x = 1 + 24xy^2 z^3$$
$$f_{zyx}(x, y, z) = (x + 12x^2 y^2 z^3)_x = 1 + 24xy^2 x^3 \qquad \blacksquare$$

12.3 Problem Set

 1. ■ **What Does This Say?** What is a partial derivative?

2. ■ **What Does This Say?** Describe two basic applications of the partial derivatives $f_x(x, y)$ and $f_y(x, y)$.

Find f_x, f_y, f_{xx}, and f_{yx} in Problems 3–8.

3. $f(x, y) = x^3 + x^2 y + xy^2 + y^3$

4. $f(x, y) = (x + xy + y)^3$

5. $f(x, y) = \dfrac{x}{y}$

6. $f(x, y) = xe^{xy}$

7. $f(x, y) = \ln(2x + 3y)$

8. $f(x, y) = \sin x^2 y$

Find f_x and f_y in Problems 9–16.

9. a. $f(x, y) = (\sin x^2)\cos y$ **b.** $f(x, y) = \sin(x^2 \cos y)$

10. a. $f(x, y) = (\sin \sqrt{x})\ln y^2$ **b.** $f(x, y) = \sin(\sqrt{x} \ln y^2)$

11. $f(x, y) = \sqrt{3x^2 + y^4}$ **12.** $f(x, y) = xy^2 \ln(x + y)$

13. $f(x, y) = x^2 e^{x+y} \cos y$ **14.** $f(x, y) = xy^3 \tan^{-1} y$

15. $f(x, y) = \sin^{-1}(xy)$ **16.** $f(x, y) = \cos^{-1}(xy)$

Find f_x, f_y, and f_z in Problems 17–22.

17. $f(x, y, z) = xy^2 + yz^3 + xyz$

18. $f(x, y, z) = xye^z$

19. $f(x, y, z) = \dfrac{x + y^2}{z}$

20. $f(x, y, z) = \dfrac{xy + yz}{xz}$

21. $f(x, y, z) = \ln(x + y^2 + z^3)$

22. $f(x, y, z) = \sin(xy + z)$

In Problems 23–28, find $\dfrac{\partial z}{\partial x}$ and $\dfrac{\partial z}{\partial y}$ by differentiating implicitly.

23. $\dfrac{x^2}{9} - \dfrac{y^2}{4} + \dfrac{z^2}{2} = 1$ **24.** $3x^2 + 4y^2 + 2z^2 = 5$

25. $3x^2 y + y^3 z - z^2 = 1$ **26.** $x^3 - xy^2 + yz^2 - z^3 = 4$

27. $\sqrt{x} + y^2 + \sin xz = 2$

28. $\ln(xy + yz + xz) = 5$
$(x > 0, y > 0, z > 0)$

 29. Find f_x and f_y for

$$f(x, y) = \int_x^y (t^2 + 2t + 1)\,dt$$

Hint: Review the second fundamental theorem.

30. Find f_x and f_y for

$$f(x, y) = \int_{x^2}^{2y} (e^t + 3t)\,dt$$

Hint: Review the second fundamental theorem.

31. Find the slope of the tangent line to the level curve at the point $P(1, 1, 3)$ on the surface $z = x^2 + xy^2 + y^3$.

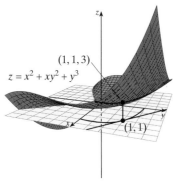

32. Find the slope of the tangent line to the level curve at the point $P\left(\dfrac{\sqrt{\pi}}{2}, \dfrac{\sqrt{\pi}}{2}, \dfrac{\pi}{4}\right)$ on the surface $z = xy + \cos(x^2 + y^2)$.

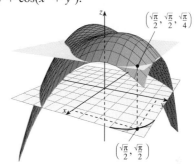

*A function $f(x, y)$ is said to be **harmonic** on the set S if f_{xx} and f_{yy} are continuous and*

$$f_{xx} + f_{yy} = 0$$

throughout S. Show that each function given in Problems 33–36 is harmonic on the prescribed set.

33. $f(x, y) = 3x^2 y - y^3$; S is the entire plane.

34. $f(x, y) = \ln(x^2 + y^2)$; S is the plane with the point $(0, 0)$ removed.

35. $f(x, y) = e^x \sin y$; S is the entire plane.

36. $f(x, y) = \sin x \cosh y$; S is the entire plane.

37. Let $f(x, y) = xy^3 + x^3 y$. Find the slope of the tangent line to the graph of f at $P(1, -1, -2)$ in the direction of
a. the x-axis.   b. the y-axis.

38. Find the slope of the tangent line at the point $P(1, -1, -2)$ on the graph of

$$f(x, y) = \dfrac{x^2 + y^2}{xy}$$

in the direction
a. parallel to the xz-plane.
b. parallel to the yz-plane.

39. For $f(x, y) = \cos xy^2$, show $f_{xy} = f_{yx}$.

40. For $f(x, y) = (\sin^2 x)(\sin y)$, show $f_{xy} = f_{yx}$.

41. Find $f_{xzy} - f_{yzz}$, where $f(x, y, z) = x^2 + y^2 - 2xy \cos z$.

42. MODELING PROBLEM   It has been determined that the flow (in cm^3/s) of blood from an artery into a small capillary can be modeled by

$$F(x, y, z) = \frac{c\pi x^2}{4}\sqrt{y - z}$$

for constant $c > 0$, where x is the diameter of the capillary, y is the pressure in the artery, and z is the pressure in the capillary. Find the rate of change of the flow of blood with respect to:

a. the diameter of the capillary.

b. the arterial pressure.

c. the capillary pressure.

43. MODELING PROBLEM   Biologists have studied the oxygen consumption of certain furry mammals. They have found that if the mammal's body temperature is T degrees Celsius, fur temperature is t degrees Celsius, and the mammal does not sweat, then its relative oxygen consumption can be modeled by

$$C(m, t, T) = \sigma(T - t)m^{-0.67}$$

(kg/h) where m is the mammal's mass (in kg) and σ is a physical constant. Find the rate (rounded to two decimal places) at which the oxygen consumption changes with respect to:

a. the mass m.

b. the body temperature T.

c. the fur temperature t.

44. MODELING PROBLEM   A gas that gathers on a surface in a condensed layer is said to be *adsorbed* in the surface, and the surface is called an *adsorbing surface*. It has been determined that the amount of gas adsorbed per unit area on an adsorbing surface can be modeled by

$$S(p, T, h) = ape^{h/(bT)}$$

where p is the gas pressure, T is the temperature of the gas, h is the heat of the adsorbed layer of gas, and a and b are physical constants. Find the rate of change of S with respect to: a. p b. h c. T

45. The ideal gas law says that $PV = kT$, where P is the pressure of a confined gas, V is the volume, T is the temperature, and k is a physical constant.

a. Find $\dfrac{\partial V}{\partial T}$ b. Find $\dfrac{\partial P}{\partial V}$

c. Show that $\dfrac{\partial P}{\partial V} \cdot \dfrac{\partial V}{\partial T} \cdot \dfrac{\partial T}{\partial P} = -1$.

46. At a certain factory, the output is given by $Q = 120K^{2/3}L^{2/5}$, where K denotes the capital investment (in units of $1,000) and L measures the size of the labor force (in worker-hours).

a. Find the *marginal productivity of capital*, $\partial Q/\partial K$, and the *marginal productivity of labor*, $\partial Q/\partial L$.

b. Determine the signs of the second partial derivatives $\partial^2 Q/\partial L^2$ and $\partial^2 Q/\partial K^2$, and give an economic interpretation.

Note: Q is an example of a Cobb–Douglas production function.

47. The temperature at a point (x, y) on a given metal plate in the xy-plane is determined according to the formula $T(x, y) = x^3 + 2xy^2 + y$ degrees. Find the rate at which the temperature changes with distance if we start at $(2, 1)$ and move

a. up (parallel to the y-axis).

b. to the right (parallel to the x-axis).

48. In physics, the **wave equation** is $\dfrac{\partial^2 z}{\partial t^2} = c^2 \dfrac{\partial^2 z}{\partial x^2}$

and the **heat equation** is $\dfrac{\partial z}{\partial t} = c^2 \dfrac{\partial^2 z}{\partial x^2}$. In each of the following cases, determine whether z satisfies the wave equation, the heat equation, or neither.

a. $z = e^{-t}\left(\sin \dfrac{x}{c} + \cos \dfrac{x}{c}\right)$ b. $z = \sin 3\,ct \sin 3x$

c. $z = \sin 5\,ct \cos 5x$

49. THINK TANK PROBLEM   Let

$$f(x, y) = \begin{cases} xy\left(\dfrac{x^2 - y^2}{x^2 + y^2}\right) & \text{if } (x, y) \neq (0, 0) \\ 0 & \text{if } (x, y) = (0, 0) \end{cases}$$

Show that $f_x(0, y) = -y$ and $f_y(x, 0) = x$, for all x and y. Then show that $f_{xy}(0, 0) = -1$ and $f_{yx}(0, 0) = 1$. Why does this not violate the equality of mixed partials theorem?

50. Show that $f_x(0, 0) = 0$ but $f_y(0, 0)$ does not exist, where

$$f(x, y) = \begin{cases} (x^2 + y) \sin\left(\dfrac{1}{x^2 + y^2}\right) & \text{if } (x, y) \neq (0, 0) \\ 0 & \text{if } (x, y) = (0, 0) \end{cases}$$

51. Suppose a substance is injected into a tube containing a liquid solvent. Suppose the tube is placed so that its axis is parallel to the x-axis, as shown in Figure 12.15.

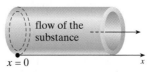

flow of the substance

$x = 0$ x

■ **FIGURE 12.15**   Problem 51

Assume that the concentration of the substance varies only in the x-direction, and let $C(x, t)$ denote the concentration at position x and time t. Because the number of molecules in this substance is very large, it is reasonable to assume that C is a continuous function whose partial derivatives exist. One model for the flow yields the **diffusion equation in one dimension,** namely,

$$\frac{\partial C}{\partial t} = \delta \frac{\partial^2 C}{\partial x^2}$$

where δ is the **diffusion constant.**

a. What must δ be for a function of the form

$$C(x, t) = e^{ax + bt}$$

(a and b are constants) to satisfy the diffusion equation?

b. Verify that

$$C(x, t) = t^{-1/2} e^{-x^2/(4\delta t)}$$

satisfies the diffusion equation.

52. The area of a triangle is $A = \frac{1}{2}ab \sin \gamma$, where γ is the angle between sides of length a and b.

a. Find $\dfrac{\partial A}{\partial a}, \dfrac{\partial A}{\partial b}$, and $\dfrac{\partial A}{\partial \gamma}$.

b. Suppose a is given as a function of b, A, and γ.

What is $\dfrac{\partial a}{\partial \gamma}$?

53. **JOURNAL PROBLEM** (*Crux*, problem by John A. Winterink.)* Prove the validity of the following simple method for finding the center of a conic: For the central conic,

$$\phi(x, y) = ax^2 + 2hxy + by^2 + 2gx + 2fy + c = 0$$

$ab - h^2 \neq 0$, show that the center is the intersection of the lines $\partial\phi/\partial x = 0$ and $\partial\phi/\partial y = 0$.

_____

*Crux, Problem 54, Vol. 6 (1980), p. 154.

12.4 Tangent Planes, Approximations, and Differentiability

IN THIS SECTION   **tangent planes, incremental approximations, the total differential, differentiability** ■

TANGENT PLANES

Suppose S is a surface with the equation $z = f(x, y)$, where f has continuous first partial derivatives f_x and f_y. Let $P(x_0, y_0, z_0)$ be a point on S, and let C_1 be the curve of intersection of S with the plane $x = x_0$ and C_2, the intersection of S with the plane $y = y_0$, as shown in Figure 12.16. The tangent lines T_1 and T_2, respectively, at P determine a unique plane, and we shall find that this plane actually contains the tangent to *every* smooth curve on S that passes through P. It is reasonable to call this plane the **tangent plane** to S at P.

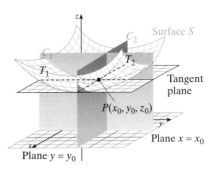

■ **FIGURE 12.16**   Tangent plane

To find an equation for the tangent plane, recall that the equation of a plane with normal $\mathbf{N} = A\mathbf{i} + B\mathbf{j} + C\mathbf{k}$ is

$$A(x - x_0) + B(y - y_0) + C(z - z_0) = 0$$

If $C \neq 0$, divide both sides by C and let $a = -A/C$ and $b = -B/C$ to obtain

$$z - z_0 = a(x - x_0) + b(y - y_0)$$

The intersection of this plane and the plane $x = x_0$ is the tangent line T_1, which we know has slope $f_y(x_0, y_0)$ from the geometric interpretation of partial derivatives. Setting $x = x_0$ in the equation for the tangent plane, we find that T_1 has the point–slope form

$$z - z_0 = b(y - y_0)$$

so we must have $b = f_y(x_0, y_0) = \dfrac{\partial z}{\partial y}$. Similarly, setting $y = y_0$, we obtain

$$z - z_0 = a(x - x_0)$$

which represents the tangent line T_2, with slope $a = f_x(x_0, y_0)$. To summarize:

Equation of the Tangent Plane

Suppose S is a surface with the equation $z = f(x, y)$ and let $P_0(x_0, y_0, z_0)$ be a point on S at which a tangent plane exists. Then the **equation of the tangent plane** to S at P_0 is

$$z - z_0 = f_x(x_0, y_0)(x - x_0) + f_y(x_0, y_0)(y - y_0)$$

EXAMPLE 1   *Equation of a tangent plane for a surface defined by $z = f(x, y)$*

Find an equation for the tangent plane to the surface $z = \tan^{-1}\dfrac{y}{x}$ at the point $P_0(1, \sqrt{3}, \frac{\pi}{3})$.

Solution $f_x(x, y) = \dfrac{-yx^{-2}}{1 + \left(\dfrac{y}{x}\right)^2} = \dfrac{-y}{x^2 + y^2};$ $f_x(1, \sqrt{3}) = \dfrac{-\sqrt{3}}{1 + 3} = \dfrac{-\sqrt{3}}{4}$

$f_y(x, y) = \dfrac{x^{-1}}{1 + \left(\dfrac{y}{x}\right)^2} = \dfrac{x}{x^2 + y^2};$ $f_y(1, \sqrt{3}) = \dfrac{1}{1 + 3} = \dfrac{1}{4}$

The equation of the tangent plane is

$$z - \frac{\pi}{3} = \left(\frac{-\sqrt{3}}{4}\right)(x - 1) + \frac{1}{4}(y - \sqrt{3})$$

or

$$3\sqrt{3}x - 3y + 12z = 4\pi$$

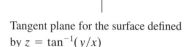

Tangent plane for the surface defined by $z = \tan^{-1}(y/x)$

INCREMENTAL APPROXIMATIONS

In Chapter 3, we observed that the tangent line to the curve $y = f(x)$ at the point $P(x_0, y_0)$ is the line that best "fits" the shape of the curve in the immediate vicinity of P. In other words, if f is differentiable at $x = x_0$ and the increment $|\Delta x|$ is sufficiently small, then

$$f(x_0 + \Delta x) \approx f(x_0) + f'(x_0)\Delta x$$

Similarly, the tangent plane at $P(x_0, y_0, z_0)$ is the plane that best fits the shape of the surface $z = f(x, y)$ near P, and the analogous **incremental** (or **linear**) approximation formula may be stated as follows.

Incremental Approximation of a Function of Two Variables

If $f(x, y)$ and its partial derivatives f_x and f_y are defined in an open region R containing the point $P(x_0, y_0)$ and f_x and f_y are continuous at P, then

$$\Delta f = f(x_0 + \Delta x, y_0 + \Delta y) - f(x_0, y_0) \approx f_x(x_0, y_0)\Delta x + f_y(x_0, y_0)\Delta y$$

so that

$$f(x_0 + \Delta x, y_0 + \Delta y) \approx f(x_0, y_0) + f_x(x_0, y_0)\Delta x + f_y(x_0, y_0)\Delta y$$

A graphical interpretation of this incremental approximation formula is shown in Figure 12.17.

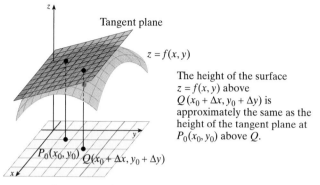

The height of the surface $z = f(x, y)$ above $Q(x_0 + \Delta x, y_0 + \Delta y)$ is approximately the same as the height of the tangent plane at $P_0(x_0, y_0)$ above Q.

■ **FIGURE 12.17**   Incremental approximation to a function of two variables

The tangent plane to the surface $z = f(x, y)$ has the equation

$$z - f(x_0, y_0) = f_x(x_0, y_0)\Delta x + f_y(x_0, y_0)\Delta y$$

As long as we are near (x_0, y_0), the height of the tangent plane is approximately the same as the height of the surface. Thus, if $|\Delta x|$ and $|\Delta y|$ are small, the point $(x_0 + \Delta x, y_0 + \Delta y)$ will be near (x_0, y_0) and we have

$$\underbrace{f(x_0 + \Delta x, y_0 + \Delta y)}_{\substack{\text{Height of } z = f(x, y) \\ \text{above } Q(x_0 + \Delta x, y_0 + \Delta y)}} \approx \underbrace{f(x_0, y_0) + f_x(x_0, y_0)\Delta x + f_y(x_0, y_0)\Delta y}_{\text{Height of the tangent plane above } Q}$$

Increments of a function of three variables $f(x, y, z)$ can be defined in a similar fashion. Suppose f has continuous partial derivatives f_x, f_y, f_z at and near the point (x_0, y_0, z_0). Then if the numbers $\Delta x, \Delta y, \Delta z$ are all sufficiently small, we have

$$\Delta f = f(x_0 + \Delta x, y_0 + \Delta y, z_0 + \Delta z) - f(x_0, y_0, z_0)$$
$$\approx f_x(x_0, y_0, z_0)\Delta x + f_y(x_0, y_0, z_0)\Delta y + f_z(x_0, y_0, z_0)\Delta z$$

EXAMPLE 2   *Using increments to estimate the change of a function*

An open box has length 3 ft, width 1 ft, and height 2 ft, and is constructed from material that costs $2/ft² for the sides and $3/ft² for the bottom. Compute the cost of constructing the box, and then use increments to estimate the change in cost if the length and width are each increased by 3 in. and the height is decreased by 4 in.

Solution   An open (no top) box with length x, width y, and height z has a surface area

$$S = xy + \underbrace{2xz + 2yz}_{}$$
$$\text{Bottom}\quad\text{Four side faces}$$

Because the sides cost $2/ft² and the bottom $3/ft², the total cost is

$$C(x, y, z) = 3xy + 2(2xz + 2yz)$$

The partial derivatives of C are

$$C_x = 3y + 4z \qquad C_y = 3x + 4z \qquad C_z = 4x + 4y$$

and the dimensions of the box change by

$$\Delta x = \tfrac{3}{12} = 0.25 \text{ ft} \qquad \Delta y = \tfrac{3}{12} = 0.25 \text{ ft} \qquad \Delta z = \tfrac{-4}{12} \approx -0.33 \text{ ft}$$

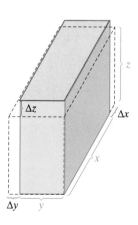

Thus, the change in the total cost is approximated by

$$\Delta C \approx C_x(3, 1, 2)\Delta x + C_y(3, 1, 2)\Delta y + C_z(3, 1, 2)\Delta z$$
$$= [3(1) + 4(2)](0.25) + [3(3) + 4(2)](0.25) + [4(3) + 4(1)](-\tfrac{4}{12})$$
$$\approx 1.67$$

That is, the cost increases by approximately $1.67. ∎

EXAMPLE 3 *Maximum percentage error using differentials*

The radius and height of a right circular cone are measured with errors of at most 3% and 2%, respectively. Use increments to approximate the maximum possible percentage error in computing the volume of the cone using these measurements and the formula $V = \tfrac{1}{3}\pi R^2 H$.

Solution We are given that

$$\left|\frac{\Delta R}{R}\right| \le 0.03 \quad \text{and} \quad \left|\frac{\Delta H}{H}\right| \le 0.02$$

The partial derivatives of V are

$$V_R = \tfrac{2}{3}\pi RH \quad \text{and} \quad V_H = \tfrac{1}{3}\pi R^2$$

so the change in V is approximated by

$$\Delta V \approx \left(\frac{2}{3}\pi RH\right)\Delta R + \left(\frac{1}{3}\pi R^2\right)\Delta H$$

Dividing by the volume $V = \tfrac{1}{3}\pi R^2 H$, we obtain

$$\frac{\Delta V}{V} \approx \frac{\tfrac{2}{3}\pi RH\Delta R + \tfrac{1}{3}\pi R^2\Delta H}{\tfrac{1}{3}\pi R^2 H} = 2\left(\frac{\Delta R}{R}\right) + \left(\frac{\Delta H}{H}\right)$$

so that

$$\left|\frac{\Delta V}{V}\right| \le 2\left|\frac{\Delta R}{R}\right| + \left|\frac{\Delta H}{H}\right| = 2(0.03) + (0.02) = 0.08$$

Thus, the maximum percentage error in computing the volume V is approximately 8%. ∎

THE TOTAL DIFFERENTIAL

For a function of one variable, $y = f(x)$, we defined the differential dy to be $dy = f'(x)\, dx$. For the two-variable case, we make the following analogous definition.

Total Differential

If $z = f(x, y)$ and Δx and Δy are increments of x and y, respectively, and if we let $dx = \Delta x$ and $dy = \Delta y$ be differentials for x and y, respectively, then the **total differential** of $f(x, y)$ is

$$df = \frac{\partial f}{\partial x}\, dx + \frac{\partial f}{\partial y}\, dy = f_x(x, y)\, dx + f_y(x, y)\, dy$$

Similarly, for a function of three variables $z = f(x, y, z)$, with $dz = \Delta z$, the **total differential** is

$$df = \frac{\partial f}{\partial x}\, dx + \frac{\partial f}{\partial y}\, dy + \frac{\partial f}{\partial z}\, dz$$

EXAMPLE 4	*Total differential*

Find the total differential of the given functions:

a. $f(x, y, z) = 2x^3 + 5y^4 - 6z$ b. $f(x, y) = x^2 \ln(3y^2 - 2x)$

Solution

a. $df = \dfrac{\partial f}{\partial x} dx + \dfrac{\partial f}{\partial y} dy + \dfrac{\partial f}{\partial z} dz = 6x^2 dx + 20y^3 dy - 6 dz$

b. $df = \dfrac{\partial f}{\partial x} dx + \dfrac{\partial f}{\partial y} dy$

$$= \left[2x \ln(3y^2 - 2x) + x^2 \dfrac{-2}{3y^2 - 2x}\right] dx + \left[x^2 \dfrac{6y}{3y^2 - 2x}\right] dy$$

$$= \left[2x \ln(3y^2 - 2x) - \dfrac{2x^2}{3y^2 - 2x}\right] dx + \dfrac{6x^2 y}{3y^2 - 2x} dy \qquad \blacksquare$$

EXAMPLE 5	*Application of the total differential*

At a certain factory, the daily output is $Q = 60K^{1/2}L^{1/3}$ units, where K denotes the capital investment (in units of $1,000) and L, the size of the labor force (in worker-hours). The current capital investment is $900,000, and 1,000 worker-hours of labor are used each day. Estimate the change in output that will result if capital investment is increased by $1,000 and labor is decreased by 2 worker-hours.

Solution   The change in output is estimated by the total differential dQ. We have $K = 900$, $L = 1,000$, $dK = \Delta K = 1$, and $dL = \Delta L = -2$. The total differential of $Q(x, y)$ is

$$dQ = \dfrac{\partial Q}{\partial K} dK + \dfrac{\partial Q}{\partial L} dL$$

$$= 60(\tfrac{1}{2})K^{-1/2}L^{1/3} dK + 60(\tfrac{1}{3})K^{1/2}L^{-2/3} dL$$

$$= 30K^{-1/2}L^{1/3} dK + 20K^{1/2}L^{-2/3} dL$$

Substituting for K, L, dK, and dL,

$$dQ = 30(900)^{-1/2}(1,000)^{1/3}(1) + 20(900)^{1/2}(1,000)^{-2/3}(-2) = -2$$

Thus, the output decreases by approximately 2 units when the capital investment is increased by $1,000 and labor is decreased by 2 worker-hours. $\blacksquare$

EXAMPLE 6	*Maximum percentage error in an electrical circuit*

When two resistances R_1 and R_2 are connected in parallel, the total resistance R is given by

$$R = \dfrac{R_1 R_2}{R_1 + R_2}$$

If R_1 is measured as 300 ohms with a maximum error of 2% and R_2 is measured as 500 ohms with a maximum error of 3%, what is the maximum percentage error in R?

Solution   We are given that

$$\left|\dfrac{dR_1}{R_1}\right| \le 0.02 \quad \text{and} \quad \left|\dfrac{dR_2}{R_2}\right| \le 0.03$$

and we wish to find the maximum value of $\left| \dfrac{dR}{R} \right|$. Because

$$\frac{\partial R}{\partial R_1} = \frac{R_2^2}{(R_1 + R_2)^2} \quad \text{and} \quad \frac{\partial R}{\partial R_2} = \frac{R_1^2}{(R_1 + R_2)^2} \qquad \textit{Quotient rule}$$

it follows that the total differential of R is

$$dR = \frac{\partial R}{\partial R_1} \, dR_1 + \frac{\partial R}{\partial R_2} \, dR_2$$

$$= \frac{R_2^2}{(R_1 + R_2)^2} \, dR_1 + \frac{R_1^2}{(R_1 + R_2)^2} \, dR_2$$

We now find $\dfrac{dR}{R}$ by dividing both sides by R; however, because $R = \dfrac{R_1 R_2}{R_1 + R_2}$, we note that $\dfrac{1}{R} = \dfrac{R_1 + R_2}{R_1 R_2}$:

$$dR \cdot \frac{1}{R} = \left[\frac{R_2^2}{(R_1 + R_2)^2} \, dR_1 + \frac{R_1^2}{(R_1 + R_2)^2} \, dR_2 \right] \cdot \frac{R_1 + R_2}{R_1 R_2}$$

$$\frac{dR}{R} = \frac{R_2}{R_1 + R_2} \cdot \frac{dR_1}{R_1} + \frac{R_1}{R_1 + R_2} \cdot \frac{dR_2}{R_2}$$

Finally, apply the triangle inequality (Table 1.1, p. 3) to this relationship:

$$\left| \frac{dR}{R} \right| \leq \left| \frac{R_2}{R_1 + R_2} \right| \left| \frac{dR_1}{R_1} \right| + \left| \frac{R_1}{R_1 + R_2} \right| \left| \frac{dR_2}{R_2} \right|$$

$$\leq \frac{500}{300 + 500} (0.02) + \frac{300}{300 + 500} (0.03) = 0.02375$$

The maximum percentage is approximately 2.4%.   ∎

DIFFERENTIABILITY

Recall from Chapter 3 that if $f(x)$ is differentiable at x_0, its *increment* is

$$\Delta f = f(x_0 + \Delta x) - f(x_0) = f'(x_0)\Delta x + \epsilon \Delta x$$

where $\epsilon \to 0$ as $\Delta x \to 0$. For a function of two variables, the increment of x is an independent variable denoted by Δx, the increment of y is an independent variable denoted by Δy, and the increment of f at (x_0, y_0) is defined as

$$\Delta f = f(x_0 + \Delta x, y_0 + \Delta y) - f(x_0, y_0)$$

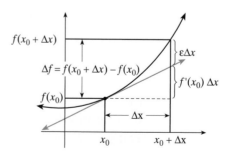

$$\Delta f = f(x_0 + \Delta x) - f(x_0) \approx f'(x_0)\, \Delta x + \underbrace{\epsilon \Delta x}$$

Error when using
df to estimate Δf

We use this increment representation to define differentiability as follows.

> **Definition of Differentiability**
>
> Suppose $f(x, y)$ is defined at each point in a circular disk that is centered at (x_0, y_0) and contains the point $(x_0 + \Delta x, y_0 + \Delta y)$. Then f is said to be **differentiable** at (x_0, y_0) if the increment of f can be expressed as
>
> $$\Delta f = f_x(x_0, y_0)\Delta x + f_y(x_0, y_0)\Delta y + \epsilon_1 \Delta x + \epsilon_2 \Delta y$$
>
> where $\epsilon_1 \to 0$ and $\epsilon_2 \to 0$ as both $\Delta x \to 0$ and $\Delta y \to 0$. Also, $f(x, y)$ is said to be **differentiable on the region R** of the plane if f is differentiable at each point in R.
>
>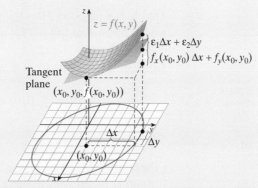

In Section 3.1, we showed that a function of one variable is continuous wherever it is differentiable. The following theorem establishes the same result for a function of two variables.

THEOREM 12.2   *Differentiability implies continuity*

If $f(x, y)$ is differentiable at (x_0, y_0), it is also continuous there.

Proof   We wish to show that $f(x, y) \to f(x_0, y_0)$ as $(x, y) \to (x_0, y_0)$ or, equivalently, that

$$\lim_{(x, y) \to (x_0, y_0)} [f(x, y) - f(x_0, y_0)] = 0$$

If we set $\Delta x = x - x_0$ and $\Delta y = y - y_0$ and let Δf denote the increment of f at (x_0, y_0), we have (by substitution)

$$f(x, y) - f(x_0, y_0) = f(x_0 + \Delta x, y_0, + \Delta y) - f(x_0, y_0) = \Delta f$$

Then, because $(\Delta x, \Delta y) \to (0, 0)$ as $(x, y) \to (x_0, y_0)$, we wish to prove that

$$\lim_{(\Delta x, \Delta y) \to (0, 0)} \Delta f = 0$$

Since f is differentiable at (x_0, y_0), we have

$$\Delta f = f_x(x_0, y_0)\Delta x + f_y(x_0, y_0)\Delta y + \epsilon_1 \Delta x + \epsilon_2 \Delta y$$

where $\epsilon_1 \to 0$ and $\epsilon_2 \to 0$ as $(\Delta x, \Delta y) \to (0, 0)$. It follows that

$$\lim_{(\Delta x, \Delta y) \to (0, 0)} \Delta f = \lim_{(\Delta x, \Delta y) \to (0, 0)} [f_x(x_0, y_0)\Delta x + f_y(x_0, y_0)\Delta y + \epsilon_1 \Delta x + \epsilon_2 \Delta y]$$

$$= [f_x(x_0, y_0)] \cdot 0 + [f_y(x_0, y_0)] \cdot 0 + 0 + 0 = 0$$

as required.

WARNING➤   Be careful about how you use the word *differentiable*. In a single-variable case, a function is differentiable at a point if its derivative exists there. However, the word is used differently for a function of two variables. In particular, the existence of the partial derivatives f_x and f_y does not guarantee that the function is differentiable, as illustrated in the following example. ⬅

EXAMPLE 7   *Think Tank example: Possible existence of partial derivatives, with a nondifferentiable function*

Let

$$f(x, y) = \begin{cases} 1 & \text{if } x > 0 \text{ and } y > 0 \\ 0 & \text{otherwise} \end{cases}$$

That is, the function f has the value 1 when (x, y) is in the first quadrant and is 0 elsewhere. Show that the partial derivatives f_x and f_y exist at the origin, but f is not differentiable there.

Solution   Since $f(0, 0) = 0$, we have

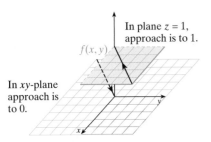

In plane $z = 1$, approach is to 1.

$f(x, y)$

In xy-plane approach is to 0.

$$f_x(0, 0) = \lim_{\Delta x \to 0} \frac{f(0 + \Delta x, 0) - f(0, 0)}{\Delta x} = 0$$

and similarly, $f_y(0, 0) = 0$. Thus, the partial derivatives both exist at the origin.

If $f(x, y)$ were differentiable at the origin, it would have to be continuous there (Theorem 12.2). Thus, we can show f is *not* differentiable by showing that it is *not* continuous at $(0, 0)$. Toward this end, note that $\lim_{(x, y) \to (0, 0)} f(x, y)$ is 1 along the line $y = x$ in the first quadrant but is 0 if the approach is along the x-axis. This means that the limit does not exist. Thus, f is not continuous at $(0, 0)$ and consequently is also not differentiable there. ∎

Although the existence of partial derivatives at $P(x_0, y_0)$ is not enough to guarantee that $f(x, y)$ is differentiable at P, we do have the following sufficient condition for differentiability.

THEOREM 12.3   *Sufficient condition for differentiability*

If f is a function of x and y and $f, f_x,$ and f_y are continuous in a disk D centered at (x_0, y_0), then f is differentiable at (x_0, y_0).

Proof   The proof is found in advanced calculus.

WARNING➤   Note that the function in Example 7 does not contradict Theorem 12.3 because there is no disk centered at $(0, 0)$ on which f is continuous. ⬅

EXAMPLE 8   *Establish differentiability*

Show that $f(x, y) = x^2y + xy^3$ is differentiable for all (x, y).

Solution   Compute the partial derivatives:

$$f_x(x, y) = \frac{\partial}{\partial x}(x^2y + xy^3) = 2xy + y^3$$

$$f_y(x, y) = \frac{\partial}{\partial y}(x^2y + xy^3) = x^2 + 3xy^2$$

Because $f, f_x,$ and f_y are all polynomials in x and y, they are continuous throughout the plane. Therefore, the sufficient condition for the differentiability theorem assures us that f must be differentiable for all x and y. ∎

12.4   Problem Set

A In Problems 1–6, find the standard-form equations for the tangent plane to the given surface at the prescribed point P_0.

1. $z = \sqrt{x^2 + y^2}$ at $P_0(3, 1, \sqrt{10})$

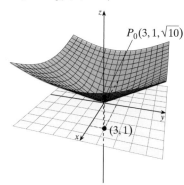

$P_0(3, 1, \sqrt{10})$

$(3, 1)$

2. $z = 10 - x^2 - y^2$ at $P_0(2, 2, 2)$

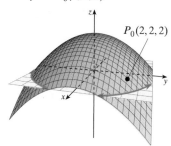

$P_0(2, 2, 2)$

3. $f(x, y) = x^2 + y^2 + \sin xy$ at $P_0 = (0, 2, 4)$

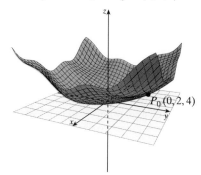

$P_0(0, 2, 4)$

4. $f(x, y) = e^{-x}\sin y$ at $P_0(0, \frac{\pi}{2}, 1)$

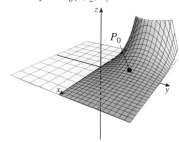

P_0

5. $z = \tan^{-1}\frac{y}{x}$ at $P_0\left(2, 2, \frac{\pi}{4}\right)$

6. $z = \ln|x + y^2|$ at $P_0(-3, -2, 0)$

Find the total differential of the functions given in Problems 7–18.

7. $f(x, y) = 5x^2y^3$
8. $f(x, y) = 8x^3y^2 - x^4y^5$
9. $f(x, y) = \sin xy$
10. $f(x, y) = \cos x^2 y$
11. $f(x, y) = \frac{y}{x}$
12. $f(x, y) = \frac{x^2}{y}$
13. $f(x, y) = ye^x$
14. $f(x, y) = e^{x^2 + y}$
15. $f(x, y, z) = 3x^3 - 2y^4 + 5z$
16. $f(x, y, z) = \sin x + \sin y + \cos z$
17. $f(x, y, z) = z^2\sin(2x - 3y)$
18. $f(x, y, z) = 3y^2z \cos x$

Show that the functions in Problems 19–22 are differentiable for all (x, y).

19. $f(x, y) = xy^3 + 3xy^2$
20. $f(x, y) = x^2 + 4x - y^2$
21. $f(x, y) = e^{2x + y^2}$
22. $f(x, y) = \sin(x^2 + 3y)$

Use an incremental approximation to estimate the functions at the values given in Problems 23–28. Check by using a calculator.

23. $f(1.01, 2.03)$, where $f(x, y) = 3x^4 + 2y^4$
24. $f(0.98, 1.03)$, where $f(x, y) = x^5 - 2y^3$
25. $f(\frac{\pi}{2} + 0.01, \frac{\pi}{2} - 0.01)$, where $f(x, y) = \sin(x + y)$
26. $f(\sqrt{\frac{\pi}{2}} + 0.01, \sqrt{\frac{\pi}{2}} - 0.01)$, where $f(x, y) = \sin(xy)$
27. $f(1.01, 0.98)$, where $f(x, y) = e^{xy}$
28. $f(1.01, 0.98)$, where $f(x, y) = e^{x^2y^2}$

B 29. Find an equation for each horizontal tangent plane to the surface

$$z = 5 - x^2 - y^2 + 4y$$

30. Find an equation for each horizontal tangent plane to the surface

$$z = 4(x - 1)^2 + 3(y + 1)^2$$

31. a. Show that if x and y are sufficiently close to zero and f is differentiable at $(0, 0)$, then

$$f(x, y) \approx f(0, 0) + xf_x(0, 0) + yf_y(0, 0)$$

  b. Use the approximation formula in part **a** to show that

$$\frac{1}{1 + x - y} \approx 1 - x + y$$

  for small x and y.

  c. If x and y are sufficiently close to zero, what is the approximate value of the expression

$$\frac{1}{(x + 1)^2 + (y + 1)^2}$$

32. When two resistors with resistances P and Q ohms are connected in parallel, the combined resistance is R, where

$$\frac{1}{R} = \frac{1}{P} + \frac{1}{Q}$$

If P and Q are measured at 6 and 10 ohms, respectively, with errors no greater than 1%, what is the maximum percentage error in the computation of R?

33. A closed box is found to have length 2 ft, width 4 ft, and height 3 ft, where the measurement of each dimension is made with a maximum possible error of ± 0.02 ft. The top of the box is made from material that costs $2/ft^2; the material for the sides and bottom costs only $1.50/ft^2. What is the maximum error involved in the computation of the cost of the box?

34. A cylindrical tank is 4 ft high and has a diameter of 2 ft. The walls of the tank are 0.2 in. thick. Approximate the volume of the interior of the tank assuming the tank has a top and a bottom that are both also 0.2 in. thick.

35. The Higrade Company sells two brands, X and Y, of a commercial soap, in thousand-pound units. If x units of brand X and y units of brand Y are sold, the unit price for brand X is

$$p(x) = 4,000 - 500x$$

and that of brand Y is

$$q(y) = 3,000 - 450y$$

a. Find an expression for the total revenue R in terms of p and q.
b. Suppose brand X sells for $500 per unit and brand Y sells for $750 per unit. Estimate the change in total revenue if the unit prices are increased by $20 for brand X and $18 for brand Y. *Hint:* Find the change in x and the change in y that correspond to $p(x)$ increasing from 500 to 520 and $q(y)$ increasing from 750 to 768.

36. The output at a certain factory is

$$Q = 150K^{2/3}L^{1/3}$$

where K is the capital investment in units of $1,000 and L is the size of the labor force, measured in worker-hours. The current capital investment is $500,000 and 1,500 worker-hours of labor are used.

a. Estimate the change in output that results when capital investment is increased by $700 and labor is increased by 6 worker-hours.
b. What if capital investment is increased by $500 and labor is decreased by 4 worker-hours?

37. According to Poiseuille's law, the resistance to the flow of blood offered by a cylindrical blood vessel of radius r and length x is

$$R(r, x) = \frac{cx}{r^4}$$

for a constant $c > 0$. A certain blood vessel in the body is 8 cm long and has a radius of 2 mm. Estimate the per-

centage change in R when x is increased by 3% and r is decreased by 2%.

38. For 1 mole of an ideal gas, the volume V, pressure P, and absolute temperature T are related by the equation $PV = RT$, where R is a certain fixed constant that depends on the gas. Suppose we know that if $T = 400$ (absolute) and $P = 3,000$ lb/ft^2, then $V = 14$ ft^3. Approximate the change in pressure if the temperature and volume are increased to 403 and 14.1 ft^3, respectively.

39. MODELING PROBLEM If x gram-moles of sulfuric acid are mixed with y gram-moles of water, the heat liberated is modeled by

$$F(x, y) = \frac{1.786\,xy}{1.798x + y}\ \text{cal}$$

Approximately how much additional heat is generated if a mixture of 5 gram-moles of acid and 4 gram-moles of water is increased to a mixture of 5.1 gram-moles of acid and 4.04 gram-moles of water?

40. MODELING PROBLEM A business analyst models the sales of a new product by the function

$$Q(x, y) = 20x^{3/2}y$$

where x thousand dollars are spent on development and y thousand dollars on promotion. Current plans call for the expenditure of $36,000 on development and $25,000 on promotion. Use the total differential of Q to estimate the change in sales that will result if the amount spent on development is increased by $500 and the amount spent on promotion is decreased by $500.

41. MODELING PROBLEM A grocer's weekly profit from the sale of two brands of orange juice is modeled by

$$P(x, y) = (x - 30)(70 - 5x + 4y) \\ + (y - 40)(80 + 6x - 7y)$$

dollars where x (cents) is the price per can of the first brand and y (cents) is the price per can of the second. Currently the first brand sells for 50¢ per can and the second for 52¢ per can. Use the total differential to estimate the change in the weekly profit that will result if the grocer raises the price of the first brand by 1¢ per can and lowers the price of the second brand by 2¢ per can.

42. A juice can is 12 cm tall and has a radius of 3 cm. A manufacturer is planning to reduce the height of the can by 0.2 cm and the radius by 0.3 cm. Use a total differential to estimate how much less the volume will be in each can after the new cans are introduced.

43. MODELING PROBLEM It is known that the period T of a simple pendulum with small oscillations is modeled by

$$T = 2\pi\sqrt{\frac{L}{g}}$$

where L is the length of the pendulum and g is the acceleration due to gravity. For a certain pendulum, it is known that $L = 4.03$ ft. It is also known that $g = 32.2$ ft/s^2.

What is the approximate error in calculating T by using $L = 4$ and $g = 32$?

44. If the weight of an object which does not float in water in the air is x pounds and its weight in water is y pounds, then the specific gravity of the object is

$$S = \frac{x}{x - y}$$

For a certain object, x and y are measured to be 1.2 lb and 0.5 lb, respectively. It is known that the measuring instrument will not register less than the true weights, but it could register more than the true weights by as much as 0.01 lb. What is the maximum possible error in the computation of the specific gravity?

45. A football has the shape of the ellipsoid

$$\frac{x^2}{9} + \frac{y^2}{36} + \frac{z^2}{9} = 1$$

where the dimensions are in inches, and is made of leather 1/8 inch thick. Use differentials to estimate the volume of the leather shell. *Hint:* The ellipsoid

$$\frac{x^2}{a^2} + \frac{y^2}{b^2} + \frac{z^2}{c^2} = 1$$

has volume $V = \frac{4}{3}\pi abc$.

46. Show that the following function is not differentiable at $(0, 0)$:

$$f(x, y) = \begin{cases} (xy)/(x^2 + y^2) & \text{if } (x, y) \neq (0, 0) \\ 0 & \text{if } (x, y) = (0, 0) \end{cases}$$

47. Compute the total differentials

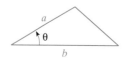

Why are these differentials alike?

48. Let A be the area of a triangle with sides a and b separated by an angle θ, as shown in Figure 12.18.

■ **FIGURE 12.18** Problem 48

Suppose $\theta = \frac{\pi}{6}$, and a is increased by 4% while b is decreased by 3%. Use differentials to estimate the percentage change in A.

49. In Problem 48, suppose that θ changes by no more than 2%. What is the maximum percentage change in A?

12.5 Chain Rules

IN THIS SECTION   chain rule for one parameter, chain rule for two parameters

CHAIN RULE FOR ONE PARAMETER

We begin with a differentiable function of two variables $f(x, y)$. If $x = x(t)$ and $y = y(t)$ are, in turn, functions of a single variable t, then $z = f[x(t), y(t)]$ is a composite function of a parameter t. In this case, the chain rule for finding the derivative with respect to one independent variable can now be stated.

THEOREM 12.4   *The chain rule for one independent parameter*

Let $f(x, y)$ be a differentiable function of x and y, and let $x = x(t)$ and $y = y(t)$ be differentiable functions of t. Then $z = f(x, y)$ is a differentiable function of t, and

$$\frac{dz}{dt} = \frac{\partial z}{\partial x}\frac{dx}{dt} + \frac{\partial z}{\partial y}\frac{dy}{dt}$$

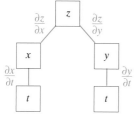

Chains from the chain rule represented schematically

■ *What This Says:* Recall the chain rule for a single variable:

$$\frac{dy}{dx} = \frac{dy}{du}\frac{du}{dx}$$

The corresponding rule for two variables is *essentially* the same except that it involves *both* variables.

Proof   Recall that because $z = f(x, y)$ is differentiable, we can write the increment Δz in the following form:

$$\Delta z = \frac{\partial z}{\partial x}\Delta x + \frac{\partial z}{\partial y}\Delta y + \epsilon_1\Delta x + \epsilon_2\Delta y$$

where $\epsilon_1 \to 0$ and $\epsilon_2 \to 0$ as both $\Delta x \to 0$ and $\Delta y \to 0$. Dividing by $\Delta t \neq 0$, we obtain

$$\frac{\Delta z}{\Delta t} = \frac{\partial z}{\partial x}\frac{\Delta x}{\Delta t} + \frac{\partial z}{\partial y}\frac{\Delta y}{\Delta t} + \epsilon_1\frac{\Delta x}{\Delta t} + \epsilon_2\frac{\Delta y}{\Delta t}$$

Because x and y are functions of t, we can write their increments as

$$\Delta x = x(t + \Delta t) - x(t) \quad \text{and} \quad \Delta y = y(t + \Delta t) - y(t)$$

We know that x and y both vary continuously with t (remember, they are differentiable), and it follows that $\Delta x \to 0$ and $\Delta y \to 0$ as $\Delta t \to 0$, so that $\epsilon_1 \to 0$ and $\epsilon_2 \to 0$ as $\Delta t \to 0$. Therefore, we have

$$\frac{dz}{dt} = \lim_{\Delta t \to 0}\frac{\Delta z}{\Delta t} = \lim_{\Delta t \to 0}\left[\frac{\partial z}{\partial x}\frac{\Delta x}{\Delta t} + \frac{\partial z}{\partial y}\frac{\Delta y}{\Delta t} + \epsilon_1\frac{\Delta x}{\Delta t} + \epsilon_2\frac{\Delta y}{\Delta t}\right]$$

$$= \frac{\partial z}{\partial x}\frac{dx}{dt} + \frac{\partial z}{\partial y}\frac{dy}{dt} + 0\frac{dx}{dt} + 0\frac{dy}{dt}$$

EXAMPLE 1   *Verifying the chain rule explicitly*

Let $z = x^2 + y^2$, where $x = \dfrac{1}{t}$ and $y = t^2$. Find $\dfrac{dz}{dt}$ in two ways:

a. by first expressing z explicitly in terms of t
b. by using the chain rule

Solution
a. By substituting $x = 1/t$ and $y = t^2$, we find that

$$z = x^2 + y^2 = \left(\frac{1}{t}\right)^2 + (t^2)^2 = t^{-2} + t^4 \quad \text{for } t \neq 0$$

Thus, $\dfrac{dz}{dt} = -2t^{-3} + 4t^3$.

b. Because $z = x^2 + y^2$ and $x = t^{-1}, y = t^2$,

$$\frac{\partial z}{\partial x} = 2x; \qquad \frac{\partial z}{\partial y} = 2y; \qquad \frac{dx}{dt} = -t^{-2}; \qquad \frac{dy}{dt} = 2t$$

Use the chain rule for one independent parameter:

$$\frac{dz}{dt} = \frac{\partial z}{\partial x}\frac{dx}{dt} + \frac{\partial z}{\partial y}\frac{dy}{dt}$$
$$= (2x)(-t^{-2}) + 2y(2t) \qquad \text{Chain rule}$$
$$= 2(t^{-1})(-t^{-2}) + 2(t^2)(2t) \qquad \text{Substitute.}$$
$$= -2t^{-3} + 4t^3$$

EXAMPLE 2   *Chain rule for one independent parameter*

Let $z = \sqrt{x^2 + 2xy}$, where $x = \cos\theta$ and $y = \sin\theta$. Find $\dfrac{dz}{d\theta}$.

Solution $\dfrac{\partial z}{\partial x} = \dfrac{1}{2}(x^2 + 2xy)^{-1/2}(2x + 2y)$ and $\dfrac{\partial z}{\partial y} = \dfrac{1}{2}(x^2 + 2xy)^{-1/2}(2x)$

Also, $\dfrac{dx}{d\theta} = -\sin\theta$ and $\dfrac{dy}{d\theta} = \cos\theta$. Use the chain rule for one independent parameter to find

$$\frac{dz}{d\theta} = \frac{\partial z}{\partial x}\frac{dx}{d\theta} + \frac{\partial z}{\partial y}\frac{dy}{d\theta}$$
$$= \tfrac{1}{2}(x^2 + 2xy)^{-1/2}(2x + 2y)(-\sin\theta) + \tfrac{1}{2}(x^2 + 2xy)^{-1/2}(2x)(\cos\theta)$$
$$= (x^2 + 2xy)^{-1/2}(x\cos\theta - x\sin\theta - y\sin\theta) \qquad\blacksquare$$

EXAMPLE 3 *Related rate application using the chain rule*

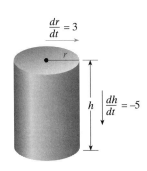

$\frac{dr}{dt} = 3$

h $\frac{dh}{dt} = -5$

A right circular cylinder is changing in such a way that its radius r is increasing at the rate of 3 in./min and its height h is decreasing at the rate of 5 in./min. At what rate is the volume of the cylinder changing when the radius is 10 in. and the height is 8 in.?

Solution The volume of the cylinder is $V = \pi r^2 h$, and we are given $\frac{dr}{dt} = 3$ and $\frac{dh}{dt} = -5$. We find that

$$\frac{\partial V}{\partial r} = \pi(2r)h \quad \text{and} \quad \frac{\partial V}{\partial h} = \pi r^2(1)$$

By the chain rule for one parameter:

$$\frac{dV}{dt} = \frac{\partial V}{\partial r}\frac{dr}{dt} + \frac{\partial V}{\partial h}\frac{dh}{dt} = 2\pi rh\frac{dr}{dt} + \pi r^2\frac{dh}{dt}$$

Thus, at the instant when $r = 10$ and $h = 8$, we have

$$\frac{dV}{dt} = 2\pi(10)(8)(3) + \pi(10)^2(-5) = -20\pi \approx -62.83185307$$

The volume is decreasing at the rate of about 62.8 in.³/min. $\qquad\blacksquare$

If $F(x, y) = 0$ defines y implicitly as a differentiable function x, then the chain rule tells us that

$$0 = \frac{\partial F}{\partial x}\frac{dx}{dx} + \frac{\partial F}{\partial y}\frac{dy}{dx} = \frac{\partial F}{\partial x} + \frac{\partial F}{\partial y}\frac{dy}{dx}$$

so

$$\frac{dy}{dx} = \frac{-\dfrac{\partial F}{\partial x}}{\dfrac{\partial F}{\partial y}} = -\frac{F_x}{F_y} \qquad \text{provided } F_y \neq 0$$

This formula provides a useful alternative to implicit differentiation; the procedure is illustrated in the following example.

EXAMPLE 4 *Implicit differentiation by formula*

If y is a differentiable function of x such that

$$\sin(x + y) + \cos(x - y) = y$$

find $\frac{dy}{dx}$.

Solution Let $F(x, y) = \sin(x+y) + \cos(x - y) - y$. Then,

$$F_x = \cos(x + y) - \sin(x - y)$$
$$F_y = \cos(x + y) + \sin(x - y) - 1$$

so

$$\frac{dy}{dx} = -\frac{F_x}{F_y} = \frac{-[\cos(x+y) + \sin(x-y) - 1]}{\cos(x+y) - \sin(x-y)}$$

■

EXAMPLE 5   *Second derivative of a function of two variables*

Let $z = f(x, y)$, where $x = at$ and $y = bt$ for constants a and b. Assuming all necessary differentiability, find d^2z/dt^2 in terms of the partial derivatives of z.

Solution   We note that $\dfrac{dx}{dt} = a$ and $\dfrac{dy}{dt} = b$, and

$$\frac{dz}{dt} = \frac{\partial z}{\partial x}\frac{dx}{dt} + \frac{\partial z}{\partial y}\frac{dy}{dt} \qquad Chain\ rule$$

We differentiate both sides with respect to t, using the chain rule again on the right:

$$\frac{d^2z}{dt^2} = \frac{\partial z}{\partial x}\frac{d^2x}{dt^2} + \frac{dx}{dt}\left[\frac{\partial^2 z}{\partial x^2}\frac{dx}{dt} + \frac{\partial^2 z}{\partial x \partial y}\frac{dy}{dt}\right] + \frac{\partial z}{\partial y}\frac{d^2y}{dt^2} + \frac{dy}{dt}\left[\frac{\partial^2 z}{\partial y \partial x}\frac{dx}{dt} + \frac{\partial^2 z}{\partial y^2}\frac{dy}{dt}\right]$$

$$= \frac{\partial z}{\partial x}(0) + a\left[\frac{\partial^2 z}{\partial x^2}a + \frac{\partial^2 z}{\partial x \partial y}b\right] + \frac{\partial z}{\partial y}(0) + b\left[\frac{\partial^2 z}{\partial y \partial x}a + \frac{\partial^2 z}{\partial y^2}b\right]$$

$$= a^2\frac{\partial^2 z}{\partial x^2} + 2ab\frac{\partial^2 z}{\partial x \partial y} + b^2\frac{\partial^2 z}{\partial y^2} \qquad Note\ \frac{\partial^2 z}{\partial x \partial y} = \frac{\partial^2 z}{\partial y \partial x}.$$

■

CHAIN RULE FOR TWO PARAMETERS

Next we shall consider the kind of composite function that occurs when x and y are both functions of *two* variables. Specifically, let $z = F(x, y)$, where $x = x(u, v)$ and $y = y(u, v)$ are both functions of two independent parameters u and v. Then $z = F[x(u, v), y(u, v)]$ is a composite function of u and v, and with suitable assumptions regarding differentiability, we can find the partial derivatives $\partial z/\partial u$ and $\partial z/\partial v$ by applying the chain rule obtained in the following theorem.

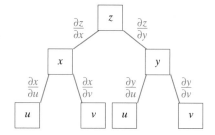

THEOREM 12.5   *The chain rule for two independent parameters*

Suppose $z = f(x, y)$ is differentiable at (x, y) and that the partial derivatives of $x = x(u, v)$ and $y = y(u, v)$ exist at (u, v). Then the composite function $z = f[x(u, v), y(u, v)]$ is differentiable at (u, v) with

$$\frac{\partial z}{\partial u} = \frac{\partial z}{\partial x}\frac{\partial x}{\partial y} + \frac{\partial z}{\partial y}\frac{\partial y}{\partial u} \quad \text{and} \quad \frac{\partial z}{\partial v} = \frac{\partial z}{\partial x}\frac{\partial x}{\partial v} + \frac{\partial z}{\partial y}\frac{\partial y}{\partial v}$$

Proof   This version of the chain rule follows immediately from the chain rule for one independent parameter. For example, if v is fixed, the composite function $z = f[x(u, v), y(u, v)]$ really depends on u alone, and we have the situation described in the chain rule of one independent variable. We apply this chain rule with a partial derivative instead of an "ordinary" derivative (because x and y are functions of more than one variable):

$$\frac{\partial z}{\partial u} = \frac{\partial z}{\partial x}\frac{\partial x}{\partial u} + \frac{\partial z}{\partial y}\frac{\partial y}{\partial u}$$

The formula for $\dfrac{\partial z}{\partial v}$ can be established in a similar fashion.

═

EXAMPLE 6   *Chain rule for two independent parameters*

Let $z = 4x - y^2$, where $x = uv^2$ and $y = u^3v$. Find $\partial z/\partial u$ and $\partial z/\partial v$.

Solution   First find the partial derivatives:

$$\frac{\partial z}{\partial x} = \frac{\partial}{\partial x}(4x - y^2) = 4 \qquad \frac{\partial z}{\partial y} = \frac{\partial}{\partial y}(4x - y^2) = -2y$$

and

$$\frac{\partial x}{\partial u} = \frac{\partial}{\partial u}(uv^2) = v^2 \qquad \frac{\partial y}{\partial u} = \frac{\partial}{\partial u}(u^3v) = 3u^2v$$

$$\frac{\partial x}{\partial v} = \frac{\partial}{\partial v}(uv^2) = 2uv \qquad \frac{\partial y}{\partial v} = \frac{\partial}{\partial v}(u^3v) = u^3$$

Therefore, the chain rule for two independent parameters gives

$$\frac{\partial z}{\partial u} = \frac{\partial z}{\partial x}\frac{\partial x}{\partial u} + \frac{\partial z}{\partial y}\frac{\partial y}{\partial u}$$

$$= (4)(v^2) + (-2y)(3u^2v) = 4v^2 - 2(u^3v)(3u^2v) = 4v^2 - 6u^5v^2$$

and

$$\frac{\partial z}{\partial v} = \frac{\partial z}{\partial x}\frac{\partial x}{\partial v} + \frac{\partial z}{\partial y}\frac{\partial y}{\partial v}$$

$$= (4)(2uv) + (-2y)(u^3) = 8uv - 2(u^3v)u^3 = 8uv - 2u^6v \qquad \blacksquare$$

The chain rules can be extended to functions of three or more variables. For instance, if $w = f(x, y, z)$ is a differentiable function of three variables and $x = x(t)$, $y = y(t), z = z(t)$ are each differentiable functions of t, then w is a differentiable composite function of t and

$$\frac{dw}{dt} = \frac{\partial w}{\partial x}\frac{dx}{dt} + \frac{\partial w}{\partial y}\frac{dy}{dt} + \frac{\partial w}{\partial z}\frac{dz}{dt}$$

In general, if $w = f(x_1, x_2, \ldots, x_n)$ is a differentiable function of the n variables $x_1, x_2, \ldots, x_n$, which in turn are differentiable functions of m variables $t_1, t_2, \ldots, t_m$, then

$$\frac{\partial w}{\partial t_1} = \frac{\partial w}{\partial x_1}\frac{\partial x_1}{\partial t_1} + \frac{\partial w}{\partial x_2}\frac{\partial x_2}{\partial t_1} + \cdots + \frac{\partial w}{\partial x_n}\frac{\partial x_n}{\partial t_1}$$

$$\vdots$$

$$\frac{\partial w}{\partial t_m} = \frac{\partial w}{\partial x_1}\frac{\partial x_1}{\partial t_m} + \frac{\partial w}{\partial x_2}\frac{\partial x_2}{\partial t_m} + \cdots + \frac{\partial w}{\partial x_n}\frac{\partial x_n}{\partial t_m}$$

EXAMPLE 7   *Chain rule for a function of several variables*

Find $\partial w/\partial s$ if $w = 4x + y^2 + z^3$, where $x = e^{rs^2}$, $y = \ln\dfrac{r + s}{t}$, and $z = rst^2$.

Solution

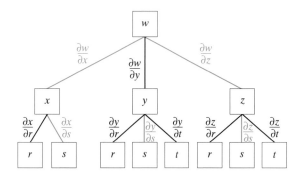

$$\frac{\partial w}{\partial s} = \frac{\partial w}{\partial x}\frac{\partial x}{\partial s} + \frac{\partial w}{\partial y}\frac{\partial y}{\partial s} + \frac{\partial w}{\partial z}\frac{\partial z}{\partial s}$$

$$= \left[\frac{\partial}{\partial x}(4x + y^2 + z^3)\right]\left[\frac{\partial}{\partial s}(e^{rs^2})\right] + \left[\frac{\partial}{\partial y}(4x + y^2 + z^3)\right]\left[\frac{\partial}{\partial s}\left(\ln\frac{r+s}{t}\right)\right]$$

$$+ \left[\frac{\partial}{\partial z}(4x + y^2 + z^3)\right]\left[\frac{\partial}{\partial s}(rst^2)\right]$$

$$= 4[e^{rs^2}(2rs)] + 2y\left(\frac{1}{(r+s)/t}\right)\left(\frac{1}{t}\right) + 3z^2(rt^2) = 8rse^{rs^2} + \frac{2y}{r+s} + 3rt^2z^2$$

In terms of r, s, and t, the partial derivative is

$$\frac{\partial w}{\partial s} = 8rse^{rs^2} + \frac{2}{r+s}\ln\frac{r+s}{t} + 3r^3s^2t^6 \qquad \blacksquare$$

12.5   Problem Set

Ⓐ 1. ▪ **What Does This Say?** Discuss the various chain rules and the need for such chain rules.

2. ▪ **What Does This Say?** Discuss the usefulness of the schematic representation for the chain rules.

3. ▪ **What Does This Say?** Write out a chain rule for three independent parameters and two variables.

In Problems 4–9, assume the given equations define y as a differentiable function of x and find dy/dx using the procedure illustrated in Example 4.

4. $x^2y + \sqrt{xy} = 4$

5. $(x^2 - y)^{3/2} + x^2y = 2$

6. $x^2y + \ln(2x + y) = 5$

7. $x\cos y + y\tan^{-1}x = x$

8. $xe^{xy} + ye^{-xy} = 3$

9. $\tan^{-1}\left(\frac{x}{y}\right) = \tan^{-1}\left(\frac{y}{x}\right)$

In Problems 10–13, the function f(x, y) depends on x and y, which in turn are each functions of t. In each case, find dz/dt two different ways:

a. *Express z explicitly in terms of t.*

b. *Use the chain rule for one parameter.*

10. $f(x, y) = 2xy + y^2$, where
$x = -3t^2$ and $y = 1 + t^3$

11. $f(x, y) = (4 + y^2)x$, where
$x = e^{2t}$ and $y = e^{3t}$

12. $f(x, y) = (1 + x^2 + y^2)^{1/2}$, where
$x = \cos 5t$ and $y = \sin 5t$

13. $f(x, y) = xy^2$, where
$x = \cos 3t$ and $y = \tan 3t$

In Problems 14–17, the function F(x, y) depends on x and y. Let $x = x(u, v)$ and $y = y(u, v)$ be given functions of u and v. Let $z = F[x(u, v), y(u, v)]$ and find the partial derivatives ∂z/∂u and ∂z/∂v in these two ways:

a. *Express z explicitly in terms of u and v.*

b. *Apply the chain rule for two independent parameters.*

14. $F(x, y) = x + y^2$, where
$x = u + v$ and $y = u - v$

15. $F(x, y) = x^2 + y^2$, where
$x = u\sin v$ and $y = u - 2v$

16. $F(x, y) = e^{xy}$, where
$x = u - v$ and $y = u + v$

17. $F(x, y) = \ln xy$, where
$x = e^{uv^2}$, $y = e^{uv}$

Write out the chain rule for the functions given in Problems 18–21.

18. $z = f(x, y)$, where $x = x(s, t), y = y(s, t)$

19. $w = f(x, y, z)$, where $x = x(s, t)$,
$y = y(s, t), z = z(s, t)$

20. $t = f(u, v)$, where $u = u(x, y, z, w)$,
$v = v(x, y, z, w)$

21. $w = f(x, y, z)$, where $x = x(s, t, u)$,
$y = y(s, t, u), z = z(s, t, u)$

Find the indicated derivatives or partial derivatives in Problems 22–27. Leave your answers in mixed form (x, y, z, t, u, v).

22. Find $\dfrac{dw}{dt}$, where $w = \ln(x + 2y - z^2)$ and

$x = 2t - 1, y = \dfrac{1}{t}, z = \sqrt{t}$.

23. Find $\dfrac{dw}{dt}$, where $w = \sin xyz$ and $x = 1 - 3t, y = e^{1-t}$,

$z = 4t$.

24. Find $\dfrac{dw}{dt}$, where $w = ze^{xy^2}$ and $x = \sin t, y = \cos t$,

$z = \tan 2t$.

25. Find $\dfrac{dw}{dt}$, where $w = e^{x^3 + yz}$ and $x = \dfrac{2}{t}$, $y = \ln(2t - 3)$, $z = t^2$.

26. Find $\dfrac{\partial w}{\partial r}$, where $w = e^{2x - y + 3z^2}$ and $x = r + s - t$, $y = 2r - 3s$, $z = \cos rst$.

27. Find $\dfrac{\partial w}{\partial r}$ and $\dfrac{\partial w}{\partial t}$, where $w = \dfrac{x + y}{2 - z}$ and $x = 2rs$, $y = \sin rt$, $z = st^2$.

Ⓑ *Find the following higher partial derivatives in Problems 28–33.*

a. $\dfrac{\partial^2 z}{\partial x \partial y}$ b. $\dfrac{\partial^2 z}{\partial x^2}$ c. $\dfrac{\partial^2 z}{\partial y^2}$

28. $x^3 + y^2 + z^2 = 5$ **29.** $xyz = 2$

30. $\ln(x + y) = y^2 + z$ **31.** $x^{-1} + y^{-1} + z^{-1} = 3$

32. $x \cos y = y + z$ **33.** $z^2 + \sin x = \tan y$

34. Let $f(x, y)$ be a differentiable function of x and y, and let $x = r \cos \theta$, $y = r \sin \theta$ for $r > 0$ and $0 < \theta < 2\pi$.

a. If $z = f[x(r, \theta), y(r, \theta)]$, find $\dfrac{\partial z}{\partial r}$ and $\dfrac{\partial z}{\partial \theta}$.

b. Show that

$$\left(\frac{\partial z}{\partial r}\right)^2 + \frac{1}{r^2}\left(\frac{\partial z}{\partial \theta}\right)^2 = \left(\frac{\partial z}{\partial x}\right)^2 + \left(\frac{\partial z}{\partial y}\right)^2$$

35. Let $z = f(x, y)$, where $x = au$ and $y = bv$, with a, b constants. Express $\partial^2 z / \partial u^2$ and $\partial^2 z / \partial v^2$ in terms of the partial derivatives of z with respect to x and y. Assume the existence and continuity of all necessary first and second partial derivatives.

36. Let (x, y, z) lie on the ellipsoid

$$\frac{x^2}{a^2} + \frac{y^2}{b^2} + \frac{z^2}{c^2} = 1$$

Without solving for z explicitly in terms of x and y, compute the higher partial derivatives

$$\frac{\partial^2 z}{\partial x^2} \quad \text{and} \quad \frac{\partial^2 z}{\partial x \partial y}$$

37. The dimensions of a rectangular box are linear functions of time, $\ell(t)$, $w(t)$, and $h(t)$. If the length and width are increasing at 2 in./sec and the height is decreasing at 3 in./sec, find the rates at which the volume V and the surface area S are changing with respect to time. If $\ell(0) = 10$, $w(0) = 8$, and $h(0) = 20$, is V increasing or decreasing when $t = 5$ sec? What about S when $t = 5$?

38. Using x hours of skilled labor and y hours of unskilled labor, a manufacturer can produce $f(x, y) = 10xy^{1/2}$ units. Currently, the manufacturer has used 30 hours of skilled labor 36 hours of unskilled labor and is planning to use 1 additional hours of skilled labor. Use calculus to estimate the corresponding change that the manufacturer should make in the level of unskilled labor so that the total output will remain the same.

39. MODELING PROBLEM   Van der Waals' equation in physical chemistry states that a gas occupying volume V at temperature T (Kelvin) exerts pressure P, where

$$\left(P + \frac{A}{V^2}\right)(V - B) = kT$$

for physical constants A, B, and k. Find the following rates:

a. The rate of change of volume with respect to temperature

b. The rate of change of pressure with respect to volume

40. MODELING PROBLEM   The concentration of a drug in the blood of a patient t hours after the drug is injected into the body intramuscularly is modeled by the Heinz function

$$C = \frac{1}{b - a}(e^{-at} - e^{-bt}) \qquad b > a$$

where a and b are parameters that depend on the patient's metabolism and the particular kind of drug being used.

a. Find the rates $\dfrac{\partial C}{\partial a}$, $\dfrac{\partial C}{\partial b}$, $\dfrac{\partial C}{\partial t}$.

b. Explore the assumption that $a = (\ln b)/t$, b constant for $t > (\ln b)/b$. In particular, what is dC/dt?

41. MODELING PROBLEM   A paint store carries two brands of latex paint. An analysis of sales figures indicates that the demand Q for the first brand is modeled by

$$Q(x, y) = 210 - 12x^2 + 18y$$

gallons/month, where x, y are the prices of the first and second brands, respectively. A separate study indicates that t months from now, the first brand will cost $x = 4 + 0.18t$ dollars/gal and the second brand will cost $y = 5 + 0.3\sqrt{t}$ dollars/gal. At what rate will the demand Q be changing with respect to time 9 months from now?

42. MODELING PROBLEM   To model the demand for the sale of bicycles, it is assumed that if 24-speed bicycles are sold for x dollars apiece and the price of gasoline is y cents per gallon, then

$$Q(x, y) = 240 - 21\sqrt{x} + 4(0.2y + 12)^{3/2}$$

bicycles will be sold each month. For this model it is furthermore assumed that, t months from now, bicycles will be selling for $x = 120 + 6t$ dollars apiece, and the price of gasoline will be $y = 80 + 10\sqrt{4t}$ cents/gal. At what rate will the monthly demand for the bicycles be changing with respect to time 4 months from now?

43. MODELING PROBLEM   At a certain factory, the amount of air pollution generated each day is modeled by the function

$$Q(E, T) = 127E^{2/3}\, T^{1/2}$$

where E is the number of employees and $T(°C)$ is the average temperature during the workday. Currently, there are 142 employees and the average temperature is 18°C. If the average daily temperature is falling at the rate of 0.23°/day, and the number of employees is increasing at the rate of 3/month, what is the corresponding effect on the rate of pollution? Express your answer in units/day. For this model, assume there are 22 workdays/month.

44. **MODELING PROBLEM**   The combined resistance R produced by three variable resistances $R_1, R_2,$ and R_3 connected in parallel is modeled by the formula

$$\frac{1}{R} = \frac{1}{R_1} + \frac{1}{R_2} + \frac{1}{R_3}$$

Suppose at a certain instant, $R_1 = 100$ ohms, $R_2 = 200$ ohms, $R_3 = 300$ ohms, and R_1 and R_3 are decreasing at the rate of 1.5 ohms/s while R_2 is increasing at the rate of 2 ohms/s. How fast is R changing with respect to time at this instant? Is it increasing or decreasing?

In Problems 45–51, assume that all functions have whatever derivatives or partial derivatives are necessary for the problem to be meaningful.

45. If $z = f(uv^2)$, show that

$$2u\frac{\partial z}{\partial u} - v\frac{\partial z}{\partial v} = 0$$

Hint: Let $w = uv^2$ and apply the chain rule.

46. If $z = u + f(u^2v^2)$, show that

$$u\frac{\partial z}{\partial u} - v\frac{\partial z}{\partial v} = u$$

Hint: Let $w = u^2v^2$ and apply the chain rule.

47. If $z = f(u - v, v - u)$, show that

$$\frac{\partial z}{\partial u} + \frac{\partial z}{\partial v} = 0$$

48. If $z = u + f(uv)$, show that $u\dfrac{\partial z}{\partial u} - v\dfrac{\partial z}{\partial v} = u$.

49. If $w = f\left(\dfrac{r - s}{s}\right)$, show that $r\dfrac{\partial w}{\partial r} + s\dfrac{\partial w}{\partial s} = 0$.

50. If $z = f(x^2 - y^2)$, evaluate $y\dfrac{\partial z}{\partial x} + x\dfrac{\partial z}{\partial y}$.

51. If $z = xy + f(x^2 - y^2)$, show that

$$y\frac{\partial z}{\partial x} - x\frac{\partial z}{\partial y} = y^2 - x^2$$

52. Let $w = f(t)$ be a differentiable function of t, where $t = (x^2 + y^2 + z^2)^{1/2}$. Show that

$$\left(\frac{dw}{dt}\right)^2 = \left(\frac{\partial w}{\partial x}\right)^2 + \left(\frac{\partial w}{\partial y}\right)^2 + \left(\frac{\partial w}{\partial z}\right)^2$$

53. Suppose f is a twice differentiable function of one variable, and let $z = f(x^2 + y^2)$. Find

a. $\dfrac{\partial^2 z}{\partial x^2}$ b. $\dfrac{\partial^2 z}{\partial y^2}$ c. $\dfrac{\partial^2 z}{\partial x\partial y}$

54. Find $\dfrac{d^2z}{d\theta^2}$ where f is a twice differentiable function of one variable and $z = f(\cos\theta, \sin\theta)$. *Hint:* Let $x = \cos\theta$ and $y = \sin\theta$. Leave your answer in terms of x and y.

55. Let f and g be twice differentiable functions of one variable, and let

$$u(x, t) = f(x + ct) + g(x - ct)$$

for a constant c. Show that $\dfrac{\partial^2 u}{\partial t^2} = c^2 \dfrac{\partial^2 u}{\partial x^2}$

Hint: Let $r = x + ct; s = x - ct$.

56. Suppose $z = f(x, y)$ has continuous second-order partial derivatives. If $x = e^r \cos\theta$ and $y = e^r \sin\theta$, show that

$$\frac{\partial^2 z}{\partial x^2} + \frac{\partial^2 z}{\partial y^2} = e^{-2r}\left[\frac{\partial^2 z}{\partial r^2} + \frac{\partial^2 z}{\partial \theta^2}\right]$$

57. Let $F(x, y, z)$ be a function of three variables with continuous partial derivatives F_x, F_y, F_z in a certain region where $F(x, y, z) = C$ for some constant C. Use the chain rule for two parameters and the fact that $\partial y/\partial x = 0$ to show that

$$\frac{\partial z}{\partial x} = \frac{-\dfrac{\partial F}{\partial x}}{\dfrac{\partial F}{\partial z}} = \frac{-F_x}{F_z}$$

58. Suppose the system

$$\begin{cases} xu + yv - uv = 0 \\ yu - xv + uv = 0 \end{cases}$$

can be solved for u and v in terms of x and y, so that $u = u(x, y)$ and $v = v(x, y)$. Use implicit differentiation to find the partial derivatives $\dfrac{\partial u}{\partial x}$ and $\dfrac{\partial v}{\partial x}$.

59. Repeat Problem 58 for the system

$$\begin{cases} xu + yv - uv = 0 \\ yu - xv + uv = 0 \end{cases}$$

for $\dfrac{\partial u}{\partial y}$ and $\dfrac{\partial v}{\partial y}$.

60. Suppose that F and G are functions of three variables and that it is possible to solve the equations $F(x, y, z) = 0$ and $G(x, y, z) = 0$ for y and z in terms of x, so that $y = y(x)$ and $z = z(x)$. Use the chain rule to express dy/dx and dz/dx in terms of the partial derivatives of F and G. Assume these partials are continuous and that

$$\frac{\partial F}{\partial y}\frac{\partial G}{\partial z} \neq \frac{\partial F}{\partial z}\frac{\partial G}{\partial y}$$

12.6   *Directional Derivatives and the Gradient*

IN THIS SECTION   **the directional derivative, directional derivative as a slope and as a rate, the gradient, maximal property of the gradient, normal property of the gradient, tangent planes, and normal lines**

In Section 12.3, we defined the partial derivatives f_x and f_y and interpreted these derivatives as slopes of tangent lines on the surface $z = f(x, y)$ in planes parallel to xz- and yz-planes, respectively, as shown in Figure 12.19.

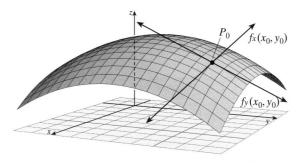

■ **FIGURE 12.19**   Partial derivatives of $z = f(x, y)$

Our next goal is to see how partial derivatives can be used to find derivatives in other directions.

THE DIRECTIONAL DERIVATIVE

In Chapter 3 we defined the *slope of a curve* at a point to be the ratio of the change in the dependent variable with respect to the change in the independent variable at the given point. To determine the slope of the tangent line at a point $P_0(x_0, y_0)$ on a surface defined by $z = f(x, y)$, we need to specify the *direction* in which we wish to measure. We do this by using vectors. In Section 12.3 we found the slope parallel to the xz-plane to be the partial derivative $f_x(x_0, y_0)$. We could have specified this direction in terms of the vector **i** (x-direction). Similarly, $f_y(x, y)$ could have been specified in terms of the **j** vector. Finally, to measure the slope of the tangent line in an *arbitrary* direction, we use a unit vector $\mathbf{u} = u_1\mathbf{i} + u_2\mathbf{j}$ in that direction.

To find the desired slope, we look at the intersection of the surface with the vertical plane passing through the point P_0 parallel to the vector **u**, as shown in Figure 12.20. This vertical plane intersects the surface to form a curve C, and we define the slope of the surface at P_0 in the direction of **u** to be the slope of the tangent line to the curve C defined by **u** at that point.

We summarize this idea of slope *in a particular direction* with the following definition.

WARNING Remember, **u** must be a *unit* vector. ←

Directional Derivative

Let f be a function of two variables, and let $\mathbf{u} = u_1\mathbf{i} + u_2\mathbf{j}$ be a unit vector. The **directional derivative of f at $P_0(x_0, y_0)$ in the direction of u** is given by

$$D_u f(x_0, y_0) = \lim_{h \to 0} \frac{f(x_0 + hu_1, y_0 + hu_2) - f(x_0, y_0)}{h}$$

provided the limit exists.

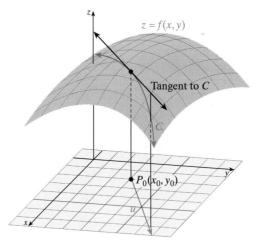

■ **FIGURE 12.20**   The directional derivative $D_u f(x_0, y_0)$ is the slope of the tangent line to the curve on the surface $z = f(x, y)$ in the direction of the unit vector **u** at $P_0(x_0, y_0)$.

EXAMPLE 1   *Computing partial derivatives as directional derivatives*

Use the definition of directional derivative to show that the directional derivatives of $f(x, y)$ in the directions of the positive x- and y-axes are f_x and f_y, respectively.

Solution   In the direction of the positive x-axis, $\mathbf{u} = \mathbf{i}$, so that $u_1 = 1$ and $u_2 = 0$.

$$D_i f(x_0, y_0) = \lim_{h \to 0} \frac{f(x_0 + h, y_0) - f(x_0, y_0)}{h}$$
Definition of directional derivative, where $u_1 = 1$ and $u_2 = 0$

$$= f_x(x_0, y_0)$$
Definition of f_x

In the direction of the positive y-axis, $\mathbf{u} = \mathbf{j}$, so that $u_1 = 0$ and $u_2 = 1$.

$$D_j f(x_0, y_0) = \lim_{h \to 0} \frac{f(x_0, y_0 + h) - f(x_0, y_0)}{h}$$
Definition of directional derivative, where $u_1 = 0$ and $u_2 = 1$

$$= f_y(x_0, y_0)$$
Definition of f_y ■

The definition of directional derivative is similar to the definition of the derivative of a function of a single variable. Just as with a single variable, it is difficult to apply the definition directly. Fortunately, the following theorem allows us to find directional derivatives more efficiently than by using the definition.

THEOREM 12.6   *Directional derivatives using partials*

Let $f(x, y)$ be a function that is differentiable at $P_0(x_0, y_0)$. Then f has a directional derivative in the direction of the unit vector $\mathbf{u} = u_1\mathbf{i} + u_2\mathbf{j}$ given by

$$D_u f(x_0, y_0) = f_x(x_0, y_0)u_1 + f_y(x_0, y_0)u_2$$

Proof   We define a function F of a single variable h by

$$F(h) = f(x_0 + hu_1, y_0 + hu_2)$$

so that

$$D_u f(x_0, y_0) = \lim_{h \to 0} \frac{f(x_0 + hu_1, y_0 + hu_2) - f(x_0, y_0)}{h}$$

$$= \lim_{h \to 0} \frac{F(h) - F(0)}{h} = F'(0)$$

Apply the chain rule with $x = x_0 + hu_1$ and $y = y_0 + hu_2$:

$$F'(h) = \frac{dF}{dh} = \frac{\partial f}{\partial x} \frac{dx}{dh} + \frac{\partial f}{\partial y} \frac{dy}{dh} = f_x(x, y)u_1 + f_y(x, y)u_2$$

When $h = 0$, we have $x = x_0$ and $y = y_0$, so that

$$D_u f(x_0, y_0) = F'(0) = \frac{\partial f}{\partial x} u_1 + \frac{\partial f}{\partial y} u_2 = f_x(x_0, y_0)u_1 + f_y(x_0, y_0)u_2 \qquad =\!=$$

| **EXAMPLE 2** | *Finding a directional derivative using partials* |

Find the directional derivative of $f(x, y) = 3 - 2x^2 + y^3$ at the point $P(1, 2)$ in the direction of the unit vector $\mathbf{u} = \dfrac{1}{2}\mathbf{i} - \dfrac{\sqrt{3}}{2}\mathbf{j}$.

Solution   First find the partial derivatives $f_x(x, y) = -4x$ and $f_y(x, y) = 3y^2$. Then because $u_1 = \dfrac{1}{2}$ and $u_2 = -\dfrac{\sqrt{3}}{2}$, we have

$$D_u f(1, 2) = f_x(1, 2)\left(\frac{1}{2}\right) + f_y(1, 2)\left(-\frac{\sqrt{3}}{2}\right)$$

$$= -4(1)\left(\frac{1}{2}\right) + 3(2)^2\left(-\frac{\sqrt{3}}{2}\right) = -2 - 6\sqrt{3} \approx -12.4 \qquad \blacksquare$$

DIRECTIONAL DERIVATIVE AS A SLOPE AND AS A RATE

Notice that the directional derivative is a number. This number can be interpreted as the slope of a tangent line to $z = f(x, y)$ or as a rate of change of the function $z = f(x, y)$.

Directional Derivative as a Slope   As an illustration, we look at the surface defined by $z = 3 - 2x^2 + y^3$ (see Figure 12.21). The intersection of this surface with a plane aligned with the vector $\mathbf{u} = \frac{1}{2}\mathbf{i} - \dfrac{\sqrt{3}}{2}\mathbf{j}$ is a curve labeled C, and the directional derivative is the slope of the tangent line to C at the point on the surface above $P(1, 2)$. This interpretation of the directional derivative is illustrated in Figure 12.21.

Directional Derivative as a Rate   The directional derivative $D_u f(x_0, y_0)$ can also be interpreted as the rate at which the function $z = f(x, y)$ changes as a point moves from $P_0(x_0, y_0)$ in the direction of the unit vector $\mathbf{u}$. Thus, in Example 2, we say that the function $f(x, y) = 3 - 2x^2 + y^3$ changes at the rate of $-2 - 6\sqrt{3}$ as a point moves from $P_0(1, 2)$ in the direction of the unit vector $\mathbf{u} = \frac{1}{2}\mathbf{i} - \dfrac{\sqrt{3}}{2}\mathbf{j}$.

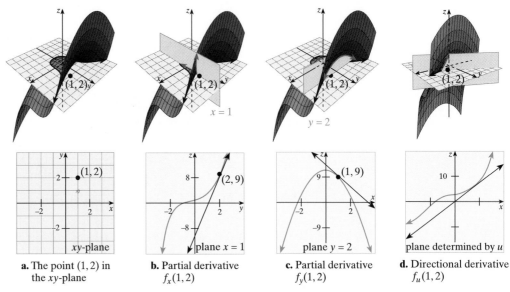

a. The point $(1, 2)$ in the xy-plane

b. Partial derivative $f_x(1, 2)$

c. Partial derivative $f_y(1, 2)$

d. Directional derivative $f_u(1, 2)$

■ **FIGURE 12.21**   Graph of $z = 3 - 2x^2 + y^3$ and some derivatives

THE GRADIENT

The directional derivative $D_u f(x, y)$ can be expressed concisely in terms of a vector function called the *gradient,* which has many important uses in mathematics. The gradient of a function of two variables may be defined as follows.

> **Gradient**
>
> Let f be a differentiable functions at (x, y) and let $f(x, y)$ have partial derivatives $f_x(x, y)$ and $f_y(x, y)$. Then the **gradient** of f, denoted by ∇f (pronounced "del eff"), is a vector given by
>
> $$\nabla f(x, y) = f_x(x, y)\mathbf{i} + f_y(x, y)\mathbf{j}$$
>
> The value of the gradient at the point $P_0(x_0, y_0)$ is denoted by
>
> $$\nabla f_0 = f_x(x_0, y_0)\mathbf{i} + f_y(x_0, y_0)\mathbf{j}$$

WARNING   Think of the symbol ∇ as an "operator" on a function that produces a vector. Another notation for ∇f is **grad** $f(x, y)$.

EXAMPLE 3   *Finding the gradient of a given function*

Find $\nabla f(x, y)$ for $f(x, y) = x^2 y + y^3$.

Solution   Begin with the partial derivatives:

$$f_x(x, y) = \frac{\partial}{\partial x}(x^2 y + y^3) = 2xy \quad \text{and} \quad f_y(x, y) = \frac{\partial}{\partial y}(x^2 y + y^3) = x^2 + 3y^2$$

Then,

$$\nabla f(x, y) = 2xy\mathbf{i} + (x^2 + 3y^2)\mathbf{j}$$ ■

The following theorem shows how the directional derivative can be expressed in terms of the gradient.

THEOREM 12.7   *The gradient formula for the directional derivative*

If f is a differentiable function of x and y, then the directional derivative at the point $P_0(x_0, y_0)$ in the direction of the unit vector **u** is

$$D_u f(x_0, y_0) = \nabla f_0 \cdot \mathbf{u}$$

Proof   Because $\nabla f_0 = f_x(x_0, y_0)\mathbf{i} + f_y(x_0, y_0)\mathbf{j}$ and $\mathbf{u} = u_1\mathbf{i} + u_2\mathbf{j}$, we have

$$D_u f(x_0, y_0) = \nabla f_0 \cdot \mathbf{u} = f_x(x_0, y_0)u_1 + f_y(x_0, y_0)u_2 \qquad \text{Dot product}$$

EXAMPLE 4   *Using the gradient formula to compute a directional derivative*

Find the directional derivative of $f(x, y) = \ln(x^2 + y^3)$ at $P_0(1, -3)$ in the direction of $\mathbf{v} = 2\mathbf{i} - 3\mathbf{j}$.

Solution
$$f_x(x, y) = \frac{2x}{x^2 + y^3}, \quad \text{so} \quad f_x(1, -3) = -\frac{2}{26}$$

$$f_y(x, y) = \frac{3y^2}{x^2 + y^3}, \quad \text{so} \quad f_y(1, -3) = -\frac{27}{26}$$

$$\nabla f_0 = \nabla f(1, -3) = -\frac{2}{26}\mathbf{i} - \frac{27}{26}\mathbf{j}$$

A unit vector in the direction of **v** is

$$\mathbf{u} = \frac{\mathbf{v}}{\|\mathbf{v}\|} = \frac{2\mathbf{i} - 3\mathbf{j}}{\sqrt{2^2 + (-3)^2}} = \frac{1}{\sqrt{13}}(2\mathbf{i} - 3\mathbf{j})$$

Thus,

$$D_u(x, y) = \nabla f \cdot \mathbf{u} = \left(-\frac{2}{26}\right)\left(\frac{2}{\sqrt{13}}\right) + \left(-\frac{27}{26}\right)\left(-\frac{3}{\sqrt{13}}\right) = \frac{77\sqrt{13}}{338} \approx 0.82 \qquad \blacksquare$$

Although a differentiable function of one variable $f(x)$ has exactly one derivative $f'(x)$, a differentiable function of two variables $F(x, y)$ has two partial derivatives and a multitude of directional derivatives. Is there any single mathematical entity for functions of several variables that is the analogue of the derivative of a function of a single variable? The properties listed in the following theorem suggest that the gradient is the analogue we seek.

THEOREM 12.8   *Basic properties of the gradient*

Let f and g be differentiable functions. Then

Constant rule	$\nabla c = \mathbf{0}$ for any constant c
Linearity rule	$\nabla(af + bg) = a\nabla f + b\nabla g$ for constants a and b
Product rule	$\nabla(fg) = f\nabla g + g\nabla f$
Quotient rule	$\nabla\left(\dfrac{f}{g}\right) = \dfrac{g\nabla f - f\nabla g}{g^2}, \quad g \neq 0$
Power rule	$\nabla(f^n) = nf^{(n-1)}\nabla f$

Proof   **Linearity rule**

$$\nabla(af + bg) = (af + bg)_x\mathbf{i} + (af + bg)_y\mathbf{j} = (af_x + bg_x)\mathbf{i} + (af_y + bg_y)\mathbf{j}$$
$$= af_x\mathbf{i} + bg_x\mathbf{i} + af_y\mathbf{j} + bg_y\mathbf{j} = a(f_x\mathbf{i} + f_y\mathbf{j}) + b(g_x\mathbf{i} + g_y\mathbf{j})$$
$$= a\nabla f + b\nabla g$$

Power rule

$$\nabla f^n = [f^n]_x\,\mathbf{i} + [f^n]_y\,\mathbf{j} = nf^{n-1}f_x\mathbf{i} + nf^{n-1}f_y\mathbf{j}$$
$$= nf^{n-1}[f_x\mathbf{i} + f_y\mathbf{j}] = nf^{n-1}\nabla f$$

The other rules are left for the problem set (Problems 58 and 59).

MAXIMAL PROPERTY OF THE GRADIENT

In applications, it is often useful to compute the greatest rate of increase (or decrease) of a given function at a specified point. The direction in which this occurs is called the direction of **steepest ascent** (or **steepest descent**). For example, suppose the function $z = f(x, y)$ gives the altitude of a skier coming down a slope, and we want to state a theorem that will give the skier the *compass direction* of the path of steepest descent (see Figure 12.22b). We emphasize the words "compass direction" because the gradient gives direction in the *xy*-plane and does not itself point up or down the mountain. The following theorem shows how the optimal direction is determined by the gradient (see Figure 12.22).

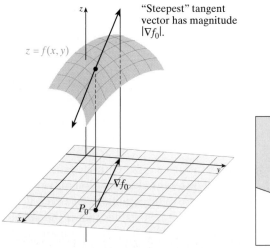

"Steepest" tangent vector has magnitude $|\nabla f_0|$.

$z = f(x, y)$

∇f_0

P_0

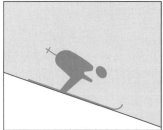

a. The optimal direction property of the gradient   **b.** Skier on a slope

■ **FIGURE 12.22** Steepest ascent or steepest descent

THEOREM 12.9 *Optimal direction property of the gradient*

Suppose *f* is differentiable and let ∇f_0 denote the gradient at P_0. Then if $\nabla f_0 \neq \mathbf{0}$:

a. The largest value of the directional derivative of $D_u f$ is $\|\nabla f_0\|$ and occurs when the unit vector **u** points in the direction of ∇f_0.
b. The smallest value of $D_u f$ is $-\|\nabla f_0\|$ and occurs when **u** points in the direction of $-\nabla f_0$.

Proof If **u** is any unit vector, then

$$D_u f = \nabla f_0 \cdot \mathbf{u} = \|\nabla f_0\|(\|\mathbf{u}\|\cos\theta) = \|\nabla f_0\|\cos\theta$$

where θ is the angle between ∇f_0 and **u**. But $\cos\theta$ assumes its largest value 1 at $\theta = 0$—that is, when **u** points in the direction ∇f_0. Thus, the largest possible value of $D_u f$ is

$$D_u f = \|\nabla f_0\|(1) = \|\nabla f_0\|$$

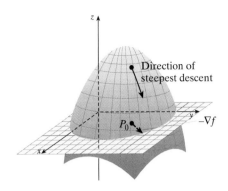

The gradient of f is the vector ∇f in the xy-plane. This vector points in the direction of steepest ascent or descent at a given point.

Statement **b** may be established in a similar fashion by noting that $\cos \theta$ assumes its smallest value -1 when $\theta = \pi$. This value occurs when **u** points toward $-\nabla f_0$, and in this direction

$$D_u f = \|\nabla f_0\|(-1) = -\|\nabla f_0\|.$$

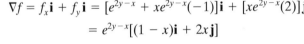

> ■ *What This Says:* The theorem states that at P_0, the function f increases most rapidly in the direction of the gradient ∇f_0 and decreases most rapidly in the opposite direction.

EXAMPLE 5   *Optimal rate of increase and decrease*

In what direction is the function defined by $f(x, y) = xe^{2y-x}$ increasing most rapidly at the point $P_0(2, 1)$, and what is the optimal rate of increase? In what direction is f decreasing most rapidly?

Solution   We begin by finding the gradient of f:

$$\nabla f = f_x \mathbf{i} + f_y \mathbf{i} = [e^{2y-x} + xe^{2y-x}(-1)]\mathbf{i} + [xe^{2y-x}(2)]\mathbf{j}$$
$$= e^{2y-x}[(1 - x)\mathbf{i} + 2x\mathbf{j}]$$

At $(2, 1)$, $\nabla f_0 = e^{2(1)-2}[(1 - 2)\mathbf{i} + 2(2)\mathbf{j}] = -\mathbf{i} + 4\mathbf{j}$. The most rapid rate of increase is $\|\nabla f_0\| = \sqrt{(-1)^2 + (4)^2} = \sqrt{17}$ and it occurs in the direction of $-\mathbf{i} + 4\mathbf{j}$. The most rapid rate of decrease occurs in the direction of $-\nabla f_0 = \mathbf{i} - 4\mathbf{j}$. ■

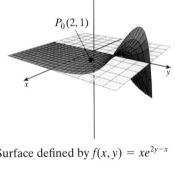

Surface defined by $f(x, y) = xe^{2y-x}$

NORMAL PROPERTY OF THE GRADIENT

Suppose S is a level surface of the function defined by $f(x, y, z)$; that is, $f(x, y, z) = K$ for some constant K. Then if $P_0(x_0, y_0, z_0)$ is a point on S, the following theorem shows that the gradient ∇f_0 at P_0 is a vector that is **normal** (that is, perpendicular) to the tangent vector of every smooth curve on S that passes through P_0 (see Figure 12.23).

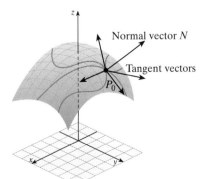

■ **FIGURE 12.23** The normal property of the gradient

========

THEOREM 12.10   *The normal property of the gradient*

Suppose the function f is differentiable at the point P_0 and that the gradient at P_0 satisfies $\nabla f_0 \neq \mathbf{0}$. If $K = f(x_0, y_0, z_0)$, then ∇f_0 is orthogonal to the level surface $f(x, y, z) = K$ at P_0.

Proof   Let C be any smooth curve on the level surface $f(x, y, z) = K$ that passes through $P_0(x_0, y_0, z_0)$, and describe the curve C by the vector function $\mathbf{R}(t) = x(t)\mathbf{i} + y(t)\mathbf{j} + z(t)\mathbf{k}$ for all t in some interval I. We shall show that the gradient ∇f_0 is orthogonal to the tangent vector $d\mathbf{R}/dt$ at P_0.

Because C lies on the level surface, any point $P(x(t), y(t), z(t))$ on C must satisfy $f[x(t), y(t), z(t)] = K$, and by applying the chain rule, we obtain

$$\frac{d}{dt}[f(x(t), y(t), z(t))] = f_x(x, y, z)\frac{dx}{dt} + f_y(x, y, z)\frac{dy}{dt} + f_z(x, y, z)\frac{dz}{dt}$$

Suppose $t = t_0$ at P_0. Then

$$\frac{d}{dt}[f(x(t), y(t), z(t))]\bigg|_{t=t_0}$$

$$= f_x(x(t_0), y(t_0), z(t_0))\frac{dx}{dt} + f_y(x(t_0), y(t_0), z(t_0))\frac{dy}{dt} + f_z(x(t_0), y(t_0), z(t_0))\frac{dz}{dt}$$

$$= \nabla f_0 \cdot \frac{d\mathbf{R}}{dt}$$

Remember that $\dfrac{d\mathbf{R}}{dt} = \dfrac{dx}{dt}\mathbf{i} + \dfrac{dy}{dt}\mathbf{j} + \dfrac{dz}{dt}\mathbf{k}$. We also know that $f(x(t), y(t), z(t)) = K$ for all t in I (because the curve C lies on the level surface $f(x, y, z) = K$). Thus, we have

$$\frac{d}{dt}\{f[x(t), y(t), z(t)]\} = \frac{d}{dt}(K) = 0$$

and it follows that $\nabla f_0 \cdot \dfrac{d\mathbf{R}}{dt} = 0$. We are given that $\nabla f_0 \neq \mathbf{0}$, and $d\mathbf{R}/dt \neq \mathbf{0}$ because the curve C is smooth. Thus, ∇f_0 is orthogonal to $d\mathbf{R}/dt$, as required.   ====

■   **What This Says:**   The gradient ∇f_0 at each point P_0 on the level surface $f(x, y, z) = K$ is normal to the surface at P_0. That is, ∇f_0 is orthogonal to all tangent lines to curves on $f(x, y, z) = K$ through P_0.

Here is an example in which f involves only two variables, so $f(x, y) = K$ is a level curve in the plane instead of a level surface in space.

EXAMPLE 6   *Finding a vector normal to a level curve*

Sketch the level curve corresponding to $C = 1$ for the function $f(x, y) = x^2 - y^2$ and find a normal vector at the point $P_0(2, \sqrt{3})$.

Solution   The level curve for $C = 1$ is a hyperbola given by $x^2 - y^2 = 1$, as shown in Figure 12.24 (along with the surface defined by $f(x, y) = x^2 - y^2$).

The gradient vector is perpendicular to the level curve, so we have

$$\nabla f = f_x\mathbf{i} + f_y\mathbf{j} = 2x\mathbf{i} - 2y\mathbf{j}$$

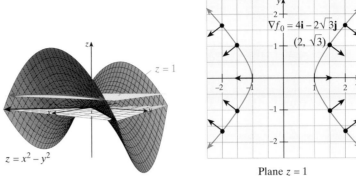

■ FIGURE 12.24   The surface $f(x, y) = x^2 - y^2$ and the level curve for which $C = 1$

At the point $(2, \sqrt{3})$, $\nabla f_0 = 4\mathbf{i} - 2\sqrt{3}\mathbf{j}$ is the required normal. This normal vector and a few others are shown in Figure 12.24.   ■

EXAMPLE 7   *Finding a vector that is normal to a level surface*

Find a vector that is normal to the level surface $x^2 + 2xy - yz + 3z^2 = 7$ at the point $P_0(1, 1, -1)$.

Solution   The gradient vector at P_0 is perpendicular to the level surface:

$$\nabla f = f_x\mathbf{i} + f_y\mathbf{j} + f_z\mathbf{k} = (2x + 2y)\mathbf{i} + (2x - z)\mathbf{j} + (6z - y)\mathbf{k}$$

At the point $(1, 1, -1)$, $\nabla f_0 = 4\mathbf{i} + 3\mathbf{j} - 7\mathbf{k}$ is the required normal.   ■

EXAMPLE 8   *Heat flow application*

The set of points (x, y) with $0 \le x \le 5$ and $0 \le y \le 5$ is a square in the first quadrant of the xy-plane. Suppose this square is heated in such a way that $T(x, y) = x^2 + y^2$ is the temperature at the point $P(x, y)$. In what direction will heat flow from the point $P_0(3, 4)$?

Solution   The flow of heat in the region is given by a vector function $\mathbf{H}(x, y)$, whose value at each point (x, y) depends on x and y. From physics it is known that $\mathbf{H}(x, y)$ will be perpendicular to the isothermal curves $T(x, y) = C$ for C constant. The gradient ∇T and all its multiples point in such a direction. Therefore, we can express the heat flow as $\mathbf{H} = -k\nabla T$, where k is a positive constant (called the **thermal conductivity**) and the negative sign is introduced to account for the fact that heat flows "downhill" (that is, toward a decreasing temperature).

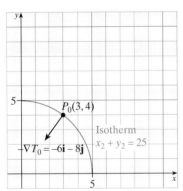

■ FIGURE 12.25   The heat flow P_0 is in the direction of $-\nabla T_0 = -6\mathbf{i} - 8\mathbf{j}$.

Because $T(3, 4) = 25$, the point $P_0(3, 4)$ lies on the isotherm $T(x, y) = 25$, which is part of the circle $x^2 + y^2 = 25$, as shown in Figure 12.25. We know that the heat flow $\mathbf{H}_0$ at P_0 will satisfy $\mathbf{H}_0 = -k\nabla T_0$, where ∇T_0 is the gradient at P_0. Because $\nabla T = 2x\mathbf{i} + 2y\mathbf{j}$, we see that $\nabla T_0 = 6\mathbf{i} + 8\mathbf{j}$. Thus, the heat flow at P_0 satisfies

$$\mathbf{H}_0 = -k\nabla T_0 = -k(6\mathbf{i} + 8\mathbf{j})$$

Because the thermal conductivity k is positive, we can say that heat flows from P_0 in the direction of the unit vector $\mathbf{u}$ given by

$$\mathbf{u} = \frac{-(6\mathbf{i} + 8\mathbf{j})}{\sqrt{(-6)^2 + (-8)^2}} = -\frac{3}{5}\mathbf{i} - \frac{4}{5}\mathbf{j}$$ ■

The directional derivative and gradient concepts can easily be extended to functions of three or more variables. For a function of three variables, $f(x, y, z)$, the gradient ∇f is defined by

$$\nabla f = f_x \mathbf{i} + f_y \mathbf{j} + f_z \mathbf{k}$$

and the directional derivative $D_u f$ of $f(x, y, z)$ at $P_0(x_0, y_0, z_0)$ in the direction of the unit vector $\mathbf{u}$ is given by

$$D_u f = \nabla f_0 \cdot \mathbf{u}$$

where, as before, ∇f_0 is the gradient ∇f evaluated at P_0. The basic properties of the gradient (Theorem 12.8) are still valid, as are the optimization properties in Theorem 12.9. Similar definitions and properties are valid for functions of more than three variables.

EXAMPLE 9   *Directional derivative of a function of three variables*

Let $f(x, y, z) = xy \sin xz$. Find ∇f_0 at the point $P_0(1, -2, \pi)$ and then compute the directional derivative of f at P_0 in the direction of the vector $\mathbf{v} = -2\mathbf{i} + 3\mathbf{j} - 5\mathbf{k}$.

Solution   Begin with the partial derivatives:

$$f_x = y \sin xz + xy(z \cos xz); \quad f_x(1, -2, \pi) = -2 \sin \pi - 2\pi \cos \pi = 2\pi$$
$$f_y = x \sin xz; \quad f_y(1, -2, \pi) = 1 \sin \pi = 0$$
$$f_z = xy(x \cos xz); \quad f_z(1, -2, \pi) = (1)(-2)(1)\cos \pi = 2$$

Thus, the gradient of f at P_0 is

$$\nabla f_0 = 2\pi \mathbf{i} + 2\mathbf{k}$$

To find $D_u f$ we need $\mathbf{u}$, the unit vector in the direction of $\mathbf{v}$:

$$\mathbf{u} = \frac{\mathbf{v}}{\|\mathbf{v}\|} = \frac{-2\mathbf{i} + 3\mathbf{j} - 5\mathbf{k}}{\sqrt{(-2)^2 + (3)^2 + (-5)^2}} = \frac{1}{\sqrt{38}}(-2\mathbf{i} + 3\mathbf{j} - 5\mathbf{k})$$

Finally,

$$D_u f(1, -2, \pi) = \nabla f_0 \cdot \mathbf{u} = \frac{1}{\sqrt{38}}(-4\pi - 10) \approx -3.66$$ ∎

TANGENT PLANES AND NORMAL LINES

Tangent planes and normal lines to a surface are the natural extensions to $\mathbb{R}^3$ of the tangent and normal lines we examined in $\mathbb{R}^2$. Suppose S is a surface and $\mathbf{N}$ is a vector normal to S at the point P_0. We would intuitively expect the normal line and the tangent plane to S at P_0 to be, respectively, the line through P_0 with the direction of $\mathbf{N}$ and the plane perpendicular to the line at P_0 (see Figure 12.26).

These observations lead us to the following definition.

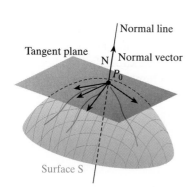

Normal line

Tangent plane $\mathbf{N}$ Normal vector

P_0

Surface S

■ **FIGURE 12.26**   Tangent plane and normal line

Normal Line to a Surface; Tangent Plane

Suppose the surface S has a nonzero normal vector $\mathbf{N}$ at the point P_0. Then the line through P_0 parallel to $\mathbf{N}$ is called the **normal line** to S at P_0, and the plane through P_0 with normal vector $\mathbf{N}$ is the **tangent plane** to S at P_0.

By analogy with the single-variable case, we would expect a surface S to have a tangent plane precisely where it can be represented by a differentiable function. In particular, if S has an equation of the form $F(x, y, z) = C$, where C is a constant and F is a function differentiable at P_0, the normal property of a gradient tells us that the gradient ∇F_0 at P_0 is normal to S (if $\nabla F_0 \neq \mathbf{0}$) and that S must therefore have a tangent plane at P_0.

> **EXAMPLE 10**   *Finding the tangent plane and normal line to a given surface*

Find equations for the tangent plane and the normal line at the point $P_0(1, -1, 2)$ on the surface S given by $x^2y + y^2z + z^2x = 5$.

Solution   We need to rewrite this problem so that the normal property of the gradient theorem applies. Let $F(x, y, z) = x^2y + y^2z + z^2x$, and consider S to be the level surface $F(x, y, z) = 5$. The gradient ∇F is normal to S at P_0. We find that

$$\nabla F(x, y, z) = (2xy + z^2)\mathbf{i} + (x^2 + 2yz)\mathbf{j} + (y^2 + 2xz)\mathbf{k}$$

so the normal vector at P_0 is

$$\mathbf{N} = \nabla F_0 = \nabla F(1, -1, 2) = 2\mathbf{i} - 3\mathbf{j} + 5\mathbf{k}$$

Hence, the required tangent plane is

$$2(x - 1) - 3(y + 1) + 5(z - 2) = 0 \quad \text{or} \quad 2x - 3y + 5z = 15$$

The normal line to the surface at P_0 is

$$x = 1 + 2t, \quad y = -1 - 3t, \quad z = 2 + 5t \qquad \blacksquare$$

By generalizing the procedure illustrated in the preceding example, we are led to the following formulas for the tangent plane and normal line. (Also see Figure 12.27.)

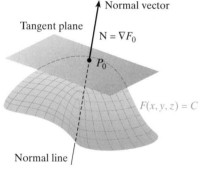

Normal vector

$\mathbf{N} = \nabla F_0$

Tangent plane

P_0

$F(x, y, z) = C$

Normal line

■ **FIGURE 12.27**   The tangent plane and normal line to a surface

---

Formulas for the Tangent Plane and Normal Line to a Surface

Suppose S is a surface with the equation $F(x, y, z) = C$ and let $P_0(x_0, y_0, z_0)$ be a point on S where F is differentiable. Then the **equation of the tangent plane** to S at P_0 is

$$F_x(x_0, y_0, z_0)(x - x_0) + F_y(x_0, y_0, z_0)(y - y_0) + F_z(x_0, y_0, z_0)(z - z_0) = 0$$

and the **equation of the normal line** to S at P_0 is

$$x = x_0 + F_x(x_0, y_0, z_0)t$$
$$y = y_0 + F_y(x_0, y_0, z_0)t$$
$$x = z_0 + F_z(x_0, y_0, z_0)t$$

provided F_x, F_y, and F_z are not all zero.

WARNING   Note that in the special case where $z = f(x, y)$, we have $F(x, y, z) = f(x, y) - z = 0$. Then $F_x = f_x, F_y = f_y$, and $F_z = -1$, and the equation of the tangent plane becomes

$$f_x(x_0, y_0, z_0)(x - x_0) + f_y(x_0, y_0, z_0)(y - y_0) - (z - z_0) = 0$$

which is equivalent to the tangent plane formula given in Section 12.4.   ▬

| **EXAMPLE 11** | *Equations of the tangent plane and the normal line* |

Find the equations for the tangent plane and the normal line to the cone $z^2 = x^2 + y^2$ at the point where $x = 3, y = 4,$ and $z > 0$.

Solution   If $P_0(x_0, y_0, z_0)$ is the point of tangency and $x_0 = 3, y_0 = 4,$ and $z_0 > 0,$ then

$$z_0 = \sqrt{x_0^2 + y_0^2} = \sqrt{9 + 16} = 5$$

If we consider $F(x, y, z) = x^2 + y^2 - z^2$, then the cone can be regarded as the level surface $F(x, y, z) = 0$. The partial derivatives of F are

$$F_x = 2x, \qquad F_y = 2y, \qquad F_z = -2z$$

so at $P_0(3, 4, 5),$

$$F_x(3, 4, 5) = 6, \qquad F_y(3, 4, 5) = 8, \qquad F_z(3, 4, 5) = -10$$

Thus the tangent plane is

$$6(x - 3) + 8(y - 4) - 10(z - 5) = 0$$

or $3x + 4y - 5z = 0,$ and the normal line is

$$x = 3 + 6t, \qquad y = 4 + 8t, \qquad z = 5 - 10t$$

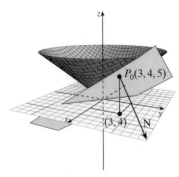

Tangent plane and normal line to the cone $z^2 = x^2 + y^2$ at $(3, 4, 5)$

12.6   Problem Set

A *Find the gradient of the functions given in Problems 1–10.*

1. $f(x, y) = x^2 - 2xy$

2. $f(x, y) = 3x + 4y^2$

3. $f(x, y) = \dfrac{y}{x} + \dfrac{x}{y}$

4. $f(x, y) = \ln(x^2 + y^2)$

5. $f(u, v) = ue^{3-v}$

6. $f(u, v) = e^{u+v}$

7. $f(x, y) = \sin(x + 2y)$

8. $f(x, y, z) = xyz^2$

9. $g(x, y, z) = xe^{y+3z}$

10. $f(x, y, z) = \dfrac{xy - 1}{z + x}$

*Compute the directional derivative of the functions given in Problems 11–16 at the point P_0 in the direction of the given vector **v**.*

Function	Point P_0	Vector **v**
11. $f(x, y) = x^2 + xy$	$(1, -2)$	$\mathbf{i} + \mathbf{j}$
12. $f(x, y) = \dfrac{e^{-x}}{y}$	$(2, -1)$	$-\mathbf{i} + \mathbf{j}$
13. $f(x, y) = \ln(x^2 + 3y)$	$(1, 1)$	$\mathbf{i} + \mathbf{j}$
14. $f(x, y) = \ln(3x + y^2)$	$(0, 1)$	$\mathbf{i} - \mathbf{j}$
15. $f(x, y) = \sec(xy - y^3)$	$(2, 0)$	$-\mathbf{i} - 3\mathbf{j}$
16. $f(x, y) = \sin xy$	$(\sqrt{\pi}, \sqrt{\pi})$	$3\pi\mathbf{i} - \pi\mathbf{j}$

Find a unit vector (if possible) that is normal to each surface given in Problems 17–24 at the prescribed point, and then find the equation of the tangent plane at the given point.

17. $x^2 + y^2 + z^2 = 3$ at $(1, -1, 1)$

18. $x^4 + y^4 + z^4 = 3$ at $(1, -1 - 1)$

19. $\cos z = \sin(x + y)$ at $(\frac{\pi}{2}, \frac{\pi}{2}, \frac{\pi}{2})$

20. $\sin(x + y) + \tan(y + z) = 1$ at $(\frac{\pi}{4}, \frac{\pi}{4}, -\frac{\pi}{4})$

21. $\ln\left(\dfrac{x}{y - z}\right) = 0$ at $(2, 5, 3)$

22. $\ln\left(\dfrac{x - y}{y + z}\right) = x - z$ at $(1, 0, 1)$

23. $ze^{x+2y} = 3$ at $(2, -1, 3)$

24. $ze^{x^2 - y^2} = 3$ at $(1, 1, 3)$

Find the direction from P_0 in which the given function f increases most rapidly and compute the magnitude of the greatest rate of increase in Problems 25–34.

25. $f(x, y) = 3x + 2y - 1$
   a. $P_0(1, -1)$ b. $P_0(1, 1)$

26. $f(x, y) = 1 - x^2 - y^2$
   a. $P_0(1, 2)$ b. $P_0(0, 0)$

27. $f(x, y) = x^3 + y^3$
   a. $P_0(3, -3)$ b. $P_0(-3, 3)$

28. $f(x, y) = ax + by + c;\ P_0(a, b)$

29. $f(x, y, z) = ax^2 + by^2 + cz^2;\ P_0(a, b, c)$
   Assume $a^4 + b^4 + c^4 \neq 0.$

30. $f(x, y) = ax^3 + by^3;\ P_0(a, b)$

31. $f(x, y) = \ln\sqrt{x^2 + y^2};\ P_0(1, 2)$

32. $f(x, y) = \sin xy;\ P_0\left(\dfrac{\sqrt{\pi}}{3}, \dfrac{\sqrt{\pi}}{2}\right)$

33. $f(x, y, z) = (x + y)^2 + (y + z)^2 + (z + x)^2; P_0(2, -1, 2)$

34. $f(x, y, z) = z \ln\left(\dfrac{y}{x}\right); P_0(1, e, -1)$

B *In Problems 35–38, find a unit vector that is normal to the given graph at the point $P_0(x_0, y_0)$ on the graph. Assume that a, b, and c are constants.*

35. the line $ax + by = c$

36. the circle $x^2 + y^2 = a^2$

37. the ellipse $\dfrac{x^2}{a^2} + \dfrac{y^2}{b^2} = 1$

38. the hyperbola $\dfrac{x^2}{a^2} - \dfrac{y^2}{b^2} = 1$

39. Find the directional derivative of $f(x, y) = x^2 + y^2$ at the point $P_0(1, 1)$ in the direction of the unit vector **u** shown in Figure 12.28.

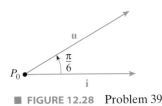

■ **FIGURE 12.28** Problem 39

40. Find the directional derivative of $f(x, y) = x^2 + xy + y^2$ at $P_0(1, -1)$ in the direction toward the origin.

41. Find the directional derivative of $f(x, y) = e^{x^2 y^2}$ at $P_0(1, -1)$ in the direction toward $Q(2, 3)$.

42. Let $f(x, y, z) = 2x^2 - y^2 + 3z^2 - 8x - 4y + 201$, and let P_0 be the point $(2, -\frac{3}{2}, \frac{1}{2})$.
   a. Find ∇f.
   b. Find $\cos \theta$, where θ is the angle between ∇f_0 and the vector toward the origin from P_0.

43. Let $f(x, y, z) = xyz$, and let **u** be a unit vector perpendicular to both $\mathbf{v} = \mathbf{i} - 2\mathbf{j} + 3\mathbf{k}$ and $\mathbf{w} = 2\mathbf{i} + \mathbf{j} - \mathbf{k}$. Find the directional derivative of f at $P_0(1, -1, 2)$ in the direction of **u**.

44. Let $f(x, y, z) = ye^{x+z} + ze^{y-x}$. At the point $P(2, 2, -2)$, find the unit vector pointing in the direction of most rapid increase of f.

45. MODELING PROBLEM   Suppose a box in space given by $0 \le x \le 2, 0 \le y \le 2, 0 \le z \le 2$ is temperature controlled so that the temperature at a point $P(x, y, z)$ in the box is modeled by $T(x, y, z) = xy + yz + xz$. A heat-seeking missile is located at $P_0(1, 1, 1)$. In what direction will the missile move for the temperature to increase as quickly as possible? What is the maximum rate of change of the temperature at the point P_0?

46. MODELING PROBLEM   A metal plate covering the rectangular region $0 < x \le 6, 0 < y \le 5$ is charged electrically in such a way that the potential at each point (x, y) is inversely proportional to the square of its

distance from the origin. If an object is at the point $(3, 4)$, in which direction should it move to increase the potential most rapidly?

47. MODELING PROBLEM   A hiker is walking on a mountain path when it begins to rain. If the surface of the mountain is modeled by $z = 1 - 3x^2 - \frac{5}{2}y^2$ (where x, y, and z are in miles) and the rain begins when the hiker is at the point $P_0(\frac{1}{4}, -\frac{1}{2})$, in what direction should she head to descend the mountainside most rapidly?

48. Let f have continuous partial derivatives, and assume the maximal directional derivative of f at $(0, 0)$ is equal to 100 and is attained in the direction of the vector $(0, 0)$ toward $(3, -4)$. Find the gradient ∇f at $(0, 0)$.

49. Suppose at the point $P_0(-1, 2)$, a certain function $f(x, y)$ has directional derivative 8 in the direction of $\mathbf{v}_1 = 3\mathbf{i} - 4\mathbf{j}$ when 1 in the direction of $\mathbf{v}_2 = 12\mathbf{i} + 5\mathbf{j}$. What is the directional derivative of f at P_0 in the direction of $\mathbf{v} = 3\mathbf{i} - 5\mathbf{j}$?

50. Suppose $f(x, y)$ has directional derivatives $D_{v_1} f = 2$ at $P_0(1, 5)$ in the direction of $\mathbf{v}_1 = 2\mathbf{i} + 3\mathbf{j}$ and $D_{v_2} f = -5$ in the direction of $\mathbf{v}_2 = 3\mathbf{i} - 5\mathbf{j}$. What is the directional derivative of f at P_0 in the direction of $\mathbf{v}_3 = 2\mathbf{i} - 3\mathbf{j}$?

51. Let f have continuous partial derivatives and suppose the maximal directional derivative of f at $P_0(1, 2)$ has magnitude 50 and is attained in the direction from P_0 toward $Q(3, -4)$. Use this information to find $\nabla f(1, 2)$.

52. Let $T(x, y) = 1 - x^2 - 2y^2$ be the temperature at each point $P(x, y)$ in the plane. A heat-loving bug is placed in the plane at the point $P_0(-1, 1)$. Find the path that the bug should take to stay as warm as possible. *Hint*: Assume that at each point on the bug's path, the tangent line will point in the direction for which T increases most rapidly.

53. Repeat Problem 52 for the temperature function $T(x, y) = 1 - ax^2 - by^2$ and starting at the point (x_0, y_0). Assume $a > 0, b > 0$, and $(x_0, y_0) \ne (0, 0)$.

┌───┐
│ ▬   **Technology Window**   ▲ ▼ │
├───┤
│ **54.** Write a computer program to solve Problem 53. │
└───┘

55. SPY PROBLEM   Just as the Spy is about to catch up with Scélérat (Problem 68 of Chapter 11 Supplementary Problems), the snow gives way and he falls into an ice cavern. He staggers to his feet and removes his skis. Why is it so warm? Good grief—the cave is a large roasting oven! Fortunately, he is wearing his heat-detector ring, which indicates the direction of greatest temperature decrease. Suppose the bunker is coordinatized so that the temperature at each point (x, y) on the floor of the bunker is given by

$$T(x, y) = 3(x - 6)^2 + 1.5(y - 1)^2 + 41$$

degrees Fahrenheit, where x and y are in feet. The Spy begins at the point $(1, 5)$ and stumbles across the room at the rate of 4 ft/min, always moving in the direction of maximum temperature decrease. But he can last no more than 2 minutes under these conditions! Assuming that there is an escape hole at the point where the temperature is minimal, does he make it or is the Spy toast at least?

56. Recall from precalculus that an ellipse is the set of all points $P(x, y)$ such that the sum of the distances from P to two fixed points (the foci) is constant. Let $P(x, y)$ be a point on the ellipse, and let r_1 and r_2 denote the respective distances from P to the two foci, F_1 and F_2, as shown in Figure 12.29.

SMH

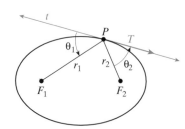

■ **FIGURE 12.29**   Unit tangent to an ellipse

a. Show that $\mathbf{T} \cdot \nabla(r_1 + r_2) = 0$, where $\mathbf{T}$ is the unit tangent to the ellipse at P.
b. Use part **a** to show that the tangent line to the ellipse at P makes equal angles with the lines joining P to the foci (that is, $\theta_1 = \theta_2$ in Figure 12.29).

C 57. **MODELING PROBLEM**   A particle P_1 with mass m_1 is located at the origin, and a particle P_2 with mass 1 unit is located at the point (x, y, z). According to Newton's law of universal gravitation, the force P_1 exerts on P_2 is modeled by

$$\mathbf{F} = \frac{-Gm_1(x\mathbf{i} + y\mathbf{j} + z\mathbf{k})}{r^3}$$

where r is the distance between P_1 and P_2, and G is the gravitational constant.

a. Starting from the fact that $r^2 = x^2 + y^2 + z^2$, show that

$$\frac{\partial}{\partial x}\left(\frac{1}{r}\right) = \frac{-x}{r^3}, \quad \frac{\partial}{\partial y}\left(\frac{1}{r}\right) = \frac{-y}{r^3}, \quad \frac{\partial}{\partial z}\left(\frac{1}{r}\right) = \frac{-z}{r^3}$$

b. The function $V = -Gm_1/r$ is called the **potential energy** function for the system. Show that $\mathbf{F} = -\nabla V$.

58. Verify each of the following properties for functions of two variables.
a. $\nabla(cf) = c(\nabla f)$ for constant c
b. $\nabla(f + g) = \nabla f + \nabla g$
c. $\nabla(f/g) = \dfrac{g\nabla f - f\nabla g}{g^2}$, $g \neq 0$

59. Verify the product rule

$$\nabla(fg) = f\nabla g + g\nabla f$$

for functions of three variables.

60. Find a general formula for the directional derivative $D_{\mathbf{u}} f$ of the function $f(x, y)$ at the point $P(x_0, y_0)$ in the direction of the unit vector $\mathbf{u} = (\cos\theta)\mathbf{i} + (\sin\theta)\mathbf{j}$. Apply your formula to obtain the directional derivative of $f(x, y) = xy^2 e^{x-2y}$ at $P_0(-1, 3)$ in the direction of the unit vector

$$\mathbf{u} = (\cos\tfrac{\pi}{6})\mathbf{i} + (\sin\tfrac{\pi}{6})\mathbf{j}$$

61. Suppose that $\mathbf{u}$ and $\mathbf{v}$ are unit vectors and that f has continuous partial derivatives. Show that

$$D_{\mathbf{u}+\mathbf{v}} f = \frac{1}{\|\mathbf{u} + \mathbf{v}\|}(D_{\mathbf{u}} f + D_{\mathbf{v}} f)$$

62. If $\mathbf{a}$ is a constant vector and $\mathbf{R} = x\mathbf{i} + y\mathbf{j} + z\mathbf{k}$, show that $\nabla(\mathbf{a} \cdot \mathbf{R}) = \mathbf{a}$.

63. Let $\mathbf{R} = x\mathbf{i} + y\mathbf{j} + z\mathbf{k}$, and let

$$r = \|\mathbf{R}\| = \sqrt{x^2 + y^2 + z^2}$$

a. Show that ∇r is a unit vector in the direction of $\mathbf{R}$.
b. Show that $\nabla(r^n) = nr^{n-2}\mathbf{R}$, for any positive integer n.

64. Suppose the surfaces $F(x, y, z) = 0$ and $G(x, y, z) = 0$ both pass through the point $P_0(x_0, y_0, z_0)$ and that the gradient ∇F_0 and ∇G_0 both exist. Show that the two surfaces are tangent at P_0 if and only if

$$\nabla F_0 \times \nabla G_0 = 0$$

12.7   Extrema of Functions of Two Variables

IN THIS SECTION   relative extrema, second partials test, absolute extrema of continuous functions   ■

There are many practical situations in which it is necessary or useful to know the largest and smallest values of a function of two variables. For example, if $T(x, y)$ is the temperature at a point (x, y) in a plate, where are the hottest and coldest points in the plate and what are these extreme temperatures? A hazardous waste dump is bounded by the curve $F(x, y) = 0$. What are the largest and smallest distances from a given interior point P_0? We begin our study of extrema with some terminology.

Absolute Extrema

The function $f(x, y)$ is said to have an **absolute maximum** at (x_0, y_0) if $f(x_0, y_0) \geq f(x, y)$ for all (x, y) in the domain D of f. Similarly, f has an **absolute minimum** at (x_0, y_0) if $f(x_0, y_0) \leq f(x, y)$ for all (x, y) in D. Collectively, absolute maxima and minima are called **absolute extrema.**

In Chapter 4, we located absolute extrema of a function of one variable by first finding *relative extrema*, those values of $f(x)$ that are larger or smaller than those at all "nearby" points. The relative extrema of a function of two variables may be defined as follows.

Relative Extrema

Let f be a function defined at (x_0, y_0). Then

$f(x_0, y_0)$ is a **relative maximum** if $f(x, y) \leq f(x_0, y_0)$ for all (x, y) in an open disk containing (x_0, y_0).

$f(x_0, y_0)$ is a **relative minimum** if $f(x, y) \geq f(x_0, y_0)$ for all (x, y) in an open disk containing (x_0, y_0).

Collectively, relative maxima and minima are called **relative extrema.**

In Chapter 4, we found that on a closed, bounded interval $[a, b]$, a continuous function f must attain both an absolute maximum and an absolute minimum.

THEOREM 12.11   *Extreme value theorem for a function of two variables*

A function of two variables $f(x, y)$ assumes an absolute extremum on any closed, bounded set S in the plane where it is continuous.

Proof   The proof of this theorem is beyond the scope of this text and is usually given in an advanced calculus course.

We also found in Chapter 4 that an absolute extremum of a function f of a single variable can occur either at an endpoint of an interval in the domain of f or at an interior critical point where it is a relative extremum—that is, at a point where either $f'(x)$ does not exist or $f'(x) = 0$. In Section 12.6, we observed that the gradient ∇f is the two-variable analogue to the derivative in the single-variable case. Thus, it is reasonable to expect extrema of a function of two variables to occur in one of the following situations:

1. either on the boundary of the domain of f, or
2. at points in the interior where the gradient ∇f does not exist, or
3. where $\nabla f = \mathbf{0}$; that is, where $f_x = f_y = 0$.

We begin by considering relative extrema, and then turn to the more general question of finding absolute extrema.

RELATIVE EXTREMA

In Chapter 4, we observed that relative extrema of the function f correspond to "peaks and valleys" on the graph of f, and the same observation can be made about

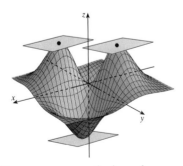

Extrema occur on the boundary where the tangent plane is horizontal, or where no tangent plane exists

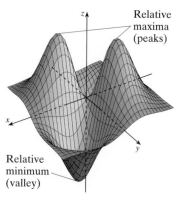

■ FIGURE 12.30   Relative extrema occur at peaks and valleys.

relative extrema in the two-variable case, as seen in Figure 12.30. For a function f of one variable, we found that the relative extrema occur where $f'(x) = 0$ or $f'(x)$ does not exist. The following theorem shows that the relative extrema of a function of two variables can be located similarly.

THEOREM 12.12   *Partial derivative criteria for relative extrema*

If f has a relative extremum (maximum or minimum) at $P_0(x_0, y_0)$ and partial derivatives f_x and f_y both exist at $f_x(x_0, y_0) = f_y(x_0, y_0)$, then

$$f_x(x_0, y_0) = f_y(x_0, y_0) = 0$$

Proof   Let $F(x) = f(x, y_0)$. Then $F(x)$ must have a relative extremum at $x = x_0$, so $F'(x_0) = 0$, which means that $f_x(x_0, y_0) = 0$. Similarly, $G(y) = f(x_0, y)$ has a relative extremum at $y = y_0$, so $G'(y_0) = 0$ and $f_y(x_0, y_0) = 0$. Thus, we must have *both* $f_x(x_0, y_0) = 0$ and $f_y(x_0, y_0) = 0$, as claimed.

WARNING   There is a horizontal tangent plane at each extreme point where the first partial derivatives exist. However, this does *not* say that whenever a horizontal tangent plane occurs at a point P, there must be an extremum there. All that can be said is that such a point P is a *possible* location for a relative extremum. ←

In single-variable calculus, we referred to a number x_0 where $f'(x_0)$ does not exist or $f'(x_0) = 0$ as a *critical number* and the point $(x_0, f(x_0))$ as a *critical point*. This terminology is extended to functions of two variables as follows.

Critical Points

A **critical point** of a function f defined on an open set S is a point (x_0, y_0) in S where either one of the following is true:

a. $f_x(x_0, y_0) = f_y(x_0, y_0) = 0$.

b. $f_x(x_0, y_0)$ or $f_y(x_0, y_0)$ does not exist (one or both).

EXAMPLE 1   *Distinguishing critical points*

Discuss the nature of the critical point $(0, 0)$ for the quadric surfaces

a. $z = x^2 + y^2$ b. $z + x^2 + y^2 = 1$ c. $z = y^2 - x^2$

Solution The graphs of these quadric surfaces are shown in Figure 12.31.

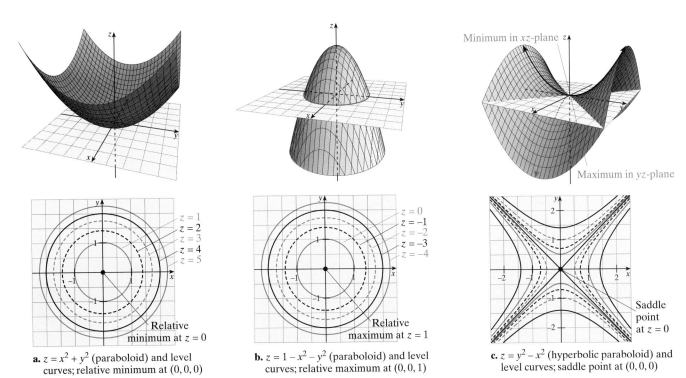

a. $z = x^2 + y^2$ (paraboloid) and level curves; relative minimum at $(0, 0, 0)$

b. $z = 1 - x^2 - y^2$ (paraboloid) and level curves; relative maximum at $(0, 0, 1)$

c. $z = y^2 - x^2$ (hyperbolic paraboloid) and level curves; saddle point at $(0, 0, 0)$

■ **FIGURE 12.31**   Classification of critical points

Let $f(x, y) = x^2 + y^2, g(x, y) = 1 - x^2 - y^2$, and $h(x, y) = y^2 - x^2$. We find the critical points:

a. $f_x(x, y) = 2x, f_y(x, y) = 2y$; the critical point is $(0, 0)$. The function f has a relative minimum at $(0, 0)$ because x^2 and y^2 are both nonnegative, yielding $x^2 + y^2 \geq 0$ for all nonzero x and y.

b. $g_x(x, y) = -2x, g_y(x, y) = -2y$; critical point $(0, 0)$. The function g has a relative maximum at $(0, 0)$ because $z = 1 - x^2 - y^2$ and x^2 and y^2 are both nonnegative, so the largest value of z occurs at $(0, 0)$.

c. $h_x(x, y) = -2x, h_y(x, y) = 2y$; critical point $(0, 0)$. The function h has neither a relative maximum nor a relative minimum at $(0, 0)$. When $z = 0, h$ is a minimum on the y-axis (where $x = 0$) and a maximum on the x-axis (where $y = 0$).   ■

If (x_0, y_0) is a critical point for $f(x, y)$ that is neither a relative maximum nor a relative minimum, it may be a **saddle point.** Such a point is a *maximum* in one direction and a *minimum* in another direction and so has a saddle shape, as shown in Figure 12.31c.

SECOND PARTIALS TEST

The previous example points to the need for some sort of a test to determine the nature of a critical point. In Chapter 4, we developed the second derivative test as a means for classifying a critical point of f as a relative maximum or minimum. According to this test, if $f'(c) = 0$, then at $x = c$, a relative maximum occurs if $f''(c) < 0$ and a relative minimum occurs if $f''(c) > 0$. If $f''(c) = 0$, the test is inconclusive. The analogous result for the two-variable case may be stated as follows.

Second Partials Test

Let $f(x, y)$ have a critical point at $P_0(x_0, y_0)$ and assume that f has continuous second-order partial derivatives in a disk centered at (x_0, y_0). Let

$$D = f_{xx}(x_0, y_0)f_{yy}(x_0, y_0) - [f_{xy}(x_0, y_0)]^2$$

Then,

A **relative maximum** occurs at P_0 if

$$D > 0 \quad \text{and} \quad f_{xx}(x_0, y_0) < 0$$
$$(\text{or } f_{yy}(x_0, y_0) < 0)$$

$z = 1 - x^2 - y^2$ (See Ex 1b)

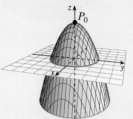

A **relative minimum** occurs at P_0 if

$$D > 0 \quad \text{and} \quad f_{xx}(x_0, y_0) > 0$$
$$(\text{or } f_{yy}(x_0, y_0) > 0)$$

$z = x^2 + y^2$ (See Ex 1a)

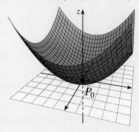

A **saddle point** occurs at P_0 if $D < 0$.

$z = y^2 - x^2$ (See Ex 1c)

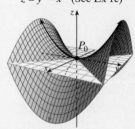

If $D = 0$, then the test is **inconclusive.**

Summary:

Critical Point	D	f_{xx}	type
(x_1, y_1)	Pos.	Neg.	Rel. max.
(x_2, y_2)	Pos.	Pos.	Rel. Min.
(x_3, y_3)	Neg.	—	Saddle pt.
(x_4, y_4)	Zero		Inconclusive

When $D(x_0, y_0) = 0$, the critical point (x_0, y_0) is said to be **degenerate.** Otherwise, it is **nondegenerate.** Note that D can be expressed in *determinant form* as

$$D = \begin{vmatrix} f_{xx} & f_{xy} \\ f_{xy} & f_{yy} \end{vmatrix}$$

where all the partials are evaluated at (x_0, y_0). Some students find this the easiest form of D to remember.

EXAMPLE 2   *Second partials test with a relative maximum*

Find all critical points on the graph of $f(x, y) = 1 - x^2 - y^2$ and use the second partials test to classify each critical point as a relative extremum or a saddle point.

Solution $f_x(x, y) = -2x, f_y(x, y) = -2y$, so the only critical point is at $(0, 0)$.
$f_{xx}(x, y) = -2, f_{xy}(x, y) = 0$, and $f_{yy}(x, y) = -2$, so

$$D = \begin{vmatrix} -2 & 0 \\ 0 & -2 \end{vmatrix} = 4$$

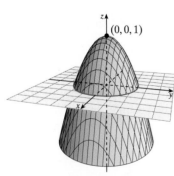

$(0, 0, 1)$

Graph of $f(x, y) = 1 - x^2 - y^2$

Because $D = 4 > 0$ for all (x, y), and because $f_{xx}(0, 0) = -2 < 0$, the second partials test tells us that a relative maximum occurs at $(0, 0)$.   ∎

| **EXAMPLE 3** | *Second partials test with a relative maximum and a saddle point* |

Find all critical points on the graph of $f(x, y) = 8x^3 - 24xy + y^3$, and use the second partials test to classify each point as a relative extremum or a saddle point.

Solution $f_x(x, y) = 24x^2 - 24y, f_y(x, y) = -24x + 3y^2$

To find the critical points, solve

$$\begin{cases} 24x^2 - 24y = 0 \\ -24x + 3y^2 = 0 \end{cases}$$

From the first equation, $y = x^2$; substitute this into the second equation to find

$$-24x + 3(x^2)^2 = 0$$
$$x^4 - 8x = 0$$
$$x(x^3 - 8) = 0$$
$$x(x - 2)(x^2 + 2x + 4) = 0$$
$$x = 0, 2 \qquad \text{The solutions of } x^2 + 2x + 4 = 0 \text{ are not real.}$$

If $x = 0$, then $y = 0$, and if $x = 2$, then $y = 4$, so the critical points are $(0, 0), (2, 4)$. To obtain D, we first find $f_{xx}(x, y) = 48x, f_{xy}(x, y) = -24$, and $f_{yy}(x, y) = 6y$ and then compute

$$D = \begin{vmatrix} f_{xx} & f_{xy} \\ f_{xy} & f_{yy} \end{vmatrix} = \begin{vmatrix} 48x & -24 \\ -24 & 6y \end{vmatrix} = 288xy - 576$$

At $(0, 0), D = -576 < 0$, so there is a saddle point at $(0, 0)$. At $(2, 4), D = 288(2)(4) - 576 = 1,728 > 0$ and $f_{xx}(2, 4) = 96 > 0$, so there is a relative minimum at $(2, 4)$.

To view the situation graphically, we calculate the coordinates of the saddle point $(0, 0, 0)$, and the relative minimum $(2, 4, -64)$, as shown in Figure 12.32.

Summary:

Critical Point	D	f_{xx}	type
$(0, 0)$	Neg.		Saddle
$(2, 4)$	Pos.	Pos.	Rel. min.

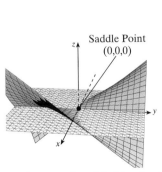

a. View near the origin (showing the saddle point)

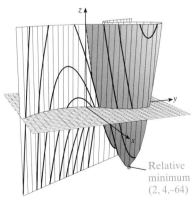

b. View away from the origin (showing the relative minimum point)

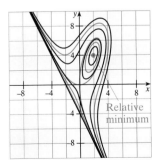

c. Level curves

■ **FIGURE 12.32**   Graph of $f(x, y) = 8x^3 - 24xy + y^3$ ∎

EXAMPLE 4   *Extrema when the second partials test fails*

Find all relative extrema and saddle points on the graph of

$$f(x, y) = x^2y^4$$

Solution $f_x(x, y) = 2xy^4, f_y(x, y) = 4x^2y^3$; we can see the critical points occur only whenever $x = 0$ or $y = 0$; that is, every point on the x- or y-axis is a critical point.

Because $f_{xx}(x, y) = 2y^4, f_{xy}(x, y) = 8xy^3, f_{yy}(x, y) = 12x^2y^2$, we have

$$D = \begin{vmatrix} f_{xx} & f_{xy} \\ f_{xy} & f_{yy} \end{vmatrix} = \begin{vmatrix} 2y^4 & 8xy^3 \\ 8xy^3 & 12x^2y^2 \end{vmatrix} = 24x^2y^6 - 64x^2y^6 = -40x^2y^6$$

For any critical point $(x_0, 0)$ or $(0, y_0)$, $D = 0$; so the second partials test fails. But $f(x, y) = 0$ for every critical point (because either $x = 0$ or $y = 0$, or both) and because $f(x, y) = x^2y^4 > 0$ when $x \neq 0$ and $y \neq 0$, it follows that each critical point must be a relative minimum. ∎

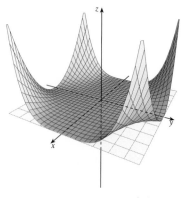

Graph of $f(x, y) = x^2y^4$

The second partials test provides explicit information about the case where f is nondegenerate at $P_0(x_0, y_0)$, but if f is degenerate, practically anything can happen, as shown in Problem 54. In that problem you are asked to show that $(0, 0)$ is a degenerate critical point for each of the following functions:

$$f(x, y) = x^4 - y^4 \qquad g(x, y) = x^2y^2 \qquad h(x, y) = x^3 + y^3$$

and the following conclusions can be made:

$f(x, y)$ has a saddle point at $(0, 0)$.

$g(x, y)$ has a relative minimum at $(0, 0)$.

$h(x, y)$ has neither kind of extremum nor a saddle point at $(0, 0)$.

ABSOLUTE EXTREMA OF CONTINUOUS FUNCTIONS

The extreme value theorem (Theorem 12.11) assures us that a continuous function f of two variables must have both an absolute maximum and an absolute minimum on any closed, bounded set S. Each extremum occurs either at a boundary point of S or at a critical point in the interior of S. Here is a procedure for finding absolute extrema:

Procedure for Determining Absolute Extrema

Given a function f that is continuous on a closed, bounded set S:

    Step 1:   Find all critical points of f in S.

    Step 2:   Find all points on the boundary of S where absolute extrema can occur (boundary critical points, endpoints, etc.).

    Step 3:   Compute the value of $f(x_0, y_0)$ for each of the points (x_0, y_0) found in steps 1 and 2.

Evaluation: The absolute maximum of f on S is the largest of the values computed in step 3, and the absolute minimum is the smallest of the computed values.

EXAMPLE 5   *Finding absolute extrema*

Find the absolute extrema of the function $f(x, y) = e^{x^2 - y^2}$ over the disk $x^2 + y^2 \leq 1$.

Solution

Step 1: $f_x(x, y) = 2xe^{x^2 - y^2}$ and $f_y(x, y) = -2ye^{x^2 - y^2}$. These partial derivatives are defined for all (x, y). Because $f_x(x, y) = f_y(x, y) = 0$ only when $x = 0$ and $y = 0$, it follows that $(0, 0)$ is the only critical point of f and it is inside the disk.

Step 2: Examine the values of f on the boundary curve $x^2 + y^2 = 1$. Because $y^2 = 1 - x^2$ on the boundary of the disk, we find that

$$f(x, y) = e^{x^2 - (1 - x^2)} = e^{2x^2 - 1}$$

We need to find the largest and smallest values of $F(x) = e^{2x^2 - 1}$ for $-1 \leq x \leq 1$. Since

$$F'(x) = 4xe^{2x^2 - 1}$$

we see that $F'(x) = 0$ only when $x = 0$ (since $e^{2x^2 - 1}$ is always positive). At $x = 0$, we have $y^2 = 1 - 0^2$, so $y = \pm 1$; and $(0, 1)$ and $(0, -1)$ are boundary critical points. At the endpoints of the interval $-1 \leq x \leq 1$, the corresponding points are $(1, 0)$ and $(-1, 0)$.

Step 3: Compute the value of f for the points found in steps 1 and 2:

Points to check	Compute $f(x_0, y_0) = e^{x_0^2 - y_0^2}$
$(0, 0)$	$f(0, 0) = e^0 = 1$
$(0, 1)$	$f(0, 1) = e^{-1}$; minimum
$(0, -1)$	$f(0, -1) = e^{-1}$; minimum
$(1, 0)$	$f(1, 0) = e$; maximum
$(-1, 0)$	$f(-1, 0) = e$; maximum

Evaluation: As indicated in the preceding table, the absolute maximum value of f on the given disk is e at $(1, 0)$ and $(-1, 0)$, and the absolute minimum value is e^{-1} at $(0, 1)$ and $(0, -1)$. ∎

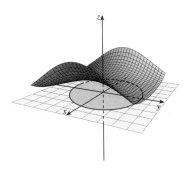

EXAMPLE 6 *Minimum distance from a point to a plane*

Find the shortest distance from the point $(0, 3, 4)$ to the plane $x + 2y + z = 5$.

Solution The distance from a point (x, y, z) to $(0, 3, 4)$ is

$$d = \sqrt{(x - 0)^2 + (y - 3)^2 + (z - 4)^2}$$

However, because (x, y, z) is on the plane $x + 2y + z = 5$, we know

$$z = 5 - x - 2y$$

so that $d = \sqrt{x^2 + (y - 3)^2 + (5 - x - 2y - 4)^2}$. Instead of minimizing d, we can minimize the expression

$$d^2 = f(x, y) = x^2 + (y - 3)^2 + (1 - x - 2y)^2$$

To find the critical values, we solve

$$f_x = 2x - 2(1 - x - 2y) = 4x + 4y - 2 = 0$$
$$f_y = 2(y - 3) - 4(1 - x - 2y) = 4x + 10y - 10 = 0$$

The only critical point is $\left(-\frac{5}{6}, \frac{4}{3}\right)$. Also, $f_{xx} = 4$, $f_{yy} = 10$, $f_{xy} = 4$, so $D > 0$, which means there is a relative minimum at $\left(-\frac{5}{6}, \frac{4}{3}\right)$. Intuitively, we see that this relative minimum must also be an absolute minimum because there must be exactly one point on the plane that is closest to the given point. We now calculate that distance:

$$d = \sqrt{(\tfrac{5}{6})^2 + (\tfrac{4}{3} - 3)^2 + [1 + \tfrac{5}{6} - 2(\tfrac{4}{3})]^2} = \sqrt{\frac{25}{6}} = \frac{5}{\sqrt{6}}$$

Check: You might want to check your work by using the formula for the distance from a point to a plane in $\mathbb{R}^3$ (Theorem 10.10):

$$d = \left| \frac{Ax_0 + By_0 + Cz_0 + D}{\sqrt{A^2 + B^2 + C^2}} \right| = \left| \frac{(1)0 + (2)3 + (1)4 + (-5)}{\sqrt{1^2 + 2^2 + 1^2}} \right| = \frac{5}{\sqrt{6}} \qquad \blacksquare$$

Comment   In general, it can be difficult to show that a relative extremum is actually an absolute extremum. In practice, however, it is often possible to make the determination using physical or geometric considerations.

In the following example, calculus is applied to justify a formula used in statistics and in many applications in the social and physical sciences.

EXAMPLE 7   *Least-squares approximation of data*

Suppose data consisting of n points $P_1, \ldots, P_n$ are known, and we wish to find a function $y = f(x)$ that fits the data reasonably well. In particular, suppose we wish to find a line $y = mx + b$ that "best fits" the data in the sense that the sum of the squares of the vertical distances from each data point to the line is minimized.

Solution   We wish to find values of m and b that minimize the sum of the squares of the differences between the y-values and the line $y = mx + b$. This line that we seek is called the **regression line**. Suppose that the point P_k has components (x_k, y_k). Now at this point the value on the regression line is $y = mx_k + b$ and the value of the data point is y_k. The "error" caused by using the point on the regression line rather than the actual data point is

$$y_k - (mx_k + b)$$

The data points may be above the regression line for some values of k and below the regression line for other values of k. We see that we need to minimize the function that represents the sum of the *squares* of all these differences:

$$F(m, b) = \sum_{k=1}^{n} [y_k - (mx_k + b)]^2$$

The situation is illustrated in Figure 12.33.

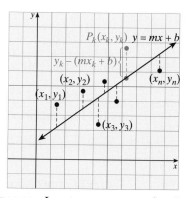

■ **FIGURE 12.33**   Least-squares approximation of data

Because F is a function of two variables, we use the second partials test:

$$F_m(m, b) = \sum_{k=1}^{n} 2[y_k - (mx_k + b)](-x_k)$$

$$= 2m \sum_{k=1}^{n} x_k^2 + 2b \sum_{k=1}^{n} x_k - 2 \sum_{k=1}^{n} x_k y_k$$

$$F_b(m, b) = \sum_{k=1}^{n} 2[y_k - (mx_k + b)](-1)$$

$$= 2m \sum_{k=1}^{n} x_k + 2b \sum_{k=1}^{n} 1 - 2 \sum_{k=1}^{n} y_k$$

$$= 2m \sum_{k=1}^{n} x_k + 2bn - 2 \sum_{k=1}^{n} y_k$$

Set each of these partial derivatives equal to 0 to find the critical values (after a great deal of algebra):

$$m = \frac{n \sum_{k=1}^{n} x_k y_k - \left(\sum_{k=1}^{n} x_k\right)\left(\sum_{k=1}^{n} y_k\right)}{n \sum_{k=1}^{n} x_k^2 - \left(\sum_{k=1}^{n} x_k\right)^2} \quad \text{and} \quad b = \frac{\sum_{k=1}^{n} x_k^2 \sum_{k=1}^{n} y_k - \left(\sum_{k=1}^{n} x_k\right)\left(\sum_{k=1}^{n} x_k y_k\right)}{n \sum_{k=1}^{n} x_k^2 - \left(\sum_{k=1}^{n} x_k\right)^2}$$

We leave it to you to complete the second partials test to verify that these values of m and b yield a minimum.   ■

Most applications of the **least-squares formula** stated in Example 7 involve using a calculator or computer software. The following technology window provides an example.

Technology Window

Many calculators will carry out the calculations required by the least-squares approximation procedure. Look at your owner's manual for specifics, but most calculators allow you to input data with keys labeled [STAT] and [DATA]. After the data are input, the m and b values are given by pressing the [Lin Reg] choice. For example, ten people are given a standard IQ test. Their scores were then compared with their high school grades:

IQ:	117	105	111	96	135	81	103	99	107	109
GPA:	3.1	2.8	2.5	2.8	3.4	1.9	2.1	3.2	2.9	2.3

A calculator output shows: $m = .0224144711$ and $b = .3173417224$. A scatter diagram with the least-squares line is shown below.

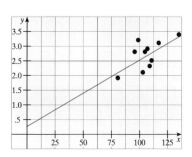

12.7 Problem Set

Ⓐ 1. ■ **What Does This Say?** Describe what is meant by a critical value.

2. ■ **What Does This Say?** Describe a procedure for determining absolute extrema.

3. ■ **What Does This Say?** Describe a procedure for determining absolute extrema on a closed, bounded set S.

Find the critical points in Problems 4–23, and classify each point as a relative maximum, a relative minimum, or a saddle point.

4. $f(x, y) = 2x^2 - 4xy + y^3 + 2$

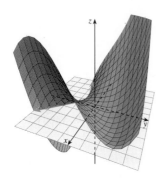

5. $f(x, y) = (x - 2)^2 + (y - 3)^4$

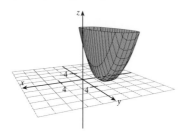

6. $f(x, y) = e^{-x}\sin y$

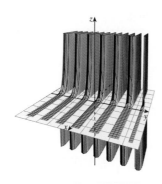

7. $f(x, y) = (1 + x^2 + y^2)e^{1-x^2-y^2}$

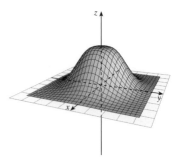

8. $f(x, y) = \dfrac{9x}{x^2 + y^2 + 1}$

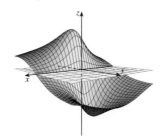

9. $f(x, y) = x^2 + xy + y^2$

10. $f(x, y) = xy - x + y$

11. $f(x, y) = -x^3 + 9x - 4y^2$

12. $f(x, y) = e^{-(x^2+y^2)}$

13. $F(x, y) = (x^2 + 2y^2)e^{1-x^2-y^2}$

14. $f(x, y) = e^{xy}$

15. $f(x, y) = x^{-1} + y^{-1} + 2xy$

16. $f(x, y) = (x - 4)\ln xy$

17. $f(x, y) = x^3 + y^3 + 3x^2 - 18y^2 + 81y + 5$

18. $f(x, y) = 2x^3 + y^3 + 3x^2 - 3y - 12x - 4$

19. $f(x, y) = x^2 + y^2 - 6xy + 9x + 5y + 2$

20. $f(x, y) = x^2 + y^2 + \dfrac{32}{xy}$

21. $f(x, y) = x^2 + y^3 + \dfrac{768}{x + y}$

22. $f(x, y) = 3xy^2 - 2x^2y + 36xy$

23. $f(x, y) = 3x^2 + 12x + 8y^3 - 12y^2 + 7$

Find the absolute extrema of f on the closed, bounded set S in the plane as described in Problems 24–30.

Ⓑ 24. $f(x, y) = 2x^2 - y^2$; S is the disk $x^2 + y^2 \le 1$.

25. $f(x, y) = xy - 2x - 5y$; S is the triangle with vertices $(0, 0)$, $(7, 0)$, and $(7, 7)$.

26. $f(x, y) = x^2 + 3y^2 - 4x + 2y - 3$; S is the square with vertices $(0, 0), (3, 0), (3, -3)$, and $(0, -3)$.

27. $f(x, y) = 2 \sin x + 5 \cos y$; S is the rectangle with vertices $(0, 0), (2, 0), (2, 5)$, and $(0, 5)$.

28. $f(x, y) = e^{x^2 + 2x + y^2}$; S is the disk $x^2 + 2x + y^2 \leq 0$.

29. $f(x, y) = x^2 + xy + y^2$; S is the disk $x^2 + y^2 \leq 1$.

30. $f(x, y) = x^2 - 4xy + y^3 + 4y$; S is the square region $0 \leq x \leq 2, 0 \leq y \leq 2$.

Find the least-squares regression line for the data points given in Problems 31–34.

31. $(-2, -3), (-1, -1), (0, 1), (1, 3), (3, 5)$

32. $(0, 1), (1, 1.6), (2.2, 3), (3.1, 3.9), (4, 5)$

33. $(3, 5.72), (4, 5.31), (6.2, 5.12), (7.52, 5.32), (8.03, 5.67)$

34. $(-4, 2), (-3, 1), (0, 0), (1, -3), (2, -1), (3, -2)$

35. Find all points on the surface $y^2 = 4 + xz$ that are closest to the origin.

36. Find all points in the plane $x + 2y + 3z = 4$ in the first octant where $f(x, y, z) = x^2 y z^3$ has a maximum value.

37. A rectangular box with no top is to have a fixed volume. What should its dimensions be if we want to use the least amount of material in its construction?

38. A wire of length L is cut into three pieces that are bent to form a circle, a square, and an equilateral triangle. How should the cuts be made to minimize the sum of the total area?

39. Find the positive numbers whose sum is 54 and whose product is as large as possible.

40. A dairy produces whole milk and skim milk in quantities x and y pints, respectively. Suppose the price (in cents) of whole milk is $p(x) = 100 - x$ and that of skim milk is $q(y) = 100 - y$, and also assume that $C(x, y) = x^2 + xy + y^2$ is the joint-cost function of the commodities. Maximize the profit

$$P(x, y) = px + qy - C(x, y)$$

41. Repeat Problem 40 for the case where $p(x) = 4 - 5x, q(y) = 4 - 2y$, and the joint-cost function is $C(x, y) = 2xy + 4$.

42. A particle of mass m in a rectangular box with dimensions x, y, z has ground state energy

$$E(x, y, z) = \frac{k^2}{8m} \left(\frac{1}{x^2} + \frac{1}{y^2} + \frac{1}{z^2} \right)$$

where k is a physical constant. If the volume of the box is fixed (say, $V_0 = xyz$), find the values of x, y, and z that minimize the ground state energy.

43. A manufacturer produces two different kinds of graphing calculators, A and B, in quantities x and y (units of 1,000), respectively. If the revenue function (in dollars) is $R(x, y) = -x^2 - 2y^2 + 2xy + 8x + 5y$, find the quantities of A and B that should be produced to maximize revenue.

44. Suppose we wish to construct a rectangular box with volume 32 ft³. Three different materials will be used in the construction. The material for the sides costs $1 per square foot, the material for the bottom costs $3 per square foot, and the material for the top costs $5 per square foot. What are the dimensions of the least expensive such box?

45. **MODELING PROBLEM** A store carries two competing brands of bottled water, one from California and the other from New York. To model this situation, assume the owner of the store can obtain both at a cost of $2/bottle. Also assume that if the California water is sold for x dollars per bottle and the New York water for y dollars per bottle, then consumers will buy approximately $40 - 50x + 40y$ bottles of California water and $20 + 60x - 70y$ bottles of the New York water each day. How should the owner price the bottled water to generate the largest possible profit?

46. **MODELING PROBLEM** The telephone company is planning to introduce two new types of executive communications systems that it hopes to sell to its largest commercial customers. To create a model to determine the maximum profit, it is assumed that if the first type of system is priced at x hundred dollars per system and the second type at y hundred dollars per system, approximately $40 - 8x + 5y$ consumers will buy the first type and $50 + 9x - 7y$ will buy the second type. If the cost of manufacturing the first type is $1,000 per system and the cost of manufacturing the second type is $3,000 per system, how should the telephone company price the systems to generate maximum profit?

47. **MODELING PROBLEM** A manufacturer with exclusive rights to a sophisticated new industrial machine is planning to sell a limited number of the machines to both foreign and domestic firms. The price the manufacturer can expect to receive for the machines will depend on the number of machines made available. For example, if only a few of the machines are placed on the market, competitive bidding among prospective purchasers will tend to drive the price up. It is estimated that if the manufacturer supplies x machines to the domestic market and y machines to the foreign market, the machines will sell for $60 - 0.2x + 0.05y$ thousand dollars apiece at home and $50 - 0.1y + 0.05x$ thousand dollars apiece abroad. If the manufacturer can produce the machines at a total cost of $10,000 apiece, how many should be supplied to each market to generate the largest possible profit?

48. **MODELING PROBLEM** A college admissions officer, Dr. Westfall, has compiled the following data relating students' high school and college GPAs:

HS GPA	2.0	2.5	3.0	3.0	3.5	3.5	4.0	4.0
College GPA	1.5	2.0	2.5	3.5	2.5	3.0	3.0	3.5

Plot the data points on a graph and find the equation of the least-squares line for these data. Then use the least-squares line to predict the college GPA of a student whose high school GPA is 3.75.

49. MODELING PROBLEM   It is known that if an ideal spring is displaced a distance y from its natural length by a force (weight) x, then $y = kx$, where k is the so-called spring constant. To compute this constant for a particular spring, a scientist obtains the following data:

x(lb)	5.2	7.3	8.4	10.12	12.37
x(in.)	11.32	15.56	17.44	21.96	26.17

Based on these data, what is the "best" choice for k?

50. ■ What Does This Say? The following table gives the approximate U.S. census figures (in millions):

Year:	1900	1910	1920	1930	1940
Population:	76.2	92.2	106.0	123.2	132.1

Year:	1950	1960	1970	1980	1990
Population:	151.3	179.3	203.3	226.5	248.7

a. Find the least-squares regression line for the given data and use this line to "predict" the population in 1997. (The actual population was about 266.5 million.)
b. Use the least-squares linear approximation to estimate the population at the present time. Check your answer by looking up the population using the Internet. Comment on the accuracy (or inaccuracy) of your prediction.

51. ■ What Does This Say? The following table gives the Dow Jones Industrial Average (DJIA) Stock Index along with the per capita consumption of wine (in gallons) for those years.*

Year:	1965	1970	1975	1980	1985	1990	1995
DJIA Index:	911	753	802	891	1,328	2,796	3,838
Consumption:	0.98	1.31	1.71	2.11	2.43	2.05	1.79

a. Plot these data on a graph, with the DJIA Index on the x-axis and consumption on the y-axis.
b. Find the equation of the least-squares line.
c. In 1998 the stock market began the year at 7,908. Use the least-squares line to predict per capita wine consumption that corresponds to this stock value.
d. Determine whether the consumption figures predicted by the least-squares line in part **c** are approximately correct. Interpret your findings.

---

*Dominick Salvatore, *Managerial Economics,* McGraw-Hill, Inc., New York, 1989. The data given in the table are on page 138

Technology Window

52. Linearizing nonlinear data. In this problem, we turn to data that do *not* tend to change linearly. Often one can "linearize" the data by taking the logarithm or exponential of the data and then doing a linear fit as described in this section.

a. Suppose we have, or suspect, a relationship $y = kx^m$. Show that by taking the natural logarithm of this equation, we obtain a linear relationship: $Y = K + mX$. Explain the new variables and constant K.

b. Below are data relating the periods of revolution t (in days) of the six inner planets and their semimajor axis a (in 10^6 km). Kepler conjectured the relationship $t = ka^m$, which is very accurate for the correct k and a. You are to "transform" the data as in part **a** (thus obtaining $T_i = \ln t_i, \dots$); and do a linear fit to the new data, thus finding k and m.

t-data: (87.97, 224.7, 365.26, 686.98, 4 332.59, 10 759.2)

a-data: (58, 108, 149, 228, 778, 1 426)

53. Below are data pertaining to a recent Olympic weight-lifting competition. The x-data are the "class data" giving eight weight classes (in kg) from featherweight to heavyweight-2. The w-data are the combined weights lifted by the winners in each class. Theoretically, we would expect a relationship $w = kx^m$, where $m = 2/3$. (Can you see why?)

a. Linearize the data as in Problem 52 and use the least-squares approximation to find k and m.

x-data: (56, 60, 67.5, 75, 82.5, 90, 100, 110)

w-data: (292.5, 342.5, 340, 375, 377.5, 412.5, 425, 455)

b. Comment on the 60 kg entry. Do you see why this participant (N. Suleymanoglu of Turkey) was referred to as the strongest man in the world?

C 54. This problem is designed to show, by example, that if f is degenerate at the critical point, then almost anything can happen.

a. Show that $f(x, y) = x^4 - y^4$ has a saddle point at $(0, 0)$.
b. Show that $g(x, y) = x^2 y^2$ has a relative minimum at $(0, 0)$.
c. Show that $h(x, y) = x^3 + y^3$ has no extremum or saddle point at $(0, 0)$.

55. THINK TANK PROBLEM   Consider the function $f(x, y) = (y - x^2)(y - 2x^2)$. Discuss the behavior of this function at $(0, 0)$.

56. THINK TANK PROBLEM Sometimes the critical points of a function can be classified by looking at the level curves. In each case shown in Figure 12.34, determine the nature of the critical point(s) of f at $(0, 0)$.

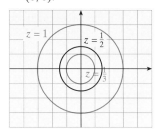
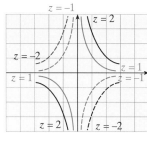

a.   **b.**

■ **FIGURE 12.34** Problem 58

57. Prove the second partials test. *Hint:* Compute the second directional derivative of f in the direction of $\mathbf{u} = h\mathbf{i} + k\mathbf{j}$ and complete the square.

58. Verify the formulas for m and b associated with the least-squares approximation.

59. This problem involves a generalization of the least-squares procedure, in which a "least-squares plane" is found to produce the best fit for a given set of data. A researcher knows that the quantity z is related to x and y by a formula of the form $z = k_1 x + k_2 y$ where k_1 and k_2 are physical constants. To determine these constants, she conducts a series of experiments, the results of which are tabulated as follows:

x	1.20	0.86	1.03	1.65	-0.95	-1.07
y	0.43	1.92	1.52	-1.03	1.22	-0.06
z	3.21	5.73	2.22	0.92	-1.11	-0.97

Modify the method of least squares to find a "best approximation" for k_1 and k_2.

12.8   Lagrange Multipliers

IN THIS SECTION method of Lagrange multipliers, constrained optimization problems, Lagrange multipliers with two parameters, a geometric interpretation

METHOD OF LAGRANGE MULTIPLIERS

In many applied problems, a function of two variables is to be optimized subject to a restriction or **constraint** on the variables. For example, consider a container heated in such a way that the temperature at the point (x, y, z) in the container is given by the function $T(x, y, z)$. Suppose that the surface $z = f(x, y)$ lies in the container, and that we wish to find the point on $z = f(x, y)$ where the temperature is the greatest. In other words, *What is the maximum value of T subject to the constraint $z = f(x, y)$, and where does this maximum value occur?* (See Figure 12.35.)

We will use the following theorem to solve such a problem.

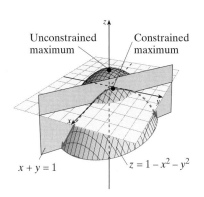

■ **FIGURE 12.35** Find the maximum value on the surface defined by $f(x, y) = 1 - x^2 - y^2$ subject to the constraint $x + y = 1$ $(x \geq 0, y \geq 0)$. See Example 1 for a solution.

THEOREM 12.13   Lagrange's theorem

Assume that f and g have continuous first partial derivatives and that f has an extremum at $P_0(x_0, y_0)$ on the smooth constraint curve $g(x, y) = c$. If $\nabla g(x_0, y_0) \neq \mathbf{0}$, there is a number λ such that

$$\nabla f(x_0, y_0) = \lambda \nabla g(x_0, y_0)$$

Proof   Denote the constraint curve $g(x, y) = c$ by C, and note that C is smooth. We represent this curve by the vector function

$$\mathbf{R}(t) = x(t)\mathbf{i} + y(t)\mathbf{j}$$

for all t in an open interval I, including t_0 corresponding to P_0, where $x'(t)$ and $y'(t)$ exist and are continuous. Let $F(t) = f[x(t), y(t)]$ for all t in I, and apply the chain rule to obtain

$$F'(t) = f_x \frac{dx}{dt} + f_y \frac{dy}{dt} = \nabla f[x(t), y(t)] \cdot \mathbf{R}'(t)$$

Because $f(x, y)$ has an extremum at P_0, we know that $F(t)$ has an extremum at t_0, the value of t that corresponds to P_0 (that is, P_0 is the point on C where $t = t_0$). Therefore, we have $F'(t_0) = 0$ and

$$F'(t_0) = \nabla f[x(t_0), y(t_0)] \cdot \mathbf{R}'(t_0) = 0$$

If $\nabla f[x(t_0), y(t_0)] = 0$, then $\lambda = 0$, and the condition $\nabla f = \lambda \nabla g$ is satisfied trivially. If $\nabla f[x(t_0), y(t_0)] \neq 0$, then $\nabla f[x(t_0), y(t_0)]$ is orthogonal to $\mathbf{R}'(t_0)$. Because $\mathbf{R}'(t_0)$ is tangent to the constraint curve C, it follows that $\nabla f(x_0, y_0)$ is normal to C. But $\nabla g(x_0, y_0)$ is also normal to C (because C is a level curve of g), and we conclude that ∇f and ∇g must be *parallel* at P_0. Thus, there is a scalar λ such that $\nabla f(x_0, y_0) = \lambda \nabla g(x_0, y_0)$, as required.

CONSTRAINED OPTIMIZATION PROBLEMS

The general procedure for the method of Lagrange multipliers may be described as follows.

Procedure for the Method of Lagrange Multipliers

Suppose f and g satisfy the hypotheses of Lagrange's theorem, and that $f(x, y)$ has an extremum subject to the constraint $g(x, y) = c$. Then to find the extreme values, proceed as follows:

1. Simultaneously solve the following three equations:

$$f_x(x, y) = \lambda g_x(x, y) \qquad f_y(x, y) = \lambda g_y(x, y) \qquad g(x, y) = c$$

2. Evaluate f at all points found in step 1. The extremum we seek must be among these values.

EXAMPLE 1   *Optimization with Lagrange multiplier*

Find the largest and smallest values of $f(x, y) = 1 - x^2 - y^2$ subject to the constraint $x + y = 1$ with $x \geq 0, y \geq 0$.

Solution   Because the constraint is $x + y = 1$, let $g(x, y) = x + y - 1$.

$$f_x(x, y) = -2x \qquad f_y(x, y) = -2y \qquad g_x(x, y) = 1 \qquad g_y(x, y) = 1$$

Form the system

$$\begin{cases} -2x = \lambda(1) \\ -2y = \lambda(1) \\ x + y = 1 \end{cases}$$

The only solution is $x = \frac{1}{2}, y = \frac{1}{2}, \lambda = -1$.

$$f\left(\frac{1}{2}, \frac{1}{2}\right) = 1 - \left(\frac{1}{2}\right)^2 - \left(\frac{1}{2}\right)^2 = \frac{1}{2}$$

The endpoints of the line segment $x + y = 1$ for $x \geq 0, y \geq 0$ are at $(1, 0)$ and $(0, 1)$, and we find that

$$f(1, 0) = 1 - 1^2 - 0^2 = 0$$
$$f(0, 1) = 1 - 0^2 - 1^2 = 0$$

Therefore, the maximum value is $\frac{1}{2}$ at $(\frac{1}{2}, \frac{1}{2})$, and the minimum value is 0 at $(1, 0)$ and $(0, 1)$.   ∎

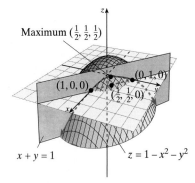

Maximum $(\frac{1}{2}, \frac{1}{2}, \frac{1}{2})$

$(1, 0, 0)$ $(0, 1, 0)$

$(\frac{1}{2}, \frac{1}{2}, 0)$

$x + y = 1$ $z = 1 - x^2 - y^2$

■ **FIGURE 12.36** The maximum is the high point of the curve of intersection of the surface and the plane

The method of Lagrange multipliers extends naturally to functions of three or more variables. If a function $f(x, y, z)$ has an extreme value subject to a constraint $g(x, y, z) = c$, then the extremum occurs at a point (x_0, y_0, z_0) such that $g(x_0, y_0, z_0) = c$ and $\nabla f(x_0, y_0, z_0) = \lambda \nabla g(x_0, y_0, z_0)$ for some number λ. Here is an example.

EXAMPLE 2   *Hottest and coldest points on a plate*

A container in $\mathbb{R}^3$ has the shape of the cube given by $0 \le x \le 1, 0 \le y \le 1, 0 \le z \le 1$. A plate is placed in the container in such a way that it occupies that portion of the plane $x + y + z = 1$ that lies in the cubical container. If the container is heated so that the temperature at each point (x, y, z) is given by

$$T(x, y, z) = 4 - 2x^2 - y^2 - z^2$$

in hundreds of degrees Celsius, what are the hottest and coldest points on the plate?

Solution   The cube and plate are shown in Figure 12.37. We shall use Lagrange multipliers to find all critical points in the interior of the plate, and then we shall examine the plate's boundary. To apply the method of Lagrange multipliers, we must solve $\nabla T = \lambda \nabla g$, where $g(x, y, z) = x + y + z - 1$. We obtain the partial derivatives.

$$T_x = -4x \qquad T_y = -2y \qquad T_z = -2z \qquad g_x = g_y = g_z = 1$$

We must solve the system

$$\begin{cases} -4x = \lambda \\ -2y = \lambda \\ -2z = \lambda \\ x + y + z = 1 \end{cases}$$

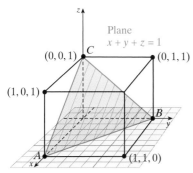

FIGURE 12.37   Find the hottest and coldest points on the plate inside the cube

Edge AC: $x + z = 1, y = 0$
Edge AB: $x + y = 1, z = 0$
Edge BC: $y + z = 1, x = 0$

The solution of this system is $\left(\frac{1}{5}, \frac{2}{5}, \frac{2}{5}\right)$. The boundary of the plate is a triangle with vertices $A(1, 0, 0)$, $B(0, 1, 0)$, and $C(0, 0, 1)$. The temperature along the edges of this triangle may be found as follows:

$$T_1(x) = 4 - 2x^2 - (0)^2 - (1 - x)^2 = 3 - 3x^2 + 2x, \quad 0 \le x \le 1$$
$$T_2(x) = 4 - 2x^2 - (1 - x)^2 - (0)^2 = 3 - 3x^2 + 2x, \quad 0 \le x \le 1$$
$$T_3(y) = 4 - 2(0)^2 - y^2 - (1 - y)^2 = 3 + 2y - 2y^2, \quad 0 \le y \le 1$$

Edge AC: Differentiating, $T_1'(x) = T_2'(x) = -6x + 2$, which equals 0 when $x = \frac{1}{3}$. If $x = \frac{1}{3}$, then $z = \frac{2}{3}$ (because $x + z = 1, y = 0$ on edge AC), so we have the critical point $\left(\frac{1}{3}, 0, \frac{2}{3}\right)$.

Edge AB: Because $T_2 = T_1$, we see $x = \frac{1}{3}$. If $x = \frac{1}{3}$, then $y = \frac{2}{3}$ (because $x + y = 1, z = 0$ on edge BC), so we have another critical point $\left(\frac{1}{3}, \frac{2}{3}, 0\right)$.

Edge BC: Differentiating, $T_3'(y) = 2 - 4y$, which equals 0 when $y = \frac{1}{2}$. Because $y + z = 1$, and $x = 0$, we have the critical point $\left(0, \frac{1}{2}, \frac{1}{2}\right)$.

Endpoints of the edges: $(1, 0, 0)$, $(0, 1, 0)$, and $(0, 0, 1)$.

The last step is to evaluate T at the critical points and the endpoints:

$$T\left(\tfrac{1}{5}, \tfrac{2}{5}, \tfrac{2}{5}\right) = 3\tfrac{3}{5};$$
$$T\left(\tfrac{1}{3}, 0, \tfrac{2}{3}\right) = 3\tfrac{1}{3}; \qquad T\left(\tfrac{1}{3}, \tfrac{2}{3}, 0\right) = 3\tfrac{1}{3}; \qquad T\left(0, \tfrac{1}{2}, \tfrac{1}{2}\right) = 3\tfrac{1}{2};$$
$$T(1, 0, 0) = 2; \qquad T(0, 1, 0) = 3; \qquad T(0, 0, 1) = 3$$

Comparing these values (remember that the temperature is in hundreds of degrees Celsius), we see that the highest temperature is 360 °C at $(\frac{1}{5}, \frac{2}{5}, \frac{2}{5})$ and the lowest temperature is 200 °C at $(1, 0, 0)$. ∎

Notice that, in the end, we do not care about the value of λ. The multiplier is just used as an intermediary device for finding the critical points. However, in some problems, we do care about the value of λ, because of the interpretation given in the following theorem.

THEOREM 12.14   *Rate of change of the extreme value*

Suppose E is an extreme value (maximum or minimum) of f subject to the constraint $g(x, y) = c$. Then the Lagrange multiplier λ is the rate of change of E with respect to c; that is, $\lambda = dE/dc$.

Proof   Note that at the extreme value (x, y) we have

$$f_x = \lambda g_x, \qquad f_y = \lambda g_y, \qquad \text{and} \qquad g(x, y) = c$$

The coordinates of the optimal ordered pair (x, y) depend on c (because different constraint levels will generally lead to different optimal combinations of x and y). Thus,

$$E = E(x, y) \quad \text{where } x \text{ and } y \text{ are functions of } c$$

By the chain rule for partial derivatives:

$$\frac{dE}{dc} = \frac{\partial E}{\partial x}\frac{dx}{dc} + \frac{\partial E}{\partial y}\frac{dy}{dc}$$

$$= f_x\frac{dx}{dc} + f_y\frac{dy}{dc} \qquad \text{Because } E = f(x, y)$$

$$= \lambda g_x\frac{dx}{dc} + \lambda g_y\frac{dy}{dc} \qquad \text{Because } f_x = \lambda g_x \text{ and } f_y = \lambda g_y$$

$$= \lambda\left(g_x\frac{dx}{dc} + g_y\frac{dy}{dc}\right)$$

$$= \lambda\frac{dg}{dc} \qquad \text{Chain rule}$$

$$= \lambda \qquad \text{Because } \frac{dg}{dc} = 1 \text{ (remember } g = c)$$

This theorem can be interpreted as saying that the multiplier gives the change in the extreme value E that results when the constraint c is increased by 1 unit. This interpretation is illustrated in the following example.

EXAMPLE 3   *Maximum output for a Cobb–Douglas production function*

If x thousand dollars is spent on labor, and y thousand dollars is spent on equipment, it is estimated that the output of a certain factory will be

$$Q(x, y) = 50x^{2/5}\,y^{3/5}$$

units. If \$150,000 is available, how should this capital be allocated between labor and equipment to generate the largest possible output? How does the maximum output change if the money available for labor and equipment is increased by \$1,000? In economics, an output function of the general form $Q(x, y) = x^{\alpha}y^{1-\alpha}$ is known as a **Cobb–Douglas production function.**

Solution   Because x and y are given in units of \$1,000, the constraint equation is $x + y = 150$. If we set $g(x, y) = x + y - 150$, we wish to maximize Q subject to $g(x, y) = 0$. To apply the method of Lagrange multipliers, we first find

$$Q_x = 20x^{-3/5}y^{3/5} \qquad Q_y = 30x^{2/5}y^{-2/5} \qquad g_x = 1 \qquad g_y = 1$$

Next, solve the system

$$\begin{cases} 20x^{-3/5}y^{3/5} = \lambda \\ 30x^{2/5}y^{-2/5} = \lambda \\ x + y - 150 = 0 \end{cases}$$

From the first two equations we have

$$20x^{-3/5}y^{3/5} = 30x^{2/5}y^{-2/5}$$
$$20y = 30x$$
$$y = 1.5x$$

Substitute $y = 1.5x$ into the equation $x + y = 150$ to find $x = 60$. This leads to the solution $y = 90$, so that the maximum output is

$$Q(60, 90) = 50(60)^{2/5}(90)^{3/5} \approx 3,826.273502 \text{ units}$$

We also find that

$$\lambda = 20(60)^{-3/5}(90)^{3/5} \approx 25.50849001$$

Thus, the maximum output is about 3,826 units and occurs when \$60,000 is allocated to labor and \$90,000 to equipment. We also note that an increase of \$1,000 (1 unit) in the available funds will increase the maximum output by approximately $\lambda \approx 25.5$ units (from 3,826.27 to 3,851.78 units).   ◼

LAGRANGE MULTIPLIERS WITH TWO PARAMETERS

The method of Lagrange multipliers can also be applied in situations with more than one constraint equation. Suppose we wish to locate an extremum of a function defined by $f(x, y, z)$ subject to *two* constraints, $g(x, y, z) = c_1$ and $h(x, y, z) = c_2$, where g and h are also differentiable and ∇g and ∇h are not parallel. By generalizing Lagrange's theorem, it can be shown that if (x_0, y_0, z_0) is the desired extremum, then there are numbers λ and μ such that $g(x_0, y_0, z_0) = c_1, h(x_0, y_0, z_0) = c_2$, and

$$\nabla f(x_0, y_0, z_0) = \lambda \nabla g(x_0, y_0, z_0) + \mu \nabla h(x_0, y_0, z_0)$$

As in the case of one constraint, we proceed by first solving this system of equations simultaneously to find $\lambda, \mu, x_0, y_0, z_0$, then evaluating $f(x, y, z)$ at each solution, and comparing to find the required extremum. This approach is illustrated in our final example of this section.

EXAMPLE 4   *Optimization with two constraints*

Find the point on the intersection of the plane $x + 2y + z = 10$ and the paraboloid $z = x^2 + y^2$ that is closest to the origin.

Solution   The distance from a point (x, y, z) to the origin is $s = \sqrt{x^2 + y^2 + z^2}$, but instead of minimizing this quantity, it is easier to minimize its square. That is, we shall minimize

$$f(x, y, z) = x^2 + y^2 + z^2$$

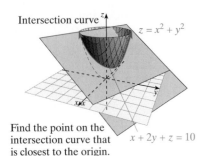

Intersection curve

$z = x^2 + y^2$

Find the point on the intersection curve that is closest to the origin.

$x + 2y + z = 10$

Graphic representation of Example 4

subject to the joint constraints

$$g(x, y, z) = x + 2y + z = 10 \quad \text{and} \quad h(x, y, z) = x^2 + y^2 - z = 0$$

Compute the partial derivatives of $f, g,$ and h:

$$
\begin{array}{ccc}
f_x = 2x & f_y = 2y & f_z = 2z \\
g_x = 1 & g_y = 2 & g_z = 1 \\
h_x = 2x & h_y = 2y & h_z = -1
\end{array}
$$

To apply the method of Lagrange multipliers, we use the formula

$$\nabla f(x_0, y_0, z_0) = \lambda \nabla g(x_0, y_0, z_0) + \mu \nabla h(x_0, y_0, z_0)$$

which leads to the following system of equations:

$$
\begin{cases}
2x = \lambda(1) + \mu(2x) \\
2y = \lambda(2) + \mu(2y) \\
2z = \lambda(1) + \mu(-1) \\
x + 2y + z = 10 \\
z = x^2 + y^2
\end{cases}
$$

This is not a linear system, so solving it requires ingenuity.

Multiply the first equation by 2 and subtract the second equation to obtain

$$4x - 2y = (4x - 2y)\mu$$
$$(4x - 2y) - (4x - 2y)\mu = 0$$
$$(4x - 2y)(1 - \mu) = 0$$
$$4x - 2y = 0 \quad \text{or} \quad 1 - \mu = 0$$

CASE I   **If $4x - 2y = 0$,** then $y = 2x$. Substitute this into the two constraint equations:

$x + 2y + z = 10$	$x^2 + y^2 - z = 0$
$x + 2(2x) + z = 10$	$x^2 + (2x)^2 - z = 0$
$z = 10 - 5x$	$z = 5x^2$

By substitution we have $5x^2 = 10 - 5x$, which has solutions $x = 1$ and $x = -2$. This implies

$x = 1$	$x = -2$
$y = 2x = 2(1) = 2$	$y = 2x = 2(-2) = -4$
$z = 5x^2 = 5(1)^2 = 5$	$z = 5x^2 = 5(-2)^2 = 20$

Thus, the points $(1, 2, 5)$ and $(-2, -4, 20)$ are candidates for the minimal distance.

CASE II   **If $1 - \mu = 0$,** then $\mu = 1$, and we look at the system of equations involving $x, y, z, \lambda,$ and μ:

$$
\begin{cases}
2x = \lambda(1) + \mu(2x) \\
2y = \lambda(2) + \mu(2y) \\
2z = \lambda(1) + \mu(-1) \\
x + 2y + z = 10 \\
z = x^2 + y^2
\end{cases}
$$

The top equation becomes $2x = \lambda + 2x$, so that $\lambda = 0$. We now find z from the third equation:

$$2z = -1 \quad \text{or} \quad z = -\tfrac{1}{2}$$

Next, turn to the constraint equations:

$x + 2y + z = 10$	$x^2 + y^2 - z = 0$
$x + 2y - \frac{1}{2} = 10$	$x^2 + y^2 + \frac{1}{2} = 0$
$x + 2y = 10 + \frac{1}{2}$	$x^2 + y^2 = -\frac{1}{2}$

There is no solution because $x^2 + y^2$ cannot equal a negative number.

We check the candidates for the minimal distance:

$$f(x, y, z) = x^2 + y^2 + z^2 \quad \text{so that}$$
$$f(1, 2, 5) = 1^2 + 2^2 + 5^2 = 30$$
$$f(-2, -4, 20) = (-2)^2 + (-4)^2 + 20^2 = 420$$

Because $f(x, y, z)$ represents the square of the distance, the minimal distance is $\sqrt{30}$ and the point on the intersection of the two surfaces nearest to the origin is $(1, 2, 5)$.   ∎

A GEOMETRIC INTERPRETATION

Lagrange's theorem can be interpreted geometrically. Suppose the constraint curve $g(x, y) = c$ and the level curves $f(x, y) = k$ are drawn in the xy-plane, as shown in Figure 12.38.

To maximize $f(x, y)$ subject to the constraint $g(x, y) = c$, we must find the "highest" (rightmost, actually) level curve of f that intersects the constraint curve. As the sketch in Figure 12.38 suggests, this critical intersection occurs at a point where the constraint curve is tangent to a level curve—that is, where the slope of the constraint curve $g(x, y) = c$ is equal to the slope of a level curve $f(x, y) = k$. According to the formula derived in Section 12.5 (p. 829).

$$\text{Slope of constraint curve } g(x, y) = c \text{ is } \frac{-g_x}{g_y}$$

$$\text{Slope of each level curve is } \frac{-f_x}{f_y}$$

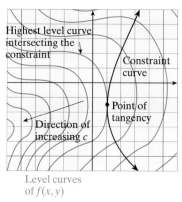

Highest level curve intersecting the constraint

Constraint curve

Point of tangency

Direction of increasing c

Level curves of $f(x, y)$

■ **FIGURE 12.38** Increasing level curves and the constraint curve

The condition that the slopes are equal can be expressed by

$$\frac{-f_x}{f_y} = \frac{-g_x}{g_y}, \quad \text{or, equivalently,} \quad \frac{f_x}{g_x} = \frac{f_y}{g_y}$$

Let λ equal this common ratio,

$$\lambda = \frac{f_x}{g_x} \quad \text{and} \quad \lambda = \frac{f_y}{g_y}$$

so that

$$f_x = \lambda g_x \quad \text{and} \quad f_y = \lambda g_y$$

and $\nabla f = f_x \mathbf{i} + f_y \mathbf{j} = \lambda(g_x \mathbf{i} + g_y \mathbf{j}) = \lambda \nabla g$.

Because the point in question must lie on the constraint curve, we also have $g(x, y) = c$. If these equations are satisfied at a certain point (a, b), then f will reach its constrained *maximum* at (a, b) if the *highest* level curve that intersects the constraint curve does so at this highest point. On the other hand, if the *lowest* level curve that intersects the constraint curve does so at (a, b), then f achieves its constrained *minimum* at this point.

12.8 Problem Set

Ⓐ *Use the method of Lagrange multipliers to find the required constrained extrema in Problems 1–14.*

1. Maximize $f(x, y) = xy$ subject to $2x + 2y = 5$.
2. Maximize $f(x, y) = xy$ subject to $x + y = 20$.
3. Maximize $f(x, y) = 16 - x^2 - y^2$ subject to $x + 2y = 6$.
4. Minimize $f(x, y) = x^2 + y^2$ subject to $x + y = 24$.
5. Minimize $f(x, y) = x^2 + y^2$ subject to $xy = 1$.
6. Minimize $f(x, y) = x^2 - xy + 2y^2$ subject to $2x + y = 22$.
7. Minimize $f(x, y) = x^2 - y^2$ subject to $x^2 + y^2 = 4$.
8. Maximize $f(x, y) = x^2 - 2y - y^2$ subject to $x^2 + y^2 = 1$.
9. Maximize $f(x, y) = \cos x + \cos y$ subject to $y = x + \frac{\pi}{4}$.
10. Maximize $f(x, y) = e^{xy}$ subject to $x^2 + y^2 = 3$.
11. Maximize $f(x, y) = \ln(xy^2)$ subject to $2x^2 + 3y^2 = 8$ for $x > 0, y > 0$.
12. Maximize $f(x, y, z) = xyz$ subject to $3x + 2y + z = 6$.
13. Minimize $f(x, y, z) = x^2 + y^2 + z^2$ subject to $x - 2y + 3z = 4$.
14. Minimize $f(x, y, z) = x^2 + y^2 + z^2$ subject to $4x^2 + 2y^2 + z^2 = 4$.

Ⓑ 15. Find the smallest value of $f(x, y, z) = 2x^2 + 4y^2 + z^2$ subject to $4x - 8y + 2z = 10$. What, if anything, can be said about the largest value of f subject to this constraint?
16. Let $f(x, y, z) = x^2y^2z^2$. Show that the maximum value of f on the sphere $x^2 + y^2 + z^2 = R^2$ is $R^6/27$.
17. Find the maximum and minimum values of $f(x, y, z) = x - y + z$ on the sphere $x^2 + y^2 + z^2 = 100$.
18. Find the maximum and minimum values $f(x, y, z) = 4x - 2y - 3z$ on the sphere $x^2 + y^2 + z^2 = 100$.
19. Use Lagrange multipliers to find the distance from the origin to the plane $Ax + By + Cz = D$ where at least one of A, B, C is nonzero.
20. Find the maximum and minimum distance from the origin to the ellipse $5x^2 - 6xy + 5y^2 = 4$.
21. Find the point on the plane

$$2x + y + z = 1$$

that is nearest to the origin.
22. Find the largest product of positive numbers x, y, and z such that their sum is 24.
23. Write the number 12 as the sum of three positive numbers x, y, z in such a way that the product xy^2z is a maximum.
24. A rectangular box with no top is to be constructed from 96 ft² of material. What should be the dimensions of the box if it is to enclose maximum volume?

25. The temperature T at point (x, y, z) in a region of space is given by the formula $T = 100 - xy - xz - yz$. Find the lowest temperature on the plane $x + y + z = 10$.
26. A farmer wishes to fence off a rectangular pasture along the bank of a river. The area of the pasture is to be 3,200 yd², and no fencing is needed along the river bank. Find the dimensions of the pasture that will require the least amount of fencing.
27. There are 320 yd of fencing available to enclose a rectangular field. How should the fencing be used so that the enclosed area is as large as possible?
28. Use the fact that 12 fl oz is approximately 6.89π in.³ to find the dimensions of the 12-oz soda can that can be constructed using the least amount of metal. Compare your answer with an actual can of Pepsi.
29. A cylindrical can is to hold 4π in.³ of orange juice. The cost per square inch of constructing the metal top and bottom is twice the cost per square inch of constructing the cardboard side. What are the dimensions of the least expensive can?
30. Find the volume of the largest rectangular parallelepiped that can be inscribed in the ellipsoid

$$x^2 + \frac{y^2}{4} + \frac{z^2}{9} = 1$$

31. A manufacturer has $8,000 to spend on the development and promotion of a new product. It is estimated that if x thousand dollars is spent on development and y thousand is spent on promotion, sales will be approximately $f(x, y) = 50x^{1/2}y^{3/2}$ units. How much money should the manufacturer allocate to development and how much to promotion to maximize sales?
32. **MODELING PROBLEM**   If x thousand dollars is spent on labor and y thousand dollars is spent on equipment, the output at a certain factory may be modeled by

$$Q(x, y) = 60x^{1/3}y^{2/3}$$

units. Assume $120,000 is available.
   a. How should money be allocated between labor and equipment to generate the largest possible output?
   b. Use the Lagrange multiplier λ to estimate the change in the maximum output of the factory that would result if the money available for labor and equipment is increased by $1,000.
33. **MODELING PROBLEM**   An architect decides to model the usable living space in a building by the volume of space that can be used comfortably by a person 6 feet tall—that is, by the largest 6-foot-high rectangular box that can be inscribed in the building. Find the dimensions

of an A-frame building y ft long with equilateral triangular ends x ft on a side that maximizes usable living space if the exterior surface area of the building cannot exceed 500 ft².

34. Find the radius of the largest cylinder of height 6 in. that can be inscribed in an inverted cone of height H, radius R, and surface area 250 in.².

35. In Problem 42 of Problem Set 12.7, you were asked to minimize the ground state energy

$$E(x, y, z) = \frac{k^2}{8m}\left(\frac{1}{x^2} + \frac{1}{y^2} + \frac{1}{z^2}\right)$$

subject to the volume constraint $V = xyz = C$. Solve the problem using Lagrange multipliers.

36. A university extension agricultural service concludes that, on a particular farm, the yield of wheat per acre is a function of water and fertilizer. Let x be the number of acre-feet of water applied, and y the number of pounds of fertilizer applied during the growing season. The agricultural service then concluded that the yield (measured in bushels), represented by f, can be defined by the formula $f(x, y) = 500 + x^2 + 2y^2$. Suppose that water costs $20 per acre-foot, fertilizer costs $12 per pound, and the farmer will invest $236 per acre for water and fertilizer. How much water and fertilizer should the farmer buy to maximize the yield?

37. How would the farmer of Problem 36 maximize the yield if the amount spent is $100 instead of $236?

38. Present post office regulations specify that a box (that is, a package in the form of a rectangular parallelepiped) can be mailed parcel post only if the sum of its length and girth does not exceed 108 inches, as shown in Figure 12.39. Find the maximum volume of such a package. (Compare your solution here with the one you might have given to Problem 17, Section 4.6, page 280.)

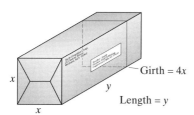

■ **FIGURE 12.39** Maximum volume for a box mailed by U.S. parcel post

39. Heron's formula says that the area of a triangle with sides a, b, c is

$$A = \sqrt{s(s - a)(s - b)(s - c)}$$

where $s = \frac{1}{2}(a + b + c)$ is the semi-perimeter of the triangle. Use this result and the method of Lagrange multipliers to show that of all triangles with a given fixed perimeter P, the equilateral triangle has the largest area.

40. If x, y, z are the angles of a triangle, what is the maximum value of the product
$P(x, y, z) = \sin x \sin y \sin z$? What about
$Q(x, y, z) = \cos x \cos y \cos z$?

Use the method of Lagrange multipliers in Problems 41–44 to find the required extrema for the two given constraints.

41. Find the minimum of $f(x, y, z) = x^2 + y^2 + z^2$ subject to $x + y = 4$ and $y + z = 6$.

42. Find the maximum of $f(x, y, z) = xyz$ subject to $x^2 + y^2 = 3$ and $y = 2z$.

43. Maximize $f(x, y, z) = xy + xz$ subject to $2x + 3z = 5$ and $xy = 4$.

44. Minimize $f(x, y, z) = 2x^2 + 3y^2 + 4z^2$ subject to $x + y + z = 4$ and $x - 2y + 5z = 3$.

45. MODELING PROBLEM A manufacturer is planning to sell a new product at the price of $150 per unit and estimates that if x thousand dollars is spent on development and y thousand dollars on promotion, then approximately

$$\frac{320y}{y + 2} + \frac{160x}{x + 4}$$

units of the product will be sold. The cost of manufacturing the product is $50 per unit.
   a. If the manufacturer has a total of $8,000 to spend on the development and promotion, how should this money be allocated to generate the largest possible profit?
   b. Suppose the manufacturer decides to spend $8,100 instead of $8,000 on the development and promotion of the new product. Estimate how this change will affect the maximum possible profit.
   c. If unlimited funds are available, how much should the manufacturer spend on development and promotion to maximize profit?
   d. What is the Lagrange multiplier in part **c**? Your answer should suggest another method for solving the problem in part **c**. Solve the problem using this alternative approach.

46. MODELING PROBLEM A jewelry box with a square base has an interior partition and is required to have volume 800 cm³ (see Figure 12.40).

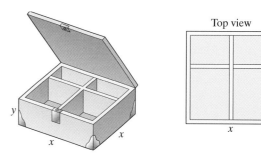

■ **FIGURE 12.40** Constructing a jewelry box

a. The material in the top costs twice as much as the material in the side and in the bottom, which in turn, costs twice as much as the material in the partitions. Find the dimensions of the box that minimize the total cost of construction. *Drat! We forgot to tell you where the partitions are located. Does that matter?*

b. Suppose the volume constraint changes from 800 cm³ to 801 cm³. Estimate the appropriate effect on the minimal cost.

C 47. A farmer wants to build a metal silo in the shape of a right circular cylinder with a right circular cone on the top (the bottom of the silo will be a concrete slab). What is the least amount of metal that can be used if the silo is to have a fixed volume V_0?

48. Find the volume of the largest rectangular parallelepiped (box) that can be inscribed in the ellipsoid

$$\frac{x^2}{a^2} + \frac{y^2}{b^2} + \frac{z^2}{c^2} = 1$$

(See Problem 30.)

49. ■ **What Does This Say?** The method of Lagrange multipliers gives a constrained extremum only if one exists. Apply the method to the problem: Optimize $f(x, y) = x + y$ subject to $xy = 1$. The method yields two candidates for an extremum. Is one a maximum and the other a minimum? Explain.

In Problems 50–53, let $Q(x, y)$ be a production function in which x and y represent units of labor and capital, respectively. If p and q represent unit costs of labor and capital, respectively, then $C = px + qy$ represents the total cost of production.

50. Use Lagrange multipliers to show that, subject to a fixed production level Q_0, the total cost is smallest when

$$\frac{Q_x}{p} = \frac{Q_y}{q} \quad \text{and} \quad Q(x, y) = Q_0$$

(provided $\nabla f \neq 0$, and $p \neq 0, q \neq 0$). This is often referred to as the **minimum cost problem,** and its solution is called the **least-cost combination of inputs.**

51. Show that the inputs x, y that maximize the production level $Q = f(x, y)$ subject to a fixed cost k satisfy

$$\frac{f_x}{p} = \frac{f_y}{q} \quad \text{with } px + qy = k$$

(assume $p \neq 0, q \neq 0$). This is called a **fixed-budget problem.**

52. A Cobb–Douglas production function is an output function of the form $Q(x, y) = cx^\alpha y^\beta$, with $\alpha + \beta = 1$. Show that such a function is maximized with respect to the fixed cost $px + qy = k$ when $x = \alpha k/p$ and $y = \beta k/q$.

Where does the maximum occur if we drop the condition $\alpha + \beta = 1$? How does the maximum output change if k is increased by 1 unit?

53. Show that the cost function

$$C(x, y) = px + qy$$

is minimized subject to the fixed production level $Ax^\alpha y^\beta = k$, with $\alpha + \beta = 1$, when

$$x = \frac{k}{A}\left(\frac{\alpha q}{\beta p}\right)^\beta \qquad y = \frac{k}{A}\left(\frac{\beta p}{\alpha q}\right)^\alpha$$

54. **HISTORICAL QUEST** A discussion of Lagrange multipliers would not be complete without mention of Joseph Lagrange, generally acknowledged as one of the two greatest mathematicians of the 18th century, Leonhard Euler being the other (see Historical Quest Problem 84 of the supplementary problems for Chapter 4). There is a distinct difference in style between Lagrange and Euler. Lagrange has been characterized as the first true analyst in the sense that he attempted to write concisely and with rigor. On the other hand, Euler wrote using intuition

JOSEPH LAGRANGE
1736–1813

and with an abundance of detail. Lagrange was described by Napoleon Bonaparte as "the lofty pyramid of the mathematical sciences" and followed Euler as the court mathematician for Frederick the Great. He was the first to use the notation $f'(x)$ and $f''(x)$ for derivatives. In this section, we were introduced to the method of Lagrange multipliers, which provides a procedure for constrained optimization. This method was contained in a paper on mechanics that Lagrange wrote when he was only 19 years old.*

For this Quest, we consider Lagrange's work with solving *algebraic* equations. You are familiar with the quadratic formula, which provides a general solution for any second-degree equation $ax^2 + bx + c = 0, a \neq 0$. Lagrange made an exhaustive study of the general solution for the first four degrees. Here is what he did. Suppose you are given a general algebraic expression involving letters $a, b, c, \ldots$; how many *different* expressions can be derived from the given one if the letters are interchanged in all possible ways? For example, from $ab + cd$ we obtain $ad + cb$ by interchanging b and d.

This problem suggests another closely related problem, so part of Lagrange's approach. Lagrange solved general algebraic equations of degree 2, 3, and 4. It was proved later (not by Lagrange, but by Galois and Abel), that no general solution for equations greater than 5 can be found. Do some research and find the general solution for equations of degree 1, 2, 3, and 4.

———

*From *Men of Mathematics* by E. T. Bell, Simon and Schuster, New York, 1937, p. 165.

Chapter 12   Review

Proficiency Examination

Concept Problems

1. What is a function of two variables?
2. What is the domain of a function of two variables?
3. What is a level curve of the function defined by $f(x, y)$?
4. What do we mean by the limit of a function of two variables?
5. State the following properties of a limit of functions of two variables.
   a. scalar rule
   b. sum rule
   c. product rule
   d. quotient rule
6. Define the continuity of a function defined by $f(x, y)$ at a point (x_0, y_0) in its domain.
7. If $z = f(x, y)$, define the first partial derivatives of f with respect to x and y.
8. What is the slope of a tangent line to the surface defined by $z = f(x, y)$ that is parallel to the xy-plane?
9. If $z = f(x, y)$, find the second partial derivatives.
10. If $z = f(x, y)$, what are the increments of x, y, and z?
11. What does it mean for a function of two variables to be differentiable at (x_0, y_0)?
12. State the incremental approximation of $f(x, y)$.
13. Define the total differential of $z = f(x, y)$.
14. State the chain rule for one parameter.
15. State the chain rule for two parameters.
16. Define the directional derivative of a function defined by $z = f(x, y)$.
17. Define the gradient $\nabla f(x, y)$.
18. State the following basic properties of the gradient.
    a. constant rule
    b. linearity rule
    c. product rule
    d. quotient rule
    e. power rule
19. Express the directional derivative in terms of the gradient.
20. State the optimal direction property of the gradient (that is, the steepest ascent and steepest descent).
21. State the normal property of the gradient.
22. Define the normal line and tangent plane to a surface S at a point P_0.
23. Define the absolute extrema of a function of two variables.
24. Define the relative extrema of a function of two variables.
25. What is a critical point of a function of two variables?
26. State the second partials test.
27. State the extreme value theorem for a function of two variables.
28. What is the least-squares approximation of data, and what is a regression line?
29. State Lagrange's theorem.
30. State the procedure for the method of Lagrange multipliers.

Practice Problems

31. If $f(x, y) = \sin^{-1} xy$, verify that $f_{xy} = f_{yx}$.
32. Let $w = x^2 y + y^2 z$, where $x = t \sin t, y = t \cos t$, and $z = 2t$. Use the chain rule to find $\dfrac{dw}{dt}$, where $t = \pi$.
33. Let $f(x, y, z) = xy + yz + xz$, and let P_0 denote the point $(1, 2, -1)$.
    a. Find the gradient of f at P_0.
    b. Find the directional derivative of f in the direction from P_0 toward the point $Q(-1, 1, -1)$.
    c. Find the direction from P_0 in which the directional derivative has its largest value. What is the magnitude of the largest directional derivative at P_0?
34. Show that the function defined by
$$f(x, y) = \begin{cases} \dfrac{x^2 y}{x^3 + y^3} & \text{if } (x, y) \neq (0, 0) \\ 0 & \text{if } (x, y) = (0, 0) \end{cases}$$
is not continuous at $(0, 0)$.
35. If $f(x, y) = \ln\left(\dfrac{y}{x}\right)$, find f_x, f_y, f_{yy}, and f_{xy}.
36. Show that if $f(x, y, z) = x^2 y + y^2 z + z^2 x$, then
$$\frac{\partial f}{\partial x} + \frac{\partial f}{\partial y} + \frac{\partial f}{\partial z} = (x + y + z)^2$$
37. Let $f(x, y) = (x^2 + y^2)^2$. Find the directional derivative of f at $(2, -2)$ in the direction that makes an angle of $\frac{2\pi}{3}$ with the positive x-axis.
38. Find all critical points of $f(x, y) = 12xy - 2x^2 - y^4$ and classify them using the second partials test.
39. Use the method of Lagrange multipliers to find the maximum and minimum values of the function $f(x, y) = x^2 + 2y^2 + 2x + 3$ subject to the constraint $x^2 + y^2 = 4$. You may assume these extreme values exist.
40. Find the largest and smallest values of the function $f(x, y) = x^2 - 4y^2 + 3x + 6y$ on the region defined by $-2 \leq x \leq 2, 0 \leq y \leq 1$.

Supplementary Problems

Describe the domain of each function given in Problems 1–4.

1. $f(x, y) = \sqrt{16 - x^2 - y^2}$ **2.** $f(x, y) = \dfrac{x^2 - y^2}{x - y}$

3. $f(x, y) = \sin^{-1} x + \cos^{-1} y$ **4.** $f(x, y) = e^{x+y} \tan^{-1}\left(\dfrac{y}{x}\right)$

Find the partial derivatives f_x and f_y for the functions defined in Problems 5–10.

5. $f(x, y) = \dfrac{x^2 - y^2}{x + y}$ **6.** $f(x, y) = x^3 e^{3y/(2x)}$

7. $f(x, y) = x^2 y + \sin\dfrac{y}{x}$ **8.** $f(x, y) = \ln\left(\dfrac{xy}{x + 2y}\right)$

9. $f(x, y) = 2x^3 y + 3xy^2 + \dfrac{y}{x}$ **10.** $f(x, y) = xye^{xy}$

For each function given in Problems 11–15, describe the level curve or level surface $f = c$ for the given values of the constant c.

11. $f(x, y) = x^2 - y; c = 2, c = -2$

12. $f(x, y) = 6x + 2y; c = 0, c = 1, c = 2$

13. $f(x, y) = \begin{cases} \sqrt{x^2 + y^2} & \text{if } x \geq 0 \\ |y| & \text{if } x < 0 \end{cases}$ $c = 0, c = 1, c = -1$

14. $f(x, y, z) = x^2 + y^2 + z^2; c = 16, c = 0, c = -25$

15. $f(x, y, z) = x^2 + \dfrac{y^2}{2} + \dfrac{z^2}{9}; c = 1, c = 2$

Evaluate the limits in Problems 16 and 17, assuming they exist.

16. $\displaystyle\lim_{(x, y)\to(1, 1)} \dfrac{xy}{x^2 + y^2}$ **17.** $\displaystyle\lim_{(x, y)\to(0, 0)} \dfrac{x + ye^{-x}}{1 + x^2}$

Show that each limit in Problems 18 and 19 does not exist.

18. $\displaystyle\lim_{(x, y)\to(0, 0)} \dfrac{x^3 - y^3}{x^3 + y^3}$

19. $\displaystyle\lim_{(x, y)\to(0, 0)} \dfrac{x^3 y^2}{x^6 + y^4}$

Find the derivatives in Problems 20–23 using the chain rule. You may leave your answers in terms of x, y, t, u, and v.

20. Find $\dfrac{dz}{dt}$ where $z = -xy + y^3$, and $x = -3t^2, y = 1 + t^3$.

21. Find $\dfrac{dz}{dt}$ where $z = xy + y^2$, and $x = e^t t^{-1}, y = \tan t$.

22. Find $\dfrac{\partial z}{\partial u}$ and $\dfrac{\partial z}{\partial v}$ where $z = x^2 - y^2$, and $x = u + 2v$, $y = u - 2v$.

23. Find $\dfrac{\partial z}{\partial u}$ and $\dfrac{\partial z}{\partial v}$ where $z = x \tan \dfrac{x}{y}$, and $x = uv, y = \dfrac{u}{v}$.

Use implicit differentiation to find $\dfrac{\partial z}{\partial x}$ and $\dfrac{\partial z}{\partial y}$ in Problems 24–27.

24. $x^2 + 6y^2 + 2z^2 = 5$ **25.** $e^x + e^y + e^z = 3$

26. $x^3 + 2xz - yz^2 - z^3 = 1$ **27.** $x + 2y - 3z = \ln z$

In Problems 28–33, find f_{xx}, and f_{yx}.

28. $f(x, y) = \tan^{-1} xy$ **29.** $f(x, y) = \sin^{-1} xy$

30. $f(x, y) = x^2 + y^3 - 2xy^2$ **31.** $f(x, y) = e^{x^2 + y^2}$

32. $f(x, y) = x \ln y$ **33.** $f(x, y) = \displaystyle\int_x^y \sin(\cos t)\, dt$

Find equations for the tangent plane and normal line to the surfaces given in Problems 34–36 at the prescribed point.

34. $x^2 y^3 z = 8$ at $P_0(2, -1, -2)$

35. $x^3 + 2xy^2 - 7x^3 + 3y + 1 = 0$ at $P_0(1, 1, 1)$

36. $z = \dfrac{-4}{2 + x^2 + y^2}$ at $P_0(1, 1, -1)$

Find all critical points of $f(x, y)$ in Problems 37–42 and classify each as a relative maximum, a relative minimum, or a saddle point.

37. $f(x, y) = x^2 - 6x + 2y^2 + 4y - 2$

38. $f(x, y) = x^3 + y^3 - 6xy$

39. $f(x, y) = (x - 1)(y - 1)(x + y - 1)$

40. $f(x, y) = x^2 + y^3 + 6xy - 7x - 6y$

41. $f(x, y) = x^3 + y^3 + 3x^2 - 18y^2 + 81y + 5$

42. $f(x, y) = \sin(x + y) + \sin x + \sin y$ for $0 < x < \pi$, $0 < y < \pi$ (See Figure 12.41.)

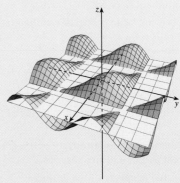

■ **FIGURE 12.41** Graph of $f(x, y) = \sin(x + y) + \sin x + \sin y$

In Problems 43–46, find the largest and smallest values of the function f on the specified closed, bounded set S.

43. $f(x, y) = xy - 2y$; S is the rectangular region $0 \leq x \leq 3$, $-1 \leq y \leq 1$

44. $f(x, y) = x^2 + 2y^2 - x - 2y$; S is the triangular region with vertices $(0, 0), (2, 0), (2, 2)$

45. $f(x, y) = x^2 + y^2 - 3y$; S is the disk $x^2 + y^2 \leq 4$

46. $f(x, y) = 6x - x^2 + 2xy - y^4$; S is the square $0 \leq x \leq 3$, $0 \leq y \leq 3$

47. Use the chain rule to find $\dfrac{dz}{dt}$ if $z = x^2 - 3xy^2$; $x = 2t$, $y = t^2$.

48. Use the chain rule to find $\dfrac{dz}{dt}$ if $z = x \ln y$; $x = 2t$, $y = e^t$.

49. Let $z = ue^{u^2 - v^2}$, where $u = 2x^2 + 3y^2$ and $v = 3x^2 - 2y^2$. Use the chain rule to find $\dfrac{\partial z}{\partial x}$ and $\dfrac{\partial z}{\partial y}$.

50. Use implicit differentiation to find $\dfrac{\partial z}{\partial x}$ and $\dfrac{\partial z}{\partial y}$, where x, y, and z are related by the equation $x^3 + 2xz - yz^2 - z^3 = 1$.

51. Find the slope of the level curve of $x^2 + y^2 = 2$ where $x = 1, y = 1$.

52. Find the slope of the level curve of $xe^y = 2$ where $x = 2$.

53. Find the equations for the tangent plane and normal line to the surface $z = \sin x + e^{xy} + 2y$ at the point $P_0(0, 1, 3)$.

54. The electric potential at each point (x, y) in the disk $x^2 + y^2 < 4$ is $V = 2(4 - x^2 - y^2)^{-1/2}$ volts. Draw the equipotential curves $V = c$ for $c = \sqrt{2}, \dfrac{2}{\sqrt{3}}$, and 8.

55. Let $f(x, y, z) = x^3y + y^3z + z^3x$. Find a function $g(x, y, z)$ such that $\dfrac{\partial f}{\partial x} + \dfrac{\partial f}{\partial y} + \dfrac{\partial f}{\partial z} = x^3 + y^3 + z^3 + 3g(x, y, z)$.

56. Let $u = \sin \dfrac{x}{y} + \ln \dfrac{y}{x}$. Show that $y\dfrac{\partial u}{\partial y} + x\dfrac{\partial u}{\partial x} = 0$.

57. Let $w = \ln(1 + x^2 + y^2) - 2\tan^{-1}y$ where $x = \ln(1 + t^2)$ and $y = e^t$. Use the chain rule to find $\dfrac{dw}{dt}$.

58. Let $f(x, y) = \tan^{-1}\dfrac{y}{x}$. Find the directional derivative of f at $(1, 2)$ in the direction that makes an angle of $\frac{\pi}{3}$ with the positive x-axis.

59. Let $f(x, y) = y^x$. Find the directional derivative of f at $P_0(3, 2)$ in the direction toward the point $Q(1, 1)$.

60. According to postal regulations, the largest cylindrical can that can be sent has a girth $(2\pi r)$ plus length ℓ of 108 inches. What is the largest volume cylindrical can that can be sent?

61. Let $f(x, y, z) = z(x - y)^5 + xy^2z^3$.

    a. Find the directional derivative of f at $(2, 1, -1)$ in the direction of the outward normal to the sphere $x^2 + y^2 + z^2 = 6$.

    b. In what direction is the directional derivative at $(2, 1, -1)$ largest?

62. Find positive numbers x and y for which xyz is a maximum, given that $x + y + z = 1$.

63. Maximize $f(x, y, z) = x^2yz$ given that x, y, and z are all positive numbers and $x + y + z = 12$.

64. Find the shortest distance from the origin to the surface $y^2 - z^2 = 10$.

65. Find the shortest distance from the origin to the surface $z^2 = 3 + xy$.

66. MODELING PROBLEM   A plate is heated in such a way that its temperature at a point (x, y) measured in centimeters on the plate is given in degrees Celsius by

$$T(x, y) = \frac{64}{x^2 + y^2 + 4}$$

    a. Find the rate of change in temperature at the point $(3, 4)$ in the direction $2\mathbf{i} + \mathbf{j}$.

    b. Find the direction and the magnitude of the greatest rate of change of the temperature at the point $(3, 4)$.

67. MODELING PROBLEM   The beautiful patterns on the wings of butterflies have long been a subject of curiosity and scientific study. Mathematical models used to study these patterns often focus on determining the level of morphogen (a chemical that effects change). In a model dealing with eyespot patterns, a quantity of morphogen is released from an eyespot and the morphogen concentration t days later is modeled by

$$S(r, t) = \frac{1}{\sqrt{4\pi t}} \exp\left(\gamma kt + \frac{r}{4t}\right) \qquad t > 0$$

where r measures the radius of the region on the wing affected by the morphogen, and k and γ are positive constants.*

    a. Find t_m so that $\partial S/\partial t = 0$. Show that the function $S_m(t)$ formed from $S(r, t)$ by fixing r has a relative maximum at t_m. Is this the same as saying that the function of two variables $S(r, t)$ has a relative maximum?

    b. Let $M(r)$ denote the maximum found in part **a**; that is, $M(r) = S(r, t_m)$. Find an expression for M in terms of $z = (1 + 4\gamma kr^2)^{1/2}$.

    c. Show that $\dfrac{dM}{dr} < 0$ and interpret this result.

68. MODELING PROBLEM   Certain malignant tumors that do not respond to conventional methods of treatment (surgery, chemotherapy, etc.) may be treated by *hyperthermia*, a process involving the application of extreme heat using microwave transmission. For one

---

*J. D. Murray, *Mathematical Biology,* 2nd edition, Springer-Verlag, New York, 1993, p. 464.

particular kind of microwave application used in such therapy, the temperature at each point located r units from the central axis of the tumor and h units inside it is modeled by the formula

$$T(r, h) = Ke^{-pr^2}[e^{-qh} - e^{-sh}]$$

where A, p, q, and s are positive constants that depend on the properties of the patient's blood and the heating application.*

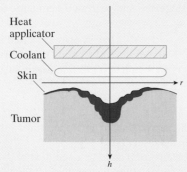

Heat applicator

Coolant

Skin

Tumor

a. At what depth inside the tumor does the maximum temperature occur? What is the maximum temperature? Express your answers in terms of K, p, q, and s.
b. The article on which this problem is based discusses the physiology of hyperthermia in addition to raising several other interesting mathematical issues. Read this article and discuss assumptions made in the model.

69. **MODELING PROBLEM**   The marketing manager for a certain company has compiled the following data relating monthly advertising expenditure and monthly sales (in units of $1,000).

Advertising	3	4	7	9	10
Sales	78	86	138	145	156

a. Plot the data on a graph and find the least-squares line.
b. Use the least-squares line to predict monthly sales if the monthly advertising expenditure is $5,000.

70. Find $f_{xx} - f_{xy} + f_{yy}$ where $f(x, y) = x^2 y^3 + x^3 y^2$.
71. Find f_{xyz} where $f(x, y, z) = \cos(x^2 + y^3 + z^4)$.
72. Let $z = f(x, y)$ where $x = t + \cos t$ and $y = e^t$.

a. Suppose $f_x(1, 1) = 4$ and $f_y(1, 1) = -3$. Find $\dfrac{dz}{dt}$ when $t = 0$.

b. Suppose $f_x(0, 2) = -1$ and $f_y(0, 2) = 3$. Find $\dfrac{\partial z}{\partial r}$ and $\dfrac{\partial z}{\partial \theta}$ at the point where $r = 2, \theta = \dfrac{\pi}{2}$, and $x = r \cos \theta, y = r \sin \theta$.

*"Heat Therapy for Tumors," by Leah Edelstein-Keshet, *UMAP Modules 1991: Tools for Teaching,* Consortium for Mathematics and Its Applications, Inc., MA, 1992, pp. 73–101.

73. Suppose f has continuous partial derivatives in some region D in the plane, and suppose $f(x, y) = 0$ for all (x, y) in D. If $(1, 2)$ is in D and $f_x(1, 2) = 4$ and $f_y(1, 2) = 6$, find dy/dx when $x = 1$ and $y = 2$.
74. Suppose $\nabla f(x, y, z)$ is parallel to the vector $x\mathbf{i} + y\mathbf{j} + z\mathbf{k}$ for all (x, y, z). Show that $f(0, 0, a) = f(0, 0, -a)$ for any a.
75. Find two unit vectors that are normal to the surface given by $z = f(x, y)$ at the point $(0, 1)$, where $f(x, y) = \sin x + e^{xy} + 2y$.
76. Let $f(x, y) = 3(x - 2)^2 - 5(y + 1)^2$. Find all points on the graph of f where the tangent plane is parallel to the plane $2x + 2y - z = 0$.
77. Let z be defined implicitly as a function of x and y by the equation $\cos(x + y) + \cos(x + z) = 1$. Find $\dfrac{\partial^2 z}{\partial y \partial x}$ in terms of x, y, and z.
78. Suppose F and F' are continuous functions of t and that $F'(t) = C$. Define f by $f(x, y) = F(x^2 + y^2)$. Show that the direction of $\nabla f(a, b)$ is the same as the direction of the line joining (a, b) to $(0, 0)$.
79. Let $f(x, y) = 12x^{-1} + 18y^{-1} + xy$, where $x > 0, y > 0$. How do you know that f must necessarily have a minimum in the region $x > 0, y > 0$? Find the maximum.
80. Let $f(x, y) = 3x^4 - 4x^2 y + y^2$. Show that f has a minimum at $(0, 0)$ on every line $y = mx$ that passes through the origin. Then show that f has no relative minimum at $(0, 0)$.
81. Find the minimum of $x^2 + y^2 + z^2$ subject to the constraint $ax + by + cz = 1$ (with $a \neq 0, b \neq 0, c \neq 0$).
82. Suppose $0 < a < 1$ and $x \geq 0, y \geq 0$. Find the maximum of $x^a y^{1-a}$ subject to the constraint $ax + (1 - a)y = 1$.
83. The **geometric mean** of three positive numbers x, y, z is $G = (xyz)^{1/3}$ and the **arithmetic mean** is $A = \frac{1}{3}(x + y + z)$. Use the method of Lagrange multipliers to show that $G(x, y, z) \leq A(x, y, z)$ for all x, y, z.
84. Liquid flows through a tube with length L centimeters and internal radius r centimeters. The total volume V of fluid that flows each second is related to the pressure P and the viscosity a of the fluid by the formula $V = \dfrac{\pi P r^4}{8aL}$.

What is the maximum error that can occur in using this formula to compute the viscosity a, if errors of $\pm 1\%$ can be made in measuring r and L, $\pm 2\%$ in measuring V, and $\pm 3\%$ in measuring P?

85. Suppose the functions f and g have continuous partial derivatives and satisfy

$$\frac{\partial f}{\partial x} = \frac{\partial g}{\partial y} \quad \text{and} \quad \frac{\partial f}{\partial y} = -\frac{\partial g}{\partial x}$$

These are called the **Cauchy–Riemann equations.**
a. Show that level curves of f and g intersect at right angles provided $\nabla f \neq 0$ and $\nabla g \neq 0$.
b. Assuming that the second partials of f and g are continuous, show that f and g satisfy

Laplace's equations

$$f_{xx} + f_{yy} = 0 \quad \text{and} \quad g_{xx} + g_{yy} = 0$$

86. Show that if $z = f(r, \theta)$, where r and θ are defined implicitly as functions of x and y by the equations $x = r \cos \theta$, $y = r \sin \theta$, then the equation $\dfrac{\partial^2 z}{\partial x^2} + \dfrac{\partial^2 z}{\partial y^2} = 0$ becomes
$$\frac{\partial^2 z}{\partial r^2} + \frac{1}{r^2}\frac{\partial^2 z}{\partial \theta^2} + \frac{1}{r}\frac{\partial z}{\partial r} = 0.$$ This is Laplace's equation in polar coordinates.

87. Suppose the angle and radius of a circular sector are allowed to vary. Use Lagrange multipliers to find the angle of the circular sector for which the perimeter of the sector is smallest, assuming the area of the sector is a fixed constant.

88. For the production function given by $Q(x, y) = x^a y^b$, where $a > 0$ and $b > 0$, show that
$$x\frac{\partial Q}{\partial x} + y\frac{\partial Q}{\partial y} = (a + b)Q$$
In particular, if $b = 1 - a$ with $0 < a < 1$, then
$$x\frac{\partial Q}{\partial x} + y\frac{\partial Q}{\partial y} = Q$$

89. The diameter of the base and the height of a right circular cylinder are measured, and the measurements are known to have errors of at most 0.5 cm. If the diameter and height are taken to be 4 cm and 8 cm, respectively, find bounds for the propagated error in
a. the volume V of the cylinder.
b. the surface area S of the cylinder.

90. A right circular cone is measured and is found to have base radius $r = 40$ cm and altitude $h = 20$ cm. If it is known that each measurement is accurate to within 2%, what is the maximum percentage error in the measurement of the volume?

91. An elastic cylindrical container is filled with air so that the radius of the base is 2.02 cm and the height is 6.04 cm. If the container is deflated so that the radius of the base reduces to 2 cm and the height to 6 cm, approximately how much air has been removed? (Ignore the thickness of the container.)

92. Suppose f is a differentiable function of two variables with f_x and f_y also differentiable, and assume that f_{xx}, f_{yy}, and f_{xy} are continuous. The **second directional derivative** (x, y) of f at the point in the direction of the unit vector $\mathbf{u} = a\mathbf{i} + b\mathbf{j}$, is defined by $D_u^2 f(x, y) = D_u[D_u f(x, y)]$. Show that
$$D_u^2 f(x, y) = a^2 f_{xx}(x, y) + 2ab f_{xy}(x, y) + b^2 f_{yy}(x, y)$$

93. A capsule is a cylinder of radius r and length ℓ, capped on each end by a hemisphere. Assume that the capsule dissolves in the stomach at a rate proportional to the ratio $R = S/V$, where V is the volume and S is the surface area of the capsule. Show that
$$\frac{\partial R}{\partial r} < 0 \quad \text{and} \quad \frac{\partial R}{\partial \ell} < 0$$

94. **SPY PROBLEM**   Gasping for breath, the Spy tumbles through the escape hole in the ice cave (Problem 55, Section 12.6) and is immediately knocked unconscious. He awakes tied to a chair, alone, except for a rather large ticking bomb less than ten feet away. "So my friend," thunders the voice of Coldfinger, "you plan to eliminate Scélérat and me for Blohardt, but it appears you are the one about to be eliminated. Still, for old times' sake, I'll give you a chance. Pick a number, S (for Spy), then I'll pick a number C (for Coldfinger), and the value of the expression
$$N = \frac{\exp[(C + S)^2 + 4S + 22]}{(C + S)^2(e^{40} + e^{8S})}$$
will be the number of minutes I'll wait before pressing this little red button." Assuming Coldfinger does his best to minimize the expression once S has been chosen, what number should the Spy pick, and how many minutes will he have to escape and defuse the bomb?

95. **THINK TANK PROBLEM**   Find the minimum distance from the origin to the paraboloid $z = 4 - x^2 - 4y^2$. The graph is shown in Figure 12.42. The distance from $P(x, y, z)$ to the origin is
$$d = \sqrt{x^2 + y^2 + z^2}$$

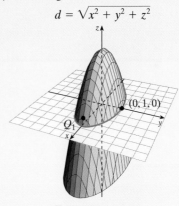

■ **FIGURE 12.42**

This distance will be minimized when d^2 is minimized. Thus, the function to be minimized (after replacing x^2 by $4 - 4y^2 - z$) is $D(x, y) = 4 - 4y^2 - z + y^2 + z^2$ $= 4 - 3y^2 + z^2 - z$. Setting the partial derivatives D_y and D_z to 0 gives the critical point $y = 0, z = 0.5$. Solving for x on the paraboloid gives the points $Q_1(\sqrt{3.5}, 0, 0.5)$ and $Q_2(-\sqrt{3.5}, 0, 0.5)$. These points are NOT minimal because $(0, 1, 0)$ is closer. Explain what is going on here. Our thanks to Herbert R. Bailey, who presented this problem in *The College Mathematics Journal* (" 'Hidden' Boundaries in Constrained Max–Min Problems," May 1991, p. 227).

96. A **minimal surface** is one that has the least surface area of any surface with a given boundary. It can be shown that if $z = f(x, y)$ is a minimal surface, then

$$(1 + z_y^2)z_{xx} - zz_{xy} + (1 + z_x^2)z_{yy} = 0$$

a. Find constants A, B so that

$$z = \ln\left(\frac{A \cos y}{B \cos x}\right)$$

is a minimal surface.

b. Is it possible to find C and D so that
$z = C \ln(\sin x) + D \ln(\sin y)$ is a minimal surface?

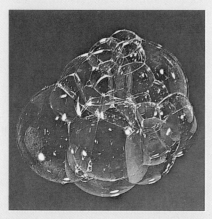

Soap bubbles form minimal surfaces. For an interesting discussion, see "The Geometry of Soap Films and Soap Bubbles," by Frederick J. Almgren, Jr. and Jean E. Taylor, *Scientific American*, July 1976, pp. 82–93.

97. **PUTNAM EXAMINATION PROBLEM**   Let f be a real-valued function having partial derivatives that is defined for $x^2 + y^2 < 1$ and that satisfies $\left|f(x, y)\right| \le 1$. Show that there exists a point (x_0, y_0) in the interior of the unit circle such that $[f_x(x_0, y_0)]^2 + [f_y(x_0, y_0)]^2 \le 16$.

98. **PUTNAM EXAMINATION PROBLEM**   Find the smallest volume bounded by the coordinate planes and a tangent plane to the ellipsoid

$$\frac{x^2}{a^2} + \frac{y^2}{b^2} + \frac{z^2}{c^2} = 1$$

99. **PUTNAM EXAMINATION PROBLEM**   Find the shortest distance between the plane $Ax + By + Cz + 1 = 0$ and the ellipsoid

$$\frac{x^2}{a^2} + \frac{y^2}{b^2} + \frac{z^2}{c^2} \le 1$$

Desertification

This project is to be done in groups of three or four students. Each group will submit a single written report.

A friend of yours named Maria is studying the causes of the continuing expansion of deserts (a process known as *desertification*). She is working on a biological index of vegetation disturbance, which she has defined. By seeing how this index and other factors change through time, she hopes to discover the role played in desertification. She is studying a huge tract of land bounded by a rectangle; this piece of land surrounds a major city but does not include it. She needs to find an economical way to calculate for this piece of land the important vegetation disturbance index $J(x, y)$.

Maria has embarked upon an ingenious approach of combining the results of photographic and radar images taken during flights over the area to calculate the index J. She is assuming that J is a smooth function. Although the flight data do not directly reveal the values of the function J, they give the rate at which the values of J change as the flights sweep over the landscape surrounding the city. Her staff has conducted numerous flights, and from the data she believes she has been able to find actual formulas for the rates at which J changes in the east–west and north–south directions. She has given these functions the names M and N. Thus, $M(x, y)$ is the rate at which J changes as one sweeps in the positive x-direction, and $N(x, y)$ is the corresponding rate in the y-direction. Maria shows you these formulas:

$$M(x, y) = 3.4e^{x(y-7.8)^2} \quad \text{and} \quad N(x, y) = 22 \sin(75 - 2xy)$$

Convince her that these two formulas cannot possibly be correct. Do this by showing her that there is a condition that the two functions M and N must satisfy if they are to be the east–west and north–south rates of change of the function J and that her formulas for M and N do not meet this condition. However, show Maria that if she can find formulas for M and N that satisfy the condition that you showed her, it is possible to find a formula for the function J from the formulas for M and N.

Mathematics in its pure form, as arithmetic, algebra, geometry, and the applications of the analytic method, as well as mathematics applied to matter and force, or statics and dynamics, furnishes the peculiar study that gives to us, whether as children or as men, the command of nature in this its quantitative aspect; mathematics furnishes the instrument, the tool of thought, which we wield in this realm.

W. T. HARRIS
PSYCHOLOGICAL FOUNDATIONS OF EDUCATION
(NEW YORK, 1898), P. 325.

*Marcus S. Cohen, Edward D. Gaughan, R. Arthur Knoebel, Douglas S. Kurtz, and David J. Pengelley, "Priming the Calculus Pump: Innovations and Resources," *MAA Notes* 17 (1991).

CONTENTS

13

Multiple Integration

PREVIEW

The *single integral*

$$\int_a^b f(x)\, dx$$

introduced in Chapter 5 has many uses, as we have seen. In this chapter, we shall generalize the single integral to define *multiple* integrals, in which the integrand is a function of several variables. We will find that multiple integration is used in much the same way as single integration, by "adding" small quantities to define and compute area, volume, surface area, moments, centroids, and probability.

PERSPECTIVE

What is the volume of a doughnut (torus)? Given the joint probability function for the amount of time a typical shopper spends shopping at a particular store and the time spent in the checkout line, how likely is it that a shopper will spend no more than 30 minutes altogether in the store? If the temperature in a solid body is given at each point (x, y, z) and time t, what is the average temperature of the body over a particular time period? Where should a security watch tower be placed in a parking lot to ensure the most comprehensive visual coverage? We shall answer these and other similar questions in this chapter using multiple integration.

13.1   Double Integration over Rectangular Regions

IN THIS SECTION **definition of the double integral, properties of double integrals, volume interpretation, iterated integration, an informal argument for Fubini's theorem** ■

DEFINITION OF THE DOUBLE INTEGRAL

Recall that in Chapter 5, we defined the single definite integral $\int_a^b f(x)\, dx$ as a special kind of limit involving Riemann sums $\sum_{k=1}^{n} f(x_k^*)\,\Delta x_k$, where $x_1, x_2, \ldots, x_n$ are points in a partition of the interval $[a, b]$, and x_k^* is a representative point in the subinterval $[x_{k-1}, x_k]$. We now apply the same ideas to define a *double* definite integral $\iint_R f(x, y)\, dA$, over the rectangle R: $a \le x \le b$, $c \le y \le d$. The definition requires the ideas and notation described in the following three steps:

Step 1: Partition the interval $a \le x \le b$ into m parts and the interval $c \le y \le d$ into n parts. Using these subdivisions, we partition the rectangle R into $N = mn$ **cells** (subrectangles), as shown in Figure 13.1. Call this partition P.

Step 2: Choose a representative point (x_k^*, y_k^*) from each cell in the partition of the rectangle. Form the sum

$$\sum_{k=1}^{N} f(x_k^*, y_k^*)\Delta A_k$$

where ΔA_k is the area of the kth representative cell. This is called the **Riemann sum** of $f(x, y)$ with respect to the partition P and cell representatives (x_k^*, y_k^*).

Step 3: We define the **norm** $\|P\|$ of the partition to be the length of the longest diagonal of any rectangle in the partition. To **refine** the partition means to subdivide the cells in such a way that the norm decreases. When this process is applied to the Riemann sum and the norm decreases indefinitely to zero, we write

$$\lim_{\|P\|\to 0} \sum_{k=1}^{N} f(x_k^*, y_k^*)\Delta A_k$$

This limit is what is called the *double integral*.

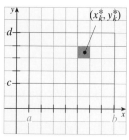

■ **FIGURE 13.1** A partition of the rectangle R into mn cells; a kth cell representative is shown

Double Integral

If f is defined on a closed, bounded rectangular region R in the xy-plane, then the **double integral of f over R** is defined by

$$\iint_R f(x, y)\, dA = \lim_{\|P\|\to 0} \sum_{k=1}^{N} f(x_k^*, y_k^*)\Delta A_k$$

provided this limit exists, in which case f is said to be **integrable** over R.

■ *What This Says:*   In considering this definition, notice:

1. The number N of cells depends on the partition P.
2. As $\|P\| \to 0$, it follows that $N \to \infty$.
3. More formally, the phrase "provided this limit exists" means:

If I is the double integral

$$I = \iint\limits_{R} f(x, y)\, dA$$

then for any $\epsilon > 0$, there exists a $\delta > 0$ such that

$$\left| I - \sum_{k=1}^{N} f(x_k^*, y_k^*)\Delta A_k \right| < \epsilon$$

whenever $\displaystyle\sum_{k=1}^{N} f(x_k^*, y_k^*)\Delta A_k$ is a Riemann sum whose norm satisfies $\|P\| < \delta$.

4. It can be shown that if $f(x, y)$ is continuous on R, then it is integrable on R.

PROPERTIES OF DOUBLE INTEGRALS

Double integrals have many of the same properties as single integrals. Three of these properties are contained in the following theorem.

THEOREM 13.1   *Properties of double integrals*

Assume that all the given integrals exist.

Linearity rule: For constants a and b,
$$\iint\limits_{D} [af(x, y) + bg(x, y)]\, dA = a \iint\limits_{D} f(x, y)\, dA + b \iint\limits_{D} g(x, y)\, dA$$

Dominance rule: If $f(x, y) \geq g(x, y)$ throughout a region D, then
$$\iint\limits_{D} f(x, y)\, dA \geq \iint\limits_{D} g(x, y)\, dA$$

Subdivision rule: If the region of integration D can be subdivided into two subregions D_1 and D_2, then
$$\iint\limits_{D} f(x, y)\, dA = \iint\limits_{D_1} f(x, y)\, dA + \iint\limits_{D_2} f(x, y)\, dA$$

Proof   To prove this, you can first show that each is true for rectangular regions of integration. Then the general case is proved by applying the definition of a double integral and various theorems for single integrals.

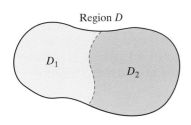

Region D

D_1

D_2

VOLUME INTERPRETATION

If $f(x) \geq 0$ on the interval $[a, b]$, the single integral $\displaystyle\int_a^b f(x)\, dx$ can be interpreted as the area under the curve $y = f(x)$ over $[a, b]$, and the double integral $\displaystyle\iint\limits_{R} f(x, y)\, dA$ has a similar interpretation in terms of volume. To see this, note that if $f(x, y) \geq 0$ on the region R and we partition R using rectangles, then the product $f(x_k^*, y_k^*)\Delta A_k$ is the volume of a parallelepiped (a box) with height $f(x_k^*, y_k^*)$ and base area ΔA_k, as shown in Figure 13.2.

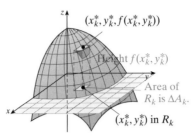

■ **FIGURE 13.2**   The approximating parallelepiped has volume $\Delta V_k = f(x_k^*, y_k^*)\Delta A_k$.

Thus, the Riemann sum

$$\sum_{k=1}^{N} f(x_k^*, y_k^*)\Delta A_k$$

provides an estimate of the total volume under the surface $z = f(x, y)$ over the rectangular domain R, and if f is continuous on R, we expect the approximation to improve if we take a more refined partition of R (that is, more rectangles with smaller norm). Thus, it is natural to *define* the total volume under the surface as the limit of Riemann sums as the norm tends to 0. That is, the volume under $z = f(x, y)$ over the domain R is given by

$$V = \lim_{\|P\|\to 0} \sum_{k=1}^{N} f(x_k^*, y_k^*)\Delta A_k = \iint_R f(x, y)\, dA$$

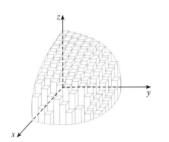

Volume approximated by rectangular parallelepipeds

EXAMPLE 1   *Evaluate a double integral by relating it to a volume*

Evaluate $\displaystyle\iint_R (2 - y)\, dA$, where R is the rectangle in the xy-plane with vertices $(0, 0), (3, 0), (3, 2),$ and $(0, 2)$.

Solution   Because $z = 2 - y$ satisfies $z \geq 0$ for all points in R, the value of the double integral is the same as the volume of the solid bounded above by the plane $z = 2 - y$ and below by the rectangle R. The solid is shown in Figure 13.3. Looking at it sideways, we see an edge with a triangular cross-section of area B and length $h = 3$. We use the formula $V = Bh$.

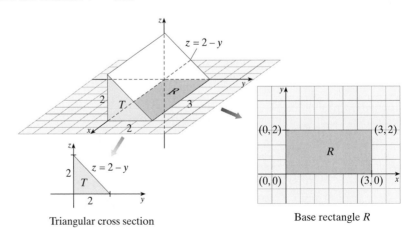

■ **FIGURE 13.3**   Evaluation of $\displaystyle\iint_R (2 - y)\, dA$ as a volume

Because the base is a triangle of side 2 and altitude 2, we have

$$V = Bh = [\tfrac{1}{2}(2)(2)](3) = 6$$

Therefore, the value of the integral is also 6; that is,

$$\iint\limits_{R} (2 - y) \, dA = 6$$ ■

ITERATED INTEGRATION

As with single integrals, it is not practical to evaluate a double integral even over a simple rectangular region by using the definition. Instead, we shall compute double integrals by a process called **successive partial integration.** To be specific, if we hold the y-variable constant (denoted by $\bar{y}$) and integrate $f(x, y)$ with respect to x, we have a single integral

$$\int f(x, \bar{y}) \, dx$$

which we shall refer to as the **partial integral of f with respect to x.** The dx tells us that the integration is with respect to x and that y is to be held constant for the integration, so that when the process is complete, the result is a function of y. Similarly, the integral

$$\int f(\bar{x}, y) \, dy$$

is the **partial integral of f with respect to y,** which means that x is held constant (denoted by $\bar{x}$) and integration is with respect to y. The result of this partial integration is a function of x.

In successive partial integration, we evaluate a double integral by integrating first with respect to one variable and then again with respect to the other variable. In partial integration, we work from the inside out, as indicated in the following notation:

$$\iint f(x, y) \, dx \, dy = \int \left[\int f(x, \bar{y}) \, dx \right] dy \qquad \text{\textit{Integrate with respect to } x \textit{ first, then with respect to } y.}$$

$$\iint f(x, y) \, dy \, dx = \int \left[\int f(\bar{x}, y) \, dy \right] dx \qquad \text{\textit{Integrate with respect to } y \textit{ first, then with respect to } x.}$$

Integrals of this form are said to be **iterated integrals.** Our next theorem tells us how iterated integrals can be used to evaluate double integrals. The theorem was first proved by the Italian mathematician Guido Fubini (1879–1943) in 1907.

THEOREM 13.2   *Fubini's theorem over a rectangular region*

If $f(x, y)$ is continuous over the rectangle $R: a \leq x \leq b, c \leq y \leq d$, then the double integral

$$\iint\limits_{R} f(x, y) \, dA$$

may be evaluated by either iterated integral; that is,

$$\iint\limits_{R} f(x, y) \, dA = \int_{c}^{d} \int_{a}^{b} f(x, y) \, dx \, dy = \int_{a}^{b} \int_{c}^{d} f(x, y) \, dy \, dx$$

■ *What This Says:* Instead of using the definition of a double integral, consider the region R, namely, $a \le x \le b$ and $c \le y \le d$, and evaluate *either* of the iterated integrals

Limits of x (variable outside brackets)   Limits of y (outside brackets)

$$\int_a^b \left[\int_c^d f(\bar{x}, y)\, dy \right] dx \quad \text{or} \quad \int_c^d \left[\int_a^b f(x, \bar{y})\, dx \right] dy$$

Limits of y (variable inside brackets)   Limits of x (inside brackets)

Proof   We shall provide an informal, geometric argument at the end of this section. The formal proof may be found in most advanced calculus textbooks.   ═══

Let us see how Fubini's theorem can be used to evaluate double integrals. We begin by taking another look at Example 1.

EXAMPLE 2   *Evaluate a double integral by using Fubini's theorem*

Use iterated integrals to compute $\iint_R (2 - y)\, dA$, where R is the rectangle with vertices $(0, 0), (3, 0), (3, 2),$ and $(0, 2)$.

Solution   The region of integration is the rectangle $0 \le x \le 3,\ 0 \le y \le 2$ (see Example 1 and Figure 13.3). Thus, by Fubini's theorem, the double integral can be evaluated as an iterated integral:

$$\iint_R (2 - y)\, dA = \int_0^3 \int_0^2 (2 - y)\, dy\, dx \qquad \textit{Integrate inner integral with respect to } y.$$

$$= \int_0^3 \left[2y - \frac{y^2}{2} \right]\Big|_0^2 dx = \int_0^3 \left[4 - \frac{4}{2} - (0) \right] dx = \int_0^3 2\, dx = 2x \big|_0^3 = 6$$

which is the same as the result obtained geometrically in Example 1.   ■

EXAMPLE 3   *Double integral using an iterated integral*

Evaluate $\iint_R x^2 y^5 dA$, where R is the rectangle $1 \le x \le 2, 0 \le y \le 1$, using an iterated integral with

a. y-integration first   b. x-integration first

Solution   The graph of the surface $z = x^2 y^5$ over the rectangle is shown in Figure 13.4a. In your work, you would usually sketch only the rectangle, as shown in Figure 13.4b.

a. $$\iint_R x^2 y^5\, dA = \int_1^2 \int_0^1 x^2 y^5 dy\, dx \qquad \textit{Read this as } \int_1^2 \left[\int_0^1 x^2 y^5\, dy \right] dx.$$

$$= \int_1^2 \left[x^2 \frac{y^6}{6} \right]\Big|_0^1 dx$$

$$= \int_1^2 \left[x^2 \left(\frac{1}{6} - \frac{0}{6} \right) \right] dx = \frac{x^3}{18}\Big|_1^2 = \frac{8}{18} - \frac{1}{18} = \frac{7}{18}$$

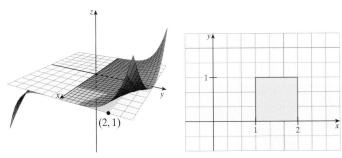

■ **FIGURE 13.4**   Graph of $z = x^2 y^5$ and defining rectangle

b. $\displaystyle\iint\limits_{R} x^2 y^5 \, dA = \int_0^1 \int_1^2 x^2 y^5 \, dx \, dy = \int_0^1 y^5 \left[\frac{x^3}{3}\right]\Bigg|_1^2 dy$

$\displaystyle = \int_0^1 \left[y^5 \left(\frac{8}{3} - \frac{1}{3}\right)\right] dy = \frac{7y^6}{18}\Bigg|_0^1 = \frac{7}{18} - \frac{0}{18} = \frac{7}{18}$ ■

Technology Window

Using technology for multiple integrals offers no special difficulties. You simply integrate with respect to one variable and then with respect to the other variable. You must be careful, however, to properly input the correct limits of integration for each integral. Notice that for $\displaystyle\int_1^2 \int_0^1 x^2 y^2 \, dy \, dx$, most calculators and software programs require the following syntax:

integrate operator, function, variable of integration, lower limit of integration, upper limit of integration.

inner limits of integration   outer limits of integration

write $\displaystyle\int$ (($\displaystyle\int \underbrace{(x^2 y^5)}_{\text{function}}$, y $\overbrace{0, 1}$,   ) x, $\overbrace{1, 2}$)

↑   ↑
inner variable of integration   outer limit of integration

$\underbrace{\hspace{6cm}}$

"inside" integral is the function for the "outside" integral

The output below (using Example 3) shows what this might look like.

a.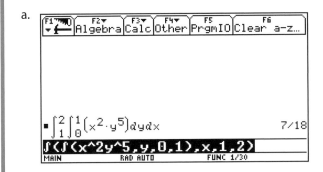

b.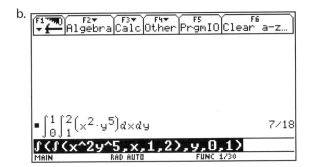

Generally speaking, when using a calculator or computer with CAS software, it does not matter much which order of integration is used. However, without technology, it can matter a great deal, as illustrated in the following example.

EXAMPLE 4   *Choosing the order of integration for a double integral*

Evaluate $\displaystyle\iint_R x \cos xy \, dA$ for $R: 0 \le x \le \dfrac{\pi}{2}, 0 \le y \le 1$.

Solution   Suppose we integrate with respect to x first:

$$\int_0^1 \left[\int_0^{\pi/2} x \cos xy \, dx \right] dy$$

The inner integral requires integration by parts. However, integrating with respect to y first is much simpler:

$$\int_0^{\pi/2} \left[\int_0^1 x \cos xy \, dy \right] dx = \int_0^{\pi/2} \left[\frac{x \sin xy}{x} \right] \Bigg|_0^1 dx = \int_0^{\pi/2} (\sin x - \sin 0) \, dx$$

$$= -\cos x \Big|_0^{\pi/2} = 1$$ ∎

AN INFORMAL ARGUMENT FOR FUBINI'S THEOREM

We can make Fubini's theorem plausible with a geometric argument in the case where $f(x, y) \ge 0$ on R. If $\int_R \int f(x, y) \, dA$ is defined on a rectangle $R: a \le x \le b, c \le y \le d$, it represents the volume of the solid S bounded above by the surface $z = f(x, y)$ and below by the rectangle R. If $A(y_k^*)$ is the cross-sectional area perpendicular to the y-axis at the point y_k^*, then $A(y_k^*)\Delta y_k$ represents the volume of a "slab" that approximates the volume of part of the solid S, as shown in Figure 13.5.

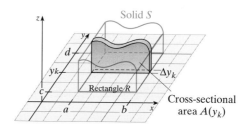

■ **FIGURE 13.5**   Cross-sectional volume parallel to the xz-plane

By using a limit to "add up" all such approximating volumes, we obtain an estimate of the volume of the entire solid S. Thus, if we let V be the exact volume of the solid S, we have

$$\iint_R f(x, y) \, dA = V = \lim_{\|P\| \to 0} \sum_{k=0}^N A(y_k^*)\Delta y_k$$

The limit on the right is just the integral of $A(y)$ over the interval $c \le y \le d$, where $A(y)$ is the area of a cross section with fixed y. In Chapter 5, we found that the area $A(y)$ can be computed by the integral

$$A(y) = \int_a^b f(x, y) \, dx \qquad \text{Integration with respect to } x \ (y \text{ is a constant})$$

We can now make this substitution for $A(y)$ to obtain

$$\iint\limits_R f(x, y)\, dA = V = \lim_{\|P\|\to 0} \sum_{k=0}^{N} A(y_k^*)\Delta y_k = \int_c^d A(y)\, dy$$

$$= \int_c^d \underbrace{\left[\int_a^b f(x, y)\, dx\right]}_{A(y)} dy \qquad \text{Substitution}$$

The fact that $\displaystyle\iint\limits_R f(x, y)\, dA = \int_a^b \int_c^d f(x, y)\, dy\, dx$

can be justified in a similar fashion (you are asked to do this in Problem 47). Thus, we have

$$\int_c^d \left[\int_a^b f(x, y)\, dx\right] dy = \iint\limits_R f(x, y)\, dA = \int_a^b \left[\int_c^d f(x, y)\, dy\right] dx$$

13.1 Problem Set

A *In Problems 1–6, evaluate the iterated integrals.*

1. $\displaystyle\int_0^2 \int_0^1 (x^2 + xy + y^2)\, dy\, dx$

2. $\displaystyle\int_1^2 \int_0^\pi x \cos y\, dy\, dx$

3. $\displaystyle\int_1^{e^2} \int_1^2 \left[\frac{1}{x} + \frac{1}{y}\right] dy\, dx$

4. $\displaystyle\int_0^{\ln 2} \int_0^1 e^{x+2y}\, dx\, dy$

5. $\displaystyle\int_3^4 \int_1^2 \frac{x}{x - y}\, dy\, dx$

6. $\displaystyle\int_2^3 \int_{-1}^2 \frac{1}{(x + y)^2}\, dy\, dx$

Use an appropriate volume formula to evaluate the double integral given in Problems 7–12.

7. $\displaystyle\iint\limits_R 4\, dA;\ R: 0 \le x \le 2;\ 0 \le y \le 4$

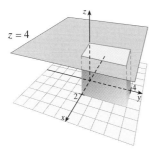

8. $\displaystyle\iint\limits_R 5\, dA;\ R: 2 \le x \le 5;\ 1 \le y \le 3$

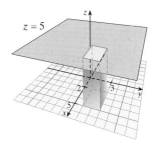

9. $\displaystyle\iint\limits_R (4 - y)\, dA;\ R: 0 \le x \le 3;\ 0 \le y \le 4$

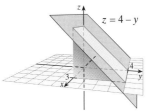

10. $\displaystyle\iint\limits_R (4 - 2y)\, dA;\ R: 0 \le x \le 4;\ 0 \le y \le 2$

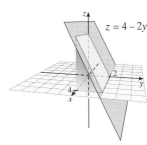

11. $\iint\limits_{R} \frac{y}{2} \, dA; R: 0 \leq x \leq 6; 0 \leq y \leq 4$

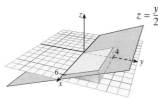

12. $\iint\limits_{R} \frac{y}{4} \, dA; R: 0 \leq x \leq 2; 0 \leq y \leq 8$

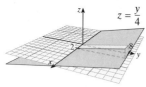

Use successive partial integration to compute the double integrals in Problems 13–20 over the specified rectangle.

13. $\iint\limits_{R} x^2 y \, dA; R: 1 \leq x \leq 2; 0 \leq y \leq 1$

14. $\iint\limits_{R} (x + 2y) \, dA; R: 2 \leq x \leq 3; -1 \leq y \leq 1$

15. $\iint\limits_{R} 2xe^y \, dA; R: -1 \leq x \leq 0; 0 \leq y \leq \ln 2$

16. $\iint\limits_{R} x^2 e^{xy} \, dA; R: 0 \leq x \leq 1; 0 \leq y \leq 1$

17. $\iint\limits_{R} \frac{2xy \, dA}{x^2 + 1}; R: 0 \leq x \leq 1; 1 \leq y \leq 3$

18. $\iint\limits_{R} y\sqrt{1 - y^2} \, dA; R: 1 \leq x \leq 5; 0 \leq y \leq 1$

19. $\iint\limits_{R} \sin(x + y) \, dA; R: 0 \leq x \leq \frac{\pi}{4}; 0 \leq y \leq \frac{\pi}{2}$

20. $\iint\limits_{R} x \sin xy \, dA; R: 0 \leq x \leq \pi; 0 \leq y \leq 1$

Find the volume of the solid bounded below by the rectangle R in the xy-plane and above by the graph of z = f(x, y) in Problems 21–31. Assume that a, b, and c are positive constants.

21. $f(x, y) = 2x + 3y; R: 0 \leq x \leq 1; 0 \leq y \leq 2$

22. $f(x, y) = 5x + 2y; R: 0 \leq x \leq 1; 0 \leq y \leq 2$

23. $f(x, y) = ax + by; R: 0 \leq x \leq a; 0 \leq y \leq b$

24. $f(x, y) = axy; R: 0 \leq x \leq b; 0 \leq y \leq c$

25. $f(x, y) = \sqrt{xy}; R: 0 \leq x \leq 1; 0 \leq y \leq 4$

26. $f(x, y) = \sqrt{xy}; R: 0 \leq x \leq a; 0 \leq y \leq b$

27. $f(x, y) = xe^y; R: 0 \leq x \leq 2; 0 \leq y \leq \ln 2$

28. $f(x, y) = xe^{xy}; R: 0 \leq x \leq 1; 0 \leq y \leq \ln 3$

29. $f(x, y) = (x + y)^5; R: 0 \leq x \leq 1; 0 \leq y \leq 1$

30. $f(x, y) = \sqrt{x + y}; R: 0 \leq x \leq 1; 0 \leq y \leq 1$

31. $f(x, y) = (x + y)^n; n \neq -1$ and $n \neq -2; R: 0 \leq x \leq 1; 0 \leq y \leq 1$

B 32. ■ **What Does This Say?** Discuss the definition of double integral.

33. ■ **What Does This Say?** Describe the implementation of Fubini's theorem.

34. MODELING PROBLEM   Suppose R is a rectangular region within the boundary of a certain national forest that contains 600,000 trees per square mile. Model the total number of trees, T, in the forest as a double integral. Assume that x and y are measured in miles.

35. MODELING PROBLEM   Suppose mass is distributed on a rectangular region R in the xy-plane so that the density (mass per unit area) at the point (x, y) is $\delta(x, y)$. Model the total mass as a double integral.

36. MODELING PROBLEM   Suppose R is the rectangular region within the boundary of a certain city, and let the city center be at the origin $(0, 0)$. For this model, assume that the population density r miles from the city center is $12e^{-.07r}$ thousand people per square mile. Model the total population of the region of the city as a double integral.

37. MODELING PROBLEM   Suppose R is a rectangular region within the boundary of a certain county, and let $f(x, y)$ denote the housing density (in homes per square mile) at the point (x, y). If the property tax on each home in the county is $1,400 per year, model the total property tax collected in the region of the county as a double integral.

In Problems 38–41, one order of integration is easier than the other. Determine the easier order and then evaluate the integral.

38. Compute $\iint\limits_{R} x\sqrt{1 - x^2} \, e^{3y} \, dA$, where R is the rectangle $0 \leq x \leq 1, 0 \leq y \leq 2$.

39. Compute $\iint\limits_{R} \frac{\ln \sqrt{y}}{xy} \, dA$, where R is the rectangle $1 \leq x \leq 4, 1 \leq y \leq e$.

40. Compute (correct to the nearest hundredth)
$$\iint\limits_{R} \frac{xy}{x^2 + y^2} \, dA, \text{ where } R$$
is the rectangle $1 \leq x \leq 3, 1 \leq y \leq 2$.

41. Evaluate $\iint\limits_{R} xe^{xy} \, dA$, where R is the rectangle $0 \leq x \leq 1, 1 \leq y \leq 2$.

42. Explain why $\iint\limits_{R} (4 - x^2 - y^2) \, dA > 2$, where R is the rectangular domain in the plane given by $0 \leq x \leq 1, 0 \leq y \leq 1$.

43. Approximate the volume of the solid lying between the surface

$$f(x, y) = 4 - x^2 - y^2$$

and the square region R given by $0 \le x \le 1$, $0 \le y \le 1$. Use a partition made up of squares whose edges have length 0.25.

44. Approximate the volume of the solid lying between the surface

$$f(x, y) = x^2 + y^2 + 1$$

and the square region R given by $0 \le x \le 1$, $0 \le y \le 1$. Use a partition made up of squares whose edges have length 0.25.

45. ℍISTORICAL ℚUEST   Guido Fubini taught at the Institute for Advanced Study in Princeton. He was nicknamed the "Little Giant," because of his small body but large mind. Even though the conclusion of Fubini's theorem was known for a long time and successfully applied in various in-stances, it was not satisfactorily proved in a general setting until 1907. His most

GUIDO FUBINI
1879–1943

important work was in differential projective geometry. In 1938 he was forced to leave Italy because of the Fascist government, and he emigrated to the United States.

For this ℍistorical ℚuest, write several paragraphs about the nature of differential projective geometry.

ⓒ 46. Let f be a function with continuous second partial derivatives on a rectangular domain R with vertices $(x_1, y_1), (x_1, y_2), (x_2, y_2),$ and (x_2, y_1), where $x_1 < x_2$ and $y_1 < y_2$. Use the fundamental theorem of calculus to show that

$$\iint\limits_R \frac{\partial^2 f}{\partial y \partial x} \, dA = f(x_1, y_1) - f(x_2, y_1)$$
$$+ f(x_2, y_2) - f(x_1, y_2)$$

47. Let f be a continuous function defined on the rectangle $R: a \le x \le b, c \le y \le d$. Use a geometric argument to show that

$$\iint\limits_R f(x, y) \, dA = \int_a^b \int_c^d f(x, y) \, dy \, dx$$

Hint: Modify the argument given in the text by taking cross-sectional areas perpendicular to the x-axis.

13.2   *Double Integration over Nonrectangular Regions*

nonrectangular regions, more on area and volume, reversing the order of integration in a double integral, properties of double integrals ■

NONRECTANGULAR REGIONS

■ **FIGURE 13.6**   The region D is bounded by a rectangle R.

Let $f(x, y)$ be a function that is continuous on the region D that can be contained in a rectangle R. (See Figure 13.6.) Define the function $F(x, y)$ on R as $f(x, y)$ if (x, y) is in D, and 0 otherwise. That is,

$$F(x, y) = \begin{cases} f(x, y) & \text{for } (x, y) \text{ in } D \\ 0 & \text{for } (x, y) \text{ not in } D \end{cases}$$

Then, if F is integrable over R, we say that f is **integrable over D,** and the **double integral of f over D** is defined as

$$\iint\limits_D f(x, y) \, dA = \iint\limits_R F(x, y) \, dA$$

In the previous section, you saw how the double integral $\int_R \int f(x, y) \, dA$ could be evaluated by successive partial integration when R is a rectangle. A modification of this procedure allows you to evaluate the double integral in the important case where R is not a rectangle, but is still bounded by vertical (or horizontal) lines on two opposite sides.

Type I Region (Vertical Strip)

A **type I region** contains points (x, y) such that for each fixed x between constants a and b, y varies from $g_1(x)$ to $g_2(x)$, where g_1 and g_2 are continuous functions. Think of a vertical strip.

Type II Region (Horizontal Strip)

A **type II region** contains points (x, y) such that for each fixed y between constants c and d, x varies from $h_1(y)$ to $h_2(y)$, where h_1 and h_2 are continuous functions. Think of a horizontal strip.

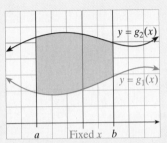

Type I region: For a fixed x between a and b, y varies from $g_1(x)$ to $g_2(x)$.

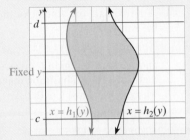

Type II region: For a fixed y between c and d, x varies from $h_1(y)$ to $h_2(y)$.

By applying Fubini's theorem for rectangular regions, we can derive the following theorem, which shows how to evaluate a double integral over a type I or type II region.

THEOREM 13.3   *Fubini's theorem for nonrectangular regions*

If D is a type I region, then

TYPE I (vertical strip):
x fixed, y varies (form $dy\, dx$)

$$\iint_D f(x, y)\, dA = \int_a^b \int_{g_1(x)}^{g_2(x)} f(x, y)\, dy\, dx$$

whenever both integrals exist. Similarly, for a type II region,

TYPE II (horizontal strip):
y fixed, x varies (form $dx\, dy$)

$$\iint_D f(x, y)\, dA = \int_c^d \int_{h_1(y)}^{h_2(y)} f(x, y)\, dx\, dy$$

Proof   This proof is found in most advanced calculus textbooks.   ═══

When using Fubini's theorem for nonrectangular regions, it helps to sketch the region of integration D and to find equations for all boundary curves of D. Such a sketch often provides the information needed to determine whether D is a type I or type II region (or neither, or both) and to set up the limits of integration of an iterated integral.

EXAMPLE 1   *Double integral over a triangular region*

Let T be the triangular region enclosed by the lines $y = 0$, $y = 2x$, and $x = 1$. Evaluate the double integral

$$\iint_T (x + y)\, dA$$

using an iterated integral with:
a. y-integration first   b. x-integration first

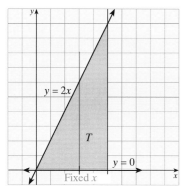

FIGURE 13.7   For each fixed x $(0 \leq x \leq 1)$, y varies from $y = 0$ to $y = 2x$.

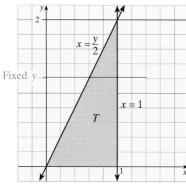

FIGURE 13.8   For each fixed y $(0 \leq y \leq 2)$, x varies from $y/2$ to 1.

Solution

a. To set up the limits of integration in the iterated integral, we draw the graph as shown in Figure 13.7 and note that for fixed x, the variable y varies from $y = 0$ (the x-axis) to the line $y = 2x$. These are the limits of integration for the inner integral (with respect to y first). The other limits of integration are the numerical limits of integration for x; that is, x varies between $x = 0$ and $x = 1$.

$$\iint_T (x + y)\, dA = \int_0^1 \int_0^{2x} (x + y)\, dy\, dx = \int_0^1 \left[xy + \frac{1}{2} y^2 \right]\Bigg|_{y=0}^{y=2x} dx$$

$$= \int_0^1 \left[x(2x) + \frac{1}{2}(2x)^2 - \left(x(0) + \frac{1}{2}(0)^2 \right) \right] dx = \int_0^1 4x^2\, dx = \frac{4}{3} x^3 \Bigg|_{x=0}^{x=1} = \frac{4}{3}$$

b. Reversing the order of integration, we see from Figure 13.8 that for each fixed y, the variable x varies (left to right) from the line $x = y/2$ to the vertical line $x = 1$. The outer limits of integration are for y as y varies from $y = 0$ to $y = 2$.

$$\iint_T (x + y)\, dA = \int_0^2 \int_{y/2}^1 (x + y)\, dx\, dy$$

$$= \int_0^2 \left[\frac{1}{2} x^2 + xy \right]\Bigg|_{x=y/2}^{x=1} dy = \int_0^2 \left[\frac{1}{2} + y - \frac{y^2}{8} - \frac{y^2}{2} \right] dy$$

$$= \left[\frac{y}{2} + \frac{y^2}{2} - \frac{5y^3}{24} \right]\Bigg|_{y=0}^{y=2} = \left[1 + 2 - \frac{5(8)}{24} \right] - [0] = \frac{4}{3} \quad ■$$

MORE ON AREA AND VOLUME

Even though we can find the area between curves with single integrals, it is often easier to compute area using a double integral. If $f(x, y) \geq 0$ over a region D in the xy-plane, then $\iint_D f(x, y)\, dA$ gives the **volume of the solid** bounded above by the surface $z = f(x, y)$ and below by the region D. In the special case where $f(x, y) = 1$, the integral gives the **area** of D.

The Double Integral as Area and Volume

The **area** of the region D in the xy-plane is given by

$$A = \iint_D dA$$

If f is continuous and $f(x, y) \geq 0$ on the region D, the **volume** of the solid under the surface $z = f(x, y)$ above the region D is given by

$$V = \iint_D f(x, y)\, dA$$

EXAMPLE 2   *Area of a region in the xy-plane using a double integral*

Find the area of the region D between $y = \cos x$ and $y = \sin x$ over the interval $0 \leq x \leq \frac{\pi}{4}$ using

a. a single integral   b. a double integral

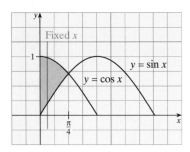

■ **FIGURE 13.9**   The area of the region between $y = \cos x$ and $y = \sin x$

Solution

a. The graph is shown in Figure 13.9.

$$\int_0^{\pi/4} (\cos x - \sin x)\, dx = [\sin x + \cos x]\Big|_0^{\pi/4} = \sqrt{2} - 1$$

b.

$$A = \iint_D dA = \int_0^{\pi/4} \int_{\sin x}^{\cos x} 1\, dy\, dx = \int_0^{\pi/4} [y]\Big|_{y=\sin x}^{y=\cos x} dx$$

$$= \int_0^{\pi/4} [\cos x - \sin x]\, dx = \sqrt{2} - 1$$

The area is $\sqrt{2} - 1 \approx 0.41$ square unit.   ■

In comparing the single and double integral solutions for area in Example 2, you might ask, "Why bother with the double integral, because it reduces to the single integral case after one step?" The answer is that it is often easier to begin with the double integral

$$A = \iint_D dA$$

and then let the *evaluation* lead to the proper form.

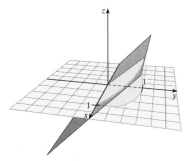

Graph of $z = y$ showing the disk in the xy-plane

| **EXAMPLE 3** | *Volume using a double integral* |

Find the volume of the solid bounded above by the plane $z = y$ and below in the xy-plane by the part of the disk $x^2 + y^2 \le 1$ in the first quadrant.

Solution   The three-dimensional solid is shown in the margin, but for your work you need be concerned only with the projection in the xy-plane, which is shown in Figure 13.10.

We can regard D as either a type I or type II region, and because we worked with a type I (vertical) region in Example 2, we shall use a type II (horizontal) region for this example. Accordingly, note that for each fixed number y between 0 and 1, x varies between $x = 0$ on the left and $x = \sqrt{1 - y^2}$ on the right. Thus,

$$V = \iint_D f(x, y)\, dA$$

$$= \int_0^1 \int_0^{\sqrt{1-y^2}} y\, dx\, dy \qquad f(x, y) = y \text{ is given; } dA = dx\, dy.$$

$$= \int_0^1 [xy]\Big|_{x=0}^{x=\sqrt{1-y^2}} dy$$

$$= \int_0^1 y\sqrt{1 - y^2}\, dy \qquad \text{Let } u = 1 - y^2 \text{ to integrate.}$$

$$= \left[-\frac{1}{3}(1 - y^2)^{3/2} \right]\Big|_{y=0}^{y=1} = \frac{1}{3}$$

The volume is $\frac{1}{3}$ cubic unit.   ■

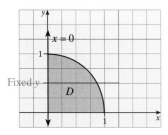

■ **FIGURE 13.10**   The quarter disk $x^2 + y^2 \le 1, x \ge 0, y \ge 0$

REVERSING THE ORDER OF INTEGRATION IN A DOUBLE INTEGRAL

It is often useful to be able to reverse the order of integration in a given iterated integral. This procedure is illustrated in the next two examples.

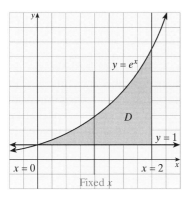

a. D as a type I region; y varies from 1 to e^x

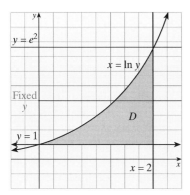

b. D as a type II region; x varies from $\ln y$ to 2

■ **FIGURE 13.11** The region of integration for Example 4

EXAMPLE 4 *Reversing order of integration in a double integral*

Reverse the order of integration in the iterated integral

$$\int_0^2 \int_1^{e^x} f(x, y) \, dy \, dx$$

Solution Begin by drawing the region D by looking at the limits of integration for both x and y in the double integral. For this example, we see that the y-integration comes first, so D is a type I region. The inner limits are

$$y = e^x \text{ (top curve)} \quad \text{and} \quad y = 1 \text{ (bottom curve)}$$

These are shown in Figure 13.11a. Next, draw the appropriate limits of integration for x. *Note:* These limits should be constants:

$$x = 0 \text{ (left point)} \quad \text{and} \quad x = 2 \text{ (right point)}$$

These vertical lines are also drawn in Figure 13.11a.

To reverse the order of integration, we need to regard D as a type II region (Figure 13.11b). Note that the region varies from $y = 1$ to $y = e^2$ (corresponding to where $y = e^x$ intersects $x = 0$ and $x = 2$, respectively). For each fixed y (horizontal strip) between 1 and e^2, the region extends from the curve $x = \ln y$ (that is, $y = e^x$) on the left to the line $x = 2$ on the right. Thus, reversing the order of integration, we find that the given integral becomes

$$\int_1^{e^2} \int_{\ln y}^2 f(x, y) \, dx \, dy$$

■ **What This Says:** There are two different ways of representing the integral, which we illustrate together for easy reference:

Type I: *x* fixed (vertical strip)	Type II: *y* fixed (horizontal strip)
y-integration first; varies from $y = 1$ to $y = e^x$. ↓	*x*-integration first; varies from $x = \ln y$ to $x = 2$. ↓
$\int_0^2 \int_1^{e^x} f(x, y) \, dy \, dx$ ↑ Constant limits for *x*; varies from 0 to 2.	$= \int_1^{e^2} \int_{\ln y}^2 f(x, y) \, dx \, dy$ ↑ Constant limits for *y*; varies from 1 to e^2.

■

EXAMPLE 5 *Reversing the order that requires a sum of integrals*

Reverse the order of integration in the integral

$$\int_{-1}^2 \int_{x^2-2}^x f(x, y) \, dy \, dx$$

Solution This is a type I form (because the y-integration comes first). To reverse the order of integration, we begin by drawing the region D, as shown in Figure 13.12a. We draw the bottom and top curves:

$$y = x^2 - 2 \text{ (bottom)} \quad \text{and} \quad y = x \text{ (top)}$$

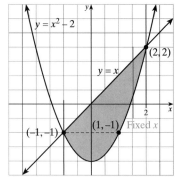

a. Type I description

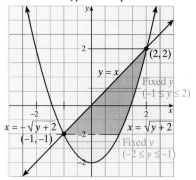

b. Type II description

■ **FIGURE 13.12** The region of integration for Example 5

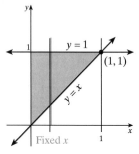

a. y-integration first

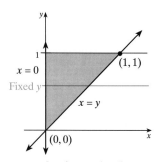

b. x-integration first

■ **FIGURE 13.13** The region of integration for Example 6

Then we draw the left and right boundaries:

$$x = -1 \quad \text{and} \quad x = 2$$

To change this to a type II form, we shift from a vertical strip (x fixed) to a horizontal strip (y fixed), as shown in Figure 13.12b. We see that for y between -2 and -1, the horizontal line is bounded by the left and right branches of the parabolic arch, whereas for y between -1 and 2, it is bounded on the left by the line $x = y$ and on the right by the parabola. This means that the integral with the order of integration reversed must be expressed as the sum of two integrals:

$$\int_{-1}^{2} \int_{x^2-2}^{x} f(x, y) \, dy \, dx \qquad \text{For a fixed } x \text{ between } -1 \text{ and } 2, y \text{ varies between } y = x^2 - 2 \text{ and } y = x.$$

$$= \int_{-2}^{-1} \int_{-\sqrt{y+2}}^{\sqrt{y+2}} f(x, y) \, dx \, dy + \int_{-1}^{2} \int_{y}^{\sqrt{y+2}} f(x, y) \, dx \, dy$$

$\underbrace{\phantom{\int_{-2}^{-1} \int_{-\sqrt{y+2}}^{\sqrt{y+2}}}}$ $\underbrace{\phantom{\int_{-1}^{2} \int_{y}^{\sqrt{y+2}}}}$
Part between left and right   Part between line and
part of parabola (blue part)   Parabola (gray part)

For a fixed y between -2 For a fixed y between -1
and -1, x varies between   and 2, x varies between
$x = -\sqrt{y+2}$ and $x = \sqrt{y+2}$ $x = y$ and $x = \sqrt{y+2}$ ■

A given iterated integral can often be evaluated in either order, although it may happen that one order is more convenient than the other. For instance, the integral in Example 5 is easier to handle by performing y-integration first, because the reverse order involves two integrations instead of just one. However, there are iterated integrals that can be evaluated *only* after reversing the order of integration. Such an integral is featured in the following example.

EXAMPLE 6 *Evaluating a double integral by reversing the order*

Evaluate $\int_{0}^{1} \int_{x}^{1} e^{y^2} \, dy \, dx$.

Solution We cannot evaluate the integral in the given order (y-integration first) because the integrand e^{y^2} has no elementary antiderivative. We shall evaluate the integral by reversing the order of integration. The region of integration is sketched in Figure 13.13a. Note that for any fixed x between 0 and 1, y varies from x to 1.

To reverse the order of integration, observe that for each fixed y between 0 and 1, x varies from 0 to y, as shown in Figure 13.13b.

$$\int_{0}^{1} \int_{x}^{1} e^{y^2} \, dy \, dx = \int_{0}^{1} \int_{0}^{y} e^{y^2} \, dx \, dy$$

$$= \int_{0}^{1} x e^{y^2} \Big|_{x=0}^{x=y} \, dy = \int_{0}^{1} y e^{y^2} \, dy \qquad \text{Let } u = y^2.$$

$$= \left[\frac{1}{2} e^{y^2} \right] \Big|_{0}^{1} = \frac{1}{2}(e^1 - e^0)$$

$$= \frac{1}{2}(e - 1) \qquad \qquad \blacksquare$$

13.2 Problem Set

A 1. ■ **What Does This Say?** Describe the process for finding volume using a double integral.

2. ■ **What Does This Say?** Describe the process of reversing the order of integration.

Sketch the region of integration in Problems 3–16, and compute the double integral (either in the order of integration given or with the order reversed).

3. $\displaystyle\int_0^4 \int_0^{4-x} xy \, dy \, dx$

4. $\displaystyle\int_0^4 \int_{x^2}^{4x} dy \, dx$

5. $\displaystyle\int_0^1 \int_{-x^2}^{x^2} dy \, dx$

6. $\displaystyle\int_1^e \int_0^{\ln x} xy \, dy \, dx$

7. $\displaystyle\int_0^{2\sqrt{2}} \int_{y^2/4}^{\sqrt{12-y^2}} dx \, dy$

8. $\displaystyle\int_{-2}^1 \int_{y^2+4y}^{3y+2} dx \, dy$

9. $\displaystyle\int_0^3 \int_1^{4-x} (x + y) \, dy \, dx$

10. $\displaystyle\int_0^1 \int_{x^3}^1 (x + y^2) \, dy \, dx$

11. $\displaystyle\int_0^1 \int_0^x (x^2 + 2y^2) \, dy \, dx$

12. $\displaystyle\int_{-1}^1 \int_{-1}^x (3x + 2y) \, dy \, dx$

13. $\displaystyle\int_0^2 \int_0^{\sin x} y \cos x \, dy \, dx$

14. $\displaystyle\int_0^{\pi/2} \int_0^{\sin x} e^y \cos x \, dy \, dx$

15. $\displaystyle\int_0^{\pi/3} \int_0^{y^2} \frac{1}{y} \sin \frac{x}{y} \, dx \, dy$

16. $\displaystyle\int_0^1 \int_0^y y^2 \, e^{xy} \, dx \, dy$

Evaluate the double integral given in Problems 17–28 for the specified region of integration D.

17. $\displaystyle\iint_D (x + y) \, dA$; D is the triangle with vertices $(0, 0), (0, 1), (1, 1)$.

18. $\displaystyle\iint_D (x + 2y) \, dA$; D is the triangle with vertices $(0, 0), (1, 0), (0, 2)$.

19. $\displaystyle\iint_D 48xy \, dA$; D is the region bounded by $y = x^3$ and $y = \sqrt{x}$.

20. $\displaystyle\iint_D (2y - x) \, dA$; D is the region bounded by $y = x^2$ and $y = 2x$.

21. $\displaystyle\iint_D y \, dA$; D is the region bounded by $y = \sqrt{x}, y = 2 - x$, and $y = 0$.

22. $\displaystyle\iint_D 4x \, dA$; D is the region bounded by $y = 4 - x^2$, $y = 3x$, and $x = 0$.

23. $\displaystyle\iint_D 4x \, dA$; D is the region in the first quadrant bounded by $y = 4 - x^2, y = 3x$, and $y = 0$.

24. $\displaystyle\iint_D (2x + 1) \, dA$; D is the triangle with vertices $(-1, 0), (1, 0)$, and $(0, 1)$.

25. $\displaystyle\iint_D 2x \, dA$; D is the region bounded by $x^2 y = 1, y = x$, $x = 2$, and $y = 0$.

26. $\displaystyle\iint_D \frac{dA}{y^2 + 1}$; D is the triangle bounded by $x = 2y$, $y = -x$, and $y = 2$.

27. $\displaystyle\iint_D 12x^2 e^{y^2} \, dA$; D is the region in the first quadrant bounded by $y = x^3$ and $y = x$.

28. $\displaystyle\iint_D \cos e^x \, dA$; D is the region bounded by $y = e^x, y = -e^x, x = 0$, and $x = \ln 2$.

Sketch the region of integration in Problems 29–36, and then compute the integral in two ways: **a.** *with the given order of integration, and* **b.** *with the order of integration reversed.*

29. $\displaystyle\int_0^4 \int_0^{4-x} xy \, dy \, dx$

30. $\displaystyle\int_0^4 \int_{x^2}^{4x} dy \, dx$

31. $\displaystyle\int_0^1 \int_x^{2x} e^{y-x} \, dy \, dx$

32. $\displaystyle\int_0^4 \int_0^{\sqrt{x}} 3x^5 \, dy \, dx$

33. $\displaystyle\int_0^1 \int_{-x^2}^{x^2} dy \, dx$

34. $\displaystyle\int_1^e \int_0^{\ln x} xy \, dy \, dx$

35. $\displaystyle\int_0^{2\sqrt{3}} \int_{y^2/6}^{\sqrt{16-y^2}} dx \, dy$

36. $\displaystyle\int_{-2}^1 \int_{y^2+4y}^{3y+2} dx \, dy$

Sketch the region of integration in Problems 37–46, and write an equivalent integral with the order of integration reversed. Do not evaluate.

37. $\displaystyle\int_0^1 \int_0^{2y} f(x, y) \, dx \, dy$

38. $\displaystyle\int_0^1 \int_{x^2}^{\sqrt{x}} f(x, y) \, dy \, dx$

39. $\displaystyle\int_0^4 \int_{y/2}^{\sqrt{y}} f(x, y) \, dx \, dy$

40. $\displaystyle\int_1^{e^2} \int_{\ln x}^2 f(x, y) \, dy \, dx$

41. $\displaystyle\int_0^3 \int_{y/3}^{\sqrt{4-y}} f(x, y) \, dx \, dy$

42. $\displaystyle\int_0^1 \int_x^{2-x} f(x, y) \, dy \, dx$

43. $\displaystyle\int_{-3}^2 \int_{x^2}^{6-x} f(x, y) \, dy \, dx$

44. $\displaystyle\int_2^4 \int_x^{16/x} f(x, y) \, dy \, dx$

45. $\displaystyle\int_0^7 \int_{x^2-6x}^x f(x, y) \, dy \, dx$

46. $\displaystyle\int_0^1 \int_{\tan^{-1}x}^{\pi/4} f(x, y) \, dy \, dx$

B *Set up a double integral for the volume of the solid region described in Problems 47–53.*

47. The tetrahedron that lies in the first octant and is bounded by the coordinate planes and the plane $z = 7 - 3x - 2y$

48. The solid bounded above by the paraboloid $z = 6 - 2x^2 - 3y^2$ and below by the plane $z = 0$

49. The solid that lies inside both the cylinder $x^2 + y^2 = 3$ and the sphere $x^2 + y^2 + z^2 = 7$

50. The solid that lies inside both the sphere $x^2 + y^2 + z^2 = 3$ and the paraboloid $2z = x^2 + y^2$

51. The ellipsoid

$$\frac{x^2}{a^2} + \frac{y^2}{b^2} + \frac{z^2}{c^2} = 1$$

52. The solid bounded above by the plane $z = 2 - 3x - 5y$ and below by the region shown in Figure 13.14.

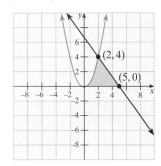

■ **FIGURE 13.14**   Region for Problem 52

53. The solid that remains when a square hole of side 2 is drilled through a sphere of radius $\sqrt{2}$

Express the area of the region D bounded by the curve given in Problems 54–55 as a double integral in two different ways (with different orders of integration). Evaluate one of these integrals to find the area.

54. Let D denote the region in the first quadrant of the xy-plane that is bounded by the curves $y = \dfrac{4}{x^2}$ and $y = 5 - x^2$.

55. Let D denote the region bounded by the ellipse $\dfrac{x^2}{a^2} + \dfrac{y^2}{b^2} = 1$.

56. Find the volume under the surface $z = x + y + 2$ above the region D bounded by the curves $y = x^2$ and $y = 2$.

57. Find the volume under the plane $z = 4x$ above the region D given by $y = x^2, y = 0$, and $x = 1$.

58. Evaluate

$$\int_0^1 \int_0^y (x^2 + y^2)\, dx\, dy + \int_1^2 \int_0^{2-y} (x^2 + y^2)\, dx\, dy$$

by reversing the order of integration.

59. Let $f(x, y)$ be a continuous function for all x and y. Reverse the order of integration in

$$\int_1^2 \int_x^{x^3} f(x, y)\, dy\, dx + \int_2^8 \int_x^8 f(x, y)\, dy\, dx$$

60. Evaluate $\displaystyle\iint_D xy\, dA$, where D is the triangular region in the xy-plane with vertices $(0, 0), (1, 0)$, and $(4, 1)$.

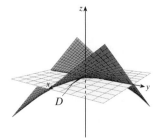

61. Compute $\displaystyle\iint_D (x^2 - xy - 1)\, dA$, where D is the triangular region bounded by the lines $x - 2y + 2 = 0$, $x + 3y - 3 = 0$, and $y = 0$.

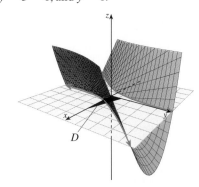

62. Find the volume under the plane $z = x + 2y + 4$ above the region D bounded by the lines $y = 2x, y = 3 - x$, and $y = 0$.

63. Find the volume under the plane $3x + y - z = 0$ above the elliptic region bounded by $4x^2 + 9y^2 \leq 36$, with $x \geq 0$ and $y \geq 0$.

64. Find the volume under the surface $z = x^2 + y^2$ above the square region bounded by $|x| \leq 1$ and $|y| \leq 1$.

Evaluate each integral in Problems 65–67 by relating it to the **C** *volume of a simple solid.*

65. Evaluate $\displaystyle\iint_R (3 - \sqrt{x^2 + y^2})\, dA$, where R is the disk $x^2 + y^2 \leq 9$ in the xy-plane.

66. Evaluate $\displaystyle\iint_R (1 - \sqrt{x^2 + z^2})\, dA$, where R is the disk $x^2 + z^2 \leq 1$ in the xz-plane.

67. Evaluate $\displaystyle\iint_R (4 - \sqrt{y^2 + z^2})\, dA$, where R is the disk $y^2 + z^2 \le 16$ in the yz-plane.

68. Show that if $f(x, y)$ is continuous on a region D and $m \le f(x, y) \le M$ for all points (x, y) in D, then

$$mA \le \iint_D f(x, y)\, dA \le MA$$

where A is in the area of D.

69. Use the result of Problem 68 to estimate the value of the double integral

$$\iint_D e^{y \sin x}\, dA$$

where D is the triangle with vertices $(-1, 0), (2, 0),$ and $(0, 1)$.

13.3   Double Integrals in Polar Coordinates

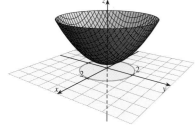

Graph of $f(x, y) = x^2 + y^2 + 1$

IN THIS SECTION   change of variables to polar form, improper double integrals in polar coordinates

In general, changing variables in a double integral is more complicated than in a single integral. In this section, we focus attention on using polar coordinates in a double integral, and in Section 13.8, we examine changing variables from a more general standpoint.

CHANGE OF VARIABLES TO POLAR FORM

Polar coordinates are used in double integrals primarily when the integrand or the region of integration (or both) have relatively simple polar descriptions. As a preview of the ideas we plan to explore, let us examine the double integral

$$\iint_R (x^2 + y^2 + 1)\, dA$$

where R is the region (disk) bounded by the circle $x^2 + y^2 = 4$.

Interpreting R as a type I (vertical strip), we see that for each fixed x between -2 and 2, y varies from the lower boundary semicircle with equation $y = -\sqrt{4 - x^2}$ to the upper semicircle $y = \sqrt{4 - x^2}$, as shown in Figure 13.15a.

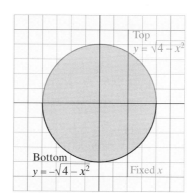

a. Type I description:
For fixed x between -2 and 2
y varies from $y = -\sqrt{4 - x^2}$
to $y = \sqrt{4 - x^2}$

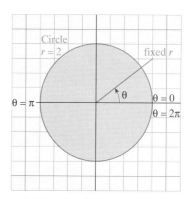

b. Polar description:
For fixed θ between 0 and 2π,
r varies from 0 to 2.

■ **FIGURE 13.15**   Two interpretations of a region R

Using the type I description, we have

$$\iint_D (x^2 + y^2 + 1)\, dA = \int_{-2}^{2} \int_{-\sqrt{4-x^2}}^{\sqrt{4-x^2}} (x^2 + y^2 + 1)\, dy\, dx$$

The iterated integral on the right is clearly difficult to evaluate, but both the integrand and the domain of integration can be represented quite simply in terms of polar coordinates. Specifically, using the polar conversion formulas

$$x = r \cos \theta \qquad y = r \sin \theta \qquad r = \sqrt{x^2 + y^2} \qquad \tan \theta = \frac{y}{x}$$

we find that the integrand $f(x, y) = x^2 + y^2 + 1$ can be rewritten

$$f(r \cos \theta, r \sin \theta) = (r \cos \theta)^2 + (r \sin \theta)^2 + 1 = r^2 + 1$$

and the region of integration D is just the interior of the circle $r = 2$. Thus, D can be described as the set of all points (r, θ) so that for each fixed angle θ between 0 and 2π, r varies from the origin ($r = 0$) to the circle $r = 2$, as shown in Figure 13.15b.

But how is the differential of integration dA changed in polar coordinates? Can we simply substitute "$dr\, d\theta$" for dA and perform the integration with respect to r and θ? The answer is no, and the correct formula for expressing a given double integral in polar form is given in the following theorem.

THEOREM 13.4   *Double integral in polar coordinates*

If f is continuous in the polar region D described by $r_1(\theta) < r < r_2(\theta)$ ($r_1(\theta) \geq 0$, $r_2(\theta) \geq 0$), $\alpha \leq \theta \leq \beta$ ($0 \leq \beta - \alpha < 2\pi$), then

$$\iint_D f(r, \theta)\, dA = \int_{\alpha}^{\beta} \int_{r_1(\theta)}^{r_2(\theta)} f(r, \theta)\, r\, dr\, d\theta$$

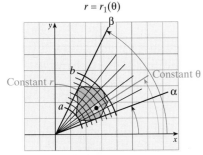

$r = r_1(\theta)$

a. A partition of a region into polar coordinates

$r = r_2(\theta)$

> ■ **What This Says:** The procedure for changing from a Cartesian integral
>
> $$\iint_R f(x, y)\, dA$$
>
> into a polar integral requires two steps. First, substitute $x = r \cos \theta$ and $y = r \sin \theta$, $dx\, dy = r\, dr\, d\theta$ into the Cartesian integral. Then convert the region of integration R to polar form D. Thus,
>
> $$\iint_R f(x, y)\, dA = \iint_D f(r \cos \theta, r \sin \theta)\, r\, dr\, d\theta$$

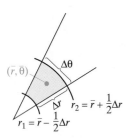

b. A typical polar rectangle in the partition

■ **FIGURE 13.16** A polar rectangle

Proof   A region described by $r_1(\theta) \leq r \leq r_2(\theta)$, $\alpha \leq \theta \leq \beta$ is called a **polar rectangle.** A typical polar rectangle is shown in Figure 13.16.

We begin by subdividing the region of integration into polar rectangles. Then we pick an arbitrary polar-form point (r_k^*, θ_k^*) in each polar rectangle in the partition and then take the limit of an appropriate Riemann sum

$$\sum_{k=1}^{N} f(r_k^*, \theta_k^*) \Delta A_k$$

where ΔA_k is the area of the kth polar rectangle.

To find the area of a typical polar rectangle, let (r_k^*, θ_k^*) be the center of the polar rectangle—that is, the point midway between the arcs and rays that form the polar rectangle, as shown in Figure 13.16b. If the circular arcs that bound the polar rectangle are Δr apart, then the arcs are given by

$$r_1 = r_k^* - \tfrac{1}{2}\Delta r_k \quad \text{and} \quad r_2 = r_k^* + \tfrac{1}{2}\Delta r_k$$

It can be shown that a circular section of radius r and central angle θ has area $\tfrac{1}{2}r^2\theta$ (see the *Student Mathematics Handbook* for details). Thus, a typical polar rectangle has area

SMH

$$\Delta A_k = \left[\underbrace{\tfrac{1}{2}(r_k^* + \tfrac{1}{2}\Delta r_k)^2}_{\text{Radius of outside arc}} - \underbrace{\tfrac{1}{2}(r_k^* - \tfrac{1}{2}\Delta r_k)^2}_{\text{Radius of inside arc}}\right]\Delta\theta = r_k^* \, \Delta r_k \, \Delta\theta_k$$

Finally, we compute the given double integral in polar form by taking the limit:

$$\iint\limits_D f(r, \theta) \, dA = \lim_{\|P\|\to 0} \sum_{k=1}^{N} f(r_k^*, \theta_k^*)\Delta A_k = \lim_{\|P\|\to 0} \sum_{k=1}^{N} f(r_k^*, \theta_k^*) \, r_k^* \, \Delta r_k \, \Delta\theta_k$$

$$= \int_\alpha^\beta \int_{r_1(\theta)}^{r_2(\theta)} f(r, \theta) \, r \, dr \, d\theta$$

PREVIEW   It can be shown that under reasonable conditions, the change of variable $x = x(u, v), y = y(u, v)$ transforms the integral $\iint f(x, y) \, dA$ into $\iint f(u, v)\,|J(u, v)| \, du \, dv$, where

$$J(u, v) = \begin{vmatrix} \dfrac{\partial x}{\partial u} & \dfrac{\partial x}{\partial v} \\[2mm] \dfrac{\partial y}{\partial u} & \dfrac{\partial y}{\partial v} \end{vmatrix}$$

This determinant is known as the **Jacobian** of the transformation. In the case of polar coordinates, we have $x = r \cos\theta$ and $y = r \sin\theta$, so that the Jacobian is

$$J(r, \theta) = \begin{vmatrix} \dfrac{\partial}{\partial r}(r \cos\theta) & \dfrac{\partial}{\partial\theta}(r \cos\theta) \\[2mm] \dfrac{\partial}{\partial r}(r \sin\theta) & \dfrac{\partial}{\partial\theta}(r \sin\theta) \end{vmatrix}$$

$$= \begin{vmatrix} \cos\theta & -r \sin\theta \\ \sin\theta & r \cos\theta \end{vmatrix} = r \cos^2\theta + r \sin^2\theta = r$$

This yields the result of Theorem 13.4.

$$\iint\limits_R f(x, y) \, dA = \iint\limits_D f(r, \theta) \, r \, dr \, d\theta = \int_\alpha^\beta \int_{r_1(\theta)}^{r_2(\theta)} f(r \cos\theta, r \sin\theta) \, r \, dr \, d\theta$$

We discuss this more completely in Section 13.8.

If you carefully compare the result in the preview box with the result of Theorem 13.4, you will see that they are not *exactly* the same. The region D in Theorem 13.4 already is in polar coordinates. The more common situation is that we are given f and a region R in rectangular coordinates that we need to change to polar coordinates.

You will need to be familiar with the graphs of many polar-form curves. These can be found in Table 9.1 on page 639, and also in the *Student Mathematics Handbook*.

We now present the example we promised in the introduction.

EXAMPLE 1   *Double integral in polar form*

Evaluate $\iint\limits_{R} (x^2 + y^2 + 1)\, dA$, where D is the region inside the circle $x^2 + y^2 = 4$.

Solution   In this example, the region R is given in rectangular form. We will describe this as a polar region D. Earlier, we observed that in D, for each fixed angle θ between 0 and 2π, r varies from $r = 0$ (the pole) to $r = 2$ (the circle). Thus,

$$\iint\limits_{R} \underbrace{(x^2 + y^2 + 1)}_{f(x,y)}\, dA = \int_0^{2\pi}\int_0^2 \underbrace{(r^2 + 1)}_{\substack{f(r\cos\theta,\, r\sin\theta)}} \underbrace{r\, dr\, d\theta}_{dA}$$

$$= \int_0^{2\pi}\int_0^2 (r^3 + r)\, dr\, d\theta = \int_0^{2\pi}\left[\frac{r^4}{4} + \frac{r^2}{2}\right]\Bigg|_0^2 d\theta = \int_0^{2\pi} 6\, d\theta = 6\theta\Big|_0^{2\pi} = 12\pi \quad \blacksquare$$

EXAMPLE 2   *Double integral in polar form*

Evaluate $\iint\limits_{D} x\, dA$, where D is the region bounded above by the line $y = x$ and below by the circle $x^2 + y^2 - 2y = 0$.

Solution   The circle $x^2 + y^2 - 2y = 0$ and the line $y = x$ are shown in Figure 13.17. The polar-form equation for the circle is $r = 2\sin\theta$, and the line $y = x$ in polar form is

$$r\sin\theta = r\cos\theta$$

This is true only when $\tan\theta = 1$ (divide both sides by $r\cos\theta$), which implies $\theta = \pi/4$. Thus, the region D may be described by saying that for each fixed angle θ between 0 and $\pi/4$, r varies from 0 (the pole) to $2\sin\theta$ (the circle), and we have

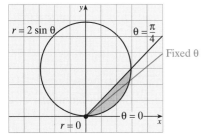

$r = 2\sin\theta$ $\theta = \dfrac{\pi}{4}$

Fixed θ

$\theta = 0$

$r = 0$

■ **FIGURE 13.17**   The region D.

$$\iint\limits_{D} x\, dA = \int_0^{\pi/4}\int_0^{2\sin\theta} \underbrace{(r\cos\theta)}_{x}\ \underbrace{r\, dr\, d\theta}_{dA} = \int_0^{\pi/4}\left[\frac{r^3}{3}\cos\theta\right]\Bigg|_{r=0}^{r=2\sin\theta} d\theta$$

$$= \int_0^{\pi/4}\left[\frac{(2\sin\theta)^3}{3}\cos\theta - 0\right] d\theta = \int_0^{\pi/4}\frac{8}{3}\sin^3\theta\cos\theta\, d\theta \qquad \text{Let } u = \sin\theta.$$

$$= \left[\frac{8}{3}\cdot\frac{1}{4}\sin^4\theta\right]\Bigg|_{\theta=0}^{\theta=\pi/4} = \frac{1}{6} \qquad\qquad \blacksquare$$

EXAMPLE 3   *Volume in polar form*

Use a polar double integral to show that a sphere of radius a has volume $\frac{4}{3}\pi a^3$.

The top hemisphere is bounded above by the sphere and below by the disk D.

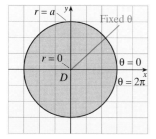

■ **FIGURE 13.18**   The disk D: For θ between 0 and 2π, r varies from 0 to a.

Solution   We shall compute the required volume by doubling the volume of the solid hemisphere $x^2 + y^2 + z^2 \leq a^2$, with $z \geq 0$. This hemisphere may be regarded as a solid bounded below by the circular disk $x^2 + y^2 \leq a^2$ and above by the spherical surface.

We need to change the equation of the hemisphere to polar form:

In rectangular form: $z = \sqrt{a^2 - x^2 - y^2}$

In polar form: $z = \sqrt{a^2 - r^2}$ *Because* $r^2 = x^2 + y^2$

Describing the disk D in polar terms, we see that for each fixed θ between 0 and 2π, r varies from the origin to the circle $x^2 + y^2 = a^2$, which has the polar equation $r = a$, as shown in Figure 13.18. Thus, the volume is given by the integral

$$V = 2 \int_D \int z \, dA \qquad \text{*Rectangular form*}$$

$$= 2 \int_0^{2\pi} \int_0^a \sqrt{a^2 - r^2} \, r \, dr \, d\theta \qquad \boxed{\text{Let } u = a^2 - r^2; \\ du = -2r \, dr.}$$

$$= 2 \int_0^{2\pi} \left[-\frac{1}{3}(a^2 - r^2)^{3/2} \right] \Big|_{r=0}^{r=a} d\theta$$

$$= -\frac{2}{3} \int_0^{2\pi} \left[(a^2 - a^2)^{3/2} - (a^2 - 0)^{3/2} \right] d\theta$$

$$= \frac{2}{3} \int_0^{2\pi} a^3 \, d\theta = \frac{2}{3} a^3 \theta \Big|_0^{2\pi} = \frac{4}{3} \pi a^3 \qquad ■$$

| EXAMPLE 4 | *Region of integration between two polar curves*

Evaluate $\int_D \int \frac{1}{x} \, dA$, where D is the region in the first quadrant that lies inside the circle $r = 3 \cos \theta$ and outside the cardioid $r = 1 + \cos \theta$.

Solution   Begin by sketching the given curves, as shown in Figure 13.19. Next, find the points of intersection:

$$3 \cos \theta = 1 + \cos \theta$$
$$2 \cos \theta = 1$$
$$\cos \theta = \tfrac{1}{2}$$
$$\theta = \tfrac{\pi}{3}$$

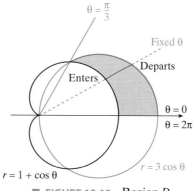

■ **FIGURE 13.19**   Region D

We see that D is the region such that, for each fixed angle θ between 0 and $\pi/3$, r varies from $1 + \cos\theta$ (the cardioid) to $3\cos\theta$ (the circle). This gives us the limits of integration. Finally, we need to write the integrand in polar form:

$$\underbrace{\frac{1}{x}}_{\text{Rectangular form}} = \underbrace{\frac{1}{r\cos\theta}}_{\text{Polar form}}$$

Thus,

$$\int_D \int \frac{1}{x}\,dA = \int_D \int \frac{1}{r\cos\theta}\, r\,dr\,d\theta$$

$$= \int_0^{\pi/3} \int_{1+\cos\theta}^{3\cos\theta} \frac{1}{\cos\theta}\,dr\,d\theta \qquad \text{Write } \frac{1}{\cos\theta} = \sec\theta.$$

$$= \int_0^{\pi/3} \left[r\sec\theta\right]\Big|_{r=1+\cos\theta}^{r=3\cos\theta}\,d\theta$$

$$= \int_0^{\pi/3} \left[3\cos\theta\sec\theta - (1+\cos\theta)\sec\theta\right]\,d\theta$$

$$= \int_0^{\pi/3} (2 - \sec\theta)\,d\theta \qquad \text{Since } \sec\theta = \frac{1}{\cos\theta}$$

$$= \left[2\theta - \ln\left|\sec\theta + \tan\theta\right|\right]\Big|_0^{\pi/3}$$

$$= \tfrac{2\pi}{3} - \ln(2 + \sqrt{3})$$ ∎

IMPROPER DOUBLE INTEGRALS IN POLAR COORDINATES

Improper double integrals play a useful role in probability theory and in certain physical applications. We shall demonstrate the basic approach to this topic by examining the special case of double integrals that are improper because the region of integration is the (unbounded) first quadrant. A general treatment of improper multiple integrals is outside the scope of this text.

Suppose the function $f(x, y)$ is continuous throughout the first quadrant Q; then define the improper integral in terms of rectangular coordinates as follows:

$$\int_Q \int f(x, y)\,dA = \lim_{n \to \infty} \int_{S_n} \int f(x, y)\,dA$$

where S_n is the square described by $0 \le x \le n, 0 \le y \le n$. In terms of polar coordinates, let C_n denote the quarter circular region described by $r \le n, 0 \le \theta \le \pi/2$, as shown in Figure 13.20. We use this quarter circular region to formulate the following definition.

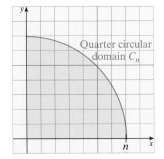

Quarter circular domain C_n

■ **FIGURE 13.20**   Quarter circular domain

Improper Double Integral in Polar Coordinates

Let Q denote the first quadrant of the Cartesian plane, and let C_n denote the quarter circular region described by $r \le n, 0 \le \theta \le \frac{\pi}{2}$.

Then the improper integral $\int_Q \int f(x, y)\,dA$ is defined in polar coordinates as

$$\lim_{n \to \infty} \int_{C_n} \int f(r\cos\theta, r\sin\theta) r\,dr\,d\theta = \lim_{n \to \infty} \int_0^{\pi/2} \int_0^n f(r\cos\theta, r\sin\theta) r\,dr\,d\theta$$

If the limit in this definition exists and is equal to L, we say that the improper integral **converges** to L. Otherwise, we say that the improper integral **diverges.**

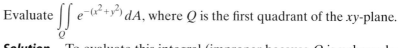

Evaluating a convergent improper double integral

Evaluate $\iint\limits_{Q} e^{-(x^2+y^2)} \, dA$, where Q is the first quadrant of the xy-plane.

Solution   To evaluate this integral (improper because Q is unbounded), we convert to polar coordinates:

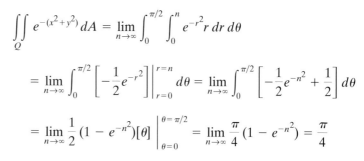

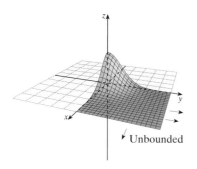

$$\iint\limits_{Q} e^{-(x^2+y^2)} \, dA = \lim_{n\to\infty} \int_0^{\pi/2} \int_0^n e^{-r^2} r \, dr \, d\theta$$

$$= \lim_{n\to\infty} \int_0^{\pi/2} \left[-\frac{1}{2} e^{-r^2} \right] \Bigg|_{r=0}^{r=n} d\theta = \lim_{n\to\infty} \int_0^{\pi/2} \left[-\frac{1}{2} e^{-n^2} + \frac{1}{2} \right] d\theta$$

$$= \lim_{n\to\infty} \frac{1}{2} (1 - e^{-n^2})[\theta] \Bigg|_{\theta=0}^{\theta=\pi/2} = \lim_{n\to\infty} \frac{\pi}{4} (1 - e^{-n^2}) = \frac{\pi}{4}$$

Thus, the improper integral converges to $\pi/4$. ∎

13.3 Problem Set

Ⓐ *Evaluate the double integral $\iint\limits_{D} f(x, y) \, dA$ in Problems 1–8, and sketch the region of integration D.*

1. $\displaystyle\int_0^{\pi/2} \int_0^{2\sin\theta} dr \, d\theta$

2. $\displaystyle\int_0^{\pi} \int_0^{1+\sin\theta} dr \, d\theta$

3. $\displaystyle\int_0^{\pi/2} \int_1^3 re^{-r^2} \, dr \, d\theta$

4. $\displaystyle\int_0^{\pi/2} \int_1^2 \sqrt{4-r^2} \, r \, dr \, d\theta$

5. $\displaystyle\int_0^{\pi} \int_0^4 r^2 \sin^2\theta \, dr \, d\theta$

6. $\displaystyle\int_0^{\pi/2} \int_1^3 r^2 \cos^2\theta \, dr \, d\theta$

7. $\displaystyle\int_0^{2\pi} \int_0^4 2r^2 \cos\theta \, dr \, d\theta$

8. $\displaystyle\int_0^{2\pi} \int_0^{1-\sin\theta} \cos\theta \, dr \, d\theta$

Use a double integral to find the area of the shaded region in Problems 9–22.

9. $r = 4$

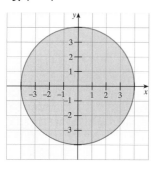

10. $r = 2\cos\theta$

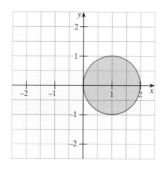

11. $r = 2(1 - \cos\theta)$

12. $r = 1 + \sin\theta$

13. $r = 4\cos 3\theta$

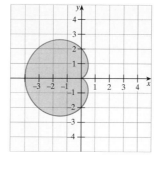

14. $r = 5\sin 2\theta$

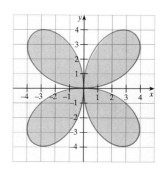

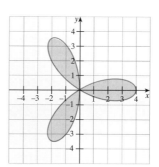

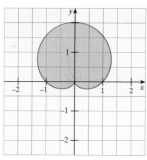

15. $r = \cos 2\theta$

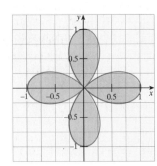

16. $r = 1$ and $r = 2 \sin \theta$
(*Hint:* Consider $0 < \theta < \frac{\pi}{6}$ and $\frac{\pi}{6} < \theta < \frac{\pi}{2}$ separately.)

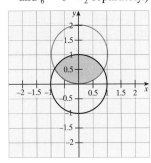

17. $r = 1$ and $r = 2 \sin \theta$

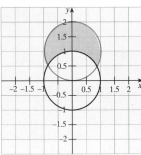

18. $r = 1$ and $r = 1 + \cos \theta$

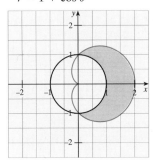

19. $r = 1$ and $r = 1 + \cos \theta$

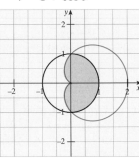

20. $r = 1$ and $r = 1 + \cos \theta$

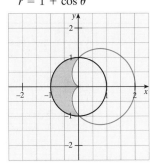

21. $r = 3 \cos \theta$ and $r = 1 + \cos \theta$

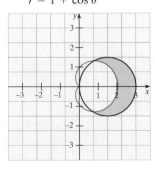

22. $r = 3 \cos \theta$ and $r = 2 - \cos \theta$

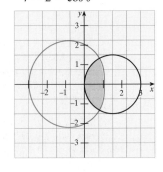

23. Use a double integral to find the area inside the inner loop of the limaçon $r = 1 + 2 \cos \theta$.

24. Use a double integral to find the area bounded by the parabola
$$r = \frac{5}{1 + \cos \theta}$$
and the lines $\theta = 0$, $\theta = \frac{\pi}{6}$, and $r = \frac{3}{4} \sec \theta$.

Use polar coordinates in Problems 25–30 to find $\iint\limits_{D} f(x, y)\, dA$, where D is the circular disk defined by $x^2 + y^2 \leq a^2$ for constant $a > 0$.

25. $f(x, y) = y^2$ **26.** $f(x, y) = x^2 + y^2$

27. $f(x, y) = \dfrac{1}{a^2 + x^2 + y^2}$ **28.** $f(x, y) = e^{-(x^2 + y^2)}$

29. $f(x, y) = \dfrac{1}{a + \sqrt{x^2 + y^2}}$ **30.** $f(x, y) = \ln(a^2 + x^2 + y^2)$

Use polar coordinates in Problems 31–36 to evaluate the given double integral.

31. $\displaystyle\iint\limits_{D} y\, dA$, where D is the disk $x^2 + y^2 \leq 4$

32. $\displaystyle\iint\limits_{D} (x^2 + y^2)\, dA$, where D is the region bounded by the x-axis, the line $y = x$, and the circle $x^2 + y^2 = 1$

33. $\displaystyle\iint\limits_{D} e^{x^2 + y^2}\, dA$, where D is the region inside the circle $x^2 + y^2 = 9$

34. $\displaystyle\iint\limits_{D} \sqrt{x^2 + y^2}\, dA$, where D is the region inside the circle $(x - 1)^2 + y^2 = 1$ in the first quadrant

35. $\displaystyle\iint\limits_{D} \ln(x^2 + y^2 + 2)\, dA$, where D is the region inside the circle $x^2 + y^2 = 4$ in the first quadrant

36. $\displaystyle\iint\limits_{D} \sin(x^2 + y^2)\, dA$, where D is the region bounded by the circles $x^2 + y^2 = 1$ and $x^2 + y^2 = 4$, and the lines $y = 0$ and $x = \sqrt{3}\, y$

37. Find the volume of the solid bounded by the paraboloid $z = 4 - x^2 - y^2$ and the xy-plane.

38. Find the volume of the solid bounded above by the cone $z = 6\sqrt{x^2 + y^2}$ and below by the circular region $x^2 + y^2 \leq a^2$ in the xy-plane, where a is a positive constant.

B **39.** Use polar coordinates to evaluate $\iint\limits_{D} xy\, dA$, where D is the intersection of the circular disks $r \leq 4 \cos \theta$ and $r \leq 4 \sin \theta$. Sketch the region of integration.

40. Let D be the region formed by intersecting the regions (in the xy-plane) described by $y \leq x$, $y \geq 0$, and $x \leq 1$.
a. Express $\iint\limits_{D} dA$ as an iterated integral in Cartesian coordinates.
b. Express $\iint\limits_{D} dA$ in terms of polar coordinates.

41. Example 5 of Section 9.3 used a single integral to find the area of the region common to the circles $r = a \cos \theta$ and $r = a \sin \theta$. Rework this example using double integrals.

42. Example 6 of Section 9.3 used a single integral to find the area between the circle $r = 5 \cos \theta$ and the limaçon $r = 2 + \cos \theta$. Use double integrals to find the same area.

In Problems 43–48, find the volume of the given solid region.

43. The solid that lies inside the sphere $x^2 + y^2 + z^2 = 25$ and outside the cylinder $x^2 + y^2 = 9$

44. The solid that bounded by the sphere $x^2 + y^2 + z^2 = 4$ and the paraboloid $3z = x^2 + y^2$

45. The solid region common to the sphere $x^2 + y^2 + z^2 = 4$ and the cylinder $r = 2 \cos \theta$

46. The solid region common to the cylinder $x^2 + y^2 = 2$ and the ellipsoid $3x^2 + 3y^2 + z^2 = 7$

47. The solid bounded above by the paraboloid $z = 1 - x^2 - y^2$, below by the plane $z = 0$, and on the sides by the cylinder $r = \cos \theta$

48. The solid region bounded above by the cone $z = x^2 + y^2$, below by the plane $z = 0$, and on the sides by the cylinder $r = \sin \theta$

Evaluate the improper integrals in Problems 49–53 over Q, the first quadrant in the xy-plane.

49. $\iint\limits_{Q} (x^2 + y^2 + 1)^{-3/2} \, dA$

50. $\iint\limits_{Q} \dfrac{dA}{(1 + x^2 + y^2)^2}$

51. $\iint\limits_{Q} \dfrac{dA}{(1 + x^2 + y^2)^3}$

52. $\iint\limits_{Q} e^{-2(x^2+y^2)} \, dA$

53. $\iint\limits_{Q} e^{-(x^2+y^2)/4} \, dA$

54. Use polar coordinates to evaluate the double integral

$$\iint\limits_{D} (x^2 + y^2) \, dA$$

over the shaded region.

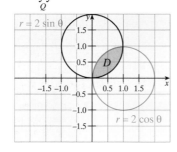

55. Use polar coordinates to evaluate the double integral $\iint\limits_{D} (x^2 + y^2) \, dA$ over the shaded region.

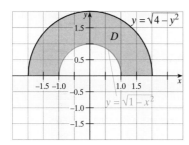

C **56.** **THINK TANK PROBLEM**   To evaluate the integral

$$\iint\limits_{R} r^2 \, dr \, d\theta$$

where R is the region in the xy-plane bounded by $r = 2 \cos \theta$, we obtain

$$\int_0^{\pi} \int_0^{2\cos\theta} r^2 \, dr \, d\theta = \int_0^{\pi} \dfrac{r^3}{3} \Big|_0^{2\cos\theta} \, d\theta$$

$$= \dfrac{8}{3} \int_0^{\pi} \cos^3\theta \, d\theta = \dfrac{8}{3}\left[\sin\theta - \dfrac{\sin^3\theta}{3}\right]_0^{\pi} = 0$$

Alternatively, we can set up the integral as

$$\int_{-\pi/2}^{\pi/2} \int_0^{2\cos\theta} r^2 \, dr \, d\theta = \int_{-\pi/2}^{\pi/2} \dfrac{r^3}{3} \Big|_0^{2\cos\theta} \, d\theta$$

$$= \dfrac{8}{3} \int_{-\pi/2}^{\pi/2} \cos^3\theta \, d\theta = \dfrac{8}{3}\left[\sin\theta - \dfrac{\sin^3\theta}{3}\right]_{-\pi/2}^{\pi/2} = \dfrac{32}{9}$$

Both of these answers cannot be correct. Which procedure (if either) is correct and why?

57. **HISTORICAL QUEST**   Newton and Leibniz have been credited with the discovery of calculus, but much of its development was due to the mathematicians Pierre-Simon Laplace, Lagrange (Historical Quest 54, Section 12.8), and Gauss (Historical Quest 61, Section 2.3). These three great mathematicians of calculus were contrasted by W. W. Rouse Ball:

PIERRE-SIMON LAPLACE 1749–1827

The great masters of modern analysis are Lagrange, Laplace, and Gauss, who were contemporaries. It is interesting to note the marked contrast in their styles. Lagrange is perfect both in form and matter, he is careful to explain his procedure, and though his arguments are general they are easy to follow. Laplace on the other hand explains nothing, is indifferent to style, and, if satisfied that his results are correct, is content to leave them either with no proof or with a faulty one. Gauss is exact and elegant as Lagrange, but even more difficult to follow than Laplace, for he removes every trace of the analysis by which he reached his results, and strives to give a proof which while rigorous shall be as concise and synthetical as possible.*

Pierre-Simon Laplace has been called the Newton of France. He taught Napoleon Bonaparte, was appointed for a time as Minister of Interior, and was at times granted favors from his powerful friend. Laplace solved problems in celestial mechanics and proved the stability of the solar system. Today, Laplace is best known as the major contributor to probability, taking it from gambling to a true branch of mathematics. He was one of the earliest to evaluate the improper

---

* *A Short Account of the History of Mathematics* as quoted in *Mathematical Circles Adieu* by Howard Eves (Boston: Prindle, Weber & Schmidt, Inc., 1977).

integral

$$I = \int_{-\infty}^{\infty} e^{-x^2}\,dx$$

which plays an important role in the theory of probability.
Show that $I = \sqrt{\pi}$. *Hint:* Note that

$$\int_{0}^{\infty} e^{-x^2}\,dx \cdot \int_{0}^{\infty} e^{-y^2}\,dy = \int_{0}^{\infty}\int_{0}^{\infty} e^{-(x^2+y^2)}\,dx\,dy$$

and use Example 5.

58. Evaluate $\int_{0}^{\infty} e^{-2x^2}\,dx$.
    Hint: See Problem 57.

59. For constant a where $0 \le a \le R$, the plane $z = R - a$ cuts off a "cap" from the sphere

$$x^2 + y^2 + z^2 = R^2$$

Use a double integral in polar coordinates to find the volume of the cap.

13.4 Surface Area

IN THIS SECTION definition of surface area, surface area projections, area of a surface defined parametrically

In chapter 6, we found that if f has a continuous derivative on the closed, bounded interval $[a, b]$, then the length L of the portion of the graph of $y = f(x)$ between the points were $x = a$ and $x = b$ may be defined by

$$L = \int_{a}^{b} \sqrt{1 + [f'(x)]^2}\,dx$$

The goal of this section is to study an analogous formula for the surface area of the graph of a differentiable function of two variables.

DEFINITION OF SURFACE AREA

Consider a surface defined by $z = f(x, y)$ defined over a region R of the xy-plane. Enclose the region R in a rectangle partitioned by a grid with lines parallel to the coordinate axes, as shown in Figure 13.21a. This creates a number of cells, and we let $R_1, R_2, \ldots, R_n$ denote those that lie entirely within R.

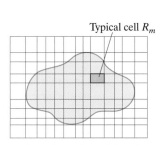

Typical cell R_m

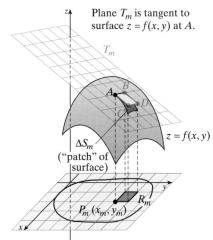

Plane T_m is tangent to surface $z = f(x, y)$ at A.

T_m

$z = f(x, y)$

ΔS_m
("patch" of surface)

$P_m(x_m, y_m)$

R_m

a. The region R is partitioned by a rectangular grid

b. The surface area above R_m is approximated by the area of the parallelogram on the tangent plane.

■ **FIGURE 13.21**   Surface area

For $m = 1, 2, \ldots, n$, let $P_m(x_m^*, y_m^*)$ be a corner of the rectangle R_m, and let T_m be the tangent plane above P_m on the surface of $z = f(x, y)$. Finally, let ΔS_m denote the area of the "patch" of surface that lies directly above R_m.

The rectangle R_m projects onto a parallelogram $ABDC$ in the tangent plane T_m, and if R_m is "small," we would expect the area of this parallelogram to approximate closely the element of surface area ΔS_m (see Figure 13.20b).

If Δx_m and Δy_m are the lengths of the sides of the rectangle R_m, the approximating parallelogram will have sides determined by the vectors

$$\mathbf{AB} = \Delta x_m \mathbf{i} + [f_x(x_m^*, y_m^*)\Delta x_m]\mathbf{k} \quad \text{and} \quad \mathbf{AC} = \Delta y_m \mathbf{j} + [f_y(x_m^*, y_m^*)\Delta y_m]\mathbf{k}$$

In Chapter 10, we showed that such a parallelogram with sides $\mathbf{AB}$ and $\mathbf{AC}$ has area

$$\|\mathbf{AB} \times \mathbf{AC}\|$$

This is shown in Figure 13.22. If K_m is the area of the approximating parallelogram, we have

$$K_m = \|\mathbf{AB} \times \mathbf{AC}\|$$

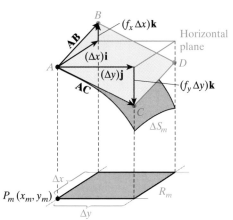

■ **FIGURE 13.22**   An element of surface area has area $\|\mathbf{AB} \times \mathbf{AC}\|$.

To compute K_m, we first find the cross product:

$$\mathbf{AB} \times \mathbf{AC} = \begin{vmatrix} \mathbf{i} & \mathbf{j} & \mathbf{k} \\ \Delta x_m & 0 & f_x(x_m^*, y_m^*)\Delta x_m \\ 0 & \Delta y_m & f_y(x_m^*, y_m^*)\Delta y_m \end{vmatrix}$$

$$= -f_x\Delta x\Delta y\mathbf{i} - f_y\Delta x\Delta y\mathbf{j} + \Delta x\Delta y\mathbf{k}$$

Then, we calculate the norm:

$$\begin{aligned} K_m &= \|\mathbf{AB} \times \mathbf{AC}\| \\ &= \sqrt{[f_x(x_m^*, y_m^*)]^2\Delta x_m^2\Delta y_m^2 + [f_y(x_m^*, y_m^*)]^2\Delta x_m^2\Delta y_m^2 + \Delta x_m^2\Delta y_m^2} \\ &= \sqrt{[f_x(x_m^*, y_m^*)]^2 + [f_y(x_m^*, y_m^*)]^2 + 1}\,\Delta x_m\Delta y_m \end{aligned}$$

Finally, summing over the entire partition, we see that the surface area over R may be approximated by the sum

$$\Delta S_n = \sum_{m=1}^{n} \sqrt{[f_x(x_m^*, y_m^*)]^2 + [f_y(x_m^*, y_m^*)]^2 + 1}\,\Delta A_m$$

where $\Delta A_m = \Delta x_m \Delta y_m$. This may be regarded as a Riemann sum, and by taking an appropriate limit (as the partition becomes more and more refined), we find that surface area, S, satisfies

$$S = \lim_{n \to \infty} \sum_{m=1}^{n} \sqrt{[f_x(x_m^*, y_m^*)]^2 + [f_y(x_m^*, y_m^*)]^2 + 1} \,\Delta A_m$$

$$= \iint_R \sqrt{[f_x(x, y)]^2 + [f_y(x, y)]^2 + 1} \, dA$$

Surface Area as a Double Integral

Assume that the function $f(x, y)$ has continuous partial derivatives f_x and f_y in a region R of the xy-plane. Then the portion of the surface $z = f(x, y)$ that lies over R has **surface area**

$$S = \iint_R \sqrt{[f_x(x, y)]^2 + [f_y(x, y)]^2 + 1} \, dA$$

■ **What This Says:**   The region R may be regarded as the projection of the surface $z = f(x, y)$ on the xy-plane. If there were a light source with rays perpendicular to the xy-plane, R would be the "shadow" of the surface on the plane. You will notice the shadows drawn in the figures shown in this section. It is also worthwhile to make the following comparisons:

Length on x-axis: **Arc length:**

$$\int_a^b dx \qquad\qquad \int_a^b ds = \int_a^b \sqrt{[f'(x)]^2 + 1} \, dx$$

Area in xy-plane: **Surface area:**

$$\iint_R dA \qquad\qquad \iint_R dS = \iint_R \sqrt{[f_x(x, y)]^2 + [f_y(x, y)]^2 + 1} \, dA$$

EXAMPLE 1   *Surface area of a plane region*

Find the surface area of the portion of the plane $x + y + z = 1$ that lies in the first octant (where $x \geq 0, y \geq 0, z \geq 0$).

Solution   The plane $x + y + z = 1$ and the triangular region T that lies beneath it in the xy-plane are shown in Figure 13.23. We begin by letting $f(x, y) = 1 - x - y$. Then $f_x(x, y) = -1$ and $f_y(x, y) = -1$, and

$$S = \iint_T \sqrt{[f_x(x, y)]^2 + [f_y(x, y)]^2 + 1} \, dA$$

$$= \iint_T \sqrt{(-1)^2 + (-1)^2 + 1} \, dA$$

$$= \iint_T \sqrt{3} \, dA$$

$$= \int_0^1 \int_0^{1-x} \sqrt{3} \, dy \, dx \qquad \text{\small Look at Figure 13.23 to see the region } T \\ \text{\small and the limits for both } x \text{ and } y.$$

$$= \sqrt{3} \int_0^1 [y]\Big|_0^{1-x} dx$$

a. The surface of the plane $x + y + z = 1$ in the first octant

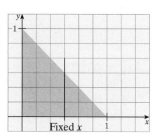

b. The surface projects onto a triangle in the xy-plane.

■ **FIGURE 13.23**   Surface area

$$= \sqrt{3} \int_0^1 (1 - x)\, dx$$

$$= \sqrt{3} \left[x - \frac{x^2}{2} \right] \Big|_0^1 = \frac{\sqrt{3}}{2}$$

The surface area is $\sqrt{3}/2 \approx 0.866$.

■

| EXAMPLE 2 | *Surface area by changing to polar coordinates* |

Find the surface area (to the nearest hundredth square unit) of that part of the paraboloid $x^2 + y^2 + z = 5$ that lies above the plane $z = 1$.

Solution Let $f(x, y) = 5 - x^2 - y^2$. Then $f_x(x, y) = -2x, f_y(x, y) = -2y$, and

$$S = \iint\limits_D \sqrt{[f_x(x, y)]^2 + [f_y(x, y)]^2 + 1}\, dA = \iint\limits_D \sqrt{4x^2 + 4y^2 + 1}\, dA$$

Now we need to determine the limits for the region D and carry out the integration by using Fubini's theorem. The paraboloid intersects the plane $x = 1$ in the circle $x^2 + y^2 = 4$. Thus, the part of the paraboloid whose surface area we seek projects onto the disk $x^2 + y^2 \leq 4$ in the xy-plane, as shown in Figure 13.24 (note the shadow).

It is easier if we convert to polar coordinates by letting $x = r \cos \theta, y = r \sin \theta$. Because the region is bounded by $0 \leq r \leq 2$ and $0 \leq \theta \leq 2\pi$, we have

$$S = \iint\limits_D \sqrt{4x^2 + 4y^2 + 1}\, dA$$

$$= \int_0^{2\pi} \int_0^2 \sqrt{4r^2 + 1}\, r\, dr\, d\theta$$

$$= \int_0^{2\pi} \int_1^{17} u^{1/2} \frac{du}{8}\, d\theta$$

Let $u = 4r^2 + 1$, so $du = 8\, r\, dr$. If $r = 2$, then $u = 17$ and if $r = 0, u = 1$

$$= \int_0^{2\pi} \frac{1}{8} \cdot \frac{2}{3} u^{3/2} \Big|_1^{17}\, d\theta$$

$$= \frac{1}{12} \int_0^{2\pi} (17^{3/2} - 1)\, d\theta$$

$$= \frac{1}{12}(17^{3/2} - 1)(2\pi - 0) \approx 36.18$$

The surface area is approximately 36.18 square units.

■

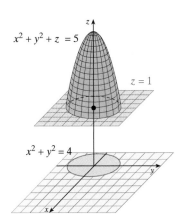

$x^2 + y^2 + z = 5$

$z = 1$

$x^2 + y^2 = 4$

Projected region

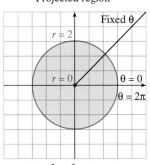

Fixed θ

$r = 2$

$r = 0$

$\theta = 0$

$\theta = 2\pi$

Disk $x^2 + y^2 \leq 4$ in polar coordinates

■ **FIGURE 13.24** The surface $x^2 + y^2 + z = 5$ above $z = 1$ projects onto a disk.

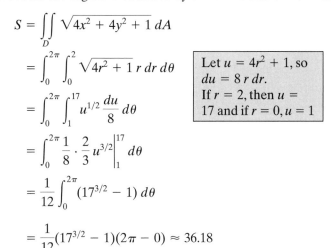

SURFACE AREA PROJECTIONS

We have been projecting the given surface onto the xy-plane, but we can easily modify our procedure to handle other cases. For instance, if the surface $y = f(x, z)$ is projected onto the region Q in the xz-plane, the formula for surface area becomes

$$S = \iint\limits_Q \sqrt{[f_x(x,z)]^2 + [f_z(x, z)]^2 + 1}\, dx\, dz$$

An analogous formula for projection onto the region T in the yz-plane would be

$$S = \iint\limits_Q \sqrt{[f_y(y,z)]^2 + [f_z(y, z)]^2 + 1}\, dy\, dz$$

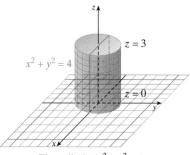

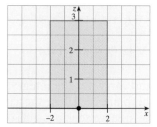

b. The projection of the cylinder on the xz-plane

■ **FIGURE 13.25**   Surface area of a cylinder

| **EXAMPLE 3** | *Surface area of a cylinder* |

Find the lateral surface area of the cylinder $x^2 + y^2 = 4, 0 \le z \le 3$.

Solution   The cylinder is shown in Figure 13.25a. We can find the lateral surface area using the formula for a rectangle by imagining the cylinder "opened up." The length of one side is the height of the cylinder (3 units), and the length of the other side of the rectangle is the circumference of the bottom, namely, $2\pi(2) = 4\pi$. Thus, the lateral surface area is 12π.

We shall illustrate our formula by using it to recompute this surface area. Specifically, we shall compute the surface area of the right half of the cylinder and then double it to obtain the required area. It does no good to project the half-cylinder onto the xy-plane, because if we do we will not be accounting for the height of the cylinder.

Instead, we will project the surface onto the xz-plane to obtain the rectangle Q bounded by $x = 2, x = -2, z = 0$, and $z = 3$, as shown in Figure 13.25b. Because we are projecting onto the xz-plane, we solve for y to express the surface as $y = f(x, z)$. Because

$$y = f(x, z) = \sqrt{4 - x^2}$$

we have $f_x(x, z) = \dfrac{-x}{\sqrt{4 - x^2}}$ and $f_z(x, z) = 0$ so that

$$[f_x(x, z)]^2 + [f_z(x, z)]^2 + 1 = \frac{x^2}{4 - x^2} + 1 = \frac{4}{4 - x^2}$$

Finally, we note the limit of integration:

$$-2 \le x \le 2 \quad \text{and} \quad 0 \le z \le 3$$

The surface area can now be calculated:

$$S = \iint\limits_Q \sqrt{\frac{4}{4 - x^2}} \, dA = \int_{-2}^{2} \int_{0}^{3} \frac{2}{\sqrt{4 - x^2}} \, dz \, dx$$

$$= \int_{-2}^{2} \frac{2z}{\sqrt{4 - x^2}} \bigg|_0^3 \, dx = \int_{-2}^{2} \frac{6}{\sqrt{4 - x^2}} \, dx = 6 \sin^{-1} \frac{x}{2} \bigg|_{-2}^{2} = 6\pi$$

The lateral surface area of the whole cylinder is $2(6\pi) = 12\pi$.   ■

AREA OF A SURFACE DEFINED PARAMETRICALLY

Suppose a surface S (see Figure 13.26) is defined parametrically by the vector function

$$\mathbf{R}(u, v) = x(u, v)\mathbf{i} + y(u, v)\mathbf{j} + z(u, v)\mathbf{k}$$

for parameters u and v.

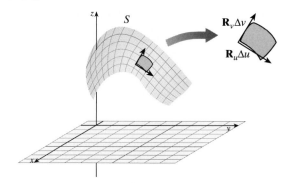

■ **FIGURE 13.26**   Area of a parametrically defined function

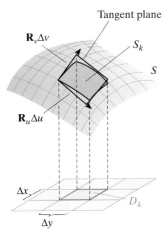

Tangent plane

$\mathbf{R}_v \Delta v$

S_k

S

$\mathbf{R}_u \Delta u$

Δx

D_k

Δy

■ **FIGURE 13.27** Area of a surface defined by a vector function

Let D be a region in the xy-plane on which x, y, and z, as well as their partial derivatives with respect to u and v, are continuous. The partial derivatives of $\mathbf{R}(u, v)$ are given by

$$\mathbf{R}_u = \frac{\partial \mathbf{R}}{\partial u} = \frac{\partial x}{\partial u}\mathbf{i} + \frac{\partial y}{\partial u}\mathbf{j} + \frac{\partial z}{\partial u}\mathbf{k} \qquad \mathbf{R}_v = \frac{\partial \mathbf{R}}{\partial v} = \frac{\partial x}{\partial v}\mathbf{i} + \frac{\partial y}{\partial v}\mathbf{j} + \frac{\partial z}{\partial v}\mathbf{k}$$

Suppose the region D is subdivided into cells, as shown in Figure 13.27. Consider a typical rectangle in this partition, with dimensions Δx and Δy, where Δx and Δy are small. If we project a typical rectangle onto the surface $\mathbf{R}(u, v)$, we obtain a **curvilinear** parallelogram with adjacent sides $\mathbf{R}_u(u, v)\Delta u$ and $\mathbf{R}_v(u, v)\Delta v$. The area of this rectangle is approximated by

$$\Delta S = \left\| \mathbf{R}_u(u, v)\Delta u \times \mathbf{R}_v(u, v) \right\| = \left\| \mathbf{R}_u(u, v) \times \mathbf{R}_v(u, v) \right\| \Delta u \, \Delta v$$

By taking an appropriate limit, we find the surface area to be a double integral.

Surface Area Defined Parametrically

Let D be a region in the xy-plane on which x, y, z and their partial derivatives with respect to u and v are continuous. Also, let S be a surface defined by a vector function

$$\mathbf{R}(u, v) = x(u, v)\mathbf{i} + y(u, v)\mathbf{j} + z(u, v)\mathbf{k}$$

Then the surface area is defined by

$$S = \iint\limits_{D} \left\| \mathbf{R}_u(u, v) \times \mathbf{R}_v(u, v) \right\| du \, dv$$

We call the quantity $\mathbf{R}_u(u, v) \times \mathbf{R}_v(u, v)$ the **fundamental cross product.**

EXAMPLE 4   *Area of a surface defined parametrically*

Find the surface area (to the nearest square unit) of the surface given parametrically by

$$\mathbf{R}(u, v) = (u \sin v)\mathbf{i} + (u \cos v)\mathbf{j} + u^2\mathbf{k} \quad \text{for } 0 \le u \le 3, 0 \le v \le 2\pi$$

Solution   We find that

$$\mathbf{R}_u = \sin v \, \mathbf{i} + \cos v \, \mathbf{j} + 2u \, \mathbf{k}$$
$$\mathbf{R}_v = u \cos v \, \mathbf{i} + (-u \sin v)\mathbf{j}$$

We begin with the fundamental cross product:

$$\mathbf{R}_u \times \mathbf{R}_v = \begin{vmatrix} \mathbf{i} & \mathbf{j} & \mathbf{k} \\ \sin v & \cos v & 2u \\ u \cos v & -u \sin v & 0 \end{vmatrix} = (2u^2 \sin v)\mathbf{i} + (2u^2 \cos v)\mathbf{j} - u\mathbf{k}$$

We find that

$$\left\| \mathbf{R}_u \times \mathbf{R}_v \right\| = \sqrt{4u^4 \sin^2 v + 4u^4 \cos^2 v + u^2} = \sqrt{4u^4 + u^2} = u\sqrt{4u^2 + 1}$$

and we can now compute the surface area:

$$S = \int_0^{2\pi} \int_0^3 u\sqrt{4u^2 + 1}\, du\, dv = \int_0^{2\pi} \left[\frac{1}{12}(4u^2 + 1)^{3/2} \right]\Bigg|_{u=0}^{u=3} dv$$

$$= \int_0^{2\pi} \left[\frac{1}{12}(37^{3/2} - 1) \right] dv = \frac{37^{3/2} - 1}{12}[2\pi - 0] \approx 117.3187007$$

The surface area is approximately 117 square units. ∎

Notice that in the special case where the surface under consideration has the explicit representation $z = f(x, y)$, it can also be represented in the vector form

$$\mathbf{R}_x = x\mathbf{i} + y\mathbf{j} + f(x, y)\mathbf{k}$$

where x and y are used as parameters ($x = u$ and $y = v$). With this vector representation, we find that

$$\mathbf{R}_x = \mathbf{i} + f_x\mathbf{k} \quad \text{and} \quad \mathbf{R}_y = \mathbf{j} + f_y\mathbf{k}$$

so the fundamental cross product is

$$\mathbf{R}_x \times \mathbf{R}_y = \begin{vmatrix} \mathbf{i} & \mathbf{j} & \mathbf{k} \\ 1 & 0 & f_x \\ 0 & 1 & f_y \end{vmatrix} = -f_x\mathbf{i} - f_y\mathbf{j} + \mathbf{k}$$

Therefore, in the case where $z = f(x, y)$, the surface area over the region D is given by

$$S = \iint_D \|\mathbf{R}_x \times \mathbf{R}_y\| dx\, dy$$

$$= \iint_D \sqrt{(-f_x)^2 + (-f_y)^2 + 1^2}\, dx\, dy$$

$$= \iint_D \sqrt{f_x^2 + f_y^2 + 1}\, dx\, dy$$

which is the formula obtained at the beginning of this section.

13.4 Problem Set

Ⓐ *Find the surface area of each surface given in Problems 1–19.*

1. The portion of the plane $2x + y + 4z = 8$ that lies in the first octant

2. The portion of the plane $4x + y + z = 9$ that lies in the first octant

3. The portion of the paraboloid
$$z = 4 - x^2 - y^2$$
that lies above the xy-plane

4. The portion of the paraboloid
$$z = x^2 + y^2 - 9$$
that lies below the xy-plane

5. The portion of the plane $3x + 6y + 2z = 12$ that is above the triangular region in the plane with vertices $(0, 0, 0)$, $(1, 0, 0)$, and $(1, 1, 0)$

6. The portion of the plane $2x + 2y - z = 0$ that is above the square region in the plane with vertices $(0, 0, 0)$, $(1, 0, 0)$, $(0, 1, 0)$, $(1, 1, 0)$

7. The portion of the surface $x^2 + z = 9$ above the square region in the plane with vertices $(0, 0, 0)$, $(2, 0, 0)$, $(0, 2, 0)$, $(2, 2, 0)$

8. The portion of the surface $z = x^2$ that lies over the triangular region in the plane with vertices $(0, 0, 0)$, $(0, 1, 0)$, and $(1, 0, 0)$

9. The portion of the surface $z = x^2$ over the square region with vertices $(0, 0, 0)$, $(0, 4, 0)$, $(4, 0, 0)$, $(4, 4, 0)$

10. The portion of the surface $z = 2x + y^2$ over the square region with vertices $(0, 0, 0)$, $(3, 0, 0)$, $(0, 3, 0)$, $(3, 3, 0)$

11. The portion of the paraboloid
$$z = x^2 + y^2$$
that lies below the plane $z = 1$

12. The portion of the sphere $x^2 + y^2 + z^2 = 25$ that lies above the plane $z = 3$

13. The part of the cylinder $x^2 + z^2 = 4$ that is in the first octant and is bounded by the plane $y = 2$

14. The portion of the plane $x + y + z = 4$ that lies inside the cylinder $x^2 + y^2 = 16$

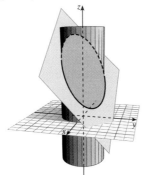

15. The portion of the sphere $x^2 + y^2 + z^2 = 4$ that lies inside the cylinder $x^2 + y^2 = 2y$

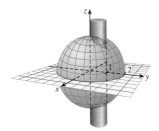

16. The portion of the sphere $x^2 + y^2 + z^2 = 8$ that is inside the elliptic cone $x^2 + y^2 - z^2 = 0$

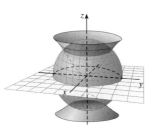

17. The portion of the surface $z = x^2 + y$ above the rectangle $0 \le x \le 2, 0 \le y \le 5$

18. The portion of the surface $z = x^2 - y^2$ that lies inside the cylinder $x^2 + y^2 = 9$

19. The portion of the surface $z = 9 - x^2 - y^2$ that lies above the xy-plane

Ⓑ 20. ■ **What Does This Say?** Describe the process for finding a surface area.

21. ■ **What Does This Say?** Describe what is meant by a surface area projection.

22. On a given map, a city parking lot is shown to be a rectangle that is 300 ft by 400 ft. However, the parking lot slopes in the 400-ft direction. It rises uniformly 1 ft for every 5 ft of horizontal displacement. What is the actual surface area of the parking lot?

23. Find the surface area of the portion of the plane $Ax + By + Cz = D$ ($A, B, C,$ and D all positive) that lies in the first octant.

24. Find the portion of the plane $x + 2y + 3z = 12$ that lies over the triangular region in the xy-plane with vertices $(0, 0, 0), (0, a, 0),$ and $(a, a, 0)$.

25. Find the surface area of that portion of the plane
$$x + y + z = a$$
that lies between the concentric cylinders
$$x^2 + y^2 = \frac{a^2}{4} \text{ and } x^2 + y^2 = a^2 (a > 0).$$

26. Find the surface area of the portion of the cylinder $x^2 + z^2 = 9$ that lies inside the cylinder $y^2 + z^2 = 9$.

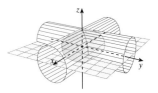

27. Find a formula for the area of the conical surface $z = \sqrt{x^2 + y^2}$ between the planes $z = 0$ and $z = h$. Express your answer in terms of h and the radius of the base of the cone.

28. Find a formula for the surface area of the frustum of the cone $z = 4\sqrt{x^2 + y^2}$ between the planes $z = h_1$ and $z = h_2, h_1 > h_2$.

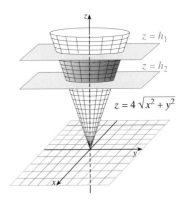

29. Find the surface area of that portion of the sphere $x^2 + y^2 + z^2 = 9z$ that lies inside the paraboloid $x^2 + y^2 = 4z$.

30. Find the surface area of that portion of the cylinder $x^2 + z^2 = 4$ that is above the triangle with vertices $(0, 0, 0), (1, 1, 0),$ and $(1, 0, 0)$.

In Problems 31–36, ***set up*** (*but do not evaluate*) *the double integral for the surface area of the given portion of surface.*

31. The surface given by $z = e^{-x} \sin y$ over the triangle with vertices $(0, 0, 0), (0, 1, 1), (0, 1, 0)$

32. The surface given by $z = x^3 - xy + y^3$ over the square $(0, 0, 0), (2, 0, 0), (0, 2, 0), (2, 2, 0)$

33. The surface given by $z = \cos(x^2 + y^2)$ over the disk $x^2 + y^2 \leq \dfrac{\pi}{2}$

34. The surface given by $z = e^{-x} \cos y$ over the disk $x^2 + y^2 \leq 2$

35. The surface given by $z = x^2 + 5xy + y^2$ over the region in the xy-plane bounded by the curve $xy = 5$ and the line $x + y = 6$

36. The surface given by $z = x^2 + 3xy + y^2$ over the region in the xy-plane bounded by $0 \leq x \leq 4, 0 \leq y \leq x$

Compute the magnitude of the fundamental cross product for the surface defined parametrically in Problems 37–40.

37. $\mathbf{R}(u, v) = (2u \sin v)\mathbf{i} + (2u \cos v)\mathbf{j} + u^2\mathbf{k}$

38. $\mathbf{R}(u, v) = (4 \sin u \cos v)\mathbf{i} + (4 \sin u \sin v)\mathbf{j} + (5 \cos u)\mathbf{k}$

39. $\mathbf{R}(u, v) = u\mathbf{i} + v^2\mathbf{j} + u^3\mathbf{k}$

40. $\mathbf{R}(u, v) = (2u \sin v)\mathbf{i} + (2u \cos v)\mathbf{j} + (u^2 \sin 2v)\mathbf{k}$

41. Find the surface area of the surface given parametrically by the equation $\mathbf{R}(u, v) = uv\mathbf{i} + (u - v)\mathbf{j} + (u + v)\mathbf{k}$ for $u^2 + v^2 \leq 1$. *Hint:* Use polar coordinates.

42. A *spiral ramp* has the vector parametric equation $\mathbf{R}(u, v) = (u \cos v)\mathbf{i} + (u \sin v)\mathbf{j} + v\mathbf{k}$ for $0 \leq u \leq 1$, $0 \leq v \leq \pi$. Find its surface area.

43. **HISTORICAL QUEST** August Möbius studied under Karl Gauss (1777–1855), as well as Gauss' own teacher Johann Pfaff (1765–1825). He was a professor at the University of Leipzig, and is best known for his work in topology, especially for his conception of the Möbius strip. A Möbius strip is a two-dimensional surface with only one side.

AUGUST MÖBIUS
1790–1868

The following parametric surface is called a **Möbius strip:**

$$x = \cos v + u \cos \dfrac{v}{2} \cos v$$

$$y = \sin v + u \cos \dfrac{v}{2} \sin v$$

$$z = u \sin \dfrac{v}{2}$$

where $-\frac{1}{2} \leq u \leq \frac{1}{2}, 0 \leq v \leq 2\pi$. Sketch the graph of the surface, and then construct a three-dimensional model of a Möbius strip. Finally, find the fundamental cross product.

44. Verify that a sphere of radius a has surface area $4\pi a^2$.

45. Verify that a cylinder of radius a and height h has surface area $2\pi ah$.

46. Find the surface area of the torus defined by

$$\mathbf{R}(u, v) = (a + b \cos v)\cos u\, \mathbf{i} + (a + b \cos v)\sin u\, \mathbf{j} + b \sin v\, \mathbf{k}$$

for $0 < b < a, 0 \leq u \leq 2\pi, 0 \leq v \leq 2\pi$.

47. Suppose a surface is given implicitly by $F(x, y, z) = 0$. If the surface can be projected onto a region D in the xy-plane, show that the surface area is given by

$$A = \iint\limits_{D} \dfrac{\sqrt{F_x^2 + F_y^2 + F_z^2}}{|F_z|}\, dA_{xy}$$

where $F_z \neq 0$. Use this formula to find the surface area of a sphere of radius R.

48. Let S be the surface defined by $f(x, y, z) = C$, and let R be the projection of S on a plane. Show that the surface area of S can be computed by the integral

$$\int_{R}\int \dfrac{\|\nabla f\|}{|\nabla f \cdot \mathbf{u}|}\, dA$$

where $\mathbf{u}$ is a unit vector normal to the plane containing R and $\nabla f \cdot \mathbf{u} \neq 0$. This is a practical formula sometimes used in calculating the surface area.

13.5 Triple Integrals

IN THIS SECTION definition of triple integral, iterated integration, volume by triple integrals ▪

DEFINITION OF TRIPLE INTEGRAL

A double integral $\iint\limits_{R} f(x, y)\, dA$ is evaluated over a closed, bounded region in the plane, and in essentially the same way, a **triple integral** $\iiint\limits_{S} f(x, y, z)\, dV$ is evaluated over a closed, bounded solid region in $\mathbb{R}^3$. Suppose $f(x, y, z)$ is defined on a closed region S, which in turn is contained in a "box" D in space. Partition D into a finite number of smaller boxes with planes parallel to the coordinate planes, as shown in Figure 13.28.

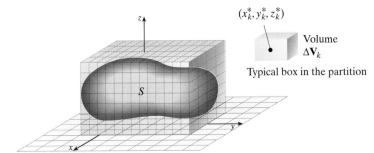

■ **FIGURE 13.28**   The region of integration S is subdivided into smaller boxes

We exclude from consideration any boxes with points outside S. Let $\Delta V_1,\ \Delta V_2,\ \ldots,$ ΔV_n denote the volumes of the boxes that remain, and define the norm $\|P\|$ of the partition to be the length of the longest diagonal of any box in the partition. Next, choose a representative point (x_k^*, y_k^*, z_k^*) from each box in the partition and form the **Riemann sum**

$$\sum_{k=1}^{n} f(x_k^*, y_k^*, z_k^*)\Delta V_k$$

If we repeat the process with more subdivisions, so that the norm approaches zero, we are led to the following definition.

Triple Integral

If f is a function defined over a closed, bounded solid region S, then the **triple integral of f over S** is defined to be the limit

$$\iiint_S f(x, y, z)\, dV = \lim_{\|P\|\to 0} \sum_{k=1}^{n} f(x_k^*, y_k^*, z_k^*)\Delta V_k$$

provided this limit exists.

In advanced calculus, it is shown that the triple integral $\iiint_S f(x, y, z)\, dV$ exists if $f(x, y, z)$ is continuous on S and S is **piecewise smooth** in the sense that it consists of a finite number of pieces, each of which has a continuously turning tangent plane (that is, the tangent plane varies continuously from point to point). It can also be shown that triple integrals have the following properties (which are analogous to those of double integrals listed in Theorem 13.1). In each case, assume the indicated integrals exist.

Linearity rule   For constants a and b

$$\iiint_S [af(x, y, z) + bg(x, y, z)]\, dV$$

$$= a \iiint_S f(x, y, z)\, dV + b \iiint_S g(x, y, z)\, dV$$

Dominance rule   If $f(x, y, z) \geq g(x, y, z)$ on S, then

$$\iiint_S f(x, y, z)\, dV \geq \iiint_S g(x, y, z)\, dV$$

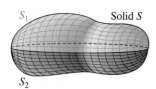

Solid S

Subdivision rule   If the solid region of integration S can be subdivided into two solid subregions S_1 and S_2, then

$$\iiint_S f(x, y, z) \, dV = \iiint_{S_1} f(x, y, z) \, dV + \iiint_{S_2} f(x, y, z) \, dV$$

ITERATED INTEGRATION

As with double integrals, we evaluate triple integrals by iterated integration. However, it is generally more difficult to set up the limits of integration in a triple integral, because the region of integration S is a solid. The relatively simple case where S is a rectangular solid (box) may be handled by applying the following theorem.

THEOREM 13.5   *Fubini's theorem over a parallelepiped in space*

If $f(x, y, z)$ is continuous over a rectangular solid $R: a \leq x \leq b, c \leq y \leq d, r \leq z \leq s$, then the triple integral may be evaluated by the iterated integral

$$\iiint_R f(x, y, z) \, dV = \int_r^s \int_c^d \int_a^b f(x, y, z) \, dx \, dy \, dz$$

The iterated integration can be performed in any order (with appropriate adjustments) to the limits of integration:

$$\begin{matrix} dx \, dy \, dz & \quad dx \, dz \, dy & \quad dz \, dx \, dy \\ dy \, dx \, dz & \quad dy \, dz \, dx & \quad dz \, dy \, dx \end{matrix}$$

Proof   The proof is beyond the scope of this course and is given in an advanced calculus course.   ▬

| **EXAMPLE 1** | *Evaluating a triple integral using Fubini's theorem* |

Evaluate $\iiint_R z^2 y e^x \, dV$, where R is the box given by

$$0 \leq x \leq 1, \quad 1 \leq y \leq 2, \quad -1 \leq z \leq 1$$

Solution   We shall evaluate the integral in the order $dx \, dy \, dz$.

$$\iiint_R f(x, y, z) \, dV = \int_{-1}^1 \int_1^2 \int_0^1 z^2 y e^x \, dx \, dy \, dz$$

$$= \int_{-1}^1 \int_1^2 z^2 y [e^x] \Big|_{x=0}^{x=1} dy \, dz = \int_{-1}^1 \int_1^2 z^2 y [e - 1] \, dy \, dz$$

$$= (e - 1) \int_{-1}^1 z^2 \left[\frac{y^2}{2}\right] \Big|_{y=1}^{y=2} dz = (e - 1) \int_{-1}^1 z^2 \left[\frac{2^2}{2} - \frac{1^2}{2}\right] dz$$

$$= \frac{3}{2}(e - 1) \int_{-1}^1 z^2 \, dz = \frac{3}{2}(e - 1) \frac{z^3}{3} \Big|_{z=-1}^{z=1}$$

$$= \frac{3}{2}(e - 1) \left[\frac{1^3}{3} - \frac{(-1)^3}{3}\right] = e - 1$$

It might be instructive for you to verify that the same result is obtained by using any other order of integration—for example, $dz \, dy \, dx$.   ■

To evaluate a triple integral over a region that is not a box, suppose we have a solid region S with a "lower" surface $z = u(x, y)$ and an "upper" suface $z = v(x, y)$ defined over a common domain A in the xy-plane. In this case, S may be described as the set of all points (x, y, z) such that for each fixed point (x, y) in A, z varies from u to v, as shown in Figure 13.29.

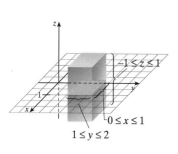

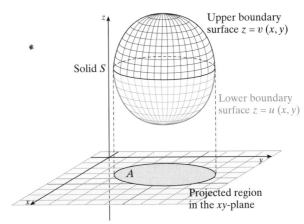

■ **FIGURE 13.29**   A solid region S bounded by the surfaces $z = u(x, y)$ and $z = v(x, y)$

The domain A is then described in a double integral. These observations are summarized in the following theorem.

THEOREM 13.6   *Triple integral over a general region*

If S is a region in space that is bounded below by the surface $z = u(x, y)$ and above by $z = v(x, y)$ as (x, y) varies over the planar region A, then

$$\iiint\limits_{S} f(x, y, z)\, dV = \iint\limits_{A} \int_{u(x, y)}^{v(x, y)} f(x, y, z)\, dz\, dA$$

Proof   This proof is omitted.   ═══

If the region of integration S has the form described in Theorem 13.6, the integral may be evaluated by integrating first with respect to z (as z varies from $z = u(x, y)$ to $z = v(x, y)$) and then computing an appropriate double integral over the planar region A.

EXAMPLE 2   *Evaluating a triple integral over a general region*

Evaluate $\iiint\limits_{S} x\, dV$, where S is the solid in the first octant bounded by the cylinder $x^2 + y^2 = 4$ and the plane $2y + z = 4$.

Solution   The solid is shown in Figure 13.30.

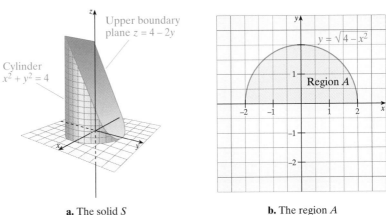

a. The solid S **b.** The region A

■ **FIGURE 13.30**   The region S bounded by the plane $2y + z = 4$ and the cylinder $x^2 + y^2 = 4$ in the first octant

The upper boundary surface of S is the plane $z = 4 - 2y$, and the lower boundary surface $z = 0$ is the xy-plane. The projection A of the solid on the xy-plane is the quarter disk $x^2 + y^2 \leq 4$ with $x \geq 0, y \geq 0$ (because S lies in the first octant). This projection may be described in type I form as the set of all (x, y) such that for each fixed x between 0 and 2, y varies from 0 to $\sqrt{4 - x^2}$. Thus, we have

$$\iiint_S x \, dV = \iint_A \int_0^{4-2y} x \, dz \, dA = \int_0^2 \int_0^{\sqrt{4-x^2}} \int_0^{4-2y} x \, dz \, dy \, dx$$

$$= \int_0^2 \int_0^{\sqrt{4-x^2}} x[(4 - 2y) - 0] \, dy \, dx = \int_0^2 \int_0^{\sqrt{4-x^2}} (4x - 2xy) \, dy \, dx$$

$$= \int_0^2 [4xy - xy^2] \Big|_{y=0}^{y=\sqrt{4-x^2}} dx = \int_0^2 [4x\sqrt{4 - x^2} - x(4 - x^2)] \, dx$$

$$= [-\tfrac{4}{3}(4 - x^2)^{3/2} - 2x^2 + \tfrac{1}{4}x^4] \Big|_0^2 = [0 - 8 + 4 + \tfrac{32}{3} + 0 - 0] = \tfrac{20}{3} \quad \blacksquare$$

VOLUME BY TRIPLE INTEGRALS

Just as a double integral can be interpreted as the area of the region of integration, a triple integral may be interpreted as the **volume** of a solid. That is, if V is the volume of the solid region S, then

$$V = \iiint_S dV$$

EXAMPLE 3   *Volume of a tetrahedron*

Find the volume of the tetrahedron T bounded by the part of the plane $2x + y + 3z = 6$ in the first octant.

Solution   The tetrahedron T is shown in Figure 13.31a.

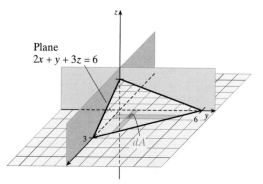

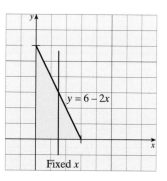

a. The tetrahedron bounded by the plane $2x + y + 3z = 6$ and the positive coordinate planes.

b. The region projected onto the xy-plane is a triangle.

■ **FIGURE 13.31**   Volume of a tetrahedron

The upper surface of T is the plane $z = \tfrac{1}{3}(6 - 2x - y)$ and its lower surface is $z = 0$. Note that T projects onto a triangle in the xy-plane, as shown in Figure 13.31b. Described in type I form, this triangle is the set of all (x, y) such that for each fixed x

between 0 and 3, y varies from 0 to $6 - 2x$. Thus,

$$V = \iiint_T dV = \iint_A \int_0^{\frac{1}{3}(6-2x-y)} dz \, dA$$

$$= \int_0^3 \int_0^{6-2x} \int_0^{\frac{1}{3}(6x-2y-y)} dz \, dy \, dx$$

$$= \int_0^3 \int_0^{6-2x} [\tfrac{1}{3}(6 - 2x - y) - 0] \, dy \, dx$$

$$= \int_0^3 \left[2y - \tfrac{2}{3}xy - \tfrac{1}{6}y^2 \right]\Big|_{y=0}^{y=6-2x} dx$$

$$= \int_0^3 [2(6 - 2x) - \tfrac{2}{3}x(6 - 2x) - \tfrac{1}{6}(6 - 2x)^2 - 0] \, dx$$

$$= \int_0^3 \tfrac{1}{6}[36 - 24x + 4x^2] \, dx = 6$$

The volume of the tetrahedron is 6 cubic units. ■

Sometimes it is easier to evaluate a triple integral by integrating first with respect to x or y instead of z. For instance, if the solid region of integration S is bounded in the back by $x = x_1(y, z)$ and in the front by $x = x_2(y, z)$, and the boundary surfaces project onto a region A in the yz-plane, denoted by A_{yz}, as shown in Figure 13.32a, then

$$\iiint_S f(x, y, z) \, dV = \iint_{A_{yz}} \int_x^{x_2} f(x, y, z) \, dx \, dA_{yz}$$

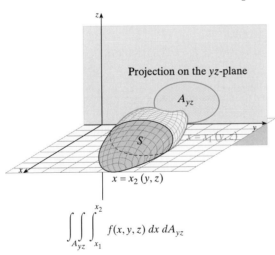

$$\iint_{A_{yz}} \int_{x_1}^{x_2} f(x, y, z) \, dx \, dA_{yz}$$

a. A solid S with "front" surface $x = x_2(y, z)$ and "back" surface $x = x_1(y, z)$

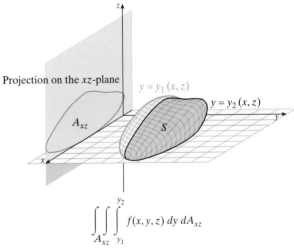

$$\iint_{A_{xz}} \int_{y_1}^{y_2} f(x, y, z) \, dy \, dA_{xz}$$

b. A solid S with "side" surfaces $y = y_2(x, z)$ and $y = y_1(x, z)$

■ **FIGURE 13.32** Iterated integration with respect to x or y first

On the other hand, if the solid region of integration S is bounded on one side by the surface $y = y_1(x, z)$ and on the other by $y = y_2(x, z)$, and the boundary surfaces project onto a region A in the xz-plane, denoted by A_{xz}, as shown in Figure 13.32b, then

$$\iiint_S f(x, y, z) \, dV = \iint_{A_{xz}} \int_{y_1}^{y_2} f(x, y, z) \, dy \, dA_{xz}$$

As an illustration, we will now rework Example 3 by projecting the tetrahedron S onto the yz-plane.

EXAMPLE 4   *Volume of a tetrahedron by changing the order of integration*

Find the volume of the tetrahedron S bounded by the coordinate planes and the plane $2x + y + 3z = 6$ in the first octant by projecting onto the yz-plane.

Solution   Note that S is bounded from "behind" by the yz-plane and "in front" by the plane $2x + y + 3z = 6$, which we express as $x = \frac{1}{2}(6 - y - 3z)$. (See Figure 13.33a.) The volume is given by

$$V = \iiint_S dV = \iint_{A_{yz}} \int_0^{\frac{1}{3}(6-y-3z)} dx \, dA_{yz}$$

where A_{yz} is the projection in the yz-plane. This projection is the triangle bounded by the lines $z = 0$, $y = 0$, and $z = \frac{1}{3}(6 - y)$, as shown in Figure 13.33b.

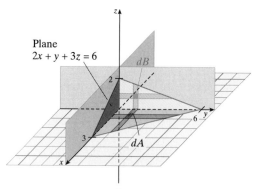

a. The tetrahedron bounded by the plane $2x + y + 3z = 6$ and the positive coordinate planes.

b. The region projected onto the yz-plane is a triangle.

■ **FIGURE 13.33**   Volume of a tetrahedron; alternative projection

Thus, for each fixed y between 0 and 6, z varies from 0 to $\frac{1}{3}(6 - y)$, and we have

$$V = \int_0^6 \int_0^{\frac{1}{3}(6-y)} \int_0^{\frac{1}{2}(6-y-3z)} dx \, dz \, dy = \int_0^6 \int_0^{\frac{1}{3}(6-y)} \tfrac{1}{2}(6 - y - 3z) \, dz \, dy$$

$$= \int_0^6 \left[3z - \tfrac{1}{2}yz - \tfrac{3}{4}z^2 \right]\Big|_{z=0}^{z=\frac{1}{3}(6-y)} dy$$

$$= \int_0^6 \left[(6 - y) - \tfrac{1}{6}y(6 - y) - \tfrac{1}{12}(6 - y)^2 - 0 \right] dy = 6$$

This is the same result we obtained in Example 3 by projecting onto the xy-plane.

■

EXAMPLE 5   *Setting up a triple integral to find a volume*

Set up (but do not evaluate) a triple integral for the volume of the solid S that is bounded above by the sphere $x^2 + y^2 + z^2 = 4$ and below by the plane $y + z = 2$.

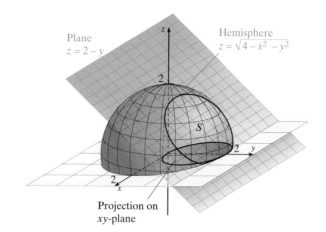

Solution First, note that the intersection of the plane and the sphere occurs above the xy-plane (where $z \geq 0$), so the sphere can be represented by the equation

$$z = \sqrt{4 - x^2 - y^2}$$

(the upper hemisphere). To find the limits of integration for x and y, we consider the projection of S onto the xy-plane. To this end, consider the intersection of the hemisphere and the plane $z = 2 - y$:

$$\sqrt{4 - x^2 - y^2} = 2 - y$$
$$4 - x^2 - y^2 = 4 - 4y + y^2$$
$$x^2 + 2y^2 - 4y = 0$$
$$x^2 + 2(y - 1)^2 = 2$$

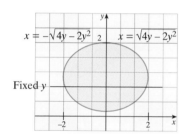

■ **FIGURE 13.34** The projection of S onto the xy-plane

Although this intersection occurs in $\mathbb{R}^3$, its equation does not contain z. Therefore, the equation serves as a projection on the xy-plane, where $z = 0$. This is an ellipse centered at $(0, 1)$. We sketch this curve in the xy-plane, as shown in Figure 13.34.

We consider this as a type II region, which means we will integrate first with respect to x, then with respect to y. Because $x^2 + 2(y - 1)^2 = 2$, we see that for fixed y between 0 and 2, x varies from $-\sqrt{2 - 2(y - 1)^2} = -\sqrt{4y - 2y^2}$ to $\sqrt{4y - 2y^2}$. However, using symmetry, we see the required volume V is twice the integral as x varies from 0 to $\sqrt{4y - 2y^2}$. This leads us to set up the following triple integral:

$$V = 2 \int_0^2 \int_0^{\sqrt{4y - 2y^2}} \int_{2-y}^{\sqrt{4 - x^2 - y^2}} dz\, dx\, dy \qquad ■$$

13.5 Problem Set

Ⓐ 1. ■ **What Does This Say?** Describe Fubini's theorem over a parallelepiped in space.

2. ■ **What Does This Say?** Set up integrals, with appropriate limits of integration, for the six possible orders of integration for $\iiint\limits_S f(x, y, z)\, dV$, where S is the solid described by

$$S: y^2 \leq x \leq 4;\ 0 \leq y \leq 2;\ 0 \leq z \leq 4 - x$$

Compute the iterated triple integrals in Problems 3–14.

3. $\int_1^4 \int_{-2}^3 \int_2^5 dx\, dy\, dz$

4. $\int_{-1}^3 \int_0^2 \int_{-2}^2 dy\, dz\, dx$

5. $\int_1^2 \int_0^1 \int_{-1}^2 8x^2yz^3\, dx\, dy\, dz$

6. $\int_4^7 \int_{-1}^2 \int_0^3 x^2y^2z^2\, dx\, dy\, dz$

7. $\int_0^2 \int_0^x \int_0^{x+y} xyz \, dz \, dy \, dx$

8. $\int_0^1 \int_{\sqrt{x}}^{\sqrt{1+x}} \int_0^{xy} y^{-1}z \, dz \, dy \, dx$

9. $\int_{-1}^2 \int_0^{\pi} \int_1^4 yz \cos xy \, dz \, dx \, dy$

10. $\int_0^{\pi} \int_0^1 \int_0^1 x^2 y \cos xyz \, dz \, dy \, dx$

11. $\int_0^1 \int_0^y \int_0^{\ln y} e^{z+2x} \, dz \, dx \, dy$

12. $\int_1^3 \int_0^{2z} \int_0^{\ln y} y \, e^{-x} \, dx \, dy \, dz$

13. $\int_1^4 \int_{-1}^{2z} \int_0^{\sqrt{3x}} \frac{x-y}{x^2+y^2} \, dy \, dx \, dz$

14. $\int_0^1 \int_{x-1}^{x^2} \int_{-x}^y (x+y) \, dz \, dy \, dx$

Evaluate the triple integrals in Problems 15–22.

15. $\iiint\limits_S (x^2y + y^2z) \, dV,$
where S is the box $1 \le x \le 3, -1 \le y \le 1, 2 \le z \le 4$

16. $\iiint\limits_S (xy + 2yz) \, dV,$
where S is the box $2 \le x \le 4, 1 \le y \le 3, -2 \le z \le 4$

17. $\iiint\limits_S xyz \, dV,$
where S is the tetrahedron with vertices $(0,0,0), (1,0,0),$ $(0,1,0),$ and $(0,0,1)$

18. $\iiint\limits_S x^2y \, dV,$
where S is the tetrahedron with vertices $(0,0,0), (3,0,0),$ $(0,2,0),$ and $(0,0,1)$

19. $\iiint\limits_S xyz \, dV,$
where S is the region given by
$x^2 + y^2 + z^2 \le 1, y \ge 0, z \ge 0$

20. $\iiint\limits_S x \, dV,$
where S is bounded by the paraboloid $z = x^2 + y^2$ and the plane $z = 1$

21. $\iiint\limits_S e^z \, dV,$
where S is the region described by the inequalities
$0 \le x \le 1, 0 \le y \le x,$ and $0 \le z \le x + y$

22. $\iiint\limits_S yz \, dV,$
where S is the solid in the first octant bounded by the hemisphere $x = \sqrt{9 - y^2 - z^2}$ and the coordinate planes

Find the volume V of the solids bounded by the graphs of the equations given in Problems 23–32 by using triple integration.

23. $x + y + z = 1$ and the coordinate planes

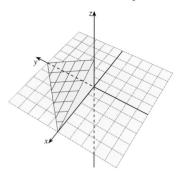

24. $y = 9 - x^2, z = 0, z = y$

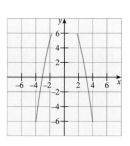
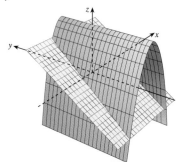

25. $(x-1)^2 + (y-2)^2 + (z-3)^2 = 1$

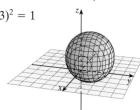

26. $z = 4 - 4x^2 - 4y^2, z = 0$

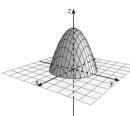

27. $x^2 + 3y^2 = z$ and the cylinder $y^2 + z = 4$

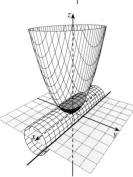

28. $x^2 + y^2 + z^3 = 9, z = 0$

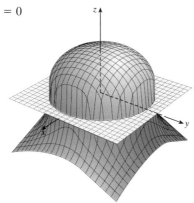

29. The solid bounded above by the paraboloid $z = 6 - x^2 - y^2$ and below by $z = 2x^2 + y^2$

30. The solid bounded by the sphere $x^2 + y^2 + z^2 = 2$ and the paraboloid $x^2 + y^2 = z$

31. The solid region common to the cylinders $x^2 + z^2 = 1$ and $y^2 + z^2 = 1$

32. The solid bounded by the cylinders $y = z^2, y = 2 - z^2$ and the planes $x = 1$ and $x = -2$

For each given iterated integral, there are five other equivalent iterated integrals. Find the one with the requested order in Problems 33–36.

33. $\int_0^1 \int_0^x \int_0^y f(x, y, z) \, dz \, dy \, dx$; change the order to $dz \, dx \, dy$.

34. $\int_1^2 \int_0^{z-1} \int_0^x f(x, y, z) \, dy \, dx \, dz$;
change the order to $dy \, dz \, dx$.

35. $\int_0^2 \int_0^{\sqrt{4-x^2}} \int_0^{\sqrt{4-x^2-y^2}} f(x, y, z) \, dz \, dy \, dx$;
change the order to $dy \, dx \, dz$.

36. $\int_0^2 \int_0^{\sqrt{4-x^2}} \int_0^{\sqrt{4-x^2}} f(x, y, z) \, dz \, dy \, dx$;
change the order to $dy \, dx \, dz$.

Ⓑ 37. Find the volume of the ellipsoid
$$\frac{x^2}{4} + \frac{y^2}{9} + \frac{z^2}{16} = 1$$

38. Find the volume of the region between the two elliptic paraboloids
$$z = \frac{x^2}{9} + y^2 - 4$$
and
$$z = -\frac{x^2}{9} - y^2 + 4$$

39. Find the volume of the region bounded by the paraboloids $z = 16 - x^2 - 2y^2$ and $z = 3x^2 + 2y^2$.

40. A wedge is cut from a right circular cylinder of radius R by a horizontal plane perpendicular to the axis of the cylinder and a second plane that meets the first on the axis at an angle of θ degrees, as shown in Figure 13.35. Set up and evaluate a triple integral for the volume of the wedge.

■ **FIGURE 13.35**

41. Find the volume of the region that is bounded above by the elliptic paraboloid
$$z = \frac{x^2}{9} + y^2$$
on the sides by the cylinder
$$\frac{x^2}{9} + y^2 = 1$$
and below by the xy-plane.

42. Find the volume of the solid region in the first octant that is bounded by the planes $z = 8 + 2x + y, y = 3 - 2x$.

Ⓒ 43. Use triple integration to find the volume of a sphere.

44. Use triple integration to find the volume of a right pyramid with height H and a square base of side S.

45. Use triple integration to find the volume of the ellipsoid
$$\frac{x^2}{a^2} + \frac{y^2}{b^2} + \frac{z^2}{c^2} = 1$$
(assume $a > 0, b > 0, c > 0$).

46. Find the volume of the solid region common to the paraboloid $z = k(x^2 + y^2)$ and the sphere
$$x^2 + y^2 + z^2 = 2k^{-2}, \quad \text{where } k > 0$$

47. Find the volume of the tetrahedron bounded by the plane
$$\frac{x}{a} + \frac{y}{b} + \frac{z}{c} = 1$$
$(a > 0, b > 0, c > 0)$ in the first octant.

48. THINK TANK PROBLEM   Let B be the box defined by $a \le x \le b, c \le y \le d, r \le z \le s$. Is it true that
$$\iiint\limits_B f(x)g(y)h(z) \, dV$$
$$= \left[\int_a^b f(x) \, dx \right] \left[\int_c^d g(y) \, dy \right] \left[\int_r^s h(z) \, dz \right]$$

Either show that this equation is generally true, or find a counterexample.

49. Change the order of integration to show that
$$\int_0^x \int_0^v f(u) \, du \, dv = \int_0^x (x - t) f(t) \, dt$$
Also, show that
$$\int_0^x \int_0^v \int_0^u f(w) \, dw \, du \, dv = \frac{1}{2} \int_0^x (x - t)^2 f(t) \, dt$$

50. Evaluate the triple integral

$$\iiint\limits_{S} \sin(\pi - z)^3 \, dz \, dx \, dy$$

where S is the solid region bounded below by the xy-plane, above by the plane $x = z$, and laterally by the planes $x = y$ and $y = \pi$. *Hint:* See Problem 49.

51. One of the following integrals has the value 0. Which is it and why?

A. $\displaystyle\int_{-2}^{2} \int_{-\sqrt{4-y^2}}^{\sqrt{4-y^2}} \int_{-\sqrt{4-x^2-y^2}}^{\sqrt{4-x^2-y^2}} (x + z^2) \, dz \, dx \, dy$

B. $\displaystyle\int_{0}^{1} \int_{x}^{2-x^2} \int_{-3}^{3} z^2 \sin xz \, dz \, dy \, dx$

Higher-dimensional multiple integrals can be defined and evaluated in essentially the same way as double integrals and triple integrals. Evaluate the given multiple integrals in Problems 52–53.

52. $\displaystyle\iiiint\limits_{H} xyz^2w^2 \, dx \, dy \, dz \, dw$,

where H is the four-dimensional "hyperbox" defined by $0 \le x \le 1, 0 \le y \le 2, -1 \le z \le 1, 1 \le w \le 2$.

53. $\displaystyle\iiiint\limits_{H} e^{x-2y+z+w} \, dw \, dz \, dy \, dx$,

where H is the four-dimensional region bounded by the hyperplane $x + y + z + w = 4$ and the coordinate spaces $x = 0, y = 0, z = 0$, and $w = 0$ in the first hyper-octant (where $x \ge 0, y \ge 0, z \ge 0, w \ge 0$).

13.6 Mass, Moments, and Probability Density Functions

IN THIS SECTION   **mass and center of mass, moments of inertia, joint probability density functions**

MASS AND CENTER OF MASS

WARNING The moment of an object about an axis is the product of its mass and the signed distance from that axis. ←

Recall (from Section 6.5) that a solid object that is sufficiently "flat" to be regarded as two-dimensional is called a **lamina.** Suppose a particular lamina occupies a bounded region R in the xy-plane, and let $\delta(x, y)$ be the density (mass per unit area) of the lamina. A **homogeneous** lamina has constant density $\delta = m/A$, where m is the mass and A is the area of the lamina. If the lamina is **nonhomogeneous,** its density $\delta(x, y)$ varies from point to point. In this case, we can partition the region R into a number of rectangles, as shown in Figure 13.36. Choose a representative point (x_k^*, y_k^*) in each rectangle of the partition, and use the formula

$$\Delta m_k = \delta(x_k^*, y_k^*)\Delta A_k$$

to approximate the mass of each rectangle. We then approximate the total mass m by the Riemann sum

$$\Delta m = \sum_{k=1}^{n} \delta(x_k^*, y_k^*)\Delta A_k$$

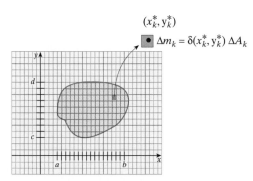

■ **FIGURE 13.36** Partition of a lamina in the plane

so that

$$m = \lim_{\|P\| \to 0} \sum_{k=1}^{n} \delta(x_k^*, y_k^*) \Delta A_k = \iint_R \delta(x, y) \, dA$$

We use these observations as the basis for the following definition.

Mass of a Planar Lamina of Variable Density

If δ is a continuous density function on the lamina corresponding to a plane region R, then the mass m of the lamina is given by

$$m = \iint_R \delta(x, y) \, dA$$

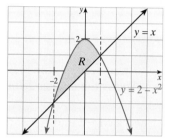

■ FIGURE 13.37   A lamina over the region R bounded by $y = 2 - x^2$ and $y = x$

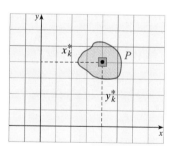

The distance from P to the x-axis is y and the distance to the y-axis is x

EXAMPLE 1   *Mass of a planar lamina*

Find the mass of the lamina of density $\delta(x, y) = x^2$ that occupies the region R bounded by the parabola $y = 2 - x^2$ and the line $y = x$.

Solution   Begin by drawing the parabola and the line, and by finding their points of intersection, as shown in Figure 13.37. By substitution,

$$x = 2 - x^2$$
$$x^2 + x - 2 = 0$$
$$x = -2, 1$$

We see that the region R is the set of all (x, y) such that for each x between -2 and 1, y varies from x to $2 - x^2$.

$$m = \iint_R x^2 \, dA = \int_{-2}^{1} \int_{x}^{2-x^2} x^2 \, dy \, dx$$

$$= \int_{-2}^{1} x^2 (2 - x^2 - x) \, dx = \int_{-2}^{1} (2x^2 - x^4 - x^3) \, dx$$

$$= \left[\frac{2x^3}{3} - \frac{x^5}{5} - \frac{x^4}{4} \right]_{-2}^{1} = \frac{63}{20} = 3.15 \qquad ■$$

The *moment* of an object about an axis is defined as the product of its mass and the signed distance from the axis. Thus, by partitioning the region R as before (with mass), we see that the moments M_x and M_y of the lamina about the x-axis and y-axis, respectively, are approximately equal to the Riemann sums

$$M_x = \sum_{k=1}^{n} \underset{\uparrow}{y_k} \, \delta(x_k^*, y_k^*) \Delta A_k \quad \text{and} \quad M_y = \sum_{k=1}^{n} \underset{\uparrow}{x_k} \, \delta(x_k^*, y_k^*) \Delta A_k$$

$$\text{Distance to } x\text{-axis} \qquad\qquad \text{Distance to } y\text{-axis}$$

By taking the limit as the norm of the partition tends to 0, we obtain

$$M_x = \iint_R y\delta(x, y) \, dA \quad \text{and} \quad M_y = \iint_R x\delta(x, y) \, dA$$

The *center of mass* of the lamina covering R is the point $(\bar{x}, \bar{y})$ where the mass m can be concentrated without affecting the moments M_x and M_y; that is,

$$m\bar{x} = M_y \quad \text{and} \quad m\bar{y} = M_x$$

For future reference, these observations are summarized in the following box.

Moments and Center of Mass of a Variable Density Planar Lamina

If $\delta(x, y)$ is a continuous density function on a lamina corresponding to a plane region R, then the **moments of mass** with respect to the x- and y-axes, respectively, are

$$M_x = \iint\limits_R y\,\delta(x, y)\,dA \quad \text{and} \quad M_y = \iint\limits_R x\,\delta(x, y)\,dA$$

Furthermore, if m is the mass of the lamina, the **center of mass** is $(\bar{x}, \bar{y})$, where

$$\bar{x} = \frac{M_y}{m} \quad \text{and} \quad \bar{y} = \frac{M_x}{m}$$

If the density δ is constant, the point $(\bar{x}, \bar{y})$ is called the **centroid** of the region.

WARNING M_x has a factor of y and M_y has a factor of x. ←

EXAMPLE 2   *Finding a center of mass*

Find the center of mass of the lamina of density $\delta(x, y) = x^2$ that occupies the region R bounded by the parabola $y = 2 - x^2$ and the line $y = x$. This is the lamina defined in Example 1.

Solution   In Example 1, we found that for each fixed x between -2 and 1, y varies from x to $2 - x^2$ (see Figure 13.37). Thus, we have

$$
\begin{aligned}
M_x &= \iint\limits_R y(x^2)\,dA & M_y &= \iint\limits_R x(x^2)\,dA \\[6pt]
&= \int_{-2}^{1} \int_{x}^{2-x^2} yx^2\,dy\,dx & &= \int_{-2}^{1} \int_{x}^{2-x^2} x^3\,dy\,dx \\[6pt]
&= \int_{-2}^{1} \left[\frac{1}{2}x^2 y^2\right]\Big|_{y=x}^{y=2-x^2}\,dx & &= \int_{-2}^{1} x^3[(2 - x^2) - x]\,dx \\[6pt]
&= \frac{1}{2}\int_{-2}^{1} x^2(x^4 - 5x^2 + 4)\,dx & &= \int_{-2}^{1}(2x^3 - x^5 - x^4)\,dx \\[6pt]
&= \frac{1}{2}\left[\frac{1}{7}x^7 - x^5 + \frac{4}{3}x^3\right]\Big|_{-2}^{1} & &= \left[\frac{2x^4}{4} - \frac{x^6}{6} - \frac{x^5}{5}\right]\Big|_{-2}^{1} \\[6pt]
&= -\frac{9}{7} & &= -\frac{18}{5}
\end{aligned}
$$

From Example 1, $m = \frac{63}{20}$, so the center of mass is $(\bar{x}, \bar{y})$, where

$$\bar{x} = \frac{M_y}{m} = \frac{-\frac{18}{5}}{\frac{63}{20}} = -\frac{8}{7} \approx -1.14 \qquad \bar{y} = \frac{M_x}{m} = \frac{-\frac{9}{7}}{\frac{63}{20}} = -\frac{20}{49} \approx -0.41 \qquad ■$$

In a completely analogous way, we can use the triple integral to find the mass and center of mass of a solid in $\mathbb{R}^3$. The density $\delta(x, y, z)$ at a point in the solid now refers to mass per unit volume, and the mass m, moments M_{yz}, M_{xz}, M_{xy} about the yz-, xz-, and xy-planes, respectively, and coordinates $\bar{x}, \bar{y}, \bar{z}$ of the center of mass are given by:

Mass $$m = \iiint\limits_R \delta(x, y, z)\,dV$$

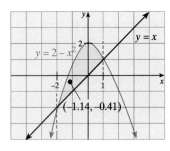

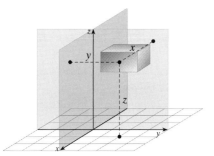

Note the distances to the coordinate planes

Moments

$$M_{yz} = \iiint\limits_{R} x\,\delta(x, y, z)\,dV$$
$$\uparrow$$
Distance to the yz-plane

$$M_{xz} = \iiint\limits_{R} y\,\delta(x, y, z)\,dV$$
$$\uparrow$$
Distance to the xz-plane

$$M_{xy} = \iiint\limits_{R} z\,\delta(x, y, z)\,dV$$
$$\uparrow$$
Distance to the xy-plane

Center of mass $$(\bar{x}, \bar{y}, \bar{z}) = \left(\frac{M_{yz}}{m}, \frac{M_{xz}}{m}, \frac{M_{xy}}{m}\right)$$

As before, if the density is constant, the center of mass is still called the **centroid.** Example 3 illustrates how this point can be found by multiple integration.

EXAMPLE 3   *Centroid of a tetrahedron*

A solid tetrahedron has vertices $(0, 0, 0)$, $(1, 0, 0)$, $(0, 1, 0)$, and $(0, 0, 1)$ and constant density $\delta = 6$. Find the centroid.

Solution   The tetrahedron may be described as the region in the first octant that lies beneath the plane $x + y + z = 1$, as shown in Figure 13.38a.

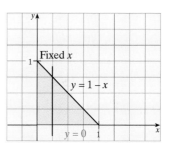

a. A tetrahedron

b. Projection on the xy-plane

■ **FIGURE 13.38**   The centroid of a tetrahedron

The boundary of the projection of the top face of the tetrahedron in the xy-plane is found by solving the equations $x + y + z = 1$ and $z = 0$ simultaneously. We find that the projection is the region bounded by the coordinate axes and the line $x + y = 1$, as shown in Figure 13.38b. This means that for each fixed x between 0 and 1, y varies from 0 to $1 - x$.

$$m = \iiint\limits_{R} \delta\,dV = \int_0^1 \int_0^{1-x} \int_0^{1-x-y} 6\,dz\,dy\,dx$$

$$= \int_0^1 \int_0^{1-x} 6(1 - x - y)\,dy\,dx = \int_0^1 \left[6y - 6xy - 3y^2\right]\Big|_{y=0}^{y=1-x}\,dx$$

$$= \int_0^1 3(x - 1)^2\,dx = \left[x^3 - 3x^2 + 3x\right]\Big|_0^1 = 1$$

Similarly, we find that

$$M_{yz} = \iiint_R 6x \, dV = \int_0^1 \int_0^{1-x} \int_0^{1-x-y} 6x \, dz \, dy \, dx = \frac{1}{4}$$

$$M_{xz} = \iiint_R 6y \, dV = \int_0^1 \int_0^{1-x} \int_0^{1-x-y} 6y \, dz \, dy \, dx = \frac{1}{4}$$

$$M_{xy} = \iiint_R 6z \, dV = \int_0^1 \int_0^{1-x} \int_0^{1-x-y} 6z \, dz \, dy \, dx = \frac{1}{4}$$

(Verify the details of these integrations.) Thus,

$$\bar{x} = \frac{M_{yz}}{m} = \frac{\frac{1}{4}}{1} = 0.25, \quad \bar{y} = \frac{M_{xz}}{m} = \frac{\frac{1}{4}}{1} = 0.25, \quad \bar{z} = \frac{M_{xy}}{m} = \frac{\frac{1}{4}}{1} = 0.25$$

The centroid is $(0.25, 0.25, 0.25)$. ∎

MOMENTS OF INERTIA

In general, a lamina of density $\delta(x, y)$ covering the region R in the first quadrant of the plane has (first) moment about a line L given by the integral

$$M_L = \iint_R s \, dm$$

where $dm = \delta(x, y) \, dA$ and $s = s(x, y)$ is the distance from the point $P(x, y)$ in R to L. Similarly, the *second* moment, or *moment of inertia*, of R about L is defined by

$$I_L = \iint_R s^2 \, dm$$

In particular, the moments of inertia about the coordinate axes are given by the integral formulas in the following box.

Moments of Inertia

The **moments of inertia** of a lamina of density δ covering the planar region R about the x-, y-, and z-axes, respectively, are given by

$$I_x = \iint_R y^2 \delta(x, y) \, dA$$

$$I_y = \iint_R x^2 \delta(x, y) \, dA$$

$$I_z = \iint_R (x^2 + y^2) \delta(x, y) \, dA$$

EXAMPLE 4   *Finding the moments of inertia*

A lamina occupies the region R in the plane that is bounded by the parabola $y = x^2$ and the lines $x = 2$ and $y = 1$. The density of the lamina at each point (x, y) is $\delta(x, y) = x^2 y$. Find the moments of inertia of the lamina about the x-axis and the y-axis.

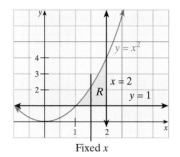

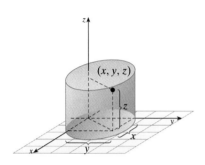

■ **FIGURE 13.39** Moment of inertia of a lamina

Solution   The graph of R is shown in Figure 13.39. We see that for each fixed x between 1 and 2, y varies from 1 to x^2.

$$I_x = \iint_R y^2 \, dm = \iint_R y^2 \delta(x, y) \, dA = \iint_R y^2(x^2y) \, dA$$

$$= \int_1^2 \int_1^{x^2} x^2 y^3 \, dy \, dx = \frac{1}{4} \int_1^2 x^2(x^8 - 1) \, dx = \frac{1{,}516}{33} \approx 45.94$$

$$I_y = \iint_R x^2 \, dm = \iint_R x^2 \delta(x, y) \, dA = \iint_R x^2(x^2y) \, dA$$

$$= \int_1^2 \int_1^{x^2} x^4 y \, dy \, dx = \frac{1}{2} \int_1^2 x^4(x^4 - 1) \, dx = \frac{1{,}138}{45} \approx 25.29$$ ■

A simple generalization enables us to compute the moment of inertia of a solid figure about an axis L outside the figure. Specifically, suppose the solid occupies a region R and that the density at each point (x, y, z) in R is given by $\delta(x, y, z)$. Because the square of the distance of a typical cell in R from the x-axis is $y^2 + z^2$, the moment of inertia about the x-axis is

$$I_x = \iiint_R \underbrace{(y^2 + z^2)}_{} \underbrace{\delta(x, y, z) \, dV}_{}$$

$$\uparrow \text{ Increment of mass}$$
Square of the distance to the x-axis

Similarly, the moments of inertia of the solid about the y-axis and the z-axis are, respectively,

$$I_y = \iiint_R \underbrace{(x^2 + z^2)}_{} \underbrace{\delta(x, y, z) \, dV}_{}$$

$$\uparrow \text{ Increment of mass}$$
Square of the distance to the y-axis

$$I_z = \iiint_R \underbrace{(x^2 + y^2)}_{} \underbrace{\delta(x, y, z) \, dV}_{}$$

$$\uparrow \text{ Increment of mass}$$
Square of the distance to the z-axis

EXAMPLE 5   *Moment of inertia of a solid*

Find the moment of inertia about the z-axis of the solid tetrahedron S with vertices $(0, 0, 0), (0, 1, 0), (1, 0, 0), (0, 0, 1)$ and density $\delta(x, y, z) = x$.

Solution   In Example 3, we observed that the solid S can be described as the set of all (x, y, z) such that for each fixed x between 0 and 1, y lies between 0 and $1 - x$, and $0 \le z \le 1 - x - y$. Thus,

$$I_z = \iiint_S (x^2 + y^2)\delta(x, y, z) \, dV = \int_0^1 \int_0^{1-x} \int_0^{1-x-y} x(x^2 + y^2) \, dz \, dy \, dx$$

$$= \int_0^1 \int_0^{1-x} x(x^2 + y^2)(1 - x - y) \, dy \, dx = \int_0^1 \left[\frac{x^3(1 - x)^2}{2} + \frac{x(1 - x)^4}{12} \right] dx = \frac{1}{90}$$ ■

Moments of inertia have a useful interpretation in physics. The **kinetic energy** of a body of mass m moving with velocity v along a straight line is defined in physics as $K = \frac{1}{2}mv^2$. Suppose a lamina covering circular disk R centered at the origin (see

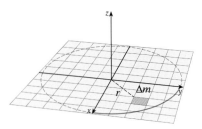

■ **FIGURE 13.40** Kinetic energy of a rotating disk

Figure 13.40) is rotating around the z-axis with angular speed ω radians/second. A cell of mass Δm located r units from the origin has linear velocity $v = r\omega$ and linear kinetic energy $K_{\text{lin}} = \frac{1}{2}(\Delta m)v^2$, and by integrating, we find that the entire disk R has kinetic energy of rotation

$$K_{\text{rot}} = \iint\limits_R \frac{1}{2}\,\omega^2\, r^2\, dm = \frac{1}{2}\,\omega^2 \iint\limits_R r^2\, dm$$

Since $r^2 = x^2 + y^2$, we see that the integral in this formula is just the moment of inertia of R about the z-axis, so the rotational kinetic energy can be expressed as

$$K_{\text{rot}} = \tfrac{1}{2} I_z \omega^2$$

Comparing this formula to the linear kinetic energy formula $K_{\text{lin}} = \frac{1}{2}mv^2$, we see that the moment of inertia may be thought of as the rotational analog of mass.

JOINT PROBABILITY DENSITY FUNCTIONS

A **continuous random variable** X is a continuous function whose domain is a set of real numbers associated with probabilities. A **probability density function** is a continuous nonnegative function $f(x)$ such that

$$P(a \leq X \leq b) = \int_a^b f(x)\, dx$$

where $P(a \leq X \leq b)$ denoted the probability that X is in the closed interval $[a, b]$.

Since $P(-\infty < X < \infty) = 1$, that is, X must be somewhere, we see that $f(x)$ must satisfy

$$\int_{-\infty}^{\infty} f(x)\, dx = 1$$

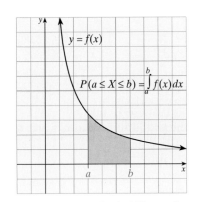

■ **FIGURE 13.41** Probability as the area under a curve

In geometric terms, the probability $P(a \leq X \leq b)$ is the area under the graph of f over the interval $a \leq x \leq b$. (See Figure 13.41.)

Similarly, we define a **joint probability density function** for two random variables X and Y to be a continuous, non-negative function $f(x, y)$ such that

$$P[(X, Y) \text{ in } R] = \iint\limits_R f(x, y)\, dA$$

where $P[(X, Y) \text{ in } R]$ denotes the probability that (X, Y) is in the region R in the xy-plane. Note that

$$P[(X, Y) \text{ in the } xy\text{-plane}] = \int_{-\infty}^{\infty} \int_{-\infty}^{\infty} f(x, y)\, dx\, dy = 1$$

Geometrically, $P[(X, Y) \text{ in } R]$ may be thought of as the volume under the surface $z = f(x, y)$ above the region R.

The techniques for constructing joint probability density functions from experimental data are outside the scope of this text and are discussed in many texts on probability and statistics. The use of a double integral to compute a probability with a given joint density function is illustrated in the next example.

EXAMPLE 6   *Compute a probability*

Suppose X measures the time (in minutes) that a customer at a particular grocery store spends shopping and Y measures the time the customer spends in the checkout

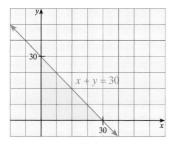

■ **FIGURE 13.42** Triangle comprised of all points (X, Y) such that $X + Y \le 30, X \ge 0, Y \ge 0$

line. A study suggests that the joint probability function for X and Y may be modeled by

$$f(x, y) = \begin{cases} \frac{1}{200} e^{-x/10} e^{-y/20} & \text{for } x \ge 0 \text{ and } y \ge 0 \\ 0 & \text{otherwise} \end{cases}$$

Find the probability that the customer's total time in the store will be no greater than 30 min.

Solution   The goal is to find the probability that $X + Y \le 30$. Stated geometrically, we wish to find the probability that a randomly selected point (x, y) lies in the region R in the first quadrant that is bounded by the coordinate axes and the line $x + y = 30$ (see Figure 13.42). This probability is given by the double integral

$$P[(X, Y) \text{ is in } R] = \iint_R f(x, y) \, dA = \int_0^{30} \int_0^{30-x} \frac{1}{200} e^{-x/10} e^{-y/20} \, dy \, dx$$

$$= \frac{1}{200} \int_0^{30} e^{-x/10} \left[\frac{e^{-y/20}}{-1/20} \right] \Big|_0^{30-x} dx$$

$$= \frac{-20}{200} \int_0^{30} e^{-x/10} [e^{-(1/20)(30-x)} - 1] \, dx$$

$$= \frac{-1}{10} \left[\frac{e^{-3/2} e^{-x/20}}{-\frac{1}{20}} - \frac{e^{-x/10}}{-\frac{1}{10}} \right] \Big|_0^{30}$$

$$= e^{-3} - 2e^{-3/2} + 1$$

$$\approx 0.6035$$

Thus, it is about 60% likely that the shopper will spend no more than 30 minutes in the store. ■

Many times we use a complementary property of probability that states

$$P(E) + P(\overline{E}) = 1$$

where E and $\overline{E}$ together constitute the set of all possible outcomes (that is, they are *complementary* probabilities). For Example 6, since the events $E = \{$the shopper will spend no more than 30 minutes in the store$\}$ and $\overline{E} = \{$the shopper will spend more than 30 minutes in the store$\}$, we can find the probability that the shopper will spend more than 30 minutes in the store by using this complementary property:

$$1 - 0.6035 = 0.3965$$

13.6 Problem Set

 1. ■ **What Does This Say?** Discuss the procedure for finding the center of mass of a lamina.

2. ■ **What Does This Say?** Discuss a procedure for finding moments in three dimensions.

3. ■ **What Does This Say?** Discuss the terms center of mass and centroid.

4. ■ **What Does This Say?** Discuss moments of inertia.

Find the centroid for the regions described in Problems 5–12.

5. A lamina with $\delta = 5$ over the rectangle with vertices $(0, 0)$, $(3, 0), (3, 4), (0, 4)$

6. A lamina with $\delta = 4$ over the region bounded by the curve $y = \sqrt{x}$ and the line $x = 4$ in the first quadrant

7. A lamina with $\delta = 2$ over the region between the line $y = 2x$ and the parabola $y = x^2$

8. A lamina with $\delta = 4$ over the region bounded by $y = \sin \frac{\pi}{2} x, x = 0, y = 0, x = \frac{1}{2}$

9. A thin homogeneous plate of density $\delta = 1$ with the shape of the region bounded above by the parabola $y = 2 - 3x^2$ and below by the line $3x + 2y = 1$

10. The part of the spherical solid with density $\delta = 2$ described by $x^2 + y^2 + z^2 \le 9, x \ge 0, y \ge 0, z \ge 0$

11. The solid tetrahedron of density $\delta = 4$ bounded by the plane $x + y + z = 4$ in the first octant.

12. The solid bounded by the surface $z = \sin x, x = 0, x = \pi$, $y = 0, z = 0$, and $y + z = 1$, where the density is $\delta = 1$.

Use double integration in Problems 13–18 to find the center of mass of a lamina covering the given region in the plane and having the specified density δ.

13. $\delta(x, y) = x^2 + y^2$ over $x^2 + y^2 \leq 9, y \geq 0$.

14. $\delta(x, y) = k(x^2 + y^2)$ over $x^2 + y^2 \leq a^2, y \geq 0$.

15. $\delta(x, y) = 7x$ over the triangle with vertices $(0, 0), (6, 5)$, and $(12, 0)$.

16. $\delta(x, y) = 3x$ over the region bounded by $y = 0, y = x^2$, and $x = 6$.

17. $\delta(x, y) = x^{-1}$ over the region bounded by $y = \ln x$, $y = 0, x = 2$.

18. $\delta(x, y) = y$ over the region bounded by $y = e^{-x}, x = 0$, $x = 2, y = 0$.

19. A lamina in the xy-plane has the shape of the semicircular region $x^2 + y^2 \leq a^2, y \geq 0$. Find the center of mass if the density at any point in the lamina is:
    a. directly proportional to the distance of the point from the origin.
    b. directly proportional to the polar angle.

20. A lamina has the shape of a semicircular region $x^2 + y^2 \leq a^2, y \geq 0$. Find the center of mass of the lamina if the density at each point is directly proportional to the square of the distance from the point to the origin.

21. Find the center of mass of a lamina that covers the region bounded by the curve $y = \ln x$ and the lines $x = e^2$ and $y = 0$ if the density at each point (x, y) is $\delta = 1$.

22. Find the center of mass of the solid bounded above by the elliptic paraboloid $z = x^2 + y^2$, on the sides by the cylinder $x^2 + y^2 = 9$ and the plane $x = 0$, and below by $z = 0$, where $\delta(x, z) = x^2 + y^2 + z^2$.

23. Find I_x, the moment of inertia about the x-axis, of the lamina that covers the region bounded by the graph of $y = 1 - x^2$ and the x-axis, if the density is $\delta(x, y) = x^2$.

24. Find I_z, the moment of inertia about the z-axis, of the lamina that covers the square in the plane with vertices $(-1, -1), (1, -1), (1, 1)$, and $(-1, 1)$, if the density is $\delta(x, y) = x^2 y^2$.

Ⓑ 25. Find the center of mass of the cardioid $r = 1 + \sin \theta$ if the density at each point (r, θ) is $\delta(r, \theta) = r$.

26. Find the center of mass of the loop of the lemniscate $r^2 = 2 \sin 2\theta$ that lies in the first quadrant, for density $\delta = 1$.

27. Find the center of mass (correct to the nearest hundredth) of the part of the large loop of the limaçon $r = 1 + 2 \cos \theta$ that does not include the small loop. Assume that $\delta = 1$.

28. Find the center of mass of the lamina that covers the triangular region with vertices $(0, 0), (a, 0), (a, b)$, if a and b

are both positive and the density at $P(x, y)$ is directly proportional to the distance of P from the y-axis.

29. **THINK TANK PROBLEM**   A solid has the shape of the rectangular parallelepiped given by $-a \leq x \leq a$, $-b \leq y \leq b, -c \leq z \leq c$, and its density is $\delta(x, y, z) = x^2 y^2 z^2$.
    a. Guess the location of the center of mass and value of the moment of inertia about the z-axis.
    b. Check your response to part **a** by direct calculation.

30. A rectangular lamina has vertices $(0, 0), (a, 0), (a, b)$, $(0, b)$, and its density at any point (x, y) is the product $\delta(x, y) = xy$. Find the center of mass of the plate.

31. A lamina of density $\delta = 1$ covers the circular disk with boundary $x^2 + y^2 = ax$. Find the moment of inertia of this circular plate about a diameter passing through the center of the lamina. *Hint:* Use polar coordinates.

32. Show that the lamina of density $\delta = 1$ that covers the circular region $x^2 + y^2 = a^2$ and has mass m, will have moment of inertia $ma^2/4$ with respect to both the x- and y-axes. What is the moment of inertia with respect to the z-axis?

33. Show that a lamina that covers the ellipse
$$\frac{x^2}{a^2} + \frac{y^2}{b^2} \leq 1$$
with mass m and density $\delta = 1$ has moment of inertia about the x-axis equal to
$$I_x = \frac{\pi a b^3 \delta}{4} = \tfrac{1}{4} m b^2$$
Note: $m = \delta(\pi a b)$.
Area of ellipse

34. Find the center of mass of the tetrahedron in the first octant bounded by the plane
$$\frac{x}{a} + \frac{y}{b} + \frac{z}{c} = 1$$
where a, b, and c are all positive constants. Assume the density is $\delta = x$.

35. A solid has the shape of the sphere $x^2 + y^2 + z^2 \leq a^2$. Find the centroid of the part of the solid in the first octant $(x \geq 0, y \geq 0, z \geq 0)$. Assume $\delta = 1$.

36. Suppose the joint probability density function for the random variables X and Y is
$$f(x, y) = \begin{cases} 2e^{-2x} e^{-y} & \text{if } x \geq 0, y \geq 0 \\ 0 & \text{otherwise} \end{cases}$$
Find the probability that $X + Y \leq 1$.

37. Suppose the joint probability density function for the random variables X and Y is
$$f(x, y) = \begin{cases} xe^{-x} e^{-y} & \text{if } x \geq 0, y \geq 0 \\ 0 & \text{otherwise} \end{cases}$$
Find the probability that $X + Y \leq 1$.

38. **MODELING PROBLEM**   Suppose X measures the length of time (in days) that a person stays in the hospital

after abdominal surgery, and Y measures the length of time (in days) that a person stays in the hospital after orthopedic surgery. On Monday, the patient in bed 107A undergoes an emergency appendectomy (abdominal surgery), while the patient's roommate in bed 107B undergoes orthopedic surgery for the repair of torn knee cartilage. If the joint probability density function for X and Y is

$$f(x, y) = \begin{cases} \frac{1}{6}e^{-x/2}e^{-y/3} & \text{if } x \geq 0, y \geq 0 \\ 0 & \text{otherwise} \end{cases}$$

find the probability (to the nearest hundredth) that both patients will be discharged from the hospital within 3 days.

39. **MODELING PROBLEM** Suppose X measures the time (in minutes) that a person stands in line at a certain bank and Y, the duration (in minutes) of a routine transaction at the teller's window. You arrive at the bank to deposit a check. If the joint probability density function for X and Y is modeled by

$$f(x, y) = \begin{cases} \frac{1}{8}e^{-x/2}e^{-y/4} & \text{if } x \geq 0, y \geq 0 \\ 0 & \text{otherwise} \end{cases}$$

find the probability that you will complete your business at the bank within 8 min.

40. **MODELING PROBLEM** Suppose X measures the time (in minutes) that a person spends with an insurance agent choosing a life insurance policy and Y, the time (in minutes) that the agent spends doing the paperwork once the client has decided. You arrange to meet with an insurance agent to buy a life insurance policy. If the joint probability density function for X and Y is

$$f(x, y) = \begin{cases} \frac{1}{300}e^{-x/30}e^{-y/10} & \text{if } x \geq 0, y \geq 0 \\ 0 & \text{otherwise} \end{cases}$$

find the probability that the entire transaction will take less than half an hour.

41. **MODELING PROBLEM** Racing yachts, such as those in the America's Cup competition, benefit from sophisticated, computer-enhanced construction techniques.* For example, define the *center of pressure* on a boat's sail as the point $(\bar{x}, \bar{y})$ where all aerodynamic forces appear to act. Suppose a sail occupies a region R in the plane, as illustrated in Figure 13.43. It can be shown that $\bar{x}$ and $\bar{y}$ are given by the formulas

$$\bar{x} = \frac{\displaystyle\iint_R xy \, dA}{\displaystyle\iint_R y \, dA} \qquad \bar{y} = \frac{\displaystyle\iint_R y^2 \, dA}{\displaystyle\iint_R y \, dA}$$

Calculate the center of pressure on the triangular sail shown in Figure 13.43.

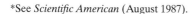

*See *Scientific American* (August 1987).

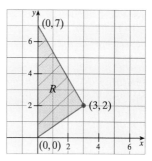

■ **FIGURE 13.43** Triangular sail

*The **average value** of the continuous function f over R is given by*

One variable: $\dfrac{1}{\text{length of segment } R} \displaystyle\int_R f(x) \, dx$

Two variables: $\dfrac{1}{\text{area of region } R} \displaystyle\iint_R f(x, y) \, dA$

Three variables: $\dfrac{1}{\text{volume of solid } R} \displaystyle\iiint_R f(x, y, z) \, dV$

Use these definitions in Problems 42–45.

42. Find the average value of $f(x, y) = e^{x^3}$, where R is the region in the first quadrant bounded by $y = x^2, y = 0$, and $x = 1$.

43. Find the average value of
$$f(x, y) = e^x y^{-1/2}$$
where R is the region in the first quadrant bounded by $y = x^2, x = 0$, and $y = 1$.

44. Find the average value of the function $f(x, y, z) = x + 2y + 3z$ over the solid region S bounded by the tetrahedron with vertices $(0, 0, 0), (1, 0, 0), (0, 1, 0)$, and $(0, 0, 1)$.

45. Find the average value of the function $f(x, y, z) = xyz$ over the solid sphere $x^2 + y^2 + z^2 \leq 1$.

46. Find the centroid of the solid bounded by the xy-plane and the surface $z = \exp(4x + 2y - x^2 - y^2)$. Note that this solid has infinite extent.

47. **MODELING PROBLEM** In a psychological experiment, x units of stimulus A and y units of stimulus B are applied to a subject, whose performance on a certain task is modeled by the function
$$f(x, y) = 10 + xye^{1-x^2-y^2}$$
Suppose the stimuli are controlled in such a way that the subject is exposed to every possible combination (x, y) with $x \geq 0, y \geq 0$, and $x + y \leq 1$. What is the subject's average response to the stimuli?

*The **radius of gyration** for revolving a region R with mass m about an axis of rotation L, with moment of inertia I, is*

$$d = \sqrt{\frac{I}{m}}$$

Note that if the entire mass m of R is located at a distance d from the axis of rotation L, then R would have the same motion of inertia. Use this definition in Problems 48–50.

48. A lamina has the shape of the right triangle in the xy-plane with vertices $(0, 0)$, $(a, 0)$, and $(0, b)$, $a > 0, b > 0$. Find the radius of gyration of the lamina about the z-axis. Assume $\delta = 1$.

49. Find the radius of gyration about the x-axis of the semicircular region $x^2 + y^2 \leq a$, $y \geq 0$, given that the density at (x, y) is directly proportional to the distance of the point from the x-axis.

50. Let R be the lamina bounded by the parabola $y = x^2$ and the lines $x = 2$ and $y = 1$, with density $\delta(x, y) = x^2 y$. What is the radius of gyration about the x-axis?

51. **MODELING PROBLEM** An industrial plant is located on a narrow river. Suppose C_0 units of pollutant are released into the river at time $t = 0$ and that the concentration of pollutant t hours later at a point x miles downstream from the plant is modeled by the diffusion function

$$C(x, t) = \frac{C_0}{\sqrt{k\pi t}} e^{-x^2/(4kt)}$$

where k is a physical constant.
   a. At what time $t_m(x_0)$ does the maximum pollution occur at point $x = x_0$ miles from the plant? What is the maximum concentration $C_m(x_0)$?
   b. Define the *danger zone* to be the portion of the riverbank such that $0 \leq x \leq x_m$ where x_m is the largest value of x such that $C_m(x_m) \geq 0.25 C_0$. Find x_m.
   c. *Set up* a double integral for the average concentration of pollutant over the set of all (x, t) such that $0 \leq t \leq t_m(x)$ for each fixed x between $x = 0$ and $x = x_m$.
   d. How would you define the "dangerous period" for the pollution spill?

52. **MODELING PROBLEM** The stiffness of a horizontal beam is modeled to be proportional to the moment of inertia of its cross section with respect to a horizontal line L through its centroid, as illustrated for three shapes shown in Figure 13.44.

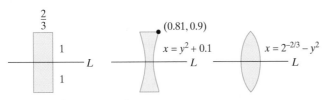

a. rectangular   **b.** concave   **c.** convex

■ **FIGURE 13.44** Cross sections of horizontal beams with area 4/3

   Which of the illustrated beams is the stiffest? For this model, assume the constant of proportionality is the same for all three cases and that $\delta = 1$.

Often the knowledge of the center of mass allows us to greatly simplify a problem, but sometimes it can lead us to an incorrect answer. We explore this in Problems 53–54.

53. **Centers of mass in physics**
   a. Suppose a volume of liquid (or granular solid) is located so that its center of mass is at the origin. The material is to be lifted to a height of h units above the center of mass. Starting with a small element of volume, ΔV, to be lifted to height h, derive the formula for the total work done:

$$\text{Work} = \delta \iiint (h - z)\, dV$$

   where δ is the weight (density) of the material.
   b. Explain why the integral in part **a** simplifies to the formula

$$\text{Work} = \delta V h = \text{Force} \times \text{distance}$$

   where h is the distance between the center of mass and the level to which the material is lifted.
   c. Picture a cylindrical tank of radius 6 ft and height 10 ft positioned so its center of mass is at the origin. Compute the work required to lift (that is, pump) the contents (of density δ) to a level of 15 ft above the center of mass. Compute this two ways: by the integral in part **a** (suggestion—do the $dx\, dy$ integral by inspection), and also by simply lifting the center of mass to the required height. These results should agree.

54. **Newton's inverse square law** Recall that if two masses m and M are each concentrated at (or nearly at) a point and are separated by a distance p, then the attractive force is expressed by $F = GmM/p^2$. What physicists and others prefer to do in the case of real-life masses (which occupy some volume) is to use p, the distance between the two centers of mass. The question you are to explore here (and again in Section 13.7) is whether, and when, this simplifying way of computing attracting forces is justified.
   a. Suppose one point mass, m, is located at $(0, 0, h)$ and the second, M, at $(0, 0, 0)$. We are interested in the total *resultant* force that M exerts on m, and we assume this to be in the z-direction. Hence we will sum the vertical components of the forces acting on m. Consider, in M, a small element of volume ΔV (at x, y, z) and argue that the magnitude of the force exerted on m and its vertical component are:

$$\Delta F_{\text{mag}} = \frac{Gm\Delta V}{p^2}$$

$$\Delta F = \frac{Gm\Delta V(h - z)^2}{p^3}$$

55. Find the moment of inertia of a rectangular lamina with dimensions h and l about an axis L through its center of mass. Assume the lamina has density $\delta = 1$.

56. Prove the following area theorem of Pappus:
Let C be a curve of length L in the plane. Then the surface obtained by rotating C about the axis L in the plane has area $2\pi Lh$, where h is the distance from the centroid of C to the axis of rotation.

57. A torus (doughnut) can be formed by rotating the circle $(x - b)^2 + y^2 = a^2$ for $b > a$ about the y-axis. Find the surface area of the torus by applying Pappus' area theorem (see Problem 56 and Figure 13.45). Compare your result with the area found parametrically in Problem 46, Section 13.4.

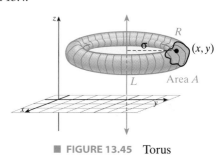

■ **FIGURE 13.45** Torus

13.7 Cylindrical and Spherical Coordinates

IN THIS SECTION cylindrical coordinates, integration with cylindrical coordinates, spherical coordinates, integration with spherical coordinates ■

CYLINDRICAL COORDINATES

Cylindrical coordinates are a generalization of polar coordinates in $\mathbb{R}^3$. Recall that the point P with Cartesian coordinates (x, y, z) is located z units above the point $Q(x, y)$ in the xy-plane (below if $z < 0$). In cylindrical coordinates, we measure the point in the xy-plane in polar coordinates, with the same z-coordinate as in the Cartesian coordinate system. These relationships are shown in Figure 13.46.

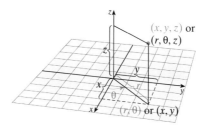

■ **FIGURE 13.46** The cylindrical coordinate system

Cylindrical coordinates are convenient for representing cylindrical surfaces and surfaces of revolution for which the z-axis is the axis of symmetry. Some examples are shown in Figure 13.47.

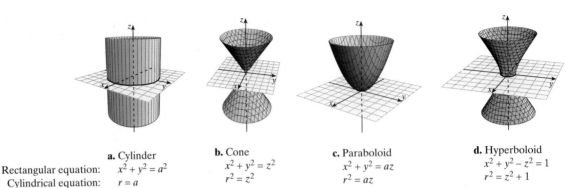

	a. Cylinder	**b.** Cone	**c.** Paraboloid	**d.** Hyperboloid
Rectangular equation:	$x^2 + y^2 = a^2$	$x^2 + y^2 = z^2$	$x^2 + y^2 = az$	$x^2 + y^2 - z^2 = 1$
Cylindrical equation:	$r = a$	$r^2 = z^2$	$r^2 = az$	$r^2 = z^2 + 1$

■ **FIGURE 13.47**   Surfaces with convenient cylindrical coordinates

We have the following conversion formulas, which follow directly from the rectangular–polar conversions.

Conversion Formulas: Rectangular–Cylindrical Coordinates

Cylindrical to rectangular: $x = r \cos \theta$
(r, θ, z) to (x, y, z) $y = r \sin \theta$
                                $z = z$

Rectangular to cylindrical: $r = \sqrt{x^2 + y^2}$

(x, y, z) to (r, θ, z) $\tan \theta = \dfrac{y}{x}$

                                $z = z$

> **EXAMPLE 1**   *Rectangular-form equation converted to cylindrical-form equation*

Find an equation in cylindrical coordinates for the elliptical paraboloid $z = x^2 + 3y^2$.

Solution   We use the conversion formulas $x = r \cos \theta$ and $y = r \sin \theta$.

$$z = x^2 + 3y^2 = (r \cos \theta)^2 + 3(r \sin \theta)^2$$
$$= r^2(\cos^2\theta + 3 \sin^2\theta)$$
$$= r^2[(1 - \sin^2\theta) + 3 \sin^2\theta]$$
$$= r^2(1 + 2 \sin^2\theta) \qquad ■$$

INTEGRATION WITH CYLINDRICAL COORDINATES

To perform triple integration with cylindrical coordinates, we simply apply our results from Section 13.3 to convert from rectangular to polar form. Even though there are six possible orders of integration, we will focus on the one for which the region is bounded below and above by $u(r, \theta) \le z \le v(r, \theta)$. In this case, we let D be the region in the xy-plane described by polar coordinates and replace dV by $dz(r\, dr\, d\theta)$, which we write as $r\, dz\, dr\, d\theta$. See Figure 13.48.

WARNING   Recall from Theorem 13.4 that in polar coordinates, $dA = r\, dr\, d\theta$.

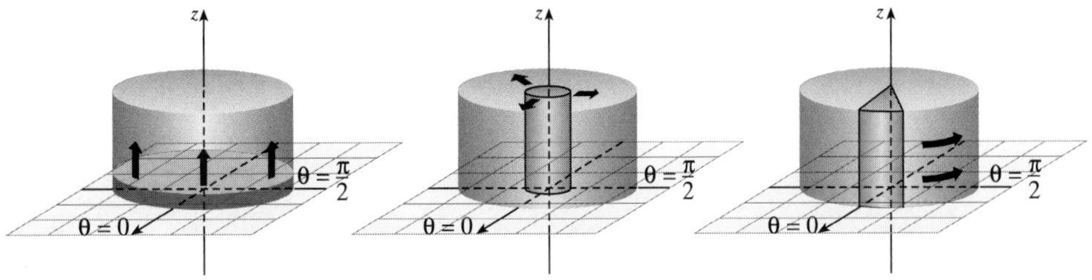

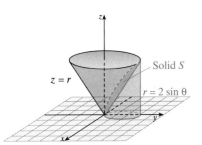

Integrate with respect to z.   Integrate with respect to r.   Integrate with respect to θ.

■ **FIGURE 13.48**   Integration with cylindrical coordinates

Triple Integral in Cylindrical Coordinates

Let R be a solid with upper surface $z = v(r, \theta)$ and lower surface $z = u(r, \theta)$, and let D be the projection of the solid on the xy-plane expressed in polar coordinates. Then, if $f(r, \theta, z)$ is continuous on R, we have

$$\iiint\limits_{R} f(r, \theta, z) \, dV = \iint\limits_{D} \int_{u(r, \theta)}^{v(r, \theta)} f(r, \theta, z) \, r \, dz \, dr \, d\theta$$

EXAMPLE 2   *Finding volume in cylindrical coordinates*

Find the volume of the solid in the first octant that is bounded by the cylinder $x^2 + y^2 = 2y$, the cone $z = \sqrt{x^2 + y^2}$, and the xy-plane.

Solution   Let S be the region occupied by the solid, as shown in Figure 13.49. This surface is most easily described in cylindrical coordinates.

Cylinder:	Cone:
$x^2 + y^2 = 2y$	$z = \sqrt{x^2 + y^2}$
$r^2 = 2r \sin \theta$	$z = r$
$z = 2 \sin \theta$	

Since the region S lies in the first octant, we have $0 \le \theta \le \frac{\pi}{2}$, so S may be described by

$$0 \le z \le r \qquad 0 \le r \le 2 \sin \theta \qquad 0 \le \theta \le \tfrac{\pi}{2}$$

$$V = \iiint\limits_{R} dV = \iint\limits_{D} \int_{u(r, \theta)}^{v(r, \theta)} r \, dz \, dr \, d\theta = \int_{0}^{\pi/2} \int_{0}^{2 \sin \theta} \int_{0}^{r} r \, dz \, dr \, d\theta$$

$$= \int_{0}^{\pi/2} \int_{0}^{2 \sin \theta} r^2 \, dr \, d\theta = \int_{0}^{\pi/2} \frac{r^3}{3} \Big|_{r=0}^{r=2 \sin \theta} d\theta = \frac{8}{3} \int_{0}^{\pi/2} \sin^3 \theta \, d\theta$$

$$= \frac{8}{3} \left[-\cos \theta + \frac{\cos^3 \theta}{3} \right] \Big|_{0}^{\pi/2} = \frac{16}{9} \qquad \text{Integration table (formula 350)} \quad ■$$

EXAMPLE 3   *Centroid in cylindrical coordinates*

A homogeneous solid S (δ constant) is bounded below by the xy-plane, on the sides by the cylinder $x^2 + y^2 = a^2 (a > 0)$, and above by the surface $z = x^2 + y^2$. Find the centroid of the solid.

■ **FIGURE 13.49**   The solid bounded by $x^2 + y^2 = 2y$, $z = \sqrt{x^2 + y^2}$, and the xy-plane

Solid S

$z = r$

$r = 2 \sin \theta$

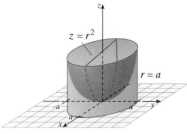

■ **FIGURE 13.50** The solid bounded by $x^2 + y^2 = a^2$ and $z = x^2 + y^2$

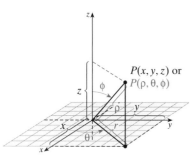

■ **FIGURE 13.51** The spherical coordinate system

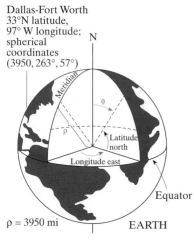

Dallas-Fort Worth 33°N latitude, 97° W longitude; spherical coordinates (3950, 263°, 57°)

$\rho = 3950$ mi   EARTH

Solution   Let S be the described solid, as shown in Figure 13.50. Because the solid is bounded by a cylinder, we will carry out the integration in cylindrical coordinates.

Cylinder:	Paraboloid:
$x^2 + y^2 = a^2$	$z = x^2 + y^2$
$r^2 = a^2$	$z = r^2$
$r = a \quad (a > 0)$	

Let $(\bar{x}, \bar{y}, \bar{z})$ denote the centroid. By symmetry, we have $\bar{x} = \bar{y} = 0$. Let m denote the mass of S. Since the projected region is $r = a$ for $0 \le \theta \le 2\pi$, we find that

$$\bar{z} = \frac{M_{xy}}{m} = \frac{\iiint\limits_S zr\, dz\, dr\, d\theta}{\iiint\limits_S r\, dz\, dr\, d\theta} = \frac{\int_0^{2\pi} \int_0^a \int_0^{r^2} zr\, dz\, dr\, d\theta}{\int_0^{2\pi} \int_0^a \int_0^{r^2} r\, dz\, dr\, d\theta} = \frac{\frac{\pi}{6}a^6}{\frac{\pi}{2}a^4} = \frac{a^2}{3}$$

Verify the details of the integration.

The centroid is $(0, 0, \frac{a^2}{3})$.   ■

SPHERICAL COORDINATES

In **spherical coordinates,** we label a point P by a triple (ρ, θ, ϕ), where ρ, θ, and ϕ are numbers determined as follows (refer to Figure 13.51):

ρ = the distance from the origin to the point P; we require $\rho \ge 0$.

θ = the polar angle (as in polar coordinates). In spherical coordinates, this is called the **azimuth** of P; we require $0 \le \theta < 2\pi$.

ϕ = the angle measured down from the positive z-axis to the ray from the origin through P. The angle ϕ is called the **colatitude** of P; we require $0 \le \phi \le \pi$.

You might recognize spherical coordinates (as well as the associated terminology) as being related to the latitude–longitude system used to identify points on the surface of the earth. In spherical coordinates, points above the xy-plane satisfy $0 \le \phi < \pi/2$ while points below satisfy $\pi/2 < \phi \le \pi$, and the equation $\phi = \pi/2$ describes the xy-plane. When this coordinate system is used to measure points on the earth, the number ρ is the distance from the center of the earth to the point with longitude θ from the prime meridian; because latitude is the angle *up* from the equator, it is measured in spherical coordinates as $\pi/2 - \phi$.

Spherical coordinates are desirable when representing spheres, cones, or certain planes. Some examples are shown in Figure 13.52.

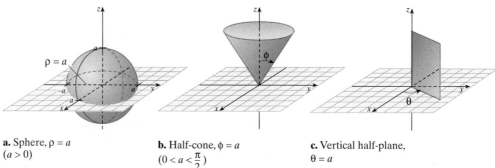

a. Sphere, $\rho = a$ $(a > 0)$

b. Half-cone, $\phi = a$ $(0 < a < \frac{\pi}{2})$

c. Vertical half-plane, $\theta = a$

■ **FIGURE 13.52** Surfaces with convenient spherical coordinates

We use the relationships in Figure 13.52 to obtain the remaining conversion formulas.

Conversion Formulas for Coordinate Systems

Spherical to
rectangular:
(ρ, θ, ϕ) to (x, y, z)

$x = \rho \sin \phi \cos \theta$
$y = \rho \sin \phi \sin \theta$
$z = \rho \cos \phi$

Rectangular to
spherical:

$\rho = \sqrt{x^2 + y^2 + z^2}$
$\tan \theta = \dfrac{y}{x}$

(x, y, z) to (ρ, θ, ϕ) $\phi = \cos^{-1}\left(\dfrac{z}{\sqrt{x^2 + y^2 + z^2}} \right)$

Spherical to
cylindrical:
(ρ, θ, ϕ) to (r, θ, z)

$r = \rho \sin \phi$
$\theta = \theta$
$z = \rho \cos \phi$

Cylindrical to
spherical:

$\rho = \sqrt{r^2 + z^2}$
$\theta = \theta$

(r, θ, z) to (ρ, θ, ϕ) $\phi = \cos^{-1}\left(\dfrac{z}{\sqrt{r^2 + z^2}} \right)$

EXAMPLE 4   *Converting rectangular-form equations to spherical-form equations*

Rewrite each of the given equations in spherical form.

a. the sphere $x^2 + y^2 + z^2 = a^2$ $(a > 0)$

b. the paraboloid $z = x^2 + 3y^2$. This is the same elliptic paraboloid we analyzed in Example 1.

Solution

a. Because $\rho = \sqrt{x^2 + y^2 + z^2}$, we see $x^2 + y^2 + z^2 = \rho^2$, so we can write

$$\rho^2 = a^2$$
$$\rho = a \qquad \textit{Because } \rho \geq 0$$

b.
$$z = x^2 + 3y^2$$
$$\rho \cos \phi = (\rho \sin \phi \cos \theta)^2 + 3(\rho \sin \phi \sin \theta)^2$$
$$\rho \cos \phi = \rho^2 \sin^2\phi \cos^2\theta + 3\rho^2 \sin^2\phi \sin^2\theta$$
$$\rho = \frac{\cos \phi}{\sin^2\phi \cos^2\theta + 3 \sin^2\phi \sin^2\theta}$$

∎

INTEGRATION WITH SPHERICAL COORDINATES

For a solid S in spherical coordinates, the fundamental element of volume is a spherical "wedge" bounded in such a way that

$$\rho_1 \leq \rho \leq \rho_1 + \Delta\rho, \qquad \phi_1 \leq \phi \leq \phi_1 + \Delta\phi, \qquad \theta_1 \leq \theta \leq \theta_1 + \Delta\theta$$

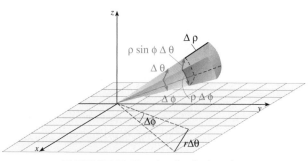

■ **FIGURE 13.53**   A spherical wedge

This "wedge" is shown in Figure 13.53. In Section 13.8, we show that the volume of the wedge is given by

$$dV = \rho^2 \sin \phi \, d\rho \, d\phi \, d\theta$$

Using this formula, we can form partitions and take a limit of a Riemann sum as the partitions are refined.

Triple Integral in Spherical Coordinates

If f is a continuous function of ρ, θ, and ϕ on a bounded, solid region S, the **triple integral of f over S** is given by

$$\iiint\limits_{S} f(\rho, \theta, \phi) \, dV = \iiint\limits_{S'} f(\rho, \theta, \phi) \, \rho^2 \sin \phi \, d\rho \, d\theta \, d\phi$$

where S' is the region S expressed in spherical coordinates.

EXAMPLE 5   *Volume of a sphere*

Find the volume of the sphere described by $x^2 + y^2 + z^2 = 9$.

Solution   It seems clear that we should work in spherical coordinates, because the equation of the sphere is $\rho = 3$ for $0 \leq \theta \leq 2\pi$ and $0 \leq \phi \leq \pi$.

$$V = \iiint\limits_{S} dV = \int_0^{\pi} \int_0^{2\pi} \int_0^3 \rho^2 \sin \phi \, d\rho \, d\theta \, d\phi$$

$$= \int_0^{\pi} \int_0^{2\pi} \frac{\rho^3}{3} \sin \phi \, \Big|_{\rho=0}^{\rho=3} \, d\theta \, d\phi = \int_0^{\pi} \int_0^{2\pi} 9 \sin \phi \, d\theta \, d\phi$$

$$= \int_0^{\pi} 18\pi \sin \phi \, d\phi = 18\pi(-\cos \phi) \Big|_0^{\pi} = 36\pi$$ ■

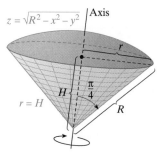

$z = \sqrt{R^2 - x^2 - y^2}$ | Axis

r

H $\dfrac{\pi}{4}$

$r = H$ R

■ **FIGURE 13.54**   Moment of inertia for a spinning top

EXAMPLE 6   *Moment of inertia using spherical coordinates*

A toy top of constant density d_0 is constructed from a portion of a solid hemisphere with a conical base, as shown in Figure 13.54. The center of the spherical cap is at the point where the top spins, and the height of the conical base is equal to its radius. Find the moment of inertia of the top about its axis of symmetry.

Solution   We use a Cartesian coordinate system in which the z-axis is the axis of symmetry of the top. Suppose the cap is part of the hemisphere $z = \sqrt{R^2 - x^2 - y^2}$.

Since the height of the conical base is equal to its radius, the cone makes an angle of $\pi/4$ radians (45°) with the z-axis. Let S denote the solid region occupied by the top. Then the moment of inertia I_z with respect to the z-axis is given by

$$I_z = \iiint_S (x^2 + y^2)\, dm = \iiint_S (x^2 + y^2) d_0\, dV$$

The shape of the top suggests that we convert to spherical coordinates, and we find that S can be described as

$$0 \le \theta < 2\pi \qquad 0 \le \phi \le \tfrac{\pi}{4} \qquad 0 \le \rho \le R$$

Also,

$$x^2 + y^2 = \rho^2 \sin^2\phi \cos^2\theta + \rho^2 \sin^2\phi \sin^2\theta = \rho^2 \sin^2\phi$$

We now evaluate the integral using spherical coordinates:

$$I_z = \iiint_S (x^2 + y^2) d_0\, dV$$

$$= d_0 \int_0^{\pi/4} \int_0^{2\pi} \int_0^{R} \underbrace{\rho^2 \sin^2\phi}_{x^2+y^2}\; \underbrace{\rho^2 \sin\phi\, d\rho\, d\theta\, d\phi}_{dV} = d_0 \int_0^{\pi/4} \int_0^{2\pi} \int_0^{R} \rho^4 \sin^3\phi\, d\rho\, d\theta\, d\phi$$

$$= d_0 \int_0^{\pi/4} \int_0^{2\pi} \frac{\rho^5}{5} \sin^3\phi \Big|_{\rho=0}^{\rho=R} d\theta\, d\phi = \frac{R^5 d_0}{5} \int_0^{\pi/4} (2\pi - 0)\sin^3\phi\, d\phi$$

$$= \frac{2R^5 d_0 \pi}{5}[-\cos\phi + \tfrac{1}{3}\cos^3\phi]\Big|_0^{\pi/4} = \frac{2R^5\, d_0\pi}{5}[-\tfrac{1}{2}\sqrt{2} + \tfrac{1}{12}\sqrt{2} + 1 - \tfrac{1}{3}]$$

$$= \frac{\pi d_0 R^5}{30}(8 - 5\sqrt{2}) \approx 0.09728 d_0 R^5 \qquad\blacksquare$$

13.7 Problem Set

Ⓐ **1.** ■ **What Does This Say?** Compare and contrast the rectangular, cylindrical, and spherical coordinate systems.

2. ■ **What Does This Say?** Suppose you need to find a particular volume. Discuss some criteria for choosing a coordinate system.

In Problems 3–16, round the coordinates to the nearest hundredth.

In Problems 3–6, convert from rectangular coordinates to
a. *cylindrical* **b.** *spherical*

3. $(0, 4, \sqrt{3}\,)$

4. $(\sqrt{2}, -2, \sqrt{3}\,)$

5. $(1, 2, 3)$

6. (π, π, π)

In Problems 7–10, convert from cylindrical coordinates to
a. *rectangular* **b.** *spherical*

7. $\left(3, \dfrac{2\pi}{3}, -3\right)$

8. $\left(4, \dfrac{\pi}{6}, -2\right)$

9. $\left(2, \dfrac{\pi}{4}, \pi\right)$

10. (π, π, π)

In Problems 11–16, convert from spherical coordinates to
a. *rectangular* **b.** *cylindrical*

11. $\left(2, \dfrac{\pi}{6}, \dfrac{2\pi}{3}\right)$

12. $\left(3, \dfrac{\pi}{4}, \dfrac{\pi}{6}\right)$

13. $\left(1, \dfrac{\pi}{6}, 0\right)$

14. $\left(2, \dfrac{\pi}{3}, \dfrac{\pi}{4}\right)$

15. $(1, 2, 3)$

16. (π, π, π)

Convert each equation in Problems 17–20 to cylindrical coordinates and sketch its graph in $\mathbb{R}^3$.

17. $z = x^2 - y^2$

18. $x^2 - y^2 = 1$

19. $\dfrac{x^2}{4} - \dfrac{y^2}{9} + z^2 = 0$

20. $z = x^2 + y^2$

Convert each equation in Problems 21–24 to spherical coordinates and sketch its graph in $\mathbb{R}^3$.

21. $z^2 = x^2 + y^2, z \ge 0$

22. $2x^2 + 2y^2 + 2z^2 = 1$

23. $4z = x^2 + 3y^2$

24. $x^2 + y^2 - 4z^2 = 1$

Convert each equation in Problems 25–30 to rectangular coordinates and sketch its graph in $\mathbb{R}^3$.

25. $z = r^2 \sin 2\theta$ **26.** $r = \sin \theta$

27. $z = r^2 \cos 2\theta$ **28.** $\rho^2 \sin^2 \phi = 1$

29. $\rho^2 \sin \phi \cos \phi \cos \theta = 1$ **30.** $\rho = \sin \phi \cos \theta$

Evaluate each iterated integral in Problems 31–38.

31. $\int_0^\pi \int_0^2 \int_0^{\sqrt{4-r^2}} r \sin \theta \, dz \, dr \, d\theta$

32. $\int_0^{\pi/4} \int_0^1 \int_0^{\sqrt{r}} r^2 \sin \theta \, dz \, dr \, d\theta$

33. $\int_0^{\pi/2} \int_0^{2\pi} \int_0^2 \cos \phi \sin \phi \, d\rho \, d\theta \, d\phi$

34. $\int_0^{\pi/2} \int_0^{\pi/4} \int_0^{\cos \phi} \rho^2 \sin \phi \, d\rho \, d\theta \, d\phi$

35. $\int_0^{2\pi} \int_0^4 \int_0^1 zr \, dz \, dr \, d\theta$

36. $\int_{-\pi/4}^{\pi/3} \int_0^{\sin \theta} \int_0^{4 \cos \theta} r \, dz \, dr \, d\theta$

37. $\int_0^{\pi/2} \int_0^{\cos \theta} \int_0^{1-r^2} r \sin \theta \, dz \, dr \, d\theta$

38. $\int_0^{\pi/3} \int_0^{\cos \theta} \int_0^{\phi} \rho^2 \sin \theta \, d\rho \, d\phi \, d\theta$

Ⓑ 39. The point (x, y, z) lies on an ellipsoid if

$$x = aR \sin \phi \cos \theta$$
$$y = bR \sin \phi \sin \theta$$
$$z = cR \cos \phi$$

Find an equation for this ellipsoid in rectangular coordinates if R is a constant.

40. Use cylindrical coordinates to compute the integral

$$\iiint_R xy \, dx \, dy \, dz$$

where R is the cylindrical solid $x^2 + y^2 \le 1$ with $0 \le z \le 1$.

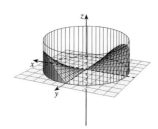

41. Use cylindrical coordinates to compute the integral

$$\iiint_R (x^4 + 2x^2y^2 + y^4) \, dx \, dy \, dz$$

where R is the cylindrical solid $x^2 + y^2 \le a^2$ with

$$0 \le z \le \frac{1}{\pi}$$

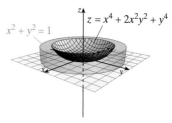

42. Use cylindrical coordinates to compute the integral

$$\iiint_S z(x^2 + y^2)^{-1/2} \, dx \, dy \, dz$$

where S is the solid bounded above by the plane $z = 2$ and below by the surface $2z = x^2 + y^2$.

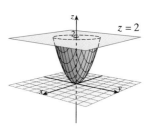

43. Find the center of mass of the solid bounded by the surface $z = \sqrt{x^2 + y^2}$ and the plane $z = 9$. Assume $\delta = 1$.

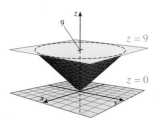

44. Find the centroid of the region bounded by the cone $z = \sqrt{x^2 + y^2}$ and the plane $z = 1$.

45. Suppose the density at each point in the hemisphere $z = \sqrt{9 - x^2 - y^2}$ is $\delta(x, y, z) = xy + z$. Set up integrals for the following quantities:
   a. the mass of the hemisphere
   b. the x-coordinate of the center of mass
   c. the moment of inertia about the z-axis

46. Find the moment of inertia about the z-axis of the portion of the homogeneous hemisphere $z = \sqrt{4 - x^2 - y^2}$ that lies between the cones $z = \sqrt{x^2 + y^2}$ and $2z = \sqrt{x^2 + y^2}$. Assume $\delta = 1$.

47. Find the average value of the function $f(x, y, z) = x + y + z$ over the sphere $x^2 + y^2 + z^2 = 4$.

48. Find the average value of θ and ϕ over the sphere $\rho \leq 3$.

49. Find the mass of the torus $\rho = 2 \sin \phi$ if the density is $\delta = \rho$.

50. Evaluate $\displaystyle\iiint\limits_{R} \sqrt{x^2 + y^2 + z^2}\, dx\, dy\, dz$ where R is defined by $x^2 + y^2 + z^2 \leq 2$.

51. Evaluate $\displaystyle\iiint\limits_{R} (x^2 + y^2 + z^2)\, dx\, dy\, dz$ where R is defined by $x^2 + y^2 + z^2 \leq 2$.

52. Evaluate $\displaystyle\iiint\limits_{S} z^2\, dx\, dy\, dz$ where S is the solid hemisphere $x^2 + y^2 + z^2 \leq 1$, $z \geq 0$.

53. Evaluate $\displaystyle\iiint\limits_{S} \frac{dx\, dy\, dz}{\sqrt{x^2 + y^2 + z^2}}$ where S is the solid sphere $x^2 + y^2 + z^2 \leq 3$.

Find the volume of the solid S given in Problems 54–58 by using integration in any convenient system of coordinates.

54. S is bounded by the surface $z = 1 - 4(x^2 + y^2)$ and the xy-plane.

55. S is bounded above by the paraboloid $z = 4 - (x^2 + y^2)$, below by the plane $z = 0$, and laterally by the cylinder $x^2 + y^2 \leq 1$.

56. S is the intersection of the solid sphere $x^2 + y^2 + z^2 \leq 9$ and the solid cylinder $x^2 + y^2 \leq 1$.

57. S is the region bounded laterally by the cylinder $r = 2 \sin \theta$, below by the plane $z = 0$, and above by the paraboloid $z = 4 - r^2$.

58. S is the region that remains in the spherical solid $\rho \leq 4$ after the solid cone $\phi \leq \dfrac{\pi}{6}$ has been removed.

59. **SPY PROBLEM**   The Spy figures out Coldfinger's puzzle (Problem 94 in the supplementary problems of Chapter 12), but just as his bonds come free, the door bursts open and in walks a group of thugs, led by . . . Purity! One of her gang starts to raise his gun. "No!" orders the erstwhile innocent maid. "I want him to think a while before he dies! Tie him again, and be quick about it. Blohardt is waiting and he is not a patient man." The Spy is tied wrist to ankle, and the thugs leave. Purity is the last to go. She pulls a lever by the door and the cave begins to fill with water. The Spy is able to pull himself into a sitting position, with his nose 3 feet above the floor. He looks around desperately and sees that the cave has a beehive shape, something like the part of the surface $\rho = 4\,(1 + \cos \phi)$ that lies above the xy-plane (where ρ is measured in ft). If the Spy needs at least 10 minutes to free himself, and if water is entering at the rate of 25 ft³/min, will he drown or survive?

60. In the previous section (Problems 53 and 54), we say that it is not always accurate to compute the attractive force between two masses by simply using the distance between their centers of mass. This approximation was applied first in astronomy, where the bodies are typically spherical in shape. As we shall see, in this case the simple calculation is very appropriate. Consider a sphere of radius a with density δ centered at the origin and a point mass m located at $(x, y, z) = (0, 0, R)$. In this problem, you are to set up the integral to compute the attractive force between the masses; and in the next problem you will do some computing.
   a. Show that the distance p between point $(x, y, z) = (0, 0, R)$ and a point in the sphere, (ρ, θ, ϕ), can be expressed $p^2 = R^2 + \rho^2 - 2R\rho \cos \phi$. *Hint:* Use the law of cosines.
   b. In the sphere, consider a small element of volume ΔV at a point (ρ, θ, ϕ). Argue that the magnitude of the force exerted on m, and the vertical component, are

$$\Delta F_{\text{mag}} = Gm(\delta \Delta V)/p^2$$

$$\Delta F = Gm(\delta \Delta V)(R - \rho \cos \phi)/p^3$$

   Note that the term $(R - \rho \cos \phi)/p$ projects the force vector onto the z-axis, giving the vertical component. Do you see why?
   c. Form the Riemann sum involving ΔF and get the following for the force acting between point mass m and the sphere of radius a:

$$F = \iiint \frac{Gm\delta(R - \rho \cos \phi)\, dV}{p^3}$$

$$= 2\pi\, Gm\delta \int_0^a \int_0^\pi p^2 \frac{(R - \rho \cos \phi) \sin \phi}{(R^2 + \rho^2 - 2R\rho \cos \phi)^{3/2}}\, d\phi\, dp$$

   Note: The integrand is independent of θ.

61. a. For an unspecified radius a and $R > a$, compute the attractive force between the masses in the previous problem using the center of mass calculation.
   b. Set a and R to numerical values and compute the attractive force by the integral in part **c** of the previous problem. Try two or three different values of R until you are satisfied that the center of mass "approximation" is exact in this case.
   c. Compare this result with the corresponding calculations of Section 13.6 and conjecture why the center of mass argument is sound in the case of the sphere and not for the rectangular region.

62. Let S be a homogeneous solid (with density 1) that has the shape of a right circular cylinder with height h and radius r. Use cylindrical coordinates to find the moment of inertia of S about its axis of symmetry.

63. Find the sum $I_x + I_y + I_z$ of the moments of inertia of the solid sphere

$$x^2 + y^2 + z^2 \le 1$$

about the coordinate axes. Assume the density is 1.

64. How much volume remains from a spherical ball of radius a when a cylindrical hole of radius b $(0 < b < a)$ is bored out of its center?

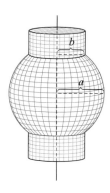

13.8 Jacobians: Change of Variables

IN THIS SECTION   change of variables in a double integral, change of variables in a triple integral

CHANGE OF VARIABLES IN A DOUBLE INTEGRAL

When the change of variable $x = g(u)$ is made in the single integral, we know

$$\int_a^b f(x)\, dx = \int_c^d f(g(u))\, g'(u)\, du$$

where the limits of integration c and d satisfy $a = g(c)$ and $b = g(d)$. To change variables in a double integral, we want to transform ("map") the region of integration D onto a region D^* that is "simpler" in some sense. In general, this process involves introducing a "mapping factor" analogous to the term $g'(u)$ in the single-variable case. This factor is called a *Jacobian* in honor of the German mathematician Karl Gustav Jacobi (1804–1851; see Historical Quest Problem 47), who made the first systematic study of change of variables in multiple integrals in the middle of the 19th century. The basic change of variable theorem may be stated as follows.

THEOREM 13.7   *Change of variables in a double integral*

Let f be a continuous function on a region D in the xy-plane, and let T be a one-to-one transformation that maps the region D^* in the uv-plane onto D under the change of variable $x = g(u, v), y = h(u, v)$, where g and h are continuously differentiable on D^*. Then

$$\iint_D f(x, y)\, dy\, dx = \iint_{D^*} f[g(u, v), h(u, v)]\, |J(u, v)|\, du\, dv$$

where $J(u, v) = \begin{vmatrix} \dfrac{\partial x}{\partial u} & \dfrac{\partial x}{\partial v} \\ \dfrac{\partial y}{\partial u} & \dfrac{\partial y}{\partial v} \end{vmatrix} = \dfrac{\partial x}{\partial u}\dfrac{\partial y}{\partial v} - \dfrac{\partial y}{\partial u}\dfrac{\partial x}{\partial v}$

is nonzero and does not change sign on D^*. The mapping factor $J(u, v)$ is called the **Jacobian** and is also denoted by $\dfrac{\partial(x, y)}{\partial(u, v)}$.

Proof   A formal proof is a matter for advanced calculus, but a geometric argument for this theorem is presented in Appendix B.   ===

EXAMPLE 1   *Finding the Jacobian*

Find the Jacobian for the change of variables from rectangular to polar coordinates, namely, $x = r \cos \theta$ and $y = r \sin \theta$.

Solution   The Jacobian of the change of variables is

$$\frac{\partial(x, y)}{\partial(r, \theta)} = \begin{vmatrix} \dfrac{\partial x}{\partial r} & \dfrac{\partial x}{\partial \theta} \\ \dfrac{\partial y}{\partial r} & \dfrac{\partial y}{\partial \theta} \end{vmatrix} = \begin{vmatrix} \cos \theta & -r \sin \theta \\ \sin \theta & r \cos \theta \end{vmatrix} = r \cos^2\theta + r \sin^2\theta = r \qquad \blacksquare$$

This result justifies the formula we have previously used:

$$\iint\limits_{D} f(x, y)\, dy\, dx = \iint\limits_{D^*} f(r \cos \theta, r \sin \theta)\, r\, dr\, d\theta$$

where D^* is the region in the (r, θ) plane that maps into region D in the (x, y) plane by the polar transform $x = r \cos \theta, y = r \sin \theta$, as illustrated in Figure 13.55.

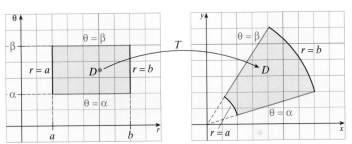

■ **FIGURE 13.55**   Transformation of a region D^* by $T: x = r \cos \theta, y = r \sin \theta$

EXAMPLE 2   *Calculating a double integral by changing variables*

Compute $\displaystyle\iint\limits_{D} \left(\frac{x - y}{x + y}\right)^4 dy\, dx$, where D is the triangular region bounded by the line $x + y = 1$ and the coordinate axes.

Solution   This is a rather difficult computation if no substitution is made. The form of the integral suggests we make the substitution

$$u = x - y \qquad v = x + y$$

Solving for x and y, we obtain

$$x = \tfrac{1}{2}(u + v) \qquad y = \tfrac{1}{2}(v - u)$$

and the Jacobian is

$$\frac{\partial(x, y)}{\partial(u, v)} = \begin{vmatrix} \dfrac{\partial}{\partial u}\left(\dfrac{u + v}{2}\right) & \dfrac{\partial}{\partial v}\left(\dfrac{u + v}{2}\right) \\ \dfrac{\partial}{\partial u}\left(\dfrac{v - u}{2}\right) & \dfrac{\partial}{\partial v}\left(\dfrac{v - u}{2}\right) \end{vmatrix} = \begin{vmatrix} \dfrac{1}{2} & \dfrac{1}{2} \\ -\dfrac{1}{2} & \dfrac{1}{2} \end{vmatrix} = \dfrac{1}{2}$$

To find the image D^* of D in the uv-plane, note that the boundary lines $x = 0$ and $y = 0$ for D map into the lines $u = -v$ and $u = v$, respectively, while $x + y = 1$ maps into $v = 1$. Therefore, the transformed region of integration D^* is the triangular region shown in Figure 13.56b, with vertices $(0, 0)$, $(1, 1)$, and $(-1, 1)$.

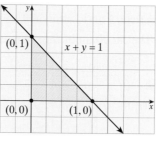

a. The region of integration D

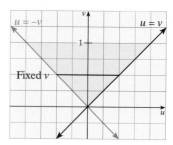

b. The transformed region D^*

■ **FIGURE 13.56**   Transformation of D to D^*

We now evaluate the given integral:

$$\iint_D \left(\frac{x - y}{x + y}\right)^4 dy\, dx = \iint_{D^*} \left(\frac{u}{v}\right)^4 \left|\frac{1}{2}\right| du\, dv = \frac{1}{2} \int_0^1 \int_{-v}^{v} u^4 v^{-4}\, du\, dv = \frac{1}{10} \qquad ■$$

In Example 2, the change of variables was chosen to simplify the integrand, but sometimes it is useful to introduce a change of variables that simplifies the region of integration.

> **EXAMPLE 3**   *Change of variable to simplify a region*

Find the area of the region E bounded by the ellipse $\dfrac{x^2}{a^2} + \dfrac{y^2}{b^2} = 1$.

Solution   The area is given by the integral

$$A = \iint_E dy\, dx$$

where E is the region shown in Figure 13.57a.

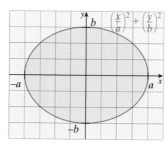

a. The elliptical region E

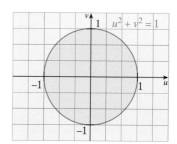

b. The transformed region C is a circular disk.

■ **FIGURE 13.57**   Transformation of the ellipse E to the circle C

Because E can be represented by

$$\left(\frac{x}{a}\right)^2 + \left(\frac{y}{b}\right)^2 \le 1$$

we consider the substitution $u = \dfrac{x}{a}$ and $v = \dfrac{y}{b}$, which will map the elliptical region E onto the circular disk

$$C: u^2 + v^2 \le 1$$

as shown in Figure 13.57b. When we solve the two equations for x and y, the Jacobian gives us the mapping (change of variable) factor:

$$\frac{\partial(x, y)}{\partial(u, v)} = \begin{vmatrix} \dfrac{\partial}{\partial u}(au) & \dfrac{\partial}{\partial v}(au) \\[2mm] \dfrac{\partial}{\partial u}(bv) & \dfrac{\partial}{\partial v}(bv) \end{vmatrix} = \begin{vmatrix} a & 0 \\ 0 & b \end{vmatrix} = ab$$

Because $ab > 0$, we have $|ab| = ab$, and the area of E is given by

$$\iint_E dy\, dx = \iint_C ab\, du\, dv = ab \iint_C du\, dv = ab[\underbrace{\pi(1)^2}] = \pi ab$$

Because C is a circle of radius 1 ∎

CHANGE OF VARIABLES IN A TRIPLE INTEGRAL

The change of variable formula for triple integrals is similar to the one given for double integrals. Let T be a change of variable that maps a region R^* in uvw-space onto a region R in xyz-space, where

$$T:\quad x = x(u, v, w) \qquad y = y(u, v, w) \qquad z = z(u, v, w)$$

Then the Jacobian of T is the determinant

$$\frac{\partial(x, y, z)}{\partial(u, v, w)} = \begin{vmatrix} \dfrac{\partial x}{\partial u} & \dfrac{\partial x}{\partial v} & \dfrac{\partial x}{\partial w} \\[2mm] \dfrac{\partial y}{\partial u} & \dfrac{\partial y}{\partial v} & \dfrac{\partial y}{\partial w} \\[2mm] \dfrac{\partial z}{\partial u} & \dfrac{\partial z}{\partial v} & \dfrac{\partial z}{\partial w} \end{vmatrix}$$

and the change of variable yields

$$\iiint_R f(x, y, z)\, dx\, dy\, dz = \iiint_{R^*} f[x(u, v, w), y(u, v, w), z(u, v, w)]\frac{\partial(x, y, z)}{\partial(u, v, w)} du\, dv\, dw$$

EXAMPLE 4   *Formula for integrating with spherical coordinates*

Obtain the formula for converting a triple integral in rectangular coordinates to one in spherical coordinates.

Solution   The conversion formulas from rectangular coordinates to spherical coordinates are:

$$x = \rho \sin \phi \cos \theta \qquad y = \rho \sin \phi \sin \theta \qquad z = \rho \cos \phi$$

The Jacobian of this transformation is

$$\frac{\partial(x, y, z)}{\partial(\rho, \theta, \phi)} = \begin{vmatrix} \dfrac{\partial}{\partial\rho}(\rho\sin\phi\cos\theta) & \dfrac{\partial}{\partial\theta}(\rho\sin\phi\cos\theta) & \dfrac{\partial}{\partial\phi}(\rho\sin\phi\cos\theta) \\[2mm] \dfrac{\partial}{\partial\rho}(\rho\sin\phi\sin\theta) & \dfrac{\partial}{\partial\theta}(\rho\sin\phi\sin\theta) & \dfrac{\partial}{\partial\phi}(\rho\sin\phi\sin\theta) \\[2mm] \dfrac{\partial}{\partial\rho}(\rho\cos\phi) & \dfrac{\partial}{\partial\theta}(\rho\cos\phi) & \dfrac{\partial}{\partial\phi}(\rho\cos\phi) \end{vmatrix}$$

$$= \begin{vmatrix} \sin\phi\cos\theta & -\rho\sin\phi\sin\theta & \rho\cos\phi\cos\theta \\ \sin\phi\sin\theta & \rho\sin\phi\cos\theta & \rho\cos\phi\sin\theta \\ \cos\phi & 0 & -\rho\sin\phi \end{vmatrix}$$

$$= -\rho^2\sin\phi \qquad \textit{After much algebra}$$

Since $0 \le \phi \le \pi$, we have $\sin\phi \ge 0$, so

$$\left|\frac{\partial(x, y, z)}{\partial(\rho, \theta, \phi)}\right| = \left|-\rho^2\sin\phi\right| = \rho^2\sin\phi$$

$$\iiint_R f(x, y, z)\, dz\, dx\, dy = \iiint_{R^*} f(\rho\sin\phi\cos\theta, \rho\sin\phi\sin\theta, \rho\cos\phi)\, \rho^2\sin\phi\, d\rho\, d\theta\, d\phi \qquad ■$$

13.8 Problem Set

Ⓐ *Find the Jacobian of the change of variables given in Problems 1–12.*

1. $x = u + v, y = uv$
2. $x = u^2, y = u + v$
3. $x = u - v, y = u + v$
4. $x = u^2 - v^2, y = 2uv$
5. $x = u^2v^2, y = v^2 - u^2$
6. $x = u\cos v, y = u\sin v$
7. $x = e^{u+v}, y = e^{u-v}$
8. $x = e^u\sin v, y = e^u\cos v$
9. $x = u + v - w; y = 2u - v + 3w;$
   $z = -u + 2v - w$
10. $x = 2u - w, y = u + 3v, z = v + 2w$
11. $x = u\cos v, y = u\sin v, z = we^{uv}$
12. $x = \dfrac{u}{v}, y = \dfrac{v}{w}, z = \dfrac{w}{u}$

A region R is given in Problems 13–16. Sketch the corresponding region R* in the uv-plane using the given transformations.

13.

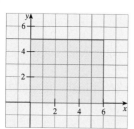

$u = x + y, v = x - y$

14.

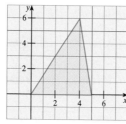

$u = 2x, v = x + y$

15.

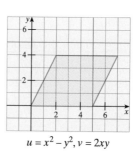

$u = x^2 - y^2, v = 2xy$

16.

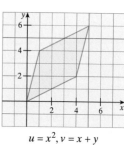

$u = x^2, v = x + y$

17. Suppose a uv-plane is mapped onto an xy-plane by the equations $x = u(1 - v), y = uv$. Express $dx\, dy$ in terms of $du\, dv$.

18. Suppose a uv-plane is mapped onto an xy-plane by the equations $x = u^2 - v^2, y = 2uv$. Express $dx\, dy$ in terms of $du\, dv$.

Use a suitable change of variable to find the area of the region specified in Problems 19 and 20.

19. The region R bounded by the hyperbolas $xy = 1$ and $xy = 4$, the lines $y = x$ and $y = 4x$.

20. The region R bounded by the parabolas $y = x^2, y = 4x^2$, $y = \sqrt{x}$, and $y = \dfrac{1}{2}\sqrt{x}$.

Let D be the region in the xy-plane that is bounded by the coordinate axes and the line $x + y = 1$. Use the change of *variable $u = x - y, v = x + y$ to compute the integrals given in Problems 21–24.*

21. $\displaystyle\iint_D \left(\frac{x-y}{x+y}\right)^5 dy\, dx$ **22.** $\displaystyle\iint_D \left(\frac{x-y}{x+y}\right)^4 dy\, dx$

23. $\displaystyle\iint_D (x-y)^5 (x+y)^3 \, dy\, dx$

24. $\displaystyle\iint_D (x-y) e^{x^2+y^2} \, dy\, dx$

Ⓑ 25. ■ **What Does This Say?** Discuss the process of changing variables in a triple integral.

26. ■ **What Does This Say?** Discuss the necessity of using the Jacobian when changing variables in a double integral.

Under the transformation

$$u = \frac{1}{5}(2x + y) \qquad v = \frac{1}{5}(x - 2y)$$

the square S in the xy-plane with vertices $(0,0), (1,-2), (3,-1), (2,1)$ is mapped onto a square in the uv-plane. Use this information to find the integrals in Problems 27–32.

27. $\displaystyle\iint_S \left(\frac{2x+y}{x-2y+5}\right)^2 dy\, dx$

28. $\displaystyle\iint_S (2x+y)(x-2y)^2 \, dy\, dx$

29. $\displaystyle\iint_S (2x+y)^2(x-2y) \, dy\, dx$

30. $\displaystyle\iint_S \sqrt{(2x+y)(x-2y)} \, dy\, dx$

31. $\displaystyle\iint_S (2x+y)\tan^{-1}(x-2y) \, dy\, dx$

32. $\displaystyle\iint_S \cos(2x+y)\sin(x-2y) \, dy\, dx$

33. Evaluate

$$\iint_R e^{(2y-x)/(y+2x)} \, dA$$

where R is the trapezoid with vertices $(0,2), (1,0), (4,0),$ and $(0,8)$.

34. Let R be the region in the xy-plane that is bounded by the parallelogram with vertices $(0,0), (1,1), (2,1),$ and $(1,0)$. Use the linear transformation $x = u + v, y = v$ to compute $\int_R \int (2x - y) \, dy\, dx$.

35. Use a suitable linear transformation $u = ax + by, v = rx + sy$ to evaluate the integral

$$\iint_R \left(\frac{x+y}{2}\right)^2 e^{(y-x)/2} \, dy\, dx$$

where R is the region inside the square with vertices $(0,0), (1,1), (0,2), (-1,1)$.

36. Under the change of variables $x = s^2 - t^2, y = 2st,$ the quarter circular region in the st-plane given by

$s^2 + t^2 \leq 1, s \geq 0, t \geq 0$ is mapped onto a certain region S of the xy-plane. Evaluate

$$\iint_S \frac{dy\, dx}{\sqrt{x^2+y^2}}$$

37. Use the change of variable $x = ar\cos\theta, y = br\sin\theta$ to evaluate

$$\iint_S \exp\left(-\frac{x^2}{a^2} - \frac{y^2}{b^2}\right) dy\, dx$$

where S is the quarter ellipse.

$$\frac{x^2}{a^2} + \frac{y^2}{b^2} \leq 1, x \geq 0, y \geq 0$$

38. A rotation of the xy-plane through the fixed angle θ is given by

$$x = u\cos\theta - v\sin\theta$$
$$y = u\sin\theta + v\cos\theta$$

(See Figure 13.58.)

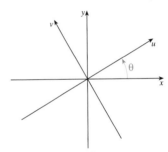

■ **FIGURE 13.58**   Rotational transformation

a. Compute the Jacobian $\dfrac{\partial(x,y)}{\partial(u,v)}$.

b. Let E denote the ellipse

$$x^2 + xy + y^2 = 3$$

Use a rotation of $\frac{\pi}{4}$ to obtain an integral that is equivalent to $\int_E \int y \, dy\, dx$. Evaluate the transformed integral.

39. Find the area of the rotated ellipse

$$5x^2 - 4xy + 2y^2 = 1$$

Note that $5x^2 - 4xy + 2y^2 = Au^2 + Bv^2,$ where A and B are constants and $u = x + 2y, v = 2x - y$.

40. **THINK TANK PROBLEM**   Let R be the region bounded by the parallelogram with vertices $(0,0), (2,0), (1,1),$ and $(-1,1)$. Show that under a suitable change of variable of the form $x = au + bv, y = cu + dv,$ we have

$$\iint_R f(x+y) \, dy\, dx = \int_0^2 f(t) \, dt$$

where f is continuous on $[0,2]$.

41. Find the Jacobian of the cylindrical coordinate transformations
$$x = r \cos \theta, \quad y = r \sin \theta, \quad z = z$$

42. Use the change of variable $x = au, y = bv, z = cw$ to find the volume of the ellipsoid
$$\frac{x^2}{a^2} + \frac{y^2}{b^2} + \frac{z^2}{c^2} = 1$$

43. Use a change of variable to find the moment of inertia about the z-axis of the ellipsoid
$$\frac{x^2}{a^2} + \frac{y^2}{b^2} + \frac{z^2}{c^2} = 1$$

44. Evaluate $\displaystyle\iint_R \ln\left(\frac{x-y}{x+y}\right) dy\, dx$, where R is the triangular region with vertices $(1, 0), (4, -3)$, and $(4, 1)$.

C 45. Show that if $ad - bc \neq 0$, the linear transformation
$$u = ax + by \quad \text{and} \quad v = cx + dy$$
can be solved for x and y to obtain
$$x = \frac{du - bv}{ad - bc} \quad \text{and} \quad y = \frac{av - cu}{ad - bc}$$

46. **THINK TANK PROBLEM** Find the volume of the solid under the surface
$$z = \frac{xy}{1 + x^2y^2}$$
over the region bounded by $xy = 1, xy = 5, x = 1$, and $x = 5$.

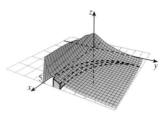

■ **FIGURE 13.59** Problem 46

47. **HISTORICAL QUEST** Karl G. Jacobi was a gifted teacher and one of Germany's most distinguished mathematicians during the first half of the 19th century. He made major contributions to the theory of elliptic functions, but his work with functional determinants is what secured his place in history. In 1841, he published a long memoir called "De determinantibus functionalibus," devoted to what we today call the Jacobian and pointing out that this determinant is in many

KARL G. JACOBI
1804–1851

ways the multivariable analogue to the differential quotient of a single variable. The memoir was published in what is usually known as *Crelle's Journal,* the first journal devoted to serious mathematics. It was begun in 1826 by August Crelle (1780–1855) with the "official" title *Journal für die reine und angewandte Mathematik.*

For this Quest, set up an integral for the circumference of the ellipse
$$\frac{x^2}{a^2} + \frac{y^2}{b^2} = 1$$
This is an example of an elliptic integral.

Chapter 13 *Review*

Proficiency Examination

Concept Problems

1. Define a double integral.
2. State Fubini's theorem over a rectangular region.
3. What is a type I region? State Fubini's theorem for a type I region.
4. What is a type II region? State Fubini's theorem for a type II region.
5. State the formula for finding an area using a double integral.
6. State the formula for finding a volume using a double integral.
7. State the following properties of a double integral:
   a. linearity rule   b. dominance rule
   c. subdivision rule

8. What is the formula for a double integral in polar coordinates?
9. Explain how to evaluate an improper integral in polar coordinates.
10. State the double integral formula for surface area.
11. State the formula for the area of a surface defined parametrically.
12. State Fubini's theorem over a parallelepiped in space.
13. Give a triple integral formula for volume.
14. Explain how to use a double integral to find the mass of a planar lamina of variable density.
15. How do you find the moment of mass with respect to the x-axis?
16. What are the formulas for the centroid of a region?

17. Define moment of inertia.

18. What is a joint probability density function?

19. Complete the following table for the conversion of coordinates.

From \ To	(r, θ, z) Cylindrical	(ρ, θ, ϕ) Spherical
(x, y, z) Rectangular	$r = \underline{\ ?\ }$ $\tan\theta = \underline{\ ?\ }$ $z = \underline{\ ?\ }$	$\rho = \underline{\ ?\ }$ $\tan\theta = \underline{\ ?\ }$ $\phi = \underline{\ ?\ }$

From \ To	(x, y, z) Rectangular	(ρ, θ, ϕ) Spherical
(r, θ, z) Cylindrical	$x = \underline{\ ?\ }$ $y = \underline{\ ?\ }$ $z = \underline{\ ?\ }$	$\rho = \underline{\ ?\ }$ $\theta = \underline{\ ?\ }$ $\phi = \underline{\ ?\ }$

From \ To	(x, y, z) Rectangular	(r, θ, z) Cylindrical
(ρ, θ, ϕ) Spherical	$x = \underline{\ ?\ }$ $y = \underline{\ ?\ }$ $z = \underline{\ ?\ }$	$r = \underline{\ ?\ }$ $\theta = \underline{\ ?\ }$ $z = \underline{\ ?\ }$

20. a. State the formula for a triple integral in cylindrical coordinates.
  b. State the formula for a triple integral in spherical coordinates.

21. Define the Jacobian for a mapping $x = g(u, v)$, $y = h(u, v)$ from u, v variables to x, y variables.

22. State the formula for the change of variables in a double integral.

Practice Problems

23. Evaluate $\displaystyle\int_0^{\pi/3} \int_0^{\sin y} e^{-x} \cos y \, dx \, dy$

24. Evaluate $\displaystyle\int_{-1}^{1} \int_0^{z} \int_y^{y-z} (x + y - z) \, dx \, dy \, dz$

25. Use a double integral to compute the area of the region R that is bounded by the x-axis and the parabola $y = 9 - x^2$.

26. Use polar coordinates to evaluate

$$\int_0^1 \int_0^{\sqrt{1-x^2}} \cos(x^2 + y^2) \, dy \, dx$$

27. MODELING PROBLEM   A certain appliance consisting of two independent electronic components will be usable as long as either one of its components is still operating. The appliance carries a warranty from the manufacturer guaranteeing replacement if the appliance becomes unusable within 1 year of the date of purchase. To model this situation, let the random variables X and Y measure the life span (in years) of the first and second components, respectively, and assume that the joint probability density function for X and Y is

$$f(x, y) = \begin{cases} \frac{1}{4} e^{-x/2} e^{-y/2} & \text{if } x \geq 0, y \geq 0 \\ 0 & \text{otherwise} \end{cases}$$

Suppose the quality assurance department selects one of these appliances at random. What is the probability that the appliance will fail during the warranty period?

28. Set up and evaluate a triple integral for the volume of the solid region in the first octant that is bounded above by the plane $z = 4x$ and below by the paraboloid $z = x^2 + 2y^2$. See Figure 13.60.

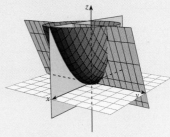

■ FIGURE 13.60   Problem 28

29. A solid S is bounded above by the plane $z = 4$ and below by the surface $z = x^2 + y^2$. Its density $\rho(x, y, z)$ at each point P is equal to the distance from P to the z-axis. Find the total mass of S.

30. Use a linear change of variables to evaluate the double integral

$$\iint\limits_R (x + y) e^{x - 2y} \, dy \, dx$$

where R is the triangular region with vertices $(0, 0)$, $(2, 0)$, $(1, 1)$.

Supplementary Problems

In Problems 1–10, sketch the region of integration, exchange the order, and evaluate the integral using either order of integration.

1. $\int_0^1 \int_0^{3x} x^2 y^2 \, dy \, dx$ **2.** $\int_0^4 \int_0^1 \sqrt{\dfrac{y}{x}} \, dx \, dy$

3. $\int_1^2 \int_0^y \dfrac{1}{x^2 + y^2} \, dx \, dy$ **4.** $\int_0^{\pi/2} \int_0^{\pi/2} \cos(x+y) \, dx \, dy$

5. $\int_0^1 \int_0^1 x \sqrt{x^2 + y} \, dx \, dy$ **6.** $\int_0^{\pi/2} \int_0^{\sqrt{\sin x}} xy \, dy \, dx$

7. $\int_0^1 \int_{-\sqrt{1-x^2}}^0 y\sqrt{x} \, dy \, dx$ **8.** $\int_0^1 \int_{\sqrt{y}}^1 \sqrt{1 - x^3} \, dx \, dy$

9. $\int_0^1 \int_{x^2}^1 x^3 \sin y^3 \, dy \, dx$

10. $\int_0^2 \int_0^{\sqrt{4-x^2}} \sqrt{4 - x^2 - y^2} \, dy \, dx$

Evaluate the integrals given in Problems 11–36.

11. $\int_0^1 \int_{\sqrt{x}}^1 e^{y^3} \, dy \, dx$

12. $\int_0^2 \int_x^2 \dfrac{y}{(x^2 + y^2)^{3/2}} \, dy \, dx$

13. $\int_0^1 \int_{-1}^4 \int_x^y z \, dz \, dy \, dx$

14. $\int_1^2 \int_0^1 \int_0^{\sqrt{1-x^2}} e\sqrt{x^2 + y^2} \, dy \, dx \, dz$

15. $\int_0^{\pi/4} \int_0^{2\pi} \int_0^\theta r^2 \sin \phi \, dr \, d\theta \, d\phi$

16. $\int_0^1 \int_0^x \int_0^y x^2 y \, dz \, dy \, dx$

17. $\int_1^2 \int_x^{x^2} \int_0^{\ln x} x e^z \, dz \, dy \, dx$

18. $\int_0^1 \int_{1-x}^{1+x} \int_0^{xy} xz \, dz \, dy \, dx$

19. $\int_{-\sqrt{3}}^{\sqrt{3}} \int_{-\sqrt{3-y^2}}^{\sqrt{3-y^2}} \int_{(x^2+y^2)^2}^9 y^2 \, dz \, dx \, dy$

Hint: Change to cylindrical coordinates.

20. $\iint_D x^2 y \, dA$, where D is the circular disk $x^2 + y^2 \le 4$

21. $\iint_D (x^2 + y^2 + 1) \, dA$, where D is the circular disk $x^2 + y^2 \le 4$

22. $\iint_D e^{x^2+y^2} \, dA$, where D is the circular disk $x^2 + y^2 \le 4$

23. $\iint_D (x^2 + y^2)^n \, dA$, where D is the circular disk $x^2 + y^2 \le 4, n \ge 0$

24. $\iint_D \dfrac{2y}{x} \, dA$, where D is the region above the line $x + y = 1$ and below the circle $x^2 + y^2 = 1$

25. $\iint_D x^3 \sqrt{4 - y^2} \, dA$, where D is the circular disk $x^2 + y^2 \le 4$

26. $\iint_D \exp\left(\dfrac{y - x}{y + x}\right) dy \, dx$, where D is the triangular region with vertices $(0,0), (2,0), (0,2)$ *Hint:* Make a suitable change of variables.

27. $\iint_D x^3 \, dA$, where D is the shaded portion of the figure

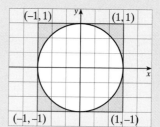

28. $\iint_D xy^2 \, dA$, where D is the shaded portion of the figure

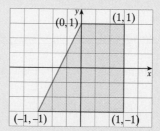

29. $\iint_P e^{-(a|x|+b|y|)} \, dA, a > 0, b > 0$, where P is the entire xy-plane. *Hint:* Use Cartesian coordinates and note that this is an improper double integral.

30. $\iint_R \dfrac{\partial^2 f}{\partial x \partial y} \, dx \, dy$, where R is the rectangle $a \le x \le b$,

$c \le y \le d$ and $\dfrac{\partial^2 f}{\partial x \partial y}$ is continuous over R with $f(a, c) = 4$, $f(a, d) = -3, f(b, c) = 1, f(b, d) = 5$

31. $\iiint\limits_{S} \sqrt{x^2 + y^2 + z^2} \, dV$, where S is the portion of the solid sphere $x^2 + y^2 + z^2 \leq 1$ that lies in the first octant

32. $\iiint\limits_{H} z^2 \, dV$, where H is the solid hemisphere $x^2 + y^2 + z^2 \leq 1$ with $z \geq 0$

33. $\iiint\limits_{H} \dfrac{dV}{\sqrt{x^2 + y^2 + z^2}}$, where H is the solid hemisphere $x^2 + y^2 + z^2 \leq 1$, with $z \geq 0$

34. $\iiint\limits_{S} \dfrac{dV}{x^2 + y^2 + z^2}$, where S is the solid region bounded below by the paraboloid $2z = x^2 + y^2$, and above by the sphere $x^2 + y^2 + z^2 = 8$

35. $\iiint\limits_{S} z(x^2 + y^2)^{-1/2} \, dV$, where S is the solid bounded by the surface $2z = x^2 + y^2$ and the plane $z = 2$

36. $\iiint\limits_{S} (x^4 + 2x^2 y^2 + y^4) \, dV$, where S is the solid cylinder given by $x^2 + y^2 \leq a^2, 0 \leq z \leq \dfrac{1}{\pi}$

37. For what values of the constant m does the improper double integral
$$\iint\limits_{R} \frac{dA}{(a^2 + x^2 + y^2)^m}$$
converge, where R is the entire xy-plane?

38. Rewrite the triple integral $\displaystyle\int_0^1 \int_0^x \int_0^{\sqrt{xy}} f(x, y, z) \, dz \, dy \, dx$ as a triple integral in the order $dy \, dx \, dz$.

39. Convert the double integral
$$\int_0^{\pi/2} \int_0^{2a\cos\theta} r \sin 2\theta \, dr \, d\theta$$
to rectangular coordinates and evaluate.

40. Reverse the order of integration
$$\int_1^2 \int_{y^2}^{y^5} e^{x/y^2} \, dx \, dy$$
and evaluate using either order.

41. Express the integral
$$\int_0^1 \int_0^y f(x, y) \, dx \, dy + \int_1^4 \int_0^{(4-y)/3} f(x, y) \, dx \, dy$$
as a double integral with the order of integration reversed.

42. Express the integral
$$\int_0^1 \int_{-\sqrt{y}}^{\sqrt{y}} f(x, y) \, dx \, dy + \int_1^3 \int_{-1}^{2-y} f(x, y) \, dx \, dy$$
as a double integral with the order of integration reversed.

43. Find numbers $a, b,$ and c so that
$$\int_{-\infty}^{\infty} \int_{-\infty}^{\infty} \exp[-(ax^2 + bxy + cy^2)] \, dA = 1$$

44. Find the area inside the circle $r = \cos\theta$ and outside the cardioid $r = 1 - \cos\theta$.

45. Find the area outside the cardioid $r = 1 + \cos\theta$ and inside the cardioid $r = 1 + \sin\theta$.

46. Find the area outside the small loop of $r = 1 + 2\sin\theta$ and inside the large loop.

47. Use a double integral to compute the volume of the tetrahedral region in the first octant that is bounded by the coordinate planes and the plane $3x + y + 2z = 6$.

48. Find the volume of the tetrahedron that is bounded by the coordinate planes and the plane
$$\frac{x}{a} + \frac{y}{b} + \frac{z}{c} = 1$$
with $a > 0, b > 0, c > 0$.

49. Find the volume of the solid that is bounded above by the paraboloid $z = x^2 + y^2$ and below by the square region $0 \leq x \leq 1, 0 \leq y \leq 1, z = 0$.

50. Find the volume of the solid that is bounded above by the surface $z = xy$, below by the xy-plane, and on the sides by the circular disk $x^2 + y^2 \leq a^2$ $(a > 0)$ that lies in the first quadrant.

51. Find the volume of the solid that is bounded above by the paraboloid $z = 4 - x^2 - y^2$ and below by the plane $z = 4 - 2x$.

52. Find the surface area of that portion of the cone $z = \sqrt{x^2 + y^2}$ that is contained in the cylinder $x^2 + y^2 = 1$.

53. Find the surface area of the portion of the paraboloid $z = x^2 + y^2$ that lies below the plane $z = 9$.

54. Find the surface area of the portion of the cylinder $x^2 + z^2 = 4$ that is bounded by the planes $x = 0, x = 1$, $y = 0$, and $y = 2$.

55. Find the volume of the region bounded by the paraboloid $y^2 + z^2 = 2x$ and the plane $x + y = 1$.

56. Find the volume of the region bounded above by the paraboloid $z = 4 - x^2 - y^2$ and below by the plane $z + 2y = 4$.

57. Find the volume of the region bounded by the ellipsoid $z^2 = 4 - 4r^2$ and the cylinder $r = \cos\theta$.

58. Find the volume of the region bounded above by the hemisphere $z = \sqrt{9 - x^2 - y^2}$, below by the xy-plane, and on the sides by the cylinder $x^2 + y^2 = 1$.

59. Show that the solid bounded below by the cone $z = \sqrt{x^2 + y^2}$ and above by the sphere $x^2 + y^2 + z^2 = 2az$ for $a > 0$ has volume $V = \pi a^3$.

60. Find the volume of the solid region bounded above by the sphere given in spherical coordinates by $\rho = a$ $(a > 0)$ and below by the cone $\phi = \phi_0$, where $0 < \phi_0 < \frac{\pi}{2}$.

61. A homogeneous solid S is bounded above by the sphere $\rho = a$ $(a > 0)$ and below by the cone $\phi = \phi_0$, where $0 < \phi_0 < \frac{\pi}{2}$. Find the moment of inertia of S about the z-axis.

62. The region between the circles $x^2 + y^2 = 1$ and $x^2 + y^2 = 4$ with $0 \le z \le 5$ forms a "washer." Find the moment of inertia of the washer about the z-axis if the density δ is given by:
a. $\delta = k$ (k a constant)   b. $\delta(x, y) = x^2 + y^2$
c. $\delta(x, y) = x^2 y^2$

63. Find $\bar{x}$, the x-coordinate of the center of mass, of a triangular lamina with vertices $(0, 0), (1, 0),$ and $(0, 1)$ if the density is $\delta(x, y) = x^2 + y^2$.

64. Find the mass of a cone of top radius R and height H if the density at each point P is proportional to the distance from P to the tip of the cone.

65. Find $\bar{z}$, the z-coordinate of the center of mass of a right circular cone with height H and base radius R.

66. Find the Jacobian $\dfrac{\partial(u, v, w)}{\partial(x, y, z)}$ of the change of variables $u = 2x - 3y + z, v = 2y - z, w = 2z$.

67. Find the Jacobian $\dfrac{\partial(u, v, w)}{\partial(x, y, z)}$ of the change of variables $u = x^2 + y^2 + z^2, v = 2y^2 + z^2, w = 2z^2$.

68. Let $u = 2x - y$ and $v = x + 2y$. Find the image of the unit square given by $0 \le x \le 1, 0 \le y \le 1$.

69. Find the image under the linear transformation $u = x - y, v = y$ for lines of the general form $y = Cx + 1$, for constant C.

70. Suppose $u = \frac{1}{2}(x^2 + y^2)$ and $v = \frac{1}{2}(x^2 - y^2)$, with $x > 0, y > 0$.
a. Find the Jacobian $\dfrac{\partial(u, v)}{\partial(x, y)}$.
b. Solve for x and y in terms of u and v, and find the Jacobian $\dfrac{\partial(x, y)}{\partial(u, v)}$.
c. Verify that $\dfrac{\partial(u, v)}{\partial(x, y)} \dfrac{\partial(x, y)}{\partial(u, v)} = 1$.

71. Let D be the region given by $u \ge 0, v \ge 0, 1 \le u + v \le 2$.
a. Show that under the transformation $u = x + y$, $v = x - y, D$ is mapped into the region R such that $-x \le y \le x, \frac{1}{2} \le x \le 1$.
b. Find the Jacobian $\dfrac{\partial(u, v)}{\partial(x, y)}$, and compute $$\iint_D (u + v)\, du\, dv.$$

72. Use a change of variables after a transformation of the form $x = Au, y = Bv, z = Cw$ to find the volume of the region bounded by the surface $\sqrt{x} + \sqrt{2y} + \sqrt{3z} = 1$ and the coordinate planes.

73. Find the centroid of a homogeneous lamina $(\delta = 1)$ that covers the part of the plane $Ax + By + Cz = 1, A > 0,$ $B > 0, C > 0$ that lies in the first quadrant.

74. A homogeneous plate $(\delta = 1)$ has the shape of the region in the first quadrant of the xy-plane that is bounded by the circle $x^2 + y^2 = 1$ and the lines $y = x$ and $x = 0$. Sketch the region and find $\bar{x}$, the x-coordinate of the centroid.

75. Let R be a lamina covering the region in the xy-plane that is bounded by the parabola $y = 1 - x^2$ and the positive coordinate axes. Assume the lamina has density $\delta(x, y) = xy$.
a. Find the center of mass of the lamina.
b. Find the moment of inertia of the plate about the z-axis.

76. A solid has the shape of the cylinder $x^2 + y^2 \le a$, with $0 \le z \le b$. Assume the solid has density $\delta(x, y, z) = x^2 + y^2$.
a. Find the mass of the solid.
b. Find the center of mass of the solid.

77. Use cylindrical coordinates to find the volume of the solid bounded by the circular cylinder $r = 2a \cos \theta$ $(a > 0)$, the cone $z = r$, and the xy-plane.

78. Let u be everywhere continuous in the plane, and define the functions f and g by
$$f(x) = \int_a^x u(x, y)\, dy \qquad g(y) = \int_y^b u(x, y)\, dx$$
Show that
$$\int_a^b f(x)\, dx = \int_a^b g(y)\, dy$$
Hint: Show that both integrals equal $\displaystyle\iint_D u(x, y)\, dA$ for a certain region D.

79. Find the volume of the solid region bounded by the surface
$$x^{2/3} + y^{2/3} + z^{2/3} = a^{2/3}$$
where a is a positive constant.

80. MODELING PROBLEM   The parking lot for a certain shopping mall has the shape shown in Figure 13.61.

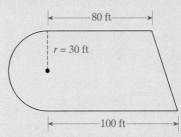

■ **FIGURE 13.61**   Shopping mall parking lot

Assuming a security observation tower can be located anywhere in the parking lot, where would you put it? State all assumptions made in setting up and analyzing your model.

81. Let f be a probability density function on a closed set R. Recall that a probability density function has the properties:

$$f(x, y) \geq 0 \quad \text{for all } (x, y) \text{ in } R$$

$$\int_{-\infty}^{\infty} \int_{-\infty}^{\infty} f(x, y) \, dA = 1$$

Show that the following functions are potential probability density functions:

a. $f(x, y) = \begin{cases} \dfrac{1}{A(R)} & \text{if } (x, y) \text{ is in } R \\ 0 & \text{if } (x, y) \text{ is not in } R \end{cases}$

b. $f(x, y) = \lambda\mu \exp(-\lambda x - \mu y)$ for $\lambda > 0$, $\mu > 0$ and for (x, y) in the first quadrant.

82. A cube of side 2 is surmounted by a hemisphere of radius 1, as shown in Figure 13.62. Suppose the origin is at the center of the hemispherical dome. If the combined solid has density $\delta = 1$, where is the center of mass?

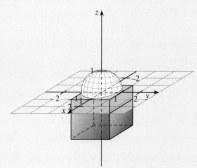

■ **FIGURE 13.62**   Center of mass of a complex solid

83. Set up (but do not evaluate) an integral for the mass of a lamina with density δ that covers the region outside the circle $r = \sqrt{2}\,a$ and inside the lemniscate $r^2 = 4a^2 \sin 2\theta$.

84. Find the mass of a lamina with density $\delta = r\theta$ that covers the region enclosed by the rose $r = \cos 3\theta$ for $0 \leq \theta \leq \frac{\pi}{6}$.

85. Find the centroid ($\delta = 1$) of the solid common to the cylindrical solids $x^2 + z^2 \leq 1$ and $y^2 + z^2 \leq 1$.

86. Show that if $z = f(r, \theta)$ is the equation of a surface S in polar coordinates, the surface area of S is given by

$$\iint_R \sqrt{1 + \left(\frac{\partial z}{\partial r}\right)^2 + \frac{1}{r^2}\left(\frac{\partial z}{\partial \theta}\right)^2} \; r \, dr \, d\theta$$

where R is the projected region in the $r\theta$-plane.

87. Find the center of mass of the solid S that lies inside the sphere $x^2 + y^2 + z^2 = 1$ in the first octant $x \geq 0$, $y \geq 0$, $z \geq 0$ if the density is $\delta(x, y, z) = (x^2 + y^2 + z^2 + 1)^{-1}$.

88. Suppose we drill a square hole of side c through the center of a sphere of radius c, as shown in Figure 13.63. What is the volume of the solid that remains?

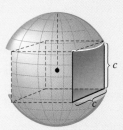

■ **FIGURE 13.63**   Sphere with a square hole drilled out

89. **PUTNAM EXAMINATION PROBLEM**   The function $K(x, y)$ is positive and continuous for $0 \leq x \leq 1$, $0 \leq y \leq 1$ and the functions $f(x)$ and $g(x)$ are positive and continuous for $0 \leq y \leq 1$. Suppose that for all $0 \leq x \leq 1$, we have

$$\int_0^1 f(y)K(x, y) \, dy = g(x) \quad \text{and} \quad \int_0^1 g(y)K(x, y) \, dy = f(x)$$

Show that $f(x) = g(x)$ for $0 \leq x \leq 1$.

90. **PUTNAM EXAMINATION PROBLEM**   A circle of radius a is revolved through 180° about a line in its plane, distant b from the center of the circle ($b > a$). For what value of the ratio b/a does the center of gravity of the solid thus generated lie on the surface of the solid?

91. **PUTNAM EXAMINATION PROBLEM**   For $f(x)$ a positive, monotone, decreasing function defined in $0 \leq x \leq 1$, prove that

$$\frac{\displaystyle\int_0^1 x[f(x)]^2 \, dx}{\displaystyle\int_0^1 x f(x) \, dx} \leq \frac{\displaystyle\int_0^1 [f(x)]^2 \, dx}{\displaystyle\int_0^1 f(x) \, dx}$$

92. **PUTNAM EXAMINATION PROBLEM**   Show that the integral equation

$$f(x, y) = 1 + \int_0^x \int_0^y f(u, v) \, du \, dv$$

has at most one solution continuous for $0 \leq x \leq 1$, $0 \leq y \leq 1$.

Space-Capsule Design

The science of mathematics has grown to such vast proportion that probably no living mathematician can claim to have achieved its mastery as a whole.

A. N. WHITEHEAD
AN INTRODUCTION TO
MATHEMATICS
(NEW YORK, 1911), P. 252

Consider the mathematics discovered from 1911 to 1998! How many times multiplied is the quotation of Whitehead?

This project is to be done in groups of three or four students. Each group will submit a single written report. Reports are to be typed, technically correct, as well as showing good grammar and style.

Suppose you are transported back in time to be part of a team of engineers designing the Apollo space capsule. The capsule is composed of two parts:

1. A cone with a height of 4 meters and a base of radius 3 meters.
2. A reentry shield in the shape of a parabola revolved about the axis of the cone, which is attached to the cone along the edge of the base of the cone. Its vertex is a distance D below the base of the cone. Find values of the design parameters D and δ so that the capsule will float with the vertex of the cone pointing up and with the waterline 2 m below the top of the cone, in order to keep the exit port 1/3 m above water.

You can make the following assumptions:

a. The capsule has uniform density δ.

b. The center of mass of the capsule should be below the center of mass of the displaced water because this will give the capsule better stability in heavy seas.

c. A body floats in a fluid at the level at which the weight of the displaced fluid equals the weight of the body (Archimedes' principle).

Your paper is not limited to the following questions but should include these concerns: Show the project director that the task is impossible; that is, there are no values of D and δ that satisfy the design specifications. However, you can solve this dilemma by incorporating a flotation collar in the shape of a torus. The collar will be made by taking hollow plastic tubing with a circular cross section of radius 1 m and wrapping it in a circular ring about the capsule, so that it fits snugly. The collar is designed to float just submerged with its top tangent to the surface of the water. Show that this flotation collar makes the capsule plus collar assembly satisfy the design specifications. Find the density δ needed to make the capsule float at the 2-meter mark. Assume the weight of the tubing is negligible compared to the weight of the capsule, that the design parameter D is equal to 1 meter, and the density of the water is 1.

MAA Notes 17 (1991), "Priming the Calculus Pump: Innovations and Resources," by Marcus S. Cohen, Edward D. Gaughan, R. Arthur Knoebel, Douglas S. Kurtz, and David J. Pengelley.

CONTENTS

14

Vector Analysis

PREVIEW

In this chapter, we draw together what we have learned about differentiation, integration, and vectors to study the calculus of vector functions defined on a set of points in $\mathbb{R}^2$ or $\mathbb{R}^3$. We introduce **line integrals,** and **surface integrals** to study such things as fluid flow and then obtain a result called **Green's theorem** that enables line integrals to be computed in terms of ordinary double integrals. This result will then be extended into $\mathbb{R}^3$ to obtain **Stokes' theorem** and the **divergence theorem,** which have extensive applications in areas such as fluid dynamics and electromagnetic theory.

PERSPECTIVE

How much work is done by a variable force acting along a given curve in space? How can the amount of heat flowing across a particular surface in unit time be measured, and is the measurement similar to measuring the flow of water or electricity? We shall use line integrals and surface integrals to answer these and other questions from physics and engineering mathematics.

14.1 *Properties of a Vector Field: Divergence and Curl*

   definition of a vector field; divergence, curl, a physical interpretation of curl

DEFINITION OF A VECTOR FIELD

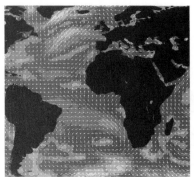

■ **FIGURE 14.1** A wind-velocity map of the Indian Ocean

The satellite photograph in Figure 14.1 shows wind measurements over the Indian Oceans. Wind direction is indicated by directed line segments, whose shading (light gray to black) indicate wind speed. This is an example of a *vector field,* in which every point in a given region of the plane or space is assigned a vector that represents a velocity or force or some other vector quantity of interest. Here is the definition of a vector field in $\mathbb{R}^3$.

> **Vector Field**
>
> A **vector field** is a collection S of points in space together with a rule that assigns to each point (x, y, z) in S exactly one vector $\mathbf{V}(x, y, z)$.

For instance, we may represent the velocity $\mathbf{V}(x, y, z)$ of a fluid flow by drawing an appropriate vector at each point (x, y, z) in the domain of the flow, and the resulting collection of vectors is called a **velocity field.** In practice, we often express a given vector field $\mathbf{V}$ in the form

$$\mathbf{V}(x, y) = u(x, y)\mathbf{i} + v(x, y)\mathbf{j} \quad \text{in } \mathbb{R}^2$$

or

$$\mathbf{V}(x, y, z) = u(x, y, z)\mathbf{i} + v(x, y, z)\mathbf{j} + w(x, y, z)\mathbf{k} \quad \text{in } \mathbb{R}^3$$

The functions $u, v,$ and w are scalar functions called the **components** of $\mathbf{V}$. For example,

$$\mathbf{V} = 2x^2y\mathbf{i} + e^{yz}\mathbf{j} + \left(\tan\frac{x}{z}\right)\mathbf{k}$$

is a vector field with **i**-component $2x^2y$, **j**-component e^{yz}, and **k**-component $\tan\dfrac{x}{z}$.

To get an idea of what a vector field "looks like," it is often helpful to draw $\mathbf{V}(x, y, z)$ as an "arrow" at selected points in S. We shall refer to such a diagram as the **graph of V**.

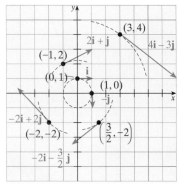

■ **FIGURE 14.2** The graph of the vector field $\mathbf{F}(x, y) = y\mathbf{i} - x\mathbf{j}$

EXAMPLE 1   *Graph of a vector field*

Sketch the graph of the vector field $\mathbf{F}(x, y) = y\mathbf{i} - x\mathbf{j}$.

Solution   We shall evaluate $\mathbf{F}$ at various points. For example,

$$\mathbf{F}(3, 4) = 4\mathbf{i} - 3\mathbf{j} \quad \text{and} \quad \mathbf{F}(-1, 2) = 2\mathbf{i} - (-1)\mathbf{j} = 2\mathbf{i} + \mathbf{j}$$

We can generate as many such vector values of $\mathbf{F}$ as we wish. Several are shown in Figure 14.2. ■

The graph of a vector field often yields useful information about the properties of the field. For instance, suppose $\mathbf{F}(x, y)$ represents the velocity of a compressible fluid (like a gas) at a point (x, y) in the plane. Then $\mathbf{F}$ assigns a velocity vector to each point in the plane, and the graph of $\mathbf{F}$ provides a picture of the fluid flow. Thus, the flow in Figure 14.3a is a constant, whereas Figure 14.3b suggests a circular flow.

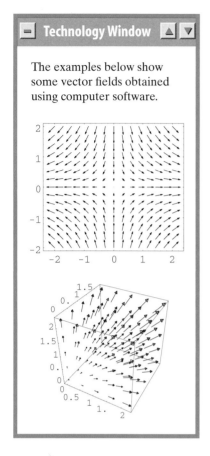
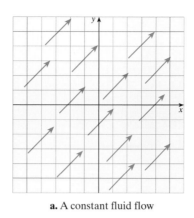

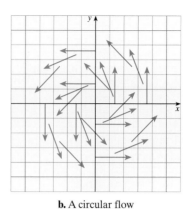

a. A constant fluid flow   **b.** A circular flow

■ **FIGURE 14.3**   Flow diagrams

Gravitational, electrical, and magnetic vector fields play an important role in physical applications. We shall discuss gravitational fields now and electrical and magnetic fields later in this section. Accordingly, we begin with Newton's law of gravitation, which says that a point mass (particle) m at the origin exerts on a unit point mass located at the point $P(x, y, z)$ a force $\mathbf{F}(x, y, z)$ given by

$$\mathbf{F}(x, y, z) = \frac{Gm}{x^2 + y^2 + z^2}\, \mathbf{u}(x, y, z)$$

where G is a constant (the universal gravitational constant) and $\mathbf{u}$ is the unit vector extending from the point P toward the origin. The vector field $\mathbf{F}(x, y, z)$ is called the **gravitational field** of the point mass m. Because

$$\mathbf{u}(x, y, z) = \frac{-1}{\sqrt{x^2 + y^2 + z^2}}\, (x\mathbf{i} + y\mathbf{j} + z\mathbf{k})$$

it follows that

$$\mathbf{F}(x, y, z) = \frac{-Gm}{(x^2 + y^2 + z^2)^{3/2}}\, (x\mathbf{i} + y\mathbf{j} + z\mathbf{k})$$

Note that the gravitational field $\mathbf{F}$ always points toward the origin and has the same magnitude for any point m located $r = \sqrt{x^2 + y^2 + z^2}$ units from the origin. Such a vector field is called a **central force field.** This force field is shown in Figure 14.4a. Some other examples of velocity fields are shown in Figures 14.4b and 14.4c.

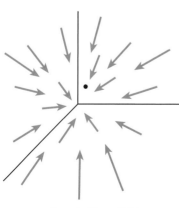

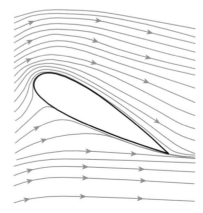

a. A central force field   **b.** Air flow vector   **c.** Wind velocity on a map

■ **FIGURE 14.4**   Examples of physical vector fields

DIVERGENCE

A vector field **V** can be differentiated in two different ways, one of which produces a derivative that is a scalar and the other, a vector. The scalar derivative may be defined as follows.

> **Divergence**
>
> The **divergence** of a vector field
>
> $$\mathbf{V}(x, y, z) = u(x, y, z)\mathbf{i} + v(x, y, z)\mathbf{j} + w(x, y, z)\mathbf{k}$$
>
> is denoted by div **V** and is given by
>
> $$\text{div } \mathbf{V} = \frac{\partial u}{\partial x}(x, y, z) + \frac{\partial v}{\partial y}(x, y, z) + \frac{\partial w}{\partial z}(x, y, z)$$

WARNING The divergence of a vector field is a scalar.

EXAMPLE 2   *Finding div* **V**

Find the divergence of each of the following vector fields.
a. $\mathbf{F}(x, y) = x^2 y \mathbf{i} + xy^3 \mathbf{j}$ b. $\mathbf{G}(x, y, z) = x\mathbf{i} + y^3 z^2 \mathbf{j} + xz^3 \mathbf{k}$

Solution

a. $\text{div } \mathbf{F} = \dfrac{\partial}{\partial x}(x^2 y) + \dfrac{\partial}{\partial y}(xy^3) = 2xy + 3xy^2$

b. $\text{div } \mathbf{G} = \dfrac{\partial}{\partial x}(x) + \dfrac{\partial}{\partial y}(y^3 z^2) + \dfrac{\partial}{\partial z}(xz^3) = 1 + 3y^2 z^2 + 3xz^2$ ∎

Suppose the vector field

$$\mathbf{V}(x, y, z) = u(x, y, z)\mathbf{i} + v(x, y, z)\mathbf{j} + w(x, y, z)\mathbf{k}$$

represents the velocity of a fluid with density $\delta(x, y, z)$ at a point (x, y, z) in a certain region R in $\mathbb{R}^3$. Then the vector field $\delta\mathbf{V}$ is called the **flux density** and is denoted by **D**. We can think of $\mathbf{D} = \delta\mathbf{V}$ as measuring the "mass flow" of the liquid.

Assuming there are no external processes acting on the fluid that would tend to create or destroy fluid, it can be shown that div **D** gives the negative of the time rate change of the density, that is,

$$\text{div } \mathbf{D} = -\frac{\partial \delta}{\partial t}$$

This is often referred to as the **continuity equation** of fluid dynamics. (A derivation is given in Section 14.7.) When div $\mathbf{D} = 0$, **D** is said to be **divergence free** or **solenoidal,** and the fluid is said to be **incompressible.** If div $\mathbf{D} > 0$ at a point (x_0, y_0, z_0), then the point is called a **source;** if div $\mathbf{D} < 0$, the point is called a **sink** (see Figure 14.5).

Div **D** > 0   Div **D** < 0   Div **D** = 0

Fluid arrives through a *source point.*

Fluid leaves through a *sink point.*

Fluid is *incompressible.*

■ **FIGURE 14.5**   Flow of a fluid across a plane region, **D**

The terms *sink, source,* and *incompressible* apply to any vector field **F** and are not reserved only for fluid applications.

A useful way to think of the divergence div **V** is in terms of the **del operator** defined by

$$\nabla = \frac{\partial}{\partial x}\mathbf{i} + \frac{\partial}{\partial y}\mathbf{j} + \frac{\partial}{\partial z}\mathbf{k}$$

The del operator *operates* on a differentiable function $f(x, y, z)$ of three variables, and for such a function, we have

$$\nabla f = \frac{\partial f}{\partial x}\mathbf{i} + \frac{\partial f}{\partial y}\mathbf{j} + \frac{\partial f}{\partial z}\mathbf{k}$$

We can also define ∇f in a totally analogous way for functions of one, two, or more than three variables. In particular, for a differentiable function $f(x, y)$ of two variables, we have

$$\nabla f = \frac{\partial f}{\partial x}\mathbf{i} + \frac{\partial f}{\partial y}\mathbf{j}$$

We recognize ∇f as the **gradient** of f, which was introduced in Chapter 12. We can think of div **V** as the *dot product* of the operator ∇ and the vector field $\mathbf{V} = u\mathbf{i} + v\mathbf{j} + w\mathbf{k}$; that is,

$$\text{div } \mathbf{V} = \frac{\partial u}{\partial x} + \frac{\partial v}{\partial y} + \frac{\partial w}{\partial z} = \left(\frac{\partial}{\partial x}\mathbf{i} + \frac{\partial}{\partial y}\mathbf{j} + \frac{\partial}{\partial z}\mathbf{k}\right) \cdot (u\mathbf{i} + v\mathbf{j} + w\mathbf{k}) = \nabla \cdot \mathbf{V}$$

CURL

The del operator may also be used to describe another derivative operation for vector fields, called the *curl*.

Curl

The **curl** of a vector field

$$\mathbf{V}(x, y, z) = u(x, y, z)\mathbf{i} + v(x, y, z)\mathbf{j} + w(x, y, z)\mathbf{k}$$

is denoted by curl **V** and is defined by

$$\text{curl } \mathbf{V} = \left(\frac{\partial w}{\partial y} - \frac{\partial v}{\partial z}\right)\mathbf{i} + \left(\frac{\partial u}{\partial z} - \frac{\partial w}{\partial x}\right)\mathbf{j} + \left(\frac{\partial v}{\partial x} - \frac{\partial u}{\partial y}\right)\mathbf{k}$$

This can be remembered in the following useful form:

$$\text{curl } \mathbf{V} = \begin{vmatrix} \mathbf{i} & \mathbf{j} & \mathbf{k} \\ \frac{\partial}{\partial x} & \frac{\partial}{\partial y} & \frac{\partial}{\partial z} \\ u & v & w \end{vmatrix}$$

We saw that $\nabla \cdot \mathbf{V} = \text{div } \mathbf{V}$, and from this definition of curl, we see that

$$\text{curl } \mathbf{V} = \left(\frac{\partial w}{\partial y} - \frac{\partial v}{\partial z}\right)\mathbf{i} + \left(\frac{\partial u}{\partial z} - \frac{\partial w}{\partial x}\right)\mathbf{j} + \left(\frac{\partial v}{\partial x} - \frac{\partial u}{\partial y}\right)\mathbf{k} = \nabla \times \mathbf{V}$$

This representation for curl and the one for divergence stated earlier are summarized in the following box.

Del Operator Forms for Divergence and Curl

Consider a vector field

$$\mathbf{V}(x, y, z) = u(x, y, z)\mathbf{i} + v(x, y, z)\mathbf{j} + w(x, y, z)\mathbf{k}$$

The divergence and curl of $\mathbf{V}$ are given by

$$\text{div } \mathbf{V} = \nabla \cdot \mathbf{V} \quad \text{and} \quad \text{curl } \mathbf{V} = \nabla \times \mathbf{V}$$

WARNING div is a scalar and curl is a vector.

EXAMPLE 3   *Curl of a vector field*

Find the curl of the vector fields

$$\mathbf{F} = x^2yz\mathbf{i} + xy^2z\mathbf{j} + xyz^2\mathbf{k} \quad \text{and} \quad \mathbf{G} = (x \cos y)\mathbf{i} + xy^2\mathbf{j}$$

Solution $\text{curl } \mathbf{F} = \begin{vmatrix} \mathbf{i} & \mathbf{j} & \mathbf{k} \\ \dfrac{\partial}{\partial x} & \dfrac{\partial}{\partial y} & \dfrac{\partial}{\partial z} \\ x^2yz & xy^2z & xyz^2 \end{vmatrix}$

$$= \left(\frac{\partial}{\partial y}xyz^2 - \frac{\partial}{\partial z}xy^2z \right)\mathbf{i} - \left(\frac{\partial}{\partial x}xyz^2 - \frac{\partial}{dz}x^2yz \right)\mathbf{j}$$

$$+ \left(\frac{\partial}{\partial x}xy^2z - \frac{\partial}{\partial y}x^2yz \right)\mathbf{k}$$

$$= (xz^2 - xy^2)\mathbf{i} + (x^2y - yz^2)\mathbf{j} + (y^2z - x^2z)\mathbf{k}$$

$\text{curl } \mathbf{G} = \begin{vmatrix} \mathbf{i} & \mathbf{j} & \mathbf{k} \\ \dfrac{\partial}{\partial x} & \dfrac{\partial}{\partial y} & \dfrac{\partial}{\partial z} \\ x \cos y & xy^2 & 0 \end{vmatrix}$

$$= \left[0 - \frac{\partial}{\partial z}xy^2 \right]\mathbf{i} - \left[0 - \frac{\partial}{\partial z}(x \cos y) \right]\mathbf{j} + \left[\frac{\partial}{\partial x}xy^2 - \frac{\partial}{\partial y}(x \cos y) \right]\mathbf{k}$$

$$= (y^2 + x \sin y)\mathbf{k}$$

■

EXAMPLE 4   *A constant vector field has divergence and curl zero*

Let $\mathbf{F}$ be a constant vector field. Show div $\mathbf{F} = 0$ and curl $\mathbf{F} = \mathbf{0}$.

Solution   Let $\mathbf{F} = a\mathbf{i} + b\mathbf{j} + c\mathbf{k}$ for constants $a, b,$ and c. Then

$$\text{div } \mathbf{F} = \frac{\partial}{\partial x}(a) + \frac{\partial}{\partial y}(b) + \frac{\partial}{\partial z}(c) = 0$$

$$\text{curl } \mathbf{F} = \begin{vmatrix} \mathbf{i} & \mathbf{j} & \mathbf{k} \\ \dfrac{\partial}{\partial x} & \dfrac{\partial}{\partial y} & \dfrac{\partial}{\partial z} \\ a & b & c \end{vmatrix} = 0\mathbf{i} - 0\mathbf{j} + 0\mathbf{k} = \mathbf{0}$$

■

WARNING   Example 4 shows that the divergence and curl of a constant vector field are zero, but this does *not* mean that if the divergence or curl vanishes, then the associated vector field is a constant. For instance, the nonconstant vector field

$$\mathbf{F}(x, y, z) = yz\mathbf{i} + x\mathbf{j} + xy^2\mathbf{k}$$

has div $\mathbf{F} = 0$.

▲ ▼

Technology Window

Derive, Maple, Mathlab, and *Mathematica* will carry out most vector operations. Some may require loading certain additional operations or subroutines. For Example 2, we define $\mathbf{G}(x, y, z) = x\mathbf{i} + y^3 z^2 \mathbf{j} + xz^3 \mathbf{k}$, and use one of these programs:

$$\text{Div}[x, y^3 z^2, xz^3] \quad \text{simplifies to} \quad 3xz^2 + 3y^2 z^2 + 1$$

Notice that the *form* of the answer differs from that shown in Example 2. You may need to enter a zero, as with $\mathbf{F}(x, y) = x^2 y \mathbf{i} + xy^3 \mathbf{j}$, as shown here:

$$\text{Div}[x^2 y, xy^3, 0] \quad \text{simplifies to} \quad x(3y^2 + 2y)$$

Finally, consider the vector field $\mathbf{F}$ from Example 3, $\mathbf{F} = x^2 yz \mathbf{i} + xy^2 z \mathbf{j} + xyz^2 \mathbf{k}$; we can find the curl:

$$\text{Curl}[x^2 yz, xy^2 z, xyz^2] \quad \text{simplifies to} \quad [x(z^2 - y^2), x^2 y - yz^2, y^2 z - x^2 z]$$

A PHYSICAL INTERPRETATION OF CURL

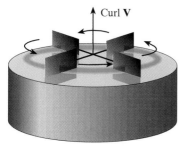

■ **FIGURE 14.6** A physical interpretation of curl $\mathbf{V}$

The terms *divergence* and *curl,* as well as other terminology in this chapter, are motived by concepts from physics. For example, we can apply the curl to the study of fluid motion. If a fluid is moving about a region in the xy-plane, the curl can be thought of as the circulation of the fluid. A good way to measure the effect (magnitude and direction) of the circulation is to place a small paddle wheel into the fluid. The curl measures the rate of the fluid's rotation at the point P where the paddle wheel is placed in the direction of the paddle wheel's axis. The curl is positive when the rotation is counterclockwise and negative when it is clockwise.

Let $\mathbf{V}(x, y, z) = u(x, y, z)\mathbf{i} + v(x, y, z)\mathbf{j} + w(x, y, z)\mathbf{k}$ be the velocity of an incompressible fluid, and suppose a paddle wheel is inserted into the fluid in such a way that its axis lies along the z-axis, as shown in Figure 14.6. Disregard the weight of the paddle. The fluid tends to swirl about the z-axis, causing the paddle to rotate, and we can analyze the fluid's swirling motion by studying the motion of the paddle. It turns out that the circulation per unit of the liquid

about the x-axis is proportional to $\left(\dfrac{\partial w}{\partial y} - \dfrac{\partial v}{\partial z}\right)$;

about the y-axis is proportional to $\left(\dfrac{\partial u}{\partial z} - \dfrac{\partial w}{\partial x}\right)$;

about the z-axis is proportional to $\left(\dfrac{\partial v}{\partial x} - \dfrac{\partial u}{\partial y}\right)$.

Thus, the tendency of the fluid to swirl is measured by curl $\mathbf{V}$. In the special case where curl $\mathbf{V} = \mathbf{0}$, the fluid has no rotational motion and is said to be **irrotational.**

Combinations of the gradient, divergence, and curl appear in a variety of applications. In particular, note that if f is a differentiable scalar function, its gradient ∇f is a vector field, and we can compute

$$\begin{aligned}
\text{div } \nabla f &= \left(\frac{\partial}{\partial x}\mathbf{i} + \frac{\partial}{\partial y}\mathbf{j} + \frac{\partial}{\partial z}\mathbf{k}\right) \cdot \left(\frac{\partial f}{\partial x}\mathbf{i} + \frac{\partial f}{\partial y}\mathbf{j} + \frac{\partial f}{\partial z}\mathbf{k}\right) \\
&= \frac{\partial^2 f}{\partial x^2} + \frac{\partial^2 f}{\partial y^2} + \frac{\partial^2 f}{\partial z^2} \\
&= \nabla \cdot \nabla f
\end{aligned}$$

We summarize this definition and introduce some terminology in the following box.

> ### The Laplacian Operator
>
> Let $f(x, y, z)$ define a function with continuous first and second partial derivatives. Then the **Laplacian of** f is
>
> $$\nabla^2 f = \nabla \cdot \nabla f = \frac{\partial^2 f}{\partial x^2} + \frac{\partial^2 f}{\partial y^2} + \frac{\partial^2 f}{\partial z^2} = f_{xx} + f_{yy} + f_{zz}$$
>
> The equation $\nabla^2 f = 0$ is called **Laplace's equation,** and a function that satisfies such an equation in a region D is said to be **harmonic** in D.

> ■ *What This Says:*   The notation $\nabla \cdot \nabla f$ is usually abbreviated by writing $\nabla^2 f$, and it is called either "del-squared f" or the *Laplacian of f*, after the French mathematician Pierre Laplace (see Historical Quest, Problem 57, Section 13.3). A function f whose first and second partials are continuous and that satisfies the equation $\nabla^2 f = 0$ is called *harmonic*.

EXAMPLE 5   *Showing that a function is harmonic*

Show that $f(x, y) = e^x \cos y$ is harmonic in the plane.

Solution $f_x(x, y) = e^x \cos y$ and $f_{xx}(x, y) = e^x \cos y$

$\qquad\qquad\qquad f_y(x, y) = -e^x \sin y$ and $f_{yy}(x, y) = -e^x \cos y$

The Laplacian of f is given by

$$f(x, y) = f_{xx}(x, y) + f_{yy}(x, y) = e^x \cos y - e^x \cos y = 0$$

Thus, f is harmonic.   ■

In many ways, the study of electricity and magnetism is analogous to that of fluid dynamics, and the curl and divergence play an important role in this study. In electromagnetic theory, it is often convenient to regard interaction between electrical charges as forces somewhat like the gravitational force between masses and then to seek quantitative measure of these forces.

One of the great scientific achievements of the 19th century was the discovery of the laws of electromagnetism by the English scientist James Clerk Maxwell (see Historical Quest, Section 14.7, Problem 30). These laws have an elegant expression in terms of the divergence and curl. It is known empirically that the force acting on a charge due to an electromagnetic field depends on the position, velocity, and amount of the particular charge, and not on the number of other charges that may be present or how those other charges are moving. Suppose a charge is located at the point (x, y, z) at time t, and consider the electric intensity field $\mathbf{E}(x, y, z, t)$ and the magnetic intensity field $\mathbf{H}(x, y, z, t)$. Then the behavior of the resulting electromagnetic field is determined by

$$\operatorname{div} \mathbf{E} = \frac{Q}{\epsilon} \qquad \operatorname{curl} \mathbf{E} = -\frac{\partial}{\partial t}(\mu \mathbf{H}) \qquad \operatorname{div}(\mu \mathbf{H}) = \mathbf{0}$$

where Q is the *electric charge density* (charge per unit volume), $\mathbf{J}$ is the *electric current density* (rate at which the charge flows through a unit area per second), and μ and ϵ are constants called the *permeability* and *permittivity,* respectively. Furthermore, if σ is a constant called the *conductivity,* then

$$(\nabla \cdot \nabla)\mathbf{E} = \mu \sigma \frac{\partial \mathbf{E}}{\partial t} + \mu \epsilon \frac{\partial^2 \mathbf{E}}{\partial t^2}$$

Working with these equations and terms is beyond the scope of this course, but if you are interested there are many references you can consult. One of the best (in spite of the fact it is 30 years old) is the classic book written by Nobel laureate Richard Feynman, Robert Leighton, and Matthew Sands, *Feynman Lectures in Physics* (Reading, MA: Addison-Wesley, 1963).

Technology Window

Derive, Maple, Mathlab, and *Mathematica* will find the Laplacian of a given function. For Example 5, we obtain

$$\text{LAPLACIAN}(e\hat{\ }x \cos(y)), \text{ which simplifies to } 0$$

If you have access to this technology, verify that $\text{LAPLACIAN}(x^2 y^3 z)$ simplifies to $6x^2 yz + 2y^3 z$.

14.1 Problem Set

A **1.** ■ **What Does This Say?** Compare and contrast divergence and curl.

2. ■ **What Does This Say?** Discuss the del operator.

Sketch several representatives of the given vector field in Problems 3–8.

3. $\mathbf{F} = x\mathbf{i} + y\mathbf{j}$
4. $\mathbf{G} = -x\mathbf{i} + y\mathbf{j}$
5. $\mathbf{V} = x^2\mathbf{i} + y\mathbf{j}$
6. $\mathbf{H} = x^2\mathbf{i} + y^2\mathbf{j}$
7. $\mathbf{F}(x, y) = y\mathbf{i} + x\mathbf{j}$
8. $\mathbf{G}(x, y) = xy\mathbf{i} - \mathbf{j}$

In Problems 9–12, find div $\mathbf{F}$ *and* curl $\mathbf{F}$ *for the given vector function.*

9. $\mathbf{F}(x, y) = x^2\mathbf{i} + xy\mathbf{j} + z^3\mathbf{k}$ **10.** $\mathbf{F}(x, y) = \mathbf{i} + (x^2 + y^2)\mathbf{j}$
11. $\mathbf{F}(x, y, z) = 2y\mathbf{j}$ **12.** $\mathbf{F}(x, y, z) = z\mathbf{i} - \mathbf{j} + 2y\mathbf{k}$

In Problems 13–18, find div $\mathbf{F}$ *and* curl $\mathbf{F}$ *for each vector field* $\mathbf{F}$ *at the given point.*

13. $\mathbf{F}(x, y, z) = \mathbf{i} + \mathbf{j} + \mathbf{k}$ at $(2, -1, 3)$
14. $\mathbf{F}(x, y, z) = xz\mathbf{i} + y^2 z\mathbf{j} + xz\mathbf{k}$ at $(1, -1, 2)$
15. $\mathbf{F}(x, y, z) = xyz\mathbf{i} + yj + x\mathbf{k}$ at $(1, 2, 3)$
16. $\mathbf{F}(x, y, z) = (\cos y)\mathbf{i} + (\sin y)\mathbf{j} + \mathbf{k}$ at $(\frac{\pi}{4}, \pi, 0)$
17. $\mathbf{F}(x, y, z) = e^{-xy}\mathbf{i} + e^{xz}\mathbf{j} + e^{yz}\mathbf{k}$ at $(3, 2, 0)$
18. $\mathbf{F}(x, y, z) = (e^{-x}\sin y)\mathbf{i} + (e^{-x}\cos y)\mathbf{j} + \mathbf{k}$ at $(1, 3, -2)$

Find div $\mathbf{F}$ *and* curl $\mathbf{F}$ *for each vector field* $\mathbf{F}$ *given in Problems 19–34.*

19. $\mathbf{F} = (\sin x)\mathbf{i} + (\cos y)\mathbf{j}$ **20.** $\mathbf{F} = (-\cos x)\mathbf{i} + (\sin y)\mathbf{j}$
21. $\mathbf{F} = x\mathbf{i} - y\mathbf{j}$ **22.** $\mathbf{F} = -x\mathbf{i} + y\mathbf{j}$
23. $\mathbf{F} = \dfrac{x}{\sqrt{x^2 + y^2}}\mathbf{i} + \dfrac{y}{\sqrt{x^2 + y^2}}\mathbf{j}$

24. $\mathbf{F} = x^2\mathbf{i} - y^2\mathbf{j}$
25. $\mathbf{F} = ax\mathbf{i} + by\mathbf{j} + c\mathbf{k}$ for constants a, b, c
26. $\mathbf{F} = (e^x \sin y)\mathbf{i} + (e^x \cos y)\mathbf{j} + \mathbf{k}$
27. $\mathbf{F} = x^2\mathbf{i} + y^2\mathbf{j} + z^2\mathbf{k}$ **28.** $\mathbf{F} = y\mathbf{i} + z\mathbf{j} + x\mathbf{k}$
29. $\mathbf{F} = xy\mathbf{i} + yz\mathbf{j} + xz\mathbf{k}$ **30.** $\mathbf{F} = 2xz\mathbf{i} + 2yz^2\mathbf{j} - \mathbf{k}$
31. $\mathbf{F} = xyz\mathbf{i} + x^2 y^2 z^2\mathbf{j} + y^2 z^3\mathbf{k}$
32. $\mathbf{F} = -z^3\mathbf{i} + 3\mathbf{j} + 2y\mathbf{k}$
33. $\mathbf{F} = (x - y)\mathbf{i} + (y - z)\mathbf{j} + (z - x)\mathbf{k}$
34. $\mathbf{F} = \dfrac{x\mathbf{i} + y\mathbf{j} + z\mathbf{k}}{\sqrt{x^2 + y^2 + z^2}}$

Determine whether each scalar function in Problems 35–38 is harmonic.

35. $u(x, y, z) = e^{-x}(\cos y - \sin y)$
36. $v(x, y, z) = (x^2 + y^2 + z^2)^{1/2}$
37. $w(x, y, z) = (x^2 + y^2 + z^2)^{-1/2}$
38. $r(x, y, z) = xyz$

B **39.** If $\mathbf{F}(x, y, z) = 2\mathbf{i} + 2x\mathbf{j} + 3y\mathbf{k}$ and $\mathbf{G}(x, y, z) = x\mathbf{i} - y\mathbf{j} + z\mathbf{k}$, find curl$(\mathbf{F} \times \mathbf{G})$.
40. If $\mathbf{F}(x, y, z) = xy\mathbf{i} + yz\mathbf{j} + z^2\mathbf{k}$ and $\mathbf{G}(x, y, z) = x\mathbf{i} + y\mathbf{j} - z\mathbf{k}$, find curl$(\mathbf{F} \times \mathbf{G})$.
41. If $\mathbf{F}(x, y, z) = 2\mathbf{i} + 2x\mathbf{j} + 3y\mathbf{k}$ and $\mathbf{G}(x, y, z) = x\mathbf{i} - y\mathbf{j} + z\mathbf{k}$, find div$(\mathbf{F} \times \mathbf{G})$.
42. If $\mathbf{F}(x, y, z) = xy\mathbf{i} + yz\mathbf{j} + z^2\mathbf{k}$ and $\mathbf{G}(x, y, z) = x\mathbf{i} + y\mathbf{j} - z\mathbf{k}$, find div$(\mathbf{F} \times \mathbf{G})$.
43. Find div $\mathbf{F}$, given that $\mathbf{F} = \nabla f$, where $f(x, y, z) = xy^3 z^2$.
44. Find div $\mathbf{F}$, given that $\mathbf{F} = \nabla f$, where $f(x, y, z) = x^2 yz^3$.
45. A magnetic field that has zero divergence everywhere is said to be *solenoidal* (because such a field can be

generated by a solenoid). Show that the field $\mathbf{B} = y^2 z\mathbf{i} + xz^3\mathbf{j} + y^2 x^2\mathbf{k}$ is solenoidal.

46. A **central force field** $\mathbf{F}$ is one that can be described as $\mathbf{F} = f(r)\mathbf{R}$, where $\mathbf{R} = x\mathbf{i} + y\mathbf{j} + z\mathbf{k}$ and $r = \|\mathbf{R}\| = (x^2 + y^2 + z^2)^{1/2}$. Show that such a field is irrotational everywhere it is defined and differentiable. That is, show curl $\mathbf{F} = \mathbf{0}$.

C 47. Let $\mathbf{A}$ be a constant vector and let $\mathbf{R} = x\mathbf{i} + y\mathbf{j} + z\mathbf{k}$. Show that div$(\mathbf{A} \times \mathbf{R}) = 0$.

48. Let $\mathbf{A}$ be a constant vector and let $\mathbf{R} = x\mathbf{i} + y\mathbf{j} + z\mathbf{k}$. Show that curl$(\mathbf{A} \times \mathbf{R}) = 2\mathbf{A}$.

49. If $\mathbf{F}(x, y) = \mathbf{u}(x, y)\mathbf{i} + \mathbf{v}(x, y)\mathbf{j}$, show that curl $\mathbf{F} = \mathbf{0}$ if and only if

$$\frac{\partial u}{\partial y} = \frac{\partial v}{\partial x}$$

50. Let $\mathbf{F} = \mathbf{R}/r^3$, where $\mathbf{R} = x\mathbf{i} + y\mathbf{j} + z\mathbf{k}$ and $r = \|\mathbf{R}\|$. Show that div $\mathbf{F} = 0$ and curl $\mathbf{F} = \mathbf{0}$.

51. Consider a rigid body that is rotating about the z-axis (counterclockwise from above) with constant angular velocity ω. If P is a point in the body located at $\mathbf{R} = x\mathbf{i} + y\mathbf{j} + z\mathbf{k}$, the velocity at P is given by the vector field $\mathbf{V} = \omega \times \mathbf{R}$.
   a. Express $\mathbf{V}$ in terms of $\mathbf{i}, \mathbf{j}$, and $\mathbf{k}$ vectors.
   b. Find div $\mathbf{V}$ and curl $\mathbf{V}$.

52. **THINK TANK PROBLEM** Find an expression for curl(curl $\mathbf{F}$) for any vector field $\mathbf{F}$.

53. Which (if any) of the following is the same as div$(\mathbf{F} \times \mathbf{G})$ for any vector fields $\mathbf{F}$ and $\mathbf{G}$?
   I. (div $\mathbf{F}$)(div $\mathbf{G}$)
   II. (curl $\mathbf{F}$) $\cdot$ $\mathbf{G}$ $-$ $\mathbf{F}$ $\cdot$ (curl $\mathbf{G}$)
   III. $\mathbf{F}$(div $\mathbf{G}$) $+$ (div $\mathbf{F}$)$\mathbf{G}$
   IV. (curl $\mathbf{F}$) $\cdot$ $\mathbf{G}$ $+$ $\mathbf{F}$ $\cdot$ (curl $\mathbf{G}$)

In Problems 54–64, prove the given property for the vector fields $\mathbf{F}$ and $\mathbf{G}$, scalar c, and scalar function f. Assume that the required partial derivatives are continuous.

54. div$(c\mathbf{F}) = c$ div $\mathbf{F}$

55. div$(\mathbf{F} + \mathbf{G}) = $ div $\mathbf{F}$ + div $\mathbf{G}$

56. curl$(\mathbf{F} + \mathbf{G}) = $ curl $\mathbf{F}$ + curl $\mathbf{G}$

57. curl$(c\mathbf{F}) = c$ curl $\mathbf{F}$

58. curl$(f\mathbf{F}) = f$ curl $\mathbf{F}$ + $(\nabla f \times \mathbf{F})$

59. div$(f\mathbf{F}) = f$ div $\mathbf{F}$ + $(\nabla f \cdot \mathbf{F})$

60. curl$(\nabla f + $ curl $\mathbf{F}) = $ curl(∇f) + curl(curl $\mathbf{F}$)

61. div$(f\nabla g) = f$ div ∇g + $\nabla f \cdot \nabla g$

62. The curl of the gradient of a function is always $\mathbf{0}$. That is, $\nabla \times (\nabla f) = \mathbf{0}$.

63. The divergence of the curl of a vector field is 0. That is, div(curl $\mathbf{F}$) = 0.

14.2 Line Integrals

IN THIS SECTION **definition of a line integral, evaluation of line integrals in parametric form, line integrals of vector fields, computing work using line integrals, evaluation of line integrals with respect to arc length** ▪

In Section 6.5, we showed that the single Riemann integral can be used to compute the work done when an object moves along a line segment against a given force, but what if the object moves along a curve in space? This is one of many situations that occur in science and engineering in which a more general definition of integration is required.

DEFINITION OF A LINE INTEGRAL

To model certain physical notions, such as work or potential, it is appropriate to generalize the original concept of integral by considering limits of sums whose summands involve partitions of a curve, called the *path of integration*. This leads us to the concept of a *line integral,* which is really integration along a curve in space. We begin by defining the line integral of a scalar function f along a curve C with respect to x. The line integral of a scalar function g along C with respect to y or the line integral of h with respect to z may be defined in an analogous fashion.

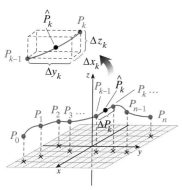

■ **FIGURE 14.7**   A partitioned smooth curve

First, we need to introduce some terminology regarding curves. Recall (Section 11.2) that a curve C with the parametric representation $R(t) = x(t)\mathbf{i} + y(t)\mathbf{j} + z(t)\mathbf{k}$ is said to be *smooth* on the interval $t_1 < t < t_2$ if the derivatives $x'(t)$, $y'(t)$, and $z'(t)$ are all continuous and not all simultaneously zero at any point t on the interval. More generally, C is **piecewise smooth** if it can be decomposed into a finite number of smooth parts. Also, C is said to be **orientable** if it is possible to describe direction along the curve for increasing t.

Assume C is a piecewise smooth, orientable curve that begins at $P = P_0$ and ends at $Q = P_n$, as shown in Figure 14.7. Let C be partitioned into n pieces by the points $P_0, P_1, \ldots, P_n$, and let (x_k, y_k, z_k) be the coordinates of the point P_k. Finally, for $k = 1, 2, \ldots, n$, let $\hat{P}_x(x_k^*, y_k^*, z_k^*)$ be a point chosen arbitrarily from the arc joining points P_{k-1} and P_k, and let $\Delta x_k = x_k - x_{k-1}$. We shall refer to the largest of the Δx_k as the **x-norm** of the partition and shall denote this norm by Δx. For a given scalar function f along the curve C with respect to x, we form the sum

$$\sum_{k=1}^{n} f(\hat{P}_k)\Delta x_k$$

and define the line integral $\int_C f \, dx$ over C with respect to x as follows.

Line Integral

The **line integral** $\int_C f \, dx$ of the scalar function f with respect to x along the piecewise smooth curve C is given by

$$\int_C f \, dx = \lim_{\Delta x \to 0} \sum_{k=1}^{n} f(\hat{P}_k)\Delta x_k$$

provided this limit exists.

The line integrals $\int_C f \, dy$ and $\int_C f \, dz$ are defined similarly.

THEOREM 14.1   *Properties of line integrals*

Let f be a given scalar function defined with respect to x on a piecewise smooth, orientable curve C. Then, for any constant k:

Constant multiple rule: $\displaystyle\int_C kf \, dx = k \int_C f \, dx$

Sum rule: $\displaystyle\int_C (f_1 + f_2) \, dx = \int_C f_1 \, dx + \int_C f_2 \, dx$

where f_1 and f_2 are scalar functions defined with respect to x on C.

Opposite direction rule: $\displaystyle\int_C f \, dx = -\int_C f \, dx$

where $-C$ denotes the curve C traversed in the opposite direction.

Subdivision rule: $\displaystyle\int_C f \, dx = \int_{C_1} f \, dx + \int_{C_2} f \, dx$

where C is subdivided into subarcs $C_1 \cup C_2$ (with $C_1 \cap C_2 = \varnothing$). This property generalizes to any finite number of subdivisions.

Similar properties hold for line integrals of the form $\int_C g \, dy$ or $\int_C h \, dz$.

Proof   The proof follows directly from the properties of limits and the definition of a line integral.   ⬜

EVALUATION OF LINE INTEGRALS IN PARAMETRIC FORM

In practice, the line integral $\int_C f\, dx$ is almost never evaluated using the definition. Instead, we note that if the integrand $f(x, y, z)$ is continuous on C and if C can be represented parametrically in vector form by $\mathbf{R}(t) = x(t)\mathbf{i} + y(t)\mathbf{j} + z(t)\mathbf{k}$, where the derivative $\mathbf{R}'(t)$ exists and is not zero for $a \leq t \leq b$, then

$$\int_C f\, dx = \int_a^b f[x(t), y(t), z(t)]\, \frac{dx}{dt}\, dt$$

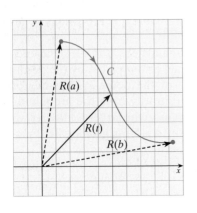

Similarly, if g and h are continuous on C, then

$$\int_C g\, dy = \int_a^b g[x(t), y(t), z(t)]\, \frac{dy}{dt}\, dt$$

$$\int_C h\, dz = \int_a^b h[x(t), y(t), z(t)]\, \frac{dz}{dt}\, dt$$

These formulas enable us to convert a line integral into "ordinary" Riemann integrals that may be evaluated by the methods developed earlier in this text. The same answer is obtained independently of the parametrization chosen.

EXAMPLE 1   *Evaluating a line integral*

Let C be the portion of the parabola $y = x^2$ between $(0, 0)$ to $(2, 4)$. Evaluate

a. $\displaystyle\int_C (x^2 + y)\, dx$ b. $\displaystyle\int_C (x^2 + y)\, dy$

Solution   First, parametrize the curve (this step is not unique). We let $x = t$ so that $y = t^2$ for $0 \leq t \leq 2$.

a. Because $\dfrac{dx}{dt} = 1$, we have

$$\int_C (x^2 + y)\, dx = \int_0^2 [(t)^2 + (t^2)]\, \frac{dx}{dt}\, dt = 2 \int_0^2 t^2\, dt = \left[\frac{2}{3} t^3 \right]_0^2 = \frac{16}{3}$$

b. Now we have $\dfrac{dy}{dt} = 2t$, and we find

$$\int_C (x^2 + y)\, dy = \int_0^2 [(t)^2 + (t^2)]\, \frac{dy}{dt}\, dt = \int_0^2 2t^2 \cdot 2t\, dt = 4 \int_0^2 t^3\, dt = [t^4]_0^2 = 16 \quad ■$$

EXAMPLE 2   *Evaluating a line integral*

Evaluate $\int_C xe^{yz}\, dz$, where C is the curve described parametrically by $x = t, y = t$, and $z = -t$ for $1 \leq t \leq 2$.

Solution
$$\int_C xe^{yz}\, dz = \int_1^2 te^{-t^2}\, \frac{dz}{dt}\, dt$$

$$= \int_1^2 (-t)e^{-t^2}\, dt \qquad \text{Because } \frac{dz}{dt} = -1$$

$$= \tfrac{1}{2}(e^{-4} - e^{-1}) \qquad\qquad ■$$

LINE INTEGRALS OF VECTOR FIELDS

We shall now discuss what it means to compute the **line integral of a vector field.**

Line Integral of a Vector Field

Let $\mathbf{F}(x, y, z) = u(x, y, z)\mathbf{i} + v(x, y, z)\mathbf{j} + w(x, y, z)\mathbf{k}$ be a vector field, and let C be a piecewise smooth curve with parametric representation

$$\mathbf{R}(t) = x(t)\mathbf{i} + y(t)\mathbf{j} + z(t)\mathbf{k} \quad \text{for } a \le t \le b$$

Using $d\mathbf{R} = dx\,\mathbf{i} + dy\,\mathbf{j} + dz\,\mathbf{k}$, we define the **line integral of F along C** by

$$\int_C \mathbf{F} \cdot d\mathbf{R} = \int_C (u\,dx + v\,dy + w\,dz)$$

$$= \int_a^b \left[u[x(t), y(t), z(t)]\frac{dx}{dt} + v[x(t), y(t), z(t)]\frac{dy}{dt} + w[x(t), y(t), z(t)]\frac{dz}{dt} \right] dt$$

EXAMPLE 3   *Line integral of a vector function*

Evaluate $\int_C \mathbf{F} \cdot d\mathbf{R}$, where $\mathbf{F} = (y^2 - z^2)\mathbf{i} + (2yz)\mathbf{j} - x^2\mathbf{k}$ and C is the curve defined parametrically by $x = t^2, y = 2t,$ and $z = t$ for $0 \le t \le 1$.

Solution   Rewrite $\mathbf{F}$ using the parameter:

$$\mathbf{F} = [(2t)^2 - (t)^2]\mathbf{i} + [2(2t)(t)]\mathbf{j} - [(t^2)^2]\mathbf{k} = 3t^2\mathbf{i} + 4t^2\mathbf{j} - t^4\mathbf{k}$$

Also, because $\mathbf{R}(t) = t^2\mathbf{i} + 2t\mathbf{j} + t\mathbf{k}$, we have $d\mathbf{R} = (2t\,dt)\mathbf{i} + (2\,dt)\mathbf{j} + dt\mathbf{k}$, so

$$\mathbf{F} \cdot d\mathbf{R} = (3t^2)(2t\,dt) + (4t^2)(2dt) + (-t^4)(dt)$$
$$= (6t^3 + 8t^2 - t^4)\,dt$$

Thus,

$$\int_C \mathbf{F} \cdot d\mathbf{R} = \int_0^1 (6t^3 + 8t^2 - t^4)\,dt = \left[\frac{3}{2}t^4 + \frac{8}{3}t^3 - \frac{1}{5}t^5 \right]_0^1 = \frac{119}{30} \qquad ∎$$

EXAMPLE 4   *Line integral of a vector defined function*

Compute the line integral $\int_C \mathbf{F} \cdot d\mathbf{R}$, where $\mathbf{F} = y\mathbf{i} + x\mathbf{j}$ and C is the top half of the circle $x^2 + y^2 = 4$ traversed counterclockwise from $(2, 0)$ to $(-2, 0)$.

Solution   First, we parametrize the curve by setting $x = 2\cos\theta, y = 2\sin\theta$ for $0 \le \theta \le \pi$. Thus,

$$\mathbf{R}(\theta) = (2\cos\theta)\mathbf{i} + (2\sin\theta)\mathbf{j} \quad \text{so that} \quad d\mathbf{R} = (-2\sin\theta\,d\theta)\mathbf{i} + (2\cos\theta\,d\theta)\mathbf{j}$$

$$\int_C \mathbf{F} \cdot d\mathbf{R} = \int_0^\pi [(2\sin\theta)\mathbf{i} + (2\cos\theta)\mathbf{j}] \cdot [(-2\sin\theta)\mathbf{i} + (2\cos\theta)\mathbf{j}]\,d\theta$$

$$= \int_0^\pi (-4\sin^2\theta + 4\cos^2\theta)\,d\theta$$

$$= 4\int_0^\pi (\cos^2\theta - \sin^2\theta)\,d\theta = 4\int_0^\pi \cos 2\theta\,d\theta = [2\sin 2\theta]_0^\pi = 0 \qquad ∎$$

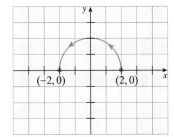

EXAMPLE 5 *Line integrals along different paths*

Let $\mathbf{F} = xy^2\mathbf{i} + x^2y\mathbf{j}$ and evaluate the line integral $\int_C \mathbf{F} \cdot d\mathbf{R}$ between the points $(0,0)$ and $(2,4)$ along the following paths:

a. the line segment connecting the points
b. the parabolic arc $y = x^2$ connecting the points

Solution The two paths we are considering are shown in Figure 14.8.

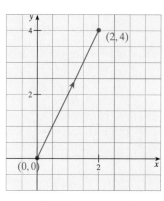
a. The line segment path

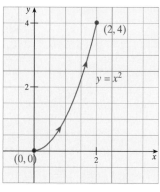
b. The parabolic path

■ **FIGURE 14.8** A line integral along different paths

a. The line joining the given points has equation $y = 2x$, which may be parametrized by setting $x = t, y = 2t$ for $0 \le t \le 2$. Thus,

$$\mathbf{R}(t) = t\mathbf{i} + 2t\mathbf{j} \quad \text{so that} \quad d\mathbf{R} = dt\,\mathbf{i} + 2\,dt\,\mathbf{j}$$

In terms of t, we find $\mathbf{F} = 4t^3\mathbf{i} + 2t^3\mathbf{j}$ and

$$\mathbf{F} \cdot d\mathbf{R} = 4t^3\,dt + 4t^3\,dt = 8t^3\,dt$$

$$\int_C \mathbf{F} \cdot d\mathbf{R} = \int_0^2 8t^3\,dt = [2t^4]_0^2 = 32$$

b. The parabola $y = x^2$ can be parametrized by setting $x = t, y = t^2$ for $0 \le t \le 2$. Thus,

$$\mathbf{R}(t) = t\mathbf{i} + t^2\mathbf{j} \quad \text{so that} \quad d\mathbf{R} = dt\,\mathbf{i} + 2t\,dt\,\mathbf{j}$$

In terms of t,

$$\mathbf{F} = xy^2\mathbf{i} + x^2y\mathbf{j} = (t)(t^2)^2\mathbf{i} + (t)^2(t^2)\mathbf{j} = t^5\mathbf{i} + t^4\mathbf{j}$$

and

$$\mathbf{F} \cdot d\mathbf{R} = t^5\,dt + 2t^5\,dt = 3t^5\,dt$$

$$\int_C \mathbf{F} \cdot d\mathbf{R} = \int_0^2 3t^5\,dt = \left[\frac{1}{2}t^6\right]_0^2 = 32 \qquad ■$$

In Example 5, we see that the value of the line integral is the same for both paths. Indeed, it can be shown that for the vector field $\mathbf{F} = xy^2\mathbf{i} + x^2y\mathbf{j}$, the line integral $\int_C \mathbf{F} \cdot d\mathbf{R}$ along *any* path C joining $(0,0)$ to $(2,4)$ has the same value. This, of course, is not true for every $\mathbf{F}$, but if it is true, we say that the line integral is **independent of path.** Path independence is an important feature of line integration and will be discussed in detail in the next section.

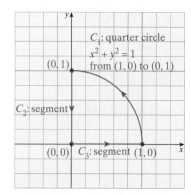
■ **FIGURE 14.9** A path C composed of three parts

EXAMPLE 6 *Line integral over a path consisting of several parts*

Evaluate $\int_C (-y\,dx + x\,dy)$, where C is the closed path shown in Figure 14.9.

Solution   Because the given closed path can be best described using three separate equations ($C_1, C_2,$ and C_3 in Figure 14.9), we perform the integration by evaluating the line integrals for each part separately, and then adding. Let $\mathbf{F} = -y\mathbf{i} + x\mathbf{j}$, so

$$\int_C (-y\,dx + x\,dy) = \int_C \mathbf{F} \cdot d\mathbf{R}$$

Then

$$\int_C \mathbf{F} \cdot d\mathbf{R} = \int_{C_1} \mathbf{F} \cdot d\mathbf{R} + \int_{C_2} \mathbf{F} \cdot d\mathbf{R} + \int_{C_3} \mathbf{F} \cdot d\mathbf{R}$$

Along C_1, a parametrization is $x = \cos t, y = \sin t$ for $0 \le t \le \frac{\pi}{2}$, so $\mathbf{R}(t) = (\cos t)\mathbf{i} + (\sin t)\mathbf{j}$ and $d\mathbf{R} = (-\sin t\,dt)\mathbf{i} + (\cos t\,dt)\mathbf{j}$. In terms of the parameter, $\mathbf{F} = (-\sin t)\mathbf{i} + (\cos t)\mathbf{j}$ and

$$\int_{C_1} \mathbf{F} \cdot d\mathbf{R} = \int_0^{\pi/2} (\sin^2 t\,dt + \cos^2 t\,dt) = \int_0^{\pi/2} dt = \frac{\pi}{2}$$

Along C_2, a parametrization is $x = 0, y = 1 - t$ for $0 \le t \le 1$, so $\mathbf{R}(t) = (1 - t)\mathbf{j}$ and $d\mathbf{R} = (-dt)\mathbf{j}$. In terms of the parameter, $\mathbf{F} = (t - 1)\mathbf{i}$ and

$$\int_{C_2} \mathbf{F} \cdot d\mathbf{R} = \int_0^1 0\,d\mathbf{R} = 0$$

Along C_3, a parametrization is $x = t, y = 0$ for $0 \le t \le 1$. $\mathbf{R}(t) = t\mathbf{i}, d\mathbf{R} = dt\mathbf{i}$, and $\mathbf{F} = t\mathbf{j}$. Because $\mathbf{F} \cdot d\mathbf{R} = 0$, we see

$$\int_{C_3} \mathbf{F} \cdot d\mathbf{R} = 0$$

Thus,

$$\int_C \mathbf{F} \cdot d\mathbf{R} = \int_{C_1} \mathbf{F} \cdot d\mathbf{R} + \int_{C_2} \mathbf{F} \cdot d\mathbf{R} + \int_{C_3} \mathbf{F} \cdot d\mathbf{R} = \frac{\pi}{2} + 0 + 0 = \frac{\pi}{2} \qquad \blacksquare$$

COMPUTING WORK USING LINE INTEGRALS

One of the most important physical applications of the line integral is in computing work. Recall from Section 10.3 that if an object moves along a straight line with displacement $\mathbf{R}$ in a constant force field $\mathbf{F}$, the work done is $\mathbf{F} \cdot \mathbf{R}$. However, the case where $\mathbf{F}$ is not constant and the object moves along a smooth curve C requires extra attention (see Figure 14.10). To analyze this case, assume that C is parametrized by $\mathbf{R}(t)$ and is oriented in the direction of motion. Partition C with subdivision points $P_0, P_1, P_2, \ldots, P_n$, as shown in Figure 14.11.

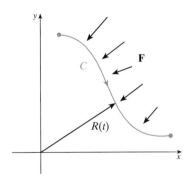

■ **FIGURE 14.10**   If $\mathbf{F}$ and $\mathbf{R}$ are constant, work $= \mathbf{F} \cdot \mathbf{R}$. If they are not, then a line integral is required

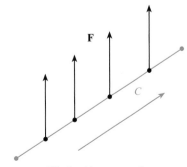

a. Work with constant force in a fixed direction

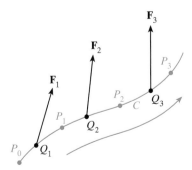

b. Work with variable force along a smooth curve

■ **FIGURE 14.11**   Computing work

For $k = 1, 2, \ldots, n$, let Q_k be a point chosen arbitrarily from the kth subarc C_k (with endpoints P_{k-1} and P_k), and let $\mathbf{F}_k$ be the value of the force field $\mathbf{F}$ at Q_k. If the length of the subarc C_k is small, $\mathbf{F}$ will be approximately constant with value $\mathbf{F}_k$ over C_k, and the object's displacement along C_k is given approximately by the secant vector $\Delta \mathbf{R}_k$ from P_{k-1} to P_k. Then, the work done by the force as the object moves along C_k may be estimated by

$$\Delta W_k = \left(\mathbf{F}_k \cdot \frac{\Delta \mathbf{R}_k}{\Delta t} \right) \Delta t$$

By adding the contributions along all n subarcs, we obtain

$$\left(\mathbf{F}_1 \cdot \frac{\Delta \mathbf{R}_1}{\Delta t} \right) \Delta t + \left(\mathbf{F}_2 \cdot \frac{\Delta \mathbf{R}_2}{\Delta t} \right) \Delta t + \cdots + \left(\mathbf{F}_n \cdot \frac{\Delta \mathbf{R}_n}{\Delta t} \right) \Delta t = \sum_{k=1}^{n} \left(\mathbf{F}_k \cdot \frac{\Delta \mathbf{R}_k}{\Delta t} \right) \Delta t$$

as an estimate of the total work done by $\mathbf{F}$ as the object moves along C. As $\Delta t \to 0$, the limiting value of this sum is the line integral of $\mathbf{F} \cdot \dfrac{d\mathbf{R}}{dt}$; that is,

$$W = \lim_{\Delta t \to 0} \sum_{k=1}^{n} \left(\mathbf{F}_k \cdot \frac{\Delta \mathbf{R}_k}{\Delta t} \right) \Delta t = \int_C \mathbf{F} \cdot \frac{d\mathbf{R}}{dt}\, dt = \int_C \mathbf{F} \cdot d\mathbf{R}$$

This leads us to the following definition.

Work as a Line Integral

Let $\mathbf{F}$ be a continuous force field over a domain D. Then the **work** W done by $\mathbf{F}$ as an object moves along a smooth curve C in D is given by the line integral

$$W = \int_C \mathbf{F} \cdot d\mathbf{R}$$

EXAMPLE 7   *Work as a line integral*

An object moves in the force field

$$\mathbf{F} = y^2 \mathbf{i} + 2(x + 1)y\mathbf{j}$$

counterclockwise from the point $(2,0)$ along the elliptical path $x^2 + 4y^2 = 4$ to $(-2,0)$, and back to the point $(2,0)$ along the x-axis. How much work is done by the force field on the object?

Solution   Let C denote the path of the object. Then the total work W done by $\mathbf{F}$ on the object as it moves along C is given by the line integral $\int_C \mathbf{F} \cdot d\mathbf{R}$. We divide the curve C into two parts so that $C = C_1 \cup C_2$. The curve $x^2 + 4y^2 = 4$ suggests the parametrization $\dfrac{x}{2} = \cos t$, $y = \sin t$ for $0 \le t \le \pi$. Thus,

$$\mathbf{R}(t) = (2\cos t)\mathbf{i} + (\sin t)\mathbf{j}$$

and

$$d\mathbf{R} = (-2\sin t\, dt)\mathbf{i} + (\cos t\, dt)\mathbf{j}$$

In parametric form,

$$\mathbf{F} = (\sin^2 t)\mathbf{i} + (4\cos t \sin t + 2\sin t)\mathbf{j}$$

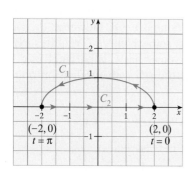

so that

$$W_1 = \int_{C_1} \mathbf{F} \cdot d\mathbf{R} = \int_0^{\pi} (-2 \sin^3 t + 4 \cos^2 t \sin t + 2 \sin t \cos t)\, dt$$

$$= \int_0^{\pi} (-2 \sin^2 t + 4 \cos^2 t + 2 \cos t) \sin t\, dt$$

$$= \int_0^{\pi} (6 \cos^2 t + 2 \cos t - 2) \sin t\, dt$$

$$= -\int_1^{-1} (6u^2 + 2u - 2)\, du = 0$$

Let $u = \cos t$;
$du = -\sin t\, dt$.
If $t = 0, u = 1$;
and if $t = \pi, u = -1$.

The curve C_2 is described by $y = 0$, so $\mathbf{F} = \mathbf{0}$ and, therefore, $W_2 = 0$. Thus, $W = W_1 + W_2 = 0$. ∎

It can be shown that the work in moving an object around *any* closed path against the force in Example 7 will always be 0. When this occurs, the force is said to be **conservative**. We shall discuss conservative force fields in Sections 14.3 and 14.4.

EVALUATION OF LINE INTEGRALS WITH RESPECT TO ARC LENGTH

Line integrals of the form $\int_C \mathbf{F} \cdot d\mathbf{R}$ can often be expressed in other forms. For example, recall from Chapter 11 that $\mathbf{T} = \dfrac{d\mathbf{R}}{ds}$ is a unit tangent vector to the curve C at the point $P(x, y, z)$, where s is the arc length parameter. We have

$$\int_C \mathbf{F} \cdot d\mathbf{R} = \int_C \mathbf{F} \cdot \frac{d\mathbf{R}}{ds}\, ds = \int_C \mathbf{F} \cdot \mathbf{T}\, ds$$

In particular, the work W done by a force field $\mathbf{F}$ on an object moving along a curve C may be expressed as

$$W = \int_C \mathbf{F} \cdot d\mathbf{R} = \int_C \mathbf{F} \cdot \mathbf{T}\, ds$$

This form of the integral is called the **line integral of the tangential component of F** and can also be written as $\int_C f(x, y, z)\, ds$. The integral will exist if f is continuous on the curve C and if C itself is piecewise smooth with finite length. A formula for computing this line integral is suggested by the following observations: If $\mathbf{R}(t) = x(t)\mathbf{i} + y(t)\mathbf{j} + z(t)\mathbf{k}$, then

$$\frac{ds}{dt} = \left\| \frac{d\mathbf{R}}{dt} \right\| = \sqrt{\left(\frac{dx}{dt}\right)^2 + \left(\frac{dy}{dt}\right)^2 + \left(\frac{dz}{dt}\right)^2}$$

so that

$$W = \int_C \mathbf{F} \cdot \mathbf{T}\, ds = \int_a^b f[x(t), y(t), z(t)] \sqrt{[x'(t)]^2 + [y'(t)]^2 + [z'(t)]^2}\, dt$$

Evaluation of a Line Integral with Respect to Arc Length

Let f be continuous on a smooth curve C. If C is defined by $\mathbf{R}(t) = x(t)\mathbf{i} + y(t)\mathbf{j} + y(t)\mathbf{k}$, where $a \le t \le b$, then

$$\int_C f(x, y, z)\, ds = \int_a^b f[x(t), y(t), z(t)] \sqrt{[x'(t)]^2 + [y'(t)]^2 + [z'(t)]^2}\, dt$$

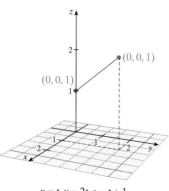

$x = t, y = 2t, z = t + 1$
for $0 \le t \le 1$

$\mathbf{R}(t) = t\mathbf{i} + 2t\mathbf{j} + (t + 1)\mathbf{k}$ for
$0 \le t \le 1$

To summarize:

| EXAMPLE 8 | *Evaluating a line integral with respect to arc length* |

Find $\displaystyle\int_C (x + y^2 - z)\, ds$, where C is the line segment described parametrically by
$x = t, y = 2t, z = t + 1$ for $0 \le t \le 1$.

Solution   Because $\dfrac{dx}{dt} = 1, \dfrac{dy}{dt} = 2$, and $\dfrac{dz}{dt} = 1$, we have

$$\int_C (x + y^2 - z)\, ds = \int_0^1 [t + (2t)^2 - (t + 1)]\sqrt{(1)^2 + (2)^2 + (1)^2}\, dt$$

$$= \sqrt{6} \int_0^1 (4t^2 - 1)\, dt = \frac{1}{3}\sqrt{6} \qquad \blacksquare$$

14.2 Problem Set

A **1.** ■ **What Does This Say?** Describe a line integral.
2. ■ **What Does This Say?** Discuss the evaluation of a
line integral.

Evaluate each line integral given in Problems 3–16.

3. $\displaystyle\int_C (-y\, dx + x\, dy)$
 C is the parabolic path
 $y = 4x^2$ from $(1, 4)$ to $(0, 0)$.

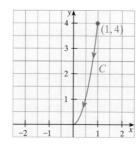

4. $\displaystyle\int_C (-y\, dx + 3x\, dy)$
 C is the parabolic path $y^2 = x$
 from $(1, 1)$ to $(9, 3)$.

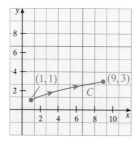

5. $\displaystyle\int_C (x\, dy - y\, dx)$
 C is the line defined
 $2x - 4y = 1$ as x varies from
 4 to 8.

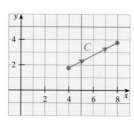

6. $\displaystyle\int_C [(y - x)\, dx + x^2 y\, dy]$
 C is the curve defined by
 $y^2 = x^3$ from $(1, -1)$ to $(1, 1)$.

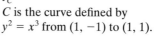

7. $\displaystyle\int_C [(x + y)^2\, dx - (x - y)^2\, dy]$
 C is the curve defined by
 $y = |2x|$ from $(-1, 2)$ to
 $(1, 2)$.

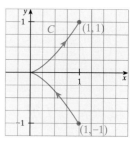

8. $\displaystyle\int_C [(y^2 - x^2)\, dx - x\, dy]$
 C is the quarter-circle
 $x^2 + y^2 = 4$ from $(0, 2)$ to
 $(2, 0)$.

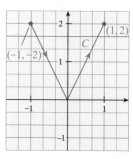

9. $\displaystyle\int_C [(x^2 + y^2)\, dx + 2xy\, dy]$ for these choices of the curve C:
 a. C is the quarter-circle $x^2 + y^2 = 1$ traversed counter-
 clockwise from $(1, 0)$ to $(0, 1)$.
 b. C is the straight line $y = 1 - x$ from $(1, 0)$ to $(0, 1)$.

10. $\int_C [x^2 y \, dx + (x^2 - y^2) \, dy]$ for these choices of the curve C:
    a. C is the arc of the parabola $y = x^2$ from $(0, 0)$ to $(2, 4)$.
    b. C is the segment of the line $y = 2x$ for $0 \leq x \leq 2$.

11. Rework Problem 10 for the path C that consists of the horizontal line segment $(0, 0)$ to $(2, 0)$, followed by the vertical segment from $(2, 0)$ to $(2, 4)$.

12. $\int_C (-xy^2 \, dx + x^2 \, dy)$ where C is the path defined by the given figure.

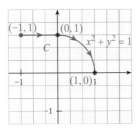

13. $\int_C (-y^2 \, dx + x^2 \, dy)$ where C is the path defined by the given figure.

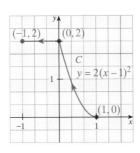

14. $\int_C [(x^2 - y^2) \, dx + x \, dy]$, where C is the closed circular path given by $x = 2 \cos \theta, y = 2 \sin \theta, 0 \leq \theta \leq 2\pi$.

15. $\int_C (x^2 y \, dx - xy \, dy)$, where C is the closed path that begins at $(0, 0)$, goes to $(1, 1)$ along the parabola $y = x^2$, and then returns to $(0, 0)$ along the line $y = x$.

16. $\int_C \dfrac{x \, dx - y \, dy}{\sqrt{x^2 + y^2}}$, where C is the quarter-circular path $x^2 + y^2 = a^2$, traversed from $(a, 0)$ to $(0, a)$.

In Problems 17–19, *evaluate* $\int_C \mathbf{F} \cdot d\mathbf{R}$, *where* $\mathbf{F} = (5x + y)\mathbf{i} + x\mathbf{j}$ *and C is the specified curve.*

17. C is the straight line segment from $(0, 0)$ to $(2, 1)$.

18. C is the curve given by $\mathbf{R}(t) = 2t\mathbf{i} + t\mathbf{j}$ for $0 \leq t \leq 1$.

19. C is the vertical line from $(0, 0)$ to $(0, 1)$, followed by the horizontal line from $(0, 1)$ to $(2, 1)$.

Evaluate the line integrals in Problems 20–35.

20. $\int_C (y \, dx - x \, dy + dz)$, where C is the helical path given by:
    a. $x = 3 \sin t, y = 3 \cos t, z = t$ for $0 \leq t \leq \frac{\pi}{2}$
    b. $x = a \sin t, y = a \cos t, z = t$ for constant a and $0 \leq t \leq \frac{\pi}{2}$

21. $\int_C (x \, dx + y \, dy + z \, dz)$, where C is the following path:
    a. the helix defined by $x = \cos t, y = \sin t, z = t$ for $0 \leq t \leq \frac{\pi}{2}$
    b. the straight line segment from $(1, 0, 0)$ to $(0, 1, \frac{\pi}{2})$

22. $\int_C (-y \, dx + x \, dy + xz \, dz)$, where C is the following path:
    a. the helix defined by $x = \cos t, y = \sin t, z = t$ for $0 \leq t \leq 2\pi$
    b. the unit circle $x^2 + y^2 = 1, z = 0$, traversed once counterclockwise as viewed from above

23. $\int_C (5 \, xy \, dx + 10 \, yz \, dy + z \, dz)$, where C is the following path:
    a. the parabolic arc $x = y^2$ from $(0, 0, 0)$ to $(1, 1, 0)$ followed by the line segment given by $x = 1, y = 1, 0 \leq z \leq 1$
    b. the straight line segment from $(0, 0, 0)$ to $(1, 1, 1)$

24. $\int_C \dfrac{dx + dy}{|x| + |y|}$, where C is the square $|x| + |y| = 1$, traversed once counterclockwise

25. $\int_C [(y + z) \, dx + (x + z) \, dy + (x + y) \, dz]$ where C is the circle of radius 1 centered on the z-axis in the plane $z = 2$, traversed once counterclockwise as viewed from above.

26. $\int_C \mathbf{F} \cdot d\mathbf{R}$, where $\mathbf{F} = (y - 3)\mathbf{i} + x\mathbf{j}$, and C is the curve given by
    $$\mathbf{R}(t) = (\sin t)\mathbf{i} - (\cos t)\mathbf{j} \quad \text{for } 0 \leq t \leq \pi$$

27. $\int_C \mathbf{F} \cdot d\mathbf{R}$, where $\mathbf{F} = (y - 2z)\mathbf{i} + x\mathbf{j} - 2xy\mathbf{k}$ and C is the path given by
    $$\mathbf{R}(t) = t\mathbf{i} + t^2\mathbf{j} - \mathbf{k} \quad \text{for } 1 \leq t \leq 2$$

28. $\int_C \mathbf{F} \cdot d\mathbf{R}$, where $\mathbf{F} = x\mathbf{i} + xy\mathbf{j} + x^2yz\mathbf{k}$, and C is the elliptical path given by $x^2 + 4y^2 - 8y + 3 = 0$ in the xy-plane, traversed once counterclockwise as viewed from above.

29. $\int_C \mathbf{F} \cdot d\mathbf{R}$, where $\mathbf{F} = y^2\mathbf{i} + x^2\mathbf{j} - (x + z)\mathbf{k}$ and C is the boundary of the triangle with vertices $(0, 0, 0), (1, 0, 0)$, $(1, 1, 0)$, traversed once clockwise, as viewed from above.

30. $\int_C \mathbf{F} \cdot \mathbf{T} \, ds$, where $\mathbf{F} = -3y\mathbf{i} + 3x\mathbf{j} + 3x\mathbf{k}$, and C is the straight line segment from $(0, 0, 1)$ to $(1, 1, 1)$.

31. $\int_C \mathbf{F} \cdot \mathbf{T} \, ds$, where $\mathbf{F} = -x\mathbf{i} + 2\mathbf{j}$, and C is the boundary of the trapezoid with vertices $(0, 0), (1, 0), (2, 1), (0, 1)$, traversed once clockwise as viewed from above.

32. $\int_C y \, ds$, where C is the curve given by
    $$\mathbf{R}(t) = t\mathbf{i} + 2t^3\mathbf{j}, \quad 0 \leq t \leq 2$$

33. $\int_C (x + y) \, ds$, where C is given by
    $$\mathbf{R}(t) = (\cos^2 t)\mathbf{i} + (\sin^2 t)\mathbf{j}, \quad -\tfrac{\pi}{4} \leq t \leq 0$$

34. $\int_C \dfrac{x^2 + xy + y^2}{z^2} \, ds$, where C is the path given by
    $$\mathbf{R}(t) = (\cos t)\mathbf{i} + (\sin t)\mathbf{j} - \mathbf{k} \text{ for } 0 \leq t \leq 2\pi.$$

35. Evaluate the line integral
    $$\int_C \frac{x \, dy - y \, dx}{x^2 + y^2}$$
    where C is the unit circle $x^2 + y^2 = 1$ traversed once counterclockwise.

Ⓑ 36. How much work is done by a constant force $\mathbf{F} = a\mathbf{i} + \mathbf{j}$ when a particle moves along the line $y = ax$ from $x = a$ to $x = 0$?

37. A force field in the plane is given by $\mathbf{F} = (x^2 - y^2)\mathbf{i} + 2xy\mathbf{j}$. Find the total work done by this force in moving a point mass counterclockwise around the square with vertices $(0,0), (2,0), (2,2), (0,2)$.

38. Find the work done by the force field $\mathbf{F} = (x^2 + y^2)\mathbf{i} + (x + y)\mathbf{j}$ as an object moves counterclockwise along the circle $x^2 + y^2 = 1$ from $(1,0)$ to $(-1, 0)$, and then back to $(1, 0)$ along the x-axis.

39. A force acting on a point mass located at (x, y) is given by $\mathbf{F} = y\mathbf{i} + 2x\mathbf{j}$. Find the work done by this force as the point mass moves along a straight line from $(1, 0)$ to $(0, 1)$.

Find the work done by the force $\mathbf{F}(x, y, z)$ on an object moving along the curve C in Problems 40–43.

40. $\mathbf{F} = (y^2 - z^2)\mathbf{i} + 2yz\mathbf{j} - x^2\mathbf{k}$, and C is the path given by $x(t) = t, y(t) = t^2, z(t) = t^3$, for $0 \le t \le 1$.

41. $\mathbf{F} = 2xy\mathbf{i} + (x^2 + 2)\mathbf{j} + y\mathbf{k}$, and C is the line segment from $(1, 0, 2)$ to $(3, 4, 1)$.

42. $\mathbf{F} = x\mathbf{i} + y\mathbf{j} + (xz - y)\mathbf{k}$, and C is the line segment from $(0, 0, 0)$ to $(2, 1, 2)$.

43. $\mathbf{F} = x\mathbf{i} + y\mathbf{j} + (xz - y)\mathbf{k}$, and C is the path given by $\mathbf{R}(t) = t^2\mathbf{i} + 2t\mathbf{j} + 4t^3\mathbf{k}$ for $0 \le t \le 1$.

44. A 180-lb laborer carries a bag of sand weighing 40 lb up a circular helical staircase on the outside of a

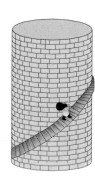

tower 50 ft high and 20 ft in diameter. How much work is done by gravity as the laborer climbs to the top in exactly five revolutions?

45. Repeat Problem 44 assuming that the bag leaks 1 lb of sand for every 10 ft of ascent. How much work is done by gravity during the laborer's climb to the top?

46. A 5,000-lb satellite orbits the earth in a circular orbit 5,000 mi from the center of the earth. How much work is done on the satellite by gravity during one complete revolution?

Ⓒ 47. Suppose a particle with charge Q and mass m moves with velocity $\mathbf{V}$ under the influence of an electric field $\mathbf{E}$ and a magnetic field $\mathbf{B}$. Then the total force on the particle is $\mathbf{F} = Q(\mathbf{E} + \mathbf{V} \times \mathbf{B})$, called the **Lorentz force**. Use Newton's second law of motion, $\mathbf{F} = m\mathbf{A}$, to show that

$$m\frac{d\mathbf{V}}{dt} \cdot \mathbf{V} = Q\mathbf{E} \cdot \frac{d\mathbf{R}}{dt}$$

and then evaluate the line integral $\displaystyle\int_C \mathbf{E} \cdot d\mathbf{R}$, where $\mathbf{C}$ is the trajectory of a particle traveling with constant speed.

48. Suppose a thin wire fits a smooth curve C in $\mathbb{R}^3$, and let $\delta(x, y, z)$ be the density of the wire. Explain why the mass m of the wire is given by the line integral

$$m = \int_C \delta(x, y, z)\, ds$$

Express the center of mass of the wire in terms of line integrals.

14.3   *Independence of Path*

IN THIS SECTION   conservative vector fields, fundamental theorem of line integrals   ■

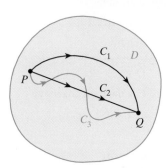

A line integral $\int_C \mathbf{F} \cdot d\mathbf{R}$ that has the same value for any curve C in D with given endpoints is said to be independent of path

In general, the value of the line integral $\displaystyle\int_C \mathbf{F} \cdot d\mathbf{R}$ depends on the path of integration C, but in certain cases, the integral will be the same for all paths in a given region D with the same initial point P and terminal point Q. In this case, we say the line integral is **independent of path** in D. In this section, we shall study path independence and characterize the kind of vector field $\mathbf{F}$ for which it occurs.

CONSERVATIVE VECTOR FIELDS

The issue of path independence for the line integral $\displaystyle\int_C \mathbf{F} \cdot d\mathbf{R}$ is closely related to determining whether $\mathbf{F}$ is the gradient of some scalar function f. It is useful to have the following terminology.

Conservative Vector Field

A vector field **F** is said to be **conservative** in a region D if it can be represented in D as the gradient of a continuously differentiable function f, which is then called a **scalar potential** of **F**. That is,

$$\mathbf{F} = \nabla f \quad \text{for } (x, y) \text{ in } D$$

$\uparrow$ Conservative vector field $\uparrow$ Scalar potential

a. A simply connected region

b. A region that is *not* simply connected

■ **FIGURE 14.12** Connected regions

■ **EXAMPLE 1**   *Conservative vector field*

Verify that the vector field $\mathbf{F} = 2xy\mathbf{i} + x^2\mathbf{j}$ is conservative, with scalar potential $f = x^2 y$.

Solution $\nabla f = 2xy\mathbf{i} + x^2\mathbf{j}$ and this is the same as **F**, so **F** is conservative.   ■

As we work this example, two questions come to mind:

1. Does there exist an easy test to determine whether **F** is conservative?

2. Once such a determination has been made, how can a scalar potential function f be found?

Before answering these questions, we need to introduce some new terminology. Specifically, a region D is **connected** if any two points in D can be joined by a piecewise smooth curve that lies entirely within D. If, in addition, every closed curve in D encloses only points that are also in D, then D is **simply connected.** Roughly speaking, a simply connected region is one with no "holes," as shown in Figure 14.12.

THEOREM 14.2   *Cross-partials test for a conservative vector field in the plane*

Consider the vector field $\mathbf{F}(x, y) = u(x, y)\mathbf{i} + v(x, y)\mathbf{j}$, where u and v have continuous first partials in the open, simply connected region D. Then $\mathbf{F}(x, y)$ is conservative in D if and only if

$$\frac{\partial u}{\partial y} = \frac{\partial v}{\partial x} \quad \text{throughout } D$$

Proof   We outline the proof in Problem Set 14.4 after our discussion of Green's theorem.   ═

In the next two examples, we use this theorem to show that a given vector field **F** is conservative. Then we "partially integrate" the components of **F** to obtain a scalar potential of **F** such that $\nabla f = \mathbf{F}$.

■ **EXAMPLE 2**   *Finding a scalar potential function*

Show that the vector field $\mathbf{F} = (e^x \sin y - y)\mathbf{i} + (e^x \cos y - x - 2)\mathbf{j}$ is conservative and then find a scalar potential function f for **F**.

Solution   Let $u(x, y) = e^x \sin y - y$ and $v(x, y) = e^x \cos y - x - 2$. Then

$$\frac{\partial u}{\partial y} = e^x \cos y - 1 \quad \text{and} \quad \frac{\partial v}{\partial x} = e^x \cos y - 1$$

Since $\dfrac{\partial u}{\partial y} = \dfrac{\partial v}{\partial x}$, it follows that **F** is conservative. To find a scalar potential function f

such that $\nabla f = \mathbf{F}$, we note that f must satisfy $u(x, y) = f_x(x, y)$ and $v(x, y) = f_y(x, y)$.

$$f(x, y) = \int u(x, y) \, dx = \underbrace{\int (e^x \sin y - y) \, dx}$$

This is the "partial integral" in the sense that y is held constant while the integration is performed with respect to x alone.

$$= e^x \sin y - yx + c(y)$$

Where $c(y)$ is a function of y alone—a "constant" as far as x-integration is concerned

Because f must also satisfy $f_y(x, y) = v(x, y)$, we compute the partial derivative of this result with respect to y:

$$f_y(x, y) = \frac{\partial}{\partial y}[e^x \sin y - yx + c(y)] = e^x \cos y - x + \frac{dc}{dy}$$

Set this equal to $v = e^x \cos y - x - 2$ and solve for $\dfrac{dc}{dy}$:

$$e^x \cos y - x + \frac{dc}{dy} = e^x \cos y - x - 2$$

$$\frac{dc}{dy} = -2$$

$$c(y) = -2y + C$$

We now have found the function $f(x, y) = e^x \sin y - xy - 2y + C$. Any such function is a scalar potential of $\mathbf{F}$ and, for simplicity, we pick $C = 0$:

$$f(x, y) = e^x \sin y - xy - 2y \qquad \blacksquare$$

In Example 2, we began by using the fact that $f_x = u$. In general, the issue of whether to start with $u = f_x$ or $v = f_y$ is a matter of personal taste and is often determined by which equation leads to the simpler integration.

EXAMPLE 3   *Testing for a conservative vector field in the plane*

Determine whether the vector field $\mathbf{F} = ye^{xy}\mathbf{i} + (xe^{xy} + x)\mathbf{j}$ is conservative; if it is, find a scalar potential.

Solution   We have $u(x, y) = ye^{xy}$ and $v(x, y) = xe^{xy} + x$.

$$\frac{\partial u}{\partial y} = xye^{xy} + e^{xy} \qquad \frac{\partial v}{\partial x} = xye^{xy} + e^{xy} + 1$$

so $\dfrac{\partial u}{\partial y} \neq \dfrac{\partial v}{\partial x}$, and $\mathbf{F}$ is not conservative. $\qquad \blacksquare$

FUNDAMENTAL THEOREM OF LINE INTEGRALS

Recall that, according to the fundamental theorem of calculus, if the function f is continuous on $[a, b]$, then

$$\int_a^b f(x) \, dx = F(b) - F(a)$$

where F is any antiderivative of f; that is, $F'(x) = f(x)$. The following is the analogous result for line integrals.

THEOREM 14.3   *Fundamental theorem of line integrals*

Let **F** be a conservative vector field on the region D and let f be a scalar potential function for **F**; that is, $\nabla f = \mathbf{F}$. Then, if C is any piecewise smooth curve lying entirely within D, with initial point P and terminal point Q, we have

$$\int_C \mathbf{F} \cdot d\mathbf{R} = f(Q) - f(P)$$

Thus, the line integral $\int_C \mathbf{F} \cdot d\mathbf{R}$ is independent of path in D.

Proof   We shall prove this theorem for the case where the curve C is smooth in D, leaving the more general case where C is piecewise smooth as an exercise. Suppose C is described by the vector function $\mathbf{R}(t) = x(t)\mathbf{i} + y(t)\mathbf{j}$, where $a \leq t \leq b$, and $P = \mathbf{R}(a), Q = \mathbf{R}(b)$. Because $\mathbf{F}(x, y) = \nabla f(x, y) = f_x(x, y)\mathbf{i} + f_y(x, y)\mathbf{j}$, we have

$$\int_C \mathbf{F} \cdot d\mathbf{R} = \int_a^b \mathbf{F} \cdot \frac{d\mathbf{R}}{dt}\, dt$$

$$= \int_a^b \left[f_x(x, y)\frac{dx}{dt} + f_y(x, y)\frac{dy}{dt} \right] dt \qquad \textit{Because } F = \nabla f$$

$$= \int_a^b \frac{d}{dt}\{f[x(t), y(t)]\}\, dt \qquad \textit{Chain rule (in reverse)}$$

$$= f[x(b), y(b)] - f[x(a), y(a)] \qquad \textit{Fundamental theorem of calculus}$$

$$= f[\mathbf{R}(b)] - f[\mathbf{R}(a)]$$

$$= f(Q) - f(P)$$

EXAMPLE 4   *Evaluating a line integral using the fundamental theorem of line integrals*

Evaluate the line integral $\int_C \mathbf{F} \cdot d\mathbf{R}$, where

$$\mathbf{F} = (e^x \sin y - y)\mathbf{i} + (e^x \cos y - x - 2)\mathbf{j} \quad \text{and } C \text{ is the path given by}$$

$$\mathbf{R}(t) = \left[t^3 \sin \frac{\pi t}{2} \right]\mathbf{i} - \left[\frac{\pi}{2}\cos\left(\frac{\pi t}{2} + \frac{\pi}{2}\right) \right]\mathbf{j} \quad \text{for } 0 \leq t \leq 1.$$

Solution   Evaluating this line integral by the parametric method would be both difficult and tedious. However, we showed in Example 2 that the vector field **F** is conservative with scalar potential

$$f(x, y) = e^x \sin y - xy - 2y$$

and according to the fundamental theorem on line integrals, the value of the line integral depends only on the value of f at the endpoints of the path C. (You should also verify that the hypotheses of the theorem are satisfied by **F** and C.)

At the endpoint where $t = 0$: $\mathbf{R}(0) = 0\mathbf{i} - [\frac{\pi}{2}\cos(\frac{\pi}{2})]\mathbf{j} = 0\mathbf{i} + 0\mathbf{j}$

$$f(0, 0) = e^0 \sin 0 - 0 - 0 = 0$$

At the endpoint where $t = 1$: $\mathbf{R}(1) = [\sin \frac{\pi}{2}]\mathbf{i} - [\frac{\pi}{2}\cos \pi]\mathbf{j} = \mathbf{i} + \frac{\pi}{2}\mathbf{j}$

$$f(1, \tfrac{\pi}{2}) = e^1 \sin \tfrac{\pi}{2} - \tfrac{\pi}{2} - 2(\tfrac{\pi}{2}) = e - \tfrac{3\pi}{2}$$

We now apply the fundamental theorem for line integrals:

$$\int_C \mathbf{F} \cdot d\mathbf{R} = f(Q) - f(P) = f(1, \tfrac{\pi}{2}) - f(0, 0)$$

$$= (e - \tfrac{3\pi}{2}) - 0 = e - \tfrac{3\pi}{2} \qquad \blacksquare$$

EXAMPLE 5 *Work along a closed path in a conservative force field*

Show that no net work is done when an object moves along a closed path back to its starting point in a connected domain where the force field is conservative.

Solution In such a force field **F**, we have $\nabla f = \mathbf{F}$, where f is a scalar potential of **F**, and because the path of motion is closed, it begins and ends at the same point P. Thus, the work is given by

$$W = \int_C \mathbf{F} \cdot d\mathbf{R} = f(P) - f(P) = 0 \qquad \blacksquare$$

This result appears as part of the proof of the following useful theorem, which ties together several equivalent conditions for path independence.

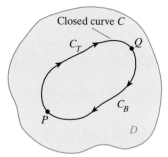

■ FIGURE 14.13 A continuous vector field **F** is conservative in the region D if and only if $\int_C \mathbf{F} \cdot d\mathbf{R} = 0$ for every closed curve C in D

THEOREM 14.4 *Closed curve theorem for a conservative force field*

The continuous vector field **F** is conservative in the open connected region D if and only if $\int_C \mathbf{F} \cdot d\mathbf{R} = 0$ for every piecewise smooth closed curve C in D.

Proof If a line integral $\int_C \mathbf{F} \cdot d\mathbf{R}$ is independent of path in the open, connected region D, then $\int_C \mathbf{F} \cdot d\mathbf{R} = 0$ for any piecewise smooth closed path C in D. Indeed, if P and Q are two points on such a path, and C_T is the path from P to Q along, say, the "top" of C, while C_B is the "bottom" path along C from Q to P (see Figure 14.13), we must have $\int_{C_B} \mathbf{F} \cdot d\mathbf{R} = -\int_{C_T} \mathbf{F} \cdot d\mathbf{R}$ and

$$\int_C \mathbf{F} \cdot d\mathbf{R} = \int_{C_T} \mathbf{F} \cdot d\mathbf{R} + \int_{C_B} \mathbf{F} \cdot d\mathbf{R} = \int_{C_T} \mathbf{F} \cdot d\mathbf{R} - \int_{C_T} \mathbf{F} \cdot d\mathbf{R} = 0$$

Conversely, if $\int_C \mathbf{F} \cdot d\mathbf{R} = 0$ for every closed curve C in a region D, it can be shown (see Problem 44) that **F** must be conservative in D. ═

We have now stated three equivalent conditions for path independence, which we summarize in the following box.

Equivalent Conditions for Path Independence

Let $\mathbf{F}(x, y)$ have continuous first partial derivatives in an open connected region D, and let C be a piecewise smooth curve in D. Then the following conditions are equivalent.

a. $\displaystyle\int_C \mathbf{F} \cdot d\mathbf{R}$ is independent of path within D.

b. **F** is conservative; that is, $\mathbf{F} = \nabla f$ for some function f defined on D.

c. $\displaystyle\int_C \mathbf{F} \cdot d\mathbf{R} = 0$ for every closed path C enclosing only points of D.

Here is an example that illustrates how these conditions can be used to evaluate line integrals.

EXAMPLE 6   *Using path independence to evaluate a line integral*

Evaluate $\int_C (xy^2\, dx + x^2 y\, dy)$ over each of the given curves:

a.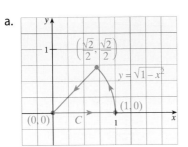

b.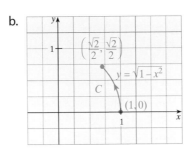

c.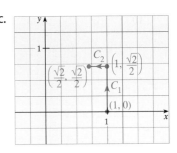

Solution   First, apply the cross-partials test to see whether $\mathbf{F}(x, y) = xy^2\mathbf{i} + x^2 y\mathbf{j}$ is conservative.

$$\frac{\partial}{\partial y}(xy^2) = 2xy \qquad \frac{\partial}{\partial x}(x^2 y) = 2xy$$

Thus, $\mathbf{F}$ is conservative.

a. Because C is a closed curve and $\mathbf{F}$ is conservative,

$$\int_C (xy^2\, dx + x^2 y\, dy) = 0$$

b. Because $y = \sqrt{1 - x^2}$ is the upper portion of the circle $x^2 + y^2 = 1$, we can parametrize C by $x = \cos\theta, y = \sin\theta$ for $0 \le \theta \le \frac{\pi}{4}$. Thus,

$$\int_C (xy^2\, dx + x^2 y\, dy) = \int_0^{\pi/4} [\cos\theta\sin^2\theta(-\sin\theta\, d\theta) + \cos^2\theta\sin\theta(\cos\theta\, d\theta)]$$

$$= \int_0^{\pi/4} [\cos^3\theta\sin\theta - \sin^3\theta\cos\theta]\, d\theta = \tfrac{1}{8}$$

c. Because $\mathbf{F}$ is conservative, the line integral is independent of path. Thus, the result is the same as for part **b.**

$$\int_C (xy^2\, dx + x^2 y\, dy) = \tfrac{1}{8}$$

It might be instructive to illustrate the power of these equivalent conditions by evaluating the line integral in part **c** without the benefit of part **b.** ■

In the next section, we develop another criterion for path independence as part of our study of an important result known as Green's theorem.

14.3 *Problem Set*

A **1.** ■ **What Does This Say?** Discuss conservative vector fields.

2. ■ **What Does This Say?** Describe the fundamental theorem of line integrals.

3. ■ **What Does This Say?** Compare and contrast the various equivalent conditions for independence of path.

Determine whether each vector field in Problems 4–9 is conservative, and if it is, find a scalar potential.

4. $y^2\mathbf{i} + 2xy\mathbf{j}$

5. $2xy^3\mathbf{i} + 3y^2x^2\mathbf{j}$

6. $(xe^{xy}\sin y)\mathbf{i} + (e^{xy}\cos y + y)\mathbf{j}$

7. $(-y + e^x\sin y)\mathbf{i} + [(x + 2)e^x\cos y]\mathbf{j}$

8. $(y - x^2)\mathbf{i} + (2x + y^2)\mathbf{j}$ **9.** $(e^{2x}\sin y)\mathbf{i} + (e^{2x}\cos y)\mathbf{j}$

Evaluate the line integrals in Problems 10–13 for each of the given paths.

10. $\displaystyle\int_C [(3x + 2y)\,dx + (2x + 3y)\,dy]$

a. 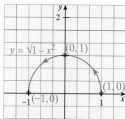   **b.**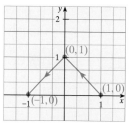

c.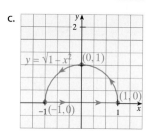

11. $\displaystyle\int_C [(3x + 2y)\,dx - (2x + 3y)\,dy]$

a. 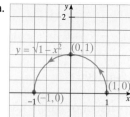   **b.**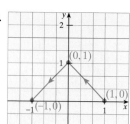

c.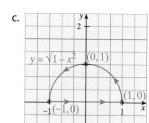

12. $\displaystyle\int_C (2x^2y\,dx + x^3\,dy)$

a. 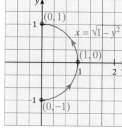   **b.**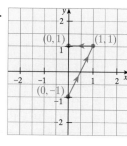

c.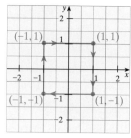

13. $\displaystyle\int_C (2xy\,dx + x^2\,dy)$

a. 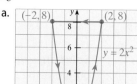   **b.**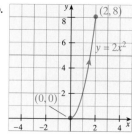

c.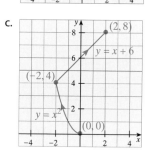

Show that the vector field $\mathbf{F}$ in Problems 14–19 is conservative, find a scalar potential f for $\mathbf{F}$, and then evaluate the line integral $\int_C \mathbf{F} \cdot d\mathbf{R}$, where C is any path connecting $A(0,0)$ to $B(1,1)$.

14. $\mathbf{F}(x, y) = (x + 2y)\mathbf{i} + (2x + y)\mathbf{j}$

15. $\mathbf{F}(x, y) = 2xy\mathbf{i} + x^2\mathbf{j}$

16. $\mathbf{F}(x, y) = (y - x^2)\mathbf{i} + (x + y^2)\mathbf{j}$

17. $\mathbf{F}(x, y) = (2x - y)\mathbf{i} + (y^2 - x)\mathbf{j}$

18. $\mathbf{F}(x, y) = e^{-y}\mathbf{i} - xe^{-y}\mathbf{j}$

19. $\mathbf{F}(x, y) = \dfrac{(y + 1)\mathbf{i} - x\mathbf{j}}{(y + 1)^2}$

B **20.** A force field

$$\mathbf{F}(x, y) = (3x^2 + 6xy^2)\mathbf{i} + (6x^2y + 4y^2)\mathbf{j}$$

acts on an object moving in the plane. Show that $\mathbf{F}$ is conservative, and find a scalar potential for $\mathbf{F}$. How much work is done as the object moves from $A(1, 0)$ to $B(0, 1)$ along any path connecting these points?

Verify that each line integral in Problems 21–26 is independent of path and then find its value.

21. $\displaystyle\int_C [(3x^2 + 2x + y^2)\,dx + (2xy + y^3)\,dy]$

where C is any path from $(0, 0)$ to $(1, 1)$

22. $\int_C [(xy \cos xy + \sin xy)dx + (x^2 \cos xy)dy]$

where C is any path from $(0, \frac{\pi}{18})$ to $(1, \frac{\pi}{6})$

23. $\int_C [(y - x^2)dx + (x + y^2)dy]$

where C is any path from $(-1, -1)$ to $(0, 3)$

24. $\int_C [(3x^2y + y^2)dx + (x^3 + 2xy)dy]$

where C is the path given parametrically by
$\mathbf{R}(t) = t\mathbf{i} + (t^2 + t - 2)\mathbf{j}$ for $0 \le t \le 2$

25. $\int_C [(\sin y)dx + (3 + x \cos y)dy]$

where C is the path given parametrically by
$\mathbf{R}(t) = 2 \sin\left(\frac{\pi t}{2}\right) \cos(\pi t)\mathbf{i} + (\sin^{-1}t)\mathbf{j}$ for $0 \le t \le 1$

26. $\int_C [(e^x \cos y)dx + (-e^x \sin y)dy]$

where C is the path given parametrically by
$\mathbf{R}(t) = (\cos t)\mathbf{i} + (\sin t)\mathbf{j}$ for $0 \le t \le \frac{\pi}{2}$

Evaluate the line integrals given in Problems 27–31 using the fundamental theorem of line integrals.

27. $\int_C (y\mathbf{i} + x\mathbf{j}) \cdot d\mathbf{R}$, where C is any path from $(0, 0)$ to $(2, 4)$

28. $\int_C (xy^2\mathbf{i} + x^2y\mathbf{j}) \cdot d\mathbf{R}$, where C is any path from $(4, 1)$ to $(0, 0)$

29. $\int_C (2y\, dx + 2x\, dy)$, where C is the line segment from $(0, 0)$ to $(4, 4)$

30. $\int_C (e^x \sin y\, dx + e^x \cos y\, dy)$, where C is any smooth curve from $(0, 0)$ to $(0, 2\pi)$

31. $\int_C \left[\tan^{-1}\frac{y}{x} - \frac{xy}{x^2 + y^2}\right]dx + \left[\frac{x^2}{x^2 + y^2} + e^{-y}(1 - y)\right]dy$

where C is any smooth curve from $(1, 1)$ to $(-1, 2)$.

32. Find a function g so $g(x)\mathbf{F}(x, y)$ is conservative where

$$\mathbf{F}(x, y) = (x^2 + y^2 + x)\mathbf{i} + xy\mathbf{j}$$

33. Find a function g so $g(x)\mathbf{F}(x, y)$ is conservative where

$$\mathbf{F}(x, y) = (x^4 + y^4)\mathbf{i} - (xy^3)\mathbf{j}$$

34. a. Over what region in the xy-plane will the line integral

$$\int_C [(-yx^{-2} + x^{-1})dx + x^{-1}dy]$$

be independent of path?

b. Evaluate the line integral in part **a** if C is defined by

$$\mathbf{R}(t) = (\cos^3 t)\mathbf{i} + (\sin 3t)\mathbf{j}$$

for $0 \le t \le \frac{\pi}{3}$.

35. **MODELING PROBLEM** The **gravitational force field F** between two particles of masses M and m separated by a distance r is modeled by

$$\mathbf{F}(x, y, z) = -\frac{KmM}{r^3}\mathbf{R}$$

where $\mathbf{R} = x\mathbf{i} + y\mathbf{j} + z\mathbf{k}$ and K is the gravitational constant.

a. Show that $\mathbf{F}$ is conservative by finding a scalar potential for $\mathbf{F}$. The scalar potential function f is often called the **Newtonian potential.**

b. Compute the amount of work done against the force field $\mathbf{F}$ in moving an object from the point $P(a_1, b_1, c_1)$ to $Q(a_2, b_2, c_2)$.

36. Let $\mathbf{F}(x, y) = \frac{-y\mathbf{i} + x\mathbf{j}}{x^2 + y^2}$.

a. Compute the line integral $\int_{C_1} \mathbf{F} \cdot d\mathbf{R}$, where C is the upper semicircle $y = \sqrt{1 - x^2}$ traversed counterclockwise. What is the value of $\int_{C_2} \mathbf{F} \cdot d\mathbf{R}$ if C_2 is the lower semicircle $y = -\sqrt{1 - x^2}$ also traversed counterclockwise?

b. Show that if $\mathbf{F} = M\mathbf{i} + N\mathbf{j}$, then

$$\frac{\partial}{\partial y}\left(\frac{-y}{x^2 + y^2}\right) = \frac{\partial}{\partial x}\left(\frac{x}{x^2 + y^2}\right)$$

but $\mathbf{F}$ is not conservative on the unit disk $x^2 + y^2 \le 1$.

37. **MODELING PROBLEM** A person whirls a bucket filled with water in a circle of radius 3 ft at the rate of 1 revolution per second. If the bucket and water weigh 30 lb, how much work is done by the force that keeps the bucket moving in a circular path?

38. **SPY PROBLEM** Holding his breath for dear life, the Spy finally escapes from Purity's watery trap (Problem 59, Section 13.7). He and the water spill into the next room, and the door slams shut. He finds himself looking across a large, rectangular room at Purity and Blohardt. He takes a step toward his nemesis, but finds his movements restricted by a force field that appears to sap his strength. "I see you have noticed the Death Force my lovely assistant has designed for you," crows Blohardt with an evil laugh. "It will do you no good to stay still," adds Purity. "The rays will eventually kill you even if you never move!" One more peal of laughter, and they leave together. Alone, the Spy quickly presses the stem on his wristwatch and the equation of the force field appears on the face:

$$\mathbf{F}(x, y) = (ye^{xy} + 2xy^3)\mathbf{i} + (xe^{xy} + 3x^2y^2 + \cos y)\mathbf{j}$$

Assuming the Spy is at $(0, 0)$ on the wristwatch's coordinate system and that the door (and safety) is at $(10, 10)$, what path should he take to minimize the work of struggling against the Death Force Field while crossing the room in the least possible time?

In Problems 39 and 40, you are to experiment with the notion of computing work along a path—with and without the benefit of independence of path. Suppose you are to power a boat of some kind from point A(0,0) to point B(2,1), and the primary consideration is the force of the wind, which generally opposes you. You are to investigate the effect of taking different paths from A to B.

39. a. Suppose the wind force is
$\mathbf{F} = \langle -a, -b \rangle$, for a and b positive. Compute the work involved along the straight-line path between A and B; then along a second path, of your choice. Does the path matter here? Why or why not?

b. Due to the effect of the harbor you are entering, suppose the wind force is $\langle -a, -ae^{-y} \rangle$. Again, compute the work along two paths as in part **a.** Does the path matter here? Why or why not?

c. Repeat part **b** for
$$\langle -a, -ae^{-y+x/9} \rangle$$

40. An important type of problem in several fields of application is, "Can we find the optimal path to minimize the work?" You are to explore this issue in regard to the wind force in Problem 39c,
$$\mathbf{F} = \langle -a, -ae^{-y+x/9} \rangle$$

a. You should have the work computed for two different paths from the previous problem. By looking at these numbers and carefully studying $\mathbf{F}$, you should see that we will be rewarded (or punished) by changing the path slightly from the straight-line path. What do your observations suggest regarding trying to minimize the work?

b. Attempt to find an (approximate) optimal path, starting with your observations in part **a** and common sense. For example, the path could be described by a parabola (or higher-degree polynomial) or by some trigonometric function. Find a "good" path for this purpose.

c. Explore the following question: "Is there a realistic optimal path from A to B?" Consider two conditions: first, there is no restriction on the path; second, suppose there is a shoreline at $y = 1$ so that one's path cannot exceed this limit. *Hint:* One approach might be to consider all parabolic paths of the form $y = x(b - ax)$, where a and b are nonnegative, and $y(2) = 1$. In this case, you can eliminate b and do a one-parameter study.

d. Assume the path in part **c** cannot go above $y = 1$. You may have discovered the optimal path. What is it? If you did the parabolic study suggested in part **c**, there is an optimal path in this case. What is it?

 41. Show that if the vector field $\mathbf{F}(x, y, z) = M(x, y, z)\mathbf{i} + N(x, y, z)\mathbf{j} + P(x, y, z)\mathbf{k}$ is conservative, then
$$\frac{\partial P}{\partial y} = \frac{\partial N}{\partial z} \qquad \frac{\partial M}{\partial z} = \frac{\partial P}{\partial x} \qquad \frac{\partial N}{\partial x} = \frac{\partial M}{\partial y}$$

42. It can be shown that $\mathbf{F} = M\mathbf{i} + N\mathbf{j} + P\mathbf{k}$ is conservative if curl $\mathbf{F} = 0$ whenever M, N, and P have continuous partial derivatives in a ball, a box, or other "simply connected" region.

a. Show that the vector field defined by
$\mathbf{F}(x, y, z) = (y^2 - 2xz)\mathbf{i} + (2xy + z)\mathbf{j} + (y - x^2)\mathbf{k}$
is conservative.

b. Note that if $\nabla f = \mathbf{F}$, we must have
$$f_x = y^2 - 2xz, \quad f_y = 2xy + z, \quad f_z = y - x^2$$
Partially integrate f_x with respect to x to express f in terms of x, y, z, and a function $c(y, z)$ that acts as a "constant" with respect to x-integration.

c. Find f_y and set it equal to $N = 2xy + z$. What can you conclude about the function $c(y, z)$? Partially integrate with respect to y to express f in terms of x, y, z and a function of z alone.

d. Find the scalar potential f.

43. Let f and g be differentiable functions of one variable. Show that the vector field $\mathbf{F} = [f(x) + y]\mathbf{i} + [g(y) + x]\mathbf{j}$ is conservative, and find the corresponding potential function.

44. Complete the proof of Theorem 14.4 by showing that if $\int_C \mathbf{F} \cdot d\mathbf{R} = 0$ for every closed curve C in a domain D where $\mathbf{F}$ is continuous, then $\mathbf{F}$ must be conservative in D. *Hint:* Let C_1 and C_2 be two curves in D with the same endpoints P and Q. Define a closed curve C so that
$$\int_C \mathbf{F} \cdot d\mathbf{R} = \int_{C_1} \mathbf{F} \cdot d\mathbf{R} - \int_{C_2} \mathbf{F} \cdot d\mathbf{R}$$
and conclude that $\int_C \mathbf{F} \cdot d\mathbf{R}$ is independent of path, so that $\mathbf{F}$ is conservative in D.

45. Let $\mathbf{R} = x\mathbf{i} + y\mathbf{j} + z\mathbf{k}$ and $r = \|\mathbf{R}\|$. Show that the work done in moving an object from a distance r_1 to a distance r_2 in the central force field $\mathbf{F} = \mathbf{R}/r^3$ is given by
$$W = \frac{1}{r_1} - \frac{1}{r_2}$$

46. An object of mass m moves along a trajectory $\mathbf{R}(t)$ with velocity $\mathbf{V}(t)$ in a conservative force field $\mathbf{F}(t_1)$. Let $\mathbf{R}(t_0) = \mathbf{Q}_0$ and $\mathbf{R}(t_1) = \mathbf{Q}_1$ be the initial and terminal points on the trajectory.

a. Show that the work done on the object is
$W = K(t_1) - K(t_0)$ where $K(t) = \frac{1}{2}m\|\mathbf{V}(t)\|^2$ is the object's *kinetic energy.*

b. Let f be a scalar potential for $\mathbf{F}$. Then $P(t) = -f(t)$ is the *potential energy* of the object. Prove the *law of conservation of energy*—namely,
$$P(t_0) + K(t_0) = P(t_1) + K(t_1)$$

14.4   Green's Theorem

IN THIS SECTION   Green's theorem, area as a line integral, Green's theorem for multiply-connected regions, alternative forms of Green's theorem

The fundamental theorem of calculus, $\int_a^b \frac{dF}{dx}\, dx = F(b) - F(a)$, can be described as saying that when the derivative $\frac{dF}{dx}$ is integrated over the closed interval $a \le x \le b$, the result is the same as that obtained by evaluating $F(x)$ at the "boundary points" a and b and forming the difference $F(b) - F(a)$. We obtained the analogous result

$$\int_C \nabla f \cdot d\mathbf{R} = f(Q) - f(P)$$

in Section 14.3, and our next goal is to obtain a different kind of analogue to the fundamental theorem of calculus called **Green's theorem** after the English mathematician George Green (Historical Quest, page 995).

GREEN'S THEOREM

Green's theorem relates the double integral of a certain differential expression over a region in the plane to a line integral over a closed boundary curve of the region. We begin with some terminology. A **Jordan curve,** named for the French mathematician Camille Jordan (1838–1922) is a closed curve C in the plane that does not intersect itself (see Figure 14.14). A **simply connected** region has the property that it is connected and the interior of every Jordan curve C in D also lies in D, as shown in Figure 14.14.

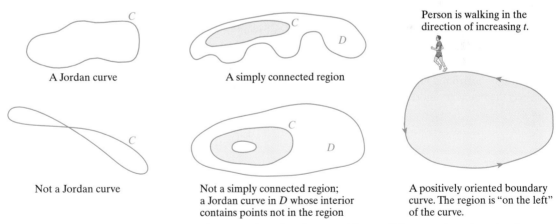

A Jordan curve

A simply connected region

Person is walking in the direction of increasing t.

Not a Jordan curve

Not a simply connected region; a Jordan curve in D whose interior contains points not in the region

A positively oriented boundary curve. The region is "on the left" of the curve.

■ **FIGURE 14.14**   A Jordan curve is a closed curve with no self-intersections. A region is simply connected if every Jordan curve has all its interior points in D

Picture yourself as a point moving along a curve. If the region D stays on your *left* as you, the point, move along the curve C with increasing t, then C is said to be **positively oriented** (see Figure 14.14). Now we are ready to state Green's theorem.

THEOREM 14.5   Green's theorem

Let D be a simply connected region with a positively oriented piecewise-smooth boundary C. Then if the vector field $\mathbf{F}(x, y) = M(x, y)\mathbf{i} + N(x, y)\mathbf{j}$ is continuously differentiable on D, we have

$$\int_C (M\, dx + N\, dy) = \iint_D \left(\frac{\partial N}{\partial x} - \frac{\partial M}{\partial y} \right) dA$$

> ■ **What This Says:** This theorem expresses an important relationship between a line integral over a simple closed curve in the plane and a double integral over the region bounded by the curve. It is one of the most important and elegant theorems in calculus. Take special note that D is required to be simply connected with a *positively oriented* boundary C. This condition is so important that sometimes the notation
>
> $$\oint_C (M\, dx + N\, dy)$$
>
> is used to indicate the line integral to emphasize the positive (or counterclockwise) orientation.

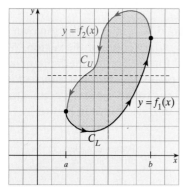

■ **FIGURE 14.15** Standard region: No vertical or horizontal line intersects the boundary more than twice

Proof   A **standard region** is one in which no vertical or horizontal line can intersect the boundary curve more than twice (see Figure 14.15). We shall prove Green's theorem for the special case where D is a standard region, and then we shall indicate how to extend the proof to more general regions.

Suppose D is a standard region with boundary curve C. We begin by showing that

$$\iint_D \frac{\partial M}{\partial y}\, dx\, dy = -\int_C M\, dx$$

Because D is a standard region, as shown in Figure 14.15, the boundary curve C is composed of a lower portion C_L and an upper portion C_U, which are the graphs of functions $f_1(x)$ and $f_2(x)$, respectively, on a certain interval $a \le x \le b$. Then we can evaluate the double integral by iterated integration:

$$\iint_D \frac{\partial M}{\partial y}\, dx\, dy = \iint_D \frac{\partial M}{\partial y}\, dy\, dx = \int_a^b \left[\int_{f_1(x)}^{f_2(x)} \frac{\partial M}{\partial y}\, dy \right] dx$$

$$= \int_a^b M[x, f_2(x)]\, dx - \int_a^b M[x, f_1(x)]\, dx = \int_{-C_U} M\, dx - \int_{C_L} M\, dx$$

$$= -\left[\int_{C_U} M\, dx + \int_{C_L} M\, dx \right] = -\int_C M\, dx$$

A similar argument shows that $\displaystyle\iint_D \frac{\partial N}{\partial x}\, dx\, dy = \int_C N\, dy$. Thus,

$$\iint_D \left(\frac{\partial N}{\partial x} - \frac{\partial M}{\partial y} \right) dA = \iint_D \frac{\partial N}{\partial x}\, dx\, dy - \iint_D \frac{\partial M}{\partial y}\, dx\, dy$$

$$= \int_C N\, dy - \int_C (-M)\, dx = \int_C (M\, dx + N\, dy)$$

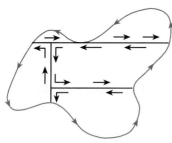

■ **FIGURE 14.16** General case: The region is decomposed into a finite number of standard regions by cuts

This completes the proof for the standard region. If D is not a standard region, it can be decomposed into a number of standard subregions by using horizontal and vertical "cuts," as shown in Figure 14.16. The proof for the standard region is then applied

Graffiti on the wall of a high school playground in Tel Aviv. Courtesy of Regev Nathansohn.

to each of these subregions, and the results are added. The line integrals along the cuts cancel in pairs, and after cancellation, the only remaining line integral is the one along the outer boundary C. Thus,

$$\int_C (M\,dx + N\,dy) = \iint_R \left(\frac{\partial N}{\partial x} - \frac{\partial M}{\partial y}\right) dA$$

The case where there is one cut is considered in Problem 37.

EXAMPLE 1   *Verifying Green's theorem*

Verify Green's theorem for the line integral
$\int_C (-y\,dx + x\,dy)$ where C is the closed path
shown in the figure.

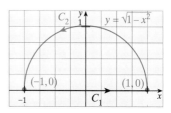

Solution   First, evaluate the line integral directly. The curve C consists of the line segment C_1 from $(-1, 0)$ to $(1, 0)$, followed by the semicircular arc C_2 from $(1, 0)$ back to $(-1, 0)$. We parametrize each of these:

$$C_1: x = t, y = 0 \quad -1 \le t \le 1 \qquad C_2: x = \cos s, y = \sin s \quad 0 \le s \le \pi$$
$$dx = dt, dy = 0 \qquad\qquad dx = -\sin s\,ds, dy = \cos s\,ds$$

$$\int_C (-y\,dx + x\,dy)$$
$$= \int_{C_1} (-y\,dx + x\,dy) + \int_{C_2} (-y\,dx + x\,dy)$$
$$= \int_{-1}^{1} [-0\,dt + t \cdot 0] + \int_0^\pi [-\sin s(-\sin s\,ds) + \cos s(\cos s\,ds)]$$
$$= \int_0^\pi (\sin^2 s + \cos^2 s)\,ds = \int_0^\pi 1\,ds = \pi$$

Next, we use Green's theorem to evaluate this integral. Note that the boundary curve C is simple and $M = -y$, $N = x$ so that $\mathbf{F}(x, y) = -y\mathbf{i} + x\mathbf{j}$ is continuously differentiable. The region D inside C is given by $0 \le y \le \sqrt{1 - x^2}$ for $-1 \le x \le 1$. We now apply Green's theorem:

$$\int_C (-y\,dx + x\,dy) = \iint_D \left(\frac{\partial}{\partial x}(x) - \frac{\partial}{\partial y}(-y)\right) dA = \int_{-1}^{1}\int_0^{\sqrt{1-x^2}} 2\,dy\,dx$$
$$= 2(\text{AREA OF SEMICIRCLE}) = 2[\tfrac{1}{2}\pi(1)^2] = \pi \qquad \blacksquare$$

EXAMPLE 2   *Computing work with Green's theorem*

A closed path C in the plane is defined by the given figure. Find the work done by an object moving along C in the force field

$$\mathbf{F}(x, y) = (x + xy^2)\mathbf{i} + 2(x^2y - y^2 \sin y)\mathbf{j}$$

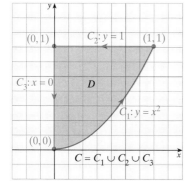

Solution   The work done, W, is given by the line integral $\int_C \mathbf{F} \cdot d\mathbf{R}$. Note that $\mathbf{F}$ is continuously differentiable on the region D enclosed by C, and since D is simply connected with a positively oriented boundary (namely, C), the hypotheses of Green's theorem are satisfied. We find that

$$
W = \int_C \mathbf{F} \cdot d\mathbf{R} = \iint_D \left[\frac{\partial}{\partial x}(2x^2y - 2y^2 \sin y) - \frac{\partial}{\partial y}(x + xy^2) \right] dA
$$

$$
= \iint_D (4xy - 2xy)\, dA = 2\int_0^1 \int_{x^2}^1 xy\, dy\, dx = 2\int_0^1 \frac{1}{2}xy^2 \Big|_{y=x^2}^{y=1} dx
$$

$$
= \int_0^1 (x - x^5)\, dx = \left[\frac{1}{2}x^2 - \frac{1}{6}x^6 \right]_0^1 = \frac{1}{3}
$$
■

AREA AS A LINE INTEGRAL

A line integral can be used to compute an area of a region in the plane by applying the following theorem.

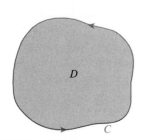

■ **FIGURE 14.17**   Area of region D

THEOREM 14.6   *Area as a line integral*

Let D be a simply connected region in the plane with piecewise smooth closed boundary C, as shown in Figure 14.17. Then the area A of region D is given by the line integral

$$
A = \frac{1}{2}\int_C (-y\, dx + x\, dy)
$$

Proof   Let $\mathbf{F}(x, y) = -y\mathbf{i} + x\mathbf{j}$. Then since $\mathbf{F}$ is continuously differentiable on D, Green's theorem applies. We have

$$
\int_C (-y\, dx + x\, dy) = \iint_D \left[\frac{\partial}{\partial x}(x) - \frac{\partial}{\partial y}(-y) \right] dA = \iint_D 2\, dA = 2A
$$

so that $A = \dfrac{1}{2}\displaystyle\int_C (-y\, dx + x\, dy)$.
=

> **WARNING**   This gives us yet another technique for finding the area of a region, especially when its boundary is specified in parametric form. In finding area, the function $\mathbf{F} = -y\mathbf{i} + x\mathbf{j}$ does not change. Do not forget the factor one-half after you finish the integration. ←

EXAMPLE 3   *Area enclosed by an ellipse*

Show that the ellipse $\dfrac{x^2}{a^2} + \dfrac{y^2}{b^2} = 1$ has area πab.

Solution   The elliptical path E is given parametrically by $x = a \cos \theta, y = b \sin \theta$ for $0 \le \theta \le 2\pi$. We find $dx = -a \sin \theta\, d\theta, dy = b \cos \theta\, d\theta$. If A is the area of this ellipse, then

$$
A = \frac{1}{2}\int_C (-y\, dx + x\, dy)
$$

$$
= \frac{1}{2}\int_0^{2\pi} [-(b \sin \theta)(-a \sin \theta\, d\theta) + (a \cos \theta)(b \cos \theta\, d\theta)]
$$

$$
= \frac{1}{2}\int_0^{2\pi} ab(\sin^2\theta + \cos^2\theta)\, d\theta
$$

$$= \frac{1}{2} \int_0^{2\pi} ab \, d\theta \qquad \textit{Because } \cos^2\theta + \sin^2\theta = 1$$

$$= \frac{1}{2} ab(2\pi - 0) = ab\pi \qquad \blacksquare$$

GREEN'S THEOREM FOR MULTIPLY-CONNECTED REGIONS

In the statement of Green's theorem, we require the region R inside the boundary curve C to be simply connected, but the theorem can be extended to multiply-connected regions—that is, regions with one or more "holes." A region with a single hole is shown in Figure 14.18a. The boundary of this region consists of an "outer" curve C_1 and an "inner" curve C_2, oriented so that the region R is always on the left as we travel around the boundary, which means that C_1 is oriented counterclockwise and C_2 clockwise.

Next, make cuts AB and CD through the region to the hole, as indicated in Figure 14.18b. Let R_1 be the simply connected region contained by the closed curve C_3 that begins at A, extends along the cut to B, and then clockwise along the bottom of the curve C_2 to C, along the cut to D, and counterclockwise along the bottom of C_1 back to A. Similarly, let R_2 be the region contained by the curve C_4 that begins at D and extends to C along the cut, to B along the top of C_2, to A along the second cut, and back to D along the top part of C_1. Then, if the vector field $\mathbf{F} = M\mathbf{i} + N\mathbf{j}$ is continuously differentiable on R, we can apply Green's theorem to show

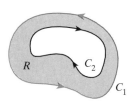

a. A doubly-connected region with oriented boundary curves C_1 and C_2.

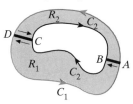

b. Two "cuts" are made through the hole

■ **FIGURE 14.18** Multiply-connected region

$$\iint\limits_R \left(\frac{\partial N}{\partial x} - \frac{\partial M}{\partial y}\right) dA = \iint\limits_{R_1} \left(\frac{\partial N}{\partial x} - \frac{\partial M}{\partial y}\right) dA + \iint\limits_{R_2} \left(\frac{\partial N}{\partial x} - \frac{\partial M}{\partial y}\right) dA$$

$$= \int_{C_3} (M \, dx + N \, dy) + \int_{C_4} (M \, dx + N \, dy)$$

But the line integrals from A to B and C to D cancel those from B to A and D to C, leaving only the line integrals along the original boundary curves C_1 and C_2, and Green's theorem for doubly-connected regions follows.

Green's Theorem for Doubly-Connected Regions

Let R be a doubly-connected region (one hole) in the plane, with outer boundary C_1 oriented counterclockwise and boundary C_2 of the hole oriented clockwise. If the boundary curves and $\mathbf{F}(x, y) = M(x, y)\mathbf{i}$ and $N(x, y)\mathbf{j}$ satisfy the hypotheses of Green's theorem then

$$\iint\limits_R \left(\frac{\partial N}{\partial x} - \frac{\partial M}{\partial y}\right) dA = \int_{C_1} (M \, dx + N \, dy) + \int_{C_2} (M \, dx + N \, dy)$$

Example 4 illustrates one way this result can be used.

EXAMPLE 4   *Green's theorem for a region containing a singular point*

Show that $\displaystyle\int_C \frac{-y \, dx + x \, dy}{x^2 + y^2} = 2\pi$, where C is any piecewise smooth Jordan curve enclosing the origin $(0, 0)$.

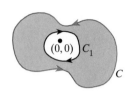

■ **FIGURE 14.19** The region R for doubly-connected regions

Solution   Let $M(x, y) = \dfrac{-y}{x^2 + y^2}$ and $N = \dfrac{x}{x^2 + y^2}$. Then

$$\frac{\partial N}{\partial x} = \frac{y^2 - x^2}{(x^2 + y^2)^2} = \frac{\partial M}{\partial y}$$

at any point (x, y) other than the origin. Next, let C_1 be a circle centered at the origin with radius r so small that the entire circle is contained in C, and let R be the region between the curve C and the circle C_1, as shown in Figure 14.19.

We know that $\dfrac{\partial N}{\partial x} = \dfrac{\partial M}{\partial y}$ throughout R (since R does not contain the origin) and Green's theorem for doubly-connected regions tells us that

$$\int_C \frac{-y\,dx + x\,dy}{x^2 + y^2} + \int_{C_1} \frac{-y\,dx + x\,dy}{x^2 + y^2} = \iint_R \left(\frac{\partial N}{\partial x} - \frac{\partial M}{\partial y} \right) dA = 0$$

so

$$\int_C \frac{-y\,dx + x\,dy}{x^2 + y^2} = -\int_{C_1} \frac{-y\,dx + x\,dy}{x^2 + y^2} = \int_{C_1^*} \frac{-y\,dx + x\,dy}{x^2 + y^2}$$

where C_1^* is the circle C_1 traversed counterclockwise instead of clockwise. In other words, we can find the value of the given line integral about the curve C by finding the line integral about the circle C^*. To do this, we parametrize C^* by

$$x = \cos\theta, \quad y = \sin\theta \quad \text{for } 0 \le \theta \le 2\pi$$

and find that

$$\int_{C_1^*} \frac{-y\,dx + x\,dy}{x^2 + y^2} = \int_0^{2\pi} \frac{-\sin\theta(-\sin\theta\,d\theta) + \cos\theta\,(\cos\theta\,d\theta)}{\cos^2\theta + \sin^2\theta}$$

$$= \int_0^{2\pi} \frac{\sin^2\theta + \cos^2\theta}{\sin^2\theta + \cos^2\theta}\,d\theta = \int_0^{2\pi} 1\,d\theta = 2\pi$$

Thus,

$$\int_C \frac{-y\,dx + x\,dy}{x^2 + y^2} = \int_{C_1^*} \frac{-y\,dx + x\,dy}{x^2 + y^2} = 2\pi \qquad\blacksquare$$

ALTERNATIVE FORMS OF GREEN'S THEOREM

Green's theorem can be expressed in a form that generalizes nicely to $\mathbb{R}^3$. In particular, note that the curl of the vector field $\mathbf{F}(x, y) = M(x, y)\mathbf{i} + N(x, y)\mathbf{j}$ is given by

$$\text{curl } \mathbf{F} = \begin{bmatrix} \mathbf{i} & \mathbf{j} & \mathbf{k} \\ \dfrac{\partial}{\partial x} & \dfrac{\partial}{\partial y} & \dfrac{\partial}{\partial z} \\ M(x, y) & N(x, y) & 0 \end{bmatrix}$$

$$= \left(-\frac{\partial N}{\partial z} \right)\mathbf{i} + \left(\frac{\partial M}{\partial z} \right)\mathbf{j} + \left[\frac{\partial N}{\partial x} - \frac{\partial M}{\partial y} \right]\mathbf{k}$$

$$= \left[\frac{\partial N}{\partial x} - \frac{\partial M}{\partial y} \right]\mathbf{k} \qquad \textit{Because } M \textit{ and } N \textit{ are functions of only } x \textit{ and } y$$

Thus, the formula in Green's theorem can be expressed as

$$\int_C \mathbf{F} \cdot d\mathbf{R} = \iint_D (\text{curl } \mathbf{F} \cdot \mathbf{k}) \, dA$$

In Section 14.6, when we extend this result to surfaces in $\mathbb{R}^3$, it will be called *Stokes' theorem.*

EXAMPLE 5   *The divergence theorem in the plane*

Suppose $\mathbf{F}(x, y) = M(x, y)\mathbf{i} + N(x, y)\mathbf{j}$ with a piecewise smooth boundary C. Show that

$$\int_C \mathbf{F} \cdot \mathbf{N} \, ds = \iint_D \text{div } \mathbf{F} \, dA$$

This is the line integral of the normal component, $\mathbf{F} \cdot \mathbf{N}$.

Solution   Let $\mathbf{R}(s) = x(s)\mathbf{i} + y(s)\mathbf{j}$ so that a unit tangent vector $\mathbf{T}$ to the curve C is $\mathbf{T} = \mathbf{R}'(s) = x'(s)\mathbf{i} + y'(s)\mathbf{j}$, which means that an outward normal vector is $\mathbf{N} = y'(s)\mathbf{i} - x'(s)\mathbf{j}$. We now apply Green's theorem to find a representation using div $\mathbf{F}$.

$$\begin{aligned}
\int_C \mathbf{F} \cdot \mathbf{N} \, ds &= \int_a^b (M\mathbf{i} + N\mathbf{j}) \cdot [y'(s)\mathbf{i} - x'(s)\mathbf{j}] \, ds \\
&= \int_a^b \left(M\frac{dy}{ds} - N\frac{dx}{ds} \right) ds = \int_C (-N \, dx + M \, dy) \\
&= \iint_D \left(\frac{\partial M}{\partial x} + \frac{\partial N}{\partial y} \right) dx \, dy = \iint_D \text{div } \mathbf{F} \, dA \qquad \blacksquare
\end{aligned}$$

When we extend this result to $\mathbb{R}^3$ in Section 14.7, it will be called the *divergence theorem.*

We repeat these alternative forms of Green's theorem for easy reference.

Alternative Forms of Green's Theorem

Let D be a simply connected region with a positively oriented boundary C. Then if the vector field $\mathbf{F} = M\mathbf{i} + N\mathbf{j}$ is continuously differentiable on D, we have

$$\int_C \underbrace{\mathbf{F} \cdot d\mathbf{R}}_{\text{tangential component of } \mathbf{F}} = \int_C (M \, dx + N \, dy) = \iint_D \left(\frac{\partial N}{\partial x} - \frac{\partial M}{\partial y} \right) dA = \iint_D (\text{curl } \mathbf{F} \cdot \mathbf{k}) \, dA$$

$$\int_C \underbrace{\mathbf{F} \cdot \mathbf{N}}_{\text{normal component of } \mathbf{F}} \, ds = \iint_D \left(\frac{\partial M}{\partial x} + \frac{\partial N}{\partial y} \right) dA = \iint_D \text{div } \mathbf{F} \, dA$$

In physics, some important applications of Green's theorem involve the so-called *normal derivative* of a scalar function f, which is defined as the directional derivative of f in the direction of the outward normal vector $\mathbf{N}$ to some curve or surface.

Normal Derivative

The **normal derivative** of f, denoted by $\dfrac{\partial f}{\partial n}$, is the directional derivative of f in the direction of the normal vector pointing to the exterior of the domain of f. In other words,

$$\frac{\partial f}{\partial n} = \nabla f \cdot \mathbf{N}$$

where $\mathbf{N}$ is an outer unit normal vector.

The following example illustrates how Green's theorem can be used in connection with the normal derivative. Additional examples are found in the problem set.

EXAMPLE 6   *Green's formula for the integral of the Laplacian*

Suppose f is a scalar function with continuous first and second partial derivatives in the simply connected region D. If the piecewise smooth curve C bounds D, show that

$$\iint_D \nabla^2 f \, dx \, dy = \int_C \frac{\partial f}{\partial n} \, ds$$

where $\nabla^2 f = f_{xx} + f_{yy}$ is the Laplacian of f and $\dfrac{\partial f}{\partial n} = \nabla f \cdot \mathbf{N}$ is the normal derivative vector.

Solution   Let $u = -\dfrac{\partial f}{\partial y}$ and $v = \dfrac{\partial f}{\partial x}$. Then we have

$$\nabla^2 f = f_{xx} + f_{yy} = \frac{\partial v}{\partial x} - \frac{\partial u}{\partial y}$$

$$
\begin{aligned}
\iint_D \nabla^2 f \, dx \, dy &= \iint_D \left(\frac{\partial v}{\partial x} - \frac{\partial u}{\partial y} \right) dx \, dy \\
&= \int_C (u \, dx + v \, dy) \qquad \text{Green's theorem} \\
&= \int_C \left(u \frac{dx}{ds} + v \frac{dy}{ds} \right) ds \\
&= \int_C \left(-\frac{\partial f}{\partial y} \frac{dx}{ds} + \frac{\partial f}{\partial x} \frac{dy}{ds} \right) ds \\
&= \int_C \left(f_x \frac{dy}{ds} - f_y \frac{dx}{ds} \right) ds \\
&= \int_C \nabla f \cdot \left(\frac{dy}{ds} \mathbf{i} - \frac{dx}{ds} \mathbf{j} \right) ds \\
&= \int_C \nabla f \cdot \mathbf{N} \, ds \qquad \text{Where } \mathbf{N} = \frac{dy}{ds}\mathbf{i} - \frac{dx}{ds}\mathbf{j} \text{ is an outward unit} \\
&\qquad\qquad\qquad\qquad\quad \text{normal vector to } C \\
&= \int_C \frac{\partial f}{\partial n} \, ds
\end{aligned}
$$

∎

14.4 Problem Set

A *Evaluate the line integrals in Problems 1–6 by applying Green's theorem. Check your answer by direct computation (that is, by parametrizing curve C).*

1. $\int_C (y^2 \, dx + x^2 \, dy)$

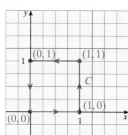

2. $\int_C (y^3 \, dx - x^3 \, dy)$

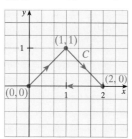

3. $\int_C [(2x^2 + 3y) \, dx - 3y^2 \, dy]$

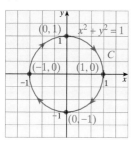

4. $\int_C (y^2 \, dx + 3xy^2 \, dy)$

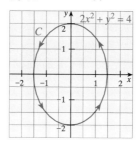

5. $\int_C 4xy \, dx$

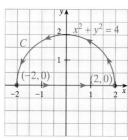

6. $\int_C (4y \, dx - 3x \, dy)$

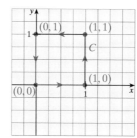

Evaluate the line integral in Problems 7–12 by applying Green's theorem.

7. $\int_C (2y \, dx - x \, dy)$

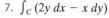

8. $\int_C (e^x \, dx - \sin x \, dy)$

9. $\int_C (x \sin x \, dx - \tan e^{y^2} \, dy)$

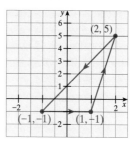

10. $\int_C [(x + y) \, dx - (3x - 2y) \, dy]$

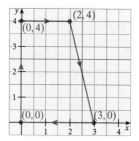

11. $\int_C [(x - y^2) \, dx + 2xy \, dy]$

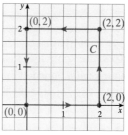

12. $\int_C (y^2 \, dx + x \, dy)$

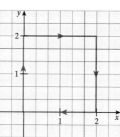

13. Use Green's theorem to find the work done by the force field

$$\mathbf{F}(x, y) = (3y - 4x)\mathbf{i} + (4x - y)\mathbf{j}$$

when an object moves once counterclockwise around the ellipse $4x^2 + y^2 = 4$.

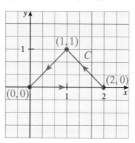
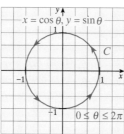

14. Find the work done when an object moves in the force field $\mathbf{F}(x, y) = y^2\mathbf{i} + x^2\mathbf{j}$ counterclockwise along the circular path $x^2 + y^2 = 2$.

Use Theorem 14.6 to find the area enclosed by the regions described in Problems 15–18, and then check by using an appropriate formula.

15. circle $x^2 + y^2 = 4$

16. triangle with vertices $(0, 0), (1, 1)$, and $(0, 2)$

17. trapezoid with vertices $(0, 0), (4, 0), (1, 3)$, and $(0, 3)$

18. semicircle $x = \sqrt{4 - y^2}$

Ⓑ 19. Evaluate the line integral

$$\int_C (x^2 y \, dx - y^2 x \, dy)$$

where C is the boundary of the region between the x-axis and the semicircle $y = \sqrt{a^2 - x^2}$, traversed counterclockwise.

20. Evaluate the line integral

$$\int_C (3y \, dx - 2x \, dy)$$

where C is the cardioid $r = 1 + \sin \theta$, traversed counterclockwise.

21. Show that

$$\int_C [(5 - xy - y^2) \, dx - (2xy - x^2) \, dy] = 3\bar{x}$$

where C is the square $0 \le x \le 1, 0 \le y \le 1$ traversed counterclockwise and $\bar{x}$ is the x-coordinate of the centroid of the square.

22. Find the work done by the force field $\mathbf{F}(x, y) = (x + 2y^2)\mathbf{j}$ as an object moves once counterclockwise about the circle $(x - 2)^2 + y^2 = 1$.

23. Let D be the region bounded by the Jordan curve C, and let A be the area of D. If $(\bar{x}, \bar{y})$ is the centroid of D, show that

$$A\bar{x} = \frac{1}{2} \int_C x^2 \, dy \quad \text{and} \quad A\bar{y} = -\frac{1}{2} \int_C y^2 \, dx$$

where C is traversed counterclockwise.

24. Use Theorem 14.6 and the polar transformation formulas $x = r \cos \theta, y = r \sin \theta$ to obtain the area formula in polar coordinates, namely,

$$A = \frac{1}{2} \int_{\theta_1}^{\theta_2} r^2 \, d\theta = \frac{1}{2} \int_{\theta_1}^{\theta_2} [g(\theta)]^2 \, d\theta$$

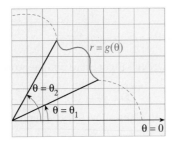

25. Evaluate $\int_C \left[\left(\frac{-y}{x^2} + \frac{1}{x} \right) dx + \frac{1}{x} \, dy \right]$, where C is any Jordan curve that does not touch or cross the y-axis, traversed counterclockwise.

26. Evaluate $\int_C \dfrac{x \, dx + y \, dy}{x^2 + y^2}$, where C is any Jordan curve whose interior does not contain the origin, traversed counterclockwise.

27. Evaluate the line integral

$$\int_C \frac{x \, dx + y \, dy}{x^2 + y^2}$$

where C is any piecewise smooth Jordan curve enclosing the origin, traversed counterclockwise.

28. Evaluate $\int_C \dfrac{-y \, dx + (x - 1) \, dy}{(x - 1)^2 + y^2}$, where C is any Jordan curve whose interior does not contain the point $(1, 0)$, traversed counterclockwise.

29. Evaluate $\int_C \dfrac{-(y + 2) \, dx + (x - 1) \, dy}{(x - 1)^2 + (y + 2)^2}$, where C is any Jordan curve whose interior does not contain the point $(1, -2)$, traversed counterclockwise.

30. Evaluate $\int_C \dfrac{\partial z}{\partial n} \, ds$, where $z(x, y) = 2x^2 + 3y^2$, and C is the circular path $x^2 + y^2 = 16$, traversed counterclockwise.

31. Evaluate $\int_C \dfrac{\partial f}{\partial n} \, ds$, where $f(x, y) = x^2 y - 2xy + y^2$, and C is the boundary of the unit square $0 \le x \le 1, 0 \le y \le 1$, traversed counterclockwise.

32. Evaluate $\int_C x \dfrac{\partial x}{\partial n} \, ds$, where C is the boundary of the unit square $0 \le x \le 1, 0 \le y \le 1$, traversed counterclockwise.

33. If C is a closed curve, show that

$$\int_C [(x - 3y) \, dx + (2x - y^2) \, dy] = 5A$$

where A is the area of the region D enclosed by C.

Ⓒ 34. Prove the following theoretical application of Green's theorem: Let $\mathbf{F}(x, y) = u(x, y)\mathbf{i} + v(x, y)\mathbf{j}$ be continuously differentiable on the simply connected region D. Then $\mathbf{F}$ is conservative if and only if

$$\frac{\partial v}{\partial x} = \frac{\partial u}{\partial y}$$

throughout D.

35. Recall that a scalar function f with continuous first and second partial derivatives is said to be *harmonic* in a region D if $\nabla^2 f = 0$ (that is, if $f_{xx} + f_{yy} = 0$). If f is such a function, show that

$$\iint_D (f_x^2 + f_y^2) \, dx \, dy = \int_C f \frac{\partial f}{\partial n} \, ds$$

36. HISTORICAL QUEST   George Green (1793–1841) was the son of a baker who worked in his father's mill and studied mathematics and physics in his spare time, using only books he obtained from the library. In 1828, he published a memoir titled "An Essay on the Application of Mathematical Analysis to the Theories of Electricity and Magnetism," which contains the result that now bears his name. Very few copies of the essay were printed and distributed, so few people knew of Green's results. In 1833, at the age of 40, he entered Cambridge University and graduated just four years before his death. His 1828 paper was discovered and publicized in 1845 by Sir William Thompson (later known as Lord Kelvin, 1824–1907), and Green finally received proper credit for his work.

In this Quest, use Green's theorem to prove the following two important results, known as **Green's formulas.**

a. **Green's first formula:**

$$\iint_D [f\nabla^2 g + \nabla f \cdot \nabla g]\, dx\, dy = \int_C f\frac{\partial g}{\partial n}\, ds$$

b. Once again start with the line integral and derive what is known as **Green's second formula:**

$$\iint_D [f\nabla^2 g - g\nabla^2 f]\, dx\, dy$$
$$= \int_C \left(f\frac{\partial g}{\partial n} - g\frac{\partial f}{\partial n}\right) ds$$

37. HISTORICAL QUEST   The result known as "Green's theorem" in the West is called "Ostrogradsky's theorem" in Russia, after Mikhail Ostrogradsky. Although Ostrogradsky published over 80 papers during a successful career as a mathematician, today he is known, even in his homeland, only for his version of Green's theorem, which appeared as part of a series of results presented to the Academy of Sciences in 1828.

MIKHAIL OSTROGRADSKY 1801–1862

In this Quest you are asked to prove Green–Ostrogradsky's theorem in the plane for the non-standard region D shown in Figure 14.20. Specifically, suppose that the line L "cuts" the region D into two standard subregions D_1 and D_2. Apply Green's theorem to D_1 and D_2, then combine the results

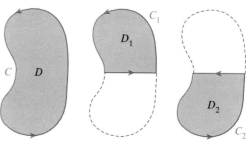

■ **FIGURE 14.20**   Green's theorem

to show that the theorem also applies to the non-standard region D. *Hint*: The key is what happens along the "cut line" L.

38. Extend Green's theorem to a "triply-connected" region (two holes), such as the one shown in Figure 14.21.

■ **FIGURE 14.21**   A triply-connected region

39. Suppose $\mathbf{F} = M(x, y)\mathbf{i} + N(x, y)\mathbf{j}$ is continuously differentiable in a doubly-connected region R and that

$$\frac{\partial N}{\partial x} = \frac{\partial M}{\partial y}$$

throughout R. How many distinct values of I are there for the integral

$$I = \int_C [M(x, y)\, dx + N(x, y)\, dy]$$

where C is a piecewise smooth Jordan curve in R?

40. Answer the question in Problem 39 for the case where R is triply-connected (two holes). See Figure 14.21.

14.5   Surface Integration

IN THIS SECTION   surface integrals, flux integrals, integrals over parametrically defined surfaces

SURFACE INTEGRALS

A *surface integral* is a generalization of the line integral in which the integration is over a surface in space rather than a curve. We shall be interested only in surfaces that are *piecewise smooth*—that is, those consisting of a finite number of pieces on which there is a continuously turning tangent plane.

We begin by defining the surface integral of a continuous scalar function $g(x, y, z)$ over a piecewise smooth surface S. Partition S into n subregions, the kth of which has

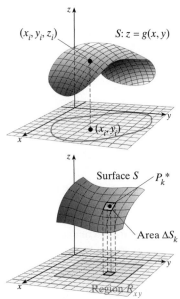

Definition of a surface integral

area ΔS_k, and let $P_k^*(x_k^*, y_k^*, z_k^*)$ be a point chosen arbitrarily from the kth subregion, for $k = 1, 2, \ldots, n$. From the sum

$$\sum_{k=1}^{n} g(P_k^*)\Delta S_k$$

and take the limit as the largest of the ΔS_k tends to 0, just as we did with surface area in Section 13.4. If this limit exists, it is called the **surface integral of g over S** and is denoted by

$$\iint_S g(x, y, z) \, dS$$

Recall from Section 13.4 that when the surface S projects onto the region R_{xy} in the xy-plane and S has the representation $z = f(x, y)$, then $dS = \sqrt{f_x^2 + f_y^2 + 1} \, dA_{xy}$, where dA_{xy} is either $dx \, dy$ or $dy \, dx$ (or $r \, dr \, d\theta$ if A_{xy} is described in terms of polar coordinates). We can now state the formula for evaluating a surface integral. (*Note:* We write R_{xy} and A_{xy} instead of the usual R and A to help you remember that these are regions in the xy-plane.)

Surface Integral

Let S be a surface defined by $z = f(x, y)$ and R_{xy} its projection on the xy-plane. If f, f_x, and f_y are continuous in R_{xy} and g is continuous on S, then the **surface integral** of g over S is

$$\iint_S g(x, y, z)dS = \int_{R_{xy}} \int g(x, y, f(x, y))\sqrt{[f_x(x, y)]^2 + [f_y(x, y)]^2 + 1} \, dA_{xy}$$

| **EXAMPLE 1** | *Evaluating a surface integral* |

Evaluate the surface integral

$$\iint_S g \, dS$$

where $g(x, y, z) = xz + 2x^2 - 3xy$ and S is that portion of the plane $2x - 3y + z = 6$ that lies over the unit square R:

$$0 \le x \le 1, \quad 0 \le y \le 1$$

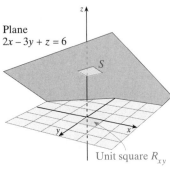

Plane
$2x - 3y + z = 6$

Unit square R_{xy}

The portion of the plane that lies above the unit square

Solution First, note that the equation of the plane can be written as $z = f(x, y)$, where $f(x, y) = 6 - 2x + 3y$. We have $f_x(x, y) = -2$ and $f_y(x, y) = 3$, so that

$$dS = \sqrt{f_x^2 + f_y^2 + 1} \, dA_{xy} = \sqrt{(-2)^2 + (3)^2 + 1} \, dA_{xy} = \sqrt{14} \, dA_{xy}$$

Consequently,

$$\begin{aligned}
\iint_S g \, dS &= \iint_S (xz + 2x^2 - 3xy)\sqrt{14} \, dA_{xy} \\
&= \iint_S [x(6 - 2x + 3y) + 2x^2 - 3xy]\sqrt{14} \, dy \, dx \\
&= \sqrt{14} \iint_S 6x \, dy \, dx \\
&= 6\sqrt{14}\int_0^1 \int_0^1 x \, dy \, dx = 6\sqrt{14} \int_0^1 x \, dx = 3\sqrt{14}
\end{aligned}$$

If the function g defined on S is simply $g(x, y, z) = 1$, then the surface integral gives the *surface area* of S.

SURFACE AREA FORMULA

$$\text{Surface area} = \iint_S dS$$

A useful application of surface integrals is to find the center of mass of a thin curved lamina whose shape is a given surface S, as shown in Figure 14.22.

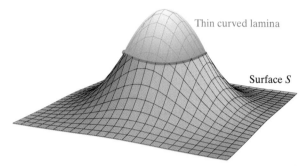

Thin curved lamina

Surface S

■ **FIGURE 14.22**   A thin lamina whose shape is the surface S

Suppose $\delta(x, y, z)$ is the density (mass per unit area) at a point (x, y, z) on the lamina. Then the *total mass, m, of the lamina* can also be represented by a surface integral.

MASS OF A LAMINA

$$m = \iint_S \delta(x, y, z)\, dS$$

and the center of mass of the surface is the point $C(\bar{x}, \bar{y}, \bar{z})$, where

CENTER OF MASS

$$\bar{x} = \frac{1}{m} \iint_S x\delta(x, y, z)\, dS, \quad \bar{y} = \frac{1}{m} \iint_S y\delta(x, y, z)\, dS, \quad \bar{z} = \frac{1}{m} \iint z\delta(x, y, z)\, dS$$

These formulas may be derived by essentially the same approach used in our previous work with moments and centroids of solid regions. (See, for example, Sections 6.5 and 13.6.)

> **EXAMPLE 2**   *Mass of a curved lamina*

Find the mass of a lamina of density $\delta(x, y, z) = z$ in the shape of the hemisphere $z = (a^2 - x^2 - y^2)^{1/2}$, as shown in Figure 14.23.

Solution   We begin by calculating dS:

$$z_x = \frac{1}{2}(a^2 - x^2 - y^2)^{-1/2}(-2x) = -x(a^2 - x^2 - y^2)^{-1/2}$$

$$z_y = \frac{1}{2}(a^2 - x^2 - y^2)^{-1/2}(-2y) = -y(a^2 - x^2 - y^2)^{-1/2}$$

$$dS = \sqrt{z_x^2 + z_y^2 + 1}\, dA_{xy} = \sqrt{\frac{x^2}{a^2 - x^2 - y^2} + \frac{y^2}{a^2 - x^2 - y^2} + 1}\, dA_{xy}$$

$$= \sqrt{\frac{a^2}{a^2 - x^2 - y^2}}\, dA_{xy} = a(a^2 - x^2 - y^2)^{-1/2}\, dA_{xy}$$

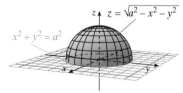

$z = \sqrt{a^2 - x^2 - y^2}$

$x^2 + y^2 = a^2$

■ **FIGURE 14.23**   Note that S projects onto the circle R in the xy-plane with the equation $x^2 + y^2 = a^2$.

The mass of the hemisphere S is given by

$$m = \iint_S \delta(x, y, z) \, dS = \iint_S z \, dS$$

$$= \iint_R (a^2 - x^2 - y^2)^{1/2} \, a(a^2 - x^2 - y^2)^{-1/2} \, dA_{xy}$$

$$= a \iint_R dA_{xy} = a(\pi a^2) = \pi a^3 \qquad \text{Since this integral represents the area of a circle of radius } a \quad \blacksquare$$

FLUX INTEGRALS

Next, we shall discuss a special kind of surface integral called a *flux integral,* which is used in physics to study fluid flow and electrostatics. To define the *flux of a vector field* **F** across a given smooth surface S, we need to know that S is **orientable** in the sense that it is possible to define a field of unit normal vectors $\mathbf{N}(x, y, z)$ on S that varies continuously as the point (x, y, z) varies over S. Most common surfaces, such as planes, spheres, ellipses, and paraboloids are orientable, but it is not difficult to construct a fairly simple surface that is not. For example, the so-called **Möbius strip,** formed by twisting a long rectangular strip before joining the ends, is not orientable (see Figure 14.24a). In advanced calculus, it is shown that for an orientable surface S, the normal vector **N** must be either **outward** (pointing toward the exterior of S) or **inward** (pointing toward the interior), as illustrated in Figure 14.24b.

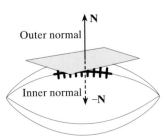

a. One-sided (non-orientable) surface: a Möbius strip

b. Two-sided (orientable) surface: a football. This is also a closed surface.

■ **FIGURE 14.24**   Surfaces in space

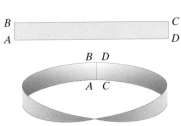

Component of **F** in the direction of **N**

Velocity field

■ **FIGURE 14.25**   Flux density is measured by $\mathbf{F} = \delta\mathbf{V}$

In Section 14.1, we defined the *flux density* as the vector field $\mathbf{F} = \delta\mathbf{V}$, where $\mathbf{V}(x, y, z)$ is the velocity of a fluid with density $\delta(x, y, z)$. The flux density measures the volume of fluid crossing surface S per unit of time and is also called the **flux of F across** S. Suppose the surface S is located in the region through which the fluid flows. If **N** is the unit normal vector to the surface S, the $\mathbf{F} \cdot \mathbf{N}$ is the component of flux in the direction normal vector to S, as shown in Figure 14.25. Then the mass of the fluid flowing through S in unit time in the direction normal vector to the surface is given by a surface integral, which is called a **flux integral.**

Flux Integral

The **flux integral** of a vector field **F** across a surface S is given by the surface integral

$$\iint_S \mathbf{F} \cdot \mathbf{N} \, dS$$

where **N** is an outward normal vector to S.

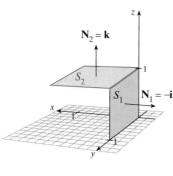

| EXAMPLE 3 | *Surface integral of a vector field* |

Compute $\iint\limits_{S} \mathbf{F} \cdot \mathbf{N} \, dS$, where $\mathbf{F} = x\mathbf{i} - 5y\mathbf{j} + 4z\mathbf{k}$, and S is the union of the two squares

$$S_1: \quad x = 0, 0 \le y \le 1, 0 \le z \le 1$$
$$S_2: \quad z = 1, 0 \le x \le 1, 0 \le y \le 1$$

Solution   On S_1, the outward unit normal vector is $\mathbf{N}_1 = -\mathbf{i}$ and

$$\mathbf{F} \cdot \mathbf{N}_1 = (x\mathbf{i} - 5y\mathbf{j} + 4z\mathbf{k}) \cdot (-\mathbf{i}) = -x = 0$$

(because $x = 0$ on S_1). Thus,

$$\iint\limits_{S_1} \mathbf{F} \cdot \mathbf{N}_1 \, dS_1 = 0$$

On S_2, $\mathbf{N}_2 = \mathbf{k}$, and $\mathbf{F} \cdot \mathbf{N}_2 = \mathbf{F} \cdot \mathbf{k} = 4z = 4$ (because $z = 1$ on S_2). Thus,

$$\iint\limits_{S_2} \mathbf{F} \cdot \mathbf{N}_2 \, dS_2 = \iint\limits_{S_2} 4 \, ds_2 = 4(\text{area of } S_2) = 4$$

Finally,

$$\iint\limits_{S} \mathbf{F} \cdot \mathbf{N} \, dS = \iint\limits_{S_1} \mathbf{F} \cdot \mathbf{N}_1 \, dS_1 + \iint\limits_{S_2} \mathbf{F} \cdot \mathbf{N}_2 \, dS_2 = 0 + 4 = 4 \qquad \blacksquare$$

The unit normal vector of a surface is not always as obvious as it was in Example 3. For an orientable surface S defined by $z = g(x, y)$, we let $G(x, y, z) = z - g(x, y)$. Then, $\mathbf{N}$ can be found as follows (see Section 12.6):

$$\mathbf{N} = \frac{\nabla G(x, y, z)}{\|\nabla G(x, y, z)\|} = \frac{-g_x\mathbf{i} - g_y\mathbf{j} + \mathbf{k}}{\sqrt{g_x^2 + g_y^2 + 1}} \qquad \textbf{Outward unit normal vector}$$

$$\mathbf{N} = \frac{-\nabla G(x, y, z)}{\|\nabla G(x, y, z)\|} = \frac{g_x\mathbf{i} + g_y\mathbf{j} - \mathbf{k}}{\sqrt{g_x^2 + g_y^2 + 1}} \qquad \textbf{Inward unit normal vector}$$

| EXAMPLE 4 | *Evaluating a flux integral* |

Compute $\iint\limits_{S} \mathbf{F} \cdot \mathbf{N} \, dS$, where $\mathbf{F} = xy\mathbf{i} + z\mathbf{j} + (x + y)\mathbf{k}$ and S is the triangular region cut off from the plane $x + y + z = 1$ by the positive coordinate axes, as shown in Figure 14.26. Assume $\mathbf{N}$ is the unit normal vector that points away from the origin.

Solution   Let $g(x, y) = z = 1 - x - y$. Then $g_x = -1, g_y = -1$, and an outward unit normal vector to the plane is

$$\mathbf{N} = \frac{-(-1)\mathbf{i} - (-1)\mathbf{j} + \mathbf{k}}{\sqrt{(-1)^2 + (-1)^2 + 1}} = \frac{1}{\sqrt{3}}(\mathbf{i} + \mathbf{j} + \mathbf{k}) \quad \text{so that}$$

$$\mathbf{F} \cdot \mathbf{N} = (xy\mathbf{i} + z\mathbf{j} + (x + y)\mathbf{k}) \cdot \left(\frac{1}{\sqrt{3}}\mathbf{i} + \frac{1}{\sqrt{3}}\mathbf{j} + \frac{1}{\sqrt{3}}\mathbf{k}\right) = \frac{1}{\sqrt{3}}(xy + z + x + y)$$

We can write this as a function of x and y because $z = 1 - x - y$:

$$\mathbf{F} \cdot \mathbf{N} = \frac{1}{\sqrt{3}}(xy + 1 - x - y + x + y) = \frac{1}{\sqrt{3}}(xy + 1)$$

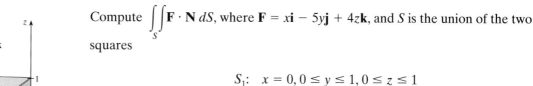

■ **FIGURE 14.26** $\mathbf{N}$ is the outward unit normal vector

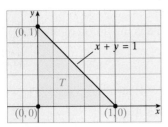

■ **FIGURE 14.27** Projected region T from Figure 14.26

We also find that

$$dS = \sqrt{g_x^2 + g_y^2 + 1}\, dA_{xy} = \sqrt{3}\, dA_{xy}$$

The final piece of the puzzle we need before turning to the evaluation of the integral is to find the projection of S onto the xy-plane. We see from Figure 14.27 that is a triangular region, which can be described as the set of all points (x, y) such that for each x between 0 and 1, y varies between $y = 0$ and $y = 1 - x$ (as shown in Figure 14.27). Finally, we find that

$$\iint_S \mathbf{F} \cdot \mathbf{N}\, dS = \iint_T \frac{1}{\sqrt{3}}(xy + 1)\sqrt{3}\, dA_{xy}$$

$$= \int_0^1 \int_0^{1-x} (xy + 1)\, dy\, dx = \int_0^1 \left[\frac{1}{2}xy^2 + y\right]_{y=0}^{y=1-x} dx$$

$$= \frac{1}{2}\int_0^1 (x^3 - 2x^2 - x + 2)\, dx = \frac{1}{2}\left[\frac{1}{4}x^4 - \frac{2}{3}x^3 - \frac{1}{2}x^2 + 2x\right]_0^1 = \frac{13}{24} \quad \blacksquare$$

INTEGRALS OVER PARAMETRICALLY DEFINED SURFACES

In Section 13.4, we showed that if a surface S is defined parametrically by the vector function

$$\mathbf{R}(u, v) = x(u, v)\mathbf{i} + y(u, v)\mathbf{j} + z(u, v)\mathbf{k}$$

over a region D in the uv-plane, the surface area of S is given by the integral

$$\iint_D \|\mathbf{R}_u \times \mathbf{R}_v\|\, du\, dv$$

Similarly, if f is continuous on D, the surface integral of f over D is given by

$$\iint_S f(x, y, z)\, dS = \iint_D f(\mathbf{R}) \|\mathbf{R}_u \times \mathbf{R}_v\|\, du\, dv$$

EXAMPLE 5   *Surface integral for a surface defined parametrically*

Evaluate $\int_S \int (x + y + z)\, dS$, where S is the surface defined parametrically by

$$\mathbf{R}(u, v) = (2u + v)\mathbf{i} + (u - 2v)\mathbf{j} + (u + 3v)\mathbf{k} \quad \text{for } 0 \le u \le 1, 0 \le v \le 2.$$

Solution

$$\iint_S (x + y + z)\, dS = \iint_D f(\mathbf{R}) \|\mathbf{R}_u \times \mathbf{R}_v\|\, du\, dv$$

Before we go on, we need to find the component parts for the integral on the right. Using $\mathbf{R} = x\mathbf{i} + y\mathbf{j} + z\mathbf{k}$, we see that for this problem $x = 2u + v$, $y = u - 2v$, and $z = u + 3v$. Because $f(x, y, z) = x + y + z$,

$$f(\mathbf{R}) = f(2u + v, u - 2v, u + 3v)$$

$$= 2u + v + u - 2v + u + 3v = 4u + 2v$$

$$\mathbf{R}_u = 2\mathbf{i} + \mathbf{j} + \mathbf{k} \text{ and } \mathbf{R}_v = \mathbf{i} - 2\mathbf{j} + 3\mathbf{k}, \text{ so}$$

$$\mathbf{R}_u \times \mathbf{R}_v = \begin{vmatrix} \mathbf{i} & \mathbf{j} & \mathbf{k} \\ 2 & 1 & 1 \\ 1 & -2 & 3 \end{vmatrix} = (3 + 2)\mathbf{i} - (6 - 1)\mathbf{j} + (-4 - 1)\mathbf{k} = 5\mathbf{i} - 5\mathbf{j} - 5\mathbf{k}$$

Thus, $\|\mathbf{R}_u \times \mathbf{R}_v\| = \sqrt{5^2 + (-5)^2 + (-5)^2} = 5\sqrt{3}$

We now substitute these values into the surface integral formula:

$$\iint_S (x + y + z) = \iint_D f(\mathbf{R}) \|\mathbf{R}_u \times \mathbf{R}_v\| \, du \, dv$$

$$= \int_0^2 \int_0^1 (4u + 2v)(5\sqrt{3}) \, du \, dv = 5\sqrt{3} \int_0^2 [2u^2 + 2u]_0^1 \, dv$$

$$= 5\sqrt{3} \int_0^2 (2 + 2v) \, dv = 5\sqrt{3} \, [2v + v^2]_0^2 = 5\sqrt{3} \, (8) = 40\sqrt{3} \qquad \blacksquare$$

14.5 Problem Set

Ⓐ *Let S be the hemisphere* $x^2 + y^2 + z^2 = 4$, *with* $z \geq 0$, *in Problems 1–6. Evaluate each surface integral.*

1. $\displaystyle\iint_S z \, dS$ **2.** $\displaystyle\iint_S z^2 \, dS$

3. $\displaystyle\iint_S (x - 2y) \, dS$ **4.** $\displaystyle\iint_S (5 - 2x) \, dS$

5. $\displaystyle\iint_S (x^2 + y^2)z \, dS$ **6.** $\displaystyle\iint_S (x^2 + y^2) \, dS$

In Problems 7–10, evaluate $\displaystyle\iint_S xy \, dS$.

7. $S: z = 2 - y, 0 \leq x \leq 2, 0 \leq y \leq 2$
8. $S: z = 4 - x - y, 0 \leq x \leq 4, 0 \leq y \leq 4$
9. $S: z = 5, x^2 + y^2 \leq 1$
10. $S: z = 10, \dfrac{x^2}{4} + \dfrac{y^2}{1} \leq 1$

In Problems 11–14, evaluate $\displaystyle\iint_S (x^2 + y^2) \, dS$.

11. $S: z = 4 - x - 2y, 0 \leq x \leq 4, 0 \leq y \leq 2$
12. $S: z = 4 - x, 0 \leq x \leq 2, 0 \leq y \leq 2$
13. $S: z = 4, x^2 + y^2 \leq 1$
14. $S: z = xy, x^2 + y^2 \leq 4, x \geq 0, y \geq 0$

In Problems 15–18, suppose S is the portion of the paraboloid $z = x^2 + y^2$ *for which* $z \leq 4$. *Evaluate the given surface integral.*

15. $\displaystyle\iint_S z \, dS$

16. $\displaystyle\iint_S (4 - z) \, dS$

17. $\displaystyle\iint_S \sqrt{1 + 4z} \, dS$

18. $\displaystyle\iint_S \dfrac{dS}{\sqrt{1 + 4z}}$

19. Evaluate $\displaystyle\iint_S (x^2 + y^2) \, dS$, where S is the surface bounded above by the hemisphere $z = \sqrt{1 - x^2 - y^2}$, and below by the plane $z = 0$.

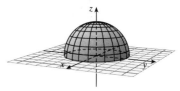

20. Evaluate $\displaystyle\iint_S 2x \, dS$, where S is the portion of the plane $x + y + z = 1$ with $x \geq 0, y \geq 0, z \geq 0$.

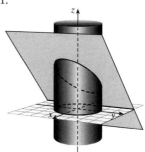

21. Evaluate $\displaystyle\iint_S (x^2 + y^2 + z^2) \, dS$, where S is the portion of the plane $z = x + 1$ that lies inside the cylinder $x^2 + y^2 = 1$.

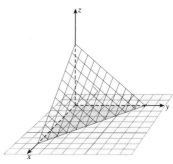

Evaluate $\iint_S \mathbf{F} \cdot \mathbf{N} \, dS$ *for the vector fields* **F** *and surfaces S given in Problems 22–27.*

22. $\mathbf{F} = x\mathbf{i} + 2y\mathbf{j} + z\mathbf{k}$, and S is the triangular region bounded by the intersection of the plane $x + 2y + z = 1$ and the positive coordinate axes.

23. $\mathbf{F} = x\mathbf{i} + 2y\mathbf{j} - 3z\mathbf{k}$, and S is that part of the plane $15x - 12y + 3z = 6$ that lies above the unit square $0 \le x \le 1, 0 \le y \le 1$.

24. $\mathbf{F} = x\mathbf{i} + y\mathbf{j} + 2z\mathbf{k}$, and S is the surface of the cube bounded by the planes $x = 0, x = 1, y = 0, y = 1, z = 0,$ and $z = 1$.

25. $\mathbf{F} = x\mathbf{i} + y\mathbf{j}$, and S is the hemisphere $z = \sqrt{1 - x^2 - y^2}$.

26. $\mathbf{F} = 2x\mathbf{i} - 3y\mathbf{j}$, and S is the part of the hemisphere given by $x^2 + y^2 + z^2 = 5$, for $z \ge 1$.

27. $\mathbf{F} = x^2\mathbf{i} + y^2\mathbf{j} + z^2\mathbf{k}$, and S is the portion of the plane $z = y + 1$ that lies inside the cylinder $x^2 + y^2 = 1$.

Ⓑ 28. Evaluate $\iint_S (3x - y + 2z) \, dS$, where S is the surface determined by $\mathbf{R}(u, v) = u\mathbf{i} + u\mathbf{j} - v\mathbf{k}, 0 \le u \le 1, 1 \le v \le 2$.

29. Evaluate $\iint_S (x - y^2 + z) \, dS$, where S is the surface defined by $\mathbf{R}(u, v) = u^2\mathbf{i} + v\mathbf{j} + u\mathbf{k}, 0 \le u \le 1, 0 \le v \le 1$.

30. Evaluate $\iint_S (\tan^{-1} x + y - z^2) \, dS$, where S is the surface defined by $\mathbf{R}(u, v) = u\mathbf{i} + v^2\mathbf{j} - v\mathbf{k}, 0 \le u \le 1, 0 \le v \le 1$.

31. Evaluate $\iint_S (x^2 + y - z) \, dS$, where S is the surface defined by $\mathbf{R}(u, v) = u\mathbf{i} - u^2\mathbf{j} + v\mathbf{k}, 0 \le u \le 2, 0 \le v \le 1$.

32. Evaluate $\iint_S (x + y + z) \, dS$, where S is the surface of the 1001cube $0 \le x \le 1, 0 \le y \le 1, 0 \le z \le 1$.

In Problems 33–38, find the mass of the homogeneous lamina that has the shape of the given surface S.

33. S is the surface $z = 4 - x - 2y$, with $z \ge 0, x \ge 0, y \ge 0$; $\delta = x$.

34. S is the surface $z = 10 - 2x - y$, with $z \ge 0, x \ge 0, y \ge 0$; $\delta = y$.

35. S is the surface $z = x^2 + y^2$, with $z \le 1; \delta = z$.

36. S is the surface $z = 1 - x^2 - y^2$, with $z \ge 0; \delta = \rho^2$.

37. S is the surface $x^2 + y^2 + z^2 = 5$, with $z \ge 1; \delta = \theta^2$.

38. S is the triangular surface with vertices $(1, 0, 0), (0, 1, 0),$ and $(0, 0, 1); \delta = x + y$.

Ⓒ 39. a. A lamina has the shape of the portion of the sphere $x^2 + y^2 + z^2 = a^2$ that lies within the cone $z = \sqrt{x^2 + y^2}$. Determine the mass of the lamina if $\delta(x, y, z) = x^2y^2z$.

b. Let S be the spherical shell centered at the origin with radius a, and let C be the right circular cone whose vertex is at the origin and whose axis of symmetry coincides with the z-axis. (This has the shape of an old-fashioned ice cream cone.) Suppose the vertex angle of the cone is ϕ_0, with $0 \le \phi_0 < \frac{\pi}{2}$. Determine the mass of that portion of the sphere that is enclosed in the intersection of S and C. Assume $\delta(x, y, z) = x^2y^2z$.

40. Recall the formula

$$I_z = \iint_S (x^2 + y^2)\delta(x, y, z) \, dS$$

for the moment of inertia about the z-axis. Show that the moment of inertia of a conical shell about its axis is $\frac{1}{2}ma^2$, where m is the mass and a is the radius of the cone. Assume $\delta(x, y, z) = 1$.

41. Recall the formula

$$I_z = \iint_S (x^2 + y^2)\delta(x, y, z) \, dS$$

for the moment of inertia about the z-axis. Show that the moment of inertia of a spherical shell of uniform density about its diameter is $\frac{2}{3}ma^2$, where m is the mass and a is the radius. Assume $\delta(x, y, z) = 1$.

42. THINK TANK PROBLEM   Show that a Möbius strip is not orientable.

14.6 Stokes' Theorem

IN THIS SECTION   Stokes' theorem, theoretical applications of Stokes' theorem, physical interpretation of Stokes' theorem ∎

In Section 14.4, we observed that Green's theorem can be written

$$\int_C \mathbf{F} \cdot d\mathbf{R} = \iint_A (\text{curl } \mathbf{F} \cdot \mathbf{k}) \, dA$$

where A is the plane region bounded by the closed curve C. Stokes' theorem is a generalization of this result to surfaces in space and their boundaries.

STOKES' THEOREM

Before stating Stokes' theorem, we need to explain what is meant by a *compatible orientation*. We say that the orientation of a closed path C on the surface S is **compatible with the orientation on S** if the positive direction is *counterclockwise* in relation to the outward normal vector of the surface (see Figure 14.28). If you point the thumb of your right hand in the direction of the outward unit normal vector, your fingers will curl in the direction of the compatibly oriented curve C.

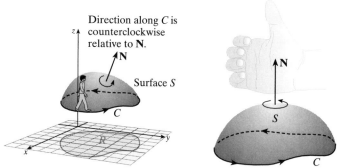

Direction along C is counterclockwise relative to **N**.

■ **FIGURE 14.28** Compatible orientation: The surface S is on the left of someone walking in a counterclockwise direction around the boundary curve C

THEOREM 14.7   *Stokes' theorem*

Let S be an oriented surface with unit normal vector **N**, and assume that S is bounded by a closed, piecewise smooth curve C whose orientation is compatible with that of S. If **F** is a vector field that is continuously differentiable on S, then

$$\int_C \mathbf{F} \cdot d\mathbf{R} = \iint_S (\text{curl } \mathbf{F} \cdot \mathbf{N}) \, dS$$

Proof   The general proof is beyond the scope of this text. However, a proof assuming **F**, S, and C are "well behaved" is given in Appendix B.

Notice that the surface S in the statement of Stokes' theorem is *not* required to be closed. That is, S does not have to bound a solid region.

EXAMPLE 1   *Verifying Stokes' theorem for a particular example*

Let $\mathbf{F} = -\frac{3}{2}y^2\mathbf{i} - 2xy\mathbf{j} + yz\mathbf{k}$, where S is that part of the surface of the plane $x + y + z = 1$ contained within triangle C with vertices $(1, 0, 0)$, $(0, 1, 0)$, and $(0, 0, 1)$, traversed counterclockwise as viewed from above. Verify Stokes' theorem.

Solution   The planar surface S and boundary curve C are shown in Figure 14.29. We shall show that the line integral $\int_C \mathbf{F} \cdot d\mathbf{R}$ and the surface integral $\iint_S (\text{curl } \mathbf{F} \cdot \mathbf{N}) \, dS$ have the same value.

I. *Evaluation of* $\displaystyle\int_C \mathbf{F} \cdot d\mathbf{R}$

The three edges of the boundary triangle C are expressed as

$$\begin{aligned} E_1:\quad & x + y = 1, \, z = 0 \\ E_2:\quad & y + z = 1, \, x = 0 \\ E_3:\quad & x + z = 1, \, y = 0 \end{aligned}$$

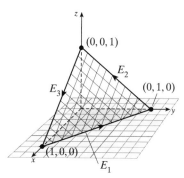

■ **FIGURE 14.29** The triangle C bounds the part of the plane $x + y + z = 1$ that lies in the first octant

We traverse C in a counterclockwise direction (see Figure 14.29).

Edge E_1: Parametrize with $x = 1 - t$, $y = t$, $z = 0$, for $0 \le t \le 1$, so $\mathbf{R}(t) = (1 - t)\mathbf{i} + t\mathbf{j}$ and $d\mathbf{R} = -dt\mathbf{i} + dt\mathbf{j}$. Finally, in terms of the parameter t, we have $\mathbf{F}(t) = -\frac{3}{2}t^2\mathbf{i} - 2t(1 - t)\mathbf{j}$.

$$\int_{E_1} \mathbf{F} \cdot d\mathbf{R} = \int_0^1 \left(\frac{3}{2}t^2 - 2t + 2t^2\right) dt = \int_0^1 \left(\frac{7}{2}t^2 - 2t\right) dt = \left(\frac{7}{6}t^3 - t^2\right)\Big|_0^1 = \frac{1}{6}$$

Edge E_2: Parametrize with $x = 0$, $y = 1 - s$, $z = s$, for $0 \le s \le 1$, so $\mathbf{R}(s) = (1 - s)\mathbf{j} + s\mathbf{k}$ and $d\mathbf{R} = -ds\mathbf{j} + ds\mathbf{k}$. In terms of the parameter s, we have $\mathbf{F}(s) = -\frac{3}{2}(1 - s)^2\mathbf{i} + (1 - s)s\mathbf{k}$.

$$\int_{E_2} \mathbf{F} \cdot d\mathbf{R} = \int_0^1 (1 - s)s\, ds = \left(\frac{s^2}{2} - \frac{s^3}{3}\right)\Big|_0^1 = \frac{1}{6}$$

Edge E_3: Parametrize with $x = r$, $y = 0$, $z = 1 - r$, for $0 \le r \le 1$, so $\mathbf{R}(r) = r\mathbf{i} + (1 - r)\mathbf{k}$ and $d\mathbf{R} = dr\mathbf{i} - dr\mathbf{k}$. In terms of the parameter r, we have $\mathbf{F}(r) = \mathbf{0}$.

$$\int_{E_3} \mathbf{F} \cdot d\mathbf{R} = 0$$

Combining these results, we find

$$\int_C \mathbf{F} \cdot d\mathbf{R} = \int_{E_1} \mathbf{F} \cdot d\mathbf{R} + \int_{E_2} \mathbf{F} \cdot d\mathbf{R} + \int_{E_3} \mathbf{F} \cdot d\mathbf{R} = \frac{1}{6} + \frac{1}{6} + 0 = \frac{1}{3}$$

II. *Evaluation of* $\displaystyle\iint_S (\text{curl } \mathbf{F} \cdot \mathbf{N})\, dS$

$$\text{curl } \mathbf{F} = \begin{vmatrix} \mathbf{i} & \mathbf{j} & \mathbf{k} \\ \dfrac{\partial}{\partial x} & \dfrac{\partial}{\partial y} & \dfrac{\partial}{\partial z} \\ -\frac{3}{2}y^2 & -2xy & yz \end{vmatrix} = z\mathbf{i} + y\mathbf{k}$$

The triangular region on the surface of the plane $x + y + z = 1$ has outward unit normal vector

$$\mathbf{N} = \frac{1}{\sqrt{3}}(\mathbf{i} + \mathbf{j} + \mathbf{k})$$

Because $z = 1 - x - y$, we find

$$\text{curl } \mathbf{F} \cdot \mathbf{N} = (z\mathbf{i} + y\mathbf{k}) \cdot \frac{1}{\sqrt{3}}(\mathbf{i} + \mathbf{j} + \mathbf{k})$$

$$= \frac{1}{\sqrt{3}}(1 - x - y + y) = \frac{1}{\sqrt{3}}(1 - x)$$

Finally,

$$dS = \sqrt{z_x^2 + z_y^2 + 1}\, dA_{xy} = \sqrt{(-1)^2 + (-1)^2 + 1}\, dA_{xy} = \sqrt{3}\, dA_{xy}$$

Combining these results, we get (see Figure 14.30)

$$\iint_S (\text{curl } \mathbf{F} \cdot \mathbf{N})\, dS = \iint_D \frac{1}{\sqrt{3}}(1 - x)\sqrt{3}\, dA_{xy} = \int_0^1 \int_0^{1-x} (1 - x)\, dy\, dx$$

$$= \int_0^1 (1 - x)[(1 - x) - 0]\, dx = -\frac{(1 - x)^3}{3}\Big|_0^1 = \frac{1}{3}$$

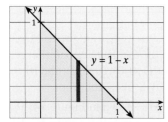

■ **FIGURE 14.30** The projected region in the xy-plane is a triangle bounded by $y = 1 - x$ and the positive coordinate axes.

We see that for this example,

$$\int_C \mathbf{F} \cdot d\mathbf{R} = \frac{1}{3} = \iint_S (\text{curl } \mathbf{F} \cdot \mathbf{N}) \, dS$$

as claimed by Stokes' theorem.

    EXAMPLE 2   *Using Stokes' theorem to evaluate a line integral*

Evaluate $\int_C (\frac{1}{2}y^2 \, dx + z \, dy + x \, dz)$ where C is the curve of intersection of the plane $x + z = 1$ and the ellipsoid $x^2 + 2y^2 + z^2 = 1$, oriented clockwise as seen from the origin. The curve C is shown in Figure 14.31.

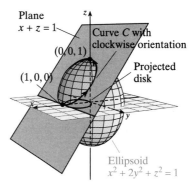

■ **FIGURE 14.31** The surface bounded by the curve of intersection projects onto a disk in the xy-plane

Solution   If we set $\mathbf{F} = \frac{1}{2}y^2\mathbf{i} + z\mathbf{j} + x\mathbf{k}$, the given line integral can be expressed as

$$\int_C \mathbf{F} \, dR$$

According to Stokes' theorem, we have

$$\int_C \mathbf{F} \, dR = \iint_S (\text{curl } \mathbf{F} \cdot \mathbf{N}) \, dS$$

We choose S to be the part of the plane $x + z = 1$ bounded by C, along with C itself, then find the required parts of the surface integral; namely, curl $\mathbf{F}$, $\mathbf{N}$, and dS.

$$\text{curl } \mathbf{F} = \begin{vmatrix} \mathbf{i} & \mathbf{j} & \mathbf{k} \\ \dfrac{\partial}{\partial x} & \dfrac{\partial}{\partial y} & \dfrac{\partial}{\partial z} \\ \frac{1}{2}y^2 & z & x \end{vmatrix} = -\mathbf{i} - \mathbf{j} - y\mathbf{k}$$

The outward unit normal vector to the plane $x + z = 1$ is $\mathbf{N} = \dfrac{1}{\sqrt{2}}(\mathbf{i} + \mathbf{k})$, so that

$$\text{curl } \mathbf{F} \cdot \mathbf{N} = \frac{1}{\sqrt{2}}(-1 - y)$$

and since $z = 1 - x$ on S, we have $z_x = -1$, $z_y = 0$, and

$$dS = \sqrt{z_x^2 + z_y^2 + 1} \, dA_{xy} = \sqrt{(-1)^2 + (0)^2 + 1} \, dA_{xy} = \sqrt{2} \, dA_{xy}$$

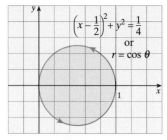

■ **FIGURE 14.32**   This is the projected disk in the xy-plane; call it S_{xy}

Finally, to describe C we substitute $z = 1 - x$ into the equation for the ellipsoid:

$$x^2 + 2y^2 + z^2 = 1$$
$$x^2 + 2y^2 + (1 - x)^2 = 1$$
$$x^2 - x + y^2 = 0$$
$$(x - \tfrac{1}{2})^2 + y^2 = \tfrac{1}{4}$$

Thus, the projection of C on the xy-plane is the circle S_{xy} shown in Figure 14.32. Notice that this circle has the polar equation $r = \cos\theta$. Note also that although C is oriented clockwise as viewed from the origin, the projected curve in the xy-plane is oriented counterclockwise as viewed from above.

We now calculate the desired line integral using Stokes' theorem:

$$\int_C (\tfrac{1}{2}y^2\, dx + z\, dy + x\, dz) = \int_C \mathbf{F} \cdot d\mathbf{R}$$

$$= \iint_S (\text{curl } \mathbf{F} \cdot \mathbf{N})\, ds \qquad \textit{Stokes' theorem}$$

$$= \int_{S_{xy}}\!\!\int \frac{1}{\sqrt{2}}(-1 - y)\,\sqrt{2}\, dA_{xy} = -\int_{S_{xy}} (1 + y)\, dA_{xy}$$

$$= -\int_{-\pi/2}^{\pi/2}\!\!\int_0^{\cos\theta} (1 + r\sin\theta)\, r\, dr\, d\theta \qquad \textit{Change to polar coordinates.}$$

$$= -\int_{-\pi/2}^{\pi/2} \left[\frac{1}{2}\cos^2\theta + \frac{1}{3}\cos^3\theta\,\sin\theta\right] d\theta = -\frac{\pi}{4} \qquad ■$$

Sometimes it is possible to exchange a particularly difficult surface integration over one surface for a less difficult integration over another surface with the same boundary curve. Suppose two surfaces S_1 and S_2 are bounded by the same curve C and induce the same orientation on C. Stokes' theorem tells us that

$$\iint_{S_1} (\text{curl } \mathbf{F} \cdot \mathbf{N}_1)\, dS_1 = \int_C \mathbf{F} \cdot d\mathbf{R} = \iint_{S_2} (\text{curl } \mathbf{F} \cdot \mathbf{N}_2)\, dS_2$$

for any function $\mathbf{F}$ whose components have continuous partial derivatives on both S_1 and S_2.

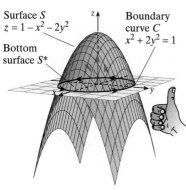

■ **FIGURE 14.33**   Surface $z = 1 - x^2 - 2y^2$ with $z \geq 0$

EXAMPLE 3   *Using Stokes' theorem to evaluate a surface integral*

Evaluate $\iint_S (\text{curl } \mathbf{F} \cdot \mathbf{N})\, dS$, where $\mathbf{F} = x\mathbf{i} + y^2\mathbf{j} + ze^{xy}\mathbf{k}$ and S is that part of the surface $z = 1 - x^2 - 2y^2$ with $z \geq 0$.

Solution   By setting $z = 0$ in the equation of the surface, we find that the boundary curve C for S is the ellipse $x^2 + 2y^2 = 1$, and the outward unit normal vector on S induces a counterclockwise orientation on C, as shown in Figure 14.33.

Let S^* be the elliptical disk defined by $x^2 + 2y^2 \leq 1$. We see that S and S^* have the same boundary, the same orientation, and the same normal vector $N_1 = \mathbf{k}$. Thus,

$$\iint_S (\text{curl } \mathbf{F} \cdot \mathbf{N})\, dS = \iint_{S^*} (\text{curl } \mathbf{F} \cdot \mathbf{k})\, dS^*$$

and we use this second integral to calculate the required integral. We obtain

$$\text{curl } \mathbf{F} = \begin{vmatrix} \mathbf{i} & \mathbf{j} & \mathbf{k} \\ \dfrac{\partial}{\partial x} & \dfrac{\partial}{\partial y} & \dfrac{\partial}{\partial z} \\ x & y^2 & ze^{xy} \end{vmatrix} = (zxe^{xy})\mathbf{i} - (zye^{xy})\mathbf{j}$$

and because curl $\mathbf{F} \cdot \mathbf{k} = 0$, we conclude

$$\iint\limits_{S} (\text{curl } \mathbf{F} \cdot \mathbf{N}) \, dS = \iint\limits_{S^*} (\text{curl } \mathbf{F} \cdot \mathbf{k}) \, dS^* = \iint\limits_{S^*} 0 \, dS^* = 0 \qquad\blacksquare$$

THEORETICAL APPLICATIONS OF STOKES' THEOREM

In physics and other applied areas, Stokes' theorem is often used as a device for establishing general properties. For instance, we can use it to prove that a vector field $\mathbf{F}$ is conservative if and only if curl $\mathbf{F} = \mathbf{0}$.

THEOREM 14.8   *Test for a vector field to be conservative*

If $\mathbf{F}$ and curl $\mathbf{F}$ are continuous in the simply connected region D, then $\mathbf{F}$ is conservative in D if and only if curl $\mathbf{F} = \mathbf{0}$ in D.

Proof   If $\mathbf{F}$ is conservative, let f be a scalar potential function, so that $\nabla f = \mathbf{F}$. Then curl $\mathbf{F} = \nabla \times \mathbf{F} = \nabla \times (\nabla f) = \mathbf{0}$. (See Problem 62, Section 14.1 for a proof of this property.)

Conversely, if curl $\mathbf{F} = \mathbf{0}$, let C be the boundary curve of the smooth surface S. Then Stokes' theorem gives

$$\int_{C} \mathbf{F} \cdot d\mathbf{R} = \iint\limits_{S} (\text{curl } \mathbf{F} \cdot \mathbf{N}) \, dS = \iint\limits_{S} 0 \, dS = 0$$

so that $\mathbf{F}$ is independent of path and $\mathbf{F}$ must be conservative.   ═══

EXAMPLE 4   *Showing that a vector field is conservative*

Show that the vector field $\mathbf{F} = yz\mathbf{i} + xz\mathbf{j} + xy\mathbf{k}$ is conservative in $\mathbb{R}^3$, then find a scalar potential function f for $\mathbf{F}$.

Solution   Since

$$\text{curl } \mathbf{F} = \begin{vmatrix} \mathbf{i} & \mathbf{j} & \mathbf{k} \\ \dfrac{\partial}{\partial x} & \dfrac{\partial}{\partial y} & \dfrac{\partial}{\partial z} \\ yz & xz & xy \end{vmatrix} = (x - x)\mathbf{i} - (y - y)\mathbf{j} + (z - z)\mathbf{k} = \mathbf{0}$$

it follows from Theorem 14.8 that $\mathbf{F}$ is conservative.

To find a scalar potential function f, we must have $\nabla f = \mathbf{F}$; that is,

$$\frac{\partial f}{\partial x} = yz \qquad \frac{\partial f}{\partial y} = xz \qquad \frac{\partial f}{\partial z} = xy$$

Partially integrating the first of these equations with respect to x, we obtain

$$f = xyz + u(y, z)$$

so
$$\frac{\partial f}{\partial y} = xz + \frac{\partial u}{\partial y} = xz \quad \text{and} \quad \frac{\partial u}{\partial y} = 0$$

Thus, $u = v(z)$ and $f = xyz + v(z)$

$$\frac{\partial f}{\partial z} = xy + v'(z) = xy \quad \text{so } v'(z) = 0 \text{ and } v(z) = C$$

We conclude that

$$f(x, y, z) = xyz$$

is a scalar potential function for $\mathbf{F}$. $\blacksquare$

EXAMPLE 5 *Maxwell's current density equation*

In physics, it is shown that if I is the current crossing any surface S bounded by the closed curve C, then

$$\int_C \mathbf{H}\cdot d\mathbf{R} = I \quad \text{and} \quad \iint_S \mathbf{J}\cdot\mathbf{N}\,dS = I$$

where $\mathbf{H}$ is the magnetic intensity and $\mathbf{J}$ is the electric current density. Use this information to derive Maxwell's current density equation curl $\mathbf{H} = \mathbf{J}$.

Solution Equating the two equations of current, we obtain

$$\int_C \mathbf{H}\cdot d\mathbf{R} = I = \iint_S \mathbf{J}\cdot\mathbf{N}\,dS$$

By Stokes' theorem, $\int_C \mathbf{H}\cdot d\mathbf{R} = \iint_S (\text{curl}\,\mathbf{H}\cdot\mathbf{N})\,dS$

Equating the two surface integrals that equal $\int_C \mathbf{H}\cdot d\mathbf{R}$, we have

$$\iint_S \mathbf{J}\cdot\mathbf{N}\,dS = \iint_S (\text{curl}\,\mathbf{H}\cdot\mathbf{N})\,dS$$

or, equivalently,

$$\iint_S (\mathbf{J} - \text{curl}\,\mathbf{H})\cdot\mathbf{N}\,dS = 0$$

Because this equation holds for *any* surface S bounded by C, it follows that

$$\mathbf{J} - \text{curl}\,\mathbf{H} = \mathbf{0}$$
$$\mathbf{J} = \text{curl}\,\mathbf{H}$$

■

PHYSICAL INTERPRETATION OF STOKES' THEOREM

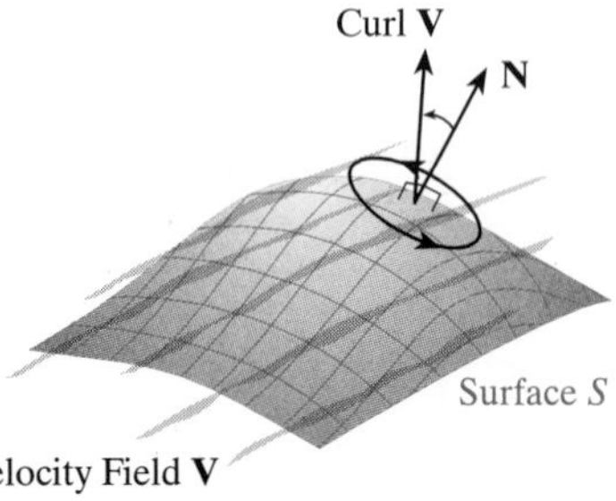

■ **FIGURE 14.34** The tendency of a fluid to swirl across the surface S is measured by curl $\mathbf{V}\cdot\mathbf{N}$

If $\mathbf{V}$ is the velocity field of a fluid flow, then curl $\mathbf{V}$ measures the tendency of the fluid to rotate, or swirl (see Figure 14.34). If the fluid flows across the surface S, the rotational tendency usually will vary from point to point on the surface, and the surface integral $\iint_S (\text{curl}\,\mathbf{V}\cdot\mathbf{N})\,dS$ provides a measure of the *cumulative* rotational tendency over the entire surface.

Stokes' theorem tells us that this cumulative measure of rotational tendency equals the line integral $\int_C \mathbf{V}\cdot d\mathbf{R}$. To interpret this line integral, recall that it can be written $\int_C \mathbf{V}\cdot\mathbf{T}\,ds$ in terms of the arc length parameter s and the unit tangent $\mathbf{T}$ to the curve. Thus, the line integral sums the *tangential component* of the velocity field $\mathbf{V}$ around the boundary C, and it is reasonable to interpret $\int_C \mathbf{V}\cdot\mathbf{T}\,ds$ as a *measure of the circulation of the fluid flow around C.*

■ ***What This Says:***

$$\underbrace{\iint_S (\text{curl}\,\mathbf{V}\cdot\mathbf{N})\,dS}_{\text{The cumulative tendency of a fluid to swirl across the surface } S} = \underbrace{\int_C \mathbf{V}\cdot\mathbf{T}\,ds}_{\text{The circulation of a fluid around the boundary curve } C}$$

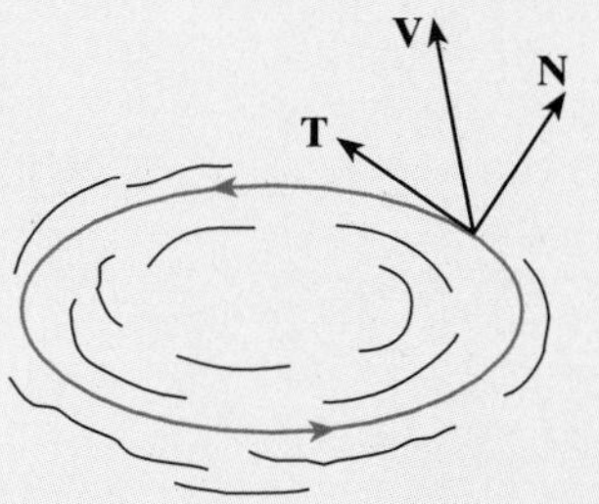

14.6   Problem Set

A *Verify Stokes' theorem for the vector functions and surfaces given in Problems 1–5.*

1. $\mathbf{F} = z\mathbf{i} + 2x\mathbf{j} + 3y\mathbf{k}$; S is the upper hemisphere $z = \sqrt{9 - x^2 - y^2}$.

2. $\mathbf{F} = (y + z)\mathbf{i} + x\mathbf{j} + (z - x)\mathbf{k}$; S is the triangular region of the plane $x + 2y + z = 3$ in the first octant.

3. $\mathbf{F} = (x + 2z)\mathbf{i} + (y - x)\mathbf{j} + (z - y)\mathbf{k}$; S is the triangular region with vertices $(3, 0, 0)$, $(0, \frac{3}{2}, 0)$, $(0, 0, 3)$.

4. $\mathbf{F} = 2xy\mathbf{i} + z^2\mathbf{k}$; S is the portion of the paraboloid $y = x^2 + z^2$, with $y \leq 4$.

5. $\mathbf{F} = 2y\mathbf{i} - 6z\mathbf{j} + 3x\mathbf{k}$; S is the portion of the paraboloid $z = 4 - x^2 - y^2$ and above the xy-plane.

Use Stokes' theorem to evaluate the line integrals given in Problems 6–13.

6. $\int_C (x^3y^2\, dx + dy + z^2\, dz)$,

    where C is the circle $x^2 + y^2 = 1$
    1 in the plane $z = 1$, coun-
    terclockwise when viewed
    from the origin.

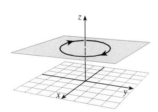

7. $\int_C (z\, dx + x\, dy + y\, dz)$,

    where C is the triangle
    with vertices $(3, 0, 0)$,
    $(0, 0, 2)$, and $(0, 6, 0)$,
    traversed in the given
    order.

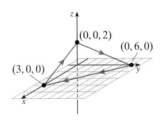

8. $\int_C (y\, dx - 2x\, dy + z\, dz)$, where C is the intersection of the surface $z = x^2 + y^2$ and the plane $x + y + z = 1$ considered counterclockwise when viewed from the origin.

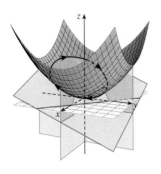

9. $\int_C [2xy^2z\, dx + 2x^2yz\, dy + (x^2y^2 - 2z)dz]$, where C is the curve given by $x = \cos t$, $y = \sin t$, $z = \sin t$, $0 \leq t \leq 2\pi$, traversed in the direction of increasing t

10. $\int_C (y\, dx + z\, dy + y\, dz)$, where C is the intersection of the sphere $x^2 + y^2 + z^2 = 4$ and the plane $x + y + z = 0$, traversed counterclockwise when viewed from above.

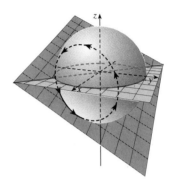

11. $\int_C (y\, dx + z\, dy + x\, dz)$, where C is the intersection of the plane $x + y = 2$ and the surface $x^2 + y^2 + z^2 = 2(x + y)$, traversed counterclockwise as viewed from the origin.

12. $\int_C [(z + \cos x)dx + (x + y^2)dy + (y + e^z)dz]$ where C is the intersection of the sphere $x^2 + y^2 + z^2 = 4$ and the cone $z = \sqrt{x^2 + y^2}$, traversed counterclockwise as viewed from above.

13. $\int_C (3y\, dx + 2z\, dy - 5x\, dz)$, where C is the intersection of the xy-plane and the hemisphere $z = \sqrt{1 - x^2 - y^2}$, tra-versed counterclockwise as viewed from above.

B *In Problems 14–19, use Stokes' theorem to evaluate $\int\int_S (\text{curl } \mathbf{F} \cdot \mathbf{N})\, dS$ for the prescribed vector fields and surfaces.*

14. $\mathbf{F} = x\mathbf{i} + y^2\mathbf{j} + xyz\mathbf{k}$ and S is the part of the paraboloid $z = 4 - x^2 - y^2$ with $z \geq 0$. Use the upward unit normal vector.

15. $\mathbf{F} = xy\mathbf{i} - z\mathbf{j}$ and S is the surface of the cube $0 \leq x \leq 1$, $0 \leq y \leq 1$, $0 \leq z \leq 1$, except for the face where $z = 0$. Use the outward unit normal vector.

16. $\mathbf{F} = y\mathbf{i} + z\mathbf{j} + x\mathbf{k}$, and S is the part of the plane $x + y + z = 1$ that lies in the first octant. Use the out-ward normal vector.

17. $\mathbf{F} = xy\mathbf{i} + x^2\mathbf{j} + z^2\mathbf{k}$, and C is the intersection of the pa-raboloid $z = x^2 + y^2$ and the plane $z = y$.

18. $\mathbf{F} = xz\mathbf{i} + y^2\mathbf{j} + x^2\mathbf{k}$, and C is the intersection of the plane $x + y + z = 3$ and the cylinder

$$x^2 + \frac{y^2}{9} = 1$$

19. $\mathbf{F} = 4y\mathbf{i} + z\mathbf{j} + 2y\mathbf{k}$, and C is the intersection of the sphere $x^2 + y^2 + z^2 = 4$ with the plane $z = 0$. Use the outward normal vector.

In Problems 20–22, use Stokes' theorem to evaluate the line integral

$$\int_C [(1 + y)z\, dx + (1 + z)x\, dy + (1 + x)y\, dz]$$

for the given path C.

20. C is the elliptic path $x = 2\cos\theta, y = \sin\theta, z = 1$ for $0 \le \theta \le 2\pi$.

21. C is the triangle with vertices $(1, 0, 0), (0, 1, 0), (0, 0, 1)$.

22. C is *any* closed path in the plane $2x - 3y + z = 1$.

In Problems 23–26, the vector field $\mathbf{V}$ represents the velocity of a fluid flow. In each case, find the circulation

$$\int_C \mathbf{V} \cdot d\mathbf{R}$$

around the boundary C, assuming a counterclockwise orientation as viewed from above.

23. $\mathbf{V} = x\mathbf{i} + (z - x)\mathbf{j} + y\mathbf{k}$, and C is the intersection of the cylinder $x^2 + y^2 = y$ and the hemisphere $z = \sqrt{1 - x^2 - y^2}$.

24. $\mathbf{V} = y\mathbf{i} + \ln(x^2 + y^2)\mathbf{j} + (x + y)\mathbf{k}$, and C is the triangle with vertices $(0, 0), (1, 0), (0, 1)$.

25. $\mathbf{V} = (e^{x^2} + z)\mathbf{i} + (x + \sin y^3)\mathbf{j} + [y + \ln(\tan^{-1}z)]\mathbf{k}$, and C is the intersection of the sphere $x^2 + y^2 + z^2 = 1$ and the cone $z = \sqrt{x^2 + y^2}$.

26. $\mathbf{V} = y^2\mathbf{i} + \tan^{-1}z\mathbf{j} + (x^2 + 1)\mathbf{k}$, and C is the intersection of the plane $z = y$ and the cylinder $x^2 + y^2 = 2x$.

27. Let $\mathbf{F} = y^2\mathbf{i} + xy\mathbf{j} + xz\mathbf{k}$, and suppose S is the hemisphere $x^2 + y^2 + z^2 = 1$ with $z \ge 0$. Use Stokes' theorem to express

$$\iint_S (\text{curl } \mathbf{F} \cdot \mathbf{N})\, dS$$

as a line integral, and then evaluate the surface integral by evaluating this line integral.

28. Let $\mathbf{F} = z\mathbf{i} + x\mathbf{j} + y\mathbf{k}$, and suppose S is a smooth surface in $\mathbb{R}^3$ whose boundary is given by $x = 2\cos\theta, y = 3\sin\theta, z = \sin\theta, 0 < \theta \le 2\pi$. Use Stokes' theorem to evaluate $\iint_S (\text{curl } \mathbf{F} \cdot \mathbf{N})\, dS$

Show that the given vector field $\mathbf{F}$ in Problems 29–34 is conservative and find a scalar potential function f for $\mathbf{F}$.

29. $yz^2\mathbf{i} + xz^2\mathbf{j} + 2xyz\mathbf{k}$

30. $e^{xy}yz\mathbf{i} + e^{xy}xz\mathbf{j} + e^{xy}\mathbf{k}$

31. $yz^{-1}\mathbf{i} + xz^{-1}\mathbf{j} - xyz^{-2}\mathbf{k}$

32. $(x^2 + y^2 + z^2)(x\mathbf{i} + y\mathbf{j} + z\mathbf{k})$

33. $(y\sin z)\mathbf{i} + (x\sin z + 2y)\mathbf{j} + (xy\cos z)\mathbf{k}$

34. $(xy^2 + yz)\mathbf{i} + (x^2y + xz + 3y^2z)\mathbf{j} + (xy + y^3)\mathbf{k}$

35. HISTORICAL QUEST   George Stokes was an English mathematical physicist who made important contributions to fluid mechanics, including the so-called Navier–Stokes equations. Most of his research was done before 1850, after which he held the Lucasian chair of mathematics at Cambridge for the better part of a half-century. William Thompson (Lord Kelvin) knew the result now known as Stokes' theorem in 1850 and sent it to Stokes as a challenge. Stokes proved the theorem, and then included it as an exam question in 1854. One of the students taking this particular examination was James Clerk Maxwell (Historical Quest #30, Section 14.7), who derived the famous electromagnetic wave equations 10 years later.

GEORGE STOKES
1819–1903

For this Quest, write a history of the Lucasian chair, which was deeded in 1663 as a gift of Henry Lucas. The first and second appointees were Isaac Barrow (Historical Quest #6, page 124) and Isaac Newton (Historical Quest #1, page 124); the chair is currently held by Stephen Hawking.

36. Let S be the ellipsoid $\frac{x^2}{4} + \frac{y^2}{9} + z^2 = 1$, and let $\mathbf{F}$ be a vector field whose component functions have continuous partial derivatives on S. Use Stokes' theorem to show that

$$\iint_S (\text{curl } \mathbf{F} \cdot \mathbf{N})\, dS = 0$$

Does it matter that S is an ellipsoid? State and prove a more general result based on what you have discovered in the first part of this problem.

37. Faraday's law of electromagnetism says that if $\mathbf{E}$ is the electric intensity vector in a system, then

$$\int_C \mathbf{E} \cdot d\mathbf{R} = -\frac{\partial \phi}{\partial t}$$

around any closed curve C where t is time and ϕ is the total magnetic flux directed outward through any surface S bounded by C. Given that

$$\phi = \iint_S \mathbf{B} \cdot \mathbf{N}\, dS$$

where $\mathbf{B}$ is the magnetic flux density, show that

$$\text{curl } \mathbf{E} = -\frac{\partial \mathbf{B}}{\partial t}$$

Hint: It can be shown that

$$\frac{\partial}{\partial t} \iint_S \mathbf{B} \cdot \mathbf{N}\, dS = \iint_S \frac{\partial \mathbf{B}}{\partial t} \cdot \mathbf{N}\, dS$$

38. The current I flowing across a surface S bounded by the closed curve C is given by

$$I = \iint_S \mathbf{J} \cdot \mathbf{N} \, dS$$

where $\mathbf{J}$ is the current density. Given that $\mu \mathbf{J} = \text{curl } \mathbf{B}$, where $\mathbf{B}$ is magnetic flux density and μ is a constant, show that

$$\oint_C \mathbf{B} \cdot d\mathbf{R} = \mu I$$

Suppose f and g are continuously differentiable functions of x, y, z, and C is a closed curve bounding the surface S. Use Stokes' theorem to verify the formulas given in Problems 39–40,

39. $\displaystyle\int_C (f \nabla g) \cdot d\mathbf{R} = \iint_S (\nabla f \times \nabla g) \cdot \mathbf{N} \, dS$

40. $\displaystyle\int_C (f \nabla g + g \nabla f) \cdot d\mathbf{R} = 0$

14.7 Divergence Theorem

IN THIS SECTION   **divergence theorem, applications of the divergence theorem, physical interpretation of divergence** ■

DIVERGENCE THEOREM

We used Green's theorem to show that $\int_C \mathbf{F} \cdot \mathbf{N} \, ds = \int_D\!\!\int \text{div } \mathbf{F} \, dA$, where D is a simply connected domain with the closed boundary curve C. The *divergence theorem* (also known as **Gauss' theorem**) is a generalization of this form of Green's theorem that relates a closed surface integral to a volume integral.

THEOREM 14.9   *The divergence theorem*

Let S be a smooth, orientable surface that encloses a solid region D in $\mathbb{R}^3$. If $\mathbf{F}$ is a continuous vector field whose components have continuous partial derivatives in D, then

$$\iint_S \mathbf{F} \cdot \mathbf{N} \, dS = \iiint_D \text{div } \mathbf{F} \, dV$$

where $\mathbf{N}$ is an outward unit normal vector to the surface S.

Proof   The general proof of this theorem is beyond the scope of this course, but an important special case is proved in Appendix B. ══

EXAMPLE 1   *Verifying the divergence theorem for a particular example*

Let $\mathbf{F} = 2x\mathbf{i} - 3y\mathbf{j} + 5z\mathbf{k}$, and let S be the hemisphere $z = \sqrt{9 - x^2 - y^2}$ together with the disk $x^2 + y^2 \le 9$ in the xy-plane. Verify the divergence theorem.

Solution   The solid is shown in Figure 14.35. We shall show that the surface integral and the triple integral have the same value.

I. *Evaluation of* $\displaystyle\iint_S \mathbf{F} \cdot \mathbf{N} \, dS$

S consists of two parts: S_1, the disk on the bottom of the hemisphere, and S_2, the hemisphere. We will consider these separately using

$$\iint_S \mathbf{F} \cdot \mathbf{N} \, dS = \iint_{S_1} \mathbf{F} \cdot \mathbf{N_1} \, dS_1 + \iint_{S_2} \mathbf{F} \cdot \mathbf{N_2} \, dS_2$$

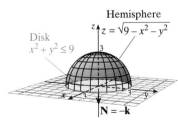

Hemisphere
$z = \sqrt{9 - x^2 - y^2}$

Disk
$x^2 + y^2 \le 9$

$\mathbf{N} = -\mathbf{k}$

■ **FIGURE 14.35**   The surface of the hemisphere

Consider S_1: Note that the disk $x^2 + y^2 \leq 9$ on the bottom of the hemisphere has outward unit normal vector $\mathbf{N} = -\mathbf{k}$. (Because $\mathbf{k}$ points into the solid, $-\mathbf{k}$ is the outward normal vector.) We have

$$\iint_{S_1} \mathbf{F} \cdot \mathbf{N} \, dS_1 = \iint_{S_1} (2x\mathbf{i} - 3y\mathbf{j} + 5z\mathbf{k}) \cdot (-\mathbf{k}) \, dS_1 = \iint_{S_1} (-5z) \, dS_1 = 0$$

Because $z = 0$ on S_1

Consider S_2: Regard the hemisphere as the level surface

$$S_2(x, y, z) = 9, \text{ where } S_2(x, y, z) = x^2 + y^2 + z^2 \text{ and } z \geq 0.$$

The outward unit (for $z \geq 0$) is given by

$$N = \frac{\nabla S_2}{\|\nabla S_2\|}$$

$$= \frac{2x\mathbf{i} + 2y\mathbf{j} + 2z\mathbf{k}}{\sqrt{4x^2 + 4y^2 + 4z^2}} \qquad \text{\em Remember that } \nabla f = \frac{\partial f}{\partial x}\mathbf{i} + \frac{\partial f}{\partial y}\mathbf{j} + \frac{\partial f}{\partial z}\mathbf{k}.$$

$$= \tfrac{1}{3}(x\mathbf{i} + y\mathbf{j} + z\mathbf{k}) \qquad \text{\em Because } x^2 + y^2 + z^2 = 9 \text{ on } S_2$$

and

$$\mathbf{F} \cdot \mathbf{N} = \tfrac{1}{3}(2x^2 - 3y^2 + 5z^2)$$

Since $z_x^2 = x^2(9 - x^2 - y^2)^{-1}$ and $z_y^2 = y^2(9 - x^2 - y^2)^{-1}$, we have

$$dS_2 = \sqrt{z_x^2 + z_y^2 + 1} \, dA_{xy}$$

$$= \sqrt{\frac{x^2}{9 - x^2 - y^2} + \frac{y^2}{9 - x^2 - y^2} + 1} \, dA_{xy}$$

$$= \sqrt{\frac{9}{9 - x^2 - y^2}} \, dA_{xy} = \frac{3}{\sqrt{9 - x^2 - y^2}} \, dA_{xy}$$

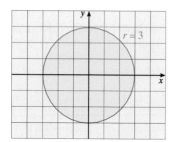

■ FIGURE 14.36 The hemisphere projects with boundary $z^2 = 9 - x^2 - y^2$ onto the disk $x^2 + y^2 \leq 9$, or, in polar form, $r \leq 3$.

and it follows that

$$\int_{S_2} \mathbf{F} \cdot \mathbf{N} \, dS_2 = \int_{S_2} \frac{1}{3}(2x^2 - 3y^2 + 5z^2) \frac{3}{\sqrt{9 - x^2 - y^2}} \, dA_{xy}$$

$$= \iint_{S_2} [2x^2 - 3y^2 + 5(9 - x^2 - y^2)][9 - x^2 - y^2]^{-1/2} \, dA_{xy}$$

$$= \int_0^{2\pi} \int_0^3 \frac{45 - 3(r \cos\theta)^2 - 8(r \sin\theta)^2}{\sqrt{9 - r^2}} \, r \, dr \, d\theta$$

The projected region is $x^2 + y^2 \leq 9$, after much computation. See Figure 14.36.

$$= \int_0^{2\pi} [135 - 54 \cos^2\theta - 144 \sin^2\theta] \, d\theta$$

$$= 72\pi$$

We can now state

$$\iint_S \mathbf{F} \cdot \mathbf{N} \, dS = \iint_{S_1} \mathbf{F} \cdot \mathbf{N} \, dS_1 + \iint_{S_2} \mathbf{F} \cdot \mathbf{N} \, dS_2 = 0 + 72\pi = 72\pi$$

II. *Evaluation of* $\iiint\limits_D \text{div } \mathbf{F} \, dV$

$$\text{div } \mathbf{F} = \frac{\partial}{\partial x}(2x) + \frac{\partial}{\partial y}(-3y) + \frac{\partial}{\partial x}(5z) = 2 - 3 + 5 = 4$$

Therefore, $\iiint \text{div } \mathbf{F} \, dV = \iiint 4 \, dV$, but $\iiint dV$ is just the volume of the hemisphere $z = \sqrt{9 - x^2 - y^2}$. A hemisphere of radius 3 has volume $\frac{1}{2}[\frac{4}{3}\pi(3)^3] = 18\pi$, so

$$\iiint\limits_D \text{div } \mathbf{F} \, dV = \iiint\limits_D 4 \, dV = 4V = 4(18\pi) = 72\pi$$

and we have $\iiint \text{div } \mathbf{F} \, dV = \iint \mathbf{F} \cdot \mathbf{N} \, dS$, as required. ∎

| **EXAMPLE 2** | *Evaluating a surface integral using the divergence theorem* |

Evaluate $\iint\limits_S \mathbf{F} \cdot \mathbf{N} \, dS$, where $\mathbf{F} = x^2\mathbf{i} + xy\mathbf{j} + x^3y^3\mathbf{k}$ and S is the surface of the tetrahedron bounded by the plane $x + y + z = 1$ and the coordinate planes, with outward unit normal vector $\mathbf{N}$.

Solution We will use the divergence theorem. Let D be the solid bounded by S, and note that

$$\text{div } \mathbf{F} = \frac{\partial}{\partial x}(x^2) + \frac{\partial}{\partial y}(xy) + \frac{\partial}{\partial z}(x^3y^3) = 2x + x + 0 = 3x$$

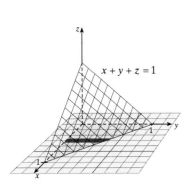

We choose to describe the solid tetrahedron as the set of all (x, y, z) such that $0 \le z \le 1 - x - y$ whenever $0 \le y \le 1 - x$ for $0 \le x \le 1$. The projection of S on the xy-plane is shown in the margin. Finally, by applying the divergence theorem, we find that

$$\iint\limits_S \mathbf{F} \cdot \mathbf{N} \, dS = \iiint\limits_D \text{div } \mathbf{F} \, dV = \int_0^1 \int_0^{1-x} \int_0^{1-x-y} 3x \, dz \, dy \, dx$$

$$= \int_0^1 \int_0^{1-x} 3x(1 - x - y) \, dy \, dx = 3 \int_0^1 \left[x(1 - x)y - \frac{1}{2}xy^2 \right]_0^{1-x} dx$$

$$= 3 \int_0^1 x(1 - x)^2 - \frac{1}{2}x(1 - x)^2 \, dx = \frac{1}{8} \quad ∎$$

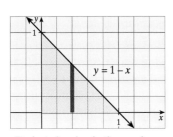

Projected region in the xy-plane

The divergence theorem applies only to closed surfaces. However, if we wish to evaluate $\int_{S_1} \int \mathbf{F} \cdot \mathbf{N} \, dS$ where S_1 is *not* closed, we may be able to find a closed surface S that is the union of S_1 and some other surface S_2. Then, if the hypotheses of the divergence theorem are satisfied by $\mathbf{F}$ and S, we have

$$\iint\limits_{S_1} \mathbf{F} \cdot \mathbf{N} \, dS + \iint\limits_{S_2} \mathbf{F} \cdot \mathbf{N} \, dS = \iint\limits_S \mathbf{F} \cdot \mathbf{N} \, dS = \iiint\limits_D \text{div } \mathbf{F} \, dV$$

where D is the solid region bounded by S. Thus, if we can compute $\iiint\limits_D \text{div } \mathbf{F} \, dV$ and $\int_{S_2} \int \mathbf{F} \cdot \mathbf{N} \, dS$, we can compute $\int_{S_1} \int \mathbf{F} \cdot \mathbf{N} \, dS$ by the equation

$$\iint\limits_{S_1} \mathbf{F} \cdot \mathbf{N} \, dS = \iiint\limits_D \text{div } \mathbf{F} \, dV - \iint\limits_{S_2} \mathbf{F} \cdot \mathbf{N} \, dS$$

This equation can also be used as a device for trading the evaluation of a difficult surface integral for that of an easier integral. Here is an example of this procedure.

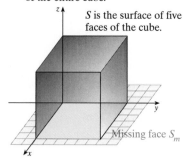

S^* is the closed surface of the entire cube.

S is the surface of five faces of the cube.

Missing face S_m

■ **FIGURE 14.37** Evaluating a surface integral

> ■ **EXAMPLE 3**   *Evaluating a surface integral over an open surface*

Evaluate $\displaystyle\iint_S \mathbf{F} \cdot \mathbf{N} \, dS$, where $\mathbf{F} = xy\mathbf{i} - z^2\mathbf{k}$ and S is the surface of the upper five faces of the unit cube $0 \le x \le 1, 0 \le y \le 1, 0 < z \le 1$, as shown in Figure 14.37.

Solution   Note that the surface S is not closed, but we can close it by adding the missing face S_m, thus forming a closed surface S^* that satisfies the conditions of the divergence theorem. The strategy is to evaluate the surface integral of S^* and then subtract the surface integral of the added face S_m.

$$\iint_{S^*} \mathbf{F} \cdot \mathbf{N} \, dS = \iiint_{\text{cube}} \operatorname{div} \mathbf{F} \, dV = \int_0^1 \int_0^1 \int_0^1 (y - 2z) \, dx \, dy \, dz$$

$$= \int_0^1 \int_0^1 (y - 2x) \, dy \, dz = \int_0^1 \left(\frac{1}{2} - 2z \right) dz = -\frac{1}{2}$$

Also, because the unit normal vector to the added face S_m is $\mathbf{N} = -\mathbf{k}$ and $z = 0$ on this face, it follows that

$$\iint_{S_m} \mathbf{F} \cdot \mathbf{N} \, dS = \iint_{S_m} (xy\mathbf{i} - z^2\mathbf{k}) \cdot (-\mathbf{k}) \, dS = \iint_{S_m} z^2 \, dS = 0$$

Therefore,

$$\iint_S \mathbf{F} \cdot \mathbf{N} \, dS = \iiint_D \operatorname{div} \mathbf{F} \, dV - \iint_{S_m} \mathbf{F} \cdot \mathbf{N} \, dS = -\frac{1}{2} - 0 = -\frac{1}{2} \qquad ■$$

APPLICATIONS OF THE DIVERGENCE THEOREM

Like Stokes' theorem, the divergence theorem is often used for theoretical purposes, especially as a tool for deriving general properties in mathematical physics. The following example deals with an important of fluid dynamics.

> ■ **EXAMPLE 4**   *Continuity equation of fluid dynamics*

Suppose a fluid with density $\delta(x, y, z, t)$ flows in some region of space with velocity $\mathbf{F}(x, y, z, t)$ at the point (x, y, z) at time t. Assuming there are no sources or sinks, show that

$$\operatorname{div} \delta\mathbf{F} = -\frac{\partial \delta}{\partial t}$$

Solution   Recall from Section 14.1 that a point is called a *source* if $\operatorname{div} \mathbf{F} > 0$, a *sink* if $\operatorname{div} \mathbf{F} < 0$, and *incompressible* if $\operatorname{div} \mathbf{F} = 0$. Let S be a smooth surface in $\mathbb{R}^3$ that encloses a solid region D. In physics, it is shown that the amount of fluid flowing out of D across S in unit time is $\iint_S \delta\mathbf{F} \cdot \mathbf{N} \, dS$. Thus, $-\iint_S \delta\mathbf{F} \cdot \mathbf{N} \, dS$ is the amount of outflow and is equal to the amount of inflow, $\iiint_D \frac{\partial \delta}{\partial t} \, dV$, because we are assuming there are no sinks or sources. Equating these two quantities, we have

$$\overbrace{-\iint_S \delta\mathbf{F} \cdot \mathbf{N} \, dS}^{\text{OUTFLOW}} = \overbrace{\iiint_D \frac{\partial \delta}{\partial t} \, dV}^{\text{INFLOW}}$$

By the divergence theorem, it follows that

$$\iint_S \delta\mathbf{F} \cdot \mathbf{N} \, dS = \iiint_D \operatorname{div} \delta\mathbf{F} \, dV$$

so that

$$\iiint\limits_{D} \text{div } \delta\mathbf{F} \, dV + \iiint\limits_{D} \frac{\partial\delta}{\partial t} \, dV = 0$$

$$\iiint\limits_{D} \left[\text{div } \delta\mathbf{F} + \frac{\partial\delta}{\partial t} \right] dV = 0$$

This equation must hold for any region D, no matter how small, which means that the integrand of the integral must be 0. That is,

$$\text{div } \delta\mathbf{F} = -\frac{\partial\delta}{\partial t} \qquad \blacksquare$$

In physics it is known that the total heat contained in a body with uniform density δ and specific heat σ is $\iiint\limits_{D} \sigma\delta T \, dV$, where T is the temperature. Thus, the amount of heat leaving D per unit of time is given by

$$-\frac{\partial}{\partial t}\left[\iiint\limits_{D} \sigma\delta T \, dV \right] = \iiint\limits_{D} -\sigma\delta\frac{\partial T}{\partial t} \, dV$$

In the following example, we use this result to obtain an important formula from mathematical physics.

EXAMPLE 5   *Derivation of the heat equation*

Let $T(x, y, z, t)$ be the temperature at each point (x, y, z) in a solid body D at time t. Given that the velocity of heat flow in the body is $\mathbf{F} = -k\nabla T$ for a positive constant k (called the **thermal conductivity**), show that

$$\frac{\partial T}{\partial t} = \frac{k}{\sigma\delta} \nabla^2 T$$

where σ is the specific heat of the body and δ is its density.

Solution   Let S be the closed surface that bounds D. Because $\mathbf{F}$ is the velocity of heat flow, the amount of heat leaving D per unit time is $\iint\limits_{S} \mathbf{F} \cdot \mathbf{N} \, dS$, and the divergence theorem applies:

$$\iint\limits_{S} \mathbf{F} \cdot \mathbf{N} \, dS = \iiint\limits_{D} \text{div } (-k\nabla T) \, dV = \iiint\limits_{D} (-k\nabla \cdot \nabla T) \, dV$$

$$= \iiint\limits_{D} -k\nabla^2 T \, dV$$

Since this is the amount of heat leaving D per unit of time, it must equal the heat integral from physics derived just before this example. Thus,

$$\iiint\limits_{D} -k\nabla^2 T \, dV = \iiint\limits_{D} -\sigma\delta\frac{\partial T}{\partial t} \, dV$$

This equation holds not only for the body as a whole, but for every part of the body, no matter how small. Thus, we can shrink the body to a single point, and it then follows that the integrands are equal; that is,

$$-k\nabla^2 T = -\sigma\delta\frac{\partial T}{\partial t}$$

$$\frac{\partial T}{\partial t} = \frac{k}{\sigma\delta} \nabla^2 T \qquad \blacksquare$$

For the concluding example, recall that the *normal derivative* $\partial g / \partial n$ of a scalar function g defined on the closed surface S is the directional derivative of f in the direction of the outward unit normal vector $\mathbf{N}$ to S; that is,

$$\frac{\partial g}{\partial n} = \nabla g \cdot \mathbf{N}$$

We will use this equation in the following example, which is a generalization of a property we first obtained for $\mathbb{R}^2$ in Section 14.4.

EXAMPLE 6   *Derivation of Green's first identity*

Show that if f and g are scalar functions such that $\mathbf{F} = f \nabla g$ is continuously differentiable in the solid domain D bounded by the closed surface S, then

$$\iiint\limits_{D} [f \nabla^2 g + \nabla f \cdot \nabla g] \, dV = \iint\limits_{S} f \frac{\partial g}{\partial n} \, dS$$

This is called **Green's first identity.**

Solution   We shall apply the divergence theorem to the vector field $\mathbf{F}$ (note that $\mathbf{F}$ is continuously differentiable), but first we need to express div $\mathbf{F}$ in a more useful form.

$$\begin{aligned}
\text{div}(f \nabla g) &= \nabla \cdot (f \nabla g) \\
&= \left[\frac{\partial}{\partial x} \mathbf{i} + \frac{\partial}{\partial y} \mathbf{j} + \frac{\partial}{\partial z} \mathbf{k} \right] \cdot \left[f \frac{\partial g}{\partial x} \mathbf{i} + f \frac{\partial g}{\partial y} \mathbf{j} + f \frac{\partial g}{\partial z} \mathbf{k} \right] \\
&= \frac{\partial}{\partial x} \left[f \frac{\partial g}{\partial x} \right] + \frac{\partial}{\partial y} \left[f \frac{\partial g}{\partial y} \right] + \frac{\partial}{\partial z} \left[f \frac{\partial g}{\partial z} \right] \\
&= \left[\frac{\partial f}{\partial x} \frac{\partial g}{\partial x} + f \frac{\partial^2 g}{\partial x^2} \right] + \left[\frac{\partial f}{\partial y} \frac{\partial g}{\partial y} + f \frac{\partial^2 g}{\partial y^2} \right] + \left[\frac{\partial f}{\partial z} \frac{\partial g}{\partial z} + f \frac{\partial^2 g}{\partial z^2} \right] \\
&= \left[\frac{\partial f}{\partial x} \frac{\partial g}{\partial x} + \frac{\partial f}{\partial y} \frac{\partial g}{\partial y} + \frac{\partial f}{\partial z} \frac{\partial g}{\partial z} \right] + f \left[\frac{\partial^2 g}{\partial x^2} + \frac{\partial^2 g}{\partial y^2} + \frac{\partial^2 g}{\partial z^2} \right] \\
&= (\nabla f) \cdot (\nabla g) + f \nabla^2 g
\end{aligned}$$

This calculation gives us the first step in the following derivation:

$$\begin{aligned}
\iiint\limits_{D} [f \nabla^2 g + \nabla f \cdot \nabla g] \, dV &= \iiint\limits_{D} \text{div}(f \nabla g) \, dV \\
&= \iint\limits_{S} (f \nabla g) \cdot \mathbf{N} \, dS \qquad \text{Divergence theorem} \\
&= \iint\limits_{S} f(\nabla g \cdot \mathbf{N}) \, dS \\
&= \iint\limits_{S} f \frac{\partial g}{\partial n} \, dS \qquad \text{Because } \nabla g \cdot \mathbf{N} = \frac{\partial g}{\partial n} \text{ by definition} \quad \blacksquare
\end{aligned}$$

PHYSICAL INTERPRETATION OF DIVERGENCE

In Section 14.1, we gave an interpretation of the curl as a measure of the tendency of a fluid to swirl. We now close this chapter with an analogous interpretation of divergence.

The following example can be interpreted as saying that the net rate of fluid mass flowing away (that is, "diverging") from point P_0 is given by div $\mathbf{F}_0$. This is the reason

P_0 is a *source* if div $\mathbf{F}_0 > 0$ (mass flowing out from P_0) and a *sink* if div $\mathbf{F}_0 < 0$ (mass flowing back into P_0).

| EXAMPLE 7 | *Physical interpretation of divergence* |

Let $\mathbf{F} = \delta \mathbf{V}$ be the flux density associated with a fluid of density δ flowing with velocity $\mathbf{V}$ and let P_0 be a point inside a solid region where the conditions of the divergence theorem are satisfied. Then,

$$\text{div } \mathbf{F}_0 = \lim_{r \to 0} \frac{1}{V(r)} \iint\limits_{S(r)} \mathbf{F} \cdot \mathbf{N} \, dS$$

where div $\mathbf{F}_0$ denotes the value of div $\mathbf{F}$ at P_0, and $S(r)$ is a sphere centered at P_0 with volume $V(r) = \frac{4}{3} \pi r^3$.

Solution   Applying the divergence theorem to the solid sphere (ball) $B(r)$ with surface $S(r)$, we obtain

$$\iint\limits_{S(r)} \mathbf{F} \cdot \mathbf{N} \, dS = \iiint\limits_{B(r)} \text{div } \mathbf{F} \, dV$$

The mean value theorem (for triple integrals) tells us that

$$\frac{1}{V(r)} \iiint\limits_{B(r)} \text{div } \mathbf{F} \, dV = \text{div } \mathbf{F}^*$$

where div $\mathbf{F}^*$ denotes the value of div $\mathbf{F}$ at some point P^* in the ball $B(r)$. Combining these results, we find that

$$\iint\limits_{S(r)} \mathbf{F} \cdot \mathbf{N} \, dS = \iiint\limits_{B(r)} \text{div } \mathbf{F} \, dV = V(r) \, \text{div } \mathbf{F}^*$$

or

$$\frac{1}{V(r)} \iint\limits_{S(r)} \mathbf{F} \cdot \mathbf{N} \, dS = \text{div } \mathbf{F}^*$$

Since the point P^* is inside the ball $B(r)$ centered at P_0, it follows that $P^* \to P$ as $r \to 0$, so div $\mathbf{F}^* \to$ div $\mathbf{F}_0$ and we have

$$\lim_{r \to 0} \frac{1}{V(r)} \iint\limits_{S(r)} \mathbf{F} \cdot \mathbf{N} \, dS = \lim_{r \to 0} \text{div } \mathbf{F}^* = \text{div } \mathbf{F}_0$$

as claimed. ∎

14.7 *Problem Set*

Ⓐ *Verify the divergence theorem for the vector function* $\mathbf{F}$ *and solid D given in Problems 1–4 Assume* $\mathbf{N}$ *is the unit normal vector pointing away from the origin.*

1. $\mathbf{F} = xz\mathbf{i} + y^2\mathbf{j} + 2z\mathbf{k}$; D is the ball $x^2 + y^2 + z^2 \leq 4$.

2. $\mathbf{F} = x\mathbf{i} - 2y\mathbf{j}$; D is the interior of the paraboloid $z = x^2 + y^2, 0 \leq z < 9$.

3. $\mathbf{F} = 2y^2\mathbf{j}$; D is the part of the surface of the plane $x + 4y + z = 8$ that lies in the first octant.

4. $\mathbf{F} = 3x\mathbf{i} + 5y\mathbf{j} + 6z\mathbf{k}$; D is the tetrahedron bounded by the coordinate planes and the plane $2x + y + z = 4$.

Use the divergence theorem in Problems 5–19 to evaluate the surface integral $\int_S \int \mathbf{F} \cdot \mathbf{N} \, dS$ *for the given choice of* $\mathbf{F}$ *and the boundary surface S. For each closed surface, assume* $\mathbf{N}$ *is the outward unit normal vector.*

5. $\mathbf{F} = x\mathbf{i} + y\mathbf{j} + z\mathbf{k}$; S is the surface of the cube $0 \leq x \leq 1$, $0 \leq y \leq 1, 0 \leq z \leq 1$. (See Problem 32, Section 14.5.)

6. $\mathbf{F} = xyz\mathbf{j}$; S is the surface of the cylinder $x^2 + y^2 = 9$, for $0 \le z \le 5$.

7. $\mathbf{F} = (\cos yz)\mathbf{i} + e^{xz}\mathbf{j} + 3z^2\mathbf{k}$; S is the surface of the hemisphere $z = \sqrt{4 - x^2 - y^2}$ together with the disk $x^2 + y^2 \le 4$ in the xy-plane.

8. $\mathbf{F} = \text{curl}[e^{xz}\mathbf{i} - 4\mathbf{j} + (\sin xyz)\mathbf{k}]$; S is the surface of the ellipsoid $2x^2 + 3y^2 + 7z^2 = 1$.

9. $\mathbf{F} = (x^2 + y^2 - x^2)\mathbf{i} + x^2y\mathbf{j} + 3z\mathbf{k}$; S is the surface of the five faces of the unit cube $0 \le x \le 1, 0 \le y \le 1, 0 \le z \le 1$, missing $z = 0$.

10. $\mathbf{F} = 2y\mathbf{i} - z\mathbf{j} + 3x\mathbf{k}$; S is the five faces of the unit cube $0 \le x \le 1, 0 \le y \le 1, 0 \le z \le 1$, missing $z = 0$.

11. $\mathbf{F} = x\mathbf{i} + y\mathbf{j} + z\mathbf{k}$; S is the surface of the paraboloid $z = x^2 + y^2$ for $0 \le z \le 9$.

12. $\mathbf{F} = \text{curl}(y\mathbf{i} + x\mathbf{j} + z\mathbf{k})$; S is the surface of the hemisphere $z = \sqrt{4 - x^2 - y^2}$ together with the disk.

13. $\mathbf{F} = x^2\mathbf{i} + y^2\mathbf{j} + z^2\mathbf{k}$; S is the surface of the sphere $x^2 + y^2 + z^2 = 4$.

14. $\mathbf{F} = xyz\mathbf{i} + xyz\mathbf{j} + xyz\mathbf{k}$; S is the surface of the box $0 \le x \le 1, 0 \le y \le 2, 0 \le z \le 3$.

15. $\mathbf{F} = x\mathbf{i} + y\mathbf{j} + (z^2 - 1)\mathbf{k}$; S is the surface of a solid bounded by the cylinder $x^2 + y^2 = 4$ and the planes $z = 0$ and $z = 1$.

16. $\mathbf{F} = (x^5 + 10xy^2z^2)\mathbf{i} + (y^5 + 10yx^2z^2)\mathbf{j} + (z^5 + 10zx^2y^2)\mathbf{k}$; S is the closed hemispherical surface $z = \sqrt{1 - x^2 - y^2}$ together with the disk $x^2 + y^2 \le 1$ in the xy-plane.

17. $\mathbf{F} = xy^2\mathbf{i} + yz^2\mathbf{j} + x^2z\mathbf{k}$; S is the surface bounded above by the sphere $\rho = 2$ and below by the cone $\varphi = \frac{\pi}{4}$ in spherical coordinates. (S is the surface of an "ice cream cone").

18. $\mathbf{F} = xy^2\mathbf{i} + yz^2\mathbf{j} + x^2z^2\mathbf{k}$; S is the surface (in spherical coordinates) with top $\rho = 2, 0 \le \phi \le \frac{\pi}{4}, 0 \le \theta \le 2\pi$, and bottom $0 \le \rho \le 2, \phi = \frac{\pi}{4}, 0 \le \theta \le 2\pi$.

19. $\mathbf{F} = x^3\mathbf{i} + y^3\mathbf{j} + 3a^2z\mathbf{k}$ (constant $a > 0$); S is the surface bounded by the cylinder $x^2 + y^2 = a^2$ and the planes $z = 0$ and $z = 1$.

B 20. Suppose that S is a closed surface that encloses a solid region D.

   a. Show that the volume of D is given by

   $$V(D) = \frac{1}{3} \iint_S (x\mathbf{i} + y\mathbf{j} + z\mathbf{k}) \cdot \mathbf{N} \, dS$$

   where N is an outward unit normal vector to S.

   b. Use the formula in part **a** to find the volume of the hemisphere

   $$z = \sqrt{R^2 - x^2 - y^2}$$

21. Use the divergence theorem to evaluate

$$\iint_S \|\mathbf{R}\|\mathbf{R} \cdot \mathbf{N} \, dS$$

where $\mathbf{R} = x\mathbf{i} + y\mathbf{j} + z\mathbf{k}$ and S is the sphere $x^2 + y^2 + z^2 = a^2$, with constant $a > 0$.

22. The moment of inertia about the z-axis of a solid D of constant density $\delta = a$ is given by

$$I_z = \iiint_T a(x^2 + y^2) \, dV$$

Express this integral as a surface integral over the surface S that bounds D.

C 23. Let u be a scalar function with continuous second partial derivatives in a region containing the solid region D, with closed boundary surface S.

   a. Show that $\displaystyle\iint_S \frac{\partial u}{\partial n} \, dS = \iiint_D \nabla^2 u \, dV$.

   b. Let $u = x + y + z$ and $v = \frac{1}{2}(x^2 + y^2 + z^2)$. Evaluate

   $$\iint_S (u\nabla v) \cdot \mathbf{N} \, dS$$

   where S is the boundary of the cube $0 \le x \le 1, 0 \le y \le 1, 0 \le z \le 1$.

24. Let f and g be scalar functions such that $\mathbf{F} = f\nabla g$ is continuously differentiable in the region D, which is bounded by the closed surface S. Prove *Green's second identity*:

$$\iiint_D (f\nabla^2 g - g\nabla^2 f) \, dV = \iint_S \left(f\frac{\partial g}{\partial n} - g\frac{\partial f}{\partial n} \right) dS$$

25. Show that if g is harmonic in the region D, then

$$\iint_S \frac{\partial g}{\partial n} \, dS = 0$$

where the closed surface S is the boundary of D. (Recall that g harmonic means $\nabla^2 g = 0$.)

26. Show that $\displaystyle\iint_S \mathbf{F} \cdot \mathbf{N} \, dS = 0$ if S is a closed surface and $\mathbf{F} = \text{curl } \mathbf{U}$ throughout the interior of S for some vector field $\mathbf{U}$. A vector field $\mathbf{U}$ with this property is said to be a *vector potential* for $\mathbf{F}$.

27. In our derivation of the heat equation in this section, we assumed that the coefficient of thermal conductivity k is constant (no sinks or sources). If $k = k(x, y, z)$ is a variable, show that the heat equation becomes

$$k\nabla^2 T + \nabla k \cdot \nabla T = \sigma\delta\frac{\partial T}{\partial t}$$

28. An electric charge q located at the origin produces the electric field

$$\mathbf{E} = \frac{q\mathbf{R}}{4\pi\epsilon\|\mathbf{R}\|^3}$$

where $\mathbf{R} = x\mathbf{i} + y\mathbf{j} + z\mathbf{k}$ and ϵ is a physical constant, called the **electric permittivity.**

   a. Show that

   $$\iint_S \mathbf{E} \cdot \mathbf{N} \, dS = 0$$

   if the closed surface S does not enclose the origin. This is **Gauss' law.**

   b. Show that

   $$\iint_S \mathbf{E} \cdot \mathbf{N} \, dS = \frac{q}{\epsilon}$$

in the case where the closed surface S encloses the origin. Note that the divergence theorem does not apply directly to this case.

29. Gauss's law can be expressed as

$$\iint_S \mathbf{D} \cdot \mathbf{N}\, dS = q$$

where $\mathbf{D} = \epsilon \mathbf{E}$ is the electric flux density, with electric intensity $\mathbf{E}$, permittivity ϵ, and q a constant. Show that div $\mathbf{D} = Q$, where Q is the charge density; that is,

$$\iiint_V Q\, dV = q$$

30. HISTORICAL QUEST   James Clerk Maxwell was one of the greatest physicists of all time. Using the experimental discoveries of Michael Faraday as a basis, he was able to express the governing rules for electrical and magnetic fields in precise mathematical form. In 1871, he published his *Theory of Heat and Magnetism*, which formed the basis for modern electromagnetic theory and contributed to quantum theory and special relativity. Maxwell was influential in convincing other mathematicians and scientists to use vectors, and was interested in areas as diverse as the behavior of light and the statistical behavior of molecular motion. He was sometimes referred to as dp/dt because in thermodynamics, $dp/dt = JCM$, a unit of measurement named for him.

JAMES CLERK MAXWELL
1831–1879

For this Quest you are to derive **Maxwell's equation for the electric intensity E:**

$$(\nabla \cdot \nabla)\mathbf{E} = \mu\sigma\frac{\partial \mathbf{E}}{\partial t} + \mu\epsilon\frac{\partial^2 \mathbf{E}}{\partial t^2}$$

To derive this equation, you need to know that

$$\text{curl } \mathbf{E} = -\frac{\partial \mathbf{B}}{\partial t} \quad \text{and} \quad \text{curl } \mathbf{H} = \sigma\mathbf{E} + \epsilon\frac{\partial \mathbf{E}}{\partial t}$$

where $\mathbf{E}$ is electric intensity, $\mathbf{B}$ is magnetic flux density, $\mathbf{H}$ is magnetic intensity, and σ, ϵ, and μ are positive constants.

a. Use the fact that $\mathbf{B} = \mu\mathbf{H}$ to show that

$$\text{curl(curl } \mathbf{E}) = -\mu\frac{\partial}{\partial t}(\text{curl } \mathbf{H})$$

b. Next, show that for any vector field $\mathbf{F} = <f, g, h>$

$$\text{curl(curl } \mathbf{F}) = -\nabla(\text{div } \mathbf{F}) - \nabla\cdot\nabla\mathbf{F}$$

c. Use the formula in b to show that

$$\nabla(\text{div } \mathbf{E}) - (\nabla \cdot \nabla)\mathbf{E} = -\mu\frac{\partial}{\partial t}\left(\sigma\mathbf{E} + \epsilon\frac{\partial \mathbf{E}}{\partial t}\right)$$

d. Complete the derivation of **Maxwell's electric intensity equation**, assuming that the charge density Q is 0 so that div $\mathbf{E} = 0$. (See Problem 29).

Chapter 14   *Review*

Chapter Checklist

In the following statements, assume that all required conditions are satisfied. Main results are collected here (without hypotheses) so that you can compare and contrast various results and conclusions.

Scalar function: $f = f(x, y, z)$

Vector field: $\mathbf{F}(x, y, z) = u(x, y, z)\mathbf{i} + v(x, y, z)\mathbf{j} + w(x, y, z)\mathbf{k}$

Notation:

Del operator: $\nabla = \dfrac{\partial}{\partial x}\mathbf{i} + \dfrac{\partial}{\partial y}\mathbf{j} + \dfrac{\partial}{\partial z}\mathbf{k}$

Gradient: $\nabla f = \dfrac{\partial f}{\partial x}\mathbf{i} + \dfrac{\partial f}{\partial y}\mathbf{j} + \dfrac{\partial f}{\partial z}\mathbf{k}$

Laplacian: $\nabla^2 f = \dfrac{\partial^2 f}{\partial x^2} + \dfrac{\partial^2 f}{\partial y^2} + \dfrac{\partial^2 f}{\partial z^2} = f_{xx} + f_{yy} + f_{zz}$

Normal derivative: $\dfrac{\partial f}{\partial n} = \nabla f \cdot \mathbf{N}$

Derivatives of a vector field:

$$\text{div } \mathbf{F} = \frac{\partial u}{\partial x} + \frac{\partial v}{\partial y} + \frac{\partial w}{\partial z} \quad \text{This is a scalar derivative.}$$

$$= \nabla \cdot \mathbf{F}$$

If P is a point (x_0, y_0, z_0), then P is a
  source if div $\mathbf{F} > 0$
  sink if div $\mathbf{F} < 0$
$\mathbf{F}$ is incompressible if div $\mathbf{F} = 0$

$$\text{curl } \mathbf{F} = \begin{vmatrix} \mathbf{i} & \mathbf{j} & \mathbf{k} \\ \dfrac{\partial}{\partial x} & \dfrac{\partial}{\partial y} & \dfrac{\partial}{\partial z} \\ u & v & w \end{vmatrix} \quad \text{This is a vector derivative.}$$

$$= \nabla \times \mathbf{F}$$

$\mathbf{F}$ is irrotational if curl $\mathbf{F} = \mathbf{0}$

Conservative vector fields:

a. **F** is *conservative* if it is the gradient of some scalar function, called the *scalar potential of* **F**; that is, $\mathbf{F} = \nabla f$.

b. **F** is conservative if and only if curl $\mathbf{F} = \mathbf{0}$.

c. Equivalent conditions:

(i) $\int_C \mathbf{F} \cdot d\mathbf{R}$ is independent of path.

(ii) **F** is conservative.

(iii) $\int_C \mathbf{F} \cdot d\mathbf{R} = 0$ for every closed path C.

Evaluation of Line Integrals: $\int_C \mathbf{F} \cdot d\mathbf{R}$

Step 1. Check to see whether **F** is conservative; if it is, then

$$\int_C \mathbf{F} \cdot d\mathbf{R} = 0 \text{ if } C \text{ is closed}$$

$$\int_C \mathbf{F} \cdot d\mathbf{R} = f(Q) - f(P) \quad \text{Initial point } P, \text{ terminal } Q$$

Step 2. If **F** is not conservative, and C is a closed curve bounding the surface S, use Stokes' theorem (or Green's theorem in $\mathbb{R}^2$) to equate the given integral to a surface integral, namely,

$$\int_C \mathbf{F} \cdot d\mathbf{R} = \iint_S (\text{curl } \mathbf{F} \cdot \mathbf{N}) \, dS$$

Step 3. If **F** is not conservative, and C is an open curve, try to add an are C_1 so that the curve formed by C and C_1 is closed. Try to choose C_1 so that $\int_{C_1} \mathbf{F} \cdot d\mathbf{R}$ is relatively easy to evaluate.

$$\int_C \mathbf{F} \cdot d\mathbf{R} = \iint_S (\text{curl } \mathbf{F} \cdot \mathbf{N}) \, dS - \int_{C_1} \mathbf{F} \cdot d\mathbf{R}$$

Step 4. As a last resort, parametrize **F** and **R**. Let $\mathbf{R}(t) = x(t)\mathbf{i} + y(t)\mathbf{j} + z(t)\mathbf{k}$ for $a \le t \le b$.

$$\int_C f(x, y, z) \, dx = \int_a^b f[x(t), y(t), z(t)] \frac{dx}{dt} \, dt$$

$$\int_C f(x, y, z) \, ds = \int_a^b f[x(t), y(t), z(t)]$$
$$\times \sqrt{[x'(t)]^2 + [y'(t)]^2 + 1} \, dt$$

Evaluation of Flux Integrals: $\iint_S \mathbf{F} \cdot \mathbf{N} \, dS$

Step 1. If the surface S is a closed surface bounding the solid region D, use the divergence theorem to write the flux integral as a triple integral.

$$\iint_S \mathbf{F} \cdot \mathbf{N} \, dS = \iiint_D \text{div } \mathbf{F} \, dV$$

Step 2. If the surface S is open, try to find a supplementary surface S_1 such that the surface formed by S and S_1 is closed. Try to choose S_1 so that $\iint_{S_1} \mathbf{F} \cdot \mathbf{N} \, dS$ is easy to evaluate.

$$\iint_S \mathbf{F} \cdot \mathbf{N} \, dS = \iiint_D \text{div } \mathbf{F} \, dV - \iint_{S_1} \mathbf{F} \cdot \mathbf{N} \, dS_1$$

Step 3. If neither step 1 nor step 2 applies, parametrize **F**, **N**, and dS. This may be quite difficult. In the special case where S has the form $z = f(x, y)$ and $g = \mathbf{F} \cdot \mathbf{N}$, we have

$$\iint_S g(x, y, z) \, dS$$
$$= \iint_R g[x, y, f(x, y)] \sqrt{[f_x(x,y)]^2 + [f_y(x, y)]^2 + 1} \, dA_{xy}$$

Applications

Fluid Mechanics

Let **V** be the velocity of an incompressible fluid. Then

div **V** measures the rate of particle flow per unit volume at a point.

curl **V** measures the tendency of a fluid to swirl.

Flux integral: $\iint_S \mathbf{V} \cdot \mathbf{N} \, dS$ measures the rate of flow across the surface S in unit time

Circulation: $\int_C \mathbf{V} \cdot \mathbf{T} \, ds$ measures the tendency of a fluid to move around C.

Stokes' theorem says that the cumulative tendency of a fluid to swirl across a surface S is equal to the circulation of a fluid around the boundary curve C.

The divergence theorem shows that the divergence of a velocity field **V** at P_0 equals the flux of flow out of P_0 per unit volume. If there are no sources or sinks, it follows that

$$\text{div } \delta \mathbf{V} = -\frac{\partial \delta}{\partial t} \quad \text{(continuity equation)}$$

where $\delta(x, y, z, t)$ is the density of the fluid.

Electromagnetism

Let **E** be the electric intensity field and **H** be the magnetic intensity field. Then

$$\text{div } \mathbf{E} = \frac{Q}{\epsilon} \quad \text{where } Q \text{ is the electric charge density and } \epsilon \text{ is the permittivity.}$$

curl $\mathbf{E} = -\dfrac{\partial(\mu\mathbf{H})}{\partial t}$ where μ is the permeability and t the time.

curl $\mathbf{H} = \mathbf{J}$ where $\mathbf{J}$ is the electric current density.

Maxwell's equation for electric intensity (assuming $Q = 0$) is

$(\nabla \cdot \nabla)\mathbf{E} = \mu\sigma\dfrac{\partial\mathbf{E}}{\partial t} + \mu\epsilon\dfrac{\partial^2\mathbf{E}}{\partial t^2}$ where σ is the conductivity.

Ampère's circuit law is $\int_C \mathbf{H} \cdot d\mathbf{R} = I$, where I is the currentcrossing any surface S bounded by the closed curve C.

Thermodynamics

The heat equation is $\dfrac{\partial T}{\partial t} = \dfrac{k}{\sigma\delta}\nabla^2 T$, where $T(x, y, z, t)$ is the temperature at (x, y, z) at time t. The constant k is the thermal conductivity, σ is the specific heat of the body, and δ is its density.

Work: $W = \displaystyle\int_C \mathbf{F} \cdot d\mathbf{R}$

Mass: $m = \displaystyle\iint_S \delta(x, y, z)\, dS$

Proficiency Examination

Concept Problems

1. What is a vector field?
2. What is the divergence of a vector field?
3. What is the curl of a vector field?
4. What is the del operator?
5. What is Laplace's equation?
6. What is the difference between the Riemann integral and a line integral? Discuss.
7. What is the formula for a line integral of a vector field?
8. How do we find work as a line integral?
9. What is the formula for a line integral in terms of arc length parameter?
10. State the fundamental theorem on line integrals.
11. Define a conservative vector field.
12. What is the scalar potential of a conservative vector field?
13. What is a Jordan curve?
14. State Green's theorem.
15. How can you use Green's theorem to find area as a line integral?
16. What is a normal derivative?
17. Define a surface integral.
18. What is the formula for a surface integral of a surface defined parametrically?
19. What is a flux integral?
20. State Stokes' theorem.
21. State the conservative vector field theorem.
22. State the divergence theorem.

Practice Problems

23. Show that $yz\mathbf{i} + xz\mathbf{j} + xy\mathbf{k}$ is conservative and find a scalar potential function.

24. Compute div $\mathbf{F}$ and curl $\mathbf{F}$ for $\mathbf{F} = x^2 y\mathbf{i} - e^{yz}\mathbf{j} + \frac{1}{2}x\mathbf{k}$.

25. Use Green's theorem to evaluate the line integral
$\displaystyle\int_C \mathbf{F} \cdot d\mathbf{R}$, where $\mathbf{F} = (2x + y)\mathbf{i} + 3y^2\mathbf{j}$ and C is the boundary of the triangle T with vertices $(-1, 2), (0, 0), (1, 2)$, traversed in the given order.

26. Use Stokes' theorem to evaluate the line integral
$\displaystyle\int_C \mathbf{F} \cdot d\mathbf{R}$, where $\mathbf{F} = 2y\mathbf{i} + z\mathbf{j} + y\mathbf{k}$ and C is the intersection of the plane $z = x + 2$ and sphere $x^2 + y^2 + z^2 = 4z$, traversed counterclockwise as viewed from above.

27. Use the divergence theorem to evaluate the surface integral $\displaystyle\iint_S \mathbf{F} \cdot \mathbf{N}\, dS$, where $\mathbf{F} = x^2\mathbf{i} + (y + z)\mathbf{j} - 2z\mathbf{k}$, and S is the surface of the unit cube $0 \le x \le 1, 0 \le y \le 1, 0 \le z \le 1$.

28. Evaluate $\displaystyle\int_C \dfrac{x\, dx + y\, dy}{(x^2 + y^2)^2}$, where C is the path shown in Figure 14.38, traversed counterclockwise.

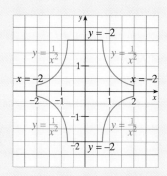

■ **FIGURE 14.38**   Problem 28

29. An object with mass m travels counterclockwise (as viewed from above) in the circular orbit $x^2 + y^2 = 9$, $z = 2$, with angular speed ω. The mass is subject to a centrifugal force $\mathbf{F} = m\omega^2\mathbf{R}$, where $\mathbf{R} = x\mathbf{i} + y\mathbf{j} + z\mathbf{k}$. Show that $\mathbf{F}$ is conservative, and find a scalar potential function for $\mathbf{F}$.

30. Find the work done by $\mathbf{F}$ from Problem 29 as the object makes a half-orbit from $(3, 0, 2)$ to $(-3, 0, 2)$.

Supplementary Problems

In Problems 1–6, determine whether the given vector field is conservative, and if it is, find a scalar potential function.

1. $\mathbf{F} = 2\mathbf{i} - 3\mathbf{j}$

2. $\mathbf{F} = xy^{-2}\mathbf{i} + x^{-2}y\mathbf{j}$

3. $\mathbf{F} = y^{-3}\mathbf{i} + (-3xy^{-4} + \cos y)\mathbf{j}$

4. $\mathbf{F} = y^2\mathbf{i} + (2xy)\mathbf{j}$

5. $\mathbf{F} = \left(\dfrac{1}{y} + \dfrac{y}{x^2}\right)\mathbf{i} - \left(\dfrac{x}{y^2} - \dfrac{1}{x}\right)\mathbf{j}$

6. $\mathbf{F} = \left[2x\tan^{-1}\left(\dfrac{y}{x}\right) - y\right]\mathbf{i} + \left[2y\tan^{-1}\left(\dfrac{y}{x}\right) + x\right]\mathbf{j}$

In Problems 7–12, find $\displaystyle\int_C \mathbf{F} \cdot d\mathbf{R}$, where C is the curve $\mathbf{R}(t) = t\mathbf{i} + t^2\mathbf{j}$, $1 \le t \le 2$. Note that these are the same as the vector fields in Problems 1–6.

7. $\mathbf{F} = 2\mathbf{i} - 3\mathbf{j}$

8. $\mathbf{F} = xy^{-2}\mathbf{i} + x^{-2}y\mathbf{j}$

9. $\mathbf{F} = y^{-3}\mathbf{i} + (-3xy^{-4} + \cos y)\mathbf{j}$

10. $\mathbf{F} = y^2\mathbf{i} + (2xy)\mathbf{j}$

11. $\mathbf{F} = \left(\dfrac{1}{y} + \dfrac{y}{x^2}\right)\mathbf{i} - \left(\dfrac{x}{y^2} - \dfrac{1}{x}\right)\mathbf{j}$

12. $\mathbf{F} = \left[2x\tan^{-1}\left(\dfrac{y}{x}\right) - y\right]\mathbf{i} + \left[2y\tan^{-1}\left(\dfrac{y}{x}\right) + x\right]\mathbf{j}$

In Problems 13–16, find div $\mathbf{F}$ and curl $\mathbf{F}$.

13. $\mathbf{F} = x\mathbf{i} + y\mathbf{j} + z\mathbf{k}$

14. $\mathbf{F} = \left(\tan^{-1}\dfrac{y}{x}\right)\mathbf{i} - 3\mathbf{j} + z^2\mathbf{k}$

15. $\mathbf{F} = \dfrac{1}{r}(x\mathbf{i} + y\mathbf{j} + z\mathbf{k})$, where $r = \sqrt{x^2 + y^2 + z^2}$, $r \ne 0$

16. $\mathbf{F} = (xy\sin z)\mathbf{i} + (x^2\cos yz)\mathbf{j} + (z\sin xy)\mathbf{k}$

Evaluate the line integrals in Problems 17–18, by parametrization.

17. $\displaystyle\int_C [(\sin \pi y)\, dx + (\cos \pi x)\, dy]$, where C is the line segment from $(1, 0)$ to $(\pi, 0)$, followed by the line segment from $(\pi, 0)$ to (π, π)

18. $\displaystyle\int_C (z\, dx - x\, dy + dz)$, where C is the arc of the helix $x = 3\sin t$, $y = 3\cos t$, $z = t$ for $0 \le t \le \frac{\pi}{4}$

In each of Problems 19–33, evaluate the line integral or the surface integral. In each case, assume the orientation of C is counterclockwise when viewed from above or that $\mathbf{N}$ is the unit outward normal vector.

19. $\displaystyle\int_C [yz\, dx + xz\, dy + (xy + 2)\, dz]$
    C is the curve $\mathbf{R}(t) = (\tan^{-1}t)\mathbf{i} + t^2\mathbf{j} - 3t\mathbf{k}$, $0 \le t \le 1$.

20. $\displaystyle\int_C [(x^2 + y)\, dx + xz\, dy - (y + z)\, dz]$
    C is the curve $\mathbf{R}(t) = t\mathbf{i} + t^2\mathbf{j} + 2\mathbf{k}$, $0 \le t \le 1$.

21. $\displaystyle\iint_S (3x^2 + y - 2z)\, dS$
    S is the surface $\mathbf{R}(u, v) = u\mathbf{i} + (u + v)\mathbf{j} + u\mathbf{k}$, $0 \le u \le 1$, $0 \le v \le 1$.

22. $\displaystyle\iint_S \mathbf{F} \cdot \mathbf{N}\, dS$
    $\mathbf{F} = 3x\mathbf{i} + z^2\mathbf{j} - 2y\mathbf{k}$ and S is the surface of the hemisphere $z = \sqrt{4 - x^2 - y^2}$.

23. $\displaystyle\int_C (x\, dx + x\, dy - y\, dz)$
    C is the curve $\mathbf{R}(t) = t\mathbf{i} + t^2\mathbf{j} + t\mathbf{k}$, $0 \le t \le 1$.

24. $\displaystyle\int_C (x^2\, dx - 3y^2\, dz)$
    C is the line segment from $(0, 1, 1)$ to $(1, 1, 2)$.

25. $\displaystyle\int_C (x^2\, dx + y\, dy)$
    C is the curve $\mathbf{R}(t) = (t\sin t)\mathbf{i} + (1 - t\cos t)\mathbf{j}$, $0 \le t \le 2\pi$.

26. $\displaystyle\int_C (xy\, dx - x^2\, dy)$
    C is the square with vertices $(1, 0), (0, 1), (-1, 0), (0, -1)$ traversed counterclockwise.

27. $\displaystyle\int_C (y\, dx + x\, dy - 2\, dz)$
    C is the curve of intersection of the cylinder $x^2 + y^2 = 2x$ and the plane $x = z$.

28. $\displaystyle\int_C [(y + z)\, dx + (x + z)\, dy + (x + y)\, dz]$

C is the curve of intersection of the sphere
$x^2 + (y - 3)^2 + z^2 = 9$ and the plane $x + 2y + z = 3$.

29. $\displaystyle\int_C (-2y\,dx + 2x\,dy + dz)$

C is the circle $x^2 + y^2 = 1$ in the plane $z = 3$.

30. $\displaystyle\iint_S (\text{curl } y\mathbf{i}) \cdot \mathbf{N}\,dS$

S is the hemisphere $z = \sqrt{1 - x^2 - y^2}$.

31. $\displaystyle\iint_S (2x^3\mathbf{i} + y^3\mathbf{j} + z^3\mathbf{k}) \cdot \mathbf{N}\,dS$

S is the surface of the ellipsoid $2x^2 + y^2 + z^2 = 1$.

32. $\displaystyle\iint_S (y^2\mathbf{i} + y^2\mathbf{j} + yz\mathbf{k}) \cdot \mathbf{N}\,dS$

S is the surface of the tetrahedron bounded by the plane $2x + 3y + z = 1$ and the coordinate planes, in the first octant.

33. $\displaystyle\iint_S \nabla\phi \cdot \mathbf{N}\,dS$, where $\phi(x, y, z) = 2x + 3y$

where S is the portion of the plane $ax + by + cz = 1$ $(a > 0, b > 0, c > 0)$ that lies in the first octant.

In Problems 34–38, find $\displaystyle\iint_S \mathbf{F} \cdot \mathbf{N}\,dS$.

34. $\mathbf{F} = x^2\mathbf{i} + y^2\mathbf{j} + x^2\mathbf{k}$, and S is the surface of the unit cube $0 \le x \le 1, 0 \le y \le 1, 0 \le z \le 1$.

35. $\mathbf{F} = 2yz\mathbf{i} + (\tan^{-1} xz)\mathbf{j} + e^{xy}\mathbf{k}$, and S is the surface of the sphere $x^2 + y^2 + z^2 = 1$.

36. $\mathbf{F} = x\mathbf{i} - 4\mathbf{j} + 3\mathbf{k}$, and S is the paraboloid $y = x^2 + z^2$ with $x^2 + z^2 \le 9$. The disk $x^2 + z^2 = 9$ is omitted; that is, the paraboloid is open on the right.

37. $\mathbf{F} = xyz\mathbf{i} + xyz\mathbf{j} + xyz\mathbf{k}$, and S is the surface of the five faces of the unit cube $0 \le x \le 1, 0 \le y \le 1, 0 \le z \le 1$ missing $z = 0$.

38. $\mathbf{F} = xy\mathbf{i} - 2z\mathbf{j}$, and S is the surface defined parametrically by $\mathbf{R}(u, v) = u\mathbf{i} + v\mathbf{j} + u\mathbf{k}$ for $0 \le u \le 1, 0 \le v \le 1$.

Find all real numbers c for which each vector field in Problems 39–41 is conservative.

39. $\mathbf{F} = (x, y) = (\sqrt{x} + 3xy)\mathbf{i} + (cx^2 + 4y)\mathbf{j}$

40. $\mathbf{F}(x, y) = \left(\dfrac{cy}{x^3} + \dfrac{y}{x^2}\right)\mathbf{i} + \left(\dfrac{1}{x^2} - \dfrac{1}{x}\right)\mathbf{j}$

41. $\mathbf{F}(x, y, z) = e^{yz/x}\left[\left(\dfrac{cyz}{x^2}\right)\mathbf{i} + \left(\dfrac{z}{x}\right)\mathbf{j} + \left(\dfrac{y}{x}\right)\mathbf{k}\right]$

42. Let $\mathbf{F} = (y^2 + x^{-2}ye^{x/y})\mathbf{i} + (2xy + z - x^{-1}e^{x/y})\mathbf{j} + y\mathbf{k}$. Is $\mathbf{F}$ conservative?

43. Find the work done when an object moves in the force field $\mathbf{F} = 2x\mathbf{i} - (x + z)\mathbf{j} + (y - x)\mathbf{k}$ along the path given by $\mathbf{R}(t) = t^2\mathbf{i} + (t^2 - t)\mathbf{j} + 3\mathbf{k}, 0 \le t \le 1$.

44. Show that the force field $\mathbf{F} = yz^2\mathbf{i} + (xz^2 - 1)\mathbf{j} + (2xyz - 1)\mathbf{k}$ is conservative and determine the work done when an object moves in the force field from the origin to the point $(1, 0, 1)$.

45. Find a region R in the plane where the vector field

$$\mathbf{F} = \frac{1}{x + y}(\mathbf{i} + \mathbf{j}) \text{ is conservative. Then evaluate}$$

$\displaystyle\int_C \mathbf{F} \cdot d\mathbf{R}$, where C is any path in R from the point $P_0(a, b)$ to $P_1(c, d)$.

46. If u is a scalar function and $\mathbf{F}$ is a continuously differentiable vector field, show that $\text{curl}(u\mathbf{F}) = u\,\text{curl }\mathbf{F} + (\nabla u \times \mathbf{F})$.

47. Show that $\text{div}(\mathbf{F} \times \mathbf{G}) = \mathbf{G} \cdot \text{curl }\mathbf{F} - \mathbf{F} \cdot \text{curl }\mathbf{G}$, for any continuously differentiable vector fields $\mathbf{F}$ and $\mathbf{G}$.

48. A vector field $\mathbf{F}$ is *incompressible* (or *solenoidal*) in a region D if $\text{div }\mathbf{F} = 0$ throughout D. If $\mathbf{F}$ and $\mathbf{G}$ are both conservative vector fields in D, show that $\mathbf{F} \times \mathbf{G}$ is solenoidal.

49. If $\mathbf{F} = \text{curl }\mathbf{G}$, show that $\mathbf{F}$ is solenoidal. (See Problem 48.)

50. Suppose $\mathbf{F} = f(x, y, z)\mathbf{A}$, where $\mathbf{A}$ is a constant vector and f is a scalar function. Show that $\text{curl }\mathbf{F}$ is orthogonal to $\mathbf{A}$ and to ∇f.

51. If $\mathbf{A}$ is a constant vector and $\mathbf{F}$ is a continuously differentiable vector field, show that $\text{div}(\mathbf{A} \times \mathbf{F}) = -\mathbf{A} \cdot \text{curl }\mathbf{F}$.

52. Evaluate the line integral $\displaystyle\int_C \frac{x\,dx - y\,dy}{x^2 - y^2}$, where C is any path in the xy-plane that is interior to the region $x > 0$, $y < x, y > -x$ and connects the point $(5, 4)$ to $(2, 0)$.

53. Evaluate $\displaystyle\int_C \left(\frac{-y}{x^2}\,dx + \frac{1}{x}\,dy\right)$, where C is the closed path $(x - 2)^2 + y^2 = 1$, traversed once counterclockwise.

54. THINK TANK PROBLEM   Let $u(x, y)$ and $v(x, y)$ be functions of two variables with continuous partial derivatives everywhere in the plane, and suppose that u and v satisfy the equation

$$\frac{\partial u}{\partial y} = \frac{\partial v}{\partial x}$$

for all (x, y). Are u and v necessarily harmonic? Either show that they are or find a counterexample.

55. a. Find a region in the plane where the vector field

$$\mathbf{F} = \left(\frac{1 + y^2}{x^3}\right)\mathbf{i} - \left(\frac{y + x^2 y}{x^2}\right)\mathbf{j}$$

is conservative, and find a scalar potential for $\mathbf{F}$.

b. Evaluate $\displaystyle\int_C \mathbf{F} \cdot d\mathbf{R}$, where C is a path from $(1, 1)$ to $(3, 4)$. Are there any limitations on the path C? Explain.

56. Determine the most general function $u(x, y)$ for which the vector field $\mathbf{F} = u(x, y)\mathbf{i} + (2y\,e^x + y^2 e^{3x})\mathbf{j}$ will be conservative.

57. Consider the line integral $\displaystyle\int_C \left(\frac{dx}{y} + \frac{dy}{x}\right)$, where C is the closed triangular path formed by the lines $y = 2x$, $x + 2y = 5$, and $x = 2$, traversed counterclockwise. First evaluate the line integral directly (by parametrizing C) and then by using Green's theorem.

58. If S is a closed surface in a region R and $\mathbf{F}$ is a continuously differentiable vector field on R, show that

$$\iint_S (\text{curl } \mathbf{F} \cdot \mathbf{N})\, dS = 0$$

where $\mathbf{N}$ is the outward unit normal vector to S.

59. A certain closed path C in the plane $2x + 2y + z = 1$ is known to project onto the unit circle $x^2 + y^2 = 1$ in the xy-plane. Let c be a constant, and let $\mathbf{R} = x\mathbf{i} + y\mathbf{j} + z\mathbf{k}$. Use Stokes' theorem to evaluate

$$\int_C (c\mathbf{k} \times \mathbf{R}) \cdot d\mathbf{R}$$

60. a. Show that if the scalar function w is harmonic, then

$$\nabla \cdot (w\,\nabla w) = \|\nabla w\|^2$$

   b. Let $w = x - y + 2z$, and let S be the surface of the sphere $x^2 + y^2 + z^2 = 9$. Evaluate

$$\iint_S w\,\frac{\partial w}{\partial n}\, dS$$

Note: The normal derivative $\dfrac{\partial w}{\partial n}$ is defined in Section 14.4.

61. A particle moves along a curve C in space that is given parametrically by $\mathbf{R}(t) = x(t)\mathbf{i} + y(t)\mathbf{j} + z(t)\mathbf{k}$ for $a \le t \le b$. A force field $\mathbf{F}$ is applied to the particle in such a way that $\mathbf{F}$ is always perpendicular to the path C. How much work is performed by the force field $\mathbf{F}$ when the particle moves from the point where $t = a$ to the point where $t = b$?

62. If $\mathbf{D}$ is the electric displacement field, then div $\mathbf{D} = \phi$, where ϕ is the **charge density.** A region of space is said to be **charge-free** if $\phi = 0$ there. Describe the charge-free regions of the electric displacement field $\mathbf{D} = 2x^2\mathbf{i} + 3y^2\mathbf{j} - 2z^2\mathbf{k}$.

63. A satellite weighing 10,000 kg travels in a circular orbit 7,000 km from the center of the earth. How much work is done by gravity on the satellite during half a revolution?

64. If $\mathbf{F}$ is a conservative force field, the scalar function f such that $\mathbf{F} = -\nabla f$ is called the *potential energy.* Suppose an

object with mass 10 g moves in the force field in such a way that its speed decreases from 3 cm/s to 2.5 cm/s. What is the corresponding change in the potential energy of the object?

65. Find the work done by the force field

$$\mathbf{F} = 2xyz\mathbf{i} + \left(x^2 z - \frac{1}{z}\tan^{-1}\frac{y}{z}\right)\mathbf{j} + \left(x^2 y + \frac{y}{z^2}\tan^{-1}\frac{y}{z}\right)\mathbf{k}$$

in moving an object along the circular helix

$$\mathbf{R}(t) = (\sin \pi t)\mathbf{i} + (\cos \pi t)\mathbf{j} + (2t + 1)\mathbf{k}$$

for $0 \le t \le \dfrac{1}{2}$.

66. Find the work done when an object moves against the force field $\mathbf{F} = 4y^2\mathbf{i} + (3x + y)\mathbf{j}$ from $(1, 0)$ to $(-1, 0)$ along the top half of the ellipse $x^2 + \dfrac{y^2}{k^2} = 1$. Which value of k minimizes the work?

67. Evaluate the line integral

$$\int_C \frac{-y\,dx + x\,dy}{x^2 + y^2}$$

where C is the limaçon given in polar coordinates by $r = 3 + 2\cos\theta$, $0 \le \theta \le 2\pi$, traversed counterclockwise.

68. Suppose f and g are both harmonic in the region R with boundary surface S. Show that

   a. $\displaystyle\iint_S f\,\frac{\partial g}{\partial n}\, dS = \iint_S g\,\frac{\partial f}{\partial n}\, dS$

   b. $\displaystyle\iint_S f\,\frac{\partial f}{\partial n}\, dS = \iiint_D \|\nabla f\|^2\, dV$

Note: $\dfrac{\partial f}{\partial n}$ denotes the normal derivative (see Section 14.4).

69. Show that curl(curl $\mathbf{F}$) $= \nabla(\text{div } \mathbf{F}) - \nabla^2 \mathbf{F}$ if the components of $\mathbf{F}$ have continuous second-order partial derivatives.

70. Evaluate the surface integral $\displaystyle\iint_S \frac{\partial f}{\partial n}\, dS$ where S is the surface of the unit sphere $x^2 + y^2 + z^2 = 1$ and f is a scalar field such that $\|\nabla f\|^2 = 3f$ and div $(f\nabla f) = 7f$. Remember that $\dfrac{\partial f}{\partial n}$ denotes the normal derivative, $\nabla f \cdot \mathbf{N}$.

71. Evaluate the surface integral $\displaystyle\iint_S dS$, where S is the torus $\mathbf{R}(u, v) = [(a + b\cos v)\cos u]\mathbf{i} + [(a + b\cos v)\sin u]\mathbf{j} + (b\sin v)\mathbf{k}$ for $0 < b < a$ and $0 \le u \le 2\pi$, $0 \le v \le 2\pi$.

72. Evaluate $\iint_S \mathbf{F} \cdot \mathbf{N}\, dS$, where $\mathbf{F} = x\mathbf{i} + y\mathbf{j} + z\mathbf{k}$ and S is a cubic surface with a corner block removed, as shown in Figure 14.39.

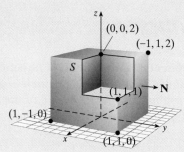

■ **FIGURE 14.39**   Cube with a corner removed

73. THINK TANK PROBLEM   Prove or disprove that it does not matter which corner is removed in Problem 72.

74. Show that a lamina that covers a standard region D in the plane with density $\delta = 1$ has moment of inertia

$$I = \frac{1}{3}\int_C (-y^3\, dx + x^3\, dy)$$

with respect to the z-axis, where C is the boundary curve of D.

75. Use vector analysis to find the centroid of the conical surface $z = \sqrt{x^2 + y^2}$ between $z = 0$ and $z = 3$.

76. JOURNAL PROBLEM by J. Chris Fischer (CRUX Problem 785).* Suppose a closed differentiable curve has exactly one tangent line parallel to every direction. More precisely, suppose that the curve has parametrization $\mathbf{V}(\theta)$ which maps the interval $[0, 2\pi]$ into $\mathbb{R}^2$ for which

i. $\dfrac{d\mathbf{V}}{d\theta} = [r(\theta)\cos\theta, r(\theta)\sin\theta]$ for some continuous real-valued function $r(\theta)$, and

ii. $\mathbf{V}(\theta) = V(\theta + \pi)$

Prove that the curve is described in the clockwise sense as θ runs from 0 to π.

77. PUTNAM EXAMINATION PROBLEM   A force acts on the element ds of a closed plane curve. The magnitude of this force is $r^{-1}\, ds$, where r is the radius of curvature at the point considered, and the direction of the force is perpendicular to the curve; it points to the convex side. Show that the system of such forces acting on all elements of the curve keeps it in equilibrium.

Continuous vs. Discrete Mathematics

William F. Lucas is a professor of mathematics and department chairman at The Claremont Graduate School. He is an Avery Fellow to Harvey Mudd College, and is known for his research in game theory, which provides a mathematical approach to the study of conflict, cooperation, and fairness. Dr. Lucas has also been active in educational reform efforts with a goal toward introducing more recently discovered topics into the mathematics curriculum.

Mathematicians often distinguish between *discrete* and *continuous* mathematics. The latter is illustrated by the calculus and its many descendants, including the subject of differential equations. Continuous mathematics deals with "solid" infinite sets such as the real number line $\mathbb{R}$ and functions defined on $\mathbb{R}$. One of the primary concerns is with the solutions of differential equations, which are typical families of such solid curves or surfaces. The many applications of such differential equations to discoveries in the physical sciences and engineering over the past three centuries has been one of the greatest intellectual success stories of all times.

Most human heads have a fixed point, in the form of a whorl, sometimes called a cowlick, from which all the hair radiates. It would be impossible to cover a sphere with hair (or with radiating lines) without at least one such fixed point.

Prior to the invention of calculus by Newton and Leibniz, most mathematics was of a discrete nature. It dealt with sets that had a finite number of elements and with infinite but "countable" sets such as the natural numbers. These sets often include continuous curves such as the conics, which can be characterized by a small number of conditions. Discrete mathematics also deals with the solutions to algebraic equations, which are usually discrete sets. The twentieth century has witnessed the creation of many additional subjects in the discrete direction. This development was spurred on by the rapidly increasing use of mathematics in the social, behavioral, decisional, and system sciences, where the items under investigation are typically finite in number and not readily approximated by some continuous idealization. Moreover, nearly all aspects of the ongoing revolution in *digital* computers involve discrete considerations.

A metal bar on a minute level is composed of discrete molecules, atoms, and elementary particles. Many of its physical properties, however, can be determined by viewing the bar as a solid continuum and employing the analytical techniques of calculus. One applies the basic laws of physics to express the local (infinitesimal) properties of the bar in terms of differential equations. The solutions of these equations, in turn, provide an excellent description of the observed global behavior of the bar.

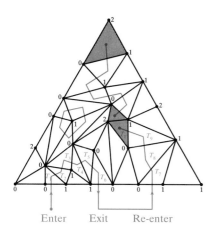

FIGURE 14.40 Arbitrary partitioning of a triangle

One of the greatest mathematical discoveries of the twentieth century in continuous mathematics is the famous fixed point theorem, published in 1912 by the Dutch mathematician L. E. J. Brouwer (1881–1966). It states that any continuous function f from a set S (with certain desirable properties) into the same set S has at least one *fixed point* x_0, that is, $f(x_0) = x_0$. For example, for any continuous function $y = f(x)$ from the set $S = \{x \in \mathbb{R} : 0 \le x \le 1\}$ into S, there is some number $x_0 \in S$ such that the curve $y = f(x)$ crosses the line $y = x$ at x_0. A head of hair must either have a seam (a parting of the hair, a discontinuity) or else a cowlick (where the hair stands straight up, a fixed point). Fixed points are realized in many physical and social phenomena such as equilibrium states in mechanics or stable prices in economics.

Like many of the outstanding results in continuous mathematics, Brouwer's theorem has an analogue in discrete mathematics, known as the labeling lemma of the German mathematician Emanuel Sperner, which appeared in 1928. Given a triangle with vertices 0, 1, and 2 as shown in Figure 14.40, one can partition the interior of this triangle into (nonoverlapping) smaller triangles in any possible way.

One can then label each of the newly created vertices with any one of the numbers 0, 1, or 2. There is only one stipulation on how the new vertices on the perimeter of the triangle can be labeled: Any vertex on a particular side of $\triangle 012$ cannot be labeled with the number appearing on the *main* vertex opposite (disjoint from) this side. For example, the side 01 of $\triangle 012$ cannot contain a vertex with the label 2. Sperner's lemma states that there must exist at least one elementary triangle inside $\triangle 012$ (or an *odd* number, in general) whose three vertices are "completely labeled" in the sense that they have all three labels 0, 1, and 2. There are three such triangles (shown in color) in Figure 14.40.

There is an elementary constructive proof of Sperner's lemma that uses the idea of a "path-following algorithm" published by Daniel I. A. Cohen in 1967. This proof is illustrated by the dashed line in Figure 14.40. One can enter into $\triangle 012$ from outside via some elementary triangle T_1 whose perimeter edge carries the labels 0 and 1. The third vertex on this elementary triangle T_1 must be either 2 (in which case T_1 is a completely labeled elementary triangle), or else it is 0 or 1. In the latter cases, one can exit T_1 via a second 01 edge. One can continue to enter and exit subsequent elementary triangles $T_2, T_3, \ldots$ through successive 01 edges until one finally reaches an elementary triangle T_n that has completely labeled vertices 0, 1, and 2. (If this path were ever to exit $\triangle 012$, then there must be still another 01 edge on the perimeter of $\triangle 012$ where one can reenter the main triangle and continue along the path, as did occur once in Figure 14.40.)

This path-following method also extends to a proof of Sperner's labeling lemma for dimensions higher than two. Furthermore, this approach is used in practical problems to arrive at a "very small" elementary triangle that can serve as an approximation to a fixed point, when it is excessively difficult or impossible to determine the precise location of the fixed point itself. In such cases, the broken path in Figure 14.40 can be viewed as the analogue in discrete mathematics of the continuous curves one arrives at when solving differential equations.

Around 1945, the great Hungarian–American mathematician John von Neumann (1903–1957) pointed out the need for approximating paths in a discrete manner when it proves too difficult to arrive at the exact continuous solutions to certain ordinary or partial differential equations. Cohen's constructive proof of Sperner's lemma, and work in the late 1960s by the mathematical economist Herbert Scarf on approximating equilibrium points, have paved the way for one very rich and extensive discrete theory that provides numerical approximations for continuous phenomena when the latter problems cannot be solved directly by the many analytic techniques of continuous mathematics.

Mathematical Essays

1. If your college offers a course on discrete mathematics, interview an instructor of the course, and using that interview as a basis, write an essay comparing continuous and discrete mathematics.

2. Draw several triangles and attempt to draw a cunterexample for Sperner's labeling lemma. In each case, show how this lemma is satisfied.

3. Consider two sheets of paper containing the numbers 1 to 120, as shown in the photograph on the left.

   If the top sheet is crumpled and dropped on the bottom sheet, the fixed point theorem tells us that one point must still be over its starting point. In the photograph on the

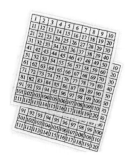

   right, it is a point in the region of the number 78. Perform this experiment several times in an attempt to find a counterexample. In each case, show how the fixed point theorem is satisfied.

4. **HISTORICAL QUEST**   William Rowan Hamilton has been called the most renowned Irish mathematician. He was a child prodigy who read Greek, Hebrew, and Latin by the time he was five, and by the age of ten he knew over a dozen languages. Many mathematical advances are credited to Hamilton. For example, he developed vector methods in analytic geometry and calculus, as well as a system of algebraic quantities called **quaternions,** which occupied his energies for the last 22 years of his life. Hamilton pursued the study of quaternions with an almost religious fervor, but by the early 20th century, the notation and terminology of vectors dominated. Much of the credit for the eventual emergence of vector methods goes not only to Hamilton, but also to the scientists James Clerk Maxwell (1831–1879), J. Willard Gibbs (1839–1903), and Oliver Heaviside (1850–1925).

WILLIAM ROWAN HAMILTON 1805–1865

   For this Quest, write a paper on quaternions.

5. Write a 500-word essay on the history of Green's theorem, Stokes' theorem, and the divergence theorem.

6. Write a report on four-dimensional geometry.

7. **Book Report**   "We often hear that mathematics consists mainly in 'proving theorems.' Is a writer's job mainly that of 'writing sentences'? A mathematician's work is mostly a tangle of guesswork, analogy, wishful thinking and frustration, and proof, far from being the core of discovery, is more often than not a way of making sure that our minds are not playing tricks. Few people, if any, had dared write this out loud before Davis and Hersh. Theorems are not to mathematics what successful courses are to a meal. The nutritional analogy is misleading. To master mathematics is to master an intangible view…" This quotation comes from the introduction to the book *The Mathematical Experience* by Philip J. Davis and Reuben Hersh (Boston: Houghton Mifflin, 1981). Read this book and prepare a book report.

8. Make up a word problem involving vector fields. Send your problem to:

   Bradley and Smith

   Prentice Hall Publishing Company

   1 Lake Street

   Upper Saddle River, NJ 07458

   The best ones submitted will appear in the next edition (along with credit to the problem poser).

15 Introduction to Differential Equations

CONTENTS

PREVIEW

We introduced and examined separable differential equations in Section 5.6 and first-order linear equations in Section 7.6. We examined applications such as orthogonal trajectories, flow of a fluid through an orifice, escape velocity of a projectile, carbon dating, the diastolic phase of blood pressure, learning curves, population models, dilution problems, and the flow of current in an RL circuit. In this chapter, we shall extend our study of first-order differential equations by examining *homogeneous* and *exact* equations and then investigate *second-order differential equations*.

PERSPECTIVE

The study of differential equations is such an extensive topic that even a brief survey of its methods and applications usually occupies a full course. Our goal in this chapter is to preview such a course by introducing some useful techniques for solving differential equations and by examining a few important applications.

15.1 *First-Order Differential Equations*

review of separable differential equations, homogeneous differential equations, review of first-order linear differential equations, exact differential equations, Euler's method

We have discussed several different kinds of first-order differential equations so far in this text. In this section, we review our previous methods and introduce two new forms, *homogeneous* and *exact* differential equations.

REVIEW OF SEPARABLE DIFFERENTIAL EQUATIONS

Recall that a differential equation is just an equation involving derivatives or differentials. In particular, an **nth-order differential equation** in the dependent variable y with respect to the independent variable x is an equation in which the highest derivative of y that appears is $d^n y/dx^n$. A **general solution** of a differential equation is an expression that completely characterizes all possible solutions of the equation, and a **particular solution** is a solution that satisfies certain specifications such as an initial value condition $y(x_0) = y_0$.

In Section 5.6, we defined a **separable differential equation** as one that can be written in the form

$$\frac{dy}{dx} = \frac{g(x)}{f(y)}$$

and observed that such an equation can be solved by separating the variables and integrating each side; that is,

$$\int f(y)\, dy = \int g(x)\, dx$$

EXAMPLE 1   *Separable differential equation*

Find the general solution of the differential equation

$$\frac{dy}{dx} = e^{-y} \sin x$$

Solution   Separate the variables and integrate:

$$\frac{dy}{dx} = e^{-y} \sin x$$

$$e^y\, dy = \sin x\, dx$$

$$\int e^y\, dy = \int \sin x\, dx$$

$$e^y = -\cos x + C \quad \textit{Combine constants of integration.}$$

This can also be written as $y = \ln|C - \cos x|$. ∎

From the standpoint of applications, one of the most important separable differential equations is

$$\frac{dy}{dx} = ky$$

which occurs in the study of exponential growth and decay. You might review Table 7.3 on page 508, which gives the solution for uninhibited growth or decay, logistic (or inhibited) growth, and limited growth functions. In this section, we consider an application of separable differential equations from chemistry.

EXAMPLE 2   *Chemical conversion*

Experiments in chemistry indicate that under certain conditions, two substances A and B will convert into a third substance C in such a way that the rate of conversion with respect to time is jointly proportional to the unconverted amounts of A and B. For simplicity, assume that one unit of C is formed from the combination of one unit of A and one unit of B, and assume that initially there are α units of A, β units of B, and no units of C present. Set up and solve a differential equation for the amount $Q(t)$ of C present at time t, assuming $\alpha \neq \beta$.

Solution   Since each unit of C is formed from one unit of A and one unit of B, it follows that at time t, $\alpha - Q(t)$ units of A and $\beta - Q(t)$ units of B remain unconverted. The specific rate condition can be expressed mathematically as

$$\frac{dQ}{dt} = k(\alpha - Q)(\beta - Q)$$

where k is a constant ($k > 0$ because $Q(t)$ is increasing).

To solve this equation, we separate the variables and integrate:

$$\int \frac{dQ}{(\alpha - Q)(\beta - Q)} = \int k \, dt$$

$$\int \frac{1}{\alpha - \beta} \left[\frac{-1}{\alpha - Q} + \frac{1}{\beta - Q} \right] dQ = \int k \, dt \qquad \text{\textit{Partial fractions decomposition}}$$

$$\frac{1}{\alpha - \beta}[\ln(\alpha - Q) - \ln(\beta - Q)] = kt + C_1$$

$$\ln \left| \frac{\alpha - Q}{\beta - Q} \right| = (\alpha - \beta)kt + C_2$$

$$\frac{\alpha - Q}{\beta - Q} = Me^{(\alpha - \beta)kt} \qquad \text{\textit{Where } } M = e^{C_2}$$

$$\alpha - Q = \beta Me^{(\alpha - \beta)kt} - QMe^{(\alpha - \beta)kt}$$

$$QMe^{(\alpha - \beta)kt} - Q = \beta Me^{(\alpha - \beta)kt} - \alpha$$

$$Q = \frac{\beta Me^{(\alpha - \beta)kt} - \alpha}{Me^{(\alpha - \beta)kt} - 1}$$

The initial condition tells us that $Q(0) = 0$, so that

$$0 = \frac{\beta Me^0 - \alpha}{Me^0 - 1}$$

$$0 = \beta M - \alpha$$

$$M = \frac{\alpha}{\beta}$$

Thus, $$Q(t) = \frac{\beta \dfrac{\alpha}{\beta} e^{(\alpha - \beta)kt} - \alpha}{\dfrac{\alpha}{\beta} e^{(\alpha - \beta)kt} - 1} = \frac{\alpha\beta[e^{(\alpha - \beta)kt} - 1]}{\alpha e^{(\alpha - \beta)kt} - \beta}$$ ■

HOMOGENEOUS DIFFERENTIAL EQUATIONS

Sometimes a first-order differential equation that is not separable can be put into separable form by a change of variables. A differential equation of the form

$$M(x, y)\, dx + N(x, y)\, dy = 0$$

is called a **homogeneous differential equation** if it can be written in the form

$$\frac{dy}{dx} = f\!\left(\frac{y}{x}\right)$$

In other words, dy/dx is isolated on one side of the equation and the other side can be expressed as a function of y/x. We can then solve the differential equation by substitution.

To see how to solve such an equation, set $v = y/x$, so that

$$vx = y$$

$$\frac{d}{dx}(vx) = \frac{d}{dx}(y) \qquad \text{Take the derivative of both sides.}$$

$$v + x\frac{dv}{dx} = \frac{dy}{dx} \qquad \text{Product rule}$$

$$v + x\frac{dv}{dx} = f(v) \qquad \text{Substitution, } \frac{dy}{dx} = f\!\left(\frac{y}{x}\right) = f(v)$$

$$x\frac{dv}{dx} = f(v) - v$$

$$\frac{dv}{f(v) - v} = \frac{dx}{x}$$

The equation can now be solved by integrating both sides; remember to express your answer in terms of the original variables x and y (use $v = y/x$).

> ### EXAMPLE 3 — *Homogeneous differential equation*

Find the general solution of the equation $2xy\, dx + (x^2 + y^2)\, dy = 0$.

Solution   First, show that the equation is homogeneous by writing it in the form $\frac{dy}{dx} = f\!\left(\frac{y}{x}\right)$:

$$2xy\, dx + (x^2 + y^2)\, dy = 0$$

$$\frac{dy}{dx} = \frac{-2xy}{x^2 + y^2} = \frac{-2\!\left(\dfrac{y}{x}\right)}{1 + \left(\dfrac{y}{x}\right)^2}$$

Let $v = \dfrac{y}{x}$ and $f(v) = \dfrac{-2v}{1 + v^2}$.

$$\frac{dv}{f(v) - v} = \frac{dx}{x}$$

$$\frac{dv}{\dfrac{-2v}{1 + v^2} - v} = \frac{dx}{x}$$

$$-\int \frac{(1 + v^2)\, dv}{v^3 + 3v} = \int x^{-1}\, dx$$

$$\int \left[\frac{\frac{1}{3}}{v} + \frac{\frac{2}{3}v}{v^2 + 3} \right] dv = -\int x^{-1} dx \qquad \text{Partial fractions decomposition}$$

$$\frac{1}{3} \ln|v| + \frac{2}{3}[\frac{1}{2} \ln|v^2 + 3|] = -\ln|x| + C_1$$

$$\frac{1}{3} \ln|v(v^2 + 3)| + \ln|x| = C_1$$

$$\ln \left| \frac{y}{x} \left[\left(\frac{y}{x} \right)^2 + 3 \right] \right| + \ln|x^3| = C_2 \qquad \text{Substituting } v = \frac{y}{x}$$

$$\ln \left| \frac{y^3 + 3x^2 y}{x^3} \cdot x^3 \right| = C_2$$

$$\qquad\qquad\qquad\qquad \text{Where } C = e^{C_2}$$

$$y^3 + 3x^2 y = C$$

This is the general solution of the given differential equation.   ∎

REVIEW OF FIRST-ORDER LINEAR DIFFERENTIAL EQUATIONS

In Section 7.6, we considered differential equations of the form

$$\frac{dy}{dx} + p(x)y = q(x)$$

Such an equation is said to be **first-order linear,** and we showed that its general solution is given by

$$y = \frac{1}{I(x)} \left[\int I(x)\, q(x)\, dx + C \right]$$

where $I(x)$ is the *integrating factor*

$$I(x) = e^{\int p(x)dx}$$

WARNING Note that the coefficient of dy/dx is 1. If it is not, then divide by that nonzero coefficient. ←

EXAMPLE 4   *First-order differential equation*

Solve $\dfrac{dy}{dx} + y \tan x = \sec x$.

Solution   Comparing the given first-order linear differential equation to the general first-order form, we see

$$p(x) = \tan x \quad \text{and} \quad q(x) = \sec x$$

The integrating factor is

$$I(x) = e^{\int \tan x\, dx} = e^{-\ln|\cos x|} = e^{\ln|(\cos x)^{-1}|} = (\cos x)^{-1} = \sec x$$

and the general solution is

$$y = \frac{1}{\sec x} \left[\int (\sec x)(\sec x)\, dx + C \right]$$

$$= \cos x[\tan x + C]$$

$$= \sin x + C \cos x \qquad\qquad ∎$$

First-order linear differential equations appear in a variety of applications. In Section 7.6, we showed how first-order linear equations may be used to model mixture (dilution) problems as well as problems involving the current in an *RL* circuit (one with only a resistor, an inductor, and an electromotive force). In this section, we consider the motion of a body that falls in a resisting medium.

EXAMPLE 5   *Motion of a body falling in a resisting medium*

Consider an object with mass m that is initially at rest and is dropped from a great height (for example, from an airplane). Suppose the body falls in a straight line and the only forces acting on it are the downward force of the earth's gravitational attraction and a resisting upward force due to air resistance in the atmosphere. Assume that the resisting force is proportional to the velocity v of the falling body. Find equations for the velocity and displacement of the body's motion. Assume the distance $s(t)$ is measured down from the drop point.

Solution   The downward force is the weight mg of the body and the upward force is $-kv$, where k is a positive constant (the negative sign indicates that the force is directed upward). According to Newton's second law, the sum of the forces acting on a body at any time equals the product ma, where a is the acceleration of the body, that is

$$\underbrace{ma}_{\substack{\text{Sum of forces} \\ \text{on the body}}} = \underbrace{mg}_{\substack{\text{Force due} \\ \text{to gravity}}} - \underbrace{kv}_{\substack{\text{Resisting} \\ \text{force}}}$$

$$m\frac{dv}{dt} = mg - kv \qquad \text{Since } a = \frac{dv}{dt}$$

$$\frac{dv}{dt} = g - \frac{k}{m}v$$

$$\frac{dv}{dt} + \frac{k}{m}v = g$$

This is a first-order linear differential equation where $p(t) = \dfrac{k}{m}$ and $q(t) = g$. The integrating factor is

$$I(t) = e^{\int k/m \, dt} = e^{kt/m}$$

so that the solution is

$$v = \frac{1}{e^{kt/m}}\left[\int e^{kt/m}(g)dt + C\right] = e^{-kt/m}\left[\frac{ge^{kt/m}}{k/m} + C\right] = \frac{mg}{k} + Ce^{-kt/m}$$

Because $v = 0$ when $t = 0$ (the body is initially at rest), it follows that

$$0 = \frac{mg}{k} + Ce^0 = \frac{mg}{k} + C$$

Thus, because $C = -\dfrac{mg}{k}$, we have

$$v = \frac{mg}{k} + \left(-\frac{mg}{k}\right)e^{-kt/m}$$

Now, to find the position $s(t)$, we use the fact that $v(t) = \dfrac{ds}{dt}$:

$$\frac{ds}{dt} = \frac{mg}{k} - \frac{mg}{k}e^{-kt/m}$$

$$\int ds = \int\left[\frac{mg}{k} - \frac{mg}{k}e^{-kt/m}\right]dt$$

$$s(t) = \frac{mg}{k}t - \frac{mg}{k}\frac{e^{-kt/m}}{-k/m} + C$$

$$= \frac{mg}{k}t + \frac{m^2g}{k^2}e^{-kt/m} + C$$

Because $s(0) = 0$ (the distance s is measured from the point where the object is dropped), we find that

$$0 = \frac{mg}{k}(0) + \frac{m^2g}{k^2} e^0 + C \quad \text{so that} \quad -\frac{m^2g}{k^2} = C$$

Thus, the displacement is

$$s(t) = \frac{mg}{k} t + \frac{m^2g}{k^2} \left(e^{-kt/m} - 1\right)$$ ∎

In the problem set, you are asked to show that no matter what the initial velocity may be, the velocity reached by the object in the long run (as $t \to +\infty$) is mg/k.

EXACT DIFFERENTIAL EQUATIONS

Sometimes a first-order differential equation can be written in the general form

$$M(x, y)dx + N(x, y)dy = 0$$

where the left side is an exact differential, namely,

$$df = M(x, y)dx + N(x, y)dy$$

In this case, the given differential equation is appropriately called **exact** and since $df = 0$, its general solution is given by $f(x, y) = C$.

But how can we tell whether a particular first-order equation is exact, and if it is, how can we find f? Since

$$df = \frac{\partial f}{\partial x} dx + \frac{\partial f}{\partial y} dy$$

for a total differential (see Section 12.4), we must have

$$df = \frac{\partial f}{\partial x} dx + \frac{\partial f}{\partial y} dy = M(x, y)dx + N(x, y)dy$$

so

$$\frac{\partial f}{\partial x} = M(x, y) \quad \text{and} \quad \frac{\partial f}{\partial y} = N(x, y)$$

This will be true if and only if f satisfies the *cross-derivative test*

$$\frac{\partial N}{\partial x} = \frac{\partial M}{\partial y}$$

and then the function $f(x, y)$ is found by partial integration, exactly as we found the potential function of a conservative vector field in Section 14.3. The procedure for identifying and then solving an exact differential equation is illustrated in Example 6.

> **EXAMPLE 6**   *Exact differential equation*

Find the general solution for $(2xy^3 + 3y)\, dx + (3x^2y^2 + 3x)\, dy = 0$.

Solution   Let $M(x, y) = 2xy^3 + 3y$ and $N(x, y) = 3x^2y^2 + 3x$, and apply the cross-derivative test

$$\frac{\partial M}{\partial y} = 6xy^2 + 3 \quad \text{and} \quad \frac{\partial N}{\partial x} = 6xy^2 + 3$$

Because $\dfrac{\partial M}{\partial y} = \dfrac{\partial N}{\partial x}$, the equation is exact. To obtain a general solution, we must find a function f such that

$$\frac{\partial f}{\partial x} = 2xy^3 + 3y \quad \text{and} \quad \frac{\partial f}{\partial y} = 3x^2y^2 + 3x$$

To find f, we integrate the first partial on the left with respect to x:

$$f(x, y) = \int (2xy^3 + 3y)\, dx = x^2y^3 + 3xy + u(y)$$

where u is a function of y. Taking the partial derivative of f with respect to y and comparing the result with $\partial f/\partial y$, we obtain

$$\frac{\partial f}{\partial y} = \frac{\partial}{\partial y}[x^2y^3 + 3xy + u(y)] = 3x^2y^2 + 3x + u'(y)$$

so that

$$3x^2y^2 + 3x = 3x^2y^2 + 3x + u'(y)$$
$$0 = u'(y)$$

This implies that u is a constant. Taking $u = 0$, we have $f = x^2y^3 + 3xy$, and the general solution to the exact differential equation is

$$x^2y^3 + 3xy = C \qquad \blacksquare$$

■ **TABLE 15.1**   **Summary of Strategies for First-Order Differential Equations**

Form of Equation	Method	Solution
$\dfrac{dy}{dx} = \dfrac{g(x)}{f(y)}$	Separate the variables.	$\displaystyle\int f(y)\, dy = \int g(x)\, dx$
$\dfrac{dy}{dx} = f\!\left(\dfrac{y}{x}\right)$	Homogeneous—use a change of variable $v = \dfrac{y}{x}$.	$\displaystyle\int \frac{dv}{f(v) - v} = \int \frac{dx}{x}$
$\dfrac{dy}{dx} + p(x)y = q(x)$	Use the integrating factor $I(x) = e^{\int p(x)\,dx}$	$y = \dfrac{1}{I(x)}\left[\displaystyle\int I(x)\, q(x)\, dx + C\right]$
$M(x, y)\, dx + N(x, y)\, dy = 0$, where $\dfrac{\partial M}{\partial y} = \dfrac{\partial N}{\partial x}$	Exact—use partial integration to find f, where $\dfrac{\partial f}{\partial x} = M$ and $\dfrac{\partial f}{\partial y} = N$	$f(x, y) = C$

EULER'S METHOD

In Section 5.6, we introduced direction fields as a means for obtaining a "picture" of various solutions to a differential equation, but sometimes we need more than a rough graph of a solution. We now consider approximating a solution by numerical means. **Euler's method** is a simple procedure for obtaining a table of approximate values for the solution of a given initial value problem*

$$\frac{dy}{dx} = f(x, y) \qquad y(x_0) = y_0$$

---

*In our discussion of Euler's method, we assume that the given initial value problem has a unique solution. It can be shown that such an initial value problem always has a unique solution if f and $\partial f/\partial y$ are both continuous in a neighborhood (x_0, y_0). The proof of this result is beyond the scope of this text but can be found in most elementary differential equations texts.

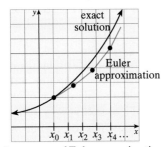

a. The first Euler approximation

b. A sequence of Euler approximations

■ **FIGURE 15.1** Graphical representation of Euler's method

The key idea in Euler's method is to increment x_0 by a small quantity h and then to approximate $y_1 = y(x_1)$ for $x_1 = x_0 + h$ by assuming x and y change by so little over the interval $[x_0, x_1]$ that $f(x, y)$ can be replaced by $f(x_0, y_0)$ for this interval. Solving the approximating initial value problem

$$\frac{dy}{dx} = f(x_0, y_0) \qquad y(x_0) = y_0$$

we obtain

$$y - y_0 = f(x_0, y_0)(x - x_0)$$

In other words, we are approximating the solution curve $y = y(x)$ near (x_0, y_0) by the tangent line to the curve at this point, as shown in Figure 15.1a.

We then repeat this process with (x_1, y_1) assuming the role of (x_0, y_0) to obtain an approximation of the solution $y = y(x)$ over the interval $x_1 \leq x \leq x_2$, where $x_2 = x_1 + h$ and

$$y_2 = y_1 + hf(x_1, y_1)$$

Continuing in this fashion, we obtain a sequence of line segments that approximates the shape of the solution curve as shown in Figure 15.1b. Euler's method is illustrated in the following example.

EXAMPLE 7   *Euler's method*

Use Euler's method with $h = 0.1$ to estimate the solution of the initial value problem

$$\frac{dy}{dx} = x + y^2 \qquad y(0) = 1$$

over the interval $0 \leq x \leq 0.5$.

Solution   Before using Euler's method, we might first look at a graphical solution. The slope field is shown in Figure 15.2a, and the particular solution through the point $(0, 1)$ is shown in Figure 15.2b.

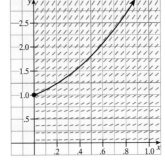

a. Slope field of $\dfrac{dy}{dx} = x + y^2$ **b.** Particular solution through $(0, 1)$

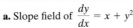

■ **FIGURE 15.2** Graphical solution using a direction field

To use Euler's method for this example, we note

$$f(x, y) = x + y^2, \quad x_0 = 0, \quad y_0 = 1, \quad \text{and} \quad h = 0.1$$

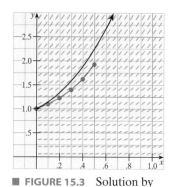

We show the calculator (or computer) solution correct to four decimal places:

$$y_0 = y(0) = 1$$
$$y_1 = y_0 + hf(x_0, y_0) = 1 + 0.1(0 + 1^2) = 1.1000$$
$$y_2 = y_1 + hf(x_1, y_1) = 1.1000 + 0.1(0.1 + 1.1000^2) = 1.2310$$
$$y_3 = y_2 + hf(x_2, y_2) = 1.2310 + 0.1(0.2 + 1.2310^2) \approx 1.4025$$
$$y_4 = y_3 + hf(x_3, y_3) = 1.4025 + 0.1(0.3 + 1.4025^2) \approx 1.6292$$
$$y_5 = y_4 + hf(x_4, y_4) = 1.6292 + 0.1(0.4 + 1.6292^2) \approx 1.9347$$

These points can be plotted to approximate the solution, as shown in Figure 15.3. Notice that we plotted these points by superimposing them on the direction field shown in Figure 15.2b.

■ **FIGURE 15.3**   Solution by Euler's method

Euler's method has educational value as the simplest numerical method for solving ordinary differential equations, and can be found in most computer-assisted programs. However, as you might guess by looking at Figure 15.3, as you move away from (x_0, y_0), the error may accumulate. The Euler method can be improved in a variety of ways, most notably by a collection of procedures known as the *Runge–Kutta* and *predictor–corrector* methods. These methods are studied in more advanced courses.

15.1 Problem Set

(A) *Find the general solution of the differential equations in Problems 1–6 by separating variables.*

1. $xy\, dx = (x - 5)\, dy$

2. $\dfrac{dy}{dx} = y \tan x$

3. $(e^{2x} + 9) \dfrac{dy}{dx} = y$

4. $y \dfrac{dy}{dx} = e^{x-3y} \cos x$

5. $9\, dx - x\sqrt{x^2 - 9}\, dy = 0$

6. $xy \dfrac{dy}{dx} = x^2 + y^2 + x^2 y^2 + 1$

In Problems 7–12, a graphical solution of a given initial value problem is shown by using a direction field. Solve the problem to find an equation for the particular solution.

7. $\dfrac{dy}{dx} + 2xy = 4$
    passing through $(0, 0)$

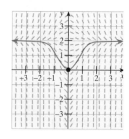

8. $\dfrac{dy}{dx} + \dfrac{y}{x} = \dfrac{\sin x}{x}$
    passing through $(-2, 0)$

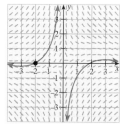

9. $\dfrac{dy}{dx} + y = \cos x$
    passing through $(0, 0)$

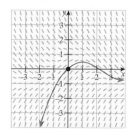

10. $\dfrac{dy}{dx} + y = \dfrac{e^x}{1 + e^{2x}}$
    passing through $(0, 2)$

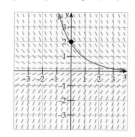

11. $x \dfrac{dy}{dx} - 2y = x^3$
    passing through $(2, -1)$

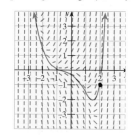

12. $\dfrac{dy}{dx} - 3xy = 5xe^{x^2}$
    passing through $(0, -3)$

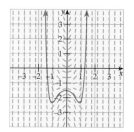

Show that each differentiable equation in Problems 13–18 is in the form $dy/dx = f(y/x)$, and then find the general solution.

13. $(3x - y) dx + (x + 3y) dy = 0$

14. $xy dx - (2x^2 + y^2) dy = 0$

15. $(3x - y) dx + (x - 3y) dy = 0$

16. $(x^2 + y^2) dx - 2xy dy = 0$

17. $(-6y^2 + 3xy + 2x^2) dx + x^2 dy = 0$

18. $x dy - (y + \sqrt{xy}) dx = 0$

Show the differential equations in Problems 19–24 are exact and find the general solution.

19. $(3x^2y + \tan y) dx + (x^3 + x \sec^2 y) dy = 0$

20. $(3x^2 - 10xy) dx + (2y - 5x^2 + 4) dy = 0$

21. $\left[\dfrac{1}{1 + x^2} + \dfrac{2x}{x^2 + y^2}\right] dx + \left[\dfrac{2y}{x^2 + y^2} - e^{-y}\right] dy = 0$

22. $(2xy^3 + 3y - 3x^2) dx + (3x^2y^2 + 3x) dy = 0$

23. $[2x \cos 2y - 3y(1 - 2x)] dx$
$\quad - [2x^2 \sin 2y + 3(2 + x - x^2)] dy = 0$

24. $[(x + xy - 3)(1 + y) - x^2\sqrt{y}\,] dx$
$\quad + \left[x^2(y + 1) - 3x - \dfrac{x^3}{6\sqrt{y}}\right] dy$

B 25. Consider the differential equation

$$\frac{dy}{dx} = x + y$$

a. Find the particular solution that contains the point $(1, 2)$.

b. Sketch isoclines for $C = 1, 3$, and 5. Then use the direction field for the given differential equation to sketch the solution through $(1, 2)$. Compare the result with part **a.**

c. Use Euler's method to approximate a solution for $x_0 = 1, y_0 = 2$, and $h = 0.2$. Compare this result with part **b.**

26. Consider the differential equation

$$\frac{dy}{dx} = x^2 - y^2$$

a. Find the particular solution that contains the point $(2, 1)$.

b. Sketch isoclines for $C = 0, 2$, and 4. Then use the direction field for the given differential equation to sketch the solution through $(2, 1)$. Compare the result with part **a.**

c. Use Euler's method to approximate a solution for $x_0 = 2, y_0 = 1$, and $h = 0.2$.

Estimate a solution for Problems 27–30 using Euler's method. For each of these problems, a direction field is given. Superimpose the segments from Euler's method on the given direction field.

27. $\dfrac{dy}{dx} = \dfrac{x + y}{y - x}$
passing through $(0, 1)$
for $0 \le x \le 0.5, h = 0.1$

28. $\dfrac{dy}{dx} = 2x(x^2 - y)$
passing through $(0, 4)$
for $0 \le x \le 1, h = 0.1$

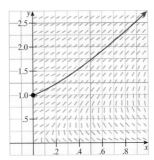

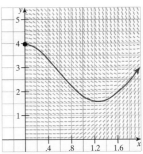

29. $\dfrac{dy}{dx} = \dfrac{5x - 3xy}{1 + x^2}$
passing through $(0, 0)$
for $0 \le x \le 0.5, h = 0.1$

30. $\dfrac{dy}{dx} = \dfrac{y^2 + 2x}{3y^2 - 2xy}$
passing through $(0, 1)$
for $0 \le x \le 0.5, h = 0.1$

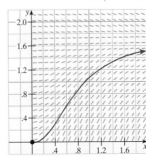

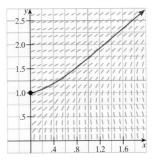

*An **integrating factor** of the differential equation*

$$M dx + N dy = 0$$

is a function $f(x, y)$ such that

$$f(x, y) M(x, y) dx + f(x, y)N(x, y) dy = 0$$

is exact. In Problems 31–32, find an integrating factor of the specified type for the given differential equations, then solve the equation.

31. $y dx + (y - x)dy = 0; f(x, y) = y^n$

32. $(x^2 + y^2)dx + (3xy)dy = 0; f(x, y) = x^n$

Identify each equation given in Problems 33–44 as separable, homogeneous, first-order linear, or exact, and then solve. It is possible for an equation to be of more than one type.

33. $(2xy^2 + 3x^2y - y^3)dx + (2x^2y + x^3 - 3xy^2) dy = 0$

34. $(1 + x)dy + \sqrt{1 - y^2}\, dx = 0$

35. $\left(\dfrac{2x}{y} - \dfrac{y^2}{x^2}\right) dx + \left(\dfrac{2y}{x} - \dfrac{x^2}{y^2} + 3\right) dy = 0$

36. $(x^2 - xy - x + y)dx - (xy - y^2) dy = 0$

37. $e^{y-x}\sin x\, dx - \csc x\, dy = 0$

38. $x^2 \dfrac{dy}{dx} + 2xy = \sin x$

39. $(3x^2 - y \sin xy)dx - (x \sin xy) dy = 0$

40. $\dfrac{dy}{dx} = \dfrac{y}{x} + x \cos \dfrac{y}{x}$

41. $(y - \sin^2 x)\,dx + (\sin x)dy = 0$

42. $(y^3 - y)dx + (x^2 + x)dy = 0$

43. $x\dfrac{dy}{dx} = y - \sqrt{x^2 + y^2}$

44. $(2x + \sin y - \cos y)dx + (x \cos y + x \sin y)dy = 0$

Find the particular solution of the differential equations in Problems 45–52 with the specified initial conditions.

45. $\left(x \sin^2 \dfrac{y}{x} - y\right) dx + x\,dy = 0; x = \dfrac{4}{\pi}, = 1$

46. $y(5x - y)dx - x(5x + 2y)dy = 0; x = 1, y = 1$

47. $x\dfrac{dy}{dx} - 3y = x^3; x = 1, y = 1$

48. $\dfrac{dy}{dx} = 1 + 3y \tan x; x = 0, y = 2$

49. $[\sin(x^2 + y) + 2x^2 \cos(x^2 + y)]dx$
   $+ [x \cos(x^2 + y)]dy = 0; x = 0, y = 0$

50. $y\dfrac{dy}{dx} = e^{x+2y}\sin x; x = 0, y = 0$

51. $ye^x\,dy = (y^2 + 2y + 2)dx; x = 0, y = -1$

52. $(2xy + y^2 + 2x)dx + (x^2 + 2xy - 1)dy = 0;$
   $x = 1, y = 3$

53. In the chemical conversion analyzed in Example 2, what happens to $Q(t)$ as $t \to +\infty$ in the case where $\alpha \neq \beta$? What if $\alpha = \beta$?

54. a. Find an equation for the velocity of the falling body in Example 5 in the case where the body has initial velocity $v_0 \neq 0$.
   b. Show that for any initial velocity (zero or nonzero) the velocity reached by the object "in the long run" (as $t \to +\infty$) is always mg/k.

55. **MODELING PROBLEM** A population of foxes grows logistically until it is decided to allow hunting at the constant rate of h foxes per month. The population $P(t)$ is then modeled by the differential equation

$$\dfrac{dP}{dt} = P(k - \ell P) - h$$

where $P(0) = P_0$.

   a. Solve this equation in terms of $k, \ell, h,$ and P_0 for the case where $h < \dfrac{k^2}{4\ell}$.
   b. If $P_0 > h$, what happens to $P(t)$ as $t \to \infty$?

56. **MODELING PROBLEM** A chemical in a solution diffuses from a compartment where the concentration is $C_0(t) = 7e^{-t}$ across a membrane with diffusion coefficient $k = 1.75$. The concentration $C(t)$ in the second compart-

ment is modeled by

$$\dfrac{dC}{dt} = 1.75[C_0(t) - C(t)]$$

Solve this equation for $C(t)$, assuming that $C(0) = 0$.

57. Our formula for the solution of a first-order linear equation requires the differential equation to be linear in x. Suppose, instead, the equation is linear in y; that is, it can be written in the form

$$\dfrac{dx}{dy} + R(y)x = S(y)$$

   a. Find a formula for the general solution of an equation that is first-order linear in y.
   b. Use the formula obtained in part **a** to solve

$$y\,dx - 2x\,dy = y^4 e^{-y}\,dy$$

58. **MODELING PROBLEM** A man is pulling a heavy sled along the ground by a rope of fixed length L. Assume the man begins walking at the origin of a coordinate plane and that the sled is initially at the point $(0, L)$. The man walks to the right (along the positive x-axis), dragging the sled behind him. When the man is at point M, the sled is at S, as shown in Figure 15.4. Find a differential equation for the path of the sled and solve this differential equation.

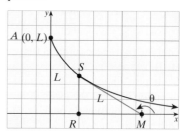

■ **FIGURE 15.4** Sled problem

59. **SPY PROBLEM** Having conserved as much energy as possible after his escape from the Death Ray (Problem 38, Section 14.3), the Spy continues his search for Purity and Blohardt. He rushes into a small room and finds no Blohardt but *two* Purities struggling ferociously on the floor. As he separates them, he notices that one is dressed in red and the other in green. The red one cries, "She's my evil twin, Rottona. She escaped from prison and has been holding me prisoner." The one in green shouts, "Liar! I'm the real Purity. I've been tied up here all along and just was freed." The Spy wishes he could remember whether the one who has been trying to kill him was wearing red or green, but it is all a big blur. Instead, he says, "I will ask a question that only the real Purity can answer. A 160-pound skydiver jumps from a plane and begins to free fall. She knows the air resistance is $0.01\,v^2$, where $v(t)$ is her velocity. She is wearing a wrist altimeter and wishes to open her parachute when she has fallen

1,000 feet. What will be her velocity at that instant?" The green Purity stamps her foot petulantly and cries, "How should I know and who cares, anyhow?" The one in red grins and answers correctly. What is her answer?

60. ALMOST HOMOGENEOUS EQUATIONS Sometimes a differential equation is not quite homogeneous but becomes homogeneous with a linear change of variable. Specifically, consider a differential equation of the form

$$\frac{dy}{dx} = f\left(\frac{ax + by + c}{rx + sy + t}\right)$$

a. Suppose $as \neq br$. Make the change of variable $x = X + A$ and $y = Y + B$ where A and B satisfy

$$\begin{cases} aA + bB + c = 0 \\ rA + sB + t = 0 \end{cases}$$

Show that with these choices for A and B, the differential equation becomes homogeneous.

b. Apply the procedure outlined in part **a** to solve the differential equation

$$\frac{dy}{dx} = \left(\frac{-3x + y + 2}{x + 3y - 5}\right)$$

61. A **Riccati equation** is a differential equation of the form

$$\frac{dy}{dx} = P(x)y^2 + Q(x)y + R(x)$$

Suppose we know that $y = u(x)$ is a solution of a given Riccati equation.

a. Change variables by setting

$$z = \frac{1}{y - u(x)}$$

Show that the given equation is transformed by this change of variables into the separable form

$$\frac{dz}{dx} + [2P(x)u(x) + Q(x)]z = -P(x)$$

b. Solve the first-order linear equation in z obtained in part **a** and explain how to find the general solution of the given Riccati equation.

c. Use the method outlined in parts **a** and **b** to solve the Riccati equation

$$\frac{dy}{dx} = \frac{1}{x^2}y^2 + \frac{2}{x}y - 2$$

Hint: There is a solution of the general form $y = Ax$.

62. JOURNAL PROBLEM* Find all solutions of the Riccati equation

$$u' = u^2 + \frac{au}{x} - b$$

$a, b \neq 0$, that are real rational functions of x.

─────────

15.2 Second-Order Homogeneous Linear Differential Equations

IN THIS SECTION linear independence, solutions of the equation $y'' + ay' + by = 0$, higher-order homogeneous linear equations, damped motion of a mass on a spring, reduction of order

LINEAR INDEPENDENCE

A **linear differential equation** is one of the general form

$$a_n(x)y^{(n)} + a_{n-1}(x)y^{(n-1)} + \cdots + a_0(x)y = R(x)$$

and if $a_n(x) \neq 0$, it is said to be of **order** n. It is *homogeneous* if $R(x) = 0$ and *nonhomogeneous* if $R(x) \neq 0$. In this section we focus attention on the homogeneous case, and we examine nonhomogeneous equations in the next section.

To characterize all solutions of the homogeneous equation

$$a_n(x)y^{(n)} + a_{n-1}y^{(n-1)} + \cdots + a_0(x)y = 0$$

we require the following definition.

Linear Dependence and Independence

The functions $y_1, y_2, \ldots, y_n$ are said to be **linearly independent** if the equation

$$C_1 y_1 + C_2 y_2 + \cdots + C_n y_n = 0 \quad \text{for constants} \quad C_1, C_2, \ldots$$

has only the trivial solution $C_1 = C_2 = \cdots = C_n = 0$ for all x in the interval I. Otherwise the y_k's are **linearly dependent.**

The functions $y_1 = \cos x$ and $y_2 = x$ are linearly independent because the only way we can have $C_1 \cos x + C_2 x = 0$ for all x is for C_1 and C_2 both to be 0. However, $y_1 = 1, y_2 = \sin^2 x$, and $y_3 = \cos 2x$ are linearly dependent, because

$$C_1(1) + C_2(\sin^2 x) + C_3(\cos 2x) = 0 \quad \text{for} \quad C_1 = 1, C_2 = -2, C_3 = -1$$

It can be quite difficult to determine whether a given collection of functions $y_1, y_2, \ldots, y_n$ is linearly independent. However, this issue can be settled by a routine computation using the following determinant, which is named after Josef Hoëné de Wronski (see the Historical Quest in Problem 31).

Wronskian

The **Wronskian** $W(y_1, y_2, \ldots, y_n)$ of n functions $y_1, y_2, \ldots, y_n$ having $n - 1$ derivatives on an interval I is defined to be the determinant function

$$W(y_1, y_2, \ldots, y_n) = \begin{vmatrix} y_1 & y_2 & \cdots & y_n \\ y_1' & y_2' & \cdots & y_n' \\ \vdots & \vdots & & \vdots \\ y_1^{(n-1)} & y_2^{(n-1)} & \cdots & y_n^{(n-1)} \end{vmatrix}$$

THEOREM 15.1   *Determining linear independence with the Wronskian*

Suppose the functions $a_n(x), a_{n-1}(x), \ldots, a_0(x)$ in the nth-order homogeneous linear differential equation

$$a_n(x)y^{(n)} + a_{n-1}(x)y^{(n-1)} + \cdots + a_0(x)y = 0$$

are all continuous on a closed interval $[c, d]$. Then solutions $y_1, y_2, \ldots, y_n$ of this differential equation are linearly independent if and only if the Wronskian is nonzero; that is,

$$W(y_1, y_2, \ldots, y_n) \neq 0$$

throughout the interval $[c, d]$.

Proof   The proof of this theorem is beyond the scope of this course but can be found in most differential equations books.

EXAMPLE 1   *Showing linear independence*

The functions $y_1 = e^{-x}, y_2 = xe^{-x}$, and $y_3 = e^{3x}$ are solutions of a certain homogeneous linear differential equation with constant coefficients. Show that these solutions are linearly independent.

Solution

$$W(e^{-x}, xe^{-x}, e^{3x}) = \begin{vmatrix} e^{-x} & xe^{-x} & e^{3x} \\ -e^{-x} & (1-x)e^{-x} & 3e^{3x} \\ e^{-x} & (x-2)e^{-x} & 9e^{3x} \end{vmatrix}$$

$$= e^{-x}[9e^{3x}(1-x)e^{-x} - 3e^{3x}(x-2)e^{-x}]$$
$$- xe^{-x}[-9e^{3x}e^{-x} - 3e^{-x}e^{3x}]$$
$$+ e^{3x}[-e^{-x}(x-2)e^{-x} - e^{-x}(1-x)e^{-x}]$$

$$= 16e^{x}$$

Because $16e^{x} \neq 0$, the functions are linearly independent. ∎

The general solution of an nth-order homogeneous linear differential equation with constant coefficients can be characterized in terms of n linearly independent solutions. Here is the theorem that applies to the second-order case.

THEOREM 15.2   ***Characterizing the general solution of $y'' + ay' + by = 0$***

If y_1 and y_2 are linearly independent solutions of the differential equation $y'' + ay' + by = 0$, then the general solution is

$$y = C_1 y_1 + C_2 y_2 \quad \text{for arbitrary constants } C_1, C_2$$

Proof   We can prove that if y_1 and y_2 are linearly independent solutions, then $y = C_1 y_1 + C_2 y_2$ is also a solution. The proof that all solutions are of this form is beyond the scope of this book. Suppose y_1 and y_2 are solutions, so that

$$y_1''(x) + ay_1'(x) + by_1(x) = 0$$
$$y_2''(x) + ay_2'(x) + by_2(x) = 0$$

If $y = C_1 y_1 + C_2 y_2$, we have

$$y'' + ay' + by = [C_1 y_1'' + C_2 y_2''] + a[C_1 y_1' + C_2 y_2'] + b[C_1 y_1 + C_2 y_2]$$
$$= C_1[y_1'' + ay_1' + by_1] + C_2[y_2'' + ay_2' + by_2]$$
$$= 0 + 0 = 0$$

Thus, $y = C_1 y_1 + C_2 y_2$ is also a solution. ═

SOLUTIONS OF THE EQUATION $y'' + ay' + by = 0$

Thanks to the characterization theorem, we now know that once we have two linearly independent solutions y_1, y_2 of the equation $y'' + ay' + by = 0$, we have them all because the general solution can be characterized as $y = C_1 y_1 + C_2 y_2$. Therefore, the whole issue of how to represent the solution of a second-order homogeneous linear equation with constant coefficients depends on finding two linearly independent solutions.

Recall that the general solution of the first-order equation $y' + ay = 0$ is $y = Ce^{-ax}$. Therefore, it is not unreasonable to expect the second-order equation $y'' + ay' + by = 0$ to have one (or more) solutions of the form $y = e^{rx}$. If $y = e^{rx}$, then $y' = re^{rx}$ and $y'' = r^2 e^{rx}$, and by substituting these derivatives into the equation $y'' + ay' + by = 0$, we obtain

$$y'' + ay' + by = 0$$
$$r^2 e^{rx} + a(re^{rx}) + be^{rx} = 0$$
$$e^{rx}(r^2 + ar + b) = 0$$
$$r^2 + ar + b = 0 \qquad e^{rx} \neq 0$$

Thus, $y = e^{rx}$ is a solution of the given second-order differential equation if and only if $r^2 + ar + b = 0$. This equation is called the **characteristic equation** of $y'' + ay' + by = 0$.

EXAMPLE 2   *Characteristic equation with distinct real roots*

Find the general solution of the differential equation $y'' + 2y' - 3y = 0$.

Solution   Begin by solving the characteristic equation:

$$r^2 + 2r - 3 = 0$$
$$(r - 1)(r + 3) = 0$$
$$r = 1, -3$$

The particular solutions are $y_1 = e^x$ and $y_2 = e^{-3x}$. Next, determine whether these equations are linearly independent by looking at the Wronskian:

$$W(e^x, e^{-3x}) = \begin{vmatrix} e^x & e^{-3x} \\ e^x & -3e^{-3x} \end{vmatrix} = e^x(-3e^{-3x}) - e^x e^{-3x} = -4e^{-2x}$$

Because $W(e^x, e^{-3x}) = -4e^{-2x} \neq 0$, the functions are linearly independent, and the characterization theorem tells us that the general solution is

$$y = C_1 y_1 + C_2 y_2 = C_1 e^x + C_2 e^{-3x} \qquad \blacksquare$$

Example 2 has a characteristic equation that factors easily. In practice, however, the quadratic formula is often required, namely,

$$r = \frac{-a \pm \sqrt{a^2 - 4b}}{2}$$

The discriminant *for this equation,* $a^2 - 4b$, figures prominently in the following theorem.

THEOREM 15.3   *Solution of $y'' + ay' + by = 0$*

If r_1 and r_2 are the roots of the characteristic equation $r^2 + ar + b = 0$, then the general solution of the homogeneous linear differential equation $y'' + ay' + by = 0$ can be expressed in one of these forms:

$a^2 - 4b > 0$:   The general solution is

$$y = C_1 e^{r_1 x} + C_2 e^{r_2 x}$$

where

$$r_1 = \frac{-a + \sqrt{a^2 - 4b}}{2} \quad \text{and} \quad r_2 = \frac{-a - \sqrt{a^2 - 4b}}{2}$$

$a^2 - 4b = 0$:   The general solution is

$$y = C_1 e^{-ax/2} + C_2 x e^{-ax/2} = (C_1 + C_2 x)e^{-ax/2}$$

$a^2 - 4b < 0$:   The general solution is

$$y = e^{-ax/2}\left[C_1 \cos\left(\frac{\sqrt{4b^2 - a}}{2}x\right) + C_2 \sin\left(\frac{\sqrt{4b^2 - a^2}}{2}x\right) \right]$$

Proof   We have just seen that the solutions of the differential equation of the form $y = e^{rx}$ correspond to the solutions of the characteristic equation $r^2 + ar + b = 0$.

The quadratic formula characterizes the three cases according to the discriminant of this equation, namely, $a^2 - 4b$. For each case, we must find two linearly independent solutions.

$a^2 - 4b > 0$: Let $y_1 = e^{r_1 x}$, and $y_2 = e^{r_2 x}$. Then

$$W(e^{r_1 x}, e^{r_2 x}) = \begin{vmatrix} e^{r_1 x} & e^{r_2 x} \\ r_1 e^{r_1 x} & r_2 e^{r_2 x} \end{vmatrix} = r_2 e^{(r_1 + r_2)x} - r_1 e^{(r_1 + r_2)x} = (r_2 - r_1)e^{(r_1 + r_2)x}$$

Because $r_2 \neq r_1$ and $e^{(r_1 + r_2)x} > 0$, we see $W(e^{r_1 x}, e^{r_2 x}) \neq 0$, so the functions are linearly independent. The characterization theorem tells us that the general solution is

$$y = C_1 y_1 + C_2 y_2 = C_1 e^{r_1 x} + C_2 e^{r_2 x}$$

$a^2 - 4b = 0$: In this case, the characteristic equation has one repeated root—namely, $r = -a/2$. The function $y_1 = e^{-ax/2}$ is one solution, and it can be shown that a second linearly independent solution is $y_2 = xe^{-ax/2}$ (see Problem 32). Thus, the general solution is

$$y = C_1 e^{-ax/2} + C_2 xe^{-ax/2}$$

$a^2 - 4b < 0$: The proof of this part is left for the reader. $=$

EXAMPLE 3   *Characteristic equation with repeated roots*

Find the general solution of the differential equation $y'' + 4y' + 4y = 0$.

Solution   Solve the characteristic equation:

$$r^2 + 4r + 4 = 0$$
$$(r + 2)^2 = 0$$
$$r = -2 \quad \text{(multiplicity 2)}$$

The roots are $r_1 = r_2 = -2$. Thus, $y_1 = e^{-2x}$ and $y_2 = xe^{-2x}$, so the general solution of the differential equation is

$$y = C_1 e^{-2x} + C_2 xe^{-2x}$$ ∎

EXAMPLE 4   *Characteristic equation with complex roots*

Find the general solution of the differential equation $2y'' + 3y' + 5y = 0$.

Solution   Solve the characteristic equation*:

$$2r^2 + 3r + 5 = 0$$
$$r = \frac{-3 \pm \sqrt{9 - 4(2)(5)}}{2(2)}$$
$$= \frac{-3 \pm \sqrt{31}\, i}{4}$$

The roots are

$$r_1 = -\frac{3}{4} + \frac{\sqrt{31}}{4}\, i, \quad r_2 = -\frac{3}{4} - \frac{\sqrt{31}}{4}\, i$$

---

*Technically, we have considered only the case where the leading coefficient of the characteristic equation is 1. Verify that you obtain the same result if you first divide both sides by 2.

Thus, the general solution is

$$y = e^{(-3/4)x}\left[C_1\cos\frac{\sqrt{31}}{4}x + C_2\sin\frac{\sqrt{31}}{4}x\right]$$ ∎

EXAMPLE 5 *Second-order initial value problem*

Solve $4y'' + 12y' + 9y = 0$ subject to $y(0) = 3$ and $y'(0) = -2$.

Solution   Solve

$$4r^2 + 12r + 9 = 0$$
$$(2r + 3)^2 = 0$$
$$r = -\tfrac{3}{2}$$

Thus, the general solution is

$$y = C_1e^{(-3/2)x} + \tfrac{3}{2}C_2xe^{(-3/2)x}$$

Because $y(0) = 3$ we have

$$3 = C_1e^0 + C_2(0)e^0$$
$$3 = C_1$$

Because $y'(0) = -2$, we find y':

$$y' = -\tfrac{3}{2}C_1e^{(-3/2)x} - \tfrac{3}{2}C_2xe^{(-3/2)x} + C_2e^{(-3/2)x}$$
$$y'(0) = -\tfrac{3}{2}C_1e^0 - \tfrac{3}{2}C_2(0)e^0 + C_2e^0$$
$$-2 = -\tfrac{3}{2}C_1 + C_2$$
$$-2 = -\tfrac{3}{2}(3) + C_2 \qquad \text{Because } C_1 = 3$$
$$\tfrac{5}{2} = C_2$$

Thus, the particular solution is

$$y = 3e^{(-3/2)x} + \tfrac{5}{2}xe^{(-3/2)x}$$ ∎

HIGHER-ORDER HOMOGENEOUS LINEAR EQUATIONS

Homogeneous linear differential equations of degree 3 or more with constant coefficients can be handled in essentially the same way as the second-order equations we have analyzed. As in the second-order case, some of the roots of the characteristic equations may be real and distinct, some may be real and repeated, and some may occur in complex conjugate pairs. But now, roots of the characteristic equation may occur more than twice, and when this happens, the linearly independent solutions are obtained by multiplying by increasing powers of x. For example, if 2 is a root of multiplicity 4 in the characteristic equation, the corresponding linearly independent solutions are e^{2x}, xe^{2x}, x^2e^{2x}, and x^3e^{2x}. The procedure for obtaining the general solution of nth-order linear homogeneous equation with constant coefficients is illustrated in the next two examples.

EXAMPLE 6 *Characteristic equation with repeated roots*

Solve $y^{(4)} - 5y''' + 6y'' + 4y' - 8y = 0$.

Solution   Solve the characteristic equation:

$$r^4 - 5r^3 + 6r^2 + 4r - 8 = 0$$

Because this is 4th-degree, we use synthetic division and the rational root theorem (or a calculator) to find the roots $-1, 2, 2,$ and $2.$ The general solution is

$$y = C_1 e^{-x} + C_2 e^{2x} + C_3 x e^{2x} + C_4 x^2 e^{2x}$$ ■

| **EXAMPLE 7** | *Characteristic equation with repeated roots (some not real)* |

Solve $y^{(7)} + 8y^{(5)} + 16y''' = 0$

Solution   Solve the characteristic equation:

$$r^7 + 8r^5 + 16r^3 = 0$$
$$r^3(r^4 + 8r^2 + 16) = 0$$
$$r^3(r^2 + 4)^2 = 0$$
$$r = 0 \quad \text{(multiplicity 3)}, \quad \pm 2i \text{ (multiplicity 2)}$$

The roots (showing multiplicity) are $0, 0, 0, 2i, 2i, -2i, -2i.$ The general solution is

$$y = C_1 + C_2 x + C_3 x^2 + C_4 \cos 2x + C_5 \sin 2x + C_6 x \cos 2x + C_7 x \sin 2x$$ ■

DAMPED MOTION OF A MASS ON A SPRING

To illustrate an application of second-order homogeneous linear differential equations, we shall consider the motion of an oscillating spring. Suppose we pull down on an object suspended at the end of a spring and then release it. Hooke's law in physics says that a spring that is stretched or compressed x units from its natural length tends to restore itself to its natural length by a force whose magnitude F is proportional to $x.$ Specifically, $F(x) = kx,$ where the constant of proportionality k is called the **spring constant** and depends on the stiffness of the spring.

Suppose the mass of the spring is negligible compared to the mass m of the object on the spring. When the object is pulled down and released, the spring begins to oscillate, and its motion is determined by two forces, the weight mg of the object and the restoring force $F(x) = k_1 x$ of the spring. According to Newton's second law of motion, the force acting on the object is $ma,$ where $a = x''(t)$ is the acceleration of the object. If there are no other external forces acting on the object, the motion is said to be **undamped,** and the motion is governed by the second-order homogeneous equation

$$mx''(t) = -k_1 x(t) \qquad k_1 > 0$$
$$mx''(t) + k_1 x(t) = 0$$

Next, suppose the object is connected to a dashpot, or a device that imposes a damping force. A good example is a shock absorber in a car, which forces a spring to move through a fluid (see Figure 15.5). Experiments indicate that the shock absorber introduces a damping force proportional to the velocity $v = x'(t).$ Thus, the total force in this case is $-k_1 x(t) - k_2 x'(t),$ and Newton's second law tells us

$$mx''(t) = -k_1 x(t) - k_2 x'(t) \qquad k_1 > 0, k_2 > 0$$

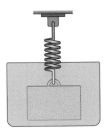

■ **FIGURE 15.5**   A damped spring system

$$mx''(t) + k_2 x'(t) + k_1 x = 0$$

The characteristic equation is

$$mr^2 + k_2 r + k_1 = 0$$

$$r = \frac{-k_2 \pm \sqrt{k_2^2 - 4k_1 m}}{2m}$$

The three cases that can occur correspond to different kinds of motion for the object on the spring.

overdamping: $k_2^2 - 4k_1 m > 0$

In this case, both roots are real and negative, and the solution is of the form

$$x(t) = C_1 e^{r_1 t} + C_2 e^{r_2 t}$$

where $r_1 = -\dfrac{k_2}{2m} + \dfrac{1}{2m}\sqrt{k_2^2 - 4k_1 m}$ and

$$r_2 = -\frac{k_2}{2m} - \frac{1}{2m}\sqrt{k_2^2 - 4k_1 m}$$

Note that the motion dies out eventually at $t \to +\infty$:

$$\lim_{t \to +\infty} x(t) = \lim_{t \to +\infty}\left(C_1 e^{r_1 t} + C_2 e^{r_2 t}\right) = 0, \quad \text{because } r_1 < 0 \text{ and } r_2 < 0$$

Overdamping is illustrated in Figure 15.6a.

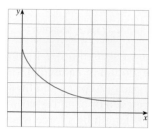

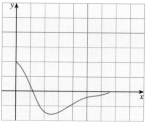

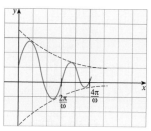

a. Overdamping $k_2^2 - 4k_1 m > 0$ **b.** Critical damping $k_2^2 - 4k_1 m = 0$ **c.** Underdamping $k_2^2 - 4k_1 m < 0$

■ **FIGURE 15.6**   Damping motion

critical damping: $k_2 - 4k_1 m = 0$

The solution has the form

$$x(t) = (C_1 + C_2 t)e^{rt}$$

where $r_1 = r_2 = r = -\dfrac{k_2}{2m}$. In this case, the motion also eventually dies out (as $t \to +\infty$), because $r < 0$. This solution is shown in Figure 15.6b.

underdamping: $k_2 - 4k_1 m < 0$

In this case, the characteristic equation has complex roots, and the solution has the form

$$x(t) = e^{-k_2 t/(2m)}\left[C_1 \cos\left(\frac{t}{2m}\sqrt{4k_1 m - k_2^2}\right) + C_2 \sin\left(\frac{t}{2m}\sqrt{4k_1 m - k_2^2}\right)\right]$$

This can be written as

$$x(t) = Ae^{\alpha t}\cos(\omega t - C)$$

where

$$A = \sqrt{C_1^2 + C_2^2}; \quad \alpha = -\frac{k_2}{2m}; \quad \omega = \frac{1}{2m}\sqrt{4k_1 m - k_2^2}; \quad C = \tan^{-1}\frac{C_2}{C_1}$$

Because k_2 and m are both positive, α must be negative and we see that $x(t) \to 0$ as $t \to +\infty$. Notice that as the motion dies out, it oscillates with frequency $2\pi/\omega$, as shown in Figure 15.6c.

REDUCTION OF ORDER

Theorem 15.2 applies even when b and c are functions of x instead of constants. In other words, the general solution of the equation.

$$y'' + b(x)y' + c(x)y = 0$$

can be expressed as

$$y = C_1 y_1 + C_2 y_2$$

where $y_1(x)$ and $y_2(x)$ are any two linearly independent solutions. Sometimes, we can find one solution y_1 by observation and then obtain a second linearly independent solution y_2 by assuming that $y_2 = vy_1$, where $v(x)$ is a twice differentiable function of x. This procedure is called **reduction of order** because it involves solving the given second-order equation by solving two related first-order equations. The basic ideas of reduction of order are illustrated in the following example.

> **EXAMPLE 8** *Reduction of order*

Show that $y = x^2$ is a solution of the equation $y'' - \dfrac{3}{x}y' + \dfrac{4}{x^2}y = 0$ for $x > 0$, then find the general solution.

Solution If $y_1 = x^2$, then $y_1' = 2x$ and $y_1'' = 2$. We have

$$y_1'' - \frac{3}{x}y_1' + \frac{4}{x^2}y_1 = 2 - \frac{3}{x}(2x) + \frac{4}{x^2}(x^2) = 0$$

and y_1 is a solution.

Next, let $y_2 = vx^2$, so

$$y_2' = 2xv + x^2 v' \quad \text{and} \quad y'' = 2v + 4xv' + x^2 v''$$

Substitute these derivatives into the given equation:

$$(2v + 4xv' + x^2 v'') - \frac{3}{x}(2xv + x^2 v') + \frac{4}{x^2}(vx^2) = 0$$

$$x^2 v'' + xv' = 0$$

$$\frac{v''}{v'} = \frac{-1}{x}$$

Integrate both sides of this equation:

$$\ln v' = -\ln x$$

$$v' = \frac{1}{x}$$

$$v = \ln x$$

Thus, $y_2 = vx^2 = x^2 \ln x$ is a second solution. To show that the solutions $y_1 = x^2$ and $y = x^2 \ln x$ are linearly independent, we compute the Wronskian (since $x > 0$):

$$W(x^2, x^2 \ln x) = \begin{vmatrix} x^2 & x^2 \ln x \\ 2x & x + 2x \ln x \end{vmatrix} = x^3 \neq 0$$

It follows that the general solution of the given differential equation is

$$y = C_1 x^2 + C_2 x^2 \ln x \qquad \blacksquare$$

15.2   Problem Set

A *Find the general solution of the second-order homogeneous linear differential equations given in Problems 1–14.*

1. $y'' + y' = 0$

2. $y'' + y' - 2y = 0$

3. $y'' + 6y' + 5y = 0$

4. $y'' + 4y = 0$

5. $y'' - y' - 6y = 0$

6. $y'' + 8y' + 16y = 0$

7. $2y'' - 5y' - 3y = 0$

8. $3y'' + 11y' - 4y = 0$

9. $y'' - y = 0$

10. $6y'' + 13y' + 6y = 0$

11. $y'' + 11y = 0$

12. $y'' - 4y' + 5y = 0$

13. $7y'' + 3y' + 5y = 0$

14. $2y'' + 5y' + 8y = 0$

Find the general solution of the given higher-order homogeneous linear differential equations in Problems 15–20.

15. $y''' + y' = 0$

16. $y''' + 4y' = 0$

17. $y^{(4)} + y''' + 2y'' = 0$

18. $y^{(4)} + 10y'' + 9y = 0$

19. $y''' + 2y'' - 5y' - 6y = 0$

20. $y^{(4)} + 2y''' + 2y'' + 2y' + y = 0$

Find the particular solution that satisfies the differential equations in Problems 21–26 subject to the specified initial conditions.

21. $y'' - 10y' + 25y = 0$; $y(0) = 1, y'(0) = -1$

22. $y'' + 6y' + 9y = 0$; $y(0) = 4, y'(0) = -3$

23. $y'' - 12y' + 11y = 0$; $y(0) = 3, y'(0) = 11$

24. $y'' + 4y' + 5y = 0$; $y(0) = -2, y'(0) = 1$

25. $y''' + 10y'' + 25y' = 0$; $y(0) = 3,$ $y'(0) = 2, y''(0) = -1$

26. $y^{(4)} - y''' = 0$; $y(0) = 3, y'(0) = 0, y''(0) = 3, y'''(0) = 4$

In Problems 27–30, find the Wronskian, W, of the given set of functions and show that $W \neq 0$.

27. $\{e^{-2x}, e^{3x}\}$

28. $\{e^{-x}, xe^{-x}\}$

29. $\{e^{-x}\cos x, e^{-x}\sin x\}$

30. $\{xe^x\cos x, xe^x\sin x\}$

31. Ηɪsᴛᴏʀɪᴄᴀʟ Qᴜᴇsᴛ   Josef Hoëné (1778–1853) adopted the name Wronski when he was 32 years old, around the time he was married. Today, he is remembered for determinants now known as Wronskians, named by Thomas Muir (1844–1934) in 1882. Wronski's main work was in the philosophy of mathematics. For years his mathematical work, which contained many errors, was dismissed as unimportant, but in recent years closer study of his work revealed that he had some significant mathematical insight.

JOSEPH HOËNÉ DE WRONSKI 1778–1853

For this Ηɪsᴛᴏʀɪᴄᴀʟ Qᴜᴇsᴛ you are asked to write a paper on one of the great philosophical issues in the history of mathematics. Here are a few quotations to get you started:

Mathematics is discovered:

"... what is physical is subject to the laws of mathematics, and what is spiritual to the laws of God, and the laws of mathematics are but the expression of the thoughts of God."

Thomas Hill, *The Uses of Mathesis; Bibliotheca Sacra*, p. 523.

"Our remote ancestors tried to interpret nature in terms of anthropomorphic concepts of their own creation and failed. The efforts of our nearer ancestors to interpret nature on engineering lines proved equally inadequate. Nature has refused to accommodate herself to either of these man-made molds. On the other hand, our efforts to interpret nature in terms of the concepts of pure mathematics have, so far, proved brilliantly successful ... from the intrinsic evidence of His creation, the Great Architect of the Universe now begins to appear as a pure mathematician."

James H. Jeans, *The Mysterious Universe*, p. 142.

Mathematics is invented:

"There is an old Armenian saying, 'He who lacks sense of the past is condemned to live in the narrow darkness of his own generation.' Mathematics without history is mathematics stripped of its greatness: for, like the other arts—and mathematics is one of the supreme arts of civilization—it derives it grandeur from the fact of being a human creation."

G. F. Simmons, *Differential Equations with Applications and Historical Notes*, 2nd edition, McGraw-Hill, Inc., 1991, p. xix.

Discuss whether the significant ideas in mathematics are *discovered* or *invented*.

B **32.** If $a^2 = b$, one solution of $y'' - 2ay' + by = 0$ is $y_1 = e^{ax}$. Use reduction of order to show that $y_2 = e^{ax}$ is a second solution, then show that y_1 and y_2 are linearly independent.

In Problems 33–38, a second-order differential equation and one solution $y_1(x)$ are given. Use reduction of order to find a second solution $y_2(x)$, then show that y_1 and y_2 are linearly independent.

33. $y'' + 6y' + 9y = 0$; $y_1 = e^{-3x}$

34. $2y'' - y' - 6y = 0$; $y_1 = e^{2x}$

35. $xy'' + 4y' = 0$; $y_1 = 1$

36. $x^2y'' + xy' - 4y = 0$; $y_1 = x^{-2}$

37. $x^2y'' + 2xy' - 12y = 0$; $y_1 = x^3$

38. $(1 - x)^2y'' - (1 - x)y' - y = 0$; $y_1 = 1 - x$

Suppose a 16-lb weight stretches a spring 8 in. from its natural length. Find a formula for the position of the weight as a function of time for the situations described in Problems 39–44.

39. The weight is pulled down an additional 6 in. and is then released with an initial upward velocity of 8 ft/s.

40. The weight is pulled down 10 in. and is released with an initial upward velocity of 6 ft/s.

41. The weight is raised 8 in. above the equilibrium point and the compressed spring is then released.

42. The weight is raised 12 in. above the equilibrium point and the compressed spring is then released.

43. The weight is pulled 6 in. below the equilibrium point and is then released. The weight is connected to a dashpot that imposes a damping force of magnitude 0.4|v| at all times.

44. The weight is pulled 12 in. below the equilibrium point and is then released. The weight is connected to a dashpot that imposes a damping force of magnitude 0.08|v| at all times.

45. Characterize the solution of the second order equation

$$y'' - 2ay' + (a^2 - b)y = 0 \qquad b > 0$$

in terms of the functions $\cosh kx$, $\sinh kx$, $\cos kx$, and $\sin kx$, where $k = \sqrt{b}$.

46. Verify that $y_1 = e^{2x}$, $y_2 = xe^{2x}$, and $y_3 = x^2 e^{2x}$ are linearly independent solutions of the third-order equation

$$y''' - 6y'' + 12y' - 8y = 0$$

47. MODELING PROBLEM   A 100-lb object is projected vertically upward from the surface of the earth with initial velocity 150 ft/s.
   a. Modeling the object's motion with negligible air resistance, how long does it take for the object to return to earth?
   b. Change the model to assume air resistance equal to half the object's velocity. Before making any computation, does your intuition tell you the object takes less or more time to return to earth this time than in part **a**? Now set up and solve a differential equation to actually determine the round-trip time. Were you right?

Ⓒ 48. MODELING PROBLEM   The motion of a pendulum subject to frictional damping proportional to its velocity is modeled by the differential equation

$$mL\frac{d^2\theta}{dt^2} + kL\frac{d\theta}{dt} + mg\sin\theta = 0$$

where L is the length of the pendulum, m is the mass of the bob at its end, and θ is the angle the

■ FIGURE 15.7

pendulum arm makes with the vertical (see Figure 15.7). Assume θ is small, so $\sin\theta$ is approximately equal to θ.
   a. Solve the resulting differential equation for the case where $k^2 \geq 4gm^2/L$. What happens to $\theta(t)$ as $t \to \infty$?
   b. If $k^2 < 4gm^2/L$, show that

$$\theta(t) = Ae^{-kt/2m}\cos\left(\sqrt{\frac{B}{L}}t + C\right)$$

for constants A, B, and C. What happens to $\theta(t)$ as $t \to \infty$ in this case?
   c. For the situation in part **b**, show that the time difference between successive vertical positions is approximately

$$T = 2\pi m\sqrt{\frac{L}{4gm^2 - k^2L}}$$

49. If there is no damping and no external forces, the motion of an object of mass m attached to a spring with spring constant k is governed by the differential equation $mx'' + kx = 0$. Show that the general solution of this equation is given by

$$x(t) = A\cos\left(\sqrt{\frac{k}{m}}t - B\right), \quad k > 0$$

where A and B are constants. This is called **simple harmonic motion** with frequency $\dfrac{1}{2\pi}\sqrt{\dfrac{k}{m}}$.

50. Consider the motion of an object of mass m on a spring with spring constant k_1 and damping constant k_2 for the case where there is critical damping.
   a. Describe the motion of the object assuming that it begins at rest: $x'(0) = 0$ at $x(0) = x_0$. How is this different from the case where the object begins at $x(0) = 0$ with initial velocity $x'(0) = v_0$? Sketch both solutions on the same graph. How do initial velocity and initial displacement affect the motion?
   b. If $x(0) = x'(0)$, what is the maximum displacement? Show that the time at which the maximum displacement occurs is independent of the initial displacement $x(0)$. Assume $x(0) \neq 0$.

51. Suppose the characteristic equation of the differential equation $y'' + ay' + by = 0$ has complex conjugate roots, $r_1 = \alpha + \beta i$ and $r_2 = \alpha - \beta i$. Show that $y_1 = e^{\alpha x}\cos\beta x$ and $y_2 = e^{\alpha x}\sin\beta x$ are both solutions and are linearly independent. This is Theorem 15.3 for the case where $a^2 - 4b < 0$.

52. PENDULUM MOTION   Suppose a ball of mass m is suspended at the end of a rod of length L and is set in motion swinging back and forth like a pendulum, as shown in Figure 15.8.

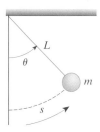

■ FIGURE 15.8   Pendulum motion

Let θ be the angle between the rod and the vertical at time t, so that the displacement of the ball from the equilibrium position is $s = L\theta$ and the acceleration of the ball's motion is

$$\frac{d^2s}{dt^2} = L\frac{d^2\theta}{dt^2}$$

a. Use Newton's second law to show that

$$mL\frac{d^2\theta}{dt^2} + mg\sin\theta = 0$$

Assume that air resistance and the mass of the rod are negligible.

b. When the displacement is small (θ close to 0), $\sin\theta$ may be replaced by θ. In this case, solve the resulting differential equation

$$\frac{d^2\theta}{dt^2} + \frac{g}{L}\theta = 0$$

How is the motion of the pendulum like the simple harmonic motion discussed in Problem 49?

53. **PATH OF A PROJECTILE WITH VARIABLE MASS**
A rocket starts from rest and moves vertically upward, along a straight line. Assume the rocket and its fuel initially weigh w kilograms and that the fuel alone weighs w_f kilograms. Further assume that the fuel is consumed at a constant rate of r kilograms per second (relative to the rocket). Finally, assume that gravitational attraction is the only external force acting on the rocket.

a. If $m(t)$ is the mass of the rocket at time t and $s(t)$ is the height above the ground at time t, it can be shown that

$$m(t)s''(t) + m'(t)v_0 + m(t)g = 0$$

where v_0 is the velocity of the exhaust gas in relation to the rocket. Express $m(t)$ in terms of w, r, v_0, and g, then integrate this differential equation to obtain the velocity $s'(t)$. Note that $s'(0) = 0$ because the rocket starts from rest.

b. Integrate the velocity $s'(t)$ to obtain $s(t)$.

c. At what time is all the fuel consumed?

d. How high is the rocket at the instant the fuel is consumed?

54. **JOURNAL PROBLEM** by Murray S. Klamkin.* Solve the differential equation

$$x^4y'' - (x^3 + 2axy)y' + 4ay^2 = 0$$

*Problem 331 in the *Canadian Mathematical Bulletin*, Vol. 26, 1983, p. 126.

15.3 Second-Order Nonhomogeneous Linear Differential Equations

IN THIS SECTION nonhomogeneous equations, method of undetermined coefficients, variation of parameters, an application to *RLC* circuits ■

NONHOMOGENEOUS EQUATIONS

Next we shall see how to solve a nonhomogeneous second-order linear equation of the general form $y'' + ay' + by = F(x)$. The key to our results is the following theorem.

THEOREM 15.4 **Characterization of the general solution of**
$$y'' + ay' + by = F(x)$$

Let y_p be a particular solution of the nonhomogeneous second-order linear equation $y'' + ay' + by = F(x)$. Let y_h be the general solution of the related homogeneous equation $y'' + ay' + by = 0$. Then the general solution of $y'' + ay' + by = F(x)$ is given by the sum

$$y = y_h + y_p$$

Proof First, the sum $y = y_h + y_p$ is a solution of the nonhomogeneous equation $y'' + ay' + by = F(x)$, because

$$
\begin{aligned}
y'' + ay' + by &= (y_h + y_p)'' + a(y_h + y_p)' + b(y_h + y_p) \\
&= y_h'' + y_p'' + ay_h' + ay_p' + by_h + by_p \\
&= (y_h'' + ay_h' + by_h) + (y_p'' + ay_p' + by_p) \\
&= 0 + F(x) \\
&= F(x)
\end{aligned}
$$

Conversely, if y is any solution of the nonhomogeneous equation, then $y - y_p$ is a solution of the related homogeneous equation because

$$(y - y_p)'' + a(y - y_p)' + b(y - y_p) = (y'' - y_p'') + a(y' - y_p') + b(y - y_p)$$
$$= (y'' + ay' + by) - (y_p'' + ay_p' + by_p)$$
$$= F(x) - F(x)$$
$$= 0$$

Thus, $y - y_p = y_h$ (because it is a solution of the homogeneous equation). Therefore, because y was *any* solution of the nonhomogeneous equation, it follows that $y = y_h + y_p$ is the general solution of the nonhomogeneous equation.

> ■ *What This Says:* We can obtain the general solution of the nonhomogeneous equation $y'' + ay' + by = F(x)$ by finding the general solution y_h of the related homogeneous equation $y'' + ay' + by = 0$ and just one particular solution y_p of the given nonhomogeneous equation.

We can use the methods of the preceding section to find the general solution of the related homogeneous equation. We now develop two methods for finding particular solutions of the nonhomogeneous equation.

METHOD OF UNDETERMINED COEFFICIENTS

Sometimes it is possible to find a particular solution y_p of the nonhomogeneous equation $y'' + ay' + by = F(x)$ by assuming a **trial solution** $\bar{y}_p$ of the same general form as $F(x)$. This procedure, called the **method of undetermined coefficients,** is illustrated in the following three examples, each of which has the related homogeneous equation $y'' + y' - 2y = 0$ with the general solution $y_h = C_1 e^x + C_2 e^{-2x}$.

EXAMPLE 1   *Method of undetermined coefficients*

Find $\bar{y}_p$ and the general solution for $y'' + y' - 2y = 2x^2 - 4x$.

Solution   The right side $F(x) = 2x^2 - 4x$ is a quadratic polynomial. Because derivatives of a polynomial are polynomials of lower degree, it seems reasonable to consider a trial solution that is also a polynomial of degree 2. That is, we "guess" that this equation has a particular solution $\bar{y}_p$ of the general form $\bar{y}_p = A_1 x^2 + A_2 x + A_3$. To find the constants A_1, A_2, and A_3, calculate

$$\bar{y}_p = 2A_1 x + A_2 \quad \text{and} \quad \bar{y}_p'' = 2A_1$$

Substitute the values for $\bar{y}_p, \bar{y}_p'$, and $\bar{y}_p''$ into the given equation:

$$y'' + y' - 2y = 2x^2 - 4x$$
$$2A_1 + 2A_1 x + A_2 - 2(A_1 x^2 + A_2 x + A_3) = 2x^2 - 4x$$
$$-2A_1 x^2 + (2A_1 - 2A_2)x + (2A_1 + A_2 - 2A_3) = 2x^2 - 4x$$

This is true only when the coefficients of each power of x on each side of the equation match, so that

$$\begin{cases} -2A_1 = 2 & (x^2 \text{ terms}) \\ 2A_1 - 2A_2 = -4 & (x \text{ terms}) \\ 2A_1 + A_2 - 2A_3 = 0 & (\text{constant terms}) \end{cases}$$

Solve this system of equations simultaneously to find $A_1 = -1, A_2 = 1$, and $A_3 = -\frac{1}{2}$. Thus, a particular solution of the given nonhomogeneous equation is

$$y_p = A_1x^2 + A_2x + A_3 = -x^2 + x - \frac{1}{2}$$

and the general solution for the nonhomogeneous equation is

$$y = y_h + y_p = C_1 e^x + C_2 e^{-2x} - x^2 + x - \frac{1}{2} \qquad \blacksquare$$

COMMENT   Notice that even though the constant term is zero in the polynomial function $F(x)$, we cannot assume that $y = A_1x^2 + A_2x$ is a suitable trial solution. In general, all terms of lower degree that could possibly lead to the given right-side function $F(x)$ must be included in the trial solution.

> **EXAMPLE 2**   *Method of undetermined coefficients*

Solve $y'' + y' - 2y = \sin x$.

Solution   Because the trial solution $\bar{y}_p$ is to be "like" the right-side function $F(x) = \sin x$, it seems that we should choose the trial solution to be $\bar{y}_p = A_1 \sin x$, but a sine function can have either a sine or a cosine in its derivatives, depending on how many derivatives are taken. Thus, to account for the $\sin x$, it is necessary to have *both* $\sin x$ and $\cos x$ in the trial solution, so we set

$$\bar{y}_p = A_1 \sin x + A_2 \cos x$$

Differentiating, we find

$$\bar{y}_p' = A_1 \cos x - A_2 \sin x \quad \text{and} \quad \bar{y}_p'' = -A_1 \sin x - A_2 \cos x$$

Substitute the values into the given equation:

$$y'' + y' - 2y = \sin x$$
$$(-A_1 \sin x - A_2 \cos x) + (A_1 \cos x - A_2 \sin x) - 2(A_1 \sin x + A_2 \cos x) = \sin x$$
$$(-3A_1 - A_2)\sin x + (A_1 - 3A_2) \cos x = \sin x$$

This gives the system

$$\begin{cases} -3A_1 - A_2 = 1 & (\sin x \text{ terms}) \\ A_1 - 3A_2 = 0 & (\cos x \text{ terms}) \end{cases}$$

with the solution $A_1 = -\frac{3}{10}, A_2 = -\frac{1}{10}$. Thus, the particular solution of the nonhomogeneous equation is $y_p = -\frac{3}{10} \sin x - \frac{1}{10} \cos x$, and the general solution is $y = y_h + y_p = C_1e^x + C_2e^{-2x} - \frac{3}{10} \sin x - \frac{1}{10} \cos x$. $\blacksquare$

COMMENT   It can be shown that if y_1 is a solution of $y'' + ay' + by = F(x)$ and y_2 is a solution of $y'' + ay' + by = G(x)$, then $y_1 + y_2$ will be a solution of $y'' + ay' + by = F(x) + G(x)$. (See Problem 55.) This is called the **principle of superposition.** For instance, by combining the results of Examples 1 and 2, we see that a particular solution of the nonhomogeneous linear equation $y'' + y' - 2y = 2x^2 - 4x + \sin x$ is

$$y_p = \overbrace{-x^2 + x - \frac{1}{2}}^{\text{Solution for } F = 2x^2 - 4x} \underbrace{-\frac{3}{10} \sin x - \frac{1}{10} \cos x}_{\text{Solution for } G = \sin x}$$

> **EXAMPLE 3**   *Method of undetermined coefficients*

Solve $y'' + y' - 2y = 4e^{-2x}$.

Solution   At first glance (looking at $F(x) = 4e^{-2x}$), it may seem that the trial solution should be $\bar{y}_p = Ae^{-2x}$. To see why this cannot be a solution, note that because $y_1 = e^{-2x}$ is a solution of the related homogeneous equation, we have

$$y_1'' + y_1' - 2y_1 = 0$$

which results in the unsolvable equation

$$0 = 4e^{-2x}$$

To deal with this situation, multiply the usual trial solution by x and consider the trial solution $\bar{y}_p = Axe^{-2x}$. Differentiating, we find

$$\bar{y}_p' = A(1 - 2x)e^{-2x} \quad \text{and} \quad \bar{y}_p'' = A(4x - 4)e^{-2x}$$

and by substituting into the given equation, we obtain

$$y'' + y' - 2y = 4e^{-2x}$$
$$A(4x - 4)e^{-2x} + A(1 - 2x)e^{-2x} - 2Axe^{-2x} = 4e^{-2x}$$
$$(4Ax - 4A + A - 2Ax - 2Ax)e^{-2x} = 4e^{-2x}$$
$$-3Ae^{-2x} = 4e^{-2x}$$
$$A = -\tfrac{4}{3}$$

Thus, $\bar{y}_p = -\tfrac{4}{3}xe^{-2x}$, so that the general solution is

$$y = y_h + y_p = C_1 e^x + C_2 e^{-2x} - \tfrac{4}{3}xe^{-2x}$$

■

The procedure illustrated in Examples 1–3 can be applied to a differential equation $y'' + ay' + by = F(x)$ only when $F(x)$ has one of the following forms:

   a. $F(x) = P_n(x)$, a polynomial of degree n
   b. $F(x) = P_n(x)e^{kx}$
   c. $F(x) = e^{kx}[P_n(x) \cos \alpha x + Q_n(x) \sin \alpha x]$, where
   $\qquad Q_n(x)$ is another polynomial of degree n

We can now describe the **method of undetermined coefficients.**

Method of Undetermined Coefficients

To solve $y'' + ay' + by = F(x)$ when $F(x)$ is one of the forms listed above:

   1. The solution is of the form $y = y_h + y_p$, where y_h is the general solution and y_p is a particular solution.

   2. Find y_h by solving the homogeneous equation

$$y'' + ay' + by = 0$$

   3. Find y_p by picking an appropriate trial solution $\bar{y}_p$:

Form of $F(x)$	Corresponding trial expression $\bar{y}_p$
a. $P_n(x) = c_n x^n + \cdots + c_1 x + c_0$	$A_n x^n + \cdots + A_1 x + A_0$
b. $P_n(x)e^{kx}$	$[A_n x^n + A_{n-1}x^{n-1} + \cdots + A_0]e^{kx}$
c. $e^{kx}[P_n(x) \cos \alpha x + Q_n(x) \sin \alpha x]$	$e^{kx}[(A_n x^n + \cdots + A_0) \cos \alpha x$ $+ (B_n x^n + \cdots + B_0) \sin \alpha x]$

> **4.** If no term in the trial expression $\bar{y}_p$ appears in the general homogeneous solution y_h, the particular solution can be found by substituting $\bar{y}_p$ into the equation $y'' + ay' + by = F(x)$ and solving for the undetermined coefficients.
>
> **5.** If any term in the trial expression $\bar{y}_p$ appears in y_h, multiply $\bar{y}_p$ by x^k, where k is the smallest integer such that no term in $x^k\bar{y}_p$ is a solution of $y'' + ay' + by = 0$. Then proceed as in step 4, using $x^k\bar{y}_p$ as the trial solution.

EXAMPLE 4   *Finding trial solutions*

Determine a suitable trial solution for undetermined coefficients in each of the given cases.

a. $y'' - 4y' + 4y = 3x^2 + 4e^{-2x}$ b. $y'' - 4y' + 4y = 5xe^{2x}$

c. $y'' + 2y' + 5y = 3e^{-x}\cos 2x$

Solution

a. The related homogeneous equation $y'' - 4y' + 4y = 0$ has the characteristic equation $r^2 - 4r + 4 = 0$, which has the root 2 of multiplicity two. Thus, the general homogeneous solution is

$$y_h = C_1 e^{2x} + C_2 x e^{2x}$$

The part of the trial solution for the nonhomogeneous equation that corresponds to $3x^2$ is $A_0 + A_1 x + A_2 x^2$ and the part that corresponds to $4e^{-2x}$ is Be^{-2x}. Since neither part includes terms in y_h, we apply the principle of superposition to conclude that

$$\bar{y}_p = A_0 + A_1 x + A_2 x^2 + Be^{-2x}$$

b. We know from part **a** that the general homogeneous solution is

$$y_h = C_1 e^{2x} + C_2 x e^{2x}$$

The normal trial solution for $5xe^{2x}$ would be $(A_0 + A_1 x)e^{2x}$, but part of this expression is contained in y_h. If we multiply by x, part of $(A_0 + A_1 x)xe^{2x}$ is still contained in y_h, so we multiply by x again to obtain

$$\bar{y}_p = (A_0 + A_1 x)x^2 e^{2x}$$

c. The related homogeneous equation $y'' + 2y' + 5y = 0$ has the characteristic equation $r^2 + 2r + 5 = 0$. This has complex conjugate roots $r = -1 \pm 2i$, so the general homogeneous solution is

$$y_h = e^{-x}[C_1\cos 2x + C_2\sin 2x]$$

Ordinarily the trial solution for the nonhomogeneous equation would be of the form $e^{-x}[A\cos 2x + B\sin 2x]$, but part of this expression is in y_h. Therefore, we multiply by x to obtain the trial solution

$$y = xe^{-x}[A\cos 2x + B\sin 2x]$$ ∎

VARIATION OF PARAMETERS

The method of undetermined coefficients applies only when the coefficients b and c are constant in the nonhomogeneous linear equation $y'' + by' + cy = F(x)$ and the driving function $F(x)$ has the same general form as a solution of a second-order homogeneous linear equation with constant coefficients. Even though many important applications are modeled by differential equations of this type, there are other situations that require a more general procedure.

Our next goal is to examine a method of J. L. Lagrange (see Historical Quest, Problem 54 in Section 12.8) called **variation of parameters,** which can be used to find a particular solution of any nonhomogeneous equation

$$y'' + P(x)y' + Q(x)y = F(x)$$

where $P, Q,$ and F are continuous.

To use variation of parameters, we must be able to find two linearly independent solutions $y_1(x)$ and $y_2(x)$ of the related homogeneous equation,

$$y'' + P(x)y' + Q(x)y = 0$$

In practice, if $P(x)$ and $Q(x)$ are not both constants, these may be difficult to find, but once we have them, we assume there is a solution of the nonhomogeneous equation of the form

$$y_p = uy_1 + vy_2$$

Differentiating this expression y_p, we obtain

$$y_p' = u'y_1 + v'y_2 + uy_1' + vy_2'$$

To simplify, assume that

$$u'y_1 + v'y_2 = 0$$

Remember, we need to find only *one* particular solution y_p, and if imposing this side condition makes it easier to find such a y_p, so much the better! With the side condition, we have

$$y_p' = uy_1' + vy_2'$$

and by differentiating again, we obtain

$$y_p'' = uy_1'' + u'y_1' + vy_2'' + v'\, y_2'$$

Next, we substitute our expressions for y_p' and y_p'' into the given differential equation:

$$F(x) = y_p'' + P(x)y_p' + Q(x)y_p$$
$$= (uy_1'' + u'\, y_1' + vy_2'' + v'\, y_2') + P(x)(uy_1' + vy_2') + Q(x)(uy_1 + vy_2)$$

This can be rewritten as

$$u[y_1'' + P(x)y_1' + Q(x)y_1] + v[y_2'' + P(x)y_2' + Q(x)y_2] + u'\, y_1' + v'\, y_2' = F(x)$$

Because y_1 and y_2 are solutions of $y'' + P(x)y' + Q(x)y = 0$, we have

$$u\underbrace{[y_1'' + P(x)y_1' + Q(x)y_1]}_{0} + v\underbrace{[y_2'' + P(x)y_2' + Q(x)y_2]}_{0} + u'y_1' + v'y_2' = F(x)$$

or

$$u'y_1' + v'y_2' = F(x)$$

Thus, the parameters u and v must satisfy the system of equations

$$\begin{cases} u'\, y_1 + v'\, y_2 = 0 \\ u'\, y_1' + v'\, y_2' = F(x) \end{cases}$$

Solve this system to obtain

$$u' = \frac{-y_2 F(x)}{y_1 y_2' - y_2 y_1'} \quad \text{and} \quad v' = \frac{y_1 F(x)}{y_1 y_2' - y_2 y_1'}$$

where in each case the denominator is not zero, because it is the Wronskian of the linearly independent solutions y_1, y_2 of the related homogeneous equation. Integrating, we find

$$u(x) = \int \frac{-y_2 F(x)}{y_1 y_2' - y_2 y_1'} \, dx \quad \text{and} \quad v(x) = \int \frac{y_1 F(x)}{y_1 y_2' - y_2 y_1'} \, dx$$

and by substituting into the expression

$$y_p = uy_1 + vy_2$$

we obtain a particular solution of the given differential equation. Here is a summary of the procedure we have described.

Variation of Parameters

To find the general solution of $y'' + P(x)y' + Q(x)y = F(x)$:

Step 1. Find the general solution, $y_h = C_1 y_1 + C_2 y_2$ to the homogeneous equation.

Step 2. Set $y_p = uy_1 + vy_2$ and substitute into the formulas:

$$u' = \frac{-y_2 F(x)}{y_1 y_2' - y_2 y_1'} \qquad v' = \frac{y_1 F(x)}{y_1 y_2' - y_2 y_1'}$$

Step 3. Integrate u' and v' to find u and v.

Step 4. A particular solution is $y_p = uy_1 + vy_2$, and the general solution is $y = y_h + y_p$.

These ideas are illustrated in the next example.

EXAMPLE 5   *Variation of parameters*

Solve $y'' + 4y = \tan 2x$.

Solution   Notice that this problem cannot be solved by undetermined coefficients, because the right-side function $F(x) = \tan 2x$ is not of the form of a solution of a homogeneous linear equation with constant coefficients.

To apply variation of parameters, begin by solving the related homogeneous equation $y'' + 4y = 0$. The characteristic equation is $r^2 + 4 = 0$ with roots $r = \pm 2i$. These complex roots have $\alpha = 0$ and $\beta = 2$ so that the general solution is

$$y = e^0[C_1 \cos 2x + C_2 \sin 2x] = C_1 \cos 2x + C_2 \sin 2x$$

This means $y_1(x) = \cos 2x$ and $y_2(x) = \sin 2x$. Set $y_p = uy_1 + vy_2$, where

$$u' = \frac{-y_2 F(x)}{y_1 y_2' - y_2 y_1'} = \frac{-\sin 2x \tan 2x}{2 \cos 2x \cos 2x + 2 \sin 2x \sin 2x} = -\frac{\sin^2 2x}{2 \cos 2x}$$

and

$$v' = \frac{-y_1 F(x)}{y_1 y_2' - y_2 y_1'} = \frac{\cos 2x \tan 2x}{2 \cos 2x \cos 2x + 2 \sin 2x \sin 2x} = \frac{1}{2} \sin 2x$$

Integrating, we obtain

$$u(x) = \int -\frac{\sin^2 2x}{2 \cos 2x} \, dx \qquad\qquad v(x) = \int \frac{1}{2} \sin 2x \, dx$$

$$= -\frac{1}{2} \int \frac{1 - \cos^2 2x}{\cos 2x} \, dx \qquad\qquad = -\frac{1}{4} \cos 2x + C_2$$

$$= -\frac{1}{2} \int (\sec 2x - \cos 2x) \, dx$$

$$= -\frac{1}{2} \left[\frac{1}{2} \ln |\sec 2x + \tan 2x| - \frac{\sin 2x}{2} \right] + C_1$$

Thus, a particular solution is

$$y_p = u y_1 + v y_2$$

$$= \left[-\frac{1}{4} \ln |\sec 2x + \tan 2x| + \frac{1}{4} \sin 2x \right] \cos 2x + \left(-\frac{1}{4} \cos 2x \right) \sin 2x$$

$$= -\frac{1}{4} (\cos 2x) \ln |\sec 2x + \tan 2x|$$

Finally, the general solution is

$$y = y_h + y_p = C_1 \cos 2x + C_2 \sin 2x - \tfrac{1}{4} (\cos 2x) \ln |\sec 2x + \tan 2x| \qquad\blacksquare$$

AN APPLICATION TO *RLC* CIRCUITS

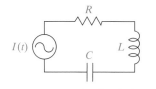

■ **FIGURE 15.9** An *RLC* circuit

An important application of second-order linear differential equations is the analysis of electric circuits. Consider a circuit with constant resistance R, inductance L, and capacitance C. Such a circuit is called an **RLC circuit** and is illustrated in Figure 15.9.

If $I(t)$ is the current in the circuit at time t and $Q(t)$ is the total charge on the capacitor, it is shown in physics that IR is the voltage drop across the resistance, Q/C is the voltage drop across the capacitor, and $L \, dI/dt$ is the voltage drop across the inductance. According to Kirchhoff's second law for circuits, the impressed voltage $E(t)$ in a circuit is the sum of the voltage drops, so that the current $I(t)$ in the circuit satisfies

$$L \frac{dI}{dt} + RI + \frac{Q}{C} = E(t)$$

By differentiating both sides of this equation and using the fact that $I = \dfrac{dQ}{dt}$, we can write

$$L \frac{d^2 I}{dt^2} + R \frac{dI}{dt} + \frac{1}{C} I = \frac{dE}{dt}$$

For instance, suppose the voltage input is sinusoidal—that is, $E(t) = A \sin \omega t$. Then we have $dE/dt = A\omega \cos \omega t$, and the second-order linear differential equation used in the analysis of the circuit is

$$L\frac{d^2I}{dt^2} + R\frac{dI}{dt} + \frac{1}{C}I = A\omega\cos\omega t$$

To solve this equation, we proceed in the usual way, solving the related homogeneous equation and then finding a particular solution of the nonhomogeneous system. The solution to the related homogeneous equation is called the **transient circuit** because the current described by this solution usually does not last very long. The part of the nonhomogeneous solution that corresponds to transient current 0 is called the **steady-state current.** Several problems dealing with RLC circuits are outlined in the problem set.

15.3 Problem Set

Ⓐ *In Problems 1–8, find a trial solution $\bar{y}_p$ for use in the method of undetermined coefficients.*

1. $y'' + 6y' = e^{2x}$

2. $y'' + 6y' + 8y = 2 + e^{-3x}$

3. $y'' + 2y' + 2y = e^{-x}$

4. $y'' + 2y' + 2y = \cos x$

5. $y'' + 2y' + 2y = e^{-x}\sin x$

6. $2y'' - y' - 6y = x^2e^{2x}$

7. $y'' + 4y' + 5y = e^{-2x}(x + \cos x)$

8. $y'' + 4y' + 5y = (e^{-x}\sin x)^2$

For the differential equation

$$y'' + 6y' + 9y = F(x)$$

find a trial solution $\bar{y}_p$ for undertermined coefficients for each choice of $F(x)$ given in Problems 9–16.

9. $3x^3 - 5x$

10. $2x^2e^{4x}$

11. $x^3\cos x$

12. $xe^{2x}\sin 5x$

13. $e^{2x} + \cos 3x$

14. $2e^{3x} + 8xe^{-3x}$

15. $4x^3 - x^2 + 5 - 3e^{-x}$

16. $(x^2 + 2x - 6)e^{-3x}\sin 3x$

Use the method of undetermined coefficients to find the general solution of the nonhomogeneous differential equations given in Problems 17–28.

17. $y'' + y' = -3x^2 + 7$

18. $y'' + 6y' + 5y = 2e^x - 3e^{-3x}$

19. $y'' + 8y' + 15y = 3e^{2x}$

20. $2y'' - 5y' - 3y = 5e^{3x}$

21. $y'' + 2y' + 2y = \cos x$

22. $y'' - 6y + 13y = e^{-3x}\sin 2x$

23. $7y'' + 6y' - y = e^{-x}(x + 1)$

24. $2y'' + 5y' = e^x\sin x$

25. $y'' - y' = x^3 - x + 5$

26. $y'' - y' = (x - 1)e^x$

27. $y'' + 2y' + y = (4 + x)e^{-x}$

28. $y'' - y' - 6y = e^{-2x} + \sin x$

Use variation of parameters to solve the differential equations given in Problems 29–38.

29. $y'' + y' = \tan x$

30. $y'' + 8y' + 16y = xe^{-2x}$

31. $y'' - y' - 6y = x^2e^{2x}$

32. $y'' - 3y' + 2y = \dfrac{e^x}{1 + e^x}$

33. $y'' + 4y = \sec 2x\tan 2x$

34. $y'' + y = \sec^2x$

35. $y'' + 2y' + y = e^{-x}\ln x$

36. $y'' - 4y' + 4y = \dfrac{e^{2x}}{1 + x}$

37. $y'' - y' = \cos^2x$

38. $y'' - y = e^{-2x}\cos e^{-x}$

Ⓑ *In each of Problems 39–45, use either undetermined coefficients or variation of parameters to find the particular solution of the differential equation that satisfies the specified initial conditions.*

39. $y'' - y = 2\cos^2x; y(0) = 0, y'(0) = 1$

40. $y'' - y = e^{-2x}\cos e^{-x}; y(0) = y'(0) = 0$

41. $y'' + 9y = 4e^{3x}; y(0) = 0, y'(0) = 2$

42. $y'' - 6y = x^2 - 3x; y(0) = 3, y'(0) = -1$

43. $y'' + 9y = x; y(0) = 0, y'(0) = 4$

44. $y'' + y' = 2\sin x; y(0) = 0, y'(0) = -4$

45. $y'' + y = \cot x; y\left(\dfrac{\pi}{2}\right) = 0, y'\left(\dfrac{\pi}{2}\right) = 5$

46. Find the general solution of the differential equation $y'' + y' - 6y = F(x)$, where F is the function whose graph is shown in Figure 15.10.

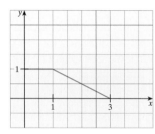

■ **FIGURE 15.10**   Graph of F for Problem 46

47. Find the general solution of the differential equation $y'' + 5y' + 6y = F(x)$, where F is the function whose graph is shown in Figure 15.11.

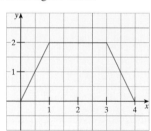

■ **FIGURE 15.11**   Graph of F for Problem 47

48. Find the steady-state current and the transient current in an RLC circuit with $L = 4$ henries, $R = 8$ ohms, $C = \dfrac{1}{8}$ farad, and $E(t) = 16\sin t$ volts. You may assume that $I(0) = 0$ and $I'(0) = 0$.

49. Find the steady-state current and the transient current in an RLC circuit with $L = 1$ henry, $R = 10$ ohms, $C = \dfrac{1}{9}$ farad, and $E(t) = 9\sin t$ volts. You may assume that $I(0) = 0$ and $I'(0) = 0$.

50. Work Problem 48 for the case where $E(t) = 10te^{-t}$.

51. Work Problem 49 for the case where $E(t) = 5t\sin t$.

C 52. Theorem 15.4 is also valid when the coefficients a and b are continuous functions of x. Solve the differential equation

$$x^2 y'' - 3xy' + 4y = x\ln x$$

by completing the following steps:

a. The related homogeneous equation

$$x^2 y'' - 3xy' + 4y = 0$$

has one solution of the form $y_1 = x^n$. Find n, then use reduction of order to find a second, linearly independent solution y_2.

b. Use variation of parameters to find a particular solution y_p of the given nonhomogeneous equation. Then use Theorem 15.4 to write the general solution. *Hint:* Divide the D. E. by x^2.

53. Repeat Problem 52 for the equation

$$(1 + x^2)y'' - 2xy' - 2y = \tan^{-1}x$$

54. **THE PRINCIPLE OF SUPERPOSITION.**   Let y_1 be a solution to the second-order linear differential equation

$$y'' + P(x)y' + Q(x)y = F_1(x)$$

and let y_2 satisfy

$$y'' + P(x)y' + Q(x)y = F_2(x)$$

Show that $y_1 + y_2$ satisfies the differential equation

$$y'' + P(x)y' + Q(x)y = F_1(x) + F_2(x)$$

Chapter 15   Review

Proficiency Examination

Concept Problems

1. What is a separable differential equation?
2. What is a homogeneous differential equation?
3. What is the form of a first-order linear differential equation?
4. What is an exact differential equation?
5. Describe Euler's method.
6. Define what it means for a set of functions to be linearly independent.
7. What is the Wronskian, and how is it used to test for linear independence?
8. a. What is the characteristic equation of $y'' + ay' + by = 0$?
   b. What is the general solution of a second-order homogeneous equation?
9. Describe the form of the general solution of a second-order nonhomogeneous equation.
10. Describe the method of undetermined coefficients.
11. Describe the method of variation of parameters.

Practice Problems

Solve the differential equations in Problems 12–18.

12. $\dfrac{dy}{dx} = \sqrt{\dfrac{1 - y^2}{1 + x^2}}$

13. $\dfrac{x}{y^2}\,dx - \dfrac{x^2}{y^3}\,dy = 0$

14. $\dfrac{dy}{dx} = \dfrac{2x + y}{3x}$

15. $xy\,dy = (x^2 - y^2)\,dx$

16. $x^2\,dy - (x^2 + y^2)\,dx = 0$

17. $y'' + 2y' + 2y = \sin x$

18. $(3x^2e^{-y} + y^{-2} + 2xy^{-3})\,dx +$
    $(-x^3e^{-y} - 2xy^{-3} - 3x^2y^{-4})\,dy = 0$

19. A spring with spring constant $k = 30$ lb/ft hangs in a vertical position with its upper end fixed and an 8-lb object attached to its lower end. The object is pulled down 4 in. from the equilibrium position of the spring and is then released. Find the displacement $x(t)$ of the object, assuming that air resistance is $0.8v$, where $v(t)$ is the velocity of the object at time t.

20. An *RLC* circuit has inductance $L = 0.1$ henry, resistance $R = 25$ ohms, and capacitance $C = 200$ microfarads (i.e., 200×10^{-6} farad). If there is a variable voltage source of $E(t) = 50 \cos 100t$ in the circuit, what is the current $I(t)$ at time t? Assume that when $t = 0$, there is no charge and no current flowing.

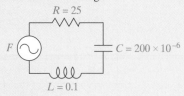

Supplementary Problems

1. a. Draw some isoclines of the differential equation
   $$\frac{dy}{dx} = 2x - 3y$$
   Use the direction field to sketch the particular solution that passes through $(0, 1)$.
   b. Use Euler's method to find a solution for $0 \le x \le 1$ and $h = 0.2$.
2. a. Draw some isoclines of the differential equation
   $$\frac{dy}{dx} = x^2y$$

   Use the direction field to sketch the particular solution that passes through $(1, 2)$.
   b. Use Euler's method to find a solution for $1 \le x \le 2$ and $h = 0.2$.

Find the general solution of the given first-order differential equations in Problems 3–20.

3. $\dfrac{dy}{dx} = \dfrac{-4x}{y^3}$

4. $x\,dy + (3y - xe^{x^2})\,dx = 0$

5. $y^2\,dy - \left(x^2 + \dfrac{y^3}{x}\right)dx = 0$

6. $\dfrac{x}{y}\,dy + \left(\dfrac{3y}{x} - 5\right)dx = 0$

7. $y\dfrac{dy}{dx} = e^{2x-y^2}$

8. $dy = (3y + e^x + \cos x)\,dx$

9. $dy - (5y + e^{5x}\text{six }x)\,dx = 0$

10. $x^2\,dy = (x^2 - y^2)\,dx$

11. $(1 - xe^y)\,dy = e^y\,dx$

12. $xy\,dx + (1 + x^2)\,dy = 0$

13. $x\,dy = (y + \sqrt{x^2 - y^2})\,dx$

14. $2xye^{x^2}dx - e^{x^2}\,dy = 0$

15. $(y - xy)\,dx + x^3\,dy = 0$

16. $\dfrac{dy}{dx} = 2y\cot 2x + 3\csc 2x$

17. $\left(4x^3y^3 + \dfrac{1}{x}\right)dx + \left(3x^4y^2 - \dfrac{1}{y}\right)dy = 0$

18. $(-y\sin xy + 2x)dx + (3y^2 - x\sin xy)\,dy = 0$

19. $\sin y\,dx + (e^x + e^{-x})\sin y\,dy = 0$

20. $\dfrac{dy}{dx} + 2y\cot x + \sin 2x = 0$

In Problems 21–28, *find*

  a. *the general solution y_h of the related homogeneous equation*

  b. *a particular solution y_p of the nonhomogeneous equation*

21. $y'' - 9y = 1 + x$ **22.** $y'' - 2y' + y = e^x$

23. $y'' - 5y' + 6y = x^2e^x$ **24.** $y'' + 2y' + y = \sinh x$

25. $y'' - 3y' + 2y = x^3e^x$ **26.** $14y'' + 29y' - 15y = \cos x$

27. $y'' + y = \tan^2 x$ **28.** $y'' + y = \sec x$

29. Solve $\dfrac{d^2y}{dx^2} = e^{-3x}$ subject to the initial conditions
$y(0) = y'(0) = 0$.

30. Use the substitution $p = dy/dx$ to solve the differential equation

$$\frac{d^2y}{dx^2} = \left(\frac{dy}{dx}\right)^3 + \frac{dy}{dx}$$

31. Solve the system of differential equations

$$\frac{dx}{dt} = -2x \qquad \frac{dy}{dt} = -3y + 2x \qquad \frac{dz}{dt} = 3y$$

subject to the initial conditions $x(0) = 1$, $y(0) = 0$, $z(0) = 0$.

32. Find a curve $y = f(x)$ that passes through $(1, 1)$ and has the property that the y-intercept of the tangent line at each point $P(x, y)$ is equal to y^2.

33. Find a curve that passes through the point $(1, 2)$ and has the property that the length of the part of the tangent line between each point $P(x, y)$ on the curve and the y-axis of the tangent is equal to the y-intercept of the tangent line at P.

34. Find a curve that passes through $(1, 2)$ and has the property that the normal line at any point $P(x, y)$ on the curve and the line joining P to the origin form an isosceles triangle with the x-axis as its base.

Orthogonal trajectories. *Recall from Section 5.6 that the orthogonal trajectories of a given family of curves are another family with the following property: Every time a member of the second family intersects a member of the given family, the tangent lines to the two curves at the common point intersect at right angles. Find the orthogonal trajectories for the families of curves in Problems 35–38.*

35. $e^x + e^{-y} = C$ **36.** $3x^2 + 5y^2 = C$

37. $x^2 - y^2 = Cx$ **38.** $y = \dfrac{Cx}{x^2 + 1}$

39. Use reduction of order to find the general solution of the Legendre equation

$$(1 - x^2)y'' - 2xy' + 2y = 0$$

given that $y_1 = x$ is one solution.

40. Find functions $x = x(t)$ and $y = y(t)$ that satisfy the linear system

$$\frac{dx}{dt} = x + y \qquad \frac{dy}{dt} = x^2 - y^2$$

Sketch the solution curve $(x(t), y(t))$ that contains the point $(0, 2)$. *Hint:* Note that $\dfrac{dy}{dx} = \dfrac{y'(t)}{x'(t)}$.

41. Solve the second-order differential equation $xy'' + 2y' = x$ by setting $p = y'$ to convert it to a first-order equation in p.

42. Find an equation for a curve for which the radius of curvature is proportional to the slope of the tangent line at each point $P(x, y)$.

43. Solve the differential equation

$$\frac{dx}{dy} - \frac{x}{y} = ye^y$$

44. MODELING PROBLEM A ship weighing 64,000 tons [mass = (64,000)(2,000)/32 slugs] starts from rest and is driven by the constant thrust of its propellers. Suppose the propellers supply 250,000 lb of thrust and the resistance of the water is $12,000v$ lb, where $v(t)$ is the velocity of the ship. Set up and solve a differential equation for $v(t)$.

45. MODELING PROBLEM   A 10-ft uniform chain (see Figure 15.12) of mass m is hanging over a peg so that 3 ft are on one side and 7 ft are on the other. Set up and solve a differential equation to find how long it takes for the chain to slide off the peg. Neglect the friction.

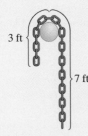

3 ft

7 ft

■ **FIGURE 15.12**   Problem 45

46. When an object weighing 10 lb is suspended from a spring, the spring is stretched 2 in. from its equilibrium position. The upper end of the spring is given a motion of $y = 2(\sin t + \cos t)$ ft. Find the displacement $x(t)$ of the object.

47. MODELING PROBLEM   A horizontal beam is freely supported at both ends, as shown in Figure 15.13.

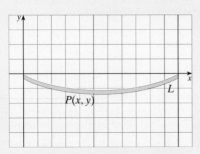

■ **FIGURE 15.13**   Problem 47

Suppose the load on the beam is W pounds per foot and one end of the beam is at the origin and the other end of the beam is at $(L, 0)$. Then the deflection y of the beam at point x is modeled by the differential equation

$$EI\frac{d^2y}{dx^2} = WLx - \frac{1}{2}Wx^2$$

for positive constants E and I.

a. Solve this equation to obtain the deflection $y(x)$. Note that $y(0) = y(L) = 0$.

b. What is the maximum deflection in the beam?

48. Consider the **Clairaut equation**

$$y = xy' + f(y')$$

a. Differentiate both sides of this equation to obtain.

$$[x + f'(y')]y'' = 0$$

Because one of the two factors in the product on the left must be 0, we have two cases to consider. What is the solution if $y'' = 0$? This is the *general solution*.

b. What is the solution if

$$x + f'(y') = 0$$

(This solution is called the *singular solution*.) *Hint:* Use the parametrization $y' = t$.

c. Find the general and singular solutions for the Clairaut equation

$$y = xy' + \sqrt{4 + (y')^2}$$

49. A differential equation of the form

$$x^2y'' + Axy' + By = F(x)$$

with $x \neq 0$ is called an **Euler equation.** To find a general solution of the related homogeneous equation

$$x^2y'' + Axy' + By = 0$$

we assume that there are solutions of the form $y = x^m$.

a. Show that m must satisfy

$$m^2 + (A - 1)m + B = 0$$

This is the characteristic equation for the Euler equation.

b. **Distinct real roots.**   Characterize the solutions of the related homogeneous equation in the case where

$$(A - 1)^2 > 4B$$

c. **Repeated real roots.**   Characterize the solutions of the related homogeneous equation in the case where

$$(A - 1)^2 = 4B$$

d. **Complex conjugate roots.**   Suppose

$$(A - 1)^2 < 4B$$

and that $\alpha \pm \beta i$ are the roots of the characteristic equation. Verify that $y_1 = x^\alpha \cos(\beta \ln|x|)$ is one solution. Use reduction of order to find a second solution y_2, then characterize the general solution.

50. Find the general solution of the Euler equation

$$x^2y'' + 7xy' + 9y = \sqrt{x}$$

51. In certain biological studies, it is important to analyze *predator–prey* relationships. Suppose $x(t)$ is the prey population and $y(t)$ is the predator population at time t. Then these populations change at rates

$$\frac{dx}{dt} = a_{11}x - a_{12}y \qquad \frac{dy}{dt} = a_{21}x - a_{22}y$$

where a_{11} is the natural growth rate of the prey, a_{12} is the predation rate, a_{21} measures the food supply of the predators, and a_{22} is the death rate of the predators. Outline a procedure for solving this system.

52. Consider the almost homogeneous differential equation

$$\frac{dy}{dx} = f\left(\frac{ax + by + c}{rx + sy + t}\right)$$

with $as = br$.

a. Let $u = \dfrac{ax + by}{a}$. Show that $u = \dfrac{rx + sy}{r}$ and $\dfrac{dy}{dx} = \dfrac{a}{b}\left(\dfrac{du}{dx} - 1\right)$.

b. Verify that by making the change of variable suggested in part **a,** you can rewrite the given differential equation in the separable form

$$\frac{du}{1 + \dfrac{b}{a}f\left(\dfrac{au + c}{ru + t}\right)} = dx$$

c. Use the procedure outlined in parts **a** and **b** to solve the almost homogeneous differential equation

$$\frac{dy}{dx} = \frac{2x + y - 3}{4x + 2y + 5}$$

53. SPY PROBLEM   "You must be Rottona!" the Spy cleverly announces to the red Purity after she correctly answers the question posed in Problem 59, Section 15.1. "Purity is too innocent to know so much physics." He starts to tie the wrists of the girl in red.

"You idiot!" she shrieks. "You ... your father wears leisure suits!" Likely there was more of the same, but a blow from behind provides the Spy with an early nap. As he's waking, it comes to him—the girl in red is Purity! He stumbles from the room into an open field where Rottona and Blohardt are both chasing Purity. His mathematical mind visualizes the field as a coordinate plane with himself at $(100, 0)$, Purity at $(0, 0)$, running along the y-axis, and Rottona and Blohardt at $(0, -18)$ and $(80, o)$, respectively (units in yards). The Spy, Rottona, and Blohardt all run directly toward Purity. If Rottona runs at 14 ft/s, Blohardt at 15 ft/s, and Purity at 12 ft/s, how fast must the Spy run to the innocent maid to get there before either of the bad guys?

54. PUTNAM EXAMINATION PROBLEM   Find all solutions of the equation

$$yy'' - 2(y')^2 = 0$$

that pass through the point $x = 1, y = 1$.

55. PUTNAM EXAMINATION PROBLEM   A coast artillery gun can fire at any angle of elevation between $0°$ and $90°$ in a fixed vertical plane. If air resistance is neglected and the muzzle velocity is constant $(v(0) = v_0)$, determine the set H of points in the plane and above the horizontal that can be hit.

56. PUTNAM EXAMINATION PROBLEM   Show that

$$x + \frac{2}{3}x^3 + \frac{2 \cdot 4}{3 \cdot 5}x^5 + \frac{2 \cdot 4 \cdot 6}{3 \cdot 5 \cdot 7}x^7 + \cdots = \frac{\sin^{-1}x}{\sqrt{1 - x^2}}$$

Save the Perch Project

This project is to be done in groups of three or four students. Each group will submit a single written report.

Happy Valley Pond is currently populated by yellow perch. A map is shown below. Water flows into the pond from two springs and evaporates from the pond, as shown by the table below.

Spring	Dry Season	Rainy Season
A	50 gal/h	60 gal/h
B	60 gal/h	75 gal/h
Evaporation:		
	110 gal/h	75 gal/h

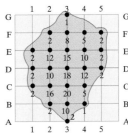

Happy Valley Pond is fed by two springs:

The overflow goes over the dam into Bubbling Brook:

Unfortunately, spring B has become contaminated with salt and is now 10% salt, which means that 10% of a gallon of water from spring B is salt. The yellow perch will start to die if the concentration of salt in the pond rises to 1%. Assume that the salt will not evaporate but will mix thoroughly with the water in the pond. There was no salt in the pond before the contamination of spring B. Your group has been called upon by the Happy Valley Bureau of Fisheries to try to save the perch.

Your paper is not limited to the following questions, but should include the number of gallons of water in the pond when the water level is exactly even with the top of

---

*This group project is courtesy of Diane Schwartz from Ithaca College, New York.

Among all the mathematical disciplines the theory of differential equations is the most important . . . . It furnishes the explanation of all those elementary manifestations of nature which involve time . . . .

SOPHUS LIE
LEIPZIGER BERICHTE, 47 (1895).

the spillover dam. The following table gives a series of measurements of the depth of the pond at the indicated points when the water level was exactly even with the top of the spillover dam.

DEPTH OF HAPPY VALLEY POND (Location/depth)

A 3/8 ft	B 2/2 ft	B 3/10 ft	B 4/1 ft	C 1/2 ft	C 2/16 ft	C 3/20 ft
C 4/5 ft	D 1/2 ft	D 2/10 ft	D 3/18 ft	D 4/12 ft	D 5/2 ft	D 3/18 ft
D 4/12 ft	D 5/2 ft	E 1/2 ft	E 2/12 ft	E 3/15 ft	E 4/10 ft	E 5/2 ft
F 2/2 ft	F 3/8 ft	F 4/5 ft	F 5/2 ft	G 3/1 ft		

Let $t = 0$ hours correspond to the time when spring B became contaminated. Assume it is the dry season and that at time $t = 0$ the water level of the pond was exactly even with the top of the spillover dam. Write a differential equation for the amount of salt in the pond after t hours. Draw a graph of the amount of salt in the pond versus time for the next 3 mo. How much salt will there be in the pond in the long run, and do the fish die? If so, when do they start to die? It is very difficult to find where the contamination of spring B originates, so the Happy Valley Bureau of Fisheries proposed to flush the pond by running 100 gal of pure water per hour through the pond. Your report should include an analysis of this plan and any modifications or improvements that could help save the perch.

Chapters 12–15 Cumulative Review

■ **TABLE 15.2 Comparison of Important Integral Theorems**

Riemann Integral (Section 5.3)	Line Integral (Section 14.2)

Riemann Integral (Section 5.3)

$$\int_a^b f(x)\, dx$$

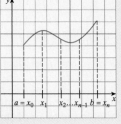

Subdivisions on
the x-axis

Line Integral (Section 14.2)

$$\int_C f(x, y, z)\, dx$$

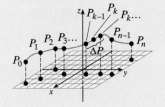

Subdivisions on a curve
C in space

Fundamental Theorem of Calculus (Section 5.4)

$$\int_a^b f(x)\, dx = F(b) - F(a)$$

F is an antiderivative of f.

Fundamental Theorem for Line Integrals (Section 14.3)

$$\int_C \mathbf{F} \cdot d\mathbf{R} = f(Q) - f(P)$$

if $\mathbf{F}$ is conservative with scalar potential f;
that is, $\nabla f = \mathbf{F}$.

Green's Theorem (Section 14.4); applies to $\mathbb{R}^2$ (two dimensions): $\mathbf{F}(x, y) = M(x, y)\mathbf{i} + N(x, y)\mathbf{j}$

$$\int_C \mathbf{F} \cdot d\mathbf{R} = \int_C (M\, dx + N\, dy) = \iint_D \left(\frac{\partial N}{\partial x} - \frac{\partial M}{\partial y}\right) dA = \iint_D (\text{curl } \mathbf{F} \cdot \mathbf{k})\, dA$$

If $\mathbf{F}$ is conservative,

$$\int_C \mathbf{F} \cdot \mathbf{N}\, ds = \int_C \nabla f \cdot \mathbf{N}\, ds = \int_C \frac{\partial f}{\partial n}\, ds = \iint_D \nabla^2 f\, dA = \iint_D \left(\frac{\partial M}{\partial x} + \frac{\partial N}{\partial y}\right) dA = \iint_D \text{div } \mathbf{F}\, dA$$

Double integral (Section 13.1)	Surface integral (Section 14.5)

Double integral (Section 13.1)

$$\iint_R f(x, y)\, dA$$

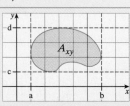

Partition of R into mn cells in the xy-plane

Surface integral (Section 14.5)

$$\iint_S g(x, y, z)\, dS = \iint_{R_{xy}} g(x, y, z)\sqrt{g_x^2 + g_y^2 + 1}\, dA_{xy}$$

where $z = f(x, y)$

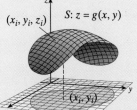

$S: z = g(x, y)$
(x_i, y_i, z_i)
(x_i, y_i)

Partition of the surface
S into n subregions

Stokes' Theorem (Section 14.6)

$$\int_C \mathbf{F} \cdot d\mathbf{R} = \iint_S (\text{curl } \mathbf{F} \cdot \mathbf{N})\, dS$$

Divergence Theorem (Section 14.7; also known
as *Gauss's theorem*)

$$\iint_S \mathbf{F} \cdot \mathbf{N}\, dS = \iiint_D \text{div } \mathbf{F}\, dV$$

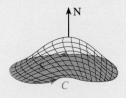

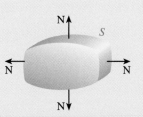

Chapters 12–15   *Cumulative Review Problems*

1. ■ **What Does This Say?** Suppose you tell a fellow college student that you are about to finish a calculus course. The student has not had any mathematics beyond high school and asks you, "What is calculus?" Answer this question using your own words.

2. ■ **What Does This Say?** There are three great ideas in calculus: the limit, the derivative, and the integral. In your own words, explain each of these concepts.

3. ■ **What Does This Say?** Chapters 12–14 were concerned with functions of several variables. This is often referred to as **multivariable calculus.** In your own words, discuss what is meant by multivariable calculus.

4. ■ **What Does This Say?** In your own words, outline a procedure for solving the types of differential equations discussed in Chapter 15.

Find f_x, f_y, and f_{xy} for the functions whose equations are given in Problems 5–10.

5. $f(x, y) = 2x^2 + xy - 5y^3$

6. $f(x, y) = x^2 e^{y/x}$

7. $f(x, y) = \dfrac{x^2 - y^2}{x - y}$

8. $f(x, y) = y \sin^2 x + \cos xy$

9. $f(x, y) = e^{x+y}$ 10. $f(x, y) = \dfrac{x^2 + y^2}{x - y}$

11. Let $f(x, y) = \dfrac{xy}{(x^2 + y^2)^2}$. Find $f_{xx} + f_{yy}$.

Evaluate the integrals in Problems 12–19.

12. $\displaystyle\int_0^1 \int_x^{2x} e^{y-x} \, dy \, dx$

13. $\displaystyle\int_0^4 \int_0^{\sqrt{x}} 3x^5 \, dy \, dx$

14. $\displaystyle\int_0^1 \int_0^z \int_y^{y-z} (x + y + z) \, dx \, dy \, dz$

15. $\displaystyle\int_0^{15\pi} \int_0^{\pi} \int_0^{\sin\phi} \rho^3 \sin\phi \, d\rho \, d\theta \, d\phi$

16. $\displaystyle\iint_R e^{x+y} \, dA; \ R: 0 \le x \le 1, 0 \le y \le 1$

17. $\displaystyle\iint_R y e^{xy} \, dA; \ R: 0 \le x \le 1, 0 \le y \le 2$

18. $\displaystyle\iint_R \sin(x + y) \, dA; \ R: 0 \le x \le \dfrac{\pi}{2}, 0 \le y \le \dfrac{\pi}{4}$

19. $\displaystyle\iint_R x \sin xy \, dA; \ R: 0 \le x \le \pi, 0 \le y \le 1$

Evaluate the line integrals in Problems 20–23.

20. $\displaystyle\int_C (y^2 z \, dx + 2xyz \, dy + xy^2 \, dz)$, where C is any path from $(0, 0, 0)$ to $(1, 1, 1)$

21. $\displaystyle\int_C (5 \, xy \, dx + 10 \, yz \, dy + z \, dz)$, where C is given by $x = t^2, y = t, z = 2t^3$ for $0 \le t \le 1$

22. $\displaystyle\int_C \mathbf{F} \cdot d\mathbf{R}$, where $\mathbf{F} = yz\mathbf{i} - x\mathbf{k}$, and C is the boundary of the triangle $(1, 1, 1)$, $(1, 0, 1)$, $(0, 0, 1)$, traversed once counterclockwise as viewed from the origin

23. $\displaystyle\int_C (x^3 + y^3) \, ds$, where C is given by $\mathbf{R}(t) = (\cos^3 t)\mathbf{i} + (\sin^3 t)\mathbf{j}$, for $0 \le t \le 2\pi$

Solve the differential equations in Problems 24–30.

24. $\dfrac{dy}{dx} + \dfrac{y}{2x} = \sqrt{x} \, e^x$

25. $x \dfrac{dy}{dx} + 3y = \sin x^3$

26. $y'' + 4y = 2 \sin x$

27. $y'' + y' - 2y = x^3 + x^2 - 2x + 5$

28. $y'' + 4y = \sin 2x$

29. $y' + xy = x + e^{-x^2/2}$ where $y = -1$ when $x = 0$

30. $xy' - 2y = x^2$, where $y = 5$ when $x = 1$

31. Find the equations for the tangent plane and the normal line to $z = x^2 + y^2 + \sin xy$ at $P = (0, 2, 4)$.

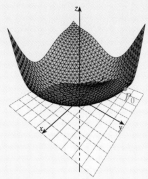

32. What are the dimensions of the rectangular box of fixed volume V_0 that has minimum surface area?

33. **MODELING PROBLEM**   A manufacturer is planning to sell a new product at the price of $350 per unit and estimates that if x thousand dollars is spent on development and y thousand dollars is spent on promotion, consumers will buy approximately

$$\frac{250y}{y + 2} + \frac{100x}{x + 5}$$

units of the product. If manufacturing costs for this product are \$150 per unit, how much should the manufacturer spend on development and how much on promotion to generate the largest possible profit, given the following circumstances?

a. Unlimited funds are available.
b. The manufacturer has only \$11,000 to spend on development and promotion of the new product.

34. **MODELING PROBLEM**   A heat-seeking missile moves in a portion of space where the temperature (in degrees Celsius) at the point (x, y, z) is given by

$$T(x, y, z) = \frac{1}{10}(x^2 + y + z^3)$$

with x, y, and z in kilometers.

a. Find the rate at which the temperature is changing as the missile moves from the point $P_0(-2, 9, 1)$ toward $Q(1, -3, 5)$.
b. If the missile is at P_0, in what direction will it travel to maximize the rate of heat increase?
c. What is the maximal rate of increase (in degrees Celsius per kilometer)?

35. Use double polar integration to find the area inside the cardioid $r = 1 - \cos\theta$ and outside the circle $r = 1$.

36. Find the volume of the solid bounded above by the surface $z = x^2 + y^2 + 1$ and below by the circular disk $x^2 + y^2 \leq 1$ in the xy-plane.

37. Find the surface area of that portion of the paraboloid $z = x^2 + y^2$ that lies below the plane $z = 16$.

38. Find the center of mass of the lamina whose shape is the region inside the circle $x^2 + y^2 = 4$ in the first quadrant, given that the density at any point (x, y) is $\delta(x, y) = x + y$. *Hint:* Reverse the order of integration.

39. A force field $\mathbf{F}(x, y) = (x - 2y)\mathbf{i} + (y - 2x)\mathbf{j}$ acts on an object moving in the plane. Show that $\mathbf{F}$ is conservative, and find a scalar potential for $\mathbf{F}$. How much work is done as the object moves from $(1, 0)$ to $(0, 1)$ along any path connecting these points?

40. A particle of weight w is acted on only by the constant gravitational force $\mathbf{F} = -w\mathbf{k}$. How much work is done in moving the weight along the helical path given by $x = \cos t$, $y = \sin t$, $z = t$ for $0 \leq t \leq 2\pi$?

Appendices

Introduction to the Theory of Limits

In Section 2.2, we defined the limit of a function as follows:

The notation

$$\lim_{x \to c} f(x) = L$$

is read "the limit of $f(x)$ as x approaches c is L" and means that the function values $f(x)$ can be made arbitrarily close to L by choosing x sufficiently close to c but not equal to c.

This informal definition was valuable because it gave you an intuitive feeling for the limit of a function and allowed you to develop a working knowledge of this fundamental concept. For theoretical work, however, this definition will not suffice, because it gives no precise, quantifiable meaning to the terms "arbitrarily close to L" and "sufficiently close to c." The following definition, derived from the work of Cauchy and Weierstrass, gives precision to the limit definition, and was also first stated in Section 2.2.

Limit of a Function (Formal definition)

The limit statement

$$\lim_{x \to c} f(x) = L$$

means that for each $\epsilon > 0$, there corresponds a number $\delta > 0$ with the property that

$$|f(x) - L| < \epsilon \quad \text{whenever} \quad 0 < |x - c| < \delta$$

Behind the formal language is a fairly straightforward idea. In particular, to establish a specific limit, say $\lim_{x \to c} f(x) = L$, a number $\epsilon > 0$ is chosen first to establish a desired degree of proximity to L, and then a number $\delta > 0$ is found that determines how close x must be to c to ensure that $f(x)$ is within ϵ units of L.

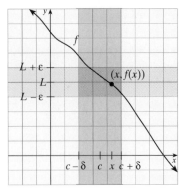

FIGURE A.1 The epsilon–delta definition of limit

The situation is illustrated in Figure A.1, which shows a function that satisfies the conditions of the definition. Notice that whenever x is within δ units of c (but not equal to c), the point $(x, f(x))$ on the graph of f must lie in the rectangle formed by the intersection of the horizontal band of width 2ϵ (blue screen) centered at L and the vertical band of width 2δ (gray region) centered at c. The smaller the ϵ-interval around the proposed limit L, generally the smaller the δ-interval will need to be for $f(x)$ to lie in the ϵ-interval. If such a δ can be found no matter how small ϵ is, then L must be the limit.

THE BELIEVER/DOUBTER FORMAT

The limit process can be thought of as a "contest" between a "believer" who claims that $\lim_{x \to c} f(x) = L$ and a "doubter" who disputes this claim. The contest begins with the doubter choosing a positive number ϵ so that whenever x is a number (other than c) within δ units of c (the gray region), the corresponding function value $f(x)$ is within ϵ units of L (the blue region). Naturally, the doubter tries to choose ϵ so small that no matter what δ the believer chooses, it will not be possible to satisfy the accuracy requirement. When will the doubter win and when will the believer win the argument? As you can see from Figure A.1, if the believer has the "correct limit" L, it will be in the intersection (the portion that is both blue and gray) no matter what ϵ the doubter chooses. On the other hand, if the believer has an "incorrect limit" as shown in Figure A.2, the limit may be in the double-screened portion for some choices of ϵ, but for other choices of ϵ the doubter can force the believer to be the loser of this contest by prohibiting the believer from making a choice within the double-screened portion.

Technology Window

The tolerance setting on a calculator often uses the $\delta - \epsilon$ notation. For example, the TI-85 allows you to set the tolerance for maximum and minimum of a function by inputting a δ-value.

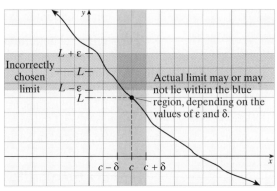

FIGURE A.2 False limit scenario

To avoid an endless series of ϵ–δ challenges, the believer usually tries to settle the issue by producing a formula relating the choice of δ to the doubter's ϵ that will satisfy the requirement no matter what ϵ is chosen. The believer/doubter format is used in Example 1 to verify a given limit statement (believer "wins") and in Example 2 to show that a certain limit does not exist (doubter "wins").

EXAMPLE 1 *Verifying a limit claim (believer wins)*

Show that $\lim_{x \to 2} (2x + 1) = 5$.

Solution To verify the given limit statement, we begin by having the doubter choose a positive number ϵ. Before picking δ, the believer might entertain the thought process shown in the following box.

> Write $|f(x) - L|$ in terms of $|x - c|$ (where $c = 2$) as follows:
> $$|f(x) - L| = |(2x + 1) - 5| = |2x - 4| = 2|x - 2|$$
> Thus, if $0 < |x - c| < \delta$ or $0 < |x - 2| < \delta$, then
> $$|f(x) - L| < 2\delta$$
> The believer *wants* $|f(x) - L| < \epsilon$, so we see that the believer should choose $2\delta = \epsilon$, or $\delta = \frac{\epsilon}{2}$.*

With the information shown in this box, we can now make the following argument, which uses the formal definition of limit:

Let $\epsilon > 0$ be given. Choose $\delta = \dfrac{\epsilon}{2} > 0$. It follows that if

$$0 < |x - 2| < \delta = \frac{\epsilon}{2}, \quad \text{then}$$

$$|(2x + 1) - 5| = 2|x - 2| < 2\left(\frac{\epsilon}{2}\right) = \epsilon$$

Thus, the conditions of the definition of limit are satisfied, and we have

$$\lim_{x \to 2} (2x + 1) = 5$$

This is shown graphically in Figure A.3a.

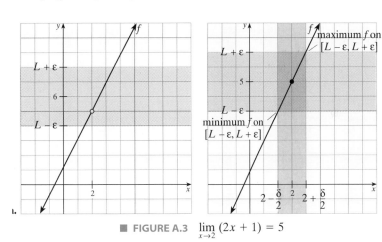

■ **FIGURE A.3** $\lim\limits_{x \to 2} (2x + 1) = 5$

Notice that no matter what ϵ is chosen by the doubter, by choosing a number that is one-half of that ϵ, the believer will force the function to stay within the double-screened portion of the graph, as shown in Figure A.3.   ■

| **EXAMPLE 2** | *Disproving a limit claim (doubter wins)* |

Determine whether $\lim\limits_{x \to 2} \dfrac{2x^2 - 3x - 2}{x - 2} = 6$.

_____

*In fact there are other choices that would work; for example, $\delta = \dfrac{\epsilon}{a}$ where $a > 2$.

Solution   Once again, the doubter will choose a positive number ϵ and the believer must respond with a δ. As before, the believer does some preliminary work with $f(x) - L$:

$$\left| \frac{2x^2 - 3x - 2}{x - 2} - 6 \right| = \left| \frac{2x^2 - 3x - 2 - 6x + 12}{x - 2} \right|$$

$$= \left| \frac{2x^2 - 9x + 10}{x - 2} \right|$$

$$= \left| \frac{(2x - 5)(x - 2)}{x - 2} \right|$$

$$= |2x - 5|$$

The believer wants to write this expression in terms of $x - c = x - 2$. This example does not seem to "fall into place" as did Example 1. Suppose the doubter chooses $\epsilon = 1.5$ (not a very small ϵ). Then the believer must find a δ so that whenever x is within this δ distance of 2, $f(x)$ will be within 1.5 units of 6. This is shown in Figure A.4a.

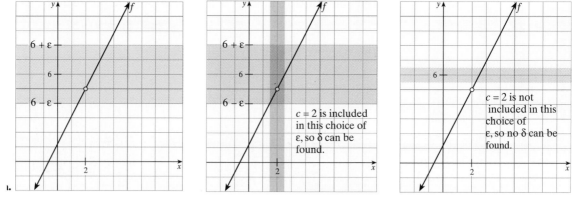

■ **FIGURE A.4**   Example of an incorrectly chosen limit

However, if the doubter chooses $\epsilon = 0.5$ (still not a very small ϵ), then the believer is defeated, because no δ can be found, as shown in Figure A.4b.   ■

The believer/doubter format is a useful device for dramatizing the way certain choices are made in epsilon–delta arguments, but it is customary to be less "chatty" in formal mathematical proofs.

EPSILON–DELTA PROOFS

The following examples illustrate epsilon–delta proofs, two in which the function has a limit and one in which it does not.

EXAMPLE 3   *An epsilon–delta proof of a limit of a linear function*

Show that $\lim\limits_{x \to 2} (4x - 3) = 5.$

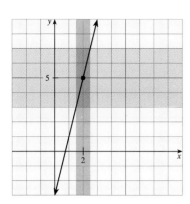

Solution   We have

$$\begin{aligned}|f(x) - L| &= |4x - 3 - 5| \\ &= |4x - 8| \\ &= \underbrace{4|x - 2|}\end{aligned}$$

This must be less than ϵ whenever $|x - 2| < \delta$.

Choose $\delta = \dfrac{\epsilon}{4}$; then

$$|f(x) - L| = 4|x - 2| < 4\delta = 4\left(\frac{\epsilon}{4}\right) = \epsilon \qquad \blacksquare$$

EXAMPLE 4   *An epsilon–delta proof of a limit of a rational function*

Show that $\displaystyle\lim_{x\to 2} \frac{x^2 - 2x + 2}{x - 4} = -1$.

Solution   We have

$$\begin{aligned}|f(x) - L| &= \left|\frac{x^2 - 2x + 2}{x - 4} - (-1)\right| \\ &= \left|\frac{x^2 - x - 2}{x - 4}\right| \\ &= \underbrace{|x - 2|\left|\frac{x + 1}{x - 4}\right|}\end{aligned}$$

This must be less than the given ϵ whenever x is near 2.

Certainly $|x - 2|$ is small if x is near 2, and the factor $\left|\dfrac{x + 1}{x - 4}\right|$ is not large (it is close to $\frac{3}{2}$). Note that if $|x - 2|$ is small, it is reasonable to assume

$$|x - 2| < 1 \quad \text{so that} \quad 1 < x < 3$$

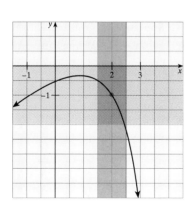

By inspection of the graph, the largest value of the fraction $\dfrac{x + 1}{x - 4}$ is

$$\frac{1 + 1}{1 - 4} = \frac{2}{-3}$$

and the smallest value of the fraction $\dfrac{x + 1}{x - 4}$ is

$$\frac{3 + 1}{3 - 4} = \frac{4}{-1} = -4$$

Thus, if $|x - 2| < 1$, we have

$$-4 < \frac{x + 1}{x - 4} < -\frac{2}{3}$$

so that

$$-4 < \frac{x + 1}{x - 4} < 4 \qquad \textit{Because } -\tfrac{2}{3} < 4$$

and

$$\left|\frac{x+1}{x-4}\right| < 4$$

Now let $\epsilon > 0$ be given. If simultaneously

$$|x-2| < \frac{\epsilon}{4} \quad \text{and} \quad \left|\frac{x+1}{x-4}\right| < 4$$

then

$$|f(x) - L| = |x-2|\left|\frac{x+1}{x-4}\right| < \frac{\epsilon}{4}(4) = \epsilon$$

Thus, we have only to take δ to be the smaller of the two numbers 1 and $\epsilon/4$ to guarantee that

$$|f(x) - L| < \epsilon$$

That is, given $\epsilon > 0$, choose δ to be the smaller of the numbers 1 and $\epsilon/4$. We write this as $\delta = \min(1, \epsilon/4)$. ∎

EXAMPLE 5   *An epsilon–delta proof that a limit does not exist*

Show that $\lim\limits_{x\to 0} \dfrac{1}{x}$ does not exist.

Solution   Let $f(x) = \dfrac{1}{x}$ and L be any number. Suppose that $\lim\limits_{x\to 0} f(x) = L$. Look at the graph of f, as shown in Figure A.5. It would seem that no matter what value of ϵ is chosen, it would be impossible to find a corresponding δ. Consider the absolute value expression required by the definition of limit: If

$$|f(x) - L| < \epsilon, \quad \text{or, for this example,} \quad \left|\frac{1}{x} - L\right| < \epsilon$$

then

$$-\epsilon < \frac{1}{x} - L < \epsilon \qquad \text{\small Property of absolute value}$$

and $\qquad\qquad\qquad\qquad\qquad\qquad\qquad$ (Table 1.1, p. 3)

$$L - \epsilon < \frac{1}{x} < L + \epsilon$$

If $\epsilon = 1$ (not a particularly small ϵ), then

$$\left|\frac{1}{x}\right| < |L| + 1$$

$$|x| > \frac{1}{|L| + 1}$$

which proves (since L was chosen arbitrarily) that $\lim\limits_{x\to 0} \dfrac{1}{x}$ does not exist. In other words,

since $|x|$ can be chosen very small, $\dfrac{1}{|x|}$ will be very large, and it will be impossible to

squeeze $\dfrac{1}{x}$ between $L - \epsilon$ and $L + \epsilon$ for any L. ∎

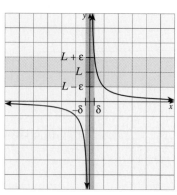

■ **FIGURE A.5** $\lim\limits_{x\to 0} \dfrac{1}{x}$

SELECTED THEOREMS WITH FORMAL PROOFS

Next, we shall prove several theoretical results using the formal definition of the limit.

The next two theorems are useful tools in the development of calculus. The first states that the points on a graph that are on or above the x-axis cannot possibly "tend toward" a point *below* the axis, as shown in Figure A.6.

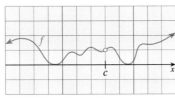

If $f(x) \geq 0$ for all x near c, then
$$\lim_{x \to c} f(x) \geq 0.$$

■ **FIGURE A.6** Limit limitation theorem

THEOREM A.1   **Limit limitation theorem**

Suppose $\displaystyle\lim_{x \to c} f(x)$ exists and $f(x) \geq 0$ throughout an open interval containing the number c, except possibly at c itself. Then

$$\lim_{x \to c} f(x) \geq 0$$

Proof   Let $L = \displaystyle\lim_{x \to c} f(x)$. To show that $L \geq 0$, assume the contrary. That is, assume $L < 0$. According to the definition of limit (with $\epsilon = -L$), there is a number $\delta > 0$ such that

$$\left| f(x) - L \right| < -L \quad \text{whenever} \quad 0 < \left| x - c \right| < \delta$$

In particular,

$$f(x) - L < -L$$

or

$$f(x) < 0$$

Thus,

$$f(x) < 0 \quad \text{whenever} \quad 0 < \left| x - c \right| < \delta$$

However, this contradicts the hypothesis that $f(x) \geq 0$ throughout an open interval containing c. Therefore, we reject the assumption that $L < 0$ and conclude that $L > 0$, as required.

Useful information about the limit of a given function f can often be obtained by examining other functions that bound f from above and below. For example, in Section 2.2 we found

$$\lim_{x \to 0} \frac{\sin x}{x} = 1$$

by using a table, and proved the limit (Theorem 2.2) by first showing that

$$\cos x \leq \frac{\sin x}{x} \leq 1$$

for all x near 0 and then noting that since $\cos x$ and 1 both tend toward 1 as x approaches 0, the function

$$\frac{\sin x}{x}$$

WARNING   It may seem reasonable to conjecture that if $f(x) > 0$ throughout an open interval containing c, then $\displaystyle\lim_{x \to c} f(x) > 0$. This is not necessarily true, and the most that can be said in this situation is that $\displaystyle\lim_{x \to c} f(x) \geq 0$, if the limit exists. For example, if

$$f(x) = \begin{cases} x^2 & \text{for } x \neq 0 \\ 1 & \text{for } x = 0 \end{cases}$$

then $f(x) > 0$ for all x, but $\displaystyle\lim_{x \to 0} f(x) = 0$. ⬤

which is "squeezed" between them, must converge to 1 as well. Theorem A.2 provides the theoretical basis for this method of proof.

THEOREM A.2   *The squeeze theorem*

If $g(x) \leq f(x) \leq h(x)$ for all x in an open interval containing c (except possibly at c itself) and if

$$\lim_{x \to c} g(x) = \lim_{x \to c} h(x) = L$$

then $\lim_{x \to c} f(x) = L$. (This is stated, without proof, in Section 2.3.)

Proof   Let $\epsilon > 0$ be given. Since $\lim_{x \to c} g(x) = L$ and $\lim_{x \to c} h(x) = L$, there are positive numbers δ_1 and δ_2 such that

$$\left| g(x) - L \right| < \epsilon \quad \text{and} \quad \left| h(x) - L \right| < \epsilon$$

whenever

$$0 < \left| x - c \right| < \delta_1 \quad \text{and} \quad 0 < \left| x - c \right| < \delta_2$$

respectively. Let δ be the smaller of the numbers δ_1 and δ_2. Then, if x is a number that satisfies $0 < \left| x - c \right| < \delta$, we have

$$-\epsilon < g(x) - L \leq f(x) - L \leq h(x) - L < \epsilon$$

and it follows that $\left| f(x) - L \right| < \epsilon$. Thus, $\lim_{x \to c} f(x) = L$, as claimed.   ▭

The geometric interpretation of the squeeze theorem is shown in Figure A.7. Notice that since $g(x) \leq f(x) \leq h(x)$, the graph of f is "squeezed" between those of g and h in the neighborhood of c. Thus, if the bounding graphs converge to a common point P as x approaches c, then the graph of f must also converge to P as well.

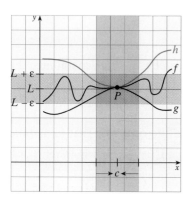

■ **FIGURE A.7**   The squeeze theorem

A.1 *Problem Set*

B *In Problems 1–6, use the believer/doubter format to prove or disprove the given limit statement.*

1. $\lim_{x \to 1} (2x - 5) = -3$

2. $\lim_{x \to -2} (3x + 7) = 1$

3. $\lim_{x \to 1} (3x + 1) = 5$

4. $\lim_{x \to 2} (x^2 - 2) = 5$

5. $\lim_{x \to 2} (x^2 + 2) = 6$

6. $\lim_{x \to 1} (x^2 - 3x + 2) = 0$

In Problems 7–12, use the formal definition of the limit to prove or disprove the given limit statement.

7. $\lim_{x \to 2} (x + 3) = 5$

8. $\lim_{t \to 0} (3t - 1) = 0$

9. $\lim_{x \to -2} (3x + 7) = 1$

10. $\lim_{x \to 1} (2x - 5) = -3$

11. $\lim_{x \to 2} (x^2 + 2) = 6$

12. $\lim_{x \to 2} \dfrac{1}{x} = \dfrac{1}{2}$

13. Prove that $f(x) = \begin{cases} \sin \dfrac{1}{x} & \text{if } x \neq 0 \\ 0 & \text{if } x = 0 \end{cases}$

is not continuous at $x = 0$. *Hint:* To show that $\lim_{x \to 0} f(x) \neq 0$, choose $\epsilon = 0.5$ and note that for any $\delta > 0$,

there exists an x of the form $x = \dfrac{2}{\pi(2n + 1)}$ with n a natural number for which $0 < \left| x \right| < \delta$.

C *In Problems 14–19, construct a formal ϵ–δ proof to show that the given limit statement is valid for any number c.*

14. If $\lim_{x \to c} f(x)$ exists and k is a constant, then

$$\lim_{x \to c} k f(x) = k \lim_{x \to c} f(x)$$

15. If $\lim_{x \to c} f(x)$ and $\lim_{x \to c} g(x)$ both exist, then

$$\lim_{x \to c} [f(x) - g(x)] = \lim_{x \to c} f(x) - \lim_{x \to c} g(x)$$

16. If $\lim_{x \to c} f(x)$ and $\lim_{x \to c} g(x)$ both exist and a, b are constants, then

$$\lim_{x \to c} [af(x) + bg(x)] = a \lim_{x \to c} f(x) + b \lim_{x \to c} g(x)$$

17. If $\lim_{x \to c} f(x) = 0$ and $\lim_{x \to c} g(x) = 0$, then

$$\lim_{x \to c} f(x)g(x) = 0$$

18. If $f(x) \geq g(x) \geq 0$ for all x and $\lim_{x \to c} f(x) = 0$, then
$$\lim_{x \to c} g(x) = 0$$

19. If $f(x) \geq g(x)$ for all x in an open interval containing the number c and $\lim_{x \to c} f(x)$ and $\lim_{x \to c} g(x)$ both exist, then $\lim_{x \to c} f(x) \geq \lim_{x \to c} g(x)$. *Hint:* Apply the limit limitation theorem (Theorem A.1) to the function $h(x) = f(x) - g(x)$.

Problems 20–23 lead to a proof of the product rule for limits.

20. If $\lim_{x \to c} f(x) = L$, show that $\lim_{x \to c} |f(x)| = |L|$. *Hint:* Note that
$$\big||f(x)| - |L|\big| \leq |f(x) - L|$$

21. If $\lim_{x \to c} f(x) = L, L \neq 0$, show that there exists a $\delta > 0$ such that
$$\tfrac{1}{2}|L| < |f(x)| < \tfrac{3}{2}|L|$$
for all x for which $0 < |x - c| < \delta$.
Hint: Use $\epsilon = \tfrac{1}{2}|L|$ in Problem 20.

22. If $\lim_{x \to c} f(x) = L$ and $L \neq 0$, show that $\lim_{x \to c} [f(x)]^2 = L^2$ by completing these steps:

a. Use Problem 21 to show that there exists a $\delta_1 > 0$ so that
$$|f(x) + L| < \tfrac{5}{2}|L|$$
whenever $0 < |x - c| < \delta_1$.

b. Given $\epsilon > 0$, show that there exists a $\delta_2 > 0$ such that
$$\big|[f(x)]^2 - L^2\big| < \tfrac{5}{2}|L|$$
whenever $0 < |x - c| < \delta_2$.

c. Complete the proof that
$$\lim_{x \to c} [f(x)]^2 = L^2$$

23. Prove the product rule for limits: If $\lim_{x \to c} f(x) = L$ and $\lim_{x \to c} g(x) = M$, then $\lim_{x \to c} f(x)g(x) = LM$. *Hint:* Use the result of Problem 22 along with the identity
$$fg = \tfrac{1}{4}[(f + g)^2 - (f - g)^2]$$

24. Show that if f is continuous at c and $f(c) > 0$, then $f(x) > 0$ throughout an open interval containing c. *Hint:* Note that
$$\lim_{x \to c} f(x) = f(c)$$
and use $\epsilon = \tfrac{1}{2}f(c)$ in the definition of limit.

25. Show that if f is continuous at L and $\lim_{x \to c} g(x) = L$, then, $\lim_{x \to c} f[g(x)] = f(L)$ by completing the following steps:

a. Explain why there exists a $\delta_1 > 0$ such that $|f(w) - f(L)| < \epsilon$ whenever $|w - L| < \delta_1$.

b. Complete the proof by setting $w = g(x)$ and using part **a.**

Technology Window

26. For linear functions, the relation between ϵ and δ is clear; but for complicated functions, it is not. However, the correct graph can illustrate this relationship. For the following function, illustrate not only that f is continuous at $x = 3$, but determine graphically how you would pick δ to accommodate a given ϵ. (For example, $\delta = K\epsilon$.)
$$f(x) = \frac{x^4 - 2x^3 + 3x^2 - 5x + 2}{x - 2}$$

APPENDIX B

Selected Proofs

CHAIN RULE (Chapter 3)

Suppose f is a differentiable function of u, and u is a differentiable function of x. Then
$$\frac{df}{dx} = \frac{df}{du}\frac{du}{dx}$$
where $f[u(x)]$.

Define an auxiliary function g by
$$g(t) = \frac{f[u(x) + t] - f[u(x)]}{t} - \frac{df}{du} \text{ if } t \neq 0 \quad \text{and} \quad g(t) = 0 \text{ if } t = 0$$

You can verify that g is continuous at $t = 0$.

Notice that for $t = \Delta u$ and $t \neq 0$,

$$g(\Delta u) = \frac{f[u(x) + \Delta u] - f[u(x)]}{\Delta u} - \frac{df}{du}$$

$$g(\Delta u) + \frac{df}{du} = \frac{f[u(x) + \Delta u] - f[u(x)]}{\Delta u}$$

$$\left[g(\Delta u) + \frac{df}{du} \right] \Delta u = f[u(x) + \Delta u] - f[u(x)]$$

We now use the definition of derivative for f.

$$\frac{df}{dx} = \lim_{\Delta x \to 0} \frac{f[u(x + \Delta x)] - f[u(x)]}{\Delta x}$$

$$= \lim_{\Delta x \to 0} \frac{f[u(x) + \Delta u] - f[u(x)]}{\Delta x} \qquad \text{Where } \Delta u = u(x + \Delta x) - u(x)$$

$$= \lim_{\Delta x \to 0} \frac{\left[g(\Delta u) + \dfrac{df}{du} \right] \Delta u}{\Delta x} \qquad \text{Substitution}$$

$$= \lim_{\Delta x \to 0} \left[g(\Delta u) + \frac{df}{du} \right] \frac{\Delta u}{\Delta x}$$

$$= \lim_{\Delta x \to 0} \left[g(\Delta u) + \frac{df}{du} \right] \lim_{\Delta x \to 0} \frac{\Delta u}{\Delta x}$$

$$= \left[\lim_{\Delta x \to 0} g(\Delta u) + \lim_{\Delta x \to 0} \frac{df}{du} \right] \lim_{\Delta x \to 0} \frac{\Delta u}{\Delta x}$$

$$= \left[0 + \frac{df}{du} \right] \frac{du}{dx} \qquad \text{Since } g \text{ is continuous at } t = 0$$

$$= \frac{df}{du} \frac{du}{dx} \qquad\qquad\qquad =$$

CAUCHY'S GENERALIZED MEAN VALUE THEOREM (Chapter 4)

Let f and g be functions that are continuous on the closed interval $[a, b]$ and differentiable on the open interval (a, b). If $g(b) \neq g(a)$ and $g'(x) \neq 0$ on (a, b), then

$$\frac{f(b) - f(a)}{g(b) - g(a)} = \frac{f'(c)}{g'(c)}$$

for at least one number c between a and b.

Proof   We begin by defining a special function, just as in the proof of the MVT as presented in Section 4.2. Specifically, let

$$F(x) = f(x) - f(a) - \frac{f(b) - f(a)}{g(b) - g(a)} [g(x) - g(a)]$$

for all x in the closed interval $[a, b]$. In the proof of the MVT in Section 4.2, we show that F satisfies the hypotheses of Rolle's theorem, which means that $F'(c) = 0$ for at least one number c in (a, b). For this number c, we have

$$0 = F'(c) = f'(c) - \frac{f(b) - f(a)}{g(b) - g(a)} g'(c)$$

and the result follows from this equation. $=$

l'HÔPITAL'S THEOREM* (Chapter 4)

For any number a, let f and g be functions that are differentiable on an open interval (a, b), where $g'(x) \neq 0$. Then if $\lim\limits_{x \to a^+} f(x) = 0$, $\lim\limits_{x \to a^+} g(x) = 0$, and $\lim\limits_{x \to a^+} \dfrac{f'(x)}{g'(x)}$ exists, then

$$\lim_{x \to a^+} \frac{f(x)}{g(x)} = \lim_{x \to a^+} \frac{f'(x)}{g'(x)}$$

Proof   First, define auxiliary functions F and G by

$F(x) = f(x)$ for $a < x \leq b$ and $F(a) = 0$;
$G(x) = g(x)$ for $a < x \leq b$ and $G(a) = 0$.

These definitions guarantee that $F(x) = f(x)$ and $G(x) = g(x)$ for $a < x \leq b$ and that $F(a) = G(a) = 0$. Thus, if w is any number between a and b, the functions F and G are continuous on the closed interval $[a, w]$ and differentiable on the open interval (a, w). According to the Cauchy generalized mean value theorem, there exists a number t between a and w for which

$$\frac{F(w) - F(a)}{G(w) - G(a)} = \frac{F'(t)}{G'(t)}$$

$$\frac{F(w)}{G(w)} = \frac{F'(t)}{G'(t)} \qquad \text{Because } F(a) = G(a) = 0$$

$$\frac{f(w)}{g(w)} = \frac{f'(t)}{g'(t)} \qquad \text{Because } F(x) = f(x) \text{ and } G(x) = g(x)$$

$$\lim_{w \to a^+} \frac{f(w)}{g(w)} = \lim_{w \to a^+} \frac{f'(t)}{g'(t)}$$

$$\lim_{w \to a^+} \frac{f(w)}{g(w)} = \lim_{t \to a^+} \frac{f'(t)}{g'(t)} \qquad \text{Because } t \text{ is "trapped" between } a \text{ and } w$$

$$\lim_{x \to a^+} \frac{f(x)}{g(x)} = \lim_{x \to a^+} \frac{f'(x)}{g'(x)} \qquad\qquad \blacksquare$$

CONTINUITY AND DIFFERENTIABILITY OF INVERSE FUNCTIONS

Let f be a one-to-one function so that it possesses an inverse.

1. If f is continuous on a domain I, then f^{-1} is continuous on $f(I)$.
2. If f is differentiable at c, and $f'(c) \neq 0$, then f^{-1} is differentiable at $f(c)$.

Proof

1. Recall that $y = f(x)$ if and only if $x = f^{-1}(y)$. Let I be the open interval (a, b), and let $y_0 = f(x_0)$ be in the open interval $(f(a), f(b))$ so that x_0 is in the open interval (a, b). To prove continuity, we must show that

$$\lim_{y \to y_0} f^{-1}(y) = f^{-1}(y_0) = x_0$$

---

*This is a special case of l'Hôpital's rule. The other cases can be found in most advanced calculus textbooks.

Consider any interval $(x_0 - \epsilon, x_0 + \epsilon)$ for $\epsilon > 0$. We must find an interval $(y_0 - \delta, y_0 + \delta)$ such that whenever y is in $(y_0 - \delta, y_0 + \delta)$, $f^{-1}(y)$ is in $(x_0 - \epsilon, x_0 + \epsilon)$.

Let $\delta_1 = y_0 - f(x_0 - \epsilon)$ and $\delta_2 = f(x_0 + \epsilon) - y_0$. If δ is the smaller of δ_1 and δ_2, it follows that if y is in $(y_0 - \delta, y_0 + \delta)$, then $f^{-1}(y)$ is in $(x_0 - \epsilon, x_0 + \epsilon)$, which is what we wanted to prove.

2. Let $g = f^{-1}$ and $f'[g(a)] \neq 0$. We want to show that g is differentiable.

$$g'(a) = \lim_{g \to a} \frac{g(x) - g(a)}{x - a} \qquad \text{From the definition of derivative}$$

$$= \lim_{x \to a} \frac{y - b}{f(y) - f(b)} \qquad \begin{array}{l} \text{If } y = g(x) \text{ then } x = f(y) \text{ and} \\ \text{if } b = g(a) \text{ then } a = f(b). \end{array}$$

$$= \lim_{y \to b} \frac{y - b}{f(y) - f(b)} \qquad \begin{array}{l} \text{Since } f \text{ is differentiable, it is} \\ \text{continuous, so } g = f^{-1} \text{ is} \\ \text{continuous by part (1). Thus, if} \\ x \to a, \text{ then } g(x) \to g(a) \text{ so} \\ \text{that } y \to b. \end{array}$$

$$= \lim_{y \to b} \frac{1}{\dfrac{f(y) - f(b)}{y - b}}$$

$$= \frac{1}{\displaystyle\lim_{y \to b} \frac{f(y) - f(b)}{y - b}}$$

$$= \frac{1}{f'(b)}$$

Since f is differentiable, we see that g is also differentiable. ▬

LIMIT COMPARISON TEST (Chapter 8)

Suppose $a_k > 0$ and $b_k > 0$ for all sufficiently large k and that

$$\lim_{k \to \infty} \frac{a_k}{b_k} = L \quad \text{where } L \text{ is finite and positive } (0 < L < \infty)$$

Then $\Sigma\, a_k$ and $\Sigma\, b_k$ either both converge or both diverge.

Proof   Assume that $\lim\limits_{k \to \infty} \dfrac{a_k}{b_k} = L$, where $L > 0$. Using $\epsilon = \dfrac{L}{2}$ in the definition of the limit of a sequence, we see that there exists a number N so that

$$\left| \frac{a_k}{b_k} - L \right| < \frac{L}{2} \qquad \text{Whenever } k > N$$

$$-\frac{L}{2} < \frac{a_k}{b_k} - L < \frac{L}{2}$$

$$\frac{L}{2} < \frac{a_k}{b_k} < \frac{3L}{2}$$

$$\frac{L}{2} b_k < a_k < \frac{3L}{2} b_k \qquad b_k > 0$$

This is true for all $k > N$. Now we can complete the proof by using the direct comparison test. Suppose Σb_k converges. Then the series $\Sigma \dfrac{3L}{2} b_k$ also converges, and the inequality

$$a_k < \frac{3L}{2} b_k$$

tells us that the series Σa_k must also converge since it is dominated by a convergent series.

Similarly if Σb_k diverges, the inequality

$$0 < \frac{L}{2} b_k < a_k$$

tells us that Σa_k dominates the divergent series $\Sigma \dfrac{L}{2} b_k$, and it follows that Σa_k also diverges.

Thus Σa_k and Σb_k either both converge or both diverge.   ∎

TAYLOR'S THEOREM (Chapter 8)

If f and all its derivatives exist in an open interval I containing c, then for each x in I,

$$f(x) = f(c) + \frac{f'(c)}{1!}(x - c) + \frac{f''(c)}{2!}(x - c)^2 + \cdots + \frac{f^{(n)}(c)}{n!}(x - c)^n + R_n(x)$$

where the remainder function $R_n(x)$ is given by

$$R_n(x) = \frac{f^{(n+1)}(z_n)}{(n + 1)!}(x - c)^{n+1}$$

for some z_n that depends on x and lies between c and x.

Proof   We shall prove Taylor's theorem by showing that if f and its first $n + 1$ derivatives are defined in an open interval I containing c, then for each fixed x in I,

$$f(x) = f(c) + \frac{f'(c)}{1!}(x - c) + \frac{f''(c)}{2!}(x - c)^2 + \cdots + \frac{f^{(n)}(c)}{n!}(x - c)^n + \frac{f^{(n+1)}(c)}{(n + 1)!}(x - c)^{n+1}$$

where z is some number between x and c. In our proof, we shall apply Cauchy's generalized mean value theorem to the auxiliary functions F and G defined for all t in I as follows:

$$F(t) = f(x) - f(t) - \frac{f'(t)}{1!}(x - t) - \cdots - \frac{f^{(n)}(t)}{n!}(x - t)^n$$

$$G(t) = \frac{(x - t)^{n+1}}{(n + 1)!}$$

Note that $F(x) = G(x) = 0$, and thus Cauchy's generalized mean value theorem tells us

$$\frac{F'(z)}{G'(z)} = \frac{F(x) - F(c)}{G(x) - G(c)} = \frac{F(c)}{G(c)}$$

for some number z between x and c. Rearranging the sides of this equation and finding the derivatives gives

$$\frac{F(c)}{G(c)} = \frac{F'(z)}{G'(z)}$$

$$= \frac{\dfrac{-f^{(n+1)}(z)}{n!}(x-z)^n}{\dfrac{-(x-z)^n}{n!}}$$

$$= f^{(n+1)}(z)$$

$$F(c) = f^{(n+1)}(z)G(c)$$

$$f(x) - f(c) - \frac{f'(c)}{1!}(x-c) - \cdots - \frac{f^{(n)}(c)}{n!}(x-c)^n = f^{(n+1)}(z)\left[\frac{(x-c)^{n+1}}{(n+1)!}\right]$$

Rearranging these terms, we obtain the required equation:

$$f(x) = f(c) + \frac{f'(c)}{1!}(x-c) + \cdots + \frac{f^{(n)}(c)}{n!}(x-c)^n + f^{(n+1)}(z)\left[\frac{(x-c)^{n+1}}{(n+1)!}\right] \quad \blacksquare$$

SUFFICIENT CONDITION FOR DIFFERENTIABILITY (Chapter 12)

If f is a function of x and y, and f, f_x, and f_y are continuous is a disk D centered at (x_0, y_0), then f is differentiable at (x_0, y_0).

Proof   If (x, y) is a point in D, we have

$$f(x, y) - f(x_0, y_0) = f(x, y) - f(x_0, y) + f(x_0, y) - f(x_0, y_0)$$

The function $f(x, y)$ with y fixed satisfies the conditions of the mean value theorem, so that

$$f(x, y) - f(x_0, y) = f_x(x_0, y)(x - x_0)$$

for some number x_1 between x and x_0, and similarly, there is a number y_1 between y and y_0 such that

$$f(x_0, y) - f(x_0, y_0) = f_y(x_0, y_1)(y - y_0)$$

Substituting these expressions, we obtain

$$f(x, y) - f(x_0, y_0) = [f(x, y) - f(x_0, y)] + [f(x_0, y) - f(x_0, y_0)]$$
$$= [f_x(x_1, y)(x - x_0)] + [f_y(x_0, y_1)(y - y_0)]$$
$$= [f_x(x_1, y)(x - x_0)] + \underbrace{f_x(x_0, y_0)(x - x_0) - f_x(x_0, y_0)(x - x_0)}_{\text{This is zero.}}$$

$$+ [f_y(x_0, y_1)(y - y_0)] + \underbrace{f_y(x_0, y_0)(y - y_0) - f_y(x_0, y_0)(y - y_0)}_{\text{This is zero.}}$$

$$= f_x(x_0, y_0)(x - x_0) + f_y(x_0, y_0)(y - y_0)$$
$$+ [f_x(x_1, y) - f_x(x_0, y_0)](x - x_0)$$
$$+ [f_y(x_0, y_1) - f_y(x_0, y_0)](y - y_0)$$

Let $\epsilon_1(x, y)$ and $\epsilon_2(x, y)$ be the functions

$$\epsilon_1(x, y) = f_x(x_1, y) - f_x(x_0, y_0) \quad \text{and} \quad \epsilon_2(x, y) = f_y(x_0, y_1) - f_y(x_0, y_0)$$

Then since x_1 is between x and x_0, and y_1 is between y and y_0, and the partial derivatives f_x and f_y are continuous at (x_0, y_0), we have

$$\lim_{(x, y) \to (x_0, y_0)} \epsilon_1(x, y) = \lim_{(x, y) \to (x_0, y_0)} [f_x(x_1, y) - f_x(x_0, y_0)] = 0$$

$$\lim_{(x, y) \to (x_0, y_0)} \epsilon_2(x, y) = \lim_{(x, y) \to (x_0, y_0)} [f_y(x_0, y_1) - f_y(x_0, y_0)] = 0$$

so that f is differentiable at (x_0, y_0), as required.

CHANGE OF VARIABLE FORMULA FOR MULTIPLE INTEGRATION (Chapter 13)

Suppose f is a continuous function of a region D, and let D^* be the image of the domain D under the change of variable $x = g(u, v), y = h(u, v)$, where g and h are continuously differentiable on D^*. Then

$$\int_D \int f(x, y) \, dy \, dx = \int_{D^*} \int f[g(u, v), h(u, v)] \underbrace{\left| \frac{\partial(x, y)}{\partial(u, v)} \right|}_{\text{absolute value of Jacobian}} dv \, du$$

Proof   A proof of this theorem is found in advanced calculus, but we can provide a geometric argument that makes this formula plausible in the special case where $f(x, y) = 1$. In particular, we shall show that in order to find the area of a region D in the xy-plane using the change of variable $x = X(u, v)$ and $y = Y(u, v)$, it is reasonable to use the formula for area, A:

$$A = \int_D \int dy \, dx = \int_{D^*} \int \left| \frac{\partial(x, y)}{\partial(u, v)} \right| dv \, du$$

where D^* is the region in the uv-plane that corresponds to D.

Suppose the given change of variable has an inverse $u = u(x, y), v = v(x, y)$ that transforms the region D in the xy-plane into a region D^* in the uv-plane. To find the area of D^* in the uv-coordinate system, it is natural to use a rectangular grid, with vertical lines $u =$ constant and horizontal lines $v =$ constant, as shown in Figure B.1a. In the xy-plane, the equations $u =$ constant and $v =$ constant will be families of parallel curves, which provide a curvilinear grid for the region D^* (Figure B.1b).

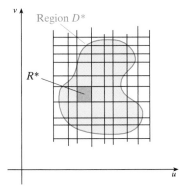

a. A rectangular grid in the uv-plane.

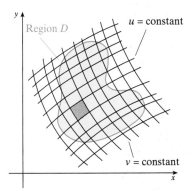

b. Corresponding grid in the xy-plane is curvilinear.

■ **FIGURE B.1**

Next, let R^* be a typical rectangular cell in the uv-grid that covers D^*, and let R be the corresponding set in the xy-plane (that is, R^* is the image of R under the given change of variable). Then, as shown in Figure B.2, if R^* has vertices $A(\bar{u}, \bar{v})$, $B(\bar{u} + \Delta u, \bar{v})$, $C(\bar{u}, \bar{v} + \Delta v)$, and $D(\bar{u} + \Delta u, \bar{v} + \Delta v)$, the set R will be the interior of a curvilinear rectangle with vertices

$$A[X(\bar{u}, \bar{v}), Y(\bar{u}, \bar{v})]$$
$$B[X(\bar{u} + \Delta u, \bar{v}), Y(\bar{u} + \Delta u, \bar{v})]$$
$$C[X(\bar{u}, \bar{v} + \Delta v), Y(\bar{u}, \bar{v} + \Delta v)]$$
$$D[X(\bar{u} + \Delta u, \bar{v} + \Delta v), Y(\bar{u} + \Delta u, \bar{v} + \Delta v)]$$

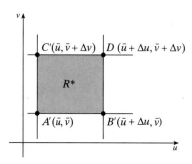

a. R^* is a typical rectangular cell in the grid covering D^*.

b. The image of R under the change of variable is a curvilinear rectangle R.

■ **FIGURE B.2**

Note that the curved side of R joining A to B may be approximated by the secant vector

$$\mathbf{AB} = [X(\bar{u} + \Delta u, \bar{v}) - X(\bar{u}, \bar{v})]\mathbf{i} + [Y(\bar{u} + \Delta u, \bar{v}) - Y(\bar{u}, \bar{v})]\mathbf{j}$$

and by applying the mean value theorem, we find that

$$\mathbf{AB} = \left[\frac{\partial x}{\partial u}(a, \bar{v}) \, \Delta u\right]\mathbf{i} + \left[\frac{\partial y}{\partial u}(b, \bar{v}) \, \Delta u\right]\mathbf{j}$$

for some numbers a, b between $\bar{u}$ and $\bar{u} + \Delta u$. If Δu is very small, a an b are approximately the same as $\bar{u}$, and we can approximate $\mathbf{AB}$ by the vector

$$\alpha = \left[\frac{\partial x}{\partial u} \Delta u\right]\mathbf{i} + \left[\frac{\partial y}{\partial u} \Delta u\right]\mathbf{j}$$

where the partials are evaluated at the point $(\bar{u}, \bar{v})$. Similarly, the curved side of R joining A and C may be approximated by the vector

$$\beta = \left[\frac{\partial x}{\partial v} \Delta v\right]\mathbf{i} + \left[\frac{\partial y}{\partial v} \Delta v\right]\mathbf{j}$$

The area of the curvilinear rectangle R is approximately the same as that of the parallelogram determined by α and β; that is,

$$\|\boldsymbol{\alpha} \times \boldsymbol{\beta}\| = \begin{vmatrix} \mathbf{i} & \mathbf{j} & \mathbf{k} \\ \dfrac{\partial x}{\partial u}\Delta u & \dfrac{\partial y}{\partial u}\Delta u & 0 \\ \dfrac{\partial x}{\partial y}\Delta v & \dfrac{\partial y}{\partial v}\Delta v & 0 \end{vmatrix}$$

$$= \left\| \begin{vmatrix} \mathbf{i} & \mathbf{j} & \mathbf{k} \\ \dfrac{\partial x}{\partial u} & \dfrac{\partial y}{\partial u} & 0 \\ \dfrac{\partial x}{\partial v} & \dfrac{\partial y}{\partial v} & 0 \end{vmatrix} \Delta v \Delta u \right\|$$

$$= \left| \dfrac{\partial x}{\partial y} \dfrac{\partial y}{\partial v} - \dfrac{\partial y}{\partial u} \dfrac{\partial x}{\partial v} \right| \Delta v \Delta u$$

$$= \left| \dfrac{\partial(x, y)}{\partial(u, v)} \right| \Delta v \Delta u$$

By adding the contributions of all cells in the partition of D, we can approximate the area of D as follows:

$$\text{APPROXIMATE AREA OF } D = \sum \text{APPROXIMATE AREA OF CURVILINEAR RECTANGLES}$$

$$= \sum \left| \dfrac{\partial(x, y)}{\partial(u, v)} \right| \Delta v \Delta u$$

Finally, using a limit to "smooth out" the approximation, we find

$$A = \iint\limits_{D} dy\, dx = \lim \sum \left| \dfrac{\partial(x, y)}{\partial(u, v)} \right| \Delta v \Delta u$$

$$= \iint\limits_{D^*} \left| \dfrac{\partial(x, y)}{\partial(u,v)} \right| dv\, du \qquad\qquad =$$

STOKES'S THEOREM (Chapter 14)

Let S be an oriented surface with unit normal vector $\mathbf{N}$, and assume that S is bounded by a closed, piecewise smooth curve C whose orientation is compatible with that of S. If $\mathbf{F}$ is a continuous vector field whose components have continuous partial derivatives on an open region containing S and C, then

$$\int_{C} \mathbf{F} \cdot d\mathbf{R} = \iint\limits_{S} (\text{curl } \mathbf{F} \cdot \mathbf{N})\, dS$$

Proof   The general proof cannot be considered until advanced calculus. However, a proof for the case where S is a graph and $\mathbf{F}$, S, and C are "well behaved" can be given. Let S be given by $z = g(x, y)$ where (x, y) is in a region D of the xy-plane. Assume g has continuous second-order partial derivatives. Let C_1 be the projection of C in the xy-plane, as shown in Figure B.3. Also let $\mathbf{F}(x, y, z) = f(x, y, z)\mathbf{i} + g(x, y, z)\mathbf{j} + h(x, y, z)\mathbf{k}$ where the partial derivatives of f, g, and h are continuous.

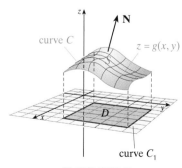

■ **FIGURE B.3**

We will evaluate each side of Stokes's theorem separately and show that the results for each are the same. If $x = x(t)$, $y = y(t)$, and $z = z(t)$ for $a \leq t \leq b$ and $\mathbf{R}(t) = x(t)\mathbf{i} + y(t)\mathbf{j} + z(t)\mathbf{k}$, then

$$
\begin{aligned}
\int_C \mathbf{F} \cdot d\mathbf{R} &= \int_a^b \left(f\frac{dx}{dt} + g\frac{dy}{dt} + h\frac{dz}{dt} \right) dt \\
&= \int_a^b \left[f\frac{dx}{dt} + g\frac{dy}{dt} + h\left(\frac{\partial z}{\partial x}\frac{dx}{dt} + \frac{\partial z}{\partial y}\frac{dy}{dt} \right) \right] dt \qquad \text{Chain rule} \\
&= \int_a^b \left[\left(f + h\frac{\partial z}{\partial x} \right)\frac{dx}{dt} + \left(g + h\frac{\partial z}{\partial y} \right)\frac{dy}{dt} \right] dt \\
&= \int_{C_1} \left(f + h\frac{\partial z}{\partial x} \right) dx + \left(g + h\frac{\partial z}{\partial y} \right) dy \\
&= \iint_D \left[\frac{\partial}{\partial x}\left(g + h\frac{\partial z}{\partial y} \right) - \frac{\partial}{\partial y}\left(f + h\frac{\partial z}{\partial x} \right) \right] dA \qquad \text{Green's theorem} \\
&= \iint_D \left[\left(\frac{\partial g}{\partial x} + \frac{\partial g}{\partial z}\frac{\partial z}{\partial x} + \frac{\partial h}{\partial x}\frac{\partial z}{\partial y} + \frac{\partial h}{\partial z}\frac{\partial z}{\partial x}\frac{\partial z}{\partial y} + h\frac{\partial^2 z}{\partial y \partial x} \right) \right. \\
&\qquad\quad \left. - \left(\frac{\partial f}{\partial y} + \frac{\partial f}{\partial z}\frac{\partial z}{\partial y} + \frac{\partial h}{\partial y}\frac{\partial z}{\partial x} + \frac{\partial h}{\partial z}\frac{\partial z}{\partial y}\frac{\partial z}{\partial x} + h\frac{\partial^2 z}{\partial y \partial x} \right) \right] dA \qquad \begin{array}{l}\text{Green's theorem}\\ \text{again}\end{array} \\
&= \iint_D \left(\frac{\partial g}{\partial x} + \frac{\partial g}{\partial z}\frac{\partial z}{\partial x} + h\frac{\partial^2 z}{\partial y \partial x} - \frac{\partial f}{\partial y} - \frac{\partial f}{\partial z}\frac{\partial z}{\partial y} - h\frac{\partial^2 z}{\partial y \partial x} \right) dA
\end{aligned}
$$

We now start over by evaluating the other side of Stokes's theorem:

$$
\begin{aligned}
\iint_S \text{curl } \mathbf{F} \cdot d\mathbf{S} &= \iint_D \left[-\left(\frac{\partial h}{\partial y} - \frac{\partial g}{\partial z} \right)\frac{\partial z}{\partial x} - \left(\frac{\partial f}{\partial z} - \frac{\partial h}{\partial x} \right)\frac{\partial z}{\partial y} + \left(\frac{\partial g}{\partial x} - \frac{\partial f}{\partial y} \right) \right] dA \\
&= \iint_D \left(\frac{\partial g}{\partial x} + \frac{\partial g}{\partial z}\frac{\partial z}{\partial x} + h\frac{\partial^2 z}{\partial y \partial x} - \frac{\partial f}{\partial y} - \frac{\partial f}{\partial z}\frac{\partial z}{\partial y} - h\frac{\partial^2 z}{\partial y \partial x} \right) dA
\end{aligned}
$$

Since these results are the same, we have

$$
\int_C \mathbf{F} \cdot d\mathbf{R} = \iint_S (\text{curl } \mathbf{F} \cdot \mathbf{N}) \, dS
$$
■

DIVERGENCE THEOREM (Chapter 14)

Let D be a region in space bounded by a smooth, orientable closed surface S. If $\mathbf{F}$ is a continuous vector field whose components have continuous partial derivatives in D, then

$$
\iint_S \mathbf{F} \cdot \mathbf{N} \, dS = \iiint_D \text{div } \mathbf{F} \, dV
$$

where $\mathbf{N}$ is an outward unit normal to the surface S.

Proof   Let $\mathbf{F}(x, y, z) = f(x, y, z)\mathbf{i} + g(x, y, z)\mathbf{j} + h(x, y, z)\mathbf{k}$. If we state the divergence theorem using this notation for $\mathbf{F}$, we have

$$
\iint_S [f(\mathbf{i} \cdot \mathbf{N}) + g(\mathbf{j} \cdot \mathbf{N}) + h(\mathbf{k} \cdot \mathbf{N})] \, dS = \iiint_D \left(\frac{\partial f}{\partial x} + \frac{\partial g}{\partial y} + \frac{\partial h}{\partial z} \right) dV
$$

$$
\iint_S f(\mathbf{i} \cdot \mathbf{N}) \, dS + \iint_S g(\mathbf{j} \cdot \mathbf{N}) \, dS + \iint_S h(\mathbf{k} \cdot \mathbf{N}) \, dS = \iiint_D \frac{\partial f}{\partial x} \, dV + \iiint_D \frac{\partial g}{\partial y} \, dV + \iiint_D \frac{\partial h}{\partial z} \, dV
$$

This result can be verified by proving

$$\iint_S f(\mathbf{i} \cdot \mathbf{N}) \, dS = \iiint_D \frac{\partial f}{\partial x} \, dV$$

$$\iint_S g(\mathbf{j} \cdot \mathbf{N}) \, dS = \iiint_D \frac{\partial g}{\partial y} \, dV$$

$$\iint_S h(\mathbf{k} \cdot \mathbf{N}) \, dS = \iiint_D \frac{\partial h}{\partial z} \, dV$$

Since the proof of each of these is virtually identical, we will show the verification for the last of these three; the other two can be done in a similar fashion. We will evaluate this third integral by separately evaluating the left and right sides to show they are the same.

We will restrict our proof to a "standard region" as described in the proof of Green's theorem in Section 14.4. The complete proof can then be completed by decomposing the general surface S into a finite number of "standard regions."

The standard solid region we shall consider has a top surface S_T with equation $z = u(x, y)$ and a bottom surface S_B with equation $z = v(x, y)$. We assume that both S_T and S_B project onto the region R in the xy-plane. The lateral surface S_L of the region is the set of all (x, y, z) such that $v(x, y) \leq z \leq u(x, y)$ on the boundary of R, as shown in Figure B.4.

We know that the outward unit normal (directed upward) to the top surface S_T is

$$\mathbf{N}_T = \frac{-u_x \mathbf{i} - u_y \mathbf{j} + \mathbf{k}}{\sqrt{u_x^2 + u_y^2 + 1}}$$

and

$$dS = \sqrt{u_x^2 + u_y^2 + 1} \, dA_{xy}$$

Thus,

$$\begin{aligned}
\iint_{S_T} h(\mathbf{k} \cdot \mathbf{N}_T) dS &= \iint_R h\left(\frac{1}{\sqrt{u_x^2 + u_y^2 + 1}}\right)\left(\sqrt{u_x^2 + u_y^2 + 1} \, dA_{xy}\right) \\
&= \iint_R h \, dA_{xy} \\
&= \iint_R h(x, y, z) \, dA_{xy} \\
&= \iint_R h[x, y, u(x, y)] \, dA_{xy}
\end{aligned}$$

Similarly, the outward unit normal $\mathbf{N}_B$ to the bottom surface S_B is directed downward so that

$$\mathbf{N}_T = \frac{v_x \mathbf{i} - v_y \mathbf{j} - \mathbf{k}}{\sqrt{u_x^2 + u_y^2 + 1}}$$

and

$$\begin{aligned}
\iint_{S_B} h(\mathbf{k} \cdot \mathbf{N}_B) \, dS &= -\iint_R h \, dA_{xy} \\
&= -\iint_R h(x, y, z) \, dA_{xy} \\
&= -\iint_R h[x, y, v(x, y)] \, dA_{xy}
\end{aligned}$$

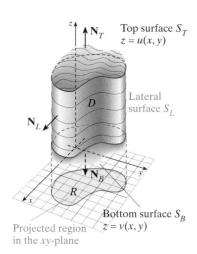

■ FIGURE B.4 A standard solid region in $\mathbb{R}^3$

Because the outward unit normal $\mathbf{N}_L$ is horizontal on the lateral surface S_L, it is perpendicular to $\mathbf{k}$, and

$$\iint_{S_L} h(\mathbf{k} \cdot \mathbf{N}_L)\, dS = 0$$

We now add for the surface S:

$$\iint_S h(\mathbf{k} \cdot \mathbf{N})\, dS = \iint_{S_T} h(\mathbf{k} \cdot \mathbf{N}_T)\, dS + \iint_{S_B} h(\mathbf{k} \cdot \mathbf{N}_B)\, dS + \iint_{S_L} h(\mathbf{k} \cdot \mathbf{N}_L)\, dS$$

$$= \iint_R h[x, y, u(x, y)]\, dA_{xy} - \iint_R h[x, y, v(x, y)] dA_{xy} + 0$$

$$= \iint_R \{h[x, y, u(x, y)] - h[x, y, v(x, y)]\}\, dA_{xy}$$

Next, we start over by looking at the triple integral on the right side. Notice that we can describe the solid S as the set of all (x, y, z) for (x, y) in R and $v(x, y) \le z \le u(x, y)$. Thus,

$$\iiint_D \frac{\partial h}{\partial z}\, dV = \iint_R \left[\int_{v(x, y)}^{u(x, y)} \frac{\partial h}{\partial z} (x, y, z)\, dz \right] dA_{xy}$$

$$= \int_R \{h[x, y, u(x, y)] - h[x, y, v(x, y)]\}\, dA_{xy}$$

We see that the left and right sides are the same, so

$$\iint_S h(\mathbf{k} \cdot \mathbf{N})\, dS = \iiint_D \frac{\partial h}{\partial z}\, dV$$

We can now conclude that (by similar arguments for the other two parts):

$$\iint_S \mathbf{F} \cdot \mathbf{N}\, dS = \iiint_D \text{div } \mathbf{F}\, dV \qquad \qquad =\!=$$

APPENDIX C

Significant Digits

Throughout this book, various technology windows appear, and in the answers to the problems you will frequently find approximate (decimal) answers. Sometimes your answer may not exactly agree with the answer found in the back of the book. This does not necessarily mean that your answer is incorrect, particularly if your answer is very close to the given answer.

To use your calculator intelligently and efficiently, you should become familiar with its functions and practice doing problems with the same calculator, whenever possible. Read the technology windows provided throughout this text, and consult the owner's manual for your particular calculator when you have questions. In addition, there are *Technology Manuals* accompanying this text, which are available for TI and HP graphic calculators, as well as for MATHLAB and for Maple.

SIGNIFICANT DIGITS

Applications involving measurements can never be exact. It is particularly important that you pay attention to the accuracy of your measurements when you use a computer or calculator, because using available technology can give you a false sense of security about the accuracy in a particular problem. For example, if you measure a triangle and find the sides are approximately 1.2 and 3.4 and then find the ratio of $1.2/3.4 \approx 0.35294117$, it appears that the result is more accurate than the original measurements! Some discussion about accuracy and significant digits is necessary.

The digits known to be correct in a number obtained by a measurement are called **significant digits.** The digits 1, 2, 3, 4, 5, 6, 7, 8, and 9 are always significant, whereas the digit 0 may or may not be significant.

1. Zeros that come between two other digits are significant, as in 203 or 10.04.

2. If the zero's only function is to place the decimal point, it is not significant, as in

$$0.00000\,23 \quad \text{or} \quad 23{,}000$$

Placeholders   Placeholders

If a zero does more than fix the decimal point, it is significant, as in

$$0.0023\,0 \qquad \text{or} \qquad 23{,}000.0\,1$$

↑   These are significant.

This digit is significant.

This second rule can, of course, result in certain ambiguities, such as 23,000 (measured to the *exact* unit). To avoid such confusion, we use scientific notation in this case:

2.3×10^4 has two significant digits;

2.3000×10^4 has five significant digits.

Numbers that come about by counting are considered to be exact and are correct to any number of significant digits.

When you compute an answer using a calculator, the answer may have 10 or more digits. In the technology windows, we generally show the 10 or 12 digits that result from the numerical calculation, but frequently the number in the answer section will have only 5 or 6 digits in the answer. It seems clear that if the first 3 or 4 nonzero digits of the answer coincide, you probably have the correct method of doing the problem.

However, you might ask why are there discrepancies, and how many digits should you use when you write down your final answer? Roughly speaking, the significant digits in a number are the digits that have meaning. To clarify the concept, we must for the moment assume that we know the exact answer. We then assume that we have been able to compute an approximation to this exact answer.

Usually we do this by some sort of iterative process, in which the answers are getting closer and closer to the exact answer. In such a process, we hope that the number of significant digits in our approximate answer is increasing at each trial. If our approximate answer is, say, 6 digits long (some of those digits might even be zero), and the difference between our answer and the exact answer is 4 units or less in the last place, then the first 5 digits are significant.

For example, if the exact answer is 3.14159 and our approximate answer is 3.14162, then our answer has 6 significant digits, but is correct to 5 significant digits. Note that saying that our answer is correct to 5 significant digits does not guarantee that all of those 5 digits exactly match the first 5 digits of the exact answer. In fact, if an exact answer is 6.001 and our computed answer is 5.997, then our answer is correct to 3 significant digits and not one of them matches the digits in the exact answer. Also note that it may be necessary for an approximation to have more digits than are actually significant for it to have a certain number of significant digits. For example, if the exact answer is 6.003 and our approximation is 5.998, then it has 3 significant digits, but only if we consider the total 4-digit number, and do not strip off the last non-significant digit.

Again, suppose you know that all digits are significant in the number 3.456; then you know that the exact number is less than 3.4565 and at least 3.4555. Some people may say that the number 3.456 is correct to 3 decimal places. This is the same as saying that it has 4 significant digits.

Why bother with significant digits? If you multiply (or divide) two numbers with the same number of significant digits, then the product will generally be at least twice as long, but will have roughly the same number of significant digits as the original factors. You can then dispense with the unneeded digits. In fact, to keep them would be misleading about the accuracy of the result.

Frequently we can make an educated guess of the number of significant digits in an answer. For example, if we compute an iterative approximation such as

$$2.3123, \quad 2.3125, \quad 2.3126, \quad 2.31261, \quad 2.31262, \ldots$$

we would generally conclude that the answer is 2.3126 to 5 significant digits. Of course, we may very well be wrong, and if we continued iterating, the answer might end up as 2.4.

ROUNDING AND RULES OF COMPUTATION USED IN THIS BOOK

In hand and calculator computations, rounding a number is done to reduce the number of digits displayed and make the number easier to comprehend. Furthermore, if you suspect that the digit in the last place is not significant, then you might be tempted to round and remove this last digit. This can lead to error. For example, if the computed value is 0.64 and the true value is known to be between 0.61 and 0.67, then the computed value has only 1 significant digit. However, if we round it to 0.6 and the true value is really 0.66, then 0.6 is not correct to even 1 significant digit. In the interest of making the text easier to read, we have used the following rounding procedure:

Rounding Procedure   To round off numbers:

1. Increase the last retained digit by 1 if the remainder is greater than or equal to 5; or

2. Retain the last digit unchanged if the remainder is less than 5.

Elaborate rules for computation of approximate data can be developed when needed (in chemistry, for example), but in this text we will use three simple rules:

Rules for Significant Digits

Addition–subtraction: Add or subtract in the usual fashion, and then round off the result so that the last digit retained is in the column farthest to the right in which both given numbers have significant digits.

Multiplication–Division: Multiply or divide in the usual fashion, and then round off the results to the smaller number of significant digits found in either of the given numbers.

Counting numbers: Numbers used to count or whole numbers used as exponents are considered to be correct to any number of significant digits.

Rounding Rule   We use the following rounding procedure in problems requiring rounding by involving several steps: *Round only once, at the end. That is, do not work with rounded results, because round-off errors can accumulate.*

CALCULATOR EXPERIMENTS

You should be aware that you are much better than your calculator at performing certain computations. For example, almost all calculators will fail to give the correct answer to

$$(10.0 \text{ EE} + 50.0) + 911.0 - (10.0 \text{ EE} + 50.0)$$

Calculators will return the value of 0, but you know at a glance that the answer is 911.0. We must reckon with this poor behavior on the part of calculators, which is called *loss of accuracy* due to *catastrophic cancellation*. In this case, it is easy to catch the error immediately, but what if the computation is so complicated (or hidden by other computations) that we do not see the error?

First, we want to point out that the order in which you perform computations can be very important. For example, most calculators will correctly conclude that

$$(10.0 \text{ EE} + 50.0) - (10.0 \text{ EE} + 50.0) + 911.0 = 911$$

There are other cases besides catastrophic cancellation where the order in which a computation is performed will substantially affect the result. For example, you may not be able to calculate

$$(10.0 \text{ EE} + 50.0)*(911.0 \text{ EE} + 73.0)/(20.0 \text{ EE} + 60.0)$$

but rearranging the factors as

$$((10.0 \text{ EE} + 50.0)/(20.0 \text{ EE} + 60.0))*(911.0 \text{ EE} + 73.0)$$

should provide the correct answer of 4.555 EE 65. So, for what do we need to watch? If you subtract two numbers that are close to each other in magnitude you *may* obtain an inaccurate result. When you have a sequence of multiplications and divisions in a string, try to arrange the factors so that the result of each intermediate calculation stays as close as possible to 1.0.

Second, since a calculator performs all computations with a finite number of digits, it is unable to do exact computations involving nonterminating decimals. This enables us to see how many digits the calculator actually uses when it computes a result. For example, the computation

$$(7.0/17.0)*(17.0)$$

should give the result 7.0, but on most calculators it does not. The size of the answer gives an indication of how many digits "Accuracy" the calculator uses internally. That is, the calculator may display decimal numbers that have 10 digits, but use 12 digits internally. If the answer to the above computation is something similar to 1.0 EE -12, then the calculator is using 12 digits internally.

TRIGONOMETRIC EVALUATIONS

In many problems you will be asked to compute the values of trigonometric functions such as the sine, cosine, or tangent. In calculus, trigonometric arguments are usually assumed to be measured in radians. You must make sure the calculator is in radian mode. If it is in radian mode, then the sine of a small number will almost be equal to that number. For example,

$$\sin(0.00001) = 1\text{E} -5 \quad (\text{which is } 0.00001)$$

If not, then you are not using radian mode. Make sure you know how to put your calculator in radian mode.

GRAPHING BLUNDERS

When you are using the graphing features, you must always be careful to choose reasonable scales for the domain (horizontal scale) and range (vertical scale). If the scale is too large, you may not see important wiggles. If the scale is too small, you may not see important behavior elsewhere in the plane. Of course, knowing the techniques of graphing discussed in Chapter 4 will prevent you from making such blunders. Some calculators may have trouble with curves that jump suddenly at a point. An example of such a curve would be

$$y = \frac{e^x}{x}$$

which jumps at the origin. Try plotting this curve with your calculator using different horizontal and vertical scales, making sure that you understand how your calculator handles such graphs.

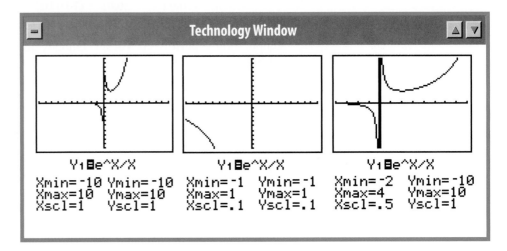

Technology Window		
Y₁■e^X/X	Y₁■e^X/X	Y₁■e^X/X
Xmin=-10 Ymin=-10	Xmin=-1 Ymin=-1	Xmin=-2 Ymin=-10
Xmax=10 Ymax=10	Xmax=1 Ymax=1	Xmax=4 Ymax=10
Xscl=1 Yscl=1	Xscl=.1 Yscl=.1	Xscl=.5 Yscl=1

APPENDIX D

Short Table of Integrals

SMH

Each formula is numbered for easy reference. The numbers in this short table are not sequential because this short table is truncated from the table of integrals found in the *Student Mathematics Handbook*.

BASIC FORMULAS

1. Constant rule $\int 0\, du = c$

2. Power rule $\int u^n\, du = \dfrac{u^{n+1}}{n+1}; \quad n \neq -1$

$\int u^n\, du = \ln|u|; \quad n = -1$

3. Exponential rule $\int e^u\, du = e^u$

4. Logarithmic rule $\int \ln|u|\, du = u \ln|u| - u$

TRIGONOMETRIC RULES

5. $\displaystyle\int \sin u\, du = -\cos u$

6. $\displaystyle\int \cos u\, du = \sin u$

7. $\displaystyle\int \tan u\, du = -\ln|\cos u|$

8. $\displaystyle\int \cot u\, du = \ln|\sin u|$

9. $\displaystyle\int \sec u\, du = \ln|\sec u + \tan u|$

10. $\displaystyle\int \csc u\, du = \ln|\csc u - \cot u|$

11. $\displaystyle\int \sec^2 u\, du = \tan u$

12. $\displaystyle\int \csc^2 u\, du = -\cot u$

13. $\displaystyle\int \sec u \tan u\, du = \sec u$

14. $\displaystyle\int \csc u \cot u\, du = -\csc u$

EXPONENTIAL RULE

15. $\displaystyle\int a^u\, du = \frac{a^u}{\ln a}; \quad a > 0, a \neq 1$

HYPERBOLIC RULES

16. $\displaystyle\int \cosh u\, du = \sinh u$

17. $\displaystyle\int \sinh u\, du = \cosh u$

18. $\displaystyle\int \tanh u\, du = \ln|\cosh u|$

19. $\displaystyle\int \coth u\, du = \ln|\sinh u|$

20. $\displaystyle\int \operatorname{sech} u\, du = \tan^{-1}(\sinh u)$

21. $\displaystyle\int \operatorname{csch} u\, du = \ln\left|\tanh \frac{u}{2}\right|$

INVERSE RULES

22. $\displaystyle\int \frac{du}{\sqrt{a^2 - u^2}} = \sin^{-1}\frac{u}{a}$

23. $\displaystyle\int \frac{du}{\sqrt{u^2 - a^2}} = \cosh^{-1}\frac{u}{a}$

24. $\displaystyle\int \frac{du}{a^2 + u^2} = \frac{1}{a}\tan^{-1}\frac{u}{a}$

25. $\displaystyle\int \frac{du}{a^2 - u^2} = \begin{cases} \dfrac{1}{a}\tanh^{-1}\dfrac{u}{a} & \text{if } \left|\dfrac{u}{a}\right| < 1 \\[2ex] \dfrac{1}{a}\coth^{-1}\dfrac{u}{a} & \text{if } \left|\dfrac{u}{a}\right| > 1 \end{cases}$

26. $\displaystyle\int \frac{du}{u\sqrt{u^2 - a^2}} = \frac{1}{a}\sec^{-1}\left|\frac{u}{a}\right|$

27. $\displaystyle\int \frac{du}{u\sqrt{a^2 - u^2}} = -\frac{1}{a}\operatorname{sech}^{-1}\left|\frac{u}{a}\right|$

28. $\displaystyle\int \frac{du}{\sqrt{1 + u^2}} = \sinh^{-1} u$

29. $\displaystyle\int \frac{du}{u\sqrt{1 + u^2}} = -\operatorname{csch}^{-1}|u|$

INTEGRALS INVOLVING $au + b$

30. $\displaystyle\int (au + b)^n\, du = \frac{(au + b)^{n+1}}{(n + 1)a}$

31. $\int u(au + b)^n du = \dfrac{(au + b)^{n+2}}{(n + 2)a^2} - \dfrac{b(au + b)^{n+1}}{(n + 1)a^2}$

32. $\int u^2(au + b)^n du = \dfrac{(au + b)^{n+3}}{(n + 3)a^3} - \dfrac{2b(au + b)^{n+2}}{(n + 2)a^3} + \dfrac{b^2(au + b)^{n+1}}{(n + 1)a^3}$

33. $\int u^m(au + b)^n du = \begin{cases} \dfrac{u^{m+1}(au + b)^n}{m + n + 1} + \dfrac{nb}{m + n + 1} \int u^m(au + b)^{n-1}\, du \\[2ex] \dfrac{u^m(au + b)^{n+1}}{(m + n + 1)a} - \dfrac{mb}{(m + n + 1)a} \int u^{m-1}(au + b)^n du \\[2ex] \dfrac{-u^{m+1}(au + b)^{n+1}}{(n + 1)b} + \dfrac{m + n + 2}{(n + 1)b} \int u^m(au + b)^{n+1} du \end{cases}$

34. $\int \dfrac{du}{au + b} = \dfrac{1}{a} \ln|au + b|$

35. $\int \dfrac{u\, du}{au + b} = \dfrac{u}{a} - \dfrac{b}{a^2} \ln|au + b|$

36. $\int \dfrac{u^2\, du}{au + b} = \dfrac{(au + b)^2}{2a^3} - \dfrac{2b(au + b)}{a^3} + \dfrac{b^2}{a^3} \ln|au + b|$

37. $\int \dfrac{u^3\, du}{au + b} = \dfrac{(au + b)^3}{3a^4} - \dfrac{3b(au + b)^2}{2a^4} + \dfrac{3b^2(au + b)}{a^4} - \dfrac{b^3}{a^4} \ln|au + b|$

INTEGRALS INVOLVING $u^2 + a^2$

55. $\int \dfrac{du}{u^2 + a^2} = \dfrac{1}{a} \tan^{-1} \dfrac{u}{a}$

56. $\int \dfrac{u\, du}{u^2 + a^2} = \dfrac{1}{2} \ln(u^2 + a^2)$

57. $\int \dfrac{u^2\, du}{u^2 + a^2} = u - a \tan^{-1} \dfrac{u}{a}$

58. $\int \dfrac{u^3\, du}{u^2 + a^2} = \dfrac{u^2}{2} - \dfrac{a^2}{2} \ln(u^2 + a^2)$

59. $\int \dfrac{du}{u(u^2 + a^2)} = \dfrac{1}{2a^2} \ln\left(\dfrac{u^2}{u^2 + a^2}\right)$

60. $\int \dfrac{du}{u^2(u^2 + a^2)} = -\dfrac{1}{a^2 u} - \dfrac{1}{a^3} \tan^{-1} \dfrac{u}{a}$

61. $\int \dfrac{du}{u^3(u^2 + a^2)} = -\dfrac{1}{2a^2 u^2} - \dfrac{1}{2a^4} \ln\left(\dfrac{u^2}{u^2 + a^2}\right)$

INTEGRALS INVOLVING $u^2 - a^2,\, u^2 > a^2$

74. $\int \dfrac{du}{u^2 - a^2} = \dfrac{1}{2a} \ln\left|\dfrac{u - a}{u + a}\right|$ or $-\dfrac{1}{a} \coth^{-1} \dfrac{u}{a}$

75. $\int \dfrac{u\, du}{u^2 - a^2} = \dfrac{1}{2} \ln|u^2 - a^2|$

76. $\int \dfrac{u^2\, du}{u^2 - a^2} = u + \dfrac{a}{2} \ln\left|\dfrac{u - a}{u + a}\right|$

77. $\int \dfrac{u^3\, du}{u^2 - a^2} = \dfrac{u^2}{2} + \dfrac{a^2}{2} \ln|u^2 - a^2|$

78. $\int \dfrac{du}{u(u^2 - a^2)} = \dfrac{1}{2a^2} \ln\left|\dfrac{u^2 - a^2}{u^2}\right|$

79. $\displaystyle\int \frac{du}{u^2(u^2 - a^2)} = \frac{1}{a^2 u} + \frac{1}{2a^3} \ln \left| \frac{u - a}{u + a} \right|$

80. $\displaystyle\int \frac{du}{u^3(u^2 - a^2)} = \frac{1}{2a^2 u^2} - \frac{1}{2a^4} \ln \left| \frac{u^2}{u^2 - a^2} \right|$

<u>INTEGRALS INVOLVING</u> $a^2 - u^2, u^2 < a^2$

93. $\displaystyle\int \frac{du}{a^2 - u^2} = \frac{1}{2a} \ln \left| \frac{a + u}{a - u} \right| \quad \text{or} \quad \frac{1}{a} \tanh^{-1} \frac{u}{a}$

94. $\displaystyle\int \frac{u\, du}{a^2 - u^2} = -\frac{1}{2} \ln \left| a^2 - u^2 \right|$

95. $\displaystyle\int \frac{u^2\, du}{a^2 - u^2} = -u + \frac{a}{2} \ln \left| \frac{a + u}{a - u} \right|$

96. $\displaystyle\int \frac{u^3\, du}{a^2 - u^2} = -\frac{u^2}{2} - \frac{a^2}{2} \ln \left| a^2 - u^2 \right|$

97. $\displaystyle\int \frac{du}{u(a^2 - u^2)} = \frac{1}{2a^2} \ln \left| \frac{u^2}{a^2 - u^2} \right|$

98. $\displaystyle\int \frac{du}{u^2(a^2 - u^2)} = -\frac{1}{a^2 u} + \frac{1}{2a^3} \ln \left| \frac{a + u}{a - u} \right|$

99. $\displaystyle\int \frac{du}{u^3(a^2 - u^2)} = -\frac{1}{2a^2 u^2} + \frac{1}{2a^4} \ln \left| \frac{u^2}{a^2 - u^2} \right|$

100. $\displaystyle\int \frac{du}{(a^2 - u^2)^2} = \frac{u}{2a^2(a^2 - u^2)} + \frac{1}{4a^3} \ln \left| \frac{a + u}{a - u} \right|$

101. $\displaystyle\int \frac{u\, du}{(a^2 - u^2)^2} = \frac{1}{2(a^2 - u^2)}$

102. $\displaystyle\int \frac{u^2\, du}{(a^2 - u^2)^2} = \frac{u}{2(a^2 - u^2)} - \frac{1}{4a} \ln \left| \frac{a + u}{a - u} \right|$

103. $\displaystyle\int \frac{u^3\, du}{(a^2 - u^2)^2} = \frac{a^2}{2(a^2 - u^2)} + \frac{1}{2} \ln \left| a^2\, u^2 \right|$

104. $\displaystyle\int \frac{du}{u(a^2 - u^2)^2} = \frac{1}{2a^2(a^2 - u^2)} + \frac{1}{2a^4} \ln \left| \frac{u^2}{a^2 - u^2} \right|$

105. $\displaystyle\int \frac{du}{u^2(a^2 - u^2)^2} = \frac{-1}{a^4 u} + \frac{u}{2a^4(a^2 - u^2)} + \frac{3}{4a^5} \ln \left| \frac{a + u}{a - u} \right|$

106. $\displaystyle\int \frac{du}{u^3(a^2 - u^2)^2} = \frac{-1}{2a^4 u^2} + \frac{1}{2a^4(a^2 - u^2)} + \frac{1}{a^6} \ln \left| \frac{u^2}{a^2 - u^2} \right|$

<u>INTEGRALS INVOLVING</u> $\sqrt{au + b}$

135. $\displaystyle\int \frac{du}{\sqrt{au + b}} = \frac{2\sqrt{au + b}}{a}$

136. $\displaystyle\int \frac{u\, du}{\sqrt{au + b}} = \frac{2(au - 2b)}{3a^2} \sqrt{au + b}$

137. $\displaystyle\int \frac{u^2\, du}{\sqrt{au + b}} = \frac{2(3a^2 u^2 - 4abu + 8b^2)}{15a^3} \sqrt{au + b}$

138. $\displaystyle\int \frac{du}{u\sqrt{au+b}} = \begin{cases} \dfrac{1}{\sqrt{b}} \ln\left|\dfrac{\sqrt{au+b}-\sqrt{b}}{\sqrt{au+b}+\sqrt{b}}\right| \\[2ex] \dfrac{2}{\sqrt{-b}} \tan^{-1}\sqrt{\dfrac{au+b}{-b}} \end{cases}$

139. $\displaystyle\int \frac{du}{u^2\sqrt{au+b}} = -\frac{\sqrt{au+b}}{bu} - \frac{a}{2b}\int \frac{du}{u\sqrt{au+b}}$

140. $\displaystyle\int \sqrt{au+b}\, du = \frac{2\sqrt{(au+b)^3}}{3a}$

141. $\displaystyle\int u\sqrt{au+b}\, du = \frac{2(3au-2b)}{15a^2}\sqrt{(au+b)^3}$

142. $\displaystyle\int u^2\sqrt{au+b}\, du = \frac{2(15a^2u^2-12abu+8b^2)}{105a^3}\sqrt{(au+b)^3}$

INTEGRALS INVOLVING $\sqrt{u^2+a^2}$

168. $\displaystyle\int \sqrt{u^2+a^2}\, du = \frac{u\sqrt{u^2+a^2}}{2} + \frac{a^2}{2}\ln\left|u+\sqrt{u^2+a^2}\right|$

169. $\displaystyle\int u\sqrt{u^2+a^2}\, du = \frac{(u^2+a^2)^{3/2}}{3}$

170. $\displaystyle\int u^2\sqrt{u^2+a^2}\, du = \frac{u(u^2+a^2)^{3/2}}{4} - \frac{a^2u\sqrt{u^2+a^2}}{8} - \frac{a^4}{8}\ln\left|u+\sqrt{u^2+a^2}\right|$

171. $\displaystyle\int u^3\sqrt{u^2+a^2}\, du = \frac{(u^2+a^2)^{5/2}}{5} - \frac{a^2(u^2+a^2)^{3/2}}{3}$

172. $\displaystyle\int \frac{du}{\sqrt{u^2+a^2}} = \ln\left|u+\sqrt{u^2+a^2}\right| \text{ or } \sinh^{-1}\frac{u}{a}$

173. $\displaystyle\int \frac{u\, du}{\sqrt{u^2+a^2}} = \sqrt{u^2+a^2}$

174. $\displaystyle\int \frac{u^2\, du}{\sqrt{u^2+a^2}} = \frac{u\sqrt{u^2+a^2}}{2} - \frac{a^2}{2}\ln\left|u+\sqrt{u^2+a^2}\right|$

175. $\displaystyle\int \frac{u^3\, du}{\sqrt{u^2+a^2}} = \frac{(u^2+a^2)^{3/2}}{3} - a^2\sqrt{u^2+a^2}$

176. $\displaystyle\int \frac{du}{u\sqrt{u^2+a^2}} = -\frac{1}{a}\ln\left|\frac{a+\sqrt{u^2+a^2}}{u}\right|$

177. $\displaystyle\int \frac{du}{u^2\sqrt{u^2+a^2}} = -\frac{\sqrt{u^2+a^2}}{a^2u}$

178. $\displaystyle\int \frac{du}{u^3\sqrt{u^2+a^2}} = -\frac{\sqrt{u^2+a^2}}{2a^2u^2} + \frac{1}{2a^3}\ln\left|\frac{a+\sqrt{u^2+a^2}}{u}\right|$

INTEGRALS INVOLVING $\sqrt{u^2-a^2}$

196. $\displaystyle\int \frac{du}{\sqrt{u^2-a^2}} = \ln\left|u+\sqrt{u^2-a^2}\right|$ **197.** $\displaystyle\int \frac{u\, du}{\sqrt{u^2-a^2}} = \sqrt{u^2-a^2}$

198. $\int \dfrac{u^2 \, du}{\sqrt{u^2 - a^2}} = \dfrac{u\sqrt{u^2 - a^2}}{2} + \dfrac{a^2}{2} \ln \left| u + \sqrt{u^2 - a^2} \right|$

199. $\int \dfrac{u^3 \, du}{\sqrt{u^2 - a^2}} = \dfrac{(u^2 - a^2)^{3/2}}{3} + a^2\sqrt{u^2 - a^2}$

200. $\int \dfrac{du}{u\sqrt{u^2 - a^2}} = \dfrac{1}{a} \sec^{-1} \left| \dfrac{u}{a} \right|$

201. $\int \dfrac{du}{u^2\sqrt{u^2 - a^2}} = \dfrac{\sqrt{u^2 - a^2}}{a^2 u}$

202. $\int \dfrac{du}{u^3\sqrt{u^2 - a^2}} = \dfrac{\sqrt{u^2 - a^2}}{2a^2 u^2} + \dfrac{1}{2a^3} \sec^{-1} \left| \dfrac{u}{a} \right|$

203. $\int \sqrt{u^2 - a^2} \, du = \dfrac{u\sqrt{u^2 - a^2}}{2} - \dfrac{a^2}{2} \ln \left| u + \sqrt{u^2 - a^2} \right|$

204. $\int u\sqrt{u^2 - a^2} \, du = \dfrac{(u^2 - a^2)^{3/2}}{3}$

205. $\int u^2\sqrt{u^2 - a^2} \, du = \dfrac{u(u^2 - a^2)^{3/2}}{4} + \dfrac{a^2 u\sqrt{u^2 - a^2}}{8} - \dfrac{a^4}{8} \ln \left| u + \sqrt{u^2 - a^2} \right|$

206. $\int u^3\sqrt{u^2 - a^2} \, du = \dfrac{(u^2 - a^2)^{5/2}}{5} + \dfrac{a^2(u^2 - a^2)^{3/2}}{3}$

INTEGRALS INVOLVING $\sqrt{a^2 - u^2}$

224. $\int \dfrac{du}{\sqrt{a^2 - u^2}} = \sin^{-1} \dfrac{u}{a}$

225. $\int \dfrac{u \, du}{\sqrt{a^2 - u^2}} = -\sqrt{a^2 - u^2}$

226. $\int \dfrac{u^2 \, du}{\sqrt{a^2 - u^2}} = -\dfrac{u\sqrt{a^2 - u^2}}{2} + \dfrac{a^2}{2} \sin^{-1} \dfrac{u}{a}$

227. $\int \dfrac{u^3 \, du}{\sqrt{a^2 - u^2}} = \dfrac{(a^2 - u^2)^{3/2}}{3} - a^2\sqrt{a^2 - u^2}$

228. $\int \dfrac{du}{u\sqrt{a^2 - u^2}} = -\dfrac{1}{a} \ln \left| \dfrac{a + \sqrt{a^2 - u^2}}{u} \right|$

229. $\int \dfrac{du}{u^2\sqrt{a^2 - u^2}} = -\dfrac{\sqrt{a^2 - u^2}}{a^2 u}$

230. $\int \dfrac{du}{u^3\sqrt{a^2 - u^2}} = -\dfrac{\sqrt{a^2 - u^2}}{2a^2 u^2} - \dfrac{1}{2a^3} \ln \left| \dfrac{a + \sqrt{a^2 - u^2}}{u} \right|$

231. $\int \sqrt{a^2 - u^2} \, du = \dfrac{u\sqrt{a^2 - u^2}}{2} + \dfrac{a^2}{2} \sin^{-1} \dfrac{u}{a}$

232. $\int u\sqrt{a^2 - u^2} \, du = -\dfrac{(a^2 - u^2)^{3/2}}{3}$

233. $\int u^2\sqrt{a^2 - u^2} \, du = -\dfrac{u(a^2 - u^2)^{3/2}}{4} + \dfrac{a^2 u\sqrt{a^2 - u^2}}{8} + \dfrac{a^4}{8} \sin^{-1} \dfrac{u}{a}$

234. $\displaystyle\int u^3\sqrt{a^2 - u^2}\, du = \frac{(a^2 - u^2)^{5/2}}{5} - \frac{a^2(a^2 - u^2)^{3/2}}{3}$

INTEGRALS INVOLVING cos au

311. $\displaystyle\int \cos au\, du = \frac{\sin au}{a}$

312. $\displaystyle\int u\cos au\, du = \frac{\cos au}{a^2} + \frac{u\sin au}{a}$

313. $\displaystyle\int u^2\cos au\, du = \frac{2u}{a^2}\cos au + \left(\frac{u^2}{a} - \frac{2}{a^3}\right)\sin au$

314. $\displaystyle\int u^3\cos au\, du = \left(\frac{3u^2}{a^2} - \frac{6}{a^4}\right)\cos au + \left(\frac{u^3}{a} - \frac{6u}{a^3}\right)\sin au$

315. $\displaystyle\int u^n\cos au\, du = \frac{u^n\sin au}{a} - \frac{n}{a}\int u^{n-1}\sin au\, du$

316. $\displaystyle\int u^n\cos au\, du = \frac{u^n\sin au}{a} + \frac{nu^{n-1}}{a^2}\cos au - \frac{n(n-1)}{a^2}\int u^{n-2}\cos au\, du$

317. $\displaystyle\int \cos^2 au\, du = \frac{u}{2} + \frac{\sin 2au}{4a}$

INTEGRALS INVOLVING sin au

342. $\displaystyle\int \sin au\, du = -\frac{\cos au}{a}$

343. $\displaystyle\int u\sin au\, du = \frac{\sin au}{a^2} - \frac{u\cos au}{a}$

344. $\displaystyle\int u^2\sin au\, du = \frac{2u}{a^2}\sin au + \left(\frac{2}{a^3} - \frac{u^2}{a}\right)\cos au$

345. $\displaystyle\int u^3\sin au\, du = \left(\frac{3u^2}{a^2} - \frac{6}{a^4}\right)\sin au + \left(\frac{6u}{a^3} - \frac{u^3}{a}\right)\cos au$

346. $\displaystyle\int u^n\sin au\, du = -\frac{u^n\cos au}{a} + \frac{n}{a}\int u^{n-1}\cos au\, du$

347. $\displaystyle\int u^n\sin au\, du = -\frac{u^n\cos au}{a} + \frac{nu^{n-1}\sin au}{a^2} - \frac{n(n-1)}{a^2}\int u^{n-2}\sin au\, du$

348. $\displaystyle\int \sin^2 au\, du = \frac{u}{2} - \frac{\sin 2au}{4a}$

INTEGRALS INVOLVING sin au and cos au

373. $\displaystyle\int \sin au\cos au\, du = \frac{\sin^2 au}{2a}$

374. $\displaystyle\int \sin pu\cos qu\, du = -\frac{\cos(p-q)u}{2(p-q)} - \frac{\cos(p+q)u}{2(p+q)}$

375. $\displaystyle\int \sin^n au\cos au\, du = \frac{\sin^{n+1} au}{(n+1)a}$

376. $\displaystyle\int \cos^n au\sin au\, du = -\frac{\cos^{n+1} au}{(n+1)a}$

377. $\displaystyle\int \sin^2 au \cos^2 au\, du = \frac{u}{8} - \frac{\sin 4au}{32a}$

403. $\displaystyle\int \tan au\, du = -\frac{1}{a}\ln|\cos au| = \frac{1}{a}\ln|\sec au|$

404. $\displaystyle\int \tan^2 au\, du = \frac{\tan au}{a} - u$

405. $\displaystyle\int \tan^3 au\, du = \frac{\tan^2 au}{2a} + \frac{1}{a}\ln|\cos au|$

406. $\displaystyle\int \tan^n au\, du = \frac{\tan^{n-1} au}{(n-1)a} - \int \tan^{n-2} au\, du$

407. $\displaystyle\int \tan^n au \sec^2 au\, du = \frac{\tan^{n+1} au}{(n+1)a}$

445. $\displaystyle\int \cos^{-1}\frac{u}{a}\, du = u\cos^{-1}\frac{u}{a} - \sqrt{a^2 - u^2}$

446. $\displaystyle\int u\cos^{-1}\frac{u}{a}\, du = \left(\frac{u^2}{2} - \frac{a^2}{4}\right)\cos^{-1}\frac{u}{a} - \frac{u\sqrt{a^2 - u^2}}{4}$

447. $\displaystyle\int u^2\cos^{-1}\frac{u}{a}\, du = \frac{u^3}{3}\cos^{-1}\frac{u}{a} - \frac{(u^2 + 2a^2)\sqrt{a^2 - u^2}}{9}$

448. $\displaystyle\int \frac{\cos^{-1}(u/a)}{u}\, du = \frac{\pi}{2}\ln|u| - \int \frac{\sin^{-1}(u/a)}{u}\, du$

449. $\displaystyle\int \frac{\cos^{-1}(u/a)}{u^2}\, du = -\frac{\cos^{-1}(u/a)}{u} + \frac{1}{a}\ln\left|\frac{a + \sqrt{a^2 - u^2}}{u}\right|$

450. $\displaystyle\int \left(\cos^{-1}\frac{u}{a}\right)^2 du = u\left(\cos^{-1}\frac{u}{a}\right)^2 - 2u - 2\sqrt{a^2 - u^2}\cos^{-1}\frac{u}{a}$

451. $\displaystyle\int \sin^{-1}\frac{u}{a}\, du = u\sin^{-1}\frac{u}{a} + \sqrt{a^2 - u^2}$

452. $\displaystyle\int u\sin^{-1}\frac{u}{a}\, du = \left(\frac{u^2}{2} - \frac{a^2}{4}\right)\sin^{-1}\frac{u}{a} + \frac{u\sqrt{a^2 - u^2}}{4}$

453. $\displaystyle\int u^2\sin^{-1}\frac{u}{a}\, du = \frac{u^3}{3}\sin^{-1}\frac{u}{a} + \frac{(u^2 + 2a^2)\sqrt{a^2 - u^2}}{9}$

454. $\displaystyle\int \frac{\sin^{-1}(u/a)}{u}\, du = \frac{u}{a} + \frac{(u/a)^3}{2\cdot 3\cdot 3} + \frac{1\cdot 3(u/a)^5}{2\cdot 4\cdot 5\cdot 5} + \frac{1\cdot 3\cdot 5(u/a)^7}{2\cdot 4\cdot 6\cdot 7\cdot 7} + \cdots$

455. $\displaystyle\int \frac{\sin^{-1}(u/a)}{u^2}\, du = -\frac{\sin^{-1}(u/a)}{u} - \frac{1}{a}\ln\left|\frac{a + \sqrt{a^2 - u^2}}{u}\right|$

456. $\displaystyle\int \left(\sin^{-1}\frac{u}{a}\right)^2 du = u\left(\sin^{-1}\frac{u}{a}\right)^2 - 2u + 2\sqrt{a^2 - u^2}\sin^{-1}\frac{u}{a}$

457. $\displaystyle\int \tan^{-1}\frac{u}{a}\, du = u\tan^{-1}\frac{u}{a} - \frac{a}{2}\ln(u^2 + a^2)$

458. $\int u \tan^{-1}\dfrac{u}{a}\,du = \tfrac{1}{2}(u^2 + a^2)\tan^{-1}\dfrac{u}{a} - \dfrac{au}{2}$

459. $\int u^2 \tan^{-1}\dfrac{u}{a}\,du = \dfrac{u^3}{3}\tan^{-1}\dfrac{u}{a} - \dfrac{au^2}{6} + \dfrac{a^3}{6}\ln(u^2 + a^2)$

INTEGRALS INVOLVING e^{au}

483. $\int e^{au}\,du = \dfrac{e^{au}}{a}$

484. $\int u e^{au}\,du = \dfrac{e^{au}}{a}\left(u - \dfrac{1}{a}\right)$

485. $\int u^2 e^{au}\,du = \dfrac{e^{au}}{a}\left(u^2 - \dfrac{2u}{a} + \dfrac{2}{a^2}\right)$

486. $\int u^n e^{au}\,du = \dfrac{u^n e^{au}}{a} - \dfrac{n}{a}\int u^{n-1}e^{au}\,du$

$$= \dfrac{e^{au}}{a}\left(u^n - \dfrac{nu^{n-1}}{a} + \dfrac{n(n-1)u^{n-2}}{a^2} - \cdots + \dfrac{(-1)^n n!}{a^n}\right)$$

if n = positive integer

487. $\int \dfrac{e^{au}}{u}\,du = \ln|u| + \dfrac{au}{1\cdot 1!} + \dfrac{(au)^2}{2\cdot 2!} + \dfrac{(au)^3}{3\cdot 3!} + \cdots$

488. $\int \dfrac{e^{au}}{u^n}\,du = \dfrac{-e^{au}}{(n-1)u^{n-1}} + \dfrac{a}{n-1}\int \dfrac{e^{au}}{u^{n-1}}\,du$

489. $\int \dfrac{du}{p + qe^{au}} = \dfrac{u}{p} - \dfrac{1}{ap}\ln|p + qe^{au}|$

490. $\int \dfrac{du}{(p + qe^{au})^2} = \dfrac{u}{p^2} + \dfrac{1}{ap(p + qe^{au})} - \dfrac{1}{ap^2}\ln|p + qe^{au}|$

491. $\int \dfrac{du}{pe^{au} + qe^{-au}} = \begin{cases} \dfrac{1}{a\sqrt{pq}}\tan^{-1}\left(\sqrt{\dfrac{p}{q}}\,e^{au}\right) \\[2ex] \dfrac{1}{2a\sqrt{-pq}}\ln\left|\dfrac{e^{au} - \sqrt{-q/p}}{e^{au} + \sqrt{-q/p}}\right| \end{cases}$

492. $\int e^{au}\sin bu\,du = \dfrac{e^{au}(a\sin bu - b\cos bu)}{a^2 + b^2}$

493. $\int e^{au}\cos bu\,du = \dfrac{e^{au}(a\cos bu + b\sin bu)}{a^2 + b^2}$

INTEGRALS INVOLVING $\ln|u|$

499. $\int \ln|u|\,du = u\ln|u| - u$

500. $\int (\ln|u|)^2\,du = u(\ln|u|)^2 - 2u\ln|u| + 2u$

501. $\int (\ln|u|)^n\,du = u(\ln|u|)^n - n\int (\ln|u|)^{n-1}\,du$

502. $\int u\ln|u|\,du = \dfrac{u^2}{2}(\ln|u| - \tfrac{1}{2})$

APPENDIX E

Answers to Selected Problems

Many problems in this book are labeled WHAT DOES THIS THIS SAY? *These problems solicit answers in your own words or a statement for you to rephrase as a given statement in your own words. For this reason, it seems inappropriate to include the answers to these questions,* Think Tank Problems, *discussion, research problems, proofs, or problems for which answers may vary.*

We also believe that an answer section should function as a check on work done, so for that reason, when an answer has both an exact answer and an approximate solution (from technology), we usually show only the approximate solution in this appendix. The exact solution (which may be the more appropriate answer) can be **checked** *by using the given approximation.*

The Student Survival and Solutions Manual offers some review, survival hints, and some added explanations for selected problems. These problems are designated in the text by a colored problem number.

CHAPTER 1: FUNCTIONS AND GRAPHS

1.1 Preliminaries (pages 12–14)

1. a. $(-3, 4)$ **b.** $3 \le x \le 5$ **c.** $-2 \le x < 1$ **d.** $(2, 7]$

3. a.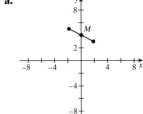

   b.

   c.

   d.

5. a.   **b.**
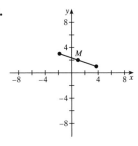

   $M = (0, 4)$ $M = (1, 2)$
   $d = 2\sqrt{5}$ $d = 2\sqrt{10}$

7. $x = 0, 1$ **9.** $y = 7, -2$

11. $x = \dfrac{b \pm \sqrt{b^2 + 12c}}{6}$ **13.** $x = 6, -10$

15. $w = -2, 5$ **17.** $\emptyset$

21. $x = \frac{3\pi}{4}, \frac{5\pi}{4}, \frac{\pi}{3}, \frac{5\pi}{3}$ **23.** $x = \frac{2\pi}{3}$

25. $\left(-\infty, -\frac{5}{3}\right)$ **27.** $\left(-\frac{5}{3}, 0\right)$

29. $(-8, -3]$ **31.** $[-1, 3]$

33. $[7.999, 8.001]$ **35.** $(x + 1)^2 + (y - 2)^2 = 9$

37. $x^2 + (y - 1.5)^2 = 0.0625$

39.   **41.**
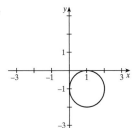   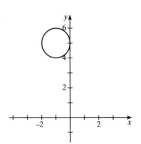

43. $\dfrac{\sqrt{2} - \sqrt{6}}{4} \approx -0.2588$ **45.** $2 - \sqrt{3} \approx 0.2679$

51. a. period 2π, amp 1   **b.** period 2π, amp 1
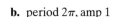
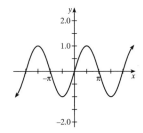

   c. period π
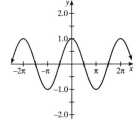
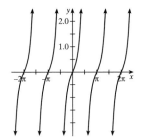

53. period $\pi/2$ **55.** period 4π

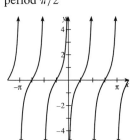   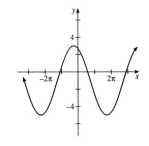

57.

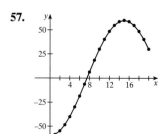

$A = 0, B = 60, C = \frac{\pi}{15}$, and $D = 7.5$

59. sun curve: $y = \cos \frac{\pi}{6}x$; moon curve: $y = 4 \cos \frac{\pi}{6}x$; combined curve: $y = 5 \cos \frac{\pi}{6} x$

61. a. The apparent depth is 3.4 m.
   b. The angle of incidence is 58°.

1.2 Lines in the Plane (pages 21–23)

3. $2x + y - 5 = 0$ **5.** $2y - 1 = 0$
7. $x + 2 = 0$ **9.** $8x - 7y - 56 = 0$
11. $3x + y - 5 = 0$
13. $3x + 4y - 1 = 0$
15. $4x + y + 3 = 0$
17. $m = -5/7$ **19.** $m = 6.001$

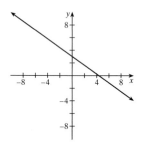   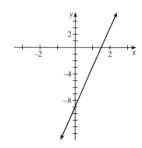

$(0, 3), (\frac{21}{5}, 0)$ $(1.50025, 0), (0, 9.003)$

21. $y = -\frac{3}{5}x - 3$ **23.** $y = \frac{3}{5}x - 0.3$
   $m = -3/5$ $m = 3/5$

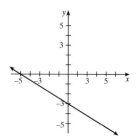   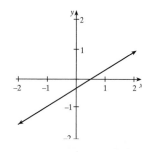

$(-5, 0), (0, -5)$ $(0.5, 0), (0, -0.3)$

25.
   $m = 3/2$

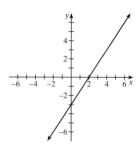

$(2, 0), (0, -3)$

27. $y = \frac{1}{5}x$
   $m = 1/5$

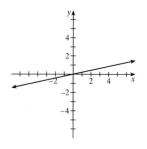

$(0, 0)$

29. no slope

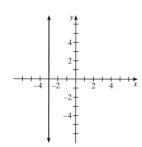

$(-3, 0)$

31. $y = 0$, and $y = 6$ **33.** $D(6, 6), E(2, 16)$
35. no values   **37.** $(4.0, -1.0)$

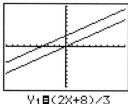
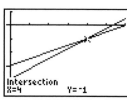

```
Y₁◘(2X+8)/3
Y₂◘(4/6)X
Xmin=-10  Ymin=-10
Xmax=10   Ymax=10
Xscl=1    Yscl=1
```

```
Y₁◘(3X-16)/4
Y₂◘(X-6)/2
Xmin=-.2  Ymin=-5
Xmax=6    Ymax=1
Xscl=1    Yscl=1
```

39. $\left(\frac{3}{4}, \frac{7}{2}\right)$ **41.** $\left(-\frac{64}{3}, \frac{100}{3}\right)$

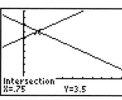
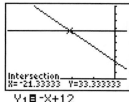

```
Y₁◘(-2X+12)/3
Y₂◘(4X+18)/6
Xmin=-1  Ymin=-1
Xmax=5   Ymax=5
Xscl=.5  Yscl=.5
```

```
Y₁◘-X+12
Y₂◘(.5/.6)(40)
Xmin=-50  Ymin=-5
Xmax=5    Ymax=50
Xscl=10   Yscl=5
```

43. $(\sqrt{2}, \sqrt{2}), (-\sqrt{2}, -\sqrt{2})$ **45.** $\left(-\dfrac{15}{8}, \dfrac{7\sqrt{15}}{8}\right),$
$\left(-\dfrac{15}{8}, -\dfrac{7\sqrt{15}}{8}\right)$

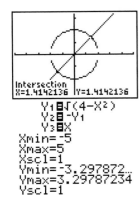 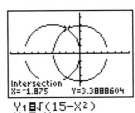

47. a. $-38.2°F$ **b.** $-17.8°C$ **c.** $-40°C$
49. $C = 60x + 5{,}000$ **51.** $V(t) = -19{,}000t + 200{,}000$

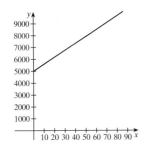 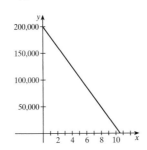

The value is 4 years is $124,000.

55. a. $B(1, 11)$ and $C(-3, -5)$
   b. The center for both triangles is $\left(\frac{1}{3}, \frac{7}{3}\right)$.
57. For the 8th month, 216 gallons
61. $x = \dfrac{a \pm \sqrt{a^2 - 4}}{2}, \ y = \dfrac{2}{a \pm \sqrt{a^2 - 4}}$

1.3 Functions (pages 32–34)

Let D represent the domain in Problems 1–12.

1. $D = (-\infty, \infty)$; $f(-2) = -1$; $f(1) = 5$; $f(0) = 3$
3. $D = (-\infty, \infty)$; $f(1) = 6$; $f(0) = -2$; $f(-2) = 0$
5. $D = (-\infty, -3) \cup (-3, \infty)$; $f(2) = 0$; $f(0) = -2$; $f(-3)$ is undefined
7. $D = (-\infty, 2)] \cup [0, \infty)$; $f(-1)$ is undefined; $f(\frac{1}{2}) = \dfrac{\sqrt{5}}{2}$; $f(1) = \sqrt{3}$
9. $D = (-\infty, \infty)$; $f(-1) = \sin 3 \approx 0.1411$; $f(\frac{1}{2}) = 0$; $f(1) = \sin(-1) \approx -0.8415$
11. $D = (-\infty, \infty)$; $f(3) = 4$; $f(1) = 2$; $f(0) = 4$
13. 9   **15.** $10x + 5h$ **17.** -1 **19.** $\dfrac{-1}{x(x + h)}$

21. 5   **23.** -1.002 **25.** $-\delta^2 - 3\delta - 3$ **27.** not equal
29. equal   **31.** not equal
33. $(f \circ g)(x) = 4x^2 + 1$; $(g \circ f)(x) = 2x^2 + 2$
35. $(f \circ g)(t) = |t|$; $(g \circ f)(t) = t$
37. $(f \circ g)(x) = \sin(2x + 3)$; $(g \circ f)(x) = 2 \sin x + 3$
39. $u(x) = 2x^2 - 1$; $g(u) = u^4$
41. $u(x) = 2x + 3$; $g(u) = |u|$
43. $u(x) = \tan x$; $g(u) = u^2$
45. $u(x) = \sqrt{x}$; $g(u) = \sin u$ **47.** $u(x) = \dfrac{x + 1}{2 - x}$; $g(u) = \sin u$
49. a. the cost is $4,500   **b.** The cost of the 20th unit is $371.
51. a. $I = \dfrac{30}{t^2(6 - t)^2}$ **b.** $I(1) = \frac{6}{5}$ candles; $I(4) = \frac{15}{32}$ candles
53. a. $625t^2 + 25t + 900$ **b.** $6,600   **c.** 4 hours
55. $(2.99, 2.99^2), (3.01, 3.01)^2$; $m = 6$
57. 27.00000001
59. $(2.9999, -25.99820003), (3.0001, -26.00180003)$; $m = -18.00135$

1.4 Functions and Graphs (pages 41–43)

1. even   **3.** neither   **5.** neither   **7.** even
9. 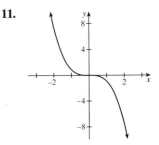   **11.**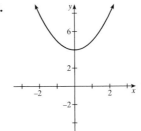
13. 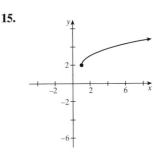   **15.**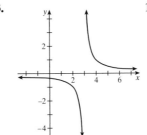
17.   **19.**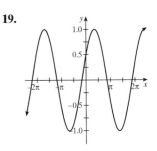

21.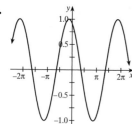

23.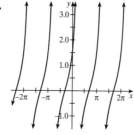

25. $P(5, f(5)); Q(x_0, f(x_0))$

27. $-\frac{1}{3}, 2$

29. $15, -\frac{25}{2}, \frac{65}{3}, -\frac{1}{4}$

31. $\frac{1 \pm \sqrt{6}}{5}$

33. 0

35. ± 1

37. ± 5.42

39. ± 2.24

41. $-3.00, 2.00, 14.00$

43. $-12.00, 18.00$

47. a.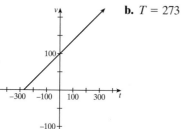

b. $T = 273$

49. $2x + 5y - 19 = 0$

51. It hits the ground approximately 1,234 ft from the firing point.

53. The basic shape is that of a parabola (standard quadratic function).

55. a. 19,400 people
b. 67 people
c. The population will tend to 20,000 people in the long run.

57. $x \approx -2.6139, 0.8031, 3.8031$

59. a. yes
b. $G(x) = (x + 2)(x^2 - 3)(x^2 + 3)$

61. Pythagorean theorem: $\triangle ABC$ with sides $a, b,$ and c is a right triangle if and only if $c^2 = a^2 + b^2$. Proofs vary.

1.5 Inverse Functions; Inverse Trigonometric Functions (pages 52–53)

3. These are inverse functions.

5. These are not inverse functions.

7. These are not inverse functions.

9. $\{(5, 4), (3, 6), (1, 7), (4, 2)\}$

11. $y = \frac{1}{2}x - \frac{3}{2}$

13. $y = \sqrt{x + 5}$

15. $y = (x - 5)^2$

17. $y = \frac{3x + 6}{2 - 3x}$

19. a. $\frac{\pi}{3}$ **b.** $-\frac{\pi}{3}$

21. a. $-\frac{\pi}{4}$ **b.** $\frac{5\pi}{6}$

23. a. $-\frac{\pi}{3}$ **b.** π

25. $\frac{\sqrt{3}}{2}$

27. 3

29. $-\frac{2\sqrt{6}}{5} \approx -0.9798$

33.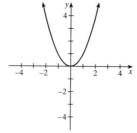

no inverse

35.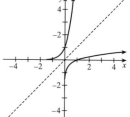

inverse exists

37.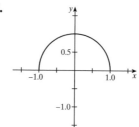

no inverse

39.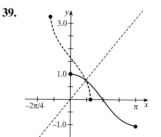

inverse exists

41. $\frac{2x}{x^2 + 1}$

43. $\frac{\sqrt{1 - x^2}}{x}$

45. 1

49. a. 0.5880 **b.** 2.5536 **c.** 2.1997 **d.** -0.5746

51. $h = \frac{d \tan \beta \tan \alpha}{\tan \alpha - \tan \beta}$

53. a. 3.141592654; conjecture is that it is π.

1.6 Exponential and Logarithm Functions (pages 65–67)

1.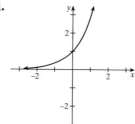

3.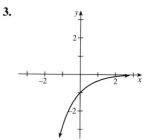

5. 31 **7.** 200.33681 **9.** 9,783.225896 **11.** 38,523.62544

13. 0 **15.** 2 **17.** -2 **19.** 3.5

21. $\frac{3}{10}$ **23.** 4 **25.** 0.23104906

27. -1.391662509 **29.** 729 **31.** $2, -1$

33. 3 **35.** $2, -\frac{5}{3}$ **37.** $-\frac{3}{2}$ **39.** $1, -\frac{3}{2}$

41. 2 **43.** 9 **45.** exponential **47.** logarithmic

49. logarithmic **51.** exponential **53.** 2.4, 0.4

55. 11 yr 202 days

57. First National offers the better deal.

59. The interest rate is approximately 5.71%.

65. a. 30.12% **b.** 77.69% **c.** 7.81%

67. a. 263.34 **b.** 1,232.72

69. Scélérat, at around 1:00 A.M. Wednesday

CHAPTER 1: REVIEW

Proficiency Examination (page 568)

21. a. $6x + 8y - 37 = 0$ **b.** $3x + 10y - 41 = 0$
c. $3x - 28y - 12 = 0$ **d.** $2x + 5y - 24 = 0$
e. $4x - 3y + 8 = 0$

22. $y = -\frac{3}{2}x + 6$

23. $y - 3 = |x + 1|$

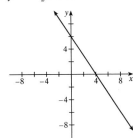

24. $y - 3 = -2(x - 1)^2$

25. $y = (x - 2)^2 - 14$

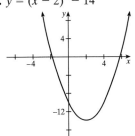

26. $y = 2\cos(x - 1)$

27. $y + 1 = \tan 2\left(x + \frac{3}{2}\right)$

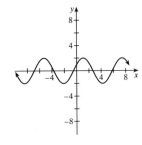

28. $y = \sin^{-1}(2x)$

29. $y = \tan^{-1}x^2$

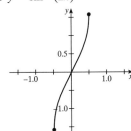

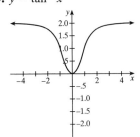

30. $y = e^{-x} + e^x$

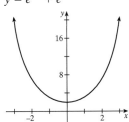

31. $y = \ln(1 - x)$

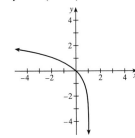

32. $y = e^{2x} + \ln x$

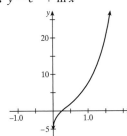

33. $y = e^x - \ln x + 15$

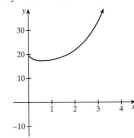

34. $-\frac{3}{2}, 1$ **35.** The two functions are not the same.

36. $f \circ g = \sin\left(\sqrt{1 - x^2}\right); g \circ f = |\cos x|$

37. 4.6286 **38.** 1.0986

39. $V = \frac{2}{3}x(12 - x^2)$

40. a. It will take 11 years, 3 quarters.
b. It will take 11 years, 6 months.
c. It will take 11 years, 166 days.

Supplementary Problems (pages 69–72)

1. $P = 30; A = 30$ **3.** $P \approx 27.1; A = 40$

5. $(x - 5)^2 + (y - 4)^2 = 16$ **7.** $5x - 3y + 3 = 0$

9. $A = -\frac{1}{6}$ and $B = \frac{1}{2}$

11. The medians meet at the point $\left(2, \frac{5}{3}\right)$.

13. 1,024 ft

15. a. 1.504077 **b.** 16.444647 **c.** 1.107149 **d.** 1.899250945

17. $\frac{3}{5}$ **19.** $\frac{5}{2}$ **21.** ± 2 **23.** 16 **25.** $\sqrt{13}$

27. $f^{-1}(x) = \sqrt[3]{\frac{1}{2}(x + 7)}$ **29.** $f^{-1}(x) = \ln(x^2 + 1)$

31. $f^{-1}(x) = \dfrac{x + a}{x - 1}$

33. The domain of f^{-1} is all real $x, x \neq 1$.

35. a. false **b.** false **c.** false **d.** true **e.** true **f.** true

37. a. $k = 0.25 \ln 2$; After 7 weeks, approximately 29.7% are burning. **b.** 0.8232 **c.** 0.07955

39. a. $P \approx \$1,075.71$ **b.** $P \approx \$1,070.52$

41. a. 15 inches **b.** 21 inches **c.** 29 inches **d.** 19 inches

43. a. The domain consists of all $x \neq 300$. **b.** x represents a percentage, so $0 \leq x \leq 100$ in order for $f(x) \geq 0$ **c.** 120 **d.** 300 **e.** The percentage of households should be 60%.

45. a. $V = \frac{256}{3}\pi; S = 64\pi$ **b.** $V = 30; S = 62$
d. $V = 15\pi; A = 3\pi\sqrt{34}$

49. $c = -\frac{4}{5}$ **51.** $\theta = \frac{\pi}{4} - \tan^{-1}\frac{5}{12}$

53. The break-even point occurs when $x = \$30,000$.

55. The glass should be taken in on the 11th or 12th day.

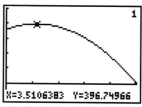

X=3.5106383   Y=396.74966

Y₁⊟-3X²+21X+360
Xmin=0   Ymin=-100
Xmax=15   Ymax=500
Xscl=3   Yscl=100

57. $C(x) = 4x^2 + 1,000x^{-1}$

59. Putnam competition problem solutions are copied with permission from *The William Lowell Putnam Mathematical Competition, Problems and Solutions: 1938–1964* by A. M. Gleason, R. E. Greenwood, and L. M. Kelly, published by The Mathematical Association of America. We will provide reference about where you can find the solution to the Putnam Problems. This is Problem 4 in the morning session for 1959.

CHAPTER 2: LIMITS AND CONTINUITY

2.1 What Is Calculus? (pages 81–84)

5. $\frac{1}{3}$ **7.** 1   **9.** $\frac{3}{11}$ **11.** π

13. a.   **b.**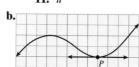

15. a.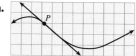

b. There is no unique tangent line.

17. 2   **19.** 1   **21.** 0

23. $\lim\limits_{n\to\infty} \dfrac{n+1}{n}$ **25.** 1   **27.** 2

29. a.   **b.**

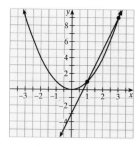   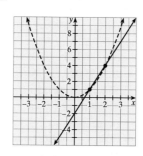

The slope of the secant line is $m = 4$.   The slope of the secant line is $m = 3$.

c.

n	x_n	point	slope
1	3	$(3, 9)$	$m = 4$
2	2	$(2, 4)$	$m = 3$
3	1.5	$(1.5, 2.25)$	$m = 2.5$
4	1.1	$(1.1, 1.21)$	$m = 2.1$

d. The slope of the tangent line is $m = 2$.

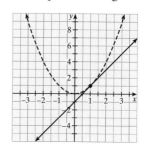

31. 2.66 square units   **33.** 6.75 square units

35. 6.28 square units

37. quadratic model   **39.** exponential model

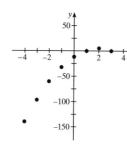   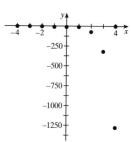

41. quadratic model   **43.** cubic model

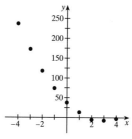   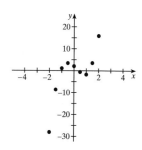

45. logarithmic model   **47.** $\pi \approx 3.16$

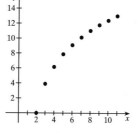

49. $A_3 \approx 1.2990$; $A_4 = 2$; $A_5 \approx 2.3776$; $A_6 \approx 2.5981$; $A_7 \approx 2.7364$; ... $A_{100} \approx 3.1395$

51. $A = 0.3984375$

57. a. Cost $= 2ax^2 + 8bx + 924ax^{-1} + 924bx^{-2}$
   b. The cost is minimized if the base is 6 in.^{2} and the height is 6.42 in.

2.2 The Limit of a Function (pages 95–97)

1. a. 0 **b.** 2 **c.** 6   **3. a.** 2 **b.** 7 **c.** 7.5
5. a. 6 **b.** 6 **c.** 6   **7.** 15
9. 10   **11.** 8
13. 2   **15.** -1
19. 1.00   **21.** 5.00
23. -0.17 **25.** does not exist
27. a. does not exist **b.** 0
29. a. 0.00 **b.** 0.64   **31. a.** 0 **b.** 0.37
33. -2.00 **35.** 0.17
37. 0.25   **39.** does not exist
41. 2.00   **43.** 1.00
45. 2.72   **47.** 0.00
49. does not exist
51. a. $-32t + 40$ **b.** 40 ft/s **c.** 3; impact velocity is -56 ft/s
   d. $t = 1.25$ seconds
53. 228   **55.** 0

2.3 Properties of Limits (pages 105–106)

1. -9 **3.** -8 **5.** $-\frac{1}{2}$ **7.** 2   **9.** $\frac{\sqrt{3}}{9}$ **11.** 4
13. -1 **15.** $\frac{1}{9}$ **7.** $\frac{1}{2}$ **19.** 2   **21.** $\frac{5}{2}$ **23.** 0
25. 1   **27.** 0   **29.** 0   **31.** $\frac{4}{3}$ **37.** -1 **39.** 0
41. the limit does not exist
43. the limit does not exist   **55.** the limit does not exist
57. 4   **59.** 8

2.4 Continuity (pages 115–117)

1. Temperature is continuous, so TEMPERATURE $= f$(time) would be a continuous function. The domain would be midnight to midnight say, $0 \le t < 24$.

3. The selling price of ATT stock is not continuous. The domain is the set of positive rational numbers that can be divided evenly by 8. At the time of this writing, there was some discussion to changing the stock quotations to the nearest cent.

5. The charges (range of the function) consist of rational numbers only (dollars and cents to the nearest cent), so the function CHARGE $= f$(MILEAGE) would be a step function (that is, not continuous). The domain would consist of the mileage from the beginning of the trip to its end.

7. No suspicious points and no points of discontinuity with a polynomial.

9. The denominator factors to $x(x - 1)$, so suspicious points would be $x = 0, 1$. There will be a hole discontinuity at $x = 0$ and a pole discontinuity at $x = 1$.

11. $x = 0$ is suspicious and is a point of discontinuity

13. $x = 1$ is a suspicious point; there are no points of discontinuity

15. The sine and cosine are continuous on the reals, but the tangent is discontinuous at $x = \pi/2 + n\pi$. Each of these values will have a pole type discontinuity.

17. a suspicious point is located at $x = 0$. There is a discontinuity at $x = 0$ (pole)

19. 3   **21.** π **23.** no value

25. a. continuous **b.** discontinuous on $[0, 1]$

27. discontinuous at $t = 0$

29. continuous

45. $a = 1; b = -18/5$ **47.** $a = 1; b = \frac{1}{2}$ **49.** $a = 5; b = 5$

51. $a = \frac{4}{3}, \frac{14}{3}, \cdots; b = \sqrt{3}$

53. It is not possible to redefine f at $x = 2$, so that it becomes continuous.

CHAPTER 2: REVIEW

Proficiency Examination (pages 117–118)

11. $\frac{3}{2}$ **12.** $\frac{1}{4}$ **13.** $-\frac{1}{4}$ **14.** 0   **15.** $\frac{9}{5}$ **16.** -1

17. We have suspicious points where the denominators are 0 at $t = 0, -1$. There are pole discontinuities at each of these points.

18. Suspicious points $x = -2$ and $x = 1$ are also points of discontinuity (since the denominator is 0).

19. $A = -1; B = 1$

Supplementary Problems (pages 118–121)

1. 5   **3.** 1   **5.** 0   **7.** $-\frac{1}{2}$ **9.** e^4 **11.** 5

13. $\frac{3}{2}$ **15.** 0.8415   **17.** 1   **19.** 3   **21.** 0   **23.** $\dfrac{1}{\sqrt{2x}}$

25. $\dfrac{-4}{x^2}$ **27.** e^x **29.** continuous on $[-5, 5]$

31. discontinuity is removable

33. not continuous anywhere

35. a. continuous on $[0, 5]$ **b.** not continuous $x = -2$
   c. not continuous at $x = -2$ **d.** continuous on $[-5, 5]$

37. a.

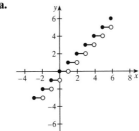

   b. $\lim\limits_{x \to 3} [\![x]\!]$ does not exist.

   c. The limit exists for all nonintegral values.

39. does not exist   **41.** $a = 2$ and $b = 1$

43. discontinuity is removable

45. $x \approx 0.6$ or 0.646944

49. a. 7.16 degrees   **b.** 1.8 second

57. a. the windchill for 20 mi/h is 3.75° and for $v \approx 25.2$
    c. at $v = 4$, $T \approx 91.4$; at $v = 45$, $T \approx 868$

61. The tangent at $x = 2$ is $y = 1.8 + 8.16(x - 2)$

63. This is Problem A1 of the morning session of the 1956 Putnam Examination.

CHAPTER 3: DIFFERENTIATION

3.1 An Introduction to the Derivative: Tangents (pages 139–142)

5.

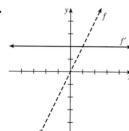

7.

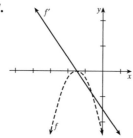

9.

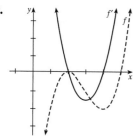

11. a. 0   **b.** 0   **13. a.** 2   **b.** 2

15. a. 0   **b.** 0   **17.** 0; differentiable for all x

19. 3; differentiable for all x **21.** $6x$; differentiable for all x

23. $2x - 1$; differentiable for all x

25. $2s - 2$; differentiable for all real s

27. $\dfrac{\sqrt{5x}}{2x}$; differentiable for $x > 0$

29. $3x - y - 7 = 0$ **31.** $3s - 4y + 1 = 0$

33. $x + 25y - 7 = 0$ **35.** $x + 3y - 9 = 0$

37. $216x - 6y - 647 = 0$ **39.** 2

41. 0   **43. a.** -3.9 **b.** -4

45.

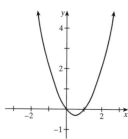

The derivative is 0 when $x = \frac{1}{2}$; the graph has a horizontal tangent at $\left(\frac{1}{2}, -\frac{1}{4}\right)$.

47. a. $-4x$ **b.** $y - 4 = 0$ **c.** $\left(\frac{2}{3}, \frac{28}{9}\right)$

---

49. yes

51. a.

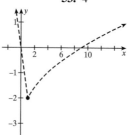

53. 4   **55.** $\frac{1}{2}$ **57.** $\frac{1}{4}$

59. The y-intercept occurs when $y = -Ac^2$.

3.2 Techniques of Differentiation (pages 150–151)

1. (11) 0   (12) 1   (13) 2   (14) 4   (15) 0   (16) -4

3. (23) $2x - 1$ (24) $-2t$ (25) $2s - 2$ (26) $-\frac{1}{2}x^{-2}$ (27) $\dfrac{\sqrt{5x}}{2x}$

5. a. $12x^3$ **b.** -1 **7. a.** $3x^2$ **b.** 1

9. $2t + 2t^{-3} - 20t^{-5}$ **11.** $-14x^{-3} + \frac{2}{3}x^{-1/3}$

13. $1 - x^{-2} + 14x^{-3}$ **15.** $-32x^3 - 12x^2 + 2$

17. $\dfrac{22}{(x + 9)^2}$ **19.** $4x^3 + 12x^2 + 8x$

21. $f'(x) = 5x^4 - 15x^2 + 1$; $f''(x) = 20x^3 - 30x$
    $f'''(x) = 60x^2 - 30$; $f^{(4)}(x) = 120x$

23. $f'(x) = 4x^{-3}$; $f''(x) = -12x^{-4}$; $f'''(x) = 48x^{-5}$;
    $f^{(4)}(x) = -240x^{-6}$

25. $\dfrac{d^2y}{dx^2} = 18x - 14$ **27.** $7x + y + 9 = 0$

29. $6x + y - 6 = 0$ **31.** $x - 6y + 5 = 0$

33. $(1, 0)$ and $\left(\frac{4}{3}, -\frac{1}{27}\right)$ **35.** $\left(\frac{29}{6}, -\frac{361}{12}\right)$

37. $(1, -2)$ **39.** no horizontal tangents

41. d. $-4x^{-3} + 9x^{-4}$ **43.** $2x - y - 2 = 0$

45. a. $4x + y - 1 = 0$ **b.** the normal line is horizontal

47. $(0, 0)$ and $(4, 64)$

49. The equation is not satisfied.

51. This function satisfies the given equation.

53. $\pi \approx \dfrac{62{,}832}{20{,}000} \approx 3.1416$

63. $A = 0$, $B = 1$, $C = 3$, so the required function is $y = x + 3$.

3.3 Derivatives of the Trigonometric, Exponential, and Logarithmic Functions (pages 158–159)

1. $\cos x - \sin x$ **3.** $2t - \sin t$

5. $\sin 2t$

7. $-\sqrt{x} \sin x + \frac{1}{2}x^{-1/2} \cos x - x\csc^2 x + \cot x$

9. $-x^2\sin x + 2x \cos x$

11. $\dfrac{x \cos x - \sin x}{x^2}$

13. $e^t\csc t(1 - \cot t)$

15. $x + 2x\ln x$ **17.** $2e^x \cos x$

19. $e^{-x}(\cos x - \sin x)$

21. $\dfrac{\sec^2 x - 2x\sec^2 x + 2\tan x}{(1 - 2x)^2}$

23. $\dfrac{t\cos t + 2\cos t - \sin t - 2}{(t + 2)^2}$

25. $\dfrac{-1}{1 - \cos x}$

27. $\dfrac{2\cos x - \sin x - 1}{(2 - \cos x)^2}$

29. $\dfrac{-2}{(\sin x - \cos x)^2}$

31. $-\sin x$

33. $-\sin\theta$

35. $2\sec^2\theta\tan\theta$

37. $\sec^3\theta + \sec\theta\tan^2\theta$

39. $-\sin x - \cos x$

41. $-2e^x\sin x$

43. $\frac{1}{4}t^{-3/2}\ln t$

45.

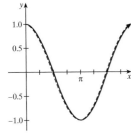

47. $4x - 2y - \pi + 2 = 0$

49. $\sqrt{3}x - 2y + \left(1 - \dfrac{\sqrt{3}\pi}{6}\right) = 0$

51. $2x - y = 0$

53. $x - y + 1 = 0$

55. a. yes **b.** yes **c.** no **d.** no

57. $A = 0, B = -\frac{3}{2}; y = -\frac{3}{2}x\sin x$

3.4 Rates of Change: Rectilinear Motion (pages 166–169)

1. 1   **3.** -3 **5.** $\frac{13}{4}$ **7.** -1

9. $\frac{1}{2}$ **11.** -1 **13.** -6

15. a. $2t - 2$ **b.** 2 **c.** 2
  d. Because $a(t) > 0$, the object is continuously accelerating.

17. a. $3t^2 - 18t + 15$ **b.** $6t - 18$ **c.** 46
  d. On $[0, 3)$ the object is decelerating, and on $(3, 6]$ it is accelerating.

19. a. $-2t^{-2} - 2t^{-3}$ **b.** $4t^{-3} + 6t^{-4}$
  c. $\frac{20}{9}$ On $[1, 3]$ the object is accelerating.

21. a. $-3\sin t$ **b.** $-3\cos t$ **c.** 12
  d. On $[0, \frac{\pi}{2})$ the object is decelerating, on $(\frac{\pi}{2}, \frac{3\pi}{2})$ the object is accelerating, and on $(\frac{3\pi}{2}, 2\pi]$ it is decelerating again.

23. a. $f'(x) = -6$
  b. The decline will be the same each year.

25. 136

27. a. 9.91 m/min **b.** 9.96 m/s

29. a. The initial velocity is 64 ft/sec.
  b. The cliff is 336 ft high.
  c. $32t + 64$ ft/sec **d.** 160 ft/sec

31. 30 ft

33. The height of the building is 144 ft.

35. The initial velocity is 24 ft/s, and the cliff is 126 ft. high.

37. a. $200t + 50\ln t + 450$ newspapers per year.
  b. 1,530 newspapers per year
  c. 1,635 newspapers

39. a. 0.2 ppm/yr **b.** 0.15 ppm **c.** 0.25 ppm

41. 91 thousand/hr

43. a. 20 persons per mo **c.** 0.39% per mo

45. 7.5% per year

47. $\dfrac{dP}{dT} = -\dfrac{4\pi\mu^2 N}{9k}T^{-2}$

49. a. $v(t) = -7\sin t; a(t) = -7\cos t$
  b. the period (one revolution) is 2π
  c. The amplitude is 7.

51. The Spy is on Mars.   **53.** They are equal.

3.5 The Chain Rule (pages 174–176)

3. $6(3x - 2)$

5. $\dfrac{-8x}{(x^2 - 9)^3}$

7. $\left[\tan u + u\sec^2 u\right]\left[3 - \dfrac{6}{x^2}\right]$

9. $e^{\sec x}\sec x\tan x$

11. a. $3u^2$ **b.** $2x$ **c.** $6x(x^2 + 1)^2$

13. a. $7u^6$ **b.** $-8 - 24x$ **c.** $-7(24x + 8)(12x^2 + 8x - 5)^6$

15. a. $\cos(\sin\theta)(\cos\theta)$ **b.** $\cos(\cos\theta)(-\sin\theta)$

17. $4\cos(4\theta + 2)$ **19.** $(1 - 2x)e^{1-2x}$

21. $2x\cos 2x^2$

23. $2x(2x^2 + 1)^3(x^2 - 2)^4(18x^2 - 11)$

25. $\dfrac{1}{2}\left(\dfrac{x^2 + 3}{x^2 - 5}\right)^{-1/2}\left[\dfrac{-16x}{(x^2 - 5)^2}\right]$

27. $\dfrac{1}{3}(x + \sqrt{2x})^{-2/3}\left(1 + \dfrac{1}{\sqrt{2x}}\right)$

29. $(2t + 1)\exp(t^2 + t + 5)$ **31.** $\dfrac{\cos x - \sin x}{\sin x + \cos x}$

33. $2x - 3y + 5 = 0$ **35.** $y - \frac{1}{16} = 0$ **37.** $y = e^2 x$

39. $\frac{2}{9}$ **41.** $1, 7$ **43.** $0, \frac{2}{3}$

45. a. 1 **b.** $\frac{3}{2}$ **c.** 1.5   **47.** 0.31 parts/million

49. The demand is decreasing by 6 lb/wk.

51. a. Illuminance is increasing by 0.035 lux/s. **b.** 7.15 m

57. a. $\dfrac{3}{(3x - 1)^2 + 1}$ **b.** $\dfrac{-1}{x^2 + 1}$

61. $\dfrac{d}{dx}f'[f(x)] = f''(x)[f(x)]f'(x)$ and

  $\dfrac{d}{dx}f[f'(x)] = f'[f'(x)]f''(x)$

3.6 Implicit Differentiation (pages 186–189)

1. $-\dfrac{x}{y}$

3. $-\dfrac{y}{x}$

5. $\dfrac{-(2x + 3y)}{3x + 2y}$

7. $-\dfrac{y^2}{x^2}$

9. $\dfrac{1 - \cos(x + y)}{\cos(x + y) + 1}$

11. $\dfrac{2x - y\sin xy}{x\sin xy}$

13. $(2e^{2x} - x^{-1})y$

15. a. $-\dfrac{2x}{3y^2}$ **b.** $\dfrac{-2x}{3(12 - x^2)^{2/3}}$

17. a. y^2 **b.** $\dfrac{1}{(x - 5)^2}$

19. $\dfrac{1}{\sqrt{-x^2 - x}}$

21. $\dfrac{x}{(x^2 + 2)\sqrt{x^2 + 1}}$

23. $\dfrac{6(\sin^{-1}2x)^2}{\sqrt{1 - 4x^2}}$

25. $\dfrac{-1}{\sqrt{e^{-2x} - 1}}$

27. $\dfrac{-1}{x^2 + 1}$

29. $\dfrac{-2}{x\sqrt{4x^4 - 1}}$

31. $\dfrac{1 - \sin^{-1}y - \dfrac{y}{1 + x^2}}{\dfrac{x}{\sqrt{1 - y^2}} + \tan^{-1}x}$

33. $2x - 3y + 13 = 0$

35. $(\pi + 1)x - y + \pi = 0$

37. $y = 0$

39. $y' = 0$

41. $y' = \frac{5}{4}$

43. $x - 1 = 0$

45. $-\dfrac{49}{100y^3}$

51. $y\left[\dfrac{10}{2x - 1} - \dfrac{1}{2(x - 9)} - \dfrac{2}{x + 3}\right]$

53. $y\left[6x - \dfrac{6x^2}{x^3 + 1} + \dfrac{8}{4x - 7}\right]$

55. $\dfrac{y \ln x}{x}$

57. $(3, -4)$ and $(-3, 4)$

63. vertical tangents at $(\sqrt{2}, 0)$, $\left(-\dfrac{2\sqrt{3}}{9}, \dfrac{\sqrt{15}}{9}\right)$ and $\left(-\dfrac{2\sqrt{3}}{9}, -\dfrac{\sqrt{15}}{9}\right)$

65. 0.44 rad/s

69. $\dfrac{d^2y}{dx^2} = -\dfrac{ac}{b^2y^3}$

3.7 Related Rates and Applications (pages 194–197)

1. -3 **3.** 1,000 **5.** 15 **7.** $\frac{4}{5}$ **9.** $\frac{30}{13}$ **11.** -3

13. $-10\sqrt{3}$ units/s **15.** 0.637 ft/s

17. The revenue will be rising at $34,000/yr.

19. -30 lb/in.²/s **23.** -3π cm³/s

25. 7.2 ft/s

27. The distance is decreasing at the rate of 7.5 ft/min.

29. The shadow is moving at 200 ft/s.

31. The angle is of the line of sight is changing at the rate of 2.78 rad/s.

33. Assume the shape of the balloon is a sphere of radius r. The volume is changing at the rate of 60.3 cm³/min.

35. The water would flow out at about 49.2 ft³/min.

37. a. $\dfrac{dH}{dt} = \dfrac{3,125t - 6,250}{\sqrt{3,125t^2 - 12,500t + 62,500}}$

b. $t = 2$ **c.** 224 mi

39. 1 rad/min **41.** 3.927 mi/h

43. At $t = 2$ P.M., 8.875 knots; at $t = 5$ P.M., 10.417 knots

45. 0.001924 ft/min or 0.0233 in./min

47. a. $\theta = \cot^{-1}\dfrac{x}{150}$

b. $d\theta/dt$ approaches 0.27 rad/s

c. as v increases so will $d\theta/dt$ and it becomes more difficult to see the seals.

3.8 Linear Approximations and Differentials (pages 206–209)

1. $6x^2\, dx$ **3.** $x^{-1/2}\, dx$

5. $(\cos x - x \sin x)dx$

7. $\dfrac{3x \sec^2 3x - \tan 3x}{2x^2}\, dx$

9. $\cot x\, dx$

11. $\dfrac{e^x}{x}(1 + x \ln x)\, dx$

13. $\dfrac{(x - 3)(x^2\sec x \tan x + 2x \sec x) - (x^2\sec x)(1)}{(x - 3)^2}\, dx$

15. $\dfrac{x + 13}{2(x + 4)^{3/2}}\, dx$

19. 0.995; by calculator, 0.9949874371

21. 217.69; by calculator 217.7155882 so we see an error of approximately 0.0255882

23. 0.06 or 6% **25.** 0.03 or 3% **27.** 0.05 parts per million

29. The output will be reduced by 12,000 units.

31. 28.37 in.³

33. a decrease of about 2 beats every 3 minutes

35. The error in S is approximately $\pm 2\%$.

37. S increases by 2% and the volume increases by 3%.

39. The length will change by about 0.0525 ft.

41. -6.93 (or about 7) particles/unit area

43. a. 472.7 **b.** 468.70

45. 1.2 units **47. b.** $x \approx 1.367$

49. 0.5183 **51.** $K \approx \frac{3}{10}$

55. $\Delta x = -3; f(97) = 9.85$; by calculator 9.848857802; if $\Delta x = 16, f(97) \approx 9.89$

CHAPTER 3: REVIEW

Proficiency Examination (pages 209–210)

21. $\dfrac{dy}{dx} = 3x^2 + \frac{3}{2}x^{1/2} - 2 \sin 2x$

22. $\dfrac{dy}{dx} = \dfrac{\sqrt{3x}}{2x} - \dfrac{6}{x^2}$

23. $\dfrac{dy}{dx} = -x[\cos (3 - x^2)][\sin (3 - x^2)]^{-1/2}$

24. $\dfrac{dy}{dx} = \dfrac{-y}{x + 3y^2}$ **25.** $y' = \frac{1}{2}xe^{-\sqrt{x}}(4 - \sqrt{x})$

26. $y' = \dfrac{\ln 1.5}{x(\ln 3x)^2}$ **27.** $y' = \dfrac{3}{\sqrt{1 - (3x + 2)^2}}$

28. $y' = \dfrac{2}{1 + 4x^2}$ **29.** $\dfrac{dy}{dx} = 0$

30. $y' = y\left[\dfrac{2x}{(x^2 - 1)\ln(x^2 - 1)} - \dfrac{1}{3x} - \dfrac{9}{3x - 1}\right]$

17. 17.3 is the constrained maximum and -17.3 is the constrained minimum

19. The minimum distance is $\dfrac{|D|}{\sqrt{A^2 + B^2 + C^2}}$

21. The nearest point is $(\frac{1}{3}, \frac{1}{6}, \frac{1}{6})$, and the minimum distance is 0.4082.

23. $x = z = 3, y = 6$

25. The lowest temperature is $\frac{200}{3}$.

27. The maximum value of A is 6,400 yd^2.

29. the radius $x = 1$ in. and the height $y = 4$ in.

31. \$2,000 to development and \$6,000 to promotion gives the maximum sales of about 1,039 units.

33. $x \approx 13.87$ and $y \approx 12.04$ **35.** $x = y = z = \sqrt[3]{C}$

37. The farmer should apply 4.24 acre-ft of water and 1.27 lb of fertilizer to maximize the yield.

39. The triangle with maximum area is equilateral.

41. $\frac{56}{3}$ **43.** $\frac{121}{24}$

45. **a.** 3 thousand dollars for development and 5 thousand dollars for promotion.
b. The actual increase in profit is \$29.69.
c. Thus, \$4,000 should be spent on development and \$6,000 should be spent on promotion to maximize profit.
d. $\lambda = 0$

47. The minimum surface area occurs when $R \approx 0.753\, V_0^{1/3}$, $L \approx 0.674\, V_0^{1/3}$, $H \approx 0.337\, V_0^{1/3}$.

49. The minimum occurs at $(1, 1)$, but there is no maximum

CHAPTER 12: REVIEW

Proficiency Examination (page 871)

32. $\dfrac{dw}{dt} = 6\pi^2$

33. **a.** $\nabla f_0 = \mathbf{i} + 3\mathbf{k}$ **b.** $D_u(f) = \dfrac{-2\sqrt{5}}{5}$

c. $\mathbf{u} = \dfrac{\mathbf{i} + 3\mathbf{k}}{\sqrt{10}}$; $\|\nabla f\| = \sqrt{10}$

35. $f_x = -\dfrac{1}{x}, f_y = \dfrac{1}{y}, f_{yy} = -\dfrac{1}{y^2}, f_{xy} = 0$

37. $D_u(f) \approx -87.4$

38. $(0, 0)$, saddle point; $(9, 3)$, relative maximum; $(-9, -3)$, relative maximum

39. maximum of 12; minimum of 3

40. The largest value of f is $49/42$ at $(2, \frac{3}{4})$ and the smallest is $-9/4$ at $(-3/2, 0)$.

Supplementary Problems (pages 872–876)

1. The domain consists of the circle with center at the origin, radius 4, and its interior.

3. $-1 \le x \le 1$ and $-1 \le y \le 1$ is the domain

5. $f_x = 1; f_y = -1$

7. $f_x = 2xy - \dfrac{y}{x^2}\cos\dfrac{y}{x}; f_y = x^2 + \dfrac{1}{x}\cos\dfrac{y}{x}$

9. $f_x = 6x^2y + 3y^2 - \dfrac{y}{x^2}; f_y = 2x^3 + 6xy + \dfrac{1}{x}$

11. For $c = 2, x^2 - y = 2$ is a parabola opening up, with vertex at $(0, -2)$. For $c = -2, x^2 - y = -2$ is a parabola opening up, with vertex at $(0, 2)$.

13. For $c = 0, \sqrt{x^2 + y^2} = 0$ is the origin only. For $c = 1, \sqrt{x^2 + y^2} = 1$ is a semicircle (to the right of the y-axis). For $c = -1, |y| = 1$ is a pair of half-lines, 1 unit above or below the x-axis, to the left of the y-axis.

15. For $c = 1, x^2 + \dfrac{y^2}{2} + \dfrac{z^2}{9} = 1$ is an ellipsoid. For $c = 2, x^2 + \dfrac{y^2}{2} + \dfrac{z^2}{9} = 2$ is an ellipsoid.

17. 0   **19.** The limit does not exist.

21. $\dfrac{dz}{dt} = ye^t(-t^{-2} + t^{-1}) + (x + 2y)\sec^2 t$

23. $\dfrac{dz}{du} = \left(\tan\dfrac{x}{y} + \dfrac{x}{y}\sec^2\dfrac{x}{y}\right)v + \left(-\dfrac{x^2}{y^2}\sec^2\dfrac{x}{y}\right)v^{-1};$

$\dfrac{dz}{dv} = \left(\tan\dfrac{x}{y} + \dfrac{x}{y}\sec^2\dfrac{x}{y}\right)u + \left(-\dfrac{x^2}{y^2}\sec^2\dfrac{x}{y}\right)(-uv^{-2})$

25. $\dfrac{\partial z}{\partial x} = -e^{x-z}; \dfrac{\partial z}{\partial y} = -e^{y-z}$

27. $\dfrac{\partial z}{\partial x} = \dfrac{z}{3z + 1}; \dfrac{\partial z}{\partial y} = \dfrac{2z}{3z + 1}$

29. $f_{xx} = \dfrac{xy^3}{(1 - x^2y^2)^{3/2}}; f_{yx} = (1 - x^2y^2)^{-3/2}$

31. $f_{xx} = 2e^{x^2+y^2}(2x^2 + 1); f_{yx} = 4xye^{x^2+y^2}$

33. $f_{xx} = \sin x \cos(\cos x); f_{yx} = 0$

35. Normal line: $\dfrac{x - 1}{16} = \dfrac{y - 1}{-7}$ and $z = 1$
Tangent plane: $-16(x - 1) + 7(y - 1) = 0$

37. relative minimum at $(3, -1)$

39. $(1, 0)$, saddle point; $(0, 1)$, saddle point; $(\frac{2}{3}, \frac{2}{3})$, relative maximum; $(1, 1)$, saddle point

41. $(0, 9)$, relative minimum; $(0, 3)$, saddle point; $(-2, 9)$ saddle point; $(-2, 3)$, relative maximum

43. The largest value is 2 at $(0, -1)$ and the smallest is -2 at $(0, 1)$.

45. The largest value of f is 10 at $(0, -2)$ and the smallest is $-9/4$ at $(0, 3/2)$.

47. $\dfrac{dz}{dt} = (2x - 3y^2)(2) + (-6xy)(2t)$

49. $\dfrac{\partial z}{\partial x} = e^{u^2 - v^2}(8xu^2 + 4x - 12uvx);$

$\dfrac{\partial z}{\partial y} = e^{u^2 - v^2}(12yu^2 + 6y + 8uvy)$

51. $\dfrac{dy}{dx} = -1$ at $(1, 1)$

53. Normal line: $\dfrac{x}{2} = \dfrac{y - 1}{2} = \dfrac{z - 3}{-1}$

Tangent plane: $2x + 2y - z + 1 = 0$

55. $g(x, y, z) = x^2y + y^2z + z^2x$

57. $\dfrac{dw}{dt} = \left[\dfrac{2x}{1 + x^2 + y^2}\right]\left(\dfrac{2t}{1 + t^2}\right) + \left[\dfrac{2y}{1 + x^2 + y^2} - \dfrac{2}{1 + y^2}\right]e^t$

59. $D_u f = \dfrac{1}{\sqrt{5}}(-16 \ln 2 - 12)$

61. a. $D_u f = -\dfrac{18}{\sqrt{6}}$ **b.** $\dfrac{1}{\sqrt{86}}(-6\mathbf{i} + \mathbf{j} + 7\mathbf{k})$

63. $x = 6, y = z = 3$; maximum is 324.

65. The minimum distance is $\sqrt{3}$.

67. a. $t_m = \dfrac{-1 + \sqrt{1 + 4k\gamma r^2}}{4k\gamma}$ **b.** $M(r) = \sqrt{\dfrac{k\gamma}{\pi}} \dfrac{e^{-z/2}}{\sqrt{z - 1}}$

69. a.

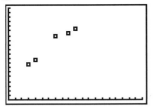

```
WINDOW FORMAT
Xmin=0
Xmax=20
Xscl=1
Ymin=0
Ymax=200
Yscl=10
```

b.

```
LinReg
y=a+bx
a=44.4516129
b=11.53763441
r=.9793293478
```

The least squares line is $y = 11.54x + 44.45$
If $x = 5,000$, the sales are $y \approx 102.15$

71. $24xy^2z^3\sin(x^2 + y^3 + z^4)$

73. $\dfrac{dy}{dx} = -\dfrac{2}{3}$

75. $\mathbf{u} = \pm\frac{1}{3}(2\mathbf{i} + 2\mathbf{j} - \mathbf{k})$

77. $\dfrac{\partial^2 z}{\partial x \partial y} = -\dfrac{\sin^2(x + z)\cos(x + y) + \sin^2(x + y)\cos(x + z)}{\sin^3(x + z)}$

79. 18 is a relative minimum.

81. $x = \dfrac{a}{a^2 + b^2 + c^2}; y = \dfrac{b}{a^2 + b^2 + c^2}; z = \dfrac{c}{a^2 + b^2 + c^2}$

87. $\theta = 2$ and $r = \sqrt{A_0}$

89. a. 10π **b.** 8π

91. Approximately 2.01 cm³ has been removed.

97. This is Putnam Problem 6 in the afternoon session of 1967.

99. This is Putnam Problem 13 in the afternoon session of 1938.

CHAPTER 13: MULTIPLE INTEGRATION

13.1 Double Integration Over Rectangular Regions (pages 887–889)

1. $\frac{13}{3}$ **3.** $(e^2 - 1)(\ln 2) + 2$

5. $\frac{15}{2}\ln 3 - 10\ln 2 + \frac{1}{2}$ **7.** 32

9. 24   **11.** 24   **13.** $\frac{7}{6}$

15. -1 **17.** $4\ln 2$ **19.** 1

21. 8   **23.** $\dfrac{a^3b}{2} + \dfrac{ab^3}{2}$ **25.** $\dfrac{32}{9}$

27. 2   **29.** 3   **31.** $\dfrac{2^{n+2} - 2}{(n + 1)(n + 2)}$

35. $M = \displaystyle\int_R\!\!\int \delta(x, y)\, dA_{xy}$

37. $T = \displaystyle\int_R\!\!\int 1{,}400\, f(x, y)\, dA_{xy}$

39. $\frac{1}{2}\ln 2$ **41.** 1.48

43. 3.55

13.2 Double Integration Over Nonrectangular Regions (pages 895–897)

3. $\frac{32}{3}$ **5.** $\frac{2}{3}$ **7.** 6.6747

9. $\frac{27}{2}$ **11.** $\frac{5}{12}$ **13.** 0.1253

15. 0.1812   **17.** $\frac{1}{2}$ **19.** 5

21. 0.4167   **23.** 13   **25.** 2.0530

27. 1.4366   **29.** $\frac{32}{3}$ **31.** $e - 2$

33. $\frac{2}{3}$ **35.** 9.5323

37. $\displaystyle\int_0^2\int_{x/2}^1 f(x, y)\, dy\, dx$ **39.** $\displaystyle\int_0^2\int_{x^2}^{2x} f(x, y)\, dy\, dx$

41. $\displaystyle\int_0^1\int_0^{3x} f(x, y)\, dy\, dx + \int_1^2\int_0^{4-x^2} f(x, y)\, dy\, dx$

43. $\displaystyle\int_0^4\int_{-\sqrt{y}}^{\sqrt{y}} f(x, y)\, dx\, dy + \int_4^9\int_{-\sqrt{y}}^{6-y} f(x, y)\, dx\, dy$

45. $\displaystyle\int_{-9}^0\int_{3-\sqrt{9+y}}^{3+\sqrt{9+y}} f(x, y)\, dx\, dy + \int_0^7\int_{y}^{3+\sqrt{9+y}} f(x, y)\, dx\, dy$

47. $V = \displaystyle\int_0^{7/3}\int_0^{(7-3x)/2} (7 - 3x - 2y)\, dy\, dx$

49. $V = 8\displaystyle\int_0^{\sqrt{3}}\int_0^{\sqrt{3-x^2}} \sqrt{7 - x^2 - y^2}\, dy\, dx$

51. $V = 4\displaystyle\int_0^a\int_0^{(b/a)\sqrt{a^2+b^2}} c\sqrt{1 - \dfrac{x^2}{a^2} - \dfrac{y^2}{b^2}}\, dy\, dx$

53. $V = \dfrac{4}{3}\pi(\sqrt{2})^3 - 8\displaystyle\int_0^1\int_0^1 \sqrt{2 - x^2 - y^2}\, dy\, dx$

55. πab **57.** 1

59. $\displaystyle\int_1^8\int_{y^{1/3}}^{y} f(x, y)\, dx\, dy$ **61.** $\frac{5}{24}$

63. 22   **65.** 18π **67.** $\frac{128\pi}{3}$

13.3 Double Integrals in Polar Coordinates (pages 903–906)

1. 2
3. 0.2888
5. 33.5103
7. 0
9. 50.2655
11. 6π
13. 4π
15. $\frac{\pi}{2}$
17. 1.9132
19. 1.9270
21. π
23. 0.5435
25. $\frac{a^4\pi}{4}$
27. $\pi \ln 2$
29. $2\pi a(1 - \ln 2)$
31. 0
33. 25,453
35. 4.2131
37. 8π
39. $\frac{8}{3}$
41. $\frac{a^2}{8}(\pi - 2)$
43. 134.0413
45. 9.6440
47. 0.9817
49. $\frac{\pi}{2}$
51. $\frac{\pi}{8}$
53. π
55. $\frac{15\pi}{4}$
57. $\sqrt{\pi}$
59. $\frac{1}{3}\pi a^2(3R - a)$

13.4 Surface Area (pages 912–914)

1. 18.3303
3. 36.1769
5. $\frac{7}{4}$
7. 9.2936
9. 67.2745
11. 5.3304
13. 2π
15. 9.1327
17. 25.6201
19. 117.3187
23. $S = \dfrac{D^2}{2ABC}\sqrt{A^2 + B^2 + C^2}$
25. $S = \dfrac{3\sqrt{3}\,\pi a^2}{4}$
27. $\sqrt{2}\,\pi h^2$
29. 36π
31. $S = \displaystyle\int_0^1 \int_0^y \sqrt{\csc^2 y + z^{-2}}\, dz\, dy$
33. $S = \displaystyle\int_0^{2\pi} \int_0^{\sqrt{\pi/2}} \sqrt{4r^2\sin^2 r^2 + 1}\; r\, dr\, d\theta$
35. $S = \displaystyle\int_1^5 \int_{5/x}^{6-x} \sqrt{(2x + 5y)^2 + (2y + 5x)^2 + 1}\; dy\, dx$
37. $4|u|\sqrt{u^2 + 1}$ **39.** $2|v|\sqrt{9u^4 + 1}$ **41.** 7.01
43. $\frac{1}{2}\left(-\frac{u}{2}\sin 2v + u\sin v - 2\cos v\sin\frac{v}{2}\right)\mathbf{i}$
$\quad +\frac{1}{2}(-u\sin^2 v - u\cos v - 2\sin v\sin\frac{v}{2})\mathbf{j}$
$\quad +\frac{1}{2}(2u\cos^2\frac{v}{2} + 2\cos\frac{v}{2})\mathbf{k}$
47. $4\pi R^2$

13.5 Triple Integrals (pages 921–924)

3. 45
5. 45
7. $\frac{68}{9}$
9. −4.7746
11. −0.2986
13. 6.3729
15. 8
17. $\frac{1}{720}$
19. 0
21. 0.9762
23. $\frac{1}{6}$
25. $\frac{4\pi}{3}$
27. 12.5664
29. 23.0859
31. $\frac{16}{3}$
33. $\displaystyle\int_0^1 \int_y^1 \int_0^y f(x, y, z)\, dz\, dx\, dy$
35. $\displaystyle\int_0^2 \int_0^{\sqrt{4-z^2}} \int_0^{\sqrt{4-y^2-z^2}} f(x, y, z)\, dx\, dy\, dz$
37. 32π
39. 32π
41. $\frac{3\pi}{2}$

43. $\frac{4}{3}\pi R^3$
45. $\dfrac{4\pi abc}{3}$
47. $\dfrac{abc}{6}$
51. B
53. 74.3197

13.6 Mass, Moments, and Probability Density Functions (pages 931–935)

5. $\left(\frac{3}{2}, 2\right)$
7. $\left(1, \frac{8}{5}\right)$
9. $\left(\frac{1}{4}, \frac{4}{5}\right)$
11. $(1, 1, 1)$
13. $\left(0, \frac{24}{5\pi}\right)$
15. $\left(7, \frac{5}{3}\right)$
17. $(1.608, 0.231)$
19. a. $\left(0, \dfrac{3a}{2\pi}\right)$ **b.** $\left(-\dfrac{8a}{3\pi^2}, \dfrac{4a}{3\pi}\right)$
21. $(4.9110, 0.7616)$
23. 0.0339
25. $(0, 1.05)$
27. $(1.4372, 0)$
31. $\dfrac{a^4\pi}{64}$
35. $\left(\dfrac{3a}{8}, \dfrac{3a}{8}, \dfrac{3a}{8}\right)$
37. 0.0803
39. The probability is roughly 75%.
41. 3.7222
43. 2.1548
45. 0
47. 10.1538
49. $\dfrac{\sqrt{10}\,a}{5}$
51. a. $\sqrt{\dfrac{2}{\pi}}\, C_0 \dfrac{e^{-1/2}}{x_0}$
b. The danger zone is approximately 1.9 miles.
c. $AV = \dfrac{1}{A} \displaystyle\int_0^{1.9358} \int_0^{x^2/(2k)} \dfrac{C_0}{\sqrt{k\pi t}} \exp\left(\dfrac{-x^2}{4kt}\right) dt\, dx$
53. a. $W = \delta \displaystyle\iiint (h - z)\, dV$ **c.** $5,400\pi\delta$
57. $4\pi^2 ab$

13.7 Cylindrical and Spherical Coordinates (pages 941–944)

3. a. $(4.00, 1.57, 1.73)$ **b.** $(4.36, 1.57, 1.16)$
5. a. $(2.24, 1.11, 3.00)$ **b.** $(3.74, 1.11, 0.64)$
7. a. $(-1.50, 2.60, -3.00)$ **b.** $(4.24, 2.09, 2.36)$
9. a. $(1.41, 1.41, 3.14)$ **b.** $(3.72, 0.79, 0.57)$
11. a. $(1.50, 0.87, -1.00)$ **b.** $(1.73, 0.52, -1.00)$
13. a. $(0.00, 0.00, 1.00)$ **b.** $(0.00, 0.52, 1.00)$
15. a. $(-0.06, 0.13, -0.99)$ **b.** $(0.14, 2.00, -0.99)$
17. $z = r^2 \cos 2\theta$
19. $r = \dfrac{6z}{\sqrt{4 - 13\cos^2\theta}}$ or
$\quad 9r^2\cos^2\theta - 4r^2\sin^2\theta + 36z^2 = 0$

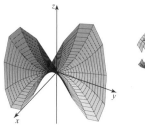
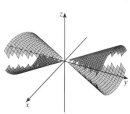

21. $\phi = \frac{\pi}{4}$

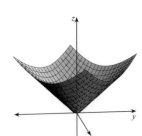

23. $\rho = \dfrac{4 \cot \phi \csc \phi}{3 - 2 \cos^2\theta}$ or

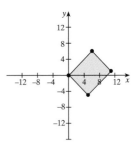

25. $z = 2xy$

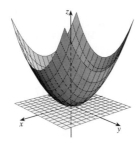

27. $z = x^2 - y^2$

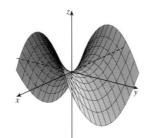

29. $xz = 1$

31. $\frac{16}{3}$ **33.** 2π **35.** 8π **37.** $\frac{7}{60}$

39. $\dfrac{x^2}{a^2} + \dfrac{y^2}{b^2} + \dfrac{z^2}{c^2} = R^2$ **41.** $\dfrac{a^6}{3}$ **43.** $(0, 0, \frac{27}{4})$

45. a. $m = \displaystyle\int_0^{2\pi} \int_0^3 \int_0^{\sqrt{9-r^2}} (r^2 \sin\theta \cos\theta + z)r\, dz\, dr\, d\theta$

b. $\bar{x} = \dfrac{1}{m} \displaystyle\int_0^{2\pi} \int_0^3 \int_0^{\sqrt{9-r^2}} r\cos\theta\,(r^2 \sin\theta \cos\theta + z)r\, dz\, dr\, d\theta$

c. $I_z = \displaystyle\int_0^{2\pi} \int_0^3 \int_0^{\sqrt{9-r^2}} r^2(r^2 \sin\theta \cos\theta + z)r\, dz\, dr\, d\theta$

47. 0   **49.** 26.8083   **51.** 14.2172

53. 18.8496   **55.** 10.9956   **57.** 7.8540

61. a. Force $= \dfrac{GmM}{R^2} = \dfrac{Gm\delta(4\pi a^3)}{3\,R^2}$

b. Force $= \dfrac{Gm\delta(9\pi)}{4}$

63. $\frac{8\pi}{5}$

13.8 Jacobians: Change of Variables (pages 948–950)

1. $u - v$ **3.** 2

5. $4uv(v^2 + u^2)$ **7.** $-2e^{2u}$

9. -9 **11.** ue^{uv}

13. $A(0, 5) \to (5, -5);$
$B(6, 5) \to (11, 1);$
$C(6, 0) \to (6, 6);$
$O(0, 0) \to (0, 0)$

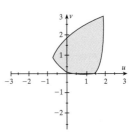

15. $A(5, 0) \to (25, 0);$
$B(7, 4) \to (33, 56);$
$C(2, 4) \to (-12, 16);$
$O(0, 0) \to (0, 0)$

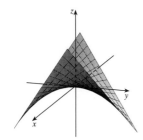

17. $dx\, dy = u\, du\, dv$

19. $3 \ln 2$

21. 0   **23.** 0

27. $\frac{5}{6}$ **29.** $\frac{625}{6}$

31. 13.0949

33. 40.6952

35. 1.1455

37. $\dfrac{ab\pi}{4}(1 - e^{-1})$

39. 1.2825

41. $dx\, dy\, dz$ becomes $r\, dr\, d\theta\, dz$

43. $\dfrac{4(a^2 + b^2)(abc)\pi}{15}$

47. $C = 4 \displaystyle\int_0^a \dfrac{1}{a\sqrt{a^2 - x^2}} \sqrt{a^4 + (b^2 - a^2)x^2}\, dx$

Chapter 13: Review

Proficiency Examination (pages 950–951)

23. 0.2866   **24.** 0

25. 36   **26.** 0.6609

27. Thus, the probability of product failure is about 15%.

28. 8.8858

29. 26.8083

30. 2.4440

Supplementary Problems (pages 952–955)

1.

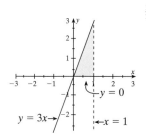

$\frac{3}{2}$

3.

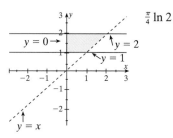

$\frac{\pi}{4}\ln 2$

5.

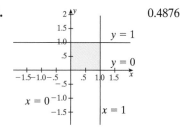

0.4876

7.

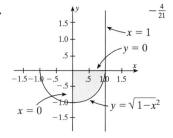

$-\frac{4}{21}$

9.

0.0383

11. $\frac{1}{3}(e-1)$　　**13.** 10　　**15.** 38.0406

17. $\frac{31}{30}$　　**19.** $\frac{81\pi}{8}$　　**21.** 12π

23. $\dfrac{2^{2n+2}\pi}{n+1}$　　**25.** 0　　**27.** 0

29. $\dfrac{4}{ab}$　　**31.** $\frac{\pi}{8}$　　**33.** π　　**35.** $\frac{32\pi}{5}$

37. The improper integral converges for $m>1$, and diverges otherwise.

39. a^2

41. $\displaystyle\int_0^1\int_x^{4-3x} f(x,y)\,dy\,dx$　　**43.** $a=b=\pi, c=0$

45. $2\sqrt{2}$　　**47.** 6　　**49.** $\frac{2}{3}$

51. $\frac{\pi}{2}$　　**53.** 117.32　　**55.** 7.0686

57. 2.4110

61. $\dfrac{2\pi a^5}{15}(2-\sin^2\phi_0\cos\phi_0-2\cos\phi_0)$

63. $\frac{2}{5}$　　**65.** $\frac{3}{4}H$　　**67.** $32xyz$

69. $y=Cx+1$ becomes $v=C(u+v)+1$

71. a. $-x\le y\le x; \frac{1}{2}\le x\le 1$

　b. $\dfrac{\partial(u,v)}{\partial(x,y)}=-2; \displaystyle\int_D\int (u+v)\,du\,dv=\frac{7}{3}$

73. $\left(\dfrac{1}{3A},\dfrac{1}{3B},\dfrac{1}{3C}\right)$

75. a. $\left(\frac{16}{35},\frac{1}{2}\right)$　**b.** $\frac{11}{240}$

77. $\dfrac{32a^3}{9}$　　**79.** $\dfrac{4a^3\pi}{35}$

81. 1　　**83.** $m=\displaystyle\int_{\pi/12}^{5\pi/12}\int_{\sqrt{2}a}^{2a\sqrt{\sin2\theta}}\delta r\,dr\,d\theta$

85. $(0,0,0)$　　**87.** $\overline{x}=\overline{y}\approx 0.3575$

89. This is Putnam Problem 4 in the afternoon session of 1993.

91. This is Putnam Problem 3 in the afternoon session of 1957.

CHAPTER 14: VECTOR ANALYSIS

14.1 Properties of a Vector Field: Divergence and Curl
(pages 965–966)

9. div $\mathbf{F}=3x+3z^2$; curl $\mathbf{F}=y\mathbf{k}$

11. div $\mathbf{F}=2$; curl $\mathbf{F}=\mathbf{0}$

13. div $\mathbf{F}=0$; curl $\mathbf{F}=\mathbf{0}$

15. At $(1,2,3)$, div $\mathbf{F}=7$; curl $\mathbf{F}=\mathbf{j}-3\mathbf{k}$

17. At $(3,2,0)$, div $\mathbf{F}=2-2e^{-6}$; curl $\mathbf{F}=-3\mathbf{i}+3e^{-6}\mathbf{k}$

19. div $\mathbf{F}=\cos x-\sin y$; curl $\mathbf{F}=\mathbf{0}$

21. div $\mathbf{F}=0$; curl $\mathbf{F}=\mathbf{0}$

23. div $\mathbf{F}=\dfrac{1}{\sqrt{x^2+y^2}}$; curl $\mathbf{F}=\mathbf{0}$

25. div $\mathbf{F}=a+b$; curl $\mathbf{F}=\mathbf{0}$

27. div $\mathbf{F}=2(x+y+z)$; curl $\mathbf{F}=\mathbf{0}$

29. div $\mathbf{F}=x+y+z$; curl $\mathbf{F}=-y\mathbf{i}-z\mathbf{j}-x\mathbf{k}$

31. div $\mathbf{F}=yz+2x^2yz^2+3y^2z^2$;
　curl $\mathbf{F}=(2yz^3-2x^2y^2z)\mathbf{i}+xy\mathbf{j}+(2xy^2z^2-xz)\mathbf{k}$

33. div $\mathbf{F}=3$; curl $\mathbf{F}=\mathbf{i}+\mathbf{j}-\mathbf{k}$

35. harmonic　　**37.** harmonic

39. $6x\mathbf{j}-3y\mathbf{k}$　　**41.** $2z+3x$

43. $6xyz^2 + 2xy^3$

51. a. $(bz - cy)\mathbf{i} + (cx - az)\mathbf{j} + (ay - bx)\mathbf{k}$
   b. div $\mathbf{V} = 0$; curl $\mathbf{V} = 2\omega$

14.2 Line Integrals (pages 974–976)

3. $-\frac{4}{3}$ **5.** 1   **7.** $\frac{26}{3}$

9. a. $-\frac{1}{3}$ **b.** $-\frac{1}{3}$ **11.** $-\frac{16}{3}$

13. $\frac{77}{15}$ **15.** $-\frac{7}{60}$ **17.** 12

19. 12   **21. a.** $\frac{\pi^2}{8}$ **b.** $\frac{\pi^2}{8}$

23. a. $\frac{5}{2}$ **b.** $\frac{11}{2}$ **25.** 0

27. 9   **29.** $-\frac{1}{3}$ **31.** 0

33. -0.7071 **35.** 2π **37.** 16

39. $\frac{1}{2}$ **41.** 42   **43.** $\frac{5}{2}$

45. The work done is 10,875 ft-lb

14.3 Independence of Path (pages 981–984)

5. $f(x, y) = x^2 y^3$; conservative

7. not conservative   **9.** not conservative

11. a. -2π **b.** -4 **c.** -2π

13. a. 0   **b.** 32   **c.** 32

15. 1   **17.** $\frac{1}{3}$ **19.** $\frac{1}{2}$

21. $\frac{13}{4}$ **23.** 8   **25.** 2.7124

27. 8   **29.** 32   **31.** 0.2245

33. $g(x) = Cx^{-5}$

35. b. $W = \dfrac{kmM}{\sqrt{a_2^2 + b_2^2 + c_2^2}} - \dfrac{kmM}{\sqrt{a_1^2 + a_2^2 + a_3^2}}$

37. 0

39. a. $W = -(2a + b)$ **b.** $W_1 = a(e^{-1} - 3)$

   c. $W_2 = \dfrac{a}{7}(9e^{-7/9} - 23)$; $W_3 = -ae^{-7/9}(2e^{7/9} + e - 1)$

14.4 Green's Theorem (pages 993–995)

1. 0   **3.** 3   **5.** 0

7. -6π **9.** 0   **11.** 16

13. 2π **15.** 4π **17.** $\frac{15}{2}$

19. $-\dfrac{\pi a^4}{4}$ **25.** 0   **27.** 0

29. 0   **31.** 3   **33.** $5A$

14.5 Surface Integrals (pages 1001–1002)

1. 8π **3.** 0   **5.** 16π

7. $4\sqrt{2}$ **9.** 0   **11.** $\dfrac{160\sqrt{6}}{3}$

13. $\frac{\pi}{2}$ **15.** 84.4635   **17.** 36π

19. $\frac{4\pi}{3}$ **21.** $\dfrac{7\pi\sqrt{2}}{4}$ **23.** -6

25. $\frac{4\pi}{3}$ **27.** π **29.** 0.9617

31. -2.3234 **33.** 13.0639   **35.** 2.9794

37. 228.5313

39. a. $\dfrac{\pi a^7}{192}$ **b.** $2\pi a^2(1 - \cos\phi)$ **41.** $I_z = \frac{2}{3}ma^2$

14.6 Stokes' Theorem (pages 1009–1011)

1. 18π **3.** $\frac{9}{2}$ **5.** -8π

7. -18 **9.** 0   **11.** $2\sqrt{2}\,\pi$

13. -3π **15.** $-\frac{1}{2}$ **17.** 0

19. -16π **21.** $\frac{3}{2}$ **23.** $2 - \pi$

25. $\frac{\pi}{2}$ **27.** 0   **29.** $f(x, y, z) = xyz^2$

31. $f(x, y, z) = \dfrac{xy}{z}$

33. $f(x, y, z) = xy \sin z + y^2$

14.7 Divergence Theorem (pages 1017–1019)

1. $\frac{64\pi}{3}$ **3.** $\frac{128}{3}$ **5.** 3

7. 24π **9.** $\frac{13}{3}$ **11.** $\frac{81\pi}{2}$

13. 0   **15.** 12π **17.** $\frac{\pi}{4}$

19. $\dfrac{9\pi a^4}{2}$ **21.** $4\pi a^4$ **23. b.** 6

CHAPTER 14: REVIEW

Proficiency Examination (pages 1021–1022)

23. $\mathbf{F}$ is conservative; $f = xyz$

24. div $\mathbf{F} = 2xy - ze^{yz}$; curl $\mathbf{F} = ye^{yz}\mathbf{i} - \frac{1}{2}\mathbf{j} - x^2\mathbf{k}$

25. -2 **26.** $-4\pi\sqrt{2}$

27. 0   **28.** 0

29. $\phi = \dfrac{m\omega^2}{2}(x^2 + y^2 + z^2)$ **30.** 0

Supplementary Problems (pages 1023–1025)

1. $\mathbf{F}$ is conservative; $f = 2x - 3y$

3. $\mathbf{F}$ is conservative; $f = xy^{-3} + \sin y$

5. $\mathbf{F}$ is not conservative

7. -7 **9.** -2.5670

11. $\frac{5}{2}$ **13.** div $\mathbf{F} = 3$; curl $\mathbf{F} = \mathbf{0}$

15. div $\mathbf{F} = \dfrac{2}{\sqrt{x^2 + y^2 + z^2}}$; curl $\mathbf{F} = \mathbf{0}$

17. $\pi \cos \pi^2$ **19.** -8.3562 **21.** $\sqrt{2}$

23. $\frac{5}{6}$ **25.** $2\pi(\pi - 1)$ **27.** 0

29. 4π **31.** $\dfrac{6\sqrt{2}\pi}{5}$ **33.** $\dfrac{2a + 3b}{2abc}$

35. 0   **37.** $\frac{3}{4}$ **39.** $\frac{3}{2}$

41. -1 **43.** $\frac{5}{6}$

45. $\mathbf{F}$ is conservative in any region of the plane where
   $x + y \neq 0$; $W = \ln\left(\dfrac{c + d}{a + b}\right)$

53. 0

55. a. C is any region that does not contain the y-axis;
$$f = -\frac{1 + y^2}{2x^2} - \frac{y^2}{2}$$
b. $\dfrac{-67}{9}$

57. $\frac{5}{4} + \ln \frac{2}{9}$ **59.** $2\pi c$ **61.** 0

63. 0 **65.** $\frac{\pi}{4} - \frac{1}{2} \ln 2$ **67.** 2π

71. $4\pi^2 ab$

73. Only the volume of the body is important, so it does not matter which corner is removed.

75. This is Putnam Problem 6i in the morning session of 1948.

CHAPTER 15: INTRODUCTION TO DIFFERENTIAL EQUATIONS

15.1 First-Order Differential Equations (pages 1038–1041)

1. $\ln|y| = x + \ln|x - 5|^5 + C$

3. $\ln|y| = -\frac{1}{18}[\ln(e^{2x} + 9) - 2x] + C$

5. $y = 3\tan^{-1}\dfrac{\sqrt{x^2 - 9}}{3} + C$ **7.** $y = 2(1 - e^{-x^2})$

9. $y = \frac{1}{2}(\cos x + \sin x) - \frac{1}{2}e^{-x}$ **11.** $y = x^3 - \frac{9}{4}x^2$

13. $-\ln\sqrt{x^2 + y^2} - \frac{1}{3}\tan^{-1}\dfrac{y}{x} = C$

15. $(y + x)^2(y - x) = C$ **17.** $\dfrac{y - x}{3y + x} = Bx^8$

19. $x^3y + x\tan y = C$

21. $\ln(x^2 + y^2) + \tan^{-1} x + e^{-y} = C$

23. $x^2\cos 2y - 3xy + 3x^2y - 6y = C$

25. a. $y = -x - 1 + 4e^{x-1}$
b. Isoclines are lines of the form $x + y = C$ **c.**

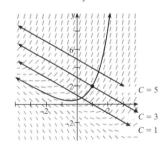

27.

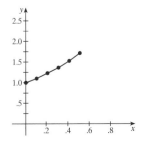

29.

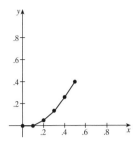

31. $\dfrac{x}{y} + \ln|y| = C$ **33.** $x^2y^2 + x^3y - xy^3 = C$

35. $\dfrac{x^2}{y} + \dfrac{y^2}{x} + 3y = C$

37. $5e^x = e^y(2\sin x\cos x + \sin^2 x + 2) + C$

39. $x^3 + xy\cos xy = C$

41. $y = (\csc x + \cot x)[(-\sin x + x) + C]$

43. $\sqrt{x^2 + y^2} + y = B$ **45.** $\cot\dfrac{y}{x} = \ln\left|\dfrac{\pi x}{4}\right| + 1$

47. $y = x^3\ln|x| + x^3$ **49.** $x\sin(x^2 + y) = 0$

51. $\frac{1}{2}\ln|y^2 + 2y + 2| - \tan^{-1}(y + 1) + e^{-x} = 1$

53. If $\alpha > \beta$, $\displaystyle\lim_{t \to +\infty} Q(t) = \beta$; if $\alpha < \beta$,
$\displaystyle\lim_{t \to +\infty} Q(t) = \alpha$, and if $\alpha = \beta$, $\displaystyle\lim_{t \to +\infty} Q(t) = \alpha$

55. a. $P(t) = \dfrac{r_1(P_0 - r_2) - r_2(P_0 - r_1)e^{-Dt}}{(P_0 - r_2) - (P_0 - r_1)e^{-Dt}}$
b. $\displaystyle\lim_{t \to +\infty} P(t) = r_1$

57. a. $x = \dfrac{1}{I}\left[\displaystyle\int I(y)\,S(y)\,dy + C\right]$ **b.** $x = y^2[(y + 1)e^{-y} + C]$

59. 125.33 ft/s   **61. c.** $y = \dfrac{3Cx + 2x^4}{3C - x^3}$

15.2 Second-Order Homogeneous Linear Differential Equations (pages 1050–1052)

1. $y = C_1 + C_2e^{-x}$ **3.** $y = C_1e^{-5x} + C_2e^{-x}$

5. $y = C_1e^{3x} + C_2e^{-2x}$ **7.** $y = C_1e^{(-1/2)x} + C_2e^{3x}$

9. $y = C_1e^x + C_2e^{-x}$

11. $y = C_1\cos\sqrt{11}x + C_2\sin\sqrt{11}x$

13. $y = e^{(-3/14)x}\left[C_1\cos\left(\dfrac{\sqrt{131}}{14}x\right) + C_2\sin\left(\dfrac{\sqrt{131}}{14}x\right)\right]$

15. $y = C_1 + C_2x + C_3e^{-x}$

17. $y = C_1 + C_2x + e^{-(1/2)x}\left[C_3\cos\left(\dfrac{\sqrt{7}}{2}x\right) + C_4\sin\left(\dfrac{\sqrt{7}}{2}x\right)\right]$

19. $y = C_1e^{2x} + C_2e^{-3x} + C_3e^{-x}$

21. $y = e^{5x}(1 - 6x)$

23. $y = \frac{11}{5}e^x + \frac{4}{5}e^{11x}$

25. $y = \frac{94}{25} - \frac{19}{25}e^{-5x} - \frac{9}{5}xe^{-5x}$ **27.** $5e^x$

29. e^{-2x} **33.** $y = C_1e^{-3x} + C_2xe^{-3x}$

35. $y = C_1 + C_2x^{-4}$

37. $y = C_1x^3 + C_2x^{-4}$

39. $y = \dfrac{1}{2}\cos(4\sqrt{3}\,t) - \dfrac{2}{\sqrt{3}}\sin(4\sqrt{3}t)$

41. $y = -\dfrac{2}{3}\cos(4\sqrt{3}\,t)$

43. $y = e^{-0.4t}[0.5\cos 6.9t + 0.03\sin 6.9t]$

47. a. 9.38 seconds

   b. It takes less time with air resistance.

53. a. $s'(t) = -v_0\ln\left(\dfrac{w - rt}{w}\right) - gt$

   b. $s(t) = \dfrac{v_0(w - rt)}{r}\ln\left(\dfrac{w - rt}{w}\right) - \dfrac{1}{2}gt^2 + v_0t$

   c. The fuel is consumed when $rt = w_f$; that is, when $t = w_f/r$.

   d. the height is $\dfrac{v_0(w - w_f)}{r}\ln\left(\dfrac{w - w_f}{w}\right) - \dfrac{1}{2}\dfrac{gw_f^2}{r^2} + \dfrac{v_0w_f}{r}$

49. The solution to this problem is found in the *Canadian Mathematical Bulletin*, Vol. 28, 1985, p. 250.

15.3 Second-Order Nonhomogeneous Linear Differential Equations (pages 1060–1061)

1. $\bar{y}_p = Ae^{2x}$

3. $\bar{y}_p = Ae^{-x}$

5. $\bar{y}_p = xe^{-x}(A\cos x + B\sin x)$

7. $\bar{y}_p = (A + Bx)e^{-2x} + xe^{-2x}(C\cos x + D\sin x)$

9. $\bar{y}_p = A_3x^3 + A_2x^2 + A_1x + A_0$

11. $\bar{y}_p = (A_3x^3 + A_2x^2 + A_1x + A_0)\cos x$
$\qquad + (B_3x^3 + B_2x^2 + B_1x + B_0)\sin x$

13. $\bar{y}_p = A_0e^{2x} + B_0\cos 3x + C_0\sin 3x$

15. $\bar{y}_p = (A_3x^3 + A_2x^2 + A_1x + A_0) + B_0e^{-x}$

17. $y = C_1 + C_2e^{-x} - x^3 + 3x^2 + x$

19. $y = C_1e^{-3x} + C_2e^{-5x} + \dfrac{3}{35}e^{2x}$

21. $y = e^{-x}(C_1\cos x + C_2\sin x) + \dfrac{1}{5}\cos x + \dfrac{2}{5}\sin x$

23. $y = C_1e^{x/7} + C_2e^{-x} + e^{-x}(-\dfrac{1}{16}x^2 - \dfrac{15}{64}x)$

25. $y = C_1 + C_2e^x - (\dfrac{1}{4}x^4 + x^3 + \dfrac{5}{2}x^2 + 10x)$

27. $y = C_1e^{-x} + C_2xe^{-x} + e^{-x}(\dfrac{1}{6}x^3 + 2x^2)$

29. $y = C_1\cos x + C_2\sin x - \cos x\ln|\sec x + \tan x|$

31. $y = C_1e^{3x} + C_2e^{-2x} - \dfrac{1}{32}(8x^2 + 12x + 13)e^{2x}$

33. $y = C_1\cos 2x + C_2\sin 2x + \dfrac{1}{2}x\cos 2x$
$\qquad - \dfrac{1}{4}\sin 2x(\ln|\cos 2x|)$

35. $y = C_1e^{-x} + C_2xe^{-x} + \dfrac{1}{4}x^2(2\ln x - 3)e^{-x}$

37. $y = C_1 + C_2e^x - \dfrac{1}{20}\sin 2x - \dfrac{1}{2}x - \dfrac{1}{5}\cos^2 x$

39. $y = \dfrac{8}{5} - \dfrac{6}{5}e^x - \dfrac{2}{10}\sin 2x + x - \dfrac{2}{5}x\cos^2 x$

41. $y = -\dfrac{2}{9}\cos 3x + \dfrac{4}{9}\sin 3x + \dfrac{2}{9}e^{3x}$

43. $y = \dfrac{35}{27}\sin 3x + \dfrac{1}{9}x$

45. $y = -4\cos x - \sin x\ln|\csc x + \cot x|$

47. $y = C_1e^{-2x} + C_2e^{-3x} + G(x)$

   where $G(x) = \begin{cases} \dfrac{1}{3}x - \dfrac{5}{18} & \text{for } 0 \le x \le 1 \\ \dfrac{1}{3} & \text{for } 1 < x < 3 \\ -\dfrac{1}{3}x + \dfrac{29}{18} & \text{for } 3 \le x \le 4 \end{cases}$

49. $I(t) = \dfrac{81}{656}e^{-9t} - \dfrac{9}{16}e^{-t} + \dfrac{18}{41}\cos t + \dfrac{45}{82}\sin t$

51. $I(t) = 0.312e^{-t} - 0.015e^{-9t} - 0.297\cos t$
$\qquad - 0.067\sin t + t(0.244\cos t + 0.305\sin t)$

53. a. $y_1 = x$; $y = C_1x + C_2(x\tan^{-1}x + 1) - \dfrac{1}{2}\tan^{-1}x + \dfrac{x}{2}$

CHAPTER 15: REVIEW

Proficiency Examination (page 1062)

12. $\sin^{-1}y = \sinh^{-1}x + C$ **13.** $y = Bx$

14. $y = x + C\sqrt[3]{x}$

15. $x^2(x^2 - 2y^2) = C$

16. $\dfrac{2}{\sqrt{3}}\tan^{-1}\left[\dfrac{2}{\sqrt{3}}\left(\dfrac{y}{x} - \dfrac{1}{2}\right)\right] = \ln|x| + C$

17. $y = e^{-x}(C_1\cos x + C_2\sin x) - \dfrac{2}{5}\cos x + \dfrac{1}{5}\sin x$

18. $x^3e^{-y} - xy^{-2} + x^2y^{-3} = C$

19. $x(t) = e^{-1.60t}[0.33\cos 10.84t + 0.05\sin 10.84t]$

20. $I(t) = e^{-125t}[-0.562\cos 185.4\,t + 2.806\sin 185.4t)]$
$\qquad + 0.562\cos 100t - 0.899\sin 100t$

Supplementary Problems (pages 1062–1065)

1. a.

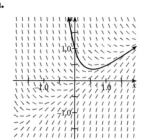

b.

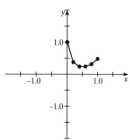

3. $\dfrac{1}{4}y^4 = -2x^2 + C$

5. $\dfrac{y^3}{3x^3} = \ln|x| + C$

7. $e^{y^2} = e^{2x} + C$

9. $y = -e^{5x}\cos x + Ce^{5x}$

11. $x = e^{-y}[y + C]$

13. $\dfrac{y}{x} = \sin(\ln|x| + C)$

15. $y = Be^{(1-2x)/(2x^2)}$

17. $x^4y^3 + \ln|x| - \ln|y| = C$

19. $y = -\tan^{-1}e^x + C$

21. a. $y_h = C_1e^{3x} + C_2e^{-3x}$ **b.** $y_p = -\dfrac{1}{9}(x + 1)$

23. a. $y_h = C_1e^{2x} + C_2e^{3x}$

   b. $y_p = (\dfrac{1}{2}x^2 + \dfrac{3}{2}x + \dfrac{7}{4})e^x$

25. a. $y_h = C_1e^x + C_2e^{2x}$

   b. $y_p = (-\dfrac{1}{4}x^4 - x^3 - 3x^2 - 6x)e^x$

27. a. $y_h = C_1\cos x + C_2\sin x$

   b. $y_p = \sin x\ln|\sec x + \tan x| - 2$

29. $y = \dfrac{1}{9}e^{-3x} + \dfrac{1}{3}x - \dfrac{1}{9}$

31. $z(t) = -3e^{-2t} + 2e^{-3t} + 1$

33. $y^2 = 5x - x^2$

35. $e^{X+Y} = 1 + Ke^X$

37. $Y^2(Y^2 + 3X^2)^2 = K$

39. $y = C_1x + C_2\left[\dfrac{x}{2}\ln\left|\dfrac{x-1}{x+1}\right| + 1\right]$

41. $\dfrac{1}{6}x^2 - \dfrac{C_1}{x} + C_2$

43. $x = y[e^y + C]$ **45.** 0.62 s

47. a. $y(t) = \dfrac{WLx^3}{6EI} - \dfrac{Wx^4}{24EI} - \dfrac{WL^3x}{8EI}$

    b. $y_{max} \approx \dfrac{WL^4}{EI}(-0.045)$

49. b. 2 real roots   **c.** $y = C_1x^{m_0} + C_2x^{m_0}\ln x$

    d. $y = x^{\alpha}[C_1\cos(\beta\ln|x|) + C_2\sin(\beta\ln|x|)]$

53. He must run at 18.8 ft/s.

55. This is Problem 4 of the morning session of the 1947 Putnam examination.

Cumulative Review for Chapters 12–15
(pages 1069–1070)

5. $f_x = 4x + y; f_y = x - 15y^2; f_{xy} = 1$

7. $f_x = 1; f_y = 1; f_{xy} = 0$

9. $f_x = e^xe^y; f_y = e^xe^y; f_{xy} = e^xe^y$

11. 0   **13.** $\dfrac{49,152}{13}$

15. $\dfrac{4\pi}{15}$ **17.** $e^2 - 3$

19. π **21.** 8

23. 0

25. $y = -\dfrac{1}{3}x^{-3}\cos x^3 + Cx^{-3}$

27. $y = C_1e^{-2x} + C_2e^x - \dfrac{1}{4}(2x^3 + 5x^2 + 7x + 18.5)$

29. $y = 1 + (x - 2)e^{-x^2/2}$

31. The equation of the tangent plane is
$$2x + 4y - z - 4 = 0$$
and the normal line is
$$\frac{x}{2} = \frac{y-2}{4} = \frac{z-4}{-1}$$

33. a. $y - x = 3$ **b.** $x = 4, y = 7$

35. $\dfrac{\pi}{4} + 2$ **37.** $\dfrac{\pi}{6}(65\sqrt{65} - 1)$

39. 0

Appendix A (pages A8–A9)

1. $\delta = \dfrac{\epsilon}{2}$ **3.** false

5. let δ = minimum of $\{1, \dfrac{\epsilon}{5}\}$ **7.** $\delta = \epsilon$

9. $\delta = \dfrac{\epsilon}{3}$ **11.** choose $\delta = \min\left(1, \dfrac{\epsilon}{5}\right)$

Credits

CHAPTER 10

p. 725   Putnam examination problem (#59) 1939, reprinted by permission from The Mathematical Association of America.

p. 725   Putnam examination problem (#60) 1959, reprinted by permission from The Mathematical Association of America.

p. 725   Putnam examination problem (#61) 1983, reprinted by permission from The Mathematical Association of America.

p. 726   Photograph of the Starship *Enterprise* from *Star Trek: The Next Generation* © Paramount Pictures, courtesy of Sygma.

p. 726   Group research project is from *MAA Notes,* Vol. 17, 1991, "Priming the Calculus Pump: Innovations and Resources," by Marcus S. Cohen, Edward D. Gaughan, R. Arthur Knoebel, Douglas S. Kurtz, and David J. Pengelley.

CHAPTER 11

p. 730   Photograph of double helix of DNA courtesy of *Scientific American,* October 1985, p. 60.

p. 752   Photograph of an earth satellite courtesy of Julian Baum, Photo Researchers, Inc.

p. 755   Diagram of Kingdome courtesy of Seattle Mariners.

p. 776   Photograph of a Ferris wheel courtesy of Marc D. Longwood.

p. 780   Putnam examination problem (#75) 1939, reprinted by permission from The Mathematical Association of America.

p. 780   Putnam examination problem (#76) 1947, reprinted by permission from The Mathematical Association of America.

p. 780   Putnam examination problem (#77) 1946, reprinted by permission from The Mathematical Association of America.

p. 781   Quotation by Johann Kepler from James R. Newman, *The World of Mathematics,* Vol. I (New York: Simon & Schuster, 1952), p. 220.

p. 781   Guest essay, "The Simulation of Science," by Howard Eves. This article is reprinted by permission from *Great Moments in Mathematics Before 1650* published by The Mathematical Association of America, 1983, p. 194.

CHAPTER 12

p. 792   Computer-generated graph of Mount St. Helens courtesy of Bill Lennox, Humboldt State University.

p. 793   Topographic map of Mount Rainier courtesy of U.S. Geological Survey.

p. 793   Map showing isotherms courtesy of Accu-Weather, Inc.

p. 860   What Does This Say? problem (#51) adapted from *Managerial Economics* by Dominick Salvatore, McGraw-Hill, Inc., New York, 1989. The data given in the table are on page 138.

p. 870   Historical Quest (#54) from *Men of Mathematics* by E. T. Bell, Simon and Schuster, New York, 1937, p. 165.

p. 873   Modeling problem (#67) from *Mathematical Biology,* 2nd ed., by J. D. Murray, Springer-Verlag, New York, 1993, p. 464.

p. 873   Modeling problem (#68) from "Heat Therapy for Tumors," Leah Edelstein-Keshet, *UMAP Modules 1991: Tools for*

Teaching, Consortium for Mathematics and Its Applications, Inc., Lexington MA, 1992, pp. 73–101.

p. 875   Think Tank Problem (#95) adapted from "Hidden Boundaries in Constrained Max-Min Problems," by Herbert R. Bailey from *The College Mathematics Journal,* May 1991, p. 227.

p. 876   Photograph of soap bubbles from *Scientific American,* "The Geometry of Soap Films and Soap Bubbles," by Frederick Almgren, Jr., and Jean Taylor.

p. 876   Putnam examination problem (#97) 1967, reprinted by permission from The Mathematical Association of America.

p. 876   Putnam examination problem (#98) 1946, reprinted by permission from The Mathematical Association of America.

p. 876   Putnam examination problem (#99) 1938, reprinted by permission from The Mathematical Association of America.

p. 877   Photograph of a desert courtesy of Scott T. Smith.

p. 877   Group research project is from *MAA Notes,* Vol. 17, 1991, "Priming the Calculus Pump: Innovations and Resources," by Marcus S. Cohen, Edward D. Gaughan, R. Arthur Knoebel, Douglas S. Kurtz, and David J. Pengelley.

CHAPTER 13

p. 905   Quotation in Historical Quest (#57) by W. W. Rouse Ball from *A Short Account of the History of Mathematics* as quoted in *Mathematical Circles Adieu* by Howard Eves, Boston: Prindle, Weber & Schmidt, Inc., 1977.

p. 955   Putnam examination problem (#89) 1994, reprinted by permission from The Mathematical Association of America.

p. 955   Putnam examination problem (#90) 1942, reprinted by permission from The Mathematical Association of America.

p. 955   Putnam examination problem (#91) 1957, reprinted by permission from The Mathematical Association of America.

p. 955   Putnam examination problem (#92) 1958, reprinted by permission from The Mathematical Association of America.

p. 956   Photograph of an Apollo spacecraft courtesy of NASA.

p. 956   Group research project is from *MAA Notes,* Vol. 17, 1991, "Priming the Calculus Pump: Innovations and Resources," by Marcus S. Cohen, Edward D. Gaughan, R. Arthur Knoebel, Douglas S. Kurtz, and David J. Pengelley.

CHAPTER 14

p. 958   Photograph of a wind-velocity map of the Indian Ocean courtesy of NASA.

p. 987   Graffiti courtesy of *The College Mathematics Journal,* Sept. 1991; photograph courtesy of Regev Nathansohn.

p. 1025   Journal Problem (#76) by J. Chris Fischer, *CRUX Mathematicorum,* problem number 785, Vol. 8, 1982, p. 277.

p. 1025   Putnam examination problem (#77) 1948, reprinted by permission from The Mathematical Association of America.

p. 1026   Guest essay, "Continuous vs Discrete Mathematics," by William Lucas.

CHAPTER 15

p. 1034   Photograph of a skydiver courtesy of Werner H. Muller/Peter Arnold, Inc.

p. 1041   Journal problem (#62) from *The American Mathematical Monthly,* Vol. 91, 1984, p. 515.

p. 1050   *Mathematics is discovered* quotation by James H. Jeans, *The Myterious Universe,* p. 142. *Mathematics is invented* quotation by G. F. Simmons, *Differential Equations with Applications and Historical Notes,* 2nd ed., McGraw-Hill, Inc., 1991, p. xix.

p. 1052   Journal problem (#54) by Murray S. Klamkin, Problem 331 in the *Canadian Mathematical Bulletin,* Vol. 26, 1983, p. 126.

p. 1065   Putnam examination problem (#55) 1938, reprinted by permission from The Mathematical Association of America.

p. 1065   Putnam examination problem (#56) 1947, reprinted by permission from The Mathematical Association of America.

p. 1065   Putnam examination problem (#57) 1948, reprinted by permission from The Mathematical Association of America.

p. 1065   Putnam examination problem (#58) 1959, reprinted by permission from The Mathematical Association of America.

p. 1066   Photograph of a dam courtesy of Tennessee Valley Authority.

p. 1066   Group research project, "Save the Perch Project," by Diane Schwartz from Ithaca College, N. Y.

Index

MISCELLANEOUS FORMULAS

SPECIAL TRIGONOMETRIC LIMITS:
$$\lim_{h \to 0} \frac{\sin h}{h} = 1 \qquad \lim_{h \to 0} \frac{\cos h - 1}{h} = 0$$

APPROXIMATION INTEGRATION METHODS: Let $\Delta x = \frac{b - a}{n}$ and, for the kth subinterval, $x_k = a + k\Delta x$; also suppose f is continuous throughout the interval $[a, b]$.

Rectangular Rule:
$$\int_a^b f(x)\, dx \approx [f(a + \Delta x) + f(a + 2\Delta x) + \; \ldots \; + f(a + n\Delta x)]\Delta x$$

Trapezoidal Rule:
$$\int_a^b f(x)\, dx \approx \tfrac{1}{2}[f(x_0) + 2f(x_1) + 2f(x_2) + \; \ldots \; + 2f(x_{n-1}) + f(x_n)]\Delta x$$

Simpson Rule:
$$\int_a^b f(x)\, dx \approx \tfrac{1}{3}[f(x_0) + 4f(x_1) + 2f(x_2) + \; \ldots \; + 4f(x_{n-1}) + f(x_n)]\Delta x$$

ARC LENGTH: Let f be a function whose derivative f' is continuous on the interval $[a, b]$. Then the **arc length**, s, of the graph of $y = f(x)$ between $x = a$ and $x = b$ is given by the integral

$$s = \int_a^b \sqrt{1 + [f'(x)]^2}\, dx$$

If C is a parametrically described curve $x = x(t)$ and $y = y(t)$, and is simple, then

$$s = \int_a^b \sqrt{\left(\frac{dx}{dt}\right)^2 + \left(\frac{dy}{dt}\right)^2}\, dt$$

SERIES Suppose there is an open interval I containing c throughout which the function f and all its derivatives exist.

Taylor Series of f
$$f(c) + \frac{f'(c)}{1!}(x - c) + \frac{f''(c)}{2!}(x - c)^2 + \frac{f'''(c)}{3!}(x - c)^3 + \cdots$$

Maclaurin Series of f
$$f(0) + \frac{f'(0)}{1!}x + \frac{f''(0)}{2!}x^2 + \frac{f'''(0)}{3!}x^3 + \cdots$$

FIRST-ORDER LINEAR D.E. The general solution of the first-order linear differential equation

$$\frac{dy}{dx} + P(x)y = Q(x)$$

is given by

$$y = \frac{1}{I(x)}\left[\int Q(x)I(x)\, dx + C\right]$$

where $I(x) = e^{\int P(x)\, dx}$ and C is an arbitrary constant.

VECTOR-VALUED FUNCTIONS

Del operator: $\nabla = \dfrac{\partial}{\partial x}\mathbf{i} + \dfrac{\partial}{\partial y}\mathbf{j} + \dfrac{\partial}{\partial z}\mathbf{k}$ 　　Gradient: $\nabla f = \dfrac{\partial f}{\partial x}\mathbf{i} + \dfrac{\partial f}{\partial y}\mathbf{j} + \dfrac{\partial f}{\partial z}\mathbf{k}$

Laplacian: $\nabla^2 f = \dfrac{\partial^2 f}{\partial x^2} + \dfrac{\partial^2 f}{\partial y^2} + \dfrac{\partial^2 f}{\partial z^2} = f_{xx} + f_{yy} + f_{zz}$

Derivatives of a vector field $\mathbf{F} = u\mathbf{i} + v\mathbf{j} + w\mathbf{k}$:

div $\mathbf{F} = \dfrac{\partial u}{\partial x} + \dfrac{\partial v}{\partial y} + \dfrac{\partial w}{\partial z} = \nabla \cdot \mathbf{F}$ 　　This is a scalar derivative.

curl $\mathbf{F} = \begin{vmatrix} \mathbf{i} & \mathbf{j} & \mathbf{k} \\ \dfrac{\partial}{\partial x} & \dfrac{\partial}{\partial y} & \dfrac{\partial}{\partial z} \\ u & v & w \end{vmatrix} = \nabla \times \mathbf{F}$ 　　This is a vector derivative.

DIFFERENTIATION FORMULAS

PROCEDURAL RULES

Constant multiple $(cf)' = cf'$

Sum rule $(f + g)' = f' + g'$

Difference rule $(f - g)' = f' - g'$

Linearity rule $(af + bg)' = af' + bg'$

Product rule $(fg)' = fg' + f'g$

Quotient rule $\left(\dfrac{f}{g}\right)' = \dfrac{gf' - fg'}{g^2}$

Chain rule $\dfrac{dy}{dx} = \dfrac{dy}{du}\dfrac{du}{dx}$

BASIC FORMULAS

Extended power rule $\dfrac{d}{dx} u^n = nu^{n-1}\dfrac{du}{dx}$

Trigonometric rules

$\dfrac{d}{dx} \cos u = -\sin u \dfrac{du}{dx}$ $\qquad$ $\dfrac{d}{dx} \sin u = \cos u \dfrac{du}{dx}$

$\dfrac{d}{dx} \tan u = \sec^2 u \dfrac{du}{dx}$ $\qquad$ $\dfrac{d}{dx} \cot u = -\csc^2 u \dfrac{du}{dx}$

$\dfrac{d}{dx} \sec u = \sec u \tan u \dfrac{du}{dx}$ $\qquad$ $\dfrac{d}{dx} \csc u = -\csc u \cot u \dfrac{du}{dx}$

Inverse trigonometric rules

$\dfrac{d}{dx} \cos^{-1} u = \dfrac{-1}{\sqrt{1 - u^2}}\dfrac{du}{dx}$ $\qquad$ $\dfrac{d}{dx} \sin^{-1} u = \dfrac{1}{\sqrt{1 - u^2}}\dfrac{du}{dx}$

$\dfrac{d}{dx} \tan^{-1} u = \dfrac{1}{1 + u^2}\dfrac{du}{dx}$ $\qquad$ $\dfrac{d}{dx} \cot^{-1} u = \dfrac{-1}{1 + u^2}\dfrac{du}{dx}$

$\dfrac{d}{dx} \sec^{-1} u = \dfrac{1}{|u|\sqrt{u^2 - 1}}\dfrac{du}{dx}$ $\qquad$ $\dfrac{d}{dx} \csc^{-1} u = \dfrac{-1}{|u|\sqrt{u^2 - 1}}\dfrac{du}{dx}$

Logarithmic rules

$\dfrac{d}{dx} \ln u = \dfrac{1}{u}\dfrac{du}{dx}$ $\qquad$ $\dfrac{d}{dx} \log_b |u| = \dfrac{\log_b e}{u}\dfrac{du}{dx} = \dfrac{1}{u \ln b}\dfrac{du}{dx}$

Exponential rules

$\dfrac{d}{dx} e^u = e^u \dfrac{du}{dx}$ $\qquad$ $\dfrac{d}{dx} b^u = b^u \ln b \dfrac{du}{dx}$

Hyperbolic rules

$\dfrac{d}{dx} \cosh u = \sinh u \dfrac{du}{dx}$ $\qquad$ $\dfrac{d}{dx} \sinh u = \cosh u \dfrac{du}{dx}$

$\dfrac{d}{dx} \tanh u = \operatorname{sech}^2 u \dfrac{du}{dx}$ $\qquad$ $\dfrac{d}{dx} \coth u = -\operatorname{csch}^2 u \dfrac{du}{dx}$

$\dfrac{d}{dx} \operatorname{sech} u = -\operatorname{sech} u \tanh u \dfrac{du}{dx}$ $\qquad$ $\dfrac{d}{dx} \operatorname{csch} u = -\operatorname{csch} u \coth u \dfrac{du}{dx}$

Inverse hyperbolic rules

$\dfrac{d}{dx} \sinh^{-1} u = \dfrac{1}{\sqrt{u^2 + 1}}\dfrac{du}{dx}$ $\qquad$ $\dfrac{d}{dx} \cosh^{-1} u = \dfrac{1}{\sqrt{u^2 - 1}}\dfrac{du}{dx}$

$\dfrac{d}{dx} \tanh^{-1} u = \dfrac{1}{1 - u^2}\dfrac{du}{dx}$ $\qquad$ $\dfrac{d}{dx} \coth^{-1} u = \dfrac{1}{1 - u^2}\dfrac{du}{dx}$

$\dfrac{d}{dx} \operatorname{sech}^{-1} u = \dfrac{-1}{|u|\sqrt{1 - u^2}}\dfrac{du}{dx}$ $\qquad$ $\dfrac{d}{dx} \operatorname{csch}^{-1} u = \dfrac{-1}{|u|\sqrt{1 + u^2}}\dfrac{du}{dx}$

INTEGRATION FORMULAS

PROCEDURAL RULES

Constant multiple		$\int cf(u)\,du = c\int f(u)\,du$

Sum rule $\quad \int [f(u) + g(u)]\,du = \int f(u)\,du + \int g(u)\,du$

Difference rule $\quad \int [f(u) - g(u)]\,du = \int f(u)\,du - \int g(u)\,du$

Linearity rule $\quad \int [af(u) + bg(u)]\,du = a\int f(u)\,du + b\int g(u)\,du$

BASIC FORMULAS

Constant rule $\quad \int 0\,du = C$

Power rules $\quad \int u^n\,du = \dfrac{u^{n+1}}{n+1} + C;\ n \neq -1$

$$\int u^{-1}\,du = \ln|u| + C$$

Exponential rules $\quad \int e^u\,du = e^u + C$

$$\int a^u\,du = \dfrac{a^u}{\ln a} + C \quad a > 0, a \neq 1$$

Logarithmic rule $\quad \int \ln u\,du = u\ln u - u + C, u > 0$

Trigonometric rules

$\int \sin u\,du = -\cos u + C$ $\qquad$ $\int \cos u\,du = \sin u + C$

$\int \tan u\,du = -\ln|\cos u| + C$ $\qquad$ $\int \cot u\,du = \ln|\sin u| + C$

$\int \sec u\,du = \ln|\sec u + \tan u| + C$ $\qquad$ $\int \csc u\,du = \ln|\csc u - \cot u| + C$

$\int \sec^2 u\,du = \tan u + C$ $\qquad$ $\int \csc^2 u\,du = -\cot u + C$

$\int \sec u \tan u\,du = \sec u + C$ $\qquad$ $\int \csc u \cot u\,du = -\csc u + C$

Cosine squared formula $\quad \int \cos^2 u\,du = \frac{1}{2}u + \frac{1}{4}\sin 2u + C$

Sine squared formula $\quad \int \sin^2 u\,du = \frac{1}{2}u - \frac{1}{4}\sin 2u + C$

Hyperbolic rules

$\int \cosh u\,du = \sinh u + C$ $\qquad$ $\int \sinh u\,du = \cosh u + C$

$\int \tanh u\,du = \ln(\cosh u) + C$ $\qquad$ $\int \coth u\,du = \ln(\sinh u) + C$

Inverse rules

$\displaystyle\int \frac{du}{\sqrt{a^2 - u^2}} = \sin^{-1}\frac{u}{a} + C$ $\qquad$ $\displaystyle\int \frac{du}{\sqrt{u^2 - a^2}} = \cosh^{-1}\frac{u}{a} + C$

$\displaystyle\int \frac{du}{a^2 + u^2} = \frac{1}{a}\tan^{-1}\frac{u}{a} + C$ $\qquad$ $\displaystyle\int \frac{du}{a^2 - u^2} = \begin{cases} \dfrac{1}{a}\tanh^{-1}\dfrac{u}{a} + C \text{ if } \left|\dfrac{u}{a}\right| < 1 \\ \dfrac{1}{a}\coth^{-1}\dfrac{u}{a} + C \text{ if } \left|\dfrac{u}{a}\right| > 1 \end{cases}$

$\displaystyle\int \frac{du}{u\sqrt{u^2 - a^2}} = \frac{1}{a}\sec^{-1}\left|\frac{u}{a}\right| + C$ $\qquad$ $\displaystyle\int \frac{du}{u\sqrt{a^2 - u^2}} = \frac{1}{a}\text{sech}^{-1}\left(\frac{u}{a}\right) + C$

$\displaystyle\int \frac{du}{u\sqrt{a^2 + u^2}} = \sinh^{-1}\left(\frac{u}{a}\right) + C$ $\qquad$ $\displaystyle\int \frac{du}{u\sqrt{a^2 + u^2}} = -\text{csch}^{-1}\left(\frac{u}{a}\right) + C$